3rd International Congress on Advances in Mechanical Sciences (ICAMS2020)

IOP Conference Series: Materials Science and Engineering
Volume 998

Hyderabad, India
27 - 28 August 2020

ISBN: 978-1-7138-4802-8
ISSN: 1757-8981

Printed from e-media with permission by:

Curran Associates, Inc.
57 Morehouse Lane
Red Hook, NY 12571

Some format issues inherent in the e-media version may also appear in this print version.

Institute of Physics
Dirac House, Temple Back
Bristol BS1 6BE UK

Phone: 44 1 17 929 7481
Fax: 44 1 17 920 0979

techtracking@iop.org

Additional copies of this publication are available from:

Curran Associates, Inc.
57 Morehouse Lane
Red Hook, NY 12571 USA
Phone: 845-758-0400
Fax: 845-758-2633
Email: curran@proceedings.com
Web: www.proceedings.com

TABLE OF CONTENTS

IOP Conf. Series: Materials Science and Engineering 998 (2020) 011001 doi:10.1088/1757-899X/998/1/011001

3rd International Congress on Advances in Mechanical Sciences

ICAMS2020

Venue: Vardhaman College of Engineering, Hyderabad

27, 28 AUGUST 2020

IOP Conf. Series: Materials Science and Engineering 998 (2020) 011001 doi:10.1088/1757-899X/998/1/011001

Summary of the Conference

The Third International Congress on Advances in Mechanical Sciences was successfully conducted on virtual mode during 27th and 28th August 2020. The delegates from all over the India have participated in the conference. The conference has provided an opportunity for researchers, academicians, industry people and students to share their knowledge and discussed the latest developments in their field of expertise. This conference provided the platform for the young researchers to interact with eminent scientists and researchers across the globe. In this conference an equal opportunity is provided to all the participants to share their ideas in the area of mechanical engineering along with interdisciplinary domains.

Around 100 numbers of delegates have participated the conference including 2 keynote speakers, 15 session chairs from the different eminent organizations and 71 participants from the different institutes, industries and R & D organizations. The two-day conference comprised 15 plenary sessions. Each presenter was given 10 minutes for presentation, followed by 5 minutes for questions.

The papers received from different premier Institutes (NIT's and Premier state government Universities etc.,), Deemed Universities (GLA University Madura, Gitam University and Hindustan University etc.,), R & D Organizations (DRDO and ISRO etc.,) and Industries (John Deere India Pvt Ltd., etc.,). It covers all the domain of mechanical engineering. Apart from that it also covers multidisciplinary areas like Structural dynamics, AI, Optimization etc.

The organizing committee has formed the International and National advisory committee by including people from Abroad and different premier

IOP Conf. Series: Materials Science and Engineering 998 (2020) 011001 doi:10.1088/1757-899X/998/1/011001

institutes of India like IITs, NITS and R& D organizations like CSIR, DRDO etc., There valuable inputs are also worked out for the event.

All the manuscripts are pearly reviewed by the technical committee members who are from the premier institutes (IITs, NITs, etc.,). The complete review processes are carried out by the external experts only.

The progress report of the Conference:

- Total number of papers submitted to ICAMS2020 – 104
- Total number of papers allowed for registration -71
- The conference conducted smoothly on virtual mode during 27th and 28th August 2020.
- Authors are requested to resubmit the revised version of manuscripts through easychair as per IOP format.

CONVENERS

Dr. S Venukumar

Dr. Shakti P Jena

Dr. M Vishnuvardhan

CONFERENCE CHAIRMAN

Dr. P. Srinivasa Rao

Head, Department of Mechanical Engineering

IOP Conf. Series: Materials Science and Engineering 998 (2020) 011002 doi:10.1088/1757-899X/998/1/011002

Peer review declaration

All papers published in this volume of *IOP Conference Series: Materials Science and Engineering* have been peer reviewed through processes administered by the Editors. Reviews were conducted by expert referees to the professional and scientific standards expected of a proceedings journal published by IOP Publishing.

- **Type of peer review: Single-blind and Double-blind**
- **Conference submission management system: EASY CHAIR**
- **Number of submissions received: 104**
- **Number of submissions sent for review: 104**
- **Number of submissions accepted:71**
- **Acceptance Rate (Number of Submissions Accepted / Number of Submissions Received X 100): 68.26**
- **Average number of reviews per paper: 2**
- **Total number of reviewers involved: 45**
- **Any additional info on review process: NA**
- **Contact person for queries:**

Dr. S Venu Kumar: venukumar24@vardhaman.org
Dr. Shakti P Jena: shaktipjena@vardhaman.org
Dr. M Vishnu Vardhan: mukkotivishnu@vardhaman.org
Conference Email: icams2020@gmail.com

IOP Conf. Series: Materials Science and Engineering 998 (2020) 012001 doi:10.1088/1757-899X/998/1/012001

Thermo-Hydraulic Performance Enhancement of Liquid He Based Cryogenic Nanofluid Flow in Laminar Region Through Rectangular Plate Fin Heat Exchangers -A Numerical Study

A Bose*[1], S Chakraborty[1] and S S Mukhopadhayay[1]

[1]Department of Mechanical Engineering, Meghnad Saha Institute of Technology, Kolkata, Pin code- 700150, West Bengal, India

*a.bose@rediffmail.com

Abstract. Cryogenic applications are increasing in recent times with the advancement of technology. This research work is focused to improve performance of cryogenic fluid in compact heat exchangers. Nanoparticles with high thermal conductivity is mixed with the cryogenic fluid to improve the thermo-hydraulic efficiency of the heat transfer fluid used in the compact heat exchangers. Liquid He has been used as base fluid and its performance has been compared using nanoparticles CuO, Al_2O_3, Fe_3O_4 and Single Wall Carbon Nanotube (SWCNT). A numerical analysis has been done in the laminar range of flows in a PFHE. Comparison of the nanofluids at cryogenic temperature is compared with heat transfer performance parameter Colburn j-factor and pressure drop performance parameter Fanning's friction factor(f). It has been found that around 15-30% enhancement of thermal performance is achieved with nanoparticles in comparison to liquid He without any nanoparticles. It has also been observed that there is an improvement of $j/f^{1/3}$ factor representing heat transfer enhancement without much increasing pumping pressure.

1. Introduction

Liquefaction of gases, refrigeration, cryogenic energy storage systems, rocket and space applications are few examples where cryogenic fluids are used to exchange heat in compact heat exchangers. Plate Fin Heat Exchangers (PFHE) are very common in exchanging heat at cryogenic temperature. It is possible to save energy if performance of the PFHE can be increased or in other words waste heat can be reduced. Performance of the PFHEs depend on geometry of the fins, properties of the fluid and flow parameters (like mass flow rate). A lot of research work has been done on geometry of the fins to improve thermal performance. But in almost all the cases it has been observed that heat transfer enhancement occurs at the cost of high flow resistant passage which ultimately increases pumping power. Same conclusion can be made to increase the convective heat transfer coefficient by maintain high mass flow rate pumping power would be increased. The idea of nanofluid is improving the thermo-hydraulic efficiency of the PFHE by increasing the thermal conductivity of the fluid. Choi [1] first coined the term "nanofluids", to refer dispersed nanoparticles (size less than 100 nm) into the base fluids. The common base fluids used were namely Water (W), oil, ethylene glycol (EG), W/EG based mixture are some of the common base fluids used for refrigeration purpose as reported by Redhwan et al. [2]. It was reported by Lee et al. [3] that nanofluids have higher thermal conductivity compared to their base fluids. They found that, Al_2O_3 nanoparticles of size 13 nm dispersed in water was increased by 30% of its thermal conductivity compared to water at 4.3% volume concentration.

PFHEs are very effective way of increasing rate of heat transfer because of very high area to volume ration. Most extensive experimental studies have been published by Kays and London [4]. It was reported that compact PFHEs of different fin shapes the pressure drop decrease in comparison with increasing the turbulence in the working fluid. provide Experimental findings of Norris and Spofford [5] reported the effect of the heat transfer coefficient and friction factor on the basis of thickness, length and pitch of the fins. Experimental studies by London and Shah [6] in 1967 concluded that a large number of fins per unit length and small offset spacing fin thickness gives better heat transfer. In1975 Wieting [7] proposed an empirical relation among the variables from earlier experiments to predict heat transfer and fluid flow friction for a PFHE of offset fin.

PFHEs performance also have been studied with cryogenic applications. Robertson [8] studied offset PFHE with subcooled liquid nitrogen at around 77K. Cao et al. [9] worked on PFHE with mixed refrigerant including CH_4, C_2H_4, C_4H_{10}, C3H8 and N_2 at about 113K. They found a large deviation in the heat transfer coefficient result from the existing correlations and experiments. A CFD simulation study done by Yang at al. [10] compared thermal performance of different cryogenic medium at gaseous and liquid state in OSF channels revealed that the fin materials with low thermal conductivity deteriorates the performance of the heat exchanger. Thermal performance of aluminum PFHEs were investigated with helium gas cooled to around 77K by Doohan et al. [11] and Goyal et al. [12]. Moreover, many cryogenic experiments were carried out to study the thermo-hydraulic performance of PFHE with helium gas at low temperature. One such report made by Wieting [13].

It has been noticed after literature survey that no work has been published comparing nanofluids performance applied in PFHE at cryogenic temperature with liquid He as base fluid. Liquid He and nanofluids with nanoparticles Al_2O_3, CuO, Fe_3O_4 and SWCNT of 5% volume fraction in each case are compared in terms of j-factor and f-factor and thermo hydraulic performance parameter $j/f^{1/3}$. In each nanofluids base fluid is considered as liquid He. A numerical simulation using Finite Volume Method (FVM) is used to analyze the thermo-hydraulic performance of the PFHE nanofluid at cryogenic temperature. Reynolds number range of the analysis has been set from 800 to 2000.

2. Numerical Analysis

2.1. Physical Description of the Model

The PFHE of rectangular cross section as shown in Fig 1 is of our interest of study for different ranges as mentioned below with liquid helium nanofluid as working fluid.

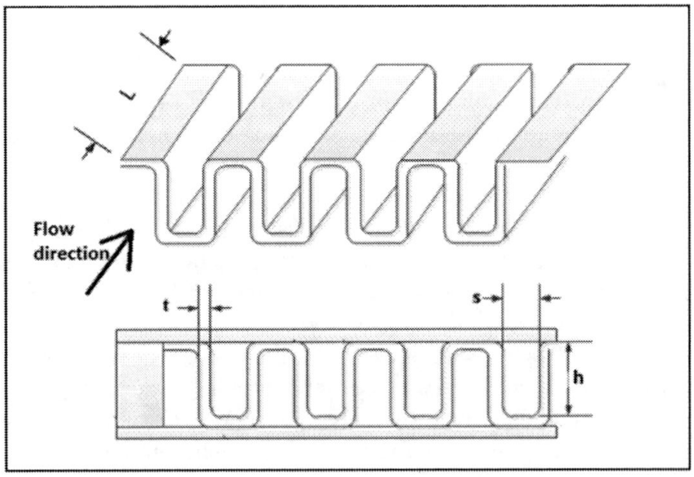

Figure 1. Geometry of the Plain Rectangular PFHE

IOP Conf. Series: Materials Science and Engineering 998 (2020) 012001 doi:10.1088/1757-899X/998/1/012001

Table 1. Geometry of the Plain Rectangular PFHE

Parameters	Working range
Reynolds number	$800 \leq Re \leq 2000$
Volume fraction of nanoparticles	$\emptyset = 5\%$
Height of the fin	$h = 9.75[mm]$
Fin spacing	$s = 4.85[mm]$
Thickness of the fin	$t = 0.25[mm]$
Length of the fin	$100[mm]$

2.2. Governing Equations

The governing equations to describe the fluid flow and heat transfer phenomena are continuity, momentum and energy equation. The mentioned equations are solved to get the velocity, pressure and temperature field. We have considered 3-D, turbulent and steady flow model, so the governing equations are as following,

Continuity equation:

$$\nabla \cdot (\rho \boldsymbol{V}) = 0 \tag{1}$$

Momentum equation:

$$\nabla \cdot (\rho \boldsymbol{V}\boldsymbol{V}) = -\nabla P + \nabla \cdot (\mu \nabla \boldsymbol{V}) \tag{2}$$

Energy equation:

$$\nabla \cdot (\rho c_p \boldsymbol{V} T) = \nabla \cdot (k \nabla T) \tag{3}$$

2.3. Thermophysical Properties of the Nanofluids

By applying the principle of mass conservation to the two species in finite control volume of the nanofluids, the nanofluid density was obtained from the relation:

$$\rho_{nf} = \varphi \rho_{np} + (1 - \varphi) \tag{4}$$

By applying the principle of calorimetry in the mixture the overall specific heat of the nanofluid was calculated from the relation:

$$c_{nf} = \frac{\varphi(\rho c)_{np} + (1-\varphi)(\rho c)_b}{\varphi(\rho)_{np} + (1-\varphi)(\rho)_b} \tag{5}$$

Maxwell model of equivalent thermal conductivity as studied by Yu et al. [14] was used for the estimation of the thermal conductivity and the Einstein model of viscosity as mentioned by Mishra et al. [15] were used to estimate the viscosity of the nanofluid.

$$\mu_{nf} = \mu_b(1 + 2.5\emptyset) \tag{6}$$

$$k_{nf} = k_b \frac{k_{np} + 2k_b + 2\varphi(k_{np} - k_b)}{k_{np} + 2k_b - \varphi(k_{np} - k_b)} \tag{7}$$

2.4. Boundary Conditions

The inlet, outlet and wall boundary conditions are as shown in Figure 2. Top, bottom and periodic side walls are assumed to be no slip boundary condition. Sidewall is kept at periodic boundary condition and bottom and top wall is kept at constant temperature boundary condition. Inlet set as velocity inlet boundary condition corresponding to the Re range from 800 to 2000 to assure the flow is laminar.

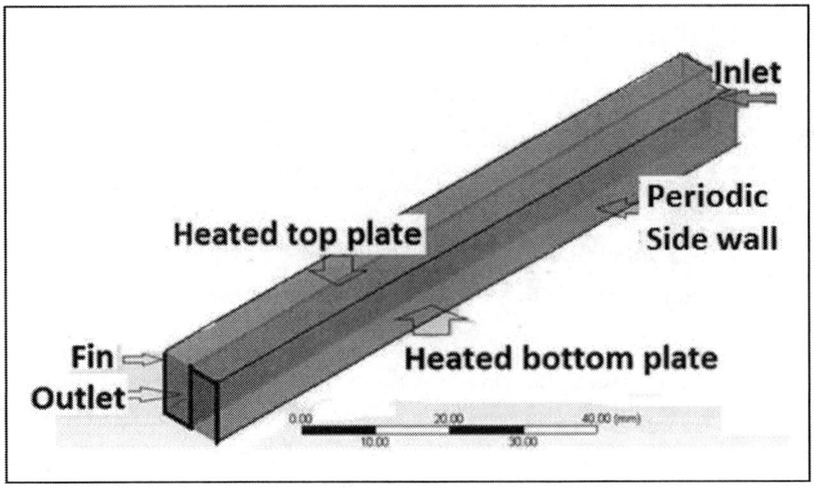

Figure 2. Geometry of the Model and Boundary Conditions

2.5. *Meshing and Model Setup*

The mesh of the computational domain was generated using a triangular patch conforming method. This mesh contains hexahedral cells having rectangular faces at the boundaries. The generated mesh consisted of 233460 elements and 245063 nodes.

A 3-D, incompressible and pressure-based solver was chosen for the computational domain. A laminar viscous model is used for numerical simulation. A 2nd order upwind interpolation formula was used for discretization of momentum and energy equation. The conventional SIMPLE (Semi-Implicit Method for Pressure-Linked Equations) algorithm was used to solve the pressure velocity coupled equations, where several iterations were performed to ensure convergence of the numerical solution was assured by monitoring the scaled residuals to a constant level of 10^{-6} for each variable.

2.6. *Grid Independency*

The solution obtained in this work is mesh independent as we have studied the mesh independency test. The result is shown in Figure 3.

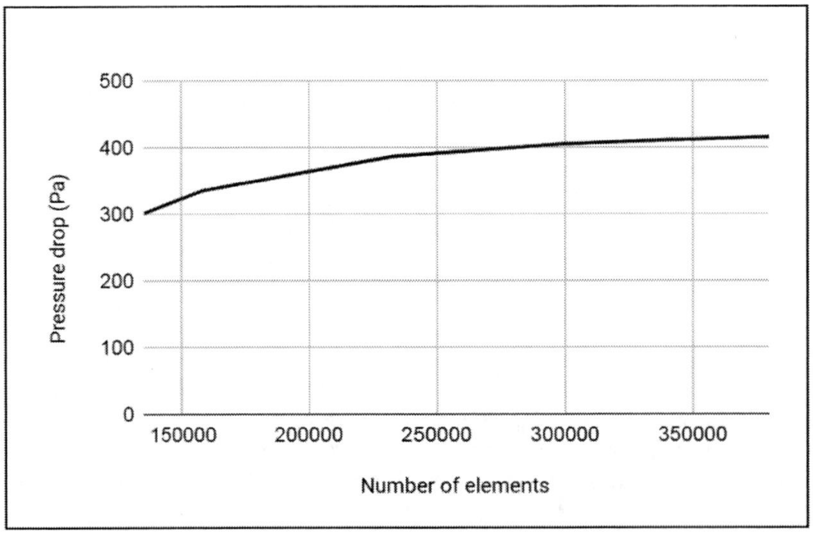

Figure 3. Grid Independence Test

IOP Conf. Series: Materials Science and Engineering 998 (2020) 012001 doi:10.1088/1757-899X/998/1/012001

2.7. Experimental Validation of the Model

Validation is done to check the authenticity of the CFD tool. We have validated the model by comparing the results of the experiment done by Kays and London [4] as shown in Figure 4.

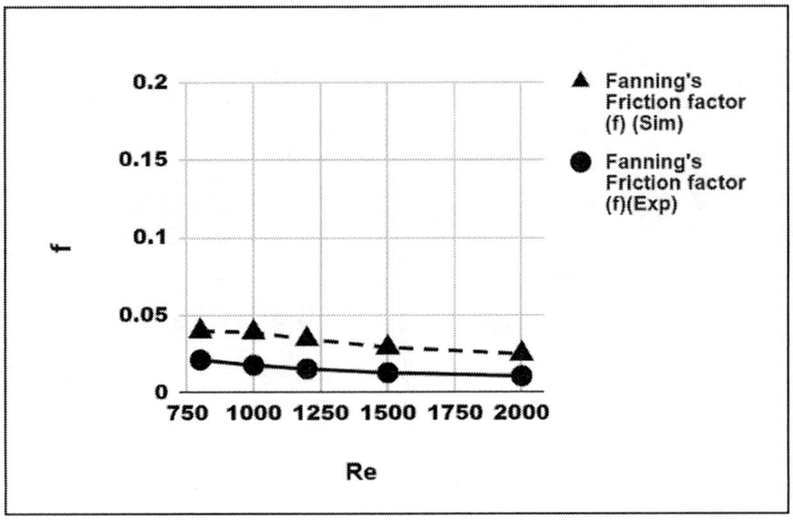

Figure 4. Validation of simulation result with the experiment of Kays and London [4]

3. Results and Discussion

This is clear from the solution of the numerical analysis as shown in Figure 5 and Figure 6 that with the increase of Re in the laminar region from 800 to 2000 friction factor and j factor decreases. These variation of f and j with Re is similar in both cryogenic nanofluids and pure liquid He. Friction factor is a non-dimensional term used to calculate the pressure drop in the PFHE. On the other hand, Colburn j factor is a non-dimensional term used to estimate the heat transfer coefficient and hence the effectiveness of the PFHE.

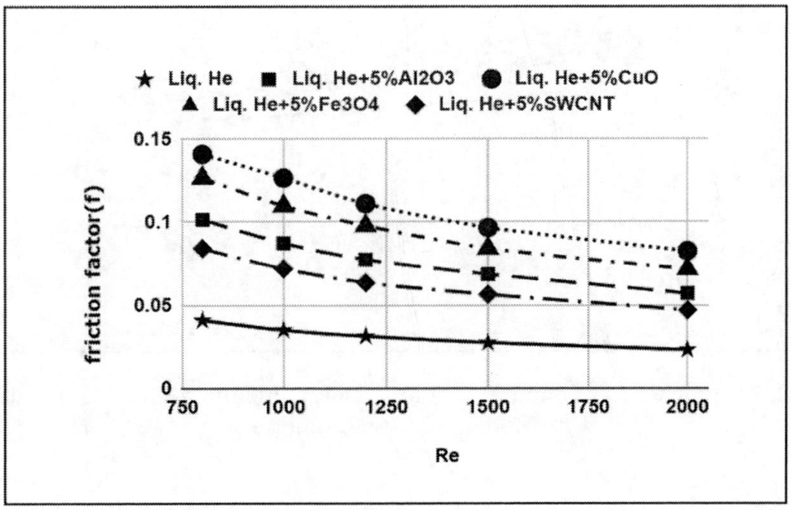

Figure 5. Comparison of friction factors among different nano-particles with liquid He

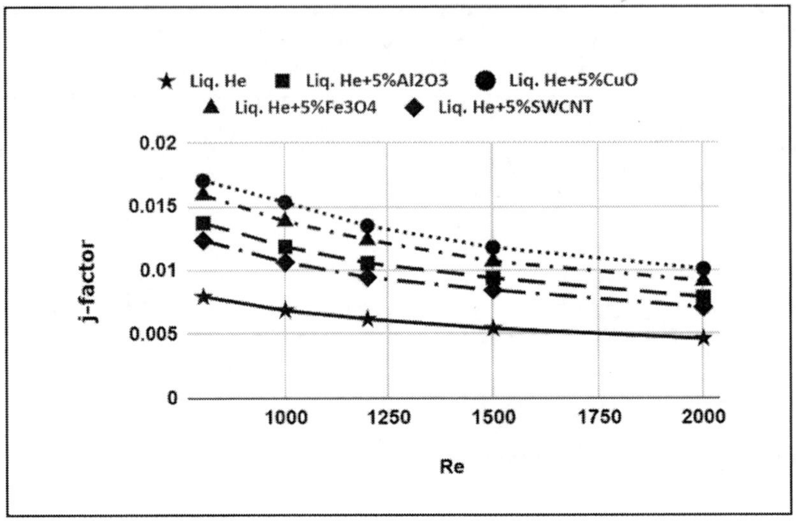

Figure 6. Comparison of Colburn j factor among different nano-particles with liquid He

It is observed in Figure 6 that CuO nanoparticles are very effective in increasing the convective heat transfer coefficient in terms of Colburn j-factor in comparison to Al_2O_3, Fe_3O_4 and SWCNT. But pressure drop in terms of Fanning's friction factor is also high in case of CuO-liquid He cryo-nanofluid in comparison to other cryo-nanofluids as shown in Figure 5. But this is quite clear from the Figure 5 and Figure 6 that in comparison to liquid He only, He based cryo-nanofluids are more effective in terms of convective heat transfer.

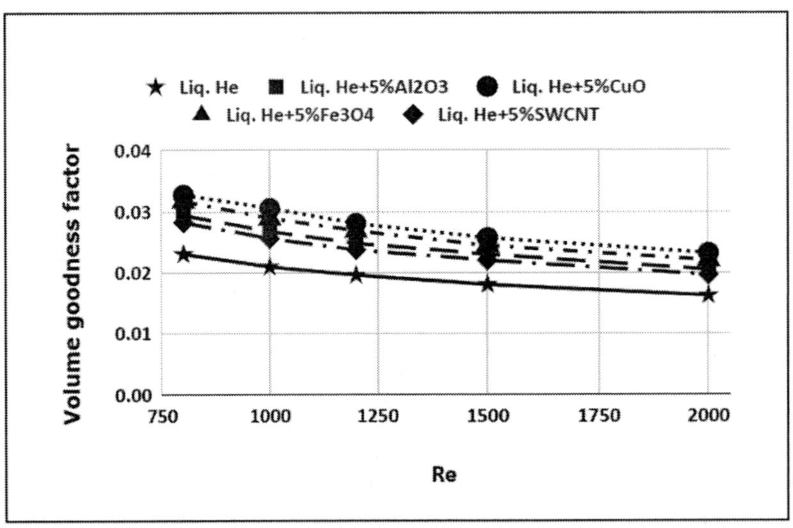

Figure 7. Comparison of volume goodness factor among different among different nano-particles with liquid He

This is very important to conclude regarding the performance of the PFHE keeping in mind both the heat exchange and pressure drop in the device. That is why volume goodness factor, defined as $j/f^{1/3}$, is compared among the different cryo-nanofluids as shown in Figure 7. It is observed that CuO nanoparticles are giving better overall performance in terms of volume goodness factor in comparison to other cryo-nanofluids. Overall performance enhancement in He based cryo-nanofluids is around 25% to 30% in comparison to only liquid He, is clearly observed in Figure 7.

4. Conclusion

In this research work performance of PFHE with one fluid as He based cryo-nanofluid is compared with respect to liquid He at laminar range. It can be concluded from the results that CuO -liq. He based cryo-nanofluid having best performance among the other cryo-nanofluids studied based on volume goodness factor. Around 25% to 30% enhancement in the overall performance considering both heat exchange and pressure drop has been found when cryo-nanofluids are used in comparison to liquid He only. But in this research certain aspects like cost of nano-particles, preparation complexity and cost of cryo-nanofluid, phase change stability of nanofluids have not been considered in evaluation of overall performance of the PFHE.

References

[1] Choi S U S & Eastman J A 1995 Enhancing thermal conductivity of fluids with nanoparticles (No. ANL/MSD/CP-84938; CONF-951135-29) *Argonne National Laboatory* IL (United States)

[2] Redhwan A A M, Azmi W H, Sharif M Z and Mamat R 2016 Development of nanorefrigerants for various types of refrigerant based: A comprehensive review on performance *International Communications in Heat and Mass Transfer* **76** 285-93

[3] Lee S, Choi S U S, Li S A and Eastman J A Measuring thermal conductivity of fluids containing oxide nanoparticles 1999 *Journal of Heat Transfer* **121** 280-89

[4] Kays W M and London A L 1984 *Compact Heat Exchangers*

[5] Norris R and Spofford W A Jul High performance fins for heat transfer 1942 *Trans. ASME* **64** 489-97

[6] Shah R K and London A L Offset rectangular plate- fin surface-heat transfer and flow friction characteristics 1968 *ASME Journal of Engineering for Power* **90** 218-28

[7] Wieting A R Empirical correlations for heat transfer and flow friction characteristics of rectangular offset-fin plate-fin heat exchangers 1975 *Trans. ASME* 488-90

[8] Robertson J M 1979 Boiling heat transfer with liquid nitrogen in brazed-aluminium plate-fin heat exchangers *National Heat Transfer Conference AIChE Symposium Series 189* **75** 151-64

[9] Cao L, Liu J, Li R, Huang S, Zhang F and Xu X Experimental study on the mixed refrigerant heat transfer performance in a plate-fin heat exchanger during a single-stage cryogenic cycle 2016 Jan 25 *Applied Thermal Engineering* **93** 1074-90

[10] Yang Y, Li Y, Si B and Zheng J Heat transfer performances of cryogenic fluids in offset strip fin-channels considering the effect of fin efficiency 2017 Nov 1 *International Journal of Heat and Mass Transfer* **114** 1114-25

[11] Doohan R S, Kush P K and Maheshwari G Exergy based optimization and experimental evaluation of plate fin heat exchanger 2016 Jun 5 *Applied Thermal Engineering* **102** 80-90

[12] Goyal M, Chakravarty A and Atrey M D Experimental investigations and validation of two dimensional model for multistream plate fin heat exchangers 2017 Mar 1 *Cryogenics* **82** 83-91

[13] Jiang Q, Zhuang M, Zhang Q, Zhu Z, Geng M, Sheng L and Zhu P Experimental study on the thermal hydraulic performance of plate-fin heat exchangers for cryogenic applications 2018 Apr 1 *Cryogenics* **91** 58-67

[14] Yu W and Choi S U The role of interfacial layers in the enhanced thermal conductivity of nanofluids: a renovated Maxwell model 2003 Apr 1 *Journal of nanoparticle research* **5** 167-71

[15] Mishra P C, Mukherjee S and Nayak S K, Panda A A brief review on viscosity of nanofluids 2014 Dec 1 *International nano letters* **4** 109-20

IOP Conf. Series: Materials Science and Engineering **998** (2020) 012002 doi:10.1088/1757-899X/998/1/012002

Thin-walled compound composite cylinders for improved specific energy absorption

V Gattineni[1] and V Nathi[2]

[1] Research Scholar, GITAM School of Technology, Hyderabad, India

[2] Director training and placement, GITAM School of Technology, Hyderabad, India

E-mail: ammu.abhi.sammi@gmail.com

Abstract. Thin tubular structures are widely used as crush cans in the automobiles to absorb energy sacrificially. As the world looks forward to greater fuel efficiency, the adoption of lightweight materials like composites becomes imperative for vehicular structural part designs. The crush cans made of composite exhibit better specific energy absorption characteristics during an impact compared to their metallic counterparts. The paper discusses how the specific energy absorption capacity of thin-walled tubes can be further enhanced using the concept of compound cylinders which induces a pre-stress in the crush can. It further compares the specific energy absorption capacity of a composite tube against metallic designs.

1. Introduction

The Automotive technology across the globe has been undergoing massive changes and the trend is likely to continue for the future. One of the notable changes has been the migration from the traditional IC engine based powertrain towards electric mobility. Whatever be the power source, the need to ensure the vehicle is of low weight for guaranteed fuel efficiency without compromising on the overall safety of the driver and passengers has been constant [1]. Composites besides being light in weight have good specific energy characteristics to absorb energy as sacrificial parts during collisions and accidents. Vehicle body made of composite materials such as CFRP has good strength to weight ratio. Additionally, it can act as a good shield against EMC hazards in electric vehicles besides being a very good contributor to eliminate or absorb acoustic noises.

Pre-stressing concepts are widely adopted by the civil engineering community in designs of varieties of structures to achieve a good overall size to strength ratio [2]. The pre-stressing of the structure is achieved using either pre-tensioning or post-tensioning techniques. In the case of pre-stress induction using post-tensioning, the component or structure is subjected to a tensioning load, after the making or manufacturing of the same. For the case of the crush can, pre-stress can be inducted into the tube after the manufacture of the tube by the application of constant internal pressure and sealing it [3]. The induction of pre-stress to improve energy absorption capacity requires additional equipment and arrangements like non-return valves [3] and/or gas generator/inflator [4]. Also, a pressurized crush can lead to catastrophic issues [4] with regards to safety around the vicinity of the vehicle but can improve energy absorption capacity drastically.

IOP Conf. Series: Materials Science and Engineering **998** (2020) 012002 doi:10.1088/1757-899X/998/1/012002

In this study, a safer method of pre-stressing which is based on the pre-tensioning technique is proposed and studied. In this approach, the pre-tensioning is introduced into the thin circular tubes during the process of manufacturing. To achieve this, instead of being a single circular tube, the crush can be made as a compound tube which is achieved by the interference of two thin-walled tubes.

2. Solid Mechanics of thin walled cylinders

Theoretically, a thin cylinder or tube is one which has thickness "t" to radius "r" ratio is less than 1/20. When such a tube is subjected to an internal pressure "P", the tube expands radially outside. When such expansion happens, the longitudinal stress is given by (P*r)/(8*t) [5]. This longitudinal stress creates an opposing force which resists the compression of the tube and thereby increases its energy absorption capacity [4].

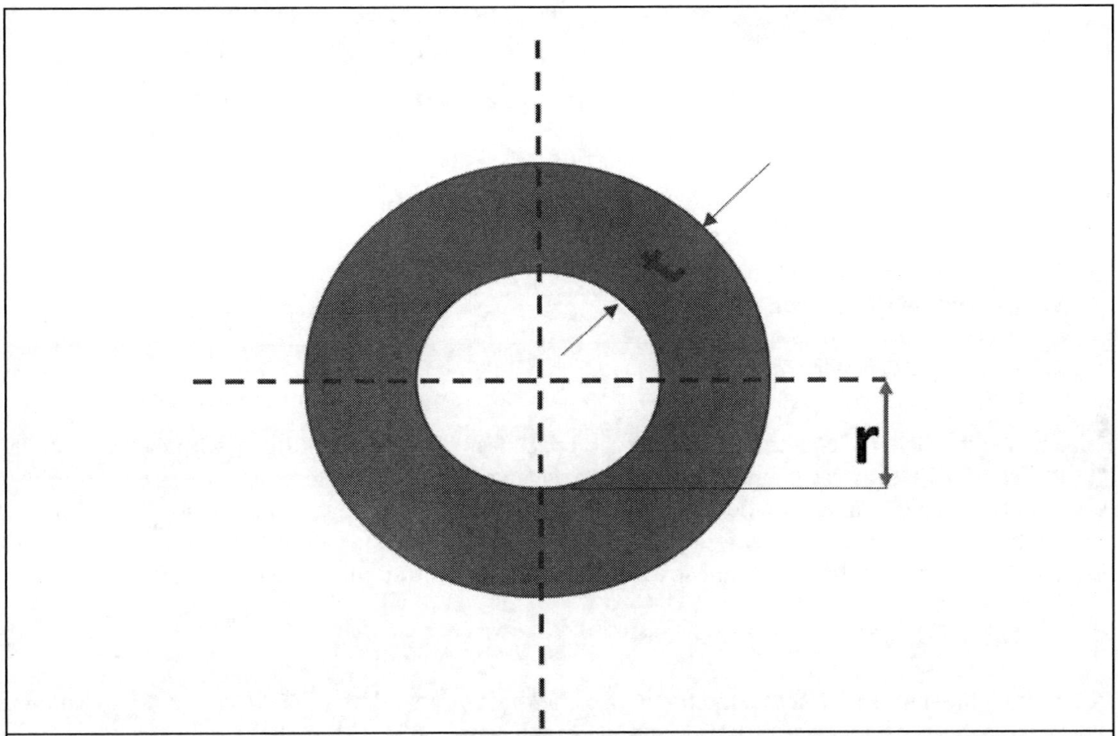

Figure 1: Schematic cross-sectional view of thin tube

2.1. *Compound Cylinders Tubes*

A compound tube is one which is formed by shrink fitting two tubes. In the case of compound cylinders, the outer radius of the inner tube shall be more than the inner radius of the outer tube. During the process of shrink fit, the outer tube exerts a compressive force on the inner tube and the inner tube exerts a tensile force on the outer tube. As a result, the combined resultant tube shall have an outer radius which shall be a little higher than the outer radius of the outer tube [5]. This means the shrink fit gives rise to a radial tension in the compound tube which is analogous to the application of internal pressure. This internal pressure also gives rise to longitudinal stress which in turn produces an opposing force that resists the tube compression [4].

IOP Conf. Series: Materials Science and Engineering **998** (2020) 012002 doi:10.1088/1757-899X/998/1/012002

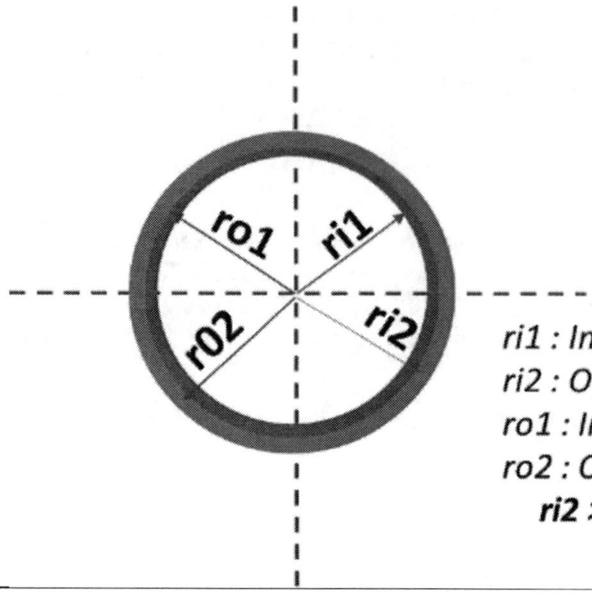

Figure 2: Schematic conceptual sketch of compound tube

3. Crushing of thin-walled tubes

3.1 Compound Cylinders Tubes

For a thin tube of length "L", Internal radius "r" and thickness "t' made of a material with Young's modulus "E", the compressive displacement for a force "P" is given as (P*L)/(E*A). Here "A" is the area of the cross-section of the tube. This relation holds good if the tube does not buckle under the load "P". For metallic tubes, the failure criteria to be considered is based on Von-Mises equivalent stress and the behavior is expected to be linear if the equivalent stress is below the yield strength of the material [5].

3.2 Tubular compression for composites

A thin tube of length "L", Internal radius "r" and thickness "t' made of a CFRP (Carbon fiber reinforced Polymer) material exhibits anisotropic behavior. The thickness "t" for the tube is achieved by making it with several layers. With the assumption that the fiber angle is oriented along the longitudinal axis of the tube, wherein the directional Young's Modulus is E1, the compressive displacement for a force "P" is given as (P*L)/(E1*A). Here "A" is the area of the cross-section of the tube. This relation also holds good if the tube does not buckle under the load "P". For CFRP tubes, the failure criteria to be considered can be based on Maximum principal stress as the material is expected to be brittle and the load is along the direction of the fiber. Here also the behavior is expected to be linear if the maximum principal stress is below the compressive strength of the material. If the same is equivalent or more than the compressive strength of the CFPR material, the damage gets initiated in the material. Post damage initiation, further application of additional load leads to complete fracture is called damage evolution [6].

3.3 Parameters of significance for energy absorbers

As can be seen from literature, thin circular tubes are mostly used as energy absorbers during an impact or accident for a vehicular frontal structure. To study the effectiveness of a component or design as an energy absorber and compare multiple designs made of multiple materials, there a host of multiple parameters [7]. But the most important parameter that is used to compare the designs made of different materials is the specific energy absorption capacity. This is defined as the energy absorbed by the tube for the unit weight for the same level of compression [7]. The typical force Vs displacement diagram for a thin tube used as energy absorber during an impact by compression is shown in figure 3. The product of force with displacement(Compression) during the impact divided by the weight of the tube indicates its specific energy absorption capacity.

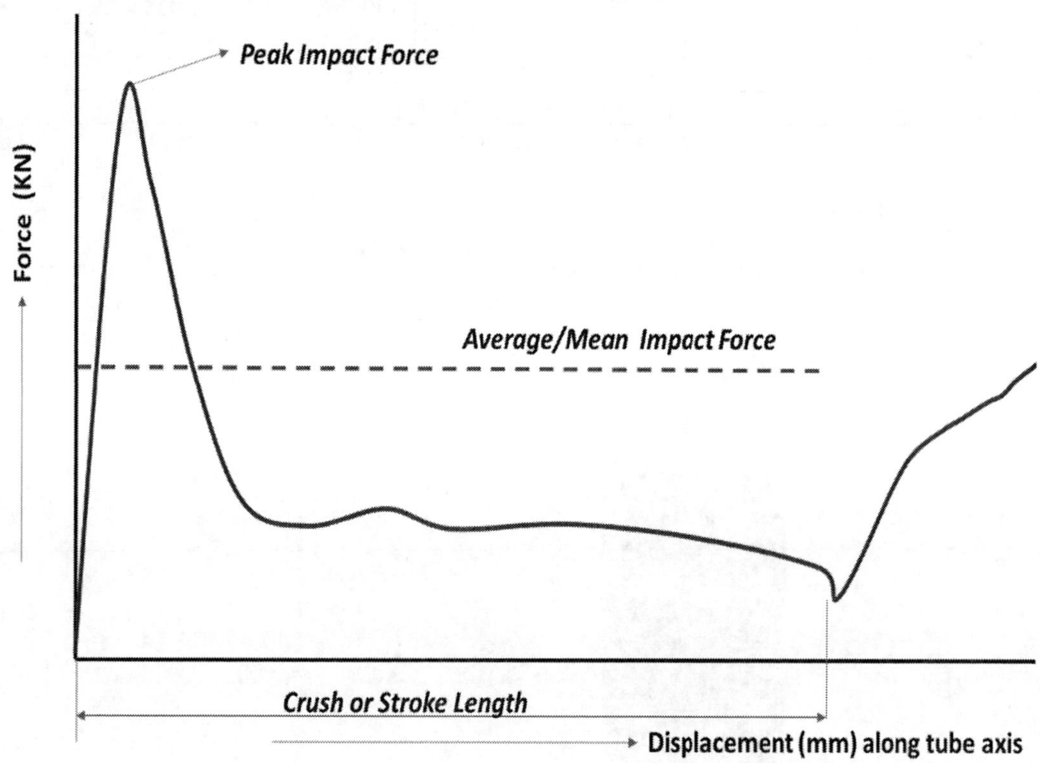

Figure 3: Typical force-displacement curve for an impact energy absorber

4. Study approach

4.1 Study methodology

As part of the study, a CFRP tube is first compressed by a small distance until the damage gets initiated in the tube and the corresponding force required is observed. and the specific energy absorption capacity is computed using finite element software Ansys [8]. Then the same amount of displacement is applied for metallic tubes of similar geometry and the specific energy absorption capacities are computed. The computed values are compared against the theoretical values [5].

The same studies are repeated for compound tubes and the specific energy absorption capacity of the compound tubes is compared against those of compound tubes. The shrink-fit process of the compound tubes (pre-tensioning) is simulated using the interference simulation approach based on finite element methods [8].

4.2 Material properties used

For the study, the material properties as listed in Table -1 were used.

Table: 1 Material properties used			
Material	Young's Modulus (MPa)		Poisson Ratio
Steel [7]	E	210000	0.3000
Aluminium [7]	E	70000	0.2800
CFRP [9]	E1	58903	0.0154
	E2	58903	0.5356
	E3	8759	0.1575
Failure Strengths of CFRP [9]			
Allowable Compressive Strength in 1/2/3 Direction(s): 555/555/500 (MPa)			

4.3 Geometric model used for metals

Figure 4 shows the geometric configuration used for the thin metal tubes. The tube has an inner radius of 25 mm and is 100 mm in length. The thickness of the tube is 3 mm.

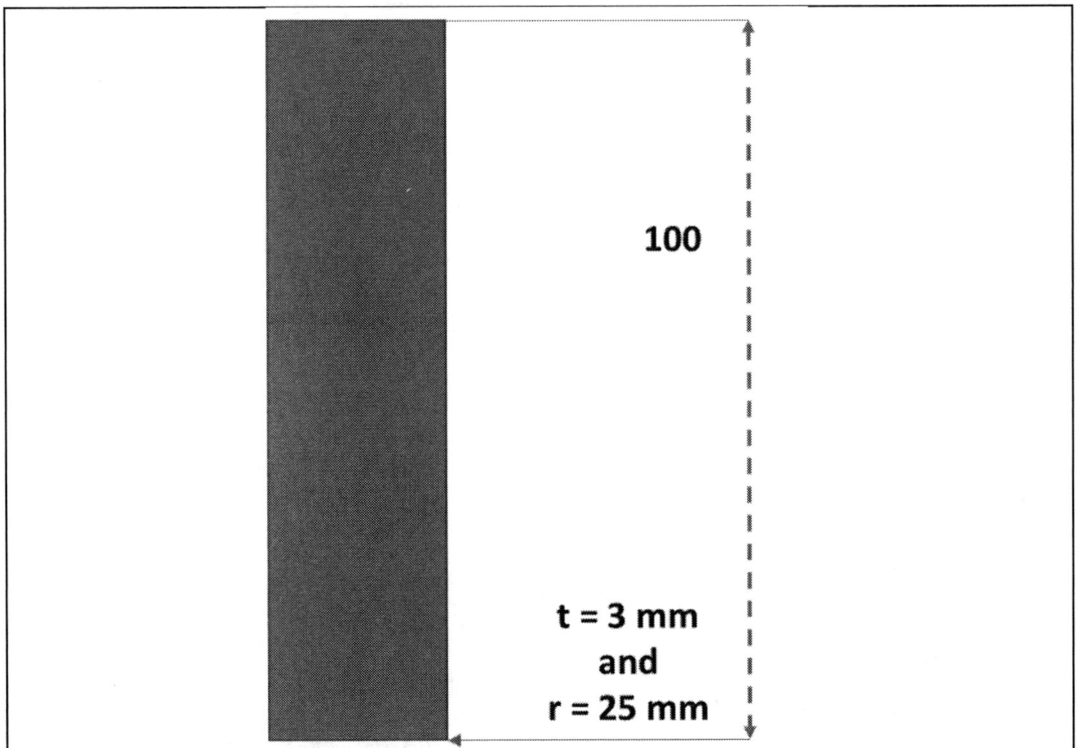

Figure 4 : Schematic cross sectional configuration of metal tube(s) used for the study

4.4 Geometric model used for composite(CFRP) tube

To achieve the trigger configuration a chamfer angle of 45o is provided [11 and 12]at the top of the tube. The normal length of the tube is 98.5 mm and the trigger has been accommodated over a length of 1.5 mm. The configuration of trigger (With trigger angle mentioned as q, Internal diameter as Di, overall thickness as t and the overall length as L) is shown in Figure 6.

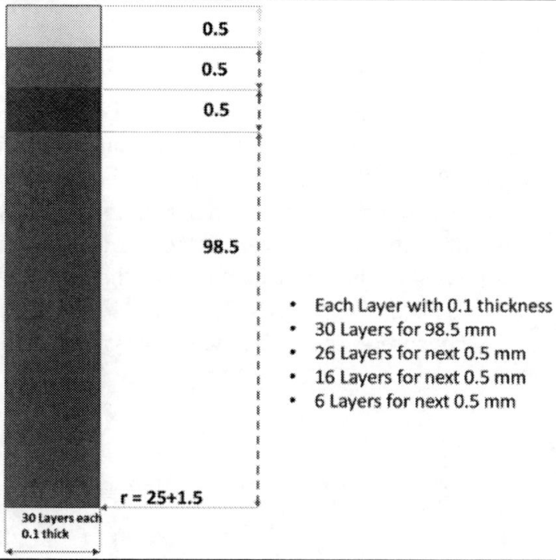

Figure 5 : Schematic cross sectional configuration of CFRP tube used for the study

4.5 Trigger configuration for composite tube

To achieve the trigger configuration a chamfer angle of 45o is provided [11 and 12]at the top of the tube. The normal length of the tube is 98.5 mm and the trigger has been accommodated over a length of 1.5 mm. The configuration of trigger (With trigger angle mentioned as θ, Internal diameter as Di, overall thickness as t and the overall length as L) is shown in Figure 6.

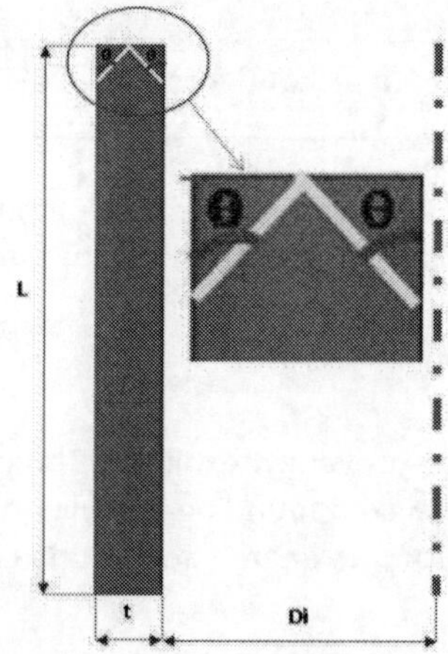

Figure 6: Schematic cross-sectional view of double bevel trigger

4.6 Interference in tubes

Since CFRP material has a very low percentage of elongation for failure, very high interference can lead to failure during shrink-fit [5]. Hence two thin CFRP tubes whose thickness is exactly half of the single CFRP tube shown in figure 7 have been chosen for the study with interference equal to 1 layer thickness(0.1mm).

4.7 Interference simulation

As described in earlier sections, two thin circular tubes each of thickness 1.5 mm (Each of 15 layers) and each with a 45o trigger whose combined thickness is equal to that of the single tube are taken. It has been assumed that the inner diameter of the inner tube is equal to that of the inner diameter of the single CFR tube with a 3-mm thickness. A contact pair is defined between the outer surface of the inner tube and the inner surface of the outer tube. The outer diameter of the inner tube has been assumed to be higher by 0.1 mm (1 layer) compared to the inner diameter of the outer tube. Solving the contact pair using the finite element method [8] calculates the stresses at the interface of the inner and outer tubes and the overall dimensions of the compound tube, with pre-stress induced. Figure 7 shows the configuration of the tubes used.

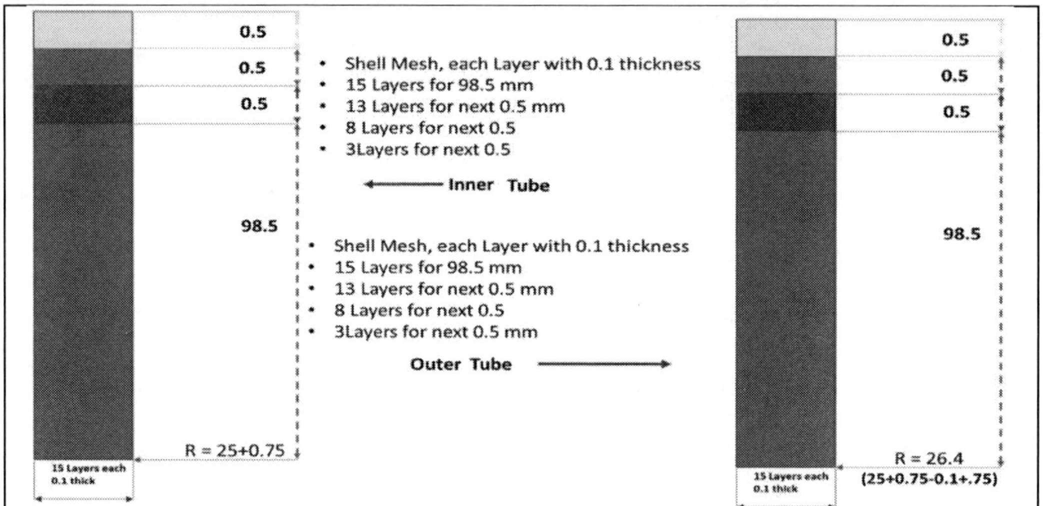

Figure 7: Schematic cross sectional configuration of CFRP tubes used for interference

In the case of metals, the geometric configuration used for compounding is the same as that of CFRP. But there is no trigger configuration provided for the metals as the same is not required for metals [7].

5. Specific energy absorption calculations

As per literature, composite tubes are proven to exhibit [13] more specific energy absorption capacity compared to metallic tubes. Also, the compounding of cylinders/tubes is expected to improve their strength. Hence the computation of specific energy absorption capacity is performed for the single and compound tubes to verify the same.

5.1 Specific energy absorption calculations for normal tube(s)

For the calculation of specific energy absorption, first, the CFRP tube with a thickness of 3 mm is initially compressed downward by an arbitrary distance of 0.3 mm. For this compression, no failure was observed when the damage contours were observed. For about a deflection of more than 0.45 mm, it was observed that the damage gets initiated in the CRFP tube with a trigger provided at the top edge. Hence a displacement slightly more than 0.45 mm and which is equal to 0.5 mm is applied and the CFRP tube is compressed downward.

Ansys software [14] which has been successfully calibrated for experimental correlation for composite and metallic material models to study damage initiation/evolution and yield/plastic behavior has been used for the study. The reaction force corresponding to compression of 0.5 mm in the respective tubes has been computed using Ansys software and tabulated.

For verification of Ansys predicted values, the values of force required for tubes made of respective materials have been computed using theory [5] and the results have been compared and tabulated as in Table-2. The finite element model used for metal tube and for CFRP with a double bevel trigger are shown in figures 8(a) and 8(b) respectively.

Figure 8(a): Mesh model of metal tube

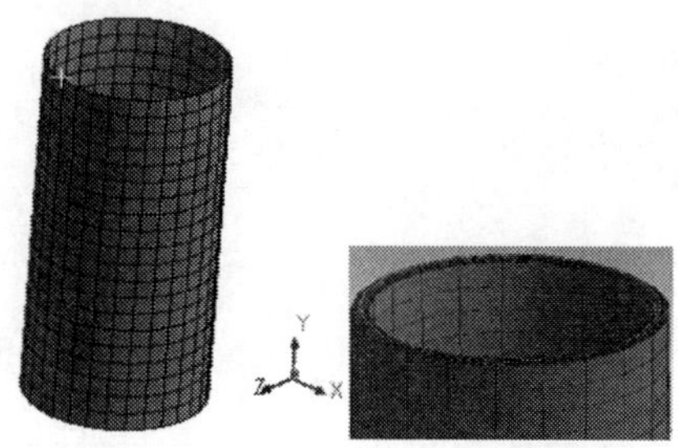

Figure 8(b) Mesh model of CFRP tube along with zoomed view of double bevel trigger

Table 2: Specific energy absorption capacity of single tube					
S. No	Material	Density of material (Kg/mm3)	Reaction force from FEM study (N)	Reaction force Calculated theoretically(N)	Specific energy absorption KJ/Kg
1	Steel	7.8 E-09	4.7192E05	4.98E05	6.31E03
2	Aluminum	2.8E-09	1.6765E05	1.66E05	6.55E03
3	CFRP	1.45E-09	1.3651E05	1.39E05	9.93E03

5.2 Specific energy absorption calculations for the compound tube

The specific energy absorption capacity for the compound tube made up of respective materials is studied using finite element software Ansys using 2 load cases. In the first load case, the inner and outer tubes are modeled with interference. A contact surface is created between the two interfering surfaces and the model is solved to predict interference stresses and hence the pre-stress. In the next load step, a compression of 0.5 mm is applied for the compound tube and the reaction force required to achieve this compression is calculated. Based on the compressive force required, the value of specific energy absorption is computed for the respective composite tubes. The results are tabulated in Table-3 for comparison. The finite element models of the compound tube (For metal and CFRP) are shown in figures 9(a) and 9(b).

Table 3: Specific energy absorption capacity for compound tube				
S. No	Material	Density of material (Kg/mm3)	Reaction force from FEM study (N)	Specific energy absorption KJ/Kg
1	Steel	7.8 E-09	4.96E05	6.63E03
2	Aluminum	2.8E-09	1.76E05	6.87E03
3	CFRP	1.45E-09	1.40E05	10.2E03

IOP Conf. Series: Materials Science and Engineering **998** (2020) 012002 doi:10.1088/1757-899X/998/1/012002

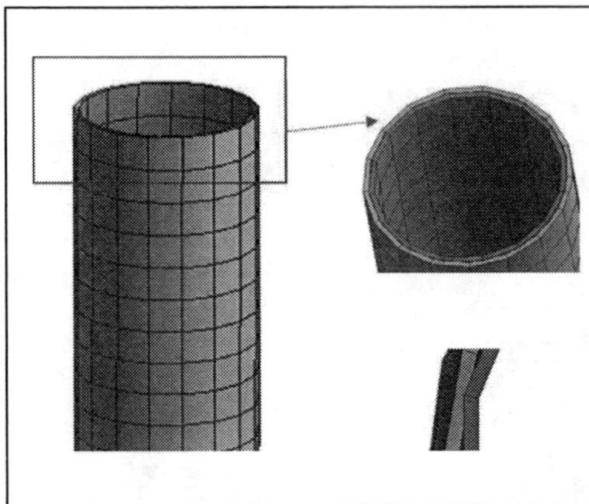

Figure 9(a): Mesh model compound metal tube

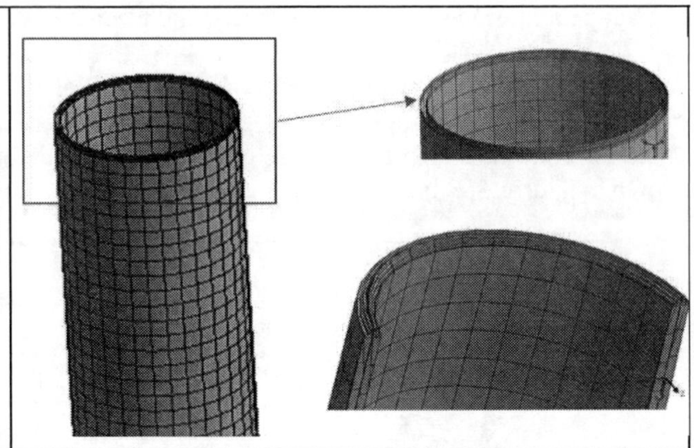

Figure 9(b) Mesh model of compound CFRP tube along with trigger at top

6. Discussion on results

For the case of single tube compression, the theoretically calculated values compare well with the values obtained through finite element calculations.

As can be seen from the results (Table-1), the composite tubes have a more specific energy absorption capacity compared to the metallic counterparts. CFRP has the highest specific energy absorption capacity followed by Aluminum. Steel owing to its high density has poor specific energy absorption capacity. Since the density of CFRP is almost half to that of Aluminum, CFRP tubes have higher specific energy absorption capacity making it suitable for lightweight vehicles. The stress results for the steel and CFRP tube for the normal tube configuration are shown in figures 10(a) and 10(b) respectively. As can be seen from the pictures, Steel being isotropic, the compression results in uniform stress along the length. For the CFRP, since the trigger forms a stress concentration, the peak stress is seen at the top (trigger vicinity).

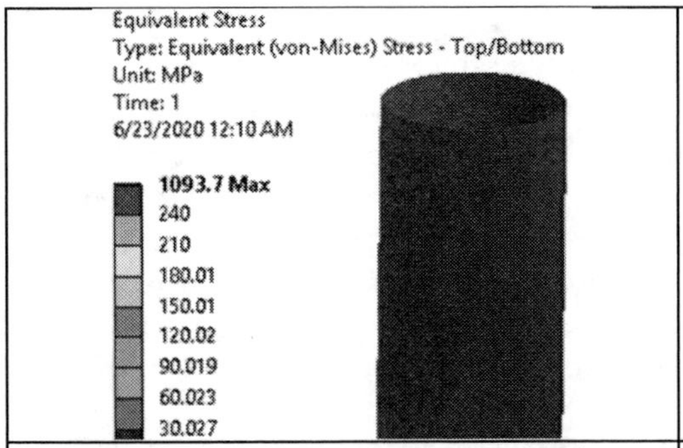

Figure 10(a): Von-mises stress for the normal tube under compression

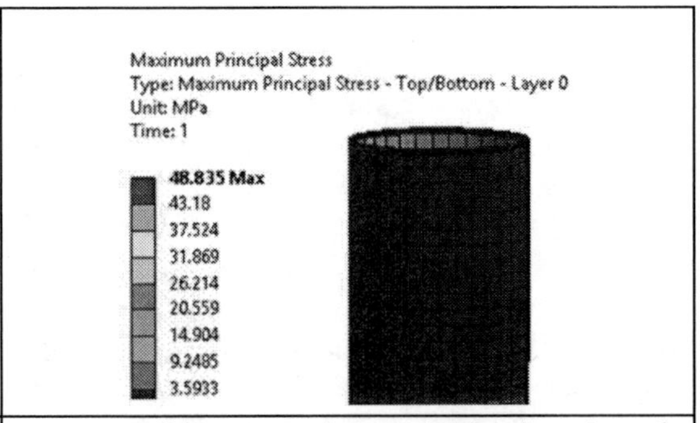

Figure 10(b): Principal stress in the direction of compression for CFRP tube

A comparison of the results between Table 2 and Table 3 reveals that compound tubes absorb more energy than their normal counterpart due to the introduction of pre-stress in the form of interference in the tubes. The percentage of improvement of specific energy between normal thin tubes and those with pre-stress are tabulated and shown in Table 4.

Table 4: Improvement in specific energy absorption capacity for compound tube				
S. No	Material	Specific energy absorption J/Kg for normal tube	Specific energy absorption J/Kg for compound tube	Percentage increase in specific energy absorption due to compounding
1	Steel	6.31E03	6.63E06	5%
2	Aluminum	6.55E03	6.87E06	5%
3	CFRP	9.93E03	10.2e06	3%

Due to the presence of trigger the improvement of compound CFRP tube is seen to be a smaller value. But when a full-pledged dynamic impact analysis is performed, the specific energy absorption capacity could be higher for CFRP tubes. Due to the nature of compounding, contact stresses are developed at the interface of the tubes that are compounded. The contact stress distribution for the steel and CFRP tubes at their respective interfaces is shown in figures 11(a) and 11(b) respectively. As can be seen from the results, steel being stronger than CFRP, the contact stress is more for steel tube than that of CFRP tube.

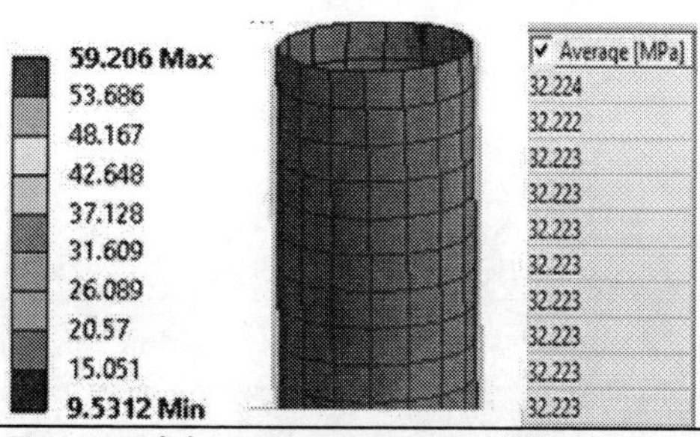

Figure 11(a): Contact stress at interface for steel metal compound tube

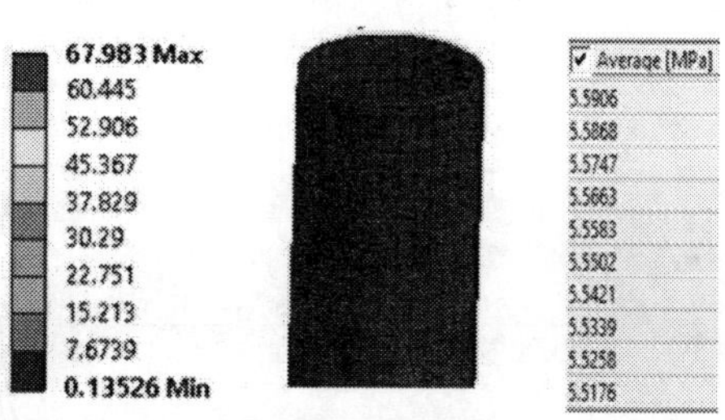

Figure 11(b): Contact stress at interface for CFRP compound tube

7. Conclusions

- A FEM based computation approach for the study of the energy absorption capacity of composite tubes subjected to pre-stress has been developed and presented.
- In place of single thin-walled tubes made of metal and CFRP material, two thin-walled tubes with half the thickness and assembled as a shrink-fit with interference have been subjected to compression to verify the feasibility of the compound cylinder concept.
- Pre-tension gets developed in the compound cylinders that shall oppose the compressive force.
- As can be seen from the results, a shrink-fit compound tube develops a pre-stress that shall improve the specific energy absorption capacity.
- Simulation based on finite element method can be effectively used to compute induced pre-strains and assess the amount of pre-strain that can be induced without damaging the tube
- The approach proposed for FEM based simulations demonstrates that thin-walled tubes can absorb more energy when made as compound tubes
- The concept proposed here does not require any special and additional equipment to induce pre-stress in the tubes and hence very advantageous compared to pre-stress induction through internal pressure
- The crash accelerations to be impacted by the people in the vehicle can be reduced by the pre-tension technique, wherein the safety rating of the vehicle can be improved without adding additional weight.

8. Future Scope

- The Current study was made for tubes made of a single material (Either metal or CFRP) and with the interference of one layer. It could be possible to achieve more interference by using a combination of materials such as FML (Fiber metal laminates) and/or MMC (Multiple metallic composites).
- The concept of pre-stress through interference and shrink fit in the form of compound cylinders has been currently performed considering only damage initiation for the CFRP material to verify the concept. A full pledged dynamic impact simulation incorporating damage evolution modeling is also suggested for the future study.

References

[1] https://www.asirt.org/safe-travel/road-safety-facts/

[2] Veerat Srilaxmi, K. Manju, M Vijaya: A case study on pre-tensioning & post tensioning systems of a pre-stressed concrete, Int. J. Eng. Technol. Manag. Res., ISSN: 2454–1907, DOI: 10.5281/zenodo/1198970

[3] T.Hou, B.G.Prusty, G.Pearce, D.Kelly and R. Thomson: Experimental investigation on energy absorbing pressurised composite tubes, The 19th International Conference on Composite Materials

[4] Venkateswarlu Gattineni, Venukumar Nathi, Pre-stressed thin tubular composite energy absorbers for improved impact energy absorption,Results in Engineering, Volume 5,2020,100102, ISSN 2590 1230, https://doi.org/10.1016/j.rineng.2020.100102.

[5] R.S. Khurmi, N. Khurmi: Strength of Materials (Mechanics of Solids) Revised Edition S Chand publishers (2018) 978-81-219-2822-2. Copyright©1968

[6] Hashin Z: Failure criteria for unidirectional fiber composites ASME J. Appl. Mech., 47 (2) (1980), pp. 329-334

[7] T. J. Reddy, Y. V. D. Rao and V. Narayanamurthy: Thin-walled structural configurations for enhanced crashworthiness, International journal of crashworthiness, 2017

[8] Ansys Theory Manual, Version 19.0

[9] Haibo Luo, Ying Yan, Xiangji Meng Can Jin: progressive failure analysis and energy-absorbing experiment of composite tubes under axial dynamic impact Composites Part B, 87 (2016)

[10] Christopher P.R. Hoppel, Steven J. De Teresa: Effect of angle-ply orientation on compression strength of composite laminates UCRL-JC-131920, 15th U.S. Army Symposium on Solid Mechanics Myrtle Beach, SC April 12-14 (1999)

[11] Deepak Siromani, Gary Henderson, Doug Mikita, Kevin Mirarchi, Ryan Park, John Smolko, Jonathan Awerbuch, Tein-Min Tan : An experimental study on the effect of failure trigger mechanisms on the energy absorption capability of CFRP tubes under axial compression, Composites: Part A 64 (2014) 25–35,http://dx.doi.org/10.1016/j.compositesa.2014.04.019

[12] Yan Ma, Shanshan Jin, Songlin Zhang: Effect of trigger on crashworthiness of unidirectional carbon fiber reinforced polyamide 6 composites, Plastics, Rubber and Composites (2018), 10.1080/14658011.2018.1466502

[13] C.T. Sun, B.J. Quinn, J. Tao, D.W. Oplinger: Report Comparative Evaluation of Failure Analysis Methods for Composite Laminates, Report Number DOT/FAA/AR-95/109, U.S. Department of Transportation Federal Aviation Administration

[14] Ever J. Barbero, Mehdi Shahbazi, Determination of material properties for ANSYS progressive damage analysis of laminated composites Compos. Struct., 176 (2017), pp. 768-779

IOP Conf. Series: Materials Science and Engineering 998 (2020) 012003 doi:10.1088/1757-899X/998/1/012003

Optimization and Experimentation of Modified Throttling Calorimeter

Sagar G. Mushan[1, a], Narayan. K. Sane[2,b], Hrushikesh Kulkarni[3,c]

[1]Assistant Professor of Mechanical Engineering, MIT Academy of Engineering, Pune, Maharashtra, India

[2]Professor of Mechanical Engineering, Walchand College of Engineering, Sangli, Maharashtra, India

[3]Project Manager, R&D, Forbes Marshall Pvt. Ltd., Pune, Maharashtra, India

[a]sgmushan@mitaoe.ac.in, [b]narayan_sane@yahoo.co.in, [c]hkulkarni@forbesmarshall.com

Abstract. Steam plays a vital role in the industrial revolution and is a major source of power generation in steam engines or steam turbine. Thus, it becomes necessary to maintain steam as dry as possible and it is necessary to monitor steam quality on a regular basis. The most commonly used method in determining steam quality is Throttling calorimeter. Here existing throttling calorimeter is designed for determining steam quality within a pressure range of 20 – 25 bar. This existing calorimeter is optimized for monitoring steam quality within a range of 4 – 12 bar gauge pressure. Pipe and orifice dimension of the calorimeter is calculated for a given flow rate of steam by using ISO norms and Bernoulli's principle. After conducting multiple trials on newly designed model test rig it is found that the 4 mm orifice diameter model gives the greater consistency of results and it is considered to be an optimized model.

Keywords: Calorimeter, Dryness fraction, Flow rate, Optimize, Steam quality.

1. Introduction

Steam is the gas formed when water changes its state from the liquid to the gaseous. In steam-using industries, steam is commonly referred to as "Dry or Wet steam." Steam plays a vital role in the industrial revolution and is the major source for power generation in steam engines or steam turbine. In case of a thermal power plant, if available steam is wet then it results in corrosion and erosion of turbine blades. Thus, it becomes necessary to maintain steam quality high enough and it should be monitored on a regular basis. Hence, steam quality plays a vital role in determining the final product quality and its outcome.

The method used in monitoring steam quality is Tank calorimeter, Separating calorimeter, Throttling calorimeter, Separating-Throttling calorimeter and Electric calorimeter. Based on the availability of inline steam quality, above calorimeters are used for determining dryness fraction of steam.

If available steam is almost dry, we make use of a throttling calorimeter. Here a sample of steam is extracted from the main steam line and passed into the calorimeter through an orifice of known size. Steam throttles at the orifice and finally expands in the expansion chamber of a calorimeter. The overall expansion process of steam takes place adiabatically.

Modified throttling calorimeter is nothing but the modification of existing calorimeter which was designed for monitoring steam quality at 25 bar pressure. Here, existing calorimeter is modified by optimizing overall design of existing calorimeter to some lower size so that it can able to monitor steam quality up to 12 bar pressure. Also, existing model is modified without using a baffle.

1.1 Literature Review

Long et al. [1] described the method for determining the steam quality used in steam flooding for the secondary recovery of petroleum. Here, the convention throttling calorimeter is modified based on available thermodynamic properties of the steam. The process of automatic continuous operation along with heat addition was also described.

Chien [2] described the method of utilizing the pressure at the entry of a nozzle and the flow rate through a nozzle in calculating the quality of steam. The condition required is critical flow through a nozzle and discharge from a nozzle is to be condensed. Amount of condensed steam is measured and using this value steam quality is determined by utilizing standard formula.

Cheung et al. [3] described in detail about the method and apparatus for determining the mass flow rate and the quality of inline steam. In this case, the total flow of steam is directed towards the liquid-vapour separator where some amount of moisture is separated and measured. The flow of one of the stream is regulated to hold the level of the separated liquid constant. Thus, by obtaining the pressure and temperature of the separated flow upstream, the mass flow rate and the steam quality is determined.

Hayes [4] described in detail about how to calculate inline steam quality by using separating calorimeter. He stated that the separating calorimeter helps in removing the moisture content present in the steam that further helps in improving steam quality before it throttles. Here for moisture separation cyclone separator is utilized.

Salunke [5] described the modified throttling calorimeter which helped in improving the performance and obtaining the correct quality of steam. In this case, the overall length of existing throttling calorimeter was reduced and modified by introducing a baffle which helps throttled steam to flow in U-path. The detailed information related to the general principle for methods of measurement and computation of the flow rate of fluid flowing in a conduit by employing pressure differential devices when they are inserted into a circular cross-section conduit running full was obtained through ISO 5167-1 [6].

1.2 Existing throttling calorimeter details

Existing throttling calorimeter shown in fig 1. features baffle plate. Since the throttled steam flow in U-path, the total length of steam travel is increased for the steady flow purpose (velocity profile will get stabilized.) On one side of the baffle, there is a jet of steam with high velocity after an orifice. After traveling some distance it flows to the other side of the baffle with lesser velocity where we can measure steady temperature, pressure.

The overall dimensions of the existing throttling calorimeter model as shown in the Fig. 1:

1) Nominal Diameter of the Cylinder body:
    a) Inner diameter = 154.051mm.
    b) Outer diameter = 168.275 mm.
2) Inlet pipe Nominal Diameter:
    a) Inner diameter = 15.8mm.
    b) Outer diameter = 21.34mm.
3) Outlet pipe Nominal Diameter:
    a) Inner diameter = 40.89mm.
    b) Outer diameter = 48.26mm

IOP Conf. Series: Materials Science and Engineering 998 (2020) 012003 doi:10.1088/1757-899X/998/1/012003

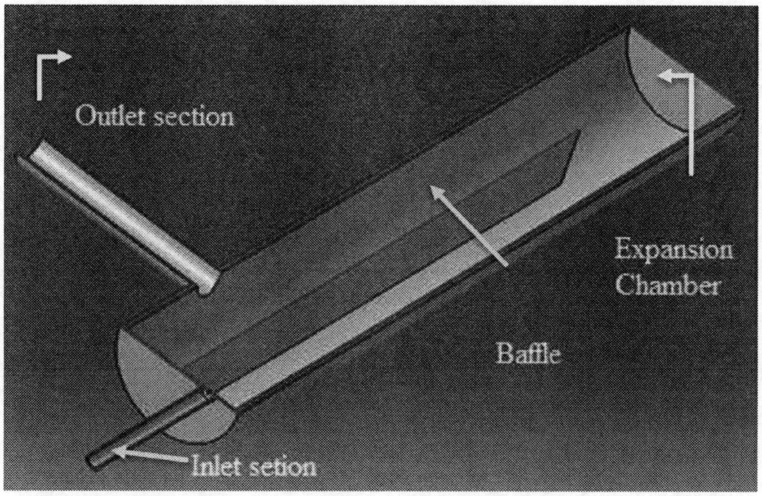

Fig 1. Existing Throttling Calorimeter Sectional View

2. Optimization of Existing Model

2.1 Design of Orifice:

Throttling Calorimeter consists of orifice plate whose purpose is to throttle inline steam. Hence a proper dimension of an orifice is calculated by applying Bernoulli's principle at the inlet section and orifice

$$\frac{P_1}{\rho} + \frac{V_1^2}{2} + gz_1 = \frac{P_2}{\rho} + \frac{V_2^2}{2} + gz_2 \qquad \text{.......................................} (1)$$

Solving above equation, it results to

$$Q_v = \frac{C \times \varepsilon \times \sqrt{\left(\frac{2\Delta p}{\rho}\right)} \times \frac{\pi D_2^2}{4}}{\sqrt{1-\beta^4}}$$

$$Q_m = Q_v \times \rho = \frac{C \times \varepsilon \times \sqrt{\left(\frac{2\Delta p}{\rho}\right)} \times \frac{\pi D_2^2}{4}}{\sqrt{1-\beta^4}} \qquad \text{.......................................} (2)$$

Where,

Q_m = Mass flow rate in kg/hr.
C_d = Discharge Coefficient.

β = Ratio of orifice diameter and internal diameter of inline steam pipe.

D_2 = orifice diameter.

ΔP = Pressure difference.

$\in$ = expansion factor $=1-(0.41+0.35\beta^4)\times\frac{\Delta p}{p\times k}$

Given condition

Mass flow rate (Q_m) = 80 kg/hr
Coefficient of Discharge (C_d) = 0.61
Inlet Pressure (P_1) = 1300000 N/m^2
Outlet Pressure (P_2) = 100000 N/m^2
Density at inlet pressure (ρ) = 6.62028 kg/m^3
Diameter of pipe (D) = 0.0158 m

Substituting above values in equation (1) and solving, value of D_2 = 4.05 mm.
Thus, orifice diameter for a given mass flow rate and pressure is 4mm and Diameter ratio (β) is 0.2563.

2.2 Calculation of pipe diameter:

For a given orifice diameter, pipe diameter is calculated by using ISO norms. As per ISO 5167 -2 formula for determining pipe diameter is [5]:

$$m_s = 2.73 \times C \times \left(\frac{d}{4.654}\right)^2 \times \left[1 - \frac{\frac{P_1 - P_2}{P_1}}{3 \times F_Y \times X_T}\right] \times \sqrt{(P_1 - P_2) \times \rho} \quad \text{......................} (3)$$

Where,

P_1= Primary pressure (kPa abs) = 1300 kPa abs P_2= Secondary pressure (kPa abs) = 100 kPa abs d = Diameter of Orifice (mm) = 4 mm

C = Discharge coefficient = 1

F_Y = Specific heat ratio factor = 1.4363

ρ = Density of steam= 6.62028 kg/

X_T= Pressure differential ratio factor = 0.72

By substituting above given values in equation (3), it results to

m_s = 126.252 kg/h = 0.03507 kg/s

Since, Specific value of steam(ϑ) = 0.151039 m^3/kg and Steam velocity (V) = 25 m/sec

Therefore, Area of pipe = $\frac{m_s \times \vartheta}{V} = \frac{(0.03507 \times 0.151039)}{25} = 2.11906 \times 10^{-4}$

Thus, diameter of pipe = $\sqrt{\frac{4 \times 2.11906 \times 10^{-4}}{\pi}}$ = 16.4 mm

Hence, nearest standard pipe dimension of 15.8mm i.e. 0.5 inch pipe size is selected.

3. CAD model of modified throttling calorimeter

Referring to the above calculation, the dimensions of orifice and pipe are

i. Orifice diameter = 4 mm.

ii. Pipe diameter = 16.4 mm.

The nearest standard pipe dimension of 15.8 mm i.e. 0.5 inch pipe size (15NB) is selected for modified calorimeter pipe. The CAD model as per above calculated dimension is shown in fig. 2 and fig. 3

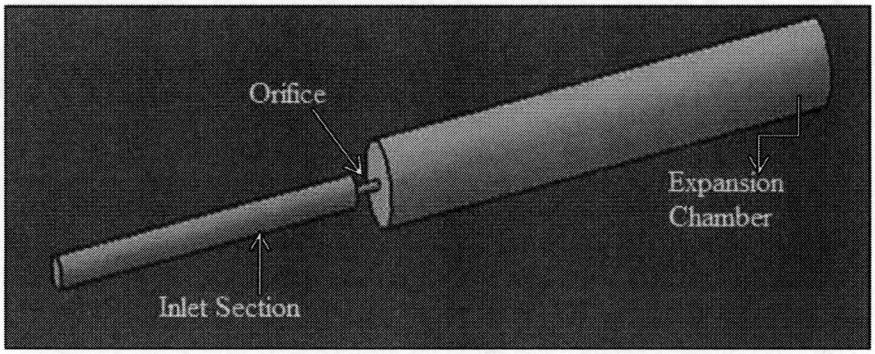

Fig 2. Isometric view of modified throttling calorimeter.

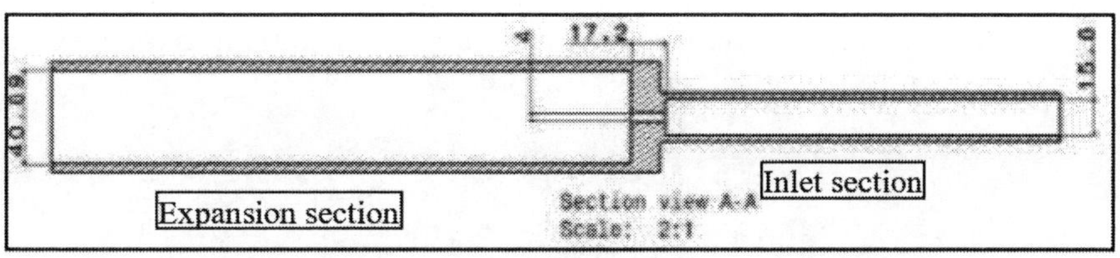

Fig 3. Sectional view of modified calorimeter.

4. Working mechanism and Experimental results

4.1 Operating procedure and layout of modified throttling calorimeter

Step 1: Steam is extracted from the main steam line and is passed to inlet section pipe.

Step 2: Steam is then passed through PRV (Pressure Reducing Valve) and ball valve (BV).

Step 3: After opening the ball valve completely, steam passes through orifice plate where it throttles to atmospheric pressure. Steady state will be observed within 5 to 10 minutes of experiment startup.

Step 4: Note down the pressure and temperature before and after throttling through pressure gauge (PG) and temperature transmitter (TT). By using the steam table, properties of steam are calculated and steam quality is determined.

Working layout shown in fig. 4 describes experimental setup of modified throttling calorimeter. Components included in experimental setup are Pressure Relief Valve (PRV), Ball valve (BV), Pressure gauge (PG 1 & PG 2) and Temperature transmitter (TT 1 & TT 2).

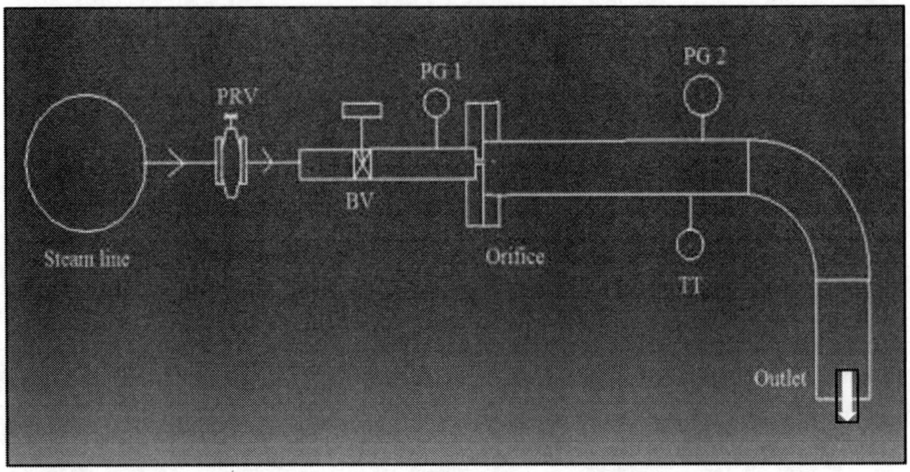

Fig 4. Layout of Modified throttling calorimeter setup.

4.2 Calculation of dryness fraction value

The amount of dryness of steam is determined by equating the enthalpy values i.e. enthalpy of extracted steam at known line pressure and enthalpy of expanded steam in the calorimeter.
After noting down the pressure before and after throttling and temperature after throttling we can able to calculate the quality of steam by using isenthalpic principle as shown below

$$h_{1v} = h_2$$

$$h_f + x \times h_{fg} = h_{sup}$$

$$x = \frac{h_{sup} - h_f}{h_{fg}} \qquad \text{...} \quad (4)$$

4.3 Experimental Results

4.3.1 Dryness fraction of steam calculated with respect to available inline pressure

Results listed in table 1 show the dryness fraction of steam obtained at different available inline steam pressure. Here, different inline pressures are obtained by regulating through pressure relief valve. Thus, through results, it is found that the value of dryness fraction is found to be within the accuracy of 0.5%.

Table 1. Results of dryness fraction value of steam with respect to available inline steam pressure.

P_1(bar)	P_2(bar)	T(°C)	H_{fb} (kJ/kg)	H_{fg}(b (kJ/kg)	H_2 (kJ/kg)	X
4.947	0.997	100.5	638.683	2109.49	2676.84	0.966
5.947	1.047	101.6	669.237	2087.1	2678.45	0.963
6.947	1.097	103.6	696.051	2066.98	2681.94	0.961
7.947	1.097	106.6	720.054	2048.6	2688.16	0.961
8.947	1.147	108.6	741.86	2031.57	2691.68	0.960
9.947	1.197	112	761.895	2015.65	2698.11	0.960
10.947	1.247	113.8	780.471	2000.64	2701.26	0.960

4.3.2 Dryness fraction of steam calculated with respect to boiler pressure

Results listed in table 2 show the dryness fraction of steam obtained with respect to available boiler pressure. Thus, through results, it is found that the value of dryness fraction is found to be within the accuracy of 0.3%.

Table 2. Results of dryness fraction value of steam with respect to boiler pressure.

P_b(bar)	P_2(bar)	T(°C)	H_{fb} (kJ/kg)	$H_{fg(b)}$ (kJ/kg)	H_2 (kJ/kg)	X_b
7.847	0.997	100.5	717.518	2049.99	2676.84	0.955
8.447	1.047	101.6	730.96	2039.55	2678.45	0.955
9.947	1.097	103.6	761.664	2015.25	2681.94	0.953
10.397	1.097	106.6	770.192	2008.39	2688.16	0.955
11.597	1.147	108.6	791.66	1990.91	2691.68	0.954
12.147	1.247	113.8	800.591	1983.24	2701.26	0.958

5. Conclusions

The following conclusions are drawn from the present work:

1. The expansion chamber diameter of calorimeter is reduced from 154.08 mm to 15.8 mm for range of 4 - 12 bar gauge pressure. Also, existing model is modified without using a baffle.
2. Value of dryness fraction obtained with respect to the available inline pressure was to be within the accuracy of 0.5%.
3. Value of dryness fraction obtained with respect to the boiler pressure was to be within the accuracy of 0.3%.
4. Thus from above experimental readings, it is found that the newly designed throttling calorimeter is an optimized and modified model for determining steam quality within the range of 4 - 12 bar gauge pressure.

References

[1] Long and Cheung: "Modified throttling calorimeter", United States Patent no. 4561785, Dec. 31, 1985.

[2] Sze-Foo Chien: "Method and apparatus for determining steam quality by measuring condensate rate of steam sample flowing through critical flow nozzle", United States Patent no. 5214956, Jun. 1, 1993.

[3] Yin L. Cheung, Sugarland; Alfred Brown, Houston, both of Tex., "Method and apparatus for determining mass flow rate and quality in a steam line", United States Patent no. 4,688,418, Aug. 25, 1987

[4] James K Hays, "Steam quality measurement using Separating calorimeter", United States Patent No. 4909067, Mar 20, 1990

[5] Sumeet Salunkhe: "Performance improvement of throttling calorimeter", M.Tech Report, 2017

[6] "Measurement of fluid flow by means of pressure differential devices inserted in circular-cross section conduits running full Part 1 – General Principles and Requirements", ISO 5167-1:2003.

[7] "Measurement of fluid flow by means of pressure differential devices inserted in circular-cross section conduits running full Part 2 – Orifice Plate", ISO 5167-2:2003.

IOP Conf. Series: Materials Science and Engineering **998** (2020) 012004 doi:10.1088/1757-899X/998/1/012004

EXPERIMENTAL INVESTIGATION ON GEO-POLYMER CONCRETE BY USING STEEL SLAG AT HIGHER TEMPERATURE

J Sanghamitra[1*], P Ramakanta[2], B Sushree Ananya[3]

[1*]Assistant Profesor, Department of Civil Engineering, VSSUT, Burla, 768018, Odisha, India
[2]Associate Professor, Department of Civil Engineering, VSSUT, Burla, 768018, Odisha, India
[3]M.Tech Dual Degree student, Department of Civil Engineering, VSSUT, Burla, 768018, Odisha, India,

Email: sanghamitrajena1987@gmail.com

Abstract. Concrete is the most durable, versatile and reliable construction material after water. Its requirement is very large so the need of ordinary portland cement is also high. Due to such large production of OPC our environment is getting polluted as carbon dioxide is emitted. Geo-polymer is the alternative to control the pollution and its an innovative construction material. It can be prepared by using fly ash, the secondary product of thermal coal and steel slag which is the by product of the production of steel in an electric arc. Another thing required to prepare the geopolymer concrete (GPC) is curing temperature. In this study mechanical properties of GPC is examined with curing temperature 70^{0}C with a replacement of steel slag as coarse aggregate as 0%, 10%, 20%, 30% and 40%. Here the strength of GPC is increasing up to 30% replacement of steel slag as coarse aggregate.

Keywords: Geopolymer concrete, Compressive Strength, Rebound Number

1. Introduction

Green house gases are the main factor for the increasing global warming and also pollution of our environment. About 65% of global warming is due to emission of CO_2 which is also a green house gas. Our cement industry contributes round six percent of carbon dioxide emission to environment. The impact of cement industry has an adverse effect on the environment. In such situation it is necessary to think an alternative to traditional concrete and geo-polymer concrete is such an innovative way so that emission of CO_2 can be controlled. GPC can be prepared by mixing activated industrial waste materials like fly-ash, coal ash and alkaline solution. Fly ash (FA) is employed as binder and almost every research on GPC has concluded that ash is the appropriate alternative to traditional method. But the only difficulty to prepare the GPC is it requires external temperature for curing. The optimum temperature for thermal curing of FAbased GPC is oven curing at 75^{0}C for 18-24 hr [1]. During preparation addition of water can improve workability around 200% and reduces the rest GPC properties up to 27% and optimum molarity of NaOH solution is 16M [2]. Here steel slag is used as aggregate i.e with partial replacement of coarse aggregate as 10, 20, 30 40%. The compressive strength of GPC with steel slag substitution was higher compared to GPC with gravel aggregate [3].

2. Experimental Procedure

2.1 Materials

Firstly the binder for GPC mixture is class F FA confirming to IS: 3812(2003) Part-I [4] which is main source of aluminosilicates. It contains CaO less than 10% and sourced from Hindalco Industries Ltd, Burla, Odisha. It has nearly 100% particles finer than 90 microns. Ground granulated blast furnace slag (GGBS) was collected from ACC cement, Raipur. Then steel slag which is used as aggregate of 10mm and 20 mm. It was collected from Bhushan Power & Steel Ltd, Jharsuguda, Odisha. Similarly natural coarse aggregate of 10mm and 20mm are used. As fine aggregate natural sand is used. Mixture of NaOH solution and Na_2SiO_3 solution is used as alkaline activator and ratio of Na_2SiO_3 to NaOH is 2.5 [5]. Chemical compositions of FA and GGBS are given in Table 1 obtained by X-ray fluorescence (XRF) analysis. Table 2 shows the properties of materials.

IOP Conf. Series: Materials Science and Engineering **998** (2020) 012004 doi:10.1088/1757-899X/998/1/012004

Table 1 Chemical compositions of Fly ash and FS were obtained by XRF

Contents	SiO_2	Al_2O_3	Fe_2O_3	CaO	MgO	Na_2O	K_2O	Cr_2O_3	SO_3	P_2O_5	MnO
FA	53.5	21.7	6.9	3.42	0.82	0.24	0.7	-	1.1	0.015	0.12
GGBS	32.36	13.4	1.3	42.9	3.7	0.21	0.32	-	3.76	0.02	-

Fig 1. Bulk Image of Fly Ash

Fig 2. NCA

Fig 3. Steel Slag

Table 2 Properties of Materials

Materials	Specific Gravity	Impact Value	Zone	Water Absorption	Los Angeles Abrasion Value
NCA	2.85	15.7%	-	0.8%	19.6%
Fly Ash	2.45	-	-	-	-
Sand	2.38	-	II	1.25%	-
Steel Slag	3.25	25%	-	1.75%	26.5%

2.2 Specimen Preparation

First 14 M Sodium hydroxide solution has to be prepared by dissolving (14 x 40) i.e 560 g of Sodium hydroxide pellets in 1L water as molecular weight of NaOH is 40gm [5]. Then according to the ration 2.5 times Sodium silicate solution was added to it. It should be prepared a day before. Alkaline solution to binder ratio was considered as 0.5. On the other hand, all the solids were mixed together up-to 3 minutes after quantified and then alkaline solution was added to it which was mixed for 4minutes. Then GPC mixture was casted and compacted to moulds of (100 ×100 × 100) mm. After completing the previous step moulds were put in oven for curing at 70^0C for 24hours. Table 3 shows the mix design proportion and slump values.

Table 3 Mix design proportion

Mix	NaOH.+ Na_2SiO_3 (kg/m^3)	Fly Ash (kg/m^3)	Steel Slag (kg/m^3)	NCA (kg/m^3)	Sand (kg)	Slump values (mm)
S-10	250	500	133.33	1200	720	74
S-20	250	500	267.67	1067.67	720	68
S-30	250	500	400	933.33	720	60
S-40	250	500	533.33	800	720	55

3. Result & Discussion

3.1 Compressive Strength

For different ratio of GPC mixtures samples were kept for 7, 14 and 28 days & after that samples were tested to check the compressive strength. Fig. 4 shows the compressive strength variation with respect to different percentages of steel slag.

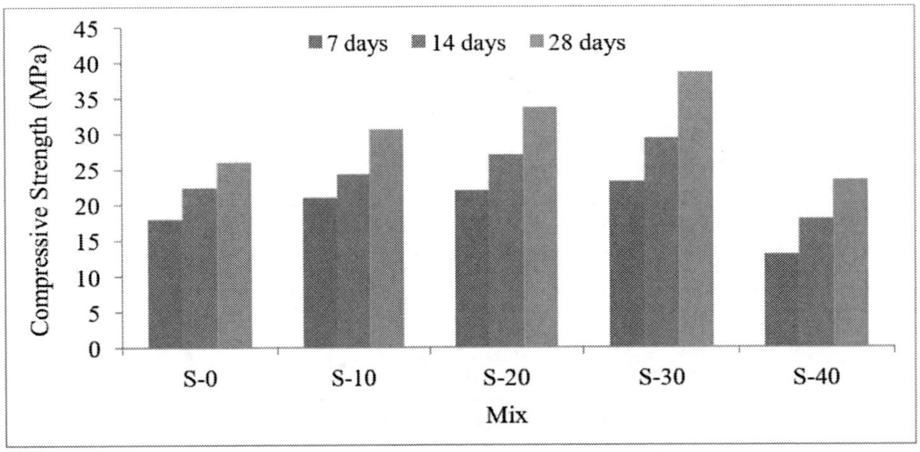

Fig. 4 Compressive strength variation

Development of high volume fly ash based GPC can show extremely good mechanical properties with higher compressive strength and longer durability [6]. Also Steel slag substitution as an aggregate can increase the strength of GPC up to 46% than natural coarse aggregate [7]. From the above result it is noted that the strength is increasing upto 30% replacement of steel slag as coarse aggregate and then it decreases.

3.2 Rebound Number

This non destructive test was carried out through the moulds of size 100 × 100 × 100 mm. The test is done to check the homogeneous microstructure of the GPC sample. Table 2 shows the rebound number values of different GPC mixes.

Table 2 Rebound number

Mix	Rebound No.
S-0	29
S-10	31
S-20	35
S-30	40
S-40	28

4. Conclusion

The study deals with the detailed investigation on partial replacement of steel slag as natural coarse aggregate which was cured at 70^0C. So it is concluded that,

- Partial use of steel slag as coarse aggregate gave high compressive strength.
- Replacement of steel slag as coarse aggregate improves 46% the strength up to 30% and accordingly rebound number also increases.

Future work,

- All Mechanical properties can be studied by using steel slag with the replacement of natural coarse aggregate.
- Durability properties can be studied with steel slag.

5. References

[1] Noushini A and Castel A 2016 *Construction and Building Materials* vol. 112 pp 464–477.

[2] Aliabdo A., Elmoaty M. A and Salem H A 2016 *Construction and Building Materials* vol. 121 pp 694-703.

[3] Ashadi H W, Aprilando B A and Astutiningsih S 2015 *International Journal of Technology* vol 2 pp 227-235.

[4] IS 3812-1 (2003): Specification for pulverized fuel ash, part 1: for use as pozzolana in cement, cement mortar and concrete

[5] Jena S and Panigrahi R 2019 *Construction and Building Materials* vol. 220 pp 525–537.

[6] Hardjito Djwantoro et. Al, On the development of fly ash based Geo-polymer concrete, ACI Materials Journal, 101 -M52, 2004

IOP Conf. Series: Materials Science and Engineering **998** (2020) 012004 doi:10.1088/1757-899X/998/1/012004

[7] Omar O M 2015 *International Journal of Civil, Environmental, Structural, Construction and Architectural Engineeri ng* Vol 9 pp 508-516.

[8] IS 13311 (1992): Indian Standard non-destructive testing of concrete Method of test: Part 2: Reboundhammer

[9] IS: 2386, Indian Standard Specification, Methods of Test for Aggregates for Concrete: Part 3,SpecificGravity, Densi ty, Voids, Absorption and Bulking, Bureau of Indian Standards, New Delhi, 1963 [Reaffirmed in 2002].

[10] IS: 2386, Indian Standard Specification, Methods of Test for Aggregates for Concrete: Part 4, Mechanical Properti es Bureau of Indian Standards, New Delhi, 1963 [Reaffirmed in 2002].

[11] IS: 383, Indian Standard Specification for coarse and fine aggregate from natural sources, Bureau of'Indian Stand ards,New Delhi,1970

IOP Conf. Series: Materials Science and Engineering 998 (2020) 012005 doi:10.1088/1757-899X/998/1/012005

Matrix Modification of Carbon Fiber Reinforced Composites with Hybrid Resin System

C.VenkateshwarReddy,[1]K.Roopa,[2]P.RameshBabu,[3]R.RamNarayanan [4]
[1]Department of Mechanical Engineering, Matrusri Engineering College, Hyderabad, India
[2]Department of Civil Engineering, M.V.S.R Engineering College, Nadergul, Hyderabad, India
[3]Department of Mechanical Engineering, UCE, Osmania University, Hyderabad, India
[4]Advanced Systems Laboratory, CPDC, Hyderabad, India
E-mail: [1]cvreddy36@gmail.com

Abstract. The most important property required for any structural composite material is high strength to weight ratio and corrosion resistance. To improve the interfacial addition between carbon fiber and epoxy resin, the epoxy matrix has been modified with hybrid resin. The unidirectional carbon/epoxy composites were prepared with different weight proportions of two resins (hybrid resin) with parts by weight(PBW).In this paper unidirectional carbon fiber T700 used as reinforcement and hybrid resin used as matrix for preparing the Uni-Directional(UD) laminate and Naval Ordnance laboratories(NOL) ring specimens. These specimens are prepared with filament winding technique, specimens were tested as per ASTM standards and test was conducted on Instron 4505 with 100KN capacity. From the test results it has been observed that the tensile strength measured by NOL ring specimens are more trustable than the results obtained with unidirectional(UD) test specimens. NOL ring test data is useful for making composite pressure vessels for aerospace applications.

1. Introduction

In recent years, composite materials derived from carbon fiber and epoxy resin are extensively employed in aircraft and space applications because of their high strength, high modulus and light weight [1,2]. Carbon fibers is a polymer having five times stronger than steel and twice as stiffness. It is made of thin strong crystalline filaments of carbon that is used to strengthen the material. The use of epoxy resins as a matrix having good engineering properties like high stiffness, strength, chemical resistance and good adhesion with fibers [4,5]. The major drawback of epoxy resin is their brittleness and having poor pot life with less resistance to crack initiation. This drawback can be eliminated by hybridizing the epoxy resin (modifying the matrix by PBW of two resins). The main objective is matrix modified in to moderate viscosity with higher elongation during the test [6,7]. This will help to design a composite pressure vessel which undergo progressive failure during testing [9]. Carbon fiber reinforced polymer (CFRP) composites have aroused considerable industrial attention because of their attractive mechanical properties. Carbon fibers are principal constituents in a fiber reinforced composite materials. Fibers occupy major volume fraction in a composite laminate and distribute the major portion of the load acting on the structure. Selection of suitable type of fiber, fiber volume fraction, fiber length and fiber orientations are important. The commercially available carbon fiber diameter is around 5-7 μm. Advanced composite materials are made with unidirectional carbon fiber T700 with epoxy resin for high strength structural applications[11]. Mechanical properties of CFRP composites depends not only on the properties of fiber and matrix, but also depends fiber matrix interfacial adhesion.

IOP Conf. Series: Materials Science and Engineering 998 (2020) 012005 doi:10.1088/1757-899X/998/1/012005

Composite material properties depend upon properties of individual constituents, distribution, orientation of fibers, fiber volume fraction, physical and chemical properties[12]. The main aim of the present work is to examine the effect of the hybrid resin on mechanical and morphological properties of carbon fiber composites. In this study two epoxy resins of equal proportions are mixed and used as hybrid resin with hardener, tests were carried out according to ASTM standards on Flat laminates and NOL ring specimens in order to compare the mechanical performance of different sample specimens[13,14].

2. Materials and Methods

The reinforcing material of carbon fiber T-700, 12K and two resins(hybrid) of diglycidal either of bisphenol-A(DGEBA) of LY 556 and LY 5052 with hardener HT 972 of solid aromatic amine of diaminodiphenyl methane(DDM) used as matrix for this investigation. The specifications of carbon fiber T700,12K as shown in table 1.and specification of Hybrid Resin LY556 and LY5052 are shown in table2.

Sl.No.	Parameter	Value
1	Grade	T700
2	Tow size	12 K
3	Carbon content	95%
4	% Elongation	2%
5	Specific gravity	1.7 – 1.8
6	Tex	800g/km(min)
7	Tensile strength	4.0 Gpa(min)
8	Tensile modulus	200 Gpa(min)
9	Fiber diameter	5-7 microns
10	Sizing	Epoxy compatible

Table 1. Carbon fiber T700,12K specifications

Sl.No.	Parameter	Value
1	Type	Hybrid
2	Grade	LY556,LY5052
3	Viscosity 25° at CPS	8000-12000
4	Specific gravity at RT	1.10 - 1.12
5	Colour	Pale Yellow/Clear liquid
6	Volatile content by weight	0.75% Max

Table 2. Hybrid Resin LY556& LY5052 specifications

When composite specimen is tested under loading condition both fiber and resin canbe able to deform in the same loading direction, but in practical it is not possible. Good adhesion between fiber and resin is required when composite is loaded under tension. This property will ensure the load transfers effectively and prevent from the fiber cracking and fiber/resin debonding.

IOP Conf. Series: Materials Science and Engineering 998 (2020) 012005 doi:10.1088/1757-899X/998/1/012005

3.Composite Specimen Preparation

Before making any structural composite product, the test samples of composite product of fiber and resin are used to prepare the laminates for preliminary characterization. In this study the matrix has been changed to hybrid resin system consisting of two resin system of LY556 and LY 5052 mixed in proportions of 50:50 ratio and hot curing hardener HT972. The proportions of hybrid resin to the hardener is in ratio of 100:27 by weight was mixed. The hybrid resin is preheated to a temperature in the range of 40-50^0 C with the help of heaters to maintain the resin viscosity during winding. Two types of samples made with filament winding technique. The flat laminate specimens are made with flat mandrel and NOL ring specimens are made with cylindrical mandrel.

The carbon fiber bands are unwind from a creel and passed through hybrid resin bath. These resin impregnated bands are passed on to a rotating (Flat and cylindrical) metal mandrel. The carbon strands are wound around the cylindrical mandrel to ensure that all fiber strands dip in the resin bath with uniform wetting of all fibers. Fiber tension is critical in filament winding because compaction of resin with fiber is achieved through the fiber tension. The fiber tension must be optimal level because high fiber tension may break the fiber completely or fiber fracture at the surface. Filament winding as shown in fig.1.Mandrel for NOL ring as shown in fig.2. Carbon roving for UD laminate preparation shown in fig.3 and Winding on NOL mandrel as shown in fig.4

Fig.1 Filament winding

Fig.2 Mandrel for NOL ring

Fig.3 Carbon roving/Epoxy UD laminate Winding by Filament Winding

Fig.4 Winding on NOL mandrel

A rectangular flat and cylindrical steel mandrels are used for specimen preparations. The surface of the flat mandrel and cylindrical mandrel is like a mirror finish. The mandrel surface is cleaned with acetone to make it free from oil, grease and dust. The surface of the mandrel is applied with Waxpol for easy removal of the samples from the mandrel. The resin impregnated fiber strands are wound around the flat/cylindrical mandrels with five hoop layers with a band width of 1.9mm and thickness of 0.4mm.. The mandrel is rotated with moderate speed, after completion of winding the mandrels is rotated for few minutes for uniform setting of the resin. Curing of specimen along with the mandrel was carried out in hot air oven with a given cure cycle.

4. Specimen Curing

Following cure cycle was maintained for NOL ring mandrel as shown in fig.5

- Oven temperature is increased from room temperature to 100^0C in 30 min. and keep the temperature for 2 hours.
- After completion of 2 hours increase the temperature 100^0C to 120^0C in 20 min. and keep the temperature for 2 hours.
- Again raise the temperature from 120^0C to 140^0C in 20 min. and keep the temperature for 6 hours.
- After completion of 6 hours cure time switch of the oven and permit the component to cool to room temperature within 4-5 hours.
- Finally remove the cured mandrel from the oven.

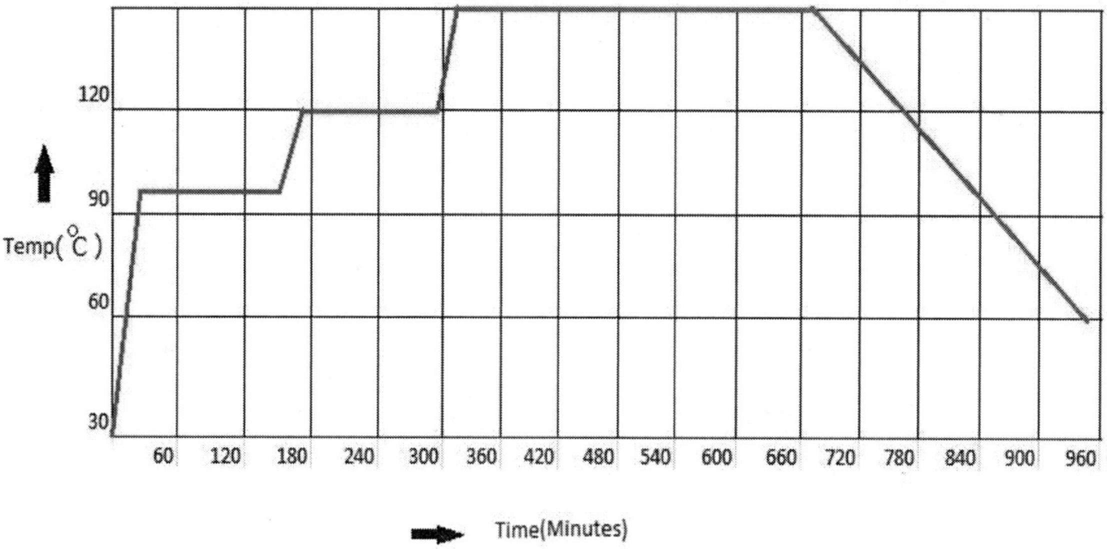

Fig.5 Cure Cycle for the Component.

The cured mandrel with filament wound cylindrical composite shell is placed on a lathe machine for separating shell in to NOL ring specimens cut with parting tool each specimen width of 6.3 mm maintained. After completion of parting the round circular specimens (NOL rings) separated from the mandrel.

5. Results and Discussions

One of the important property of Carbon fiber reinforced composite is Tensile strength. In the present study tensile strength of unidirectional (UD) flat laminates and NOL ring specimens are tested. In tensile test the UD specimen is subjected to elongate in the fiber direction by uni-axial tensile force. The standard UD flat test specimen dimensions are made according to ASTM standard is shown in fig. 6 the test is carried on electro mechanical testing machine with cross head speed of 2mm/min. is shown in fig.7 and tensile tested specimens are shown in fig 8. The output tensile test data is tabulated in table 3.

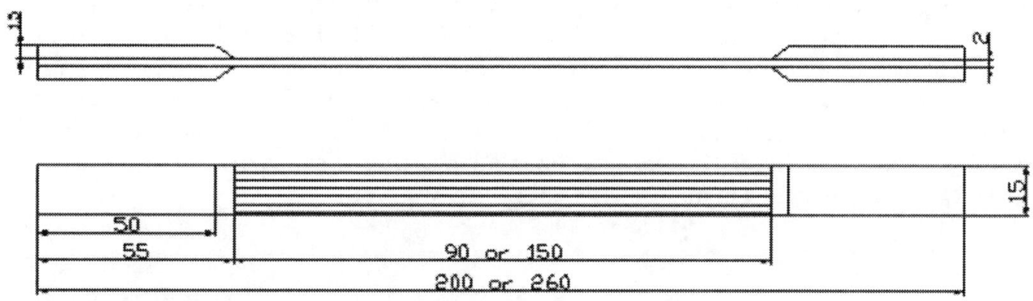

Fig.6 Dimensions of Flat specimen as per ASTM 3039

Fig.7 Tensile testing test setup

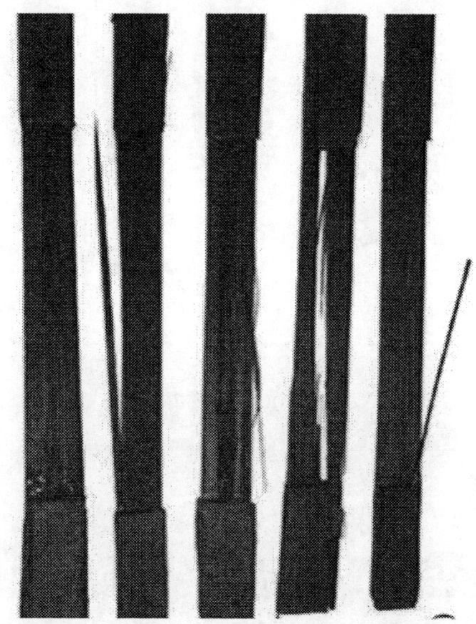

Fig.8Failure modes in UD specimens

S. No	Dimensions in MM	Max. Displace ment in MM	Max. Load in KN	Tensile Strength in MPa	Tensile Modulus in MPa	Remarks
1	14.97 x 1.94	16.72	48.87	1682	140	Explosion
2	14.95 x 1.90	16.42	44.83	1578	140	Explosion
3	14.98 x 1.90	15.80	38.71	1360	126	Splitting
4	14.97 x 1.93	16.44	47.14	1632	135	Explosion
5	14.95 x 1.93	16.30	44.80	1553	124	Explosion
6	15.02 x 1.87	15.01	37.97	1352	126	Explosion

Table.3UD laminate tensile test data with Hybrid resin 50:50 with Hardener.

National Ordnance laboratories (NOL) ring test was carried on universal testing machine (UTM) with split disc test fixture attachment as shown in fig 11. In NOL ring test four split ring is used to test the NOL ring specimens.

IOP Conf. Series: Materials Science and Engineering 998 (2020) 012005 doi:10.1088/1757-899X/998/1/012005

The internal pressure can be applied to the ring shaped specimens similarly to the burst tests of actual pressure vessels. This test gives the fracture behavior inside of the CFRP layered laminate could be observed directly whereas in actual composite pressure vessel it is highly impossible to observe the fracture behavior under the surface of pressure vessels. A NOL ring fabrication and testing techniques are used for determination of ring tensile strength has been standardized. The standard NOL ring specimen dimension is shown in fig. 9. The test specimen after testing is shown in fig 10 and 12. The tensile hoop stress values are tabulated in table 4

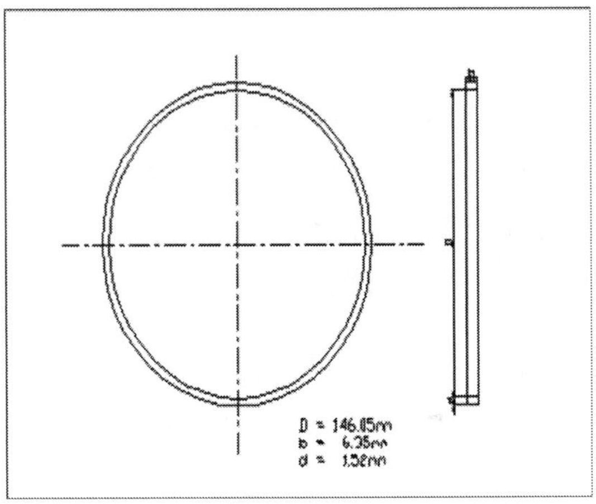

Fig.9 NOL ring specimen as per ASTM D2290

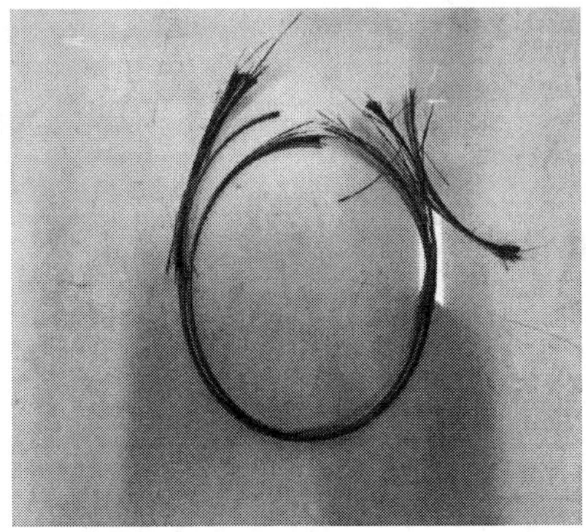

Fig.10 NOL Ring after testing (Splitting)

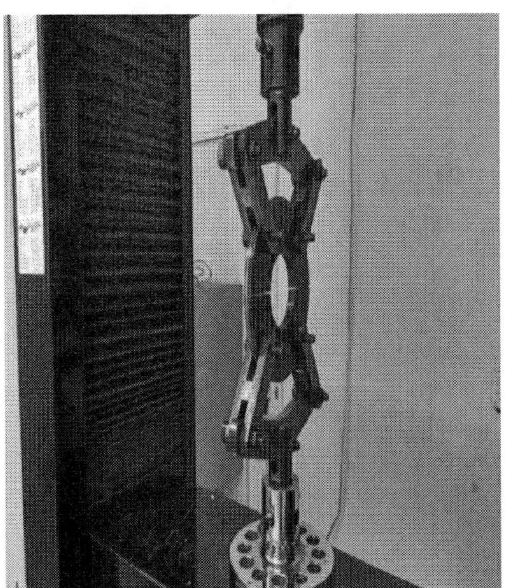

Fig.11 NOL ring Split disc test fixture

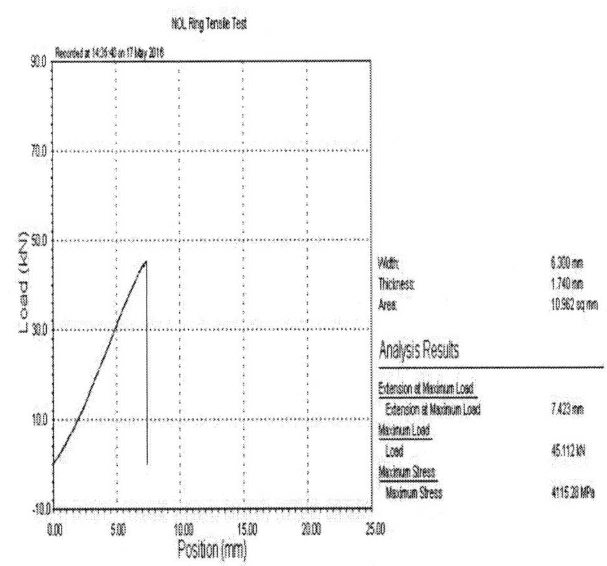

Fig.12 Hybrid resin 50:50 with Hardener

S. No	Dimensions in MM	Max. Displacement in MM	Max. Load in KN	NOL Tensile strength in MPa	Remarks
1	6.30 x 1.74	7.423	45.112	2058	Explosion
2	6.22 x 1.70	6.925	43.610	2030	Splitting
3	6.32 x 1.70	5.834	37.175	1730	Splitting
4	6.23 x 1.70	7.870	45.522	2149	Splitting
5	6.39 x 1.70	6.805	38.785	1785	Splitting
6	6.30 x 1.74	6.911	39.725	1812	Splitting

Table.4NOL ring test data with Hybrid resin 50:50 with Hardener.

Conclusion

The equal proportions of hybrid resin 50:50 is used to make the UD and NOL ring specimens. Based on the experimental result it has been found that failure stresses are more accurately measured by NOL ring test with data acquisition system compared to UD tensile test. The Unidirectional tensile test samples failure modes were observed as explosive fracture and the average ultimate tensile strength measured was 1526 MPa along fiber direction, whereas the NOL ring test data shows failure modes are in splitting and average tensile strength of NOL ring was 1927 MPa. It can be concluded that, the material could withstand with stresses of 1500MPa without any deformation and failures in UD laminates and 1900 MPa for NOL ring specimens.

References

[1] Jinshui Yang, Jiayu Xiao, Jingcheng Zeng, LipingBian, Chaoyi Peng, and Fubiao Yang "Matrix Modification with Silane Coupling Agent for Carbon Fiber Reinforced Epoxy Composites", Fibers and Polymers ,2013,Vol.14, No.5, 759-766

[2] C.Venkateshwar Reddy, P.RameshBabu, R.Ramnarayanan, Dilkeshwar Das "Mechanical properties of Unidirectional carbon and glass/epoxy reinforced composites for high strength applications", Materials Today Proceedings, volume 4, issue 2, part A, 2017, pages3166-3172

[3] Umar farooq and Peter Myler "Preparation of Aerospace grade Carbon fibrous laminated composite panels with improved performance and reduced Fabrication Process defects and flaws", ARPN Journal of Engineering and Applied Sciences, vol.12, No.4, February 2017

[4] K.Friedrich, "Polymer composites for tribological applications", advanced Industrial and Engineering Polymer Research, 1,(1) 2018, P 3-39

[5] W.M.Chen, Y.H.Yu, P.Li and X.P.Yang "Effect of new epoxy matrix for T800 carbon fiber/epoxy filament wound composites, Compos.Sci.Technol., 2007, 67, 2261-2270

[6] J.Braz. Chem. Soc. "Composite materials based on modified epoxy resin and carbon fiber", Journal of the Brazilian Chemical Society, vol.17, no.6 Sao Paulo Sept./Oct. 2006

[7] M.Nayeem Ahmed, P.Vijaya Kumar, H.K.Shivanand, Syed BasithMuzammil "A Study on flexural strength of hybrid polymer composite materials(E glass fiber-carbon, carbon fiber-graphite) on different matrix material by varying its thickness", International Journal of Mechanical Engineering and technology(IJMET), Volume 4, Issue 4, July - August (2013), pp. 274 - 286

[8] M.A. Kinna, NOL-ring Test Methods, United States naval ordnance laboratory, with oak, Maryland, NOLTR: 1964, p 64-156

[9] C.Venkateshwar reddy, Ch.Joseph s raju, P.Rameshbabu, R.Ramnarayanan "Study of void effect on tensile strength of carbon fiber reinforced composites for structural applications", International Journal for Research in Applied Science & Engineering Technology (IJRASET), volume 5, Issue viii, August 2017

[10] Sobrinho, L., V.Calado&F.Bastian (2012) Effects of rubber addition to an epoxy resin and its fiber glass reinforced composites. Polymer Composites

[11] S.Erden, K.Sever, Y.Seki&M.Sarikanat "Enhancement of the mechanical properties of glass/polyester composites via matrix modification glass/polyester composite siloxane matrix modification", Fibers and polymers, volume 11, pages 732-737 (2010)
[12] Johnsen B., Mohammed R, Taylor A, Sprenger S., "Toughening Mechanisms of particles Modified Epoxy Polymers", Journal of Polymer, Vol. 48, 530-41, 2007
[13] ASTM, Standard test method for tensile properties of polymer matrix composite materials, ASTM D3039, Annual book of ASTM standards, American Society for Testing and Materials, PA, 15(3) 2003
[14] Fang Liu, Shiqiang Deng, and Jianing Zhang "Mechanical Properties of Epoxy and Its Carbon Fiber Composites Modified by Nanoparticles", Journal of Nanomaterials, volume 2017, Article ID 8146248
[15] Fu, S.-Y. and Lauke, B. (1998) Characterisation of Tensile Behaviour of Hybrid short Glass fiber/Calcite particle/ABS composites. Composites Part A, 29A, 575-583

IOP Conf. Series: Materials Science and Engineering **998** (2020) 012006 doi:10.1088/1757-899X/998/1/012006

Finite element analysis of AISI 8620 alloy spur gear hot forging

J B Kanani[1,3*] and D I Lalwani[2,4]

[1] Research Scholar, Mechanical Engineering Department, SVNIT - Surat

[2] Associate Professor, Mechanical Engineering Department, SVNIT - Surat

(S. V. National Institute of Technology, Ichchhanath, Surat-395 007)

[3] jay.kanani85@gmail.com, [4] d.i.lalwani@gmail.com

Abstract. Among all manufacturing processes, a forging process has a remarkable position in which parts with desired mechanical properties and minimum wastage of material can be produced. Despite of mentioned advantages, forging process has few drawbacks such as high tonnage capacity machineries, expensive dies, high initial investment and die failure. The overall cost of the forging process directly depends upon the required forging load, ejection load and dies arrangements. In this paper, Finite Element Analysis (FEA) of a spur gear forging process made of low carbon alloy steel (AISI 8620) is carried out. Two load reduction methods (elastically attached die and relief hole) are simultaneously used and analyzed using DEFORM-3D software. The required forging load is compared between non-elastic die arrangements and elastic die arrangements with the relief hole. Velocity distribution is also discussed at nine different locations.

1. Introduction

Traditional manufacturing of spur gears has many disadvantages, such as more material wastage, less production, more cost and many more [1]. Over the past several decades, near-net-shape (NNS) or net-shape manufacturing has been turned into a well-known practice in the metal forming that saves time, energy and material. Therefore, forging process has tremendous potential for mass production such as automobile components and general engineering applications. Despite these favorable advantages, the economics of fully formed spur and helical gears for power transmission have not been considered to be acceptable and cannot be until a sturdy processing path is identified [2]. The overall performance of the forging operation requires an understanding of not only the flow stress of the material and frictional conditions, but also the mode of flow of the material, ejection process, forging load and die life [3].

In spur gear forging, total forging load mainly consists of deformation load and friction load. When cylindrical preform (workpiece) is forged, because of friction between die surface and cylindrical surface, the barrelling effect is produced and friction load increases. Forging load can be reduced up to 34% and better corner filling can be achieved by using the concave preform compared to simple cylindrical billet [4]. The change in friction load amount is more decisive when the coefficient of friction increases from 0.1 to 0.2. The forming load increases when the number of teeth increases and module is fix [5].

Zuo et al (2015) investigated the gear forging process. They mainly focused on corner filling, workpiece stress, die stress, and forming load. They used DEFORM-3D simulation software for FEA and AISI 8620 as workpiece material. They used relief-cavity thickness concept and observed that forging load reduced up to 20% at 3mm relief-cavity thickness.

2. Forming process and die design

In closed die forging flash formation is zero, hence billet dimension and die design is crucial. The hollow cylindrical billet is used to forge NNS spur gear and the diameter of the workpiece is equal to the dedendum diameter of the internal gear die. The detail and 3D model of forged gear is shown in figure 1. Forging of spur gear has several possibilities of die arrangements. Elastically attached die with machine frame and taper punches requires less forging load [7]. Peripheral relief and relief hole also reduce the required forging load [8]. In this paper combined effect of relief hole and elastically attached die is investigated. Die design process is complex, time consuming and requires thorough knowledge. The present die consists mainly four components: (a) bottom die, (b) counter punch, (c) middle (cavity) die and (d) top die. Bottom die and counter punch are static while middle die and top die are moving components.

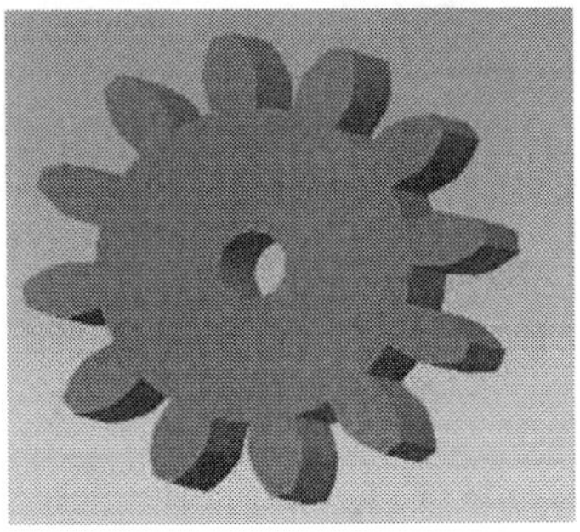

Type	Spur gear
Material	AISI 8620
Module	4
Number of teeth	12
Face width	20mm
Centre hole	8mm
Pressure angle	20°

Figure 1. Three dimentional model and specification of forged spur

Figure 2. Die assembly (Applied for patent)

3. Finite element analysis of spur gear forging process

Trial-and-error based inspection of forging tooling design is costly and time consuming. To eliminate practical experiment based evaluation of tooling designs, FEM simulation and modelling software is used. On the basis of FEM result die dimensions can be obtained for further practical experiment. DEFORM-3D has been keen in forging process finite element analysis for the last couple of decades [9]. In the present investigation, FEM based software DEFORM-3D v6.1 is used for the forging process simulations. The numerical simulation software works in three stages: (a) pre-processor, (b) solver and (c) post-processor. Modelling of the die components and the workpiece are prepared in ProEngineer Wildfire 4.0. All the geometries are converted into stl file and inserted into DEFORM-3D environment.

3.1 Numerical simulation conditions

The simulations are executed using the commercial software DEFORM-3D. The billet material is AISI 8620 and die components are AISI H13. The modulus of elasticity and poisson ratio of AISI 8620 are 206.7 MPa and 0.3, respectively [6]. The flow stress of the AISI 8620 can be defined as a function of strain, strain rate and temperature.

$$\sigma = f(\varepsilon, \dot{\varepsilon}, T) \qquad (1)$$

Here σ is flow stress, ε is strain, $\dot{\varepsilon}$ is strain rate and T is temperature of the workpiece material. The value of flow stress at various strain, strain rate and temperature can be obtained from the DEFORM-3D material library under the hot forging condition [6].

IOP Conf. Series: Materials Science and Engineering **998** (2020) 012006 doi:10.1088/1757-899X/998/1/012006

Only the workpiece is considered as plastic material and rest of the components are selected as rigid bodies (non deforming). The workpiece deforms as per von-mises flow rule. Meshing is generated as solid mesh with near about 15085 nodes and 71012 tetrahedral elements. Remeshing is required during the numerical solution when large plastic deformation occurs of the element. The automatic remeshing is chosen during simulation by providing the relative interference depth 0.7. The workpiece temperature is set to 980°C [10] and the die components are pre heated at 150°C temperature. The friction factor between die components and material contact surfaces is 0.3. Punch velocity for the simulation is set as 3mm/s and maximum stroke is 14.1mm (Including dwell time). Elastically attached die environment is given by providing relative motion between top die and middle die with scaling factor 1.

4. Results and discussion

- Figure 3 represents the relation between forging load and top die movement. Spur gear forging process comprises three stages namely, (1) primary upsetting, (2) tooth cavity filling and (3) tooth corner filling. During primary upsetting height of the billet decreases and diameter of the billet increases. Actual tooth formation starts in tooth cavity filling zone and the forging load increases almost linearly. Spur gear forging requires maximum load at the time of tooth corner filling stage. The drastic growth in forging load during tooth corner filling stage is owing to the sharp corner filling of the gear tooth.
- Figure 4 shows the effective strain distribution at full stroke and maximum strain reach to 5.14mm/mm.
- Figure 5 illustrates the effective stress distribution at full stroke and the amount of maximum effective stress is 670MPa.
- Figure 6 depicts the material flow pattern and the material is flowing towards the periphery and towards the centre. Figure 7 illustrates the velocity variation at different tracking points P1 to P9. Tracking points P4, P6 and P8 are at the dedendum or root of the gear tooth while P5, P7 and P9 are at the addendum of the gear tooth. As shown in figure 7(b, c and d) velocity increases almost linearly at the end of the stroke at point P2, P4 and P9. As shown in figure 7(b) material flow towards centre starts at the time of tooth corner filling stage. Maximum effect of barrelling (bulging) is observed at point P9.

IOP Conf. Series: Materials Science and Engineering **998** (2020) 012006 doi:10.1088/1757-899X/998/1/012006

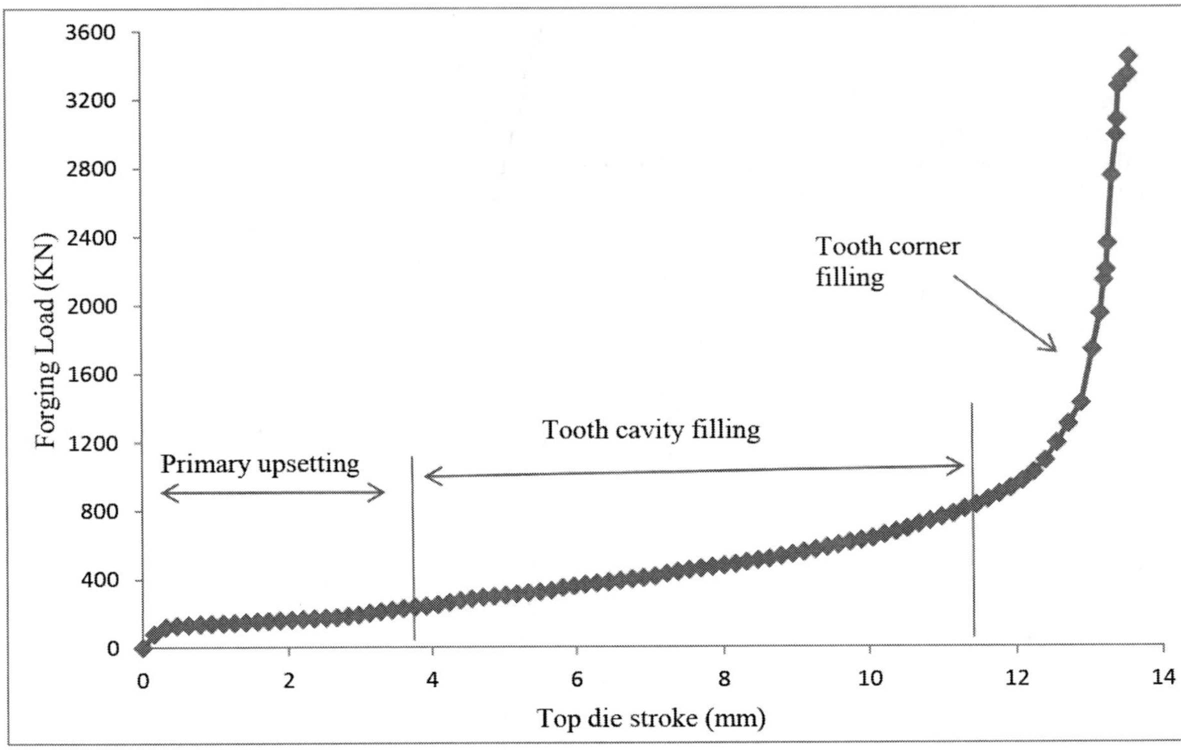

Figure 3. FEA result graph between forging load and top die stroke

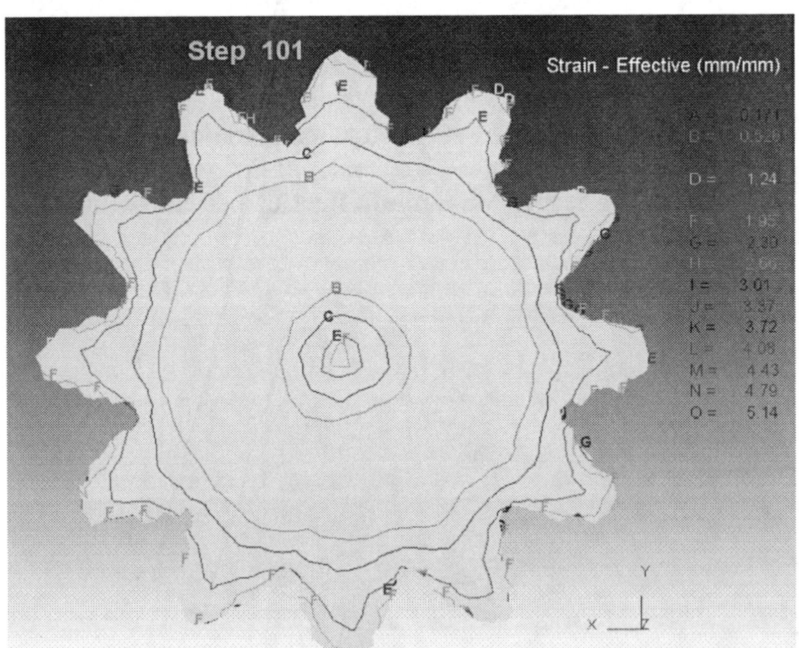

Figure 4. Effective strain

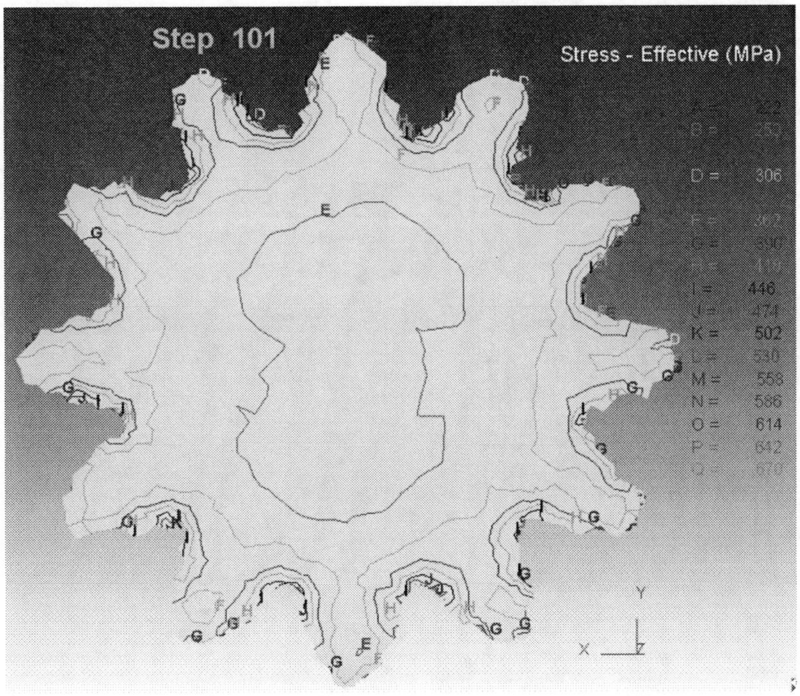

Figure 5. Effective stress

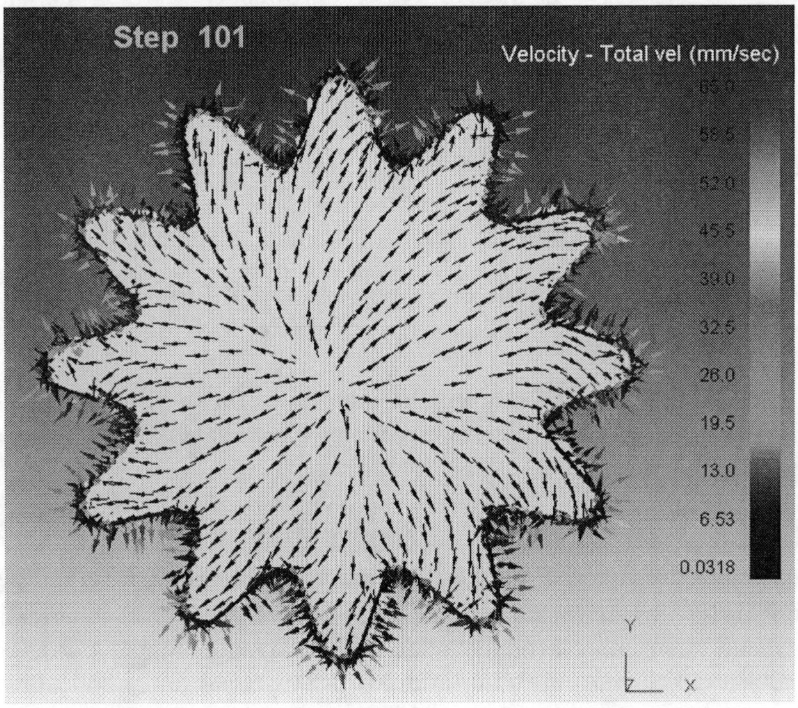

Figure 6. Velocity distribution and material flow pattern

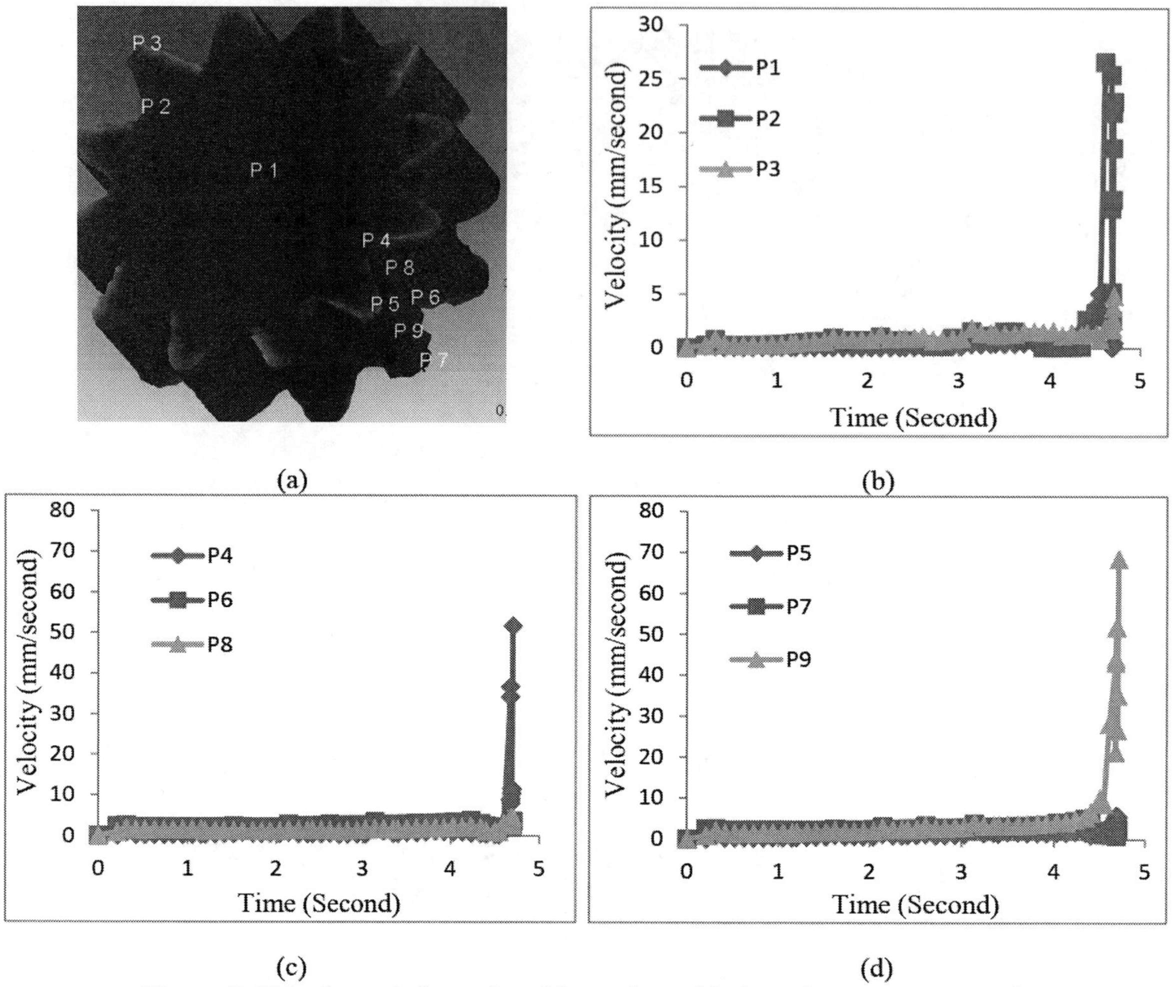

Figure 7. Velocity variations of tracking points with time along spur gear tooth

5. Conclusions

- Tooth corner filling requires the maximum amount of forging load and more than 45% load is increased during last 0.5mm top die stroke. Maximum forging load reaches to 3435 KN.
- During the same forging conditions the maximum forging load is reached up to 3648 KN in case of non-elastic die arrangments. It indicates that elastic die arrangment reduces forging load up to 213 KN.
- Strain rate also drastically increased in tooth corner filling zone therefore, flow stress increases and further deformation becomes difficult.
- Inward flow of the material reduces the centre hole size of the forged gear.
- Ejection of the forged gear will be comparatively easy in elastically attached die arrangements then non-elastically attached die arrangements. A present die design is practically tested using pure aluminium billet material and good agreement found in ejection of the forged gear.
- The grain flow lines of forged gear are continuous and the forged spur gear will have a good mechanical properties as well as dimensional stability.

References

[1] Bassan F et al 2014 Prediction of defects in multistage cold forging by using finite element method *Key Engineering Materials* **622-623** 659-663

[2] Dean T A 2000 The net-shape forming of spur gears *Materials and Design* **21** 271-278

[3] Naunit R C and Yohng J K 2001 Near net shape forg-ing of a crown gear: Some experimental results and analysis *International Journal of Machine Tools & Manufacturer* **41** 325-346

[4] Yilmaz N F and Eyercioglu O 2018 Near net shape spur gear forging using concave preform *Mechanika* **24(2)** 268-277

[5] Hosseinzade A et al 2012 Numerical investigation of the influence of friction, module and Number of teeth in forging process of a spur gear *Advanced Materials Research* **472-475** 2116-2120

[6] Zuo B et al 2015 Design of relief-cavity in closed-precision forging of gears *J. Cent. South Univ.* **22** 1287-1297

[7] Cai J et al 2004 Alternative die designs in net-shape forging of gears *Journal of Materials Processing Technology* **150** 48–55

[8] Wei F et al 2012 Finite element analysis and simulation for cold precision forging of a helical gear *J. Central South University* **19** 3369–3377

[9] Altinbalik T et al 2015 Simulation of lateral extrusion of spline form based on the software DEFORM-3D *Applied Mechanics and Materials* **729** 149-154

[10] Zhenshan G et al 2013 A study on gear tooth metal flowing law and process optimization of spiral bevel gear forging *Advanced Materials Research* **753-755** 215-220

Condition Monitoring of Induced Draft Fan in a Super Thermal Power Plant using Vibration Analysis

D Panda[1], J K Mohanty[1], S S Das[1], Anupam Mishra[2], P K Pradhan[1]*

[1]Mechanical Engineering Department,VSSUT, Burla, Odisha, India.
[2]Vedanta Limited, Jharsuguda, Odisha, India.
*Corresponding Author:prasant2001uce@gmail.com

Abstract. Timely detection of fault and taking corrective measures reduces the downtime of any industry. It also increases the reliability and efficiency of the system. Nowadays, many industries have implemented condition monitoring (or condition based maintenance) techniques as a part of maintenance strategies. Condition based maintenance monitors the health condition of machine and diagnose the defect if any. Though there are various condition monitoring techniques, vibration analysis is one of the strongest and widely used technique for fault identification. It is used to diagnose the fault of rotating machine element. In a big power plant, condition monitoring techniques play an important role to reduce the downtime of primary system and its auxiliaries, which in turn leads to reduction in productivity loss. Therefore, it is desirable to ensure system availability.

Keywords: *Super thermal power plant, condition monitoring, vibration analysis, induced draft fan.*

1. Introduction

Availability of electrical power round the clock and throughout the year is most essential for industrial and economic growth of any country. In the present era of fierce competition and financial constraints, the reliability and availability of operating power plants, are becoming a challenging issue. The thermal power plants produce larger part (i.e. almost 70%) of the electricity of world's requirement compare to other power plant. India is one of the largest producers of electricity in world and majority of the plants are coal based [1].

Rankine cycle is the basic principle of the thermal power plant [2]. In a power plant, the combustion of fuel is used to get steam from water. The generator (coupled with turbine) produces the electricity. Continuous electricity generation from a power plant depends on the availability of its components. To enhance the availability, various maintenance culture and strategies are adopted [3, 4 & 5]. These are;

- Breakdown Maintenance
- Routine Maintenance
- Precautionary Maintenance
- Condition Based Maintenance (CBM) or Condition Monitoring

In spite of high initial cost, the CBM is implemented to reduce the downtime and raise the reliability and safety of the plant. CBM watches and monitors the performance parameters of the equipment very regularly which enable to know whether any problems exists or likely to appear. There are different condition monitoring techniques, such as: Vibration Analysis [6], Noise Analysis [7], Temperature Monitoring [8], Motor Current Signature Analysis (MCSA) [9] and Wear Debris Analysis [10].

2. Methodology and Experimentation

The present work is based on a super thermal power station of Vedanta Limited, Jharsuguda, India. It has nine units and total power of 1215 MW. The plant has adopted the world class technology.

The basic components of power plant (i.e. boiler, generator and turbine) are singular and auxiliary devices like ID fans (Induced draft fans), FD fans (Forced draft fans), coal mills, Boiler feed pumps, condensate extraction pumps, cooling water pumps, ash handling units, coal handling plants etc. have inbuilt standbys. The maintenance practice at CPP of Vedanta Limited, Jharsuguda, shows that the failure rate of power components (Turbine, Induced Draft fan, Forced Draft fan etc.) is many times higher than other. In this paper, the condition monitoring of an Induced Draft fan (ID fan) by vibration analysis has been described.

3. Condition Monitoring of Induced Draft Fan (ID Fan)

An ID fan is in between dust collector and chimney. It takes the hot flue gases from furnace via a dust collector (dust separation system, Electrostatic Precipitator i.e ESP or fume extraction system) and will deliver to chimney. It produces the negative pressure in the furnace to remove the flue gases from furnace via electrostatic precipitators and to push the flue gases to chimney. The figure of ID fan is in Figure 1.

3.1 Detail Specification

Power: 1700KW, Speed: 991 RPM, Motor's NDE (Non-Drive End)/Outboard Bearing: NU240 ECM, Motor's DE (Drive End)/Inboard Bearing: 23044 CC/C3W33. Fan DE & NDE (Both Inboard & Outboard) Bearings: 22232EASMC3, Vibration velocity (RMS) limit (as per OEM); Normal: below 1.5 mm/s, Little abnormal: 1.5-4.0 mm/s, Sub critical: 4.0-6.5 mm/s, Critical: above 6.5 mm/s. Peakvue limit; Normal: less than 1.

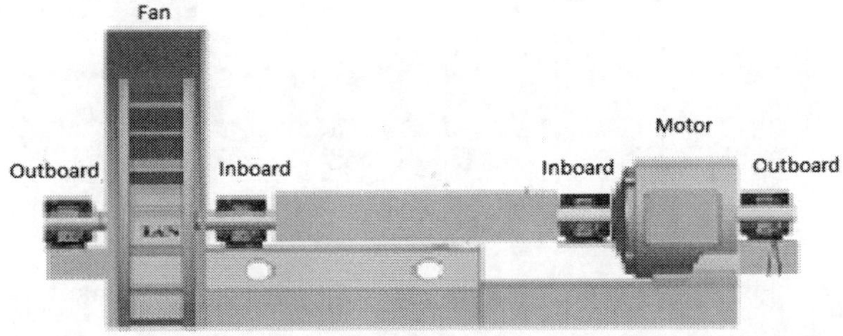

Figure 1: Schematic Diagram of a typical ID Fan

3.2 Diagnostic Symptoms:

The ID Fan is monitored through the vibration analysis. The information are recorded and analysed by using accelerometer and AMS Suite Software. A case study was observed during July 2019 – August 2019 and has been considered here. Usually, vibration readings are taken weekly. If there is some abnormality observed, then the frequency of measurement is increased. During taking readings, accelerometer is mounted on both the inboard and outboard bearings of the motor and fan. Because of unusual sound was observed on 5th August, the analysis was started to diagnose the defect. Vibration amplitude (at all bearing) in different directions; horizontal (H), vertical (V) and axial (A) were measured and analysed. The RMS values and peakvue values are presented in Table 1 and Table 2 respectively. Similarly, the vibration spectra are presented in Figure2 to Figure4.

IOP Conf. Series: Materials Science and Engineering 998 (2020) 012007 doi:10.1088/1757-899X/998/1/012007

Table 1: RMS value of vibration (in mm/s) in various days

Location	Direction	25th July	1st Aug	5th Aug	06th Aug	11th Aug	12th Aug	14th Aug
Motor Outboard	H	1.318	1.519	0.194	0.424	Bearing Replaced	0.684	0.949
	V	0.580	0.470	0.741	0.846		0.362	0.316
	A	0.522	0.396	1.167	0.953		0.350	0.350
Motor Inboard	H	1.055	1.228	1.064	1.284		0.587	0.953
	V	0.322	0.482	0.859	1.216		0.387	0.347
	A	0.798	0.443	1.314	1.214		0.332	0.411
Fan Inboard	H	1.269	1.283	2.577	2.804		0.643	1.054
	V	0.476	0.311	1.627	1.656		0.331	0.310
	A	0.602	0.630	1.725	2.726		0.392	0.469
Fan Outboard	H	1.439	1.168	1.087	1.173		0.625	0.839
	V	0.387	0.242	0.545	0.605		0.265	0.372
	A	0.611	0.324	0.636	0.885		0.493	0.443

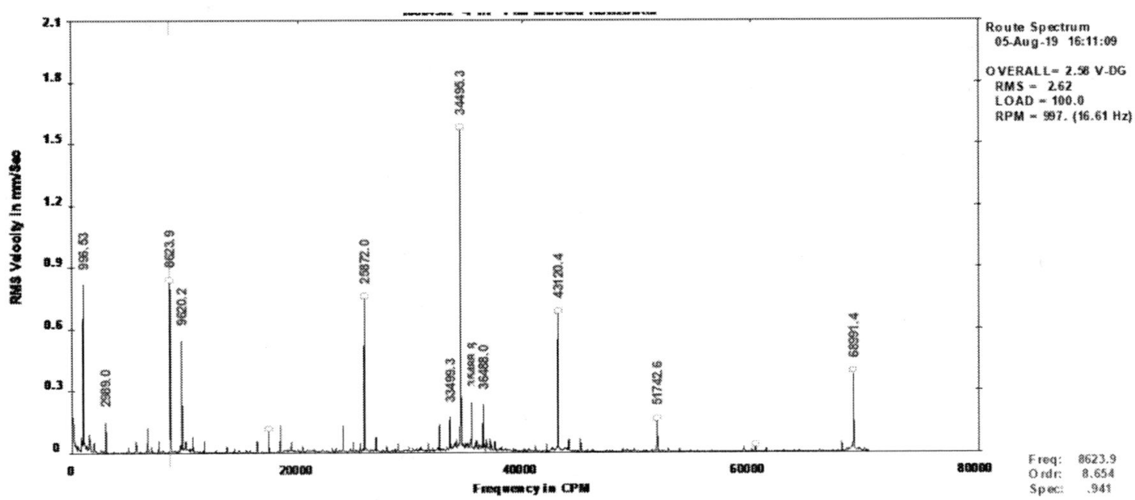

Figure 2: Vibration spectrum of ID Fan-Inboard (Horizontal).

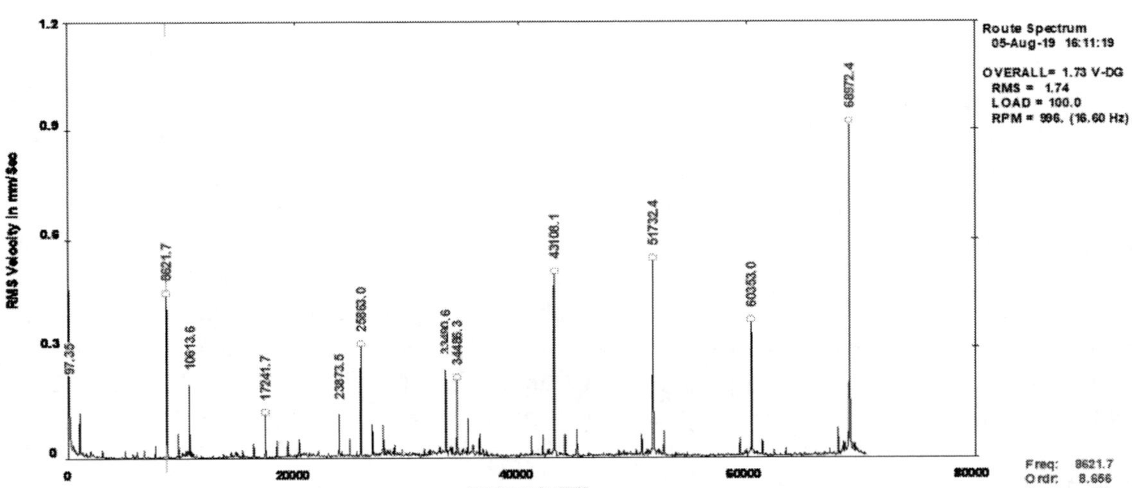

Figure 3: Vibration spectrum of ID Fan-Inboard (Vertical)

IOP Conf. Series: Materials Science and Engineering 998 (2020) 012007 doi:10.1088/1757-899X/998/1/012007

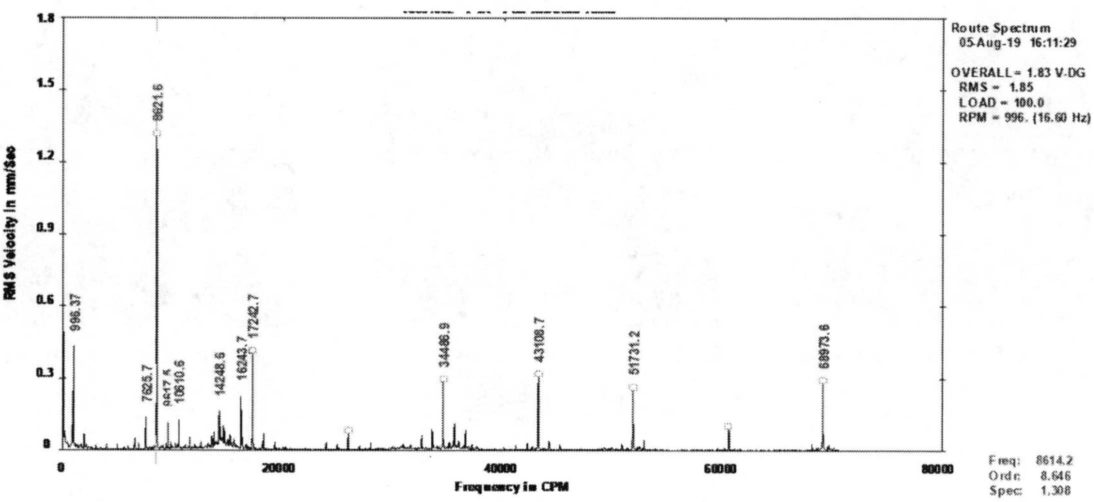

Figure 4: Vibration spectrum of ID Fan-Inboard (Axial).

Table 2: Peakvue value at Fan Inboard bearing in various days

Position	Direction	25th July	1st Aug	5th Aug	06th Aug	11th Aug	12th Aug	14th Aug
Fan Inboard	H	0.297	0.398	1.264	1.14	Bearing Replaced	0.295	0.291
	V	0.476	0.311	1.727	1.541		0.331	0.317
	A	0.602	0.630	1.825	2.594		0.392	0.384

3.3 Observations and Analysis

- The Table-1 shows that there is sudden increase of vibration amplitude (on 5th August) on the Fan DE bearing, which indicates that there is some problem in rotating element of ID fan.
- Also, the increase in Peakvue values (in table 2) signify the same thing.
- From the vibration spectrum (Fig. 2), the peak at 8623.9 cycles per minute (cpm) (which is a non-synchronous frequency) with its' harmonics indicate the abnormalities in the Fan DE bearing.
- The same frequency match with the Ball pass frequency outer race (BPFO). So it indicates that there is problem in outer race of the bearing.
- The peak at 34486 cpm and its sidebands (i. e. 4BPFO±1xRPM) indicates/ensures the outer race problem in the bearing.

As per the recommendation, the Fan-Inboard bearing was opened and it was found that there was erosion in outer race of the bearing. That is shown in Figure5 to Figure7. After replacing the bearing (on 12th August), the Fan was put back to operation. After that, vibration spectra (are shown in Figure 8 to Figure 10) shows that there is no significant peak at non synchronous frequency (i.e. at 8623.9 cpm) and the vibration amplitude (rms) is very normal.

Figure 5: Bearing housing

Figure 6: Bearing inner race

Figure 7: Eroded outer race

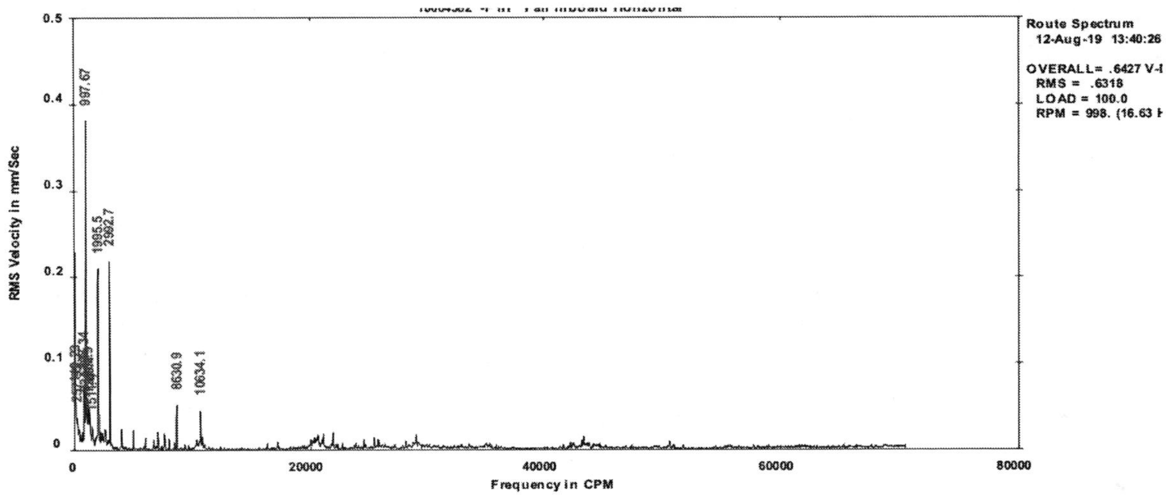

Figure 8: Vibration spectrum of ID Fan-Inboard (Horizontal) (after bearing replacement)

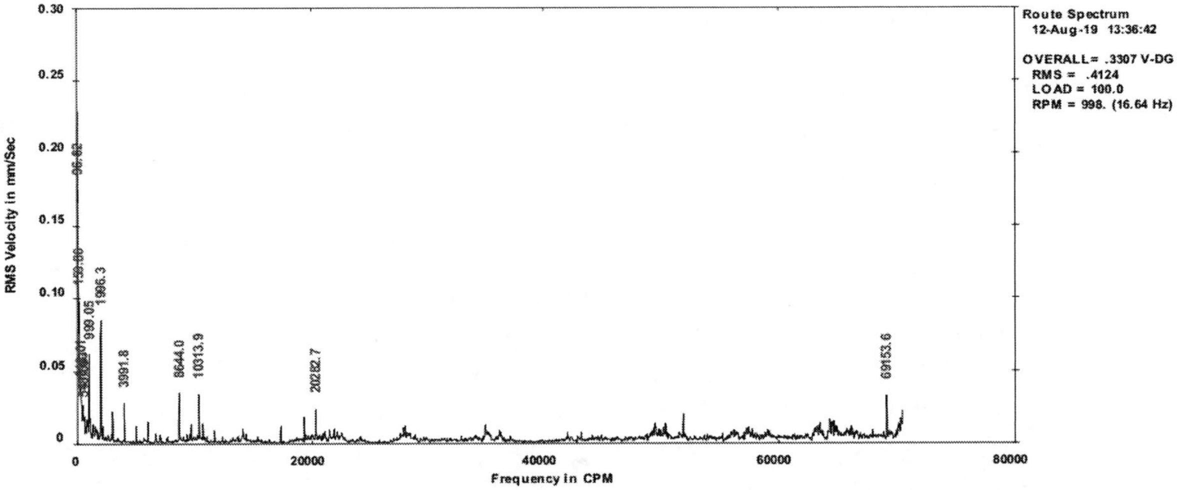

Figure 9: Vibration spectrum of ID Fan-Inboard (Vertical) (after bearing replacement)

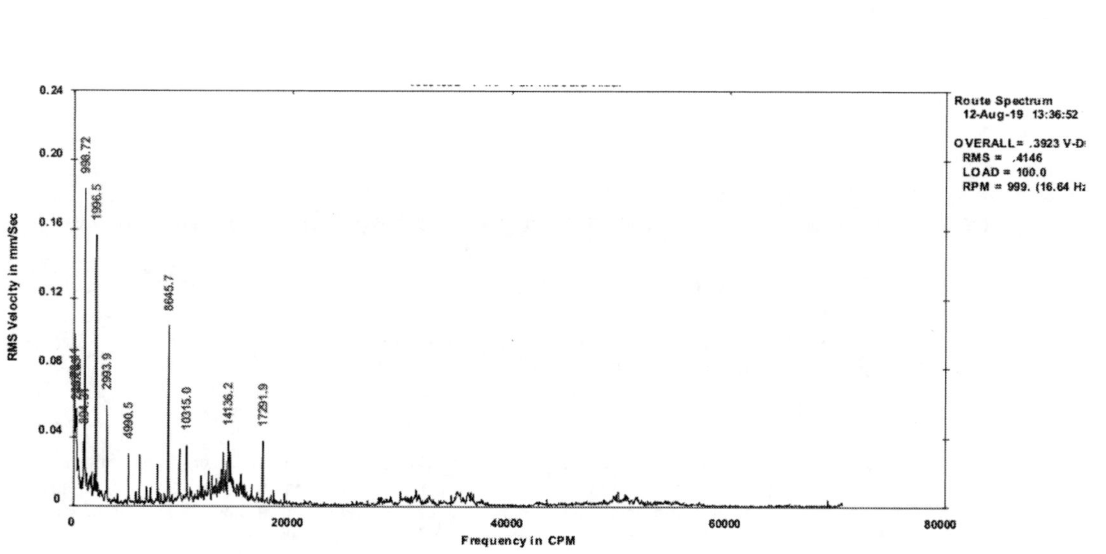

Figure 10: Vibration spectrum of ID Fan-Inboard (Axial) (after bearing replacement)

4. Summary

Vibration analysis is a very strong and sensitive technique for condition monitoring of plant machineries. It can detect the defect or any irregularity in the rotating element of the machine like; ID fan in power plant etc. Also, it can detect the defects when the machine is under operation and no need to stop it for inspection. Condition monitoring decreases the down time and improves the reliability of the plant. Also, it eliminates the sudden breakdown of the system. Equipped with proper condition monitoring technique and strategy, a power plant ensures the safe and reliable operations.

References

[1] Mohanty, J. K., Dash, P. R., & Pradhan, P. K. (2020), FMECA analysis and condition monitoring of critical equipments in super thermal power plant. *International Journal of System Assurance Engineering and Management*, 1-17, (doi:10.1007/s13198-020-00945-4).

[2] Nag, P. K. (2014), *Power Plant Engineering*, Tata Mc Graw-Hill Education, New Delhi.

[3] Mohanty, A. R..(2018), *Machinery condition monitoring: Principles and practices*. CRC Press, London.

[4] Jardine, A. K., Lin, D., & Banjevic, (2006), D. A review on machinery diagnostics and prognostics implementing condition-based maintenance. *Mechanical systems and signal processing*, *20* (7), 1483-1510.

[5] Kim, H., Na, M. G., & Heo, G. (2014), Application of monitoring, diagnosis, and prognosis in thermal performance analysis for nuclear power plants. *Nuclear Engineering and Technology*, 46 (6), 737-752.

[6] Mohanty J.K., Adarsh A., Dash P.R., Parida K. and Pradhan P.K. (2019), Integrated Condition Monitoring of Large Captive Power Plants and Aluminum Smelters, *Sound & Vibration,* 53 (5), 223-235.

[7] Collacott, R. A. (1975), Condition monitoring by sound analysis. *Non-Destructive Testing*, *8* (5), 245-248.

[8] Bagavathiappan, S., Lahiri, B. B., Saravanan, T., Philip, J., & Jayakumar, T. (2013), Infrared thermography for condition monitoring–A review. *Infrared Physics & Technology*, *60*, 35-55.

[9] Pradhan, P. K., Roy, S. K., & Mohanty, A. R.. (2019), Detection of Broken Impeller in Submersible Pump by Estimation of Rotational Frequency from Motor Current Signal. *Journal of Vibration Engineering & Technologies*, 1-8. (doi.org/10.1007/s42417-019-00165-6).

[10] Sugimura, J., Takesue, M., & Yamamoto, Y. (1998), Wear debris analysis for condition monitoring of gears. *Tribology Series,* 34, 431-442.

IOP Conf. Series: Materials Science and Engineering 998 (2020) 012008 doi:10.1088/1757-899X/998/1/012008

Health Monitoring of Induced Draft Fan in an Aluminium Smelter

S S Das[1], J K Mohanty[1], D Panda[1], Pratima Sarkar[2], P K Pradhan[1]*

[1]Mechanical Engineering Department,VSSUT, Burla, Odisha, India.
[2]Vedanta Limited, Jharsuguda, Odisha, India.
*Corresponding Author:prasant2001uce@gmail.com

Abstract. Maintenance is highly essential as to support the machine availability, product quality and plant safety. The timely repair at minimum coast is the basic criteria for good maintenance. Appropriate maintenance can sustain continuous and reliable operation with reduction in down time, high productivity, number of secondary failures, low inventory level and accident of the plant. In spite of high initial investment, most of the industries adopt condition based maintenance (CBM) to avoid sudden disruption in production. CBM is an important tool which includes observing, monitoring, diagnosis and prognosis for running a plant seamlessly, as this maintenance planning and actions have become an integral part of engineering investigation. In this paper, condition monitoring of Induced Draft (ID) fan in an Aluminium smelter has been done while it was in operation. The vibration of ID fan was measured and analyzed to detect the defect of its impeller and corrected subsequently.

Keywords: ID fan, Aluminium smelter, condition monitoring, vibration analysis.

1. Introduction

Aluminium is the most useful engineering material, next to ferrous metal. Aluminium is a good conductor of heat and electricity. It is lighter than other engineering material. The main areas of application in which aluminium is used are; electrical conductors, air craft, packaging, transportation and building materials. It is also extensively used in cooking utensils, electrical cables and telephone wires.

Aluminium is produced in mainly through two stages; mining and refining, smelting and casting. Aluminium refining is done by separating of aluminium oxide (or alumina) from the bauxite. Aluminium smelting is carried out through the electrolysis process [1]. The electrolysis is done in an iron vat, which serves as the cathode and carbon anodes are immersed for the electrolysis. Upon passage of electrical current through the anodes and electrolyte, the molten aluminium metal is deposited at the bottom of the vat/pot. The tapped metal is transferred to cast house. The molten metal is maintained at the suitable temperature of 750 ^{0}C (approx) in the furnace and cast as ingots, billets, slabs etc. in the cast house.

Besides availability of Raw materials like Alumina, Power & Carbon anodes, Continuous production of Aluminium from a Smelter also depends on the availability of its machine and machinery systems of the plant. To get maximum availability, it is essential to keep the machines and equipment in good operational condition. To achieve this, proper maintenance strategies and planning are required [2, 3, 4, 5 & 6]. The different types of maintenance strategies are as follows;

i Breakdown maintenance
ii Routine Maintenance:
iii Precautionary Maintenance

iv Condition Based Maintenance (CBM) through condition monitoring.

In spite of high initial investment, many industries adopt CBM to reduce unexpected shut down and accidents. It measure and monitors the condition of the machines using diagnostic tools and techniques, while it is in operation and the maintenance is done only when it is really necessary.

1. 1 Condition Monitoring

Condition monitoring (or CM) determines the health and condition of machines and machinery systems by observing, checking, measuring, interpreting and monitoring certain parameters. CM provides information of the machine condition before it is going to fail. It helps in improving the reliability of the machines. It is used to check the machine while it is in operation and it gives indication (in advance) of deterioration before the breakdown occurs. There are different methods and techniques of CM, described as follows;

i. *Performance monitoring*
ii. *Visual, optical, tactile and aural monitoring*
iii. *Temperature monitoring:* If there is unusual heat generation in any running equipment,-then that indicates presence of a probable defect (or abnormalities) in it [7].
iv. *Vibration monitoring:*Vibration analysis can be employed for the diagnosis of all types of defects in rotating element of the machine [8].
v. *Lubricant Monitoring &Wear debris analysis:* Lubricant Monitoring necessitates the lubricant condition in a bearing or gearbox to find out degradation in lubricant chemistry, water contamination and wear quantum. The debris, or particles, present in lubrication oil that could indicate type, size and texture of wear particles [9].
vi. *Noise monitoring:*The measurement of a machine's sound can also be employed for detecting the defects, present in the running machines. The accuracy of this method depends on the differential sound pressure level [10].
vii. *Motor Current Signature Analysis (MCSA):* It is the technique used to diagnose problems in induction motors or motor driven rotating machines [11].
viii. *Ultrasound/ultrasonic monitoring*
ix. *Leakage Monitoring*

Since, the CBM's initial investment is high; prior assessment is required before the implementation of CBM .Generally, Failure Mode, Effect and Criticality Analysis (FMECA) is done to classify the critical machines before going to execution of CBM of an industry.

1.2 FMECA through RPN

FMECA [10] analyze the probable component's failure modes, assess the risk associated with these failure modes and find out the resultant effects on system. The main steps are:

a) Identifying the machines to be analyzed.
b) Identifying the function(s), failure(s), effect(s),.
c) Evaluating the risk associated.
d) Prioritizing and assigning corrective actions.

There is a tool, used to determine the criticality of the machines. i. e. Risk Priority Numbers (RPNs).The RPN is the product of severity (S) x difficulty to detect (D) x occurrence (O). Each one is evaluated with a scale from 1 to 10 and the highest RPN is 1000 (i.e. 10x10x10) . That indicates failure is not detectable by inspection, very severe and the occurrence is almost sure. So, RPN enables to focus on the highest risks.

IOP Conf. Series: Materials Science and Engineering 998 (2020) 012008 doi:10.1088/1757-899X/998/1/012008

2. Methodology and Experimentation

The present work is based on the Aluminium Smelter in Vedanta Limited (VL), Jharsuguda, India. The plant has two smelting units (capacity of total 1.75 MTPA). The firm has collaborated with GAMI, China.It adopts rigorous health and safety measures for the machines as well as for the employees. Most of the machines are maintained through the CBM. Aluminum smelter has mainly four sub units; Green anode plant (GAP), Pot room/ pot line, Cast house, Rectifier and switchyard. In pot-line, there are four important machines; pot trending machine (PTM), Induced draft fan(ID fan), Compressor and Cooling tower. In this paper, the condition monitoring of Induced Draft fan (ID fan) using vibration analysis has been described.

2.1 FMECA of Aluminium Smelter Plant

FMECA Analysis is done for all individual machines and the same is reviewed once in a year which is a normal practice in Smelter Plant of Vedanta Limited. After considering of all three parameters, the RPN of all essential machines are evaluated and are mentioned in Table 1. The said FMECA analysis says that ID fan is the critical equipments of the Smelter Plant.

Table 1: Criticality Analysis for Aluminium Smelter's Machines

SN	Equipments	S	O	D	RPN	Remark
1	PTM	8	6	4	192	Very Critical
2	ID fan	6	7	5	210	Very Critical
3	Compressor	5	7	3	105	Critical
4	Cooling tower	5	5	3	75	Non Critical

Note: For this analysis, (i) RPN $\geq$ 150 is considered as very Critical, (ii) RPN in between 150 and 100 is considered as critical, (iii) RPN <100 is considered as Non critical.

3. Condition monitoring of ID fan of Aluminium Smelter

During the electrolysis process, the oxygen from Alumina (Al_2O_3) reacts with the carbon in the anodes, forming carbon dioxide which bubbles off. Besides these, other unwanted fumes are also being generated, which pose health hazards to the people working in Pot Line. To remove these fumes, ID fan is used. It evacuates the hot fumes from the pots to the Fume treatment plant (FTP). After treatment in FTP to remove the hazardous elements, the residual gases are discharged to atmosphere.

3.1 Specification Details:

Motor Power: 1400KW, Speed: 990 RPM, Motor's NDE (Non-Drive End)/ Outboard Bearing: NU 1032, Motor's DE (Drive End)/ Inboard Bearing: NU 1034 & 6034, Fan DE & NDE(Both Inboard & Outboard) Bearings: 22236, Vibration velocity (RMS) limit (as per OEM); Normal: below 4.5 mm/s, Sub critical: 4.5-6.5 mm/s, Critical: above 6.5 mm/s.

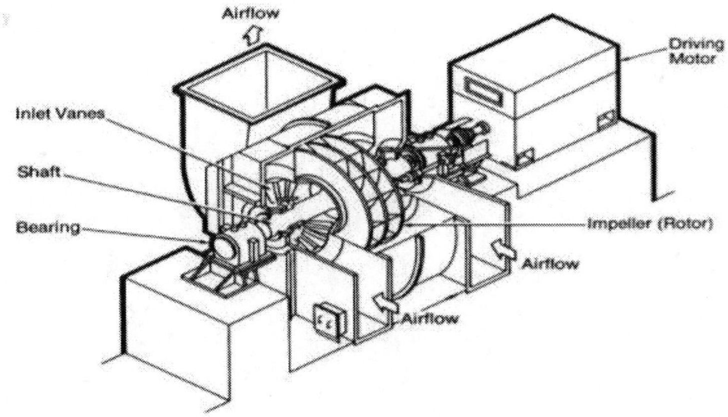

Figure 1 (a): Induced Draft (ID) Fan set up

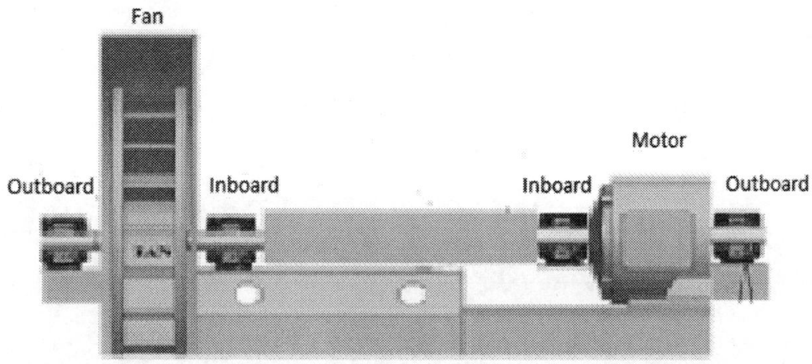

Figure 1 (b): Diagram of an ID Fan (Schematic)

3.2 Diagnostic Symptoms

The ID Fan is monitired through the vibration analysis. The data are recorded and analysed by using accelerometer and AMS Suite Software. A case study (i. e. condition monitoring of ID fan-3, Pot line-1 of Smelter-1) was carried out during February 2020 and has been considered here. Vibration measurement has been done in three different directions namely; horizontal (H), vertical (V) and axial (A).The RMS values are shown in Table 2. Similarly, the vibration spectra are shown in Figure2 and Figure3.

Table 2: RMS value of vibration (in mm/s)

Location	Direction	8th Feb	10th Feb	15th Feb	22nd Feb	27th Feb	28th Feb	1st Mar
Motor Outboard (NDE)	H	3.274	3.448	3.634	6.194	Balancing of fan impeller	2.250	2.629
	V	.419	.406	.757	.818		.390	.449
	A	1.130	.882	.603	1.230		.450	.361
Motor Inboard (DE)	H	4.760	3.004	3.036	5.646		2.237	2.579
	V	1.089	.929	.678	1.215		.572	.537
	A	.733	.812	.397	.994		.871	.427
Fan Inboard (DE)	H	4.585	3.356	2.816	5.103		2.148	2.510
	V	1.064	.417	1.119	1.182		.375	.612

	A	.916	.528	.737	.722		.577	.293
Fan Outboard (NDE)	H	2.322	1.894	3.301	4.069		1.857	1.676
	V	.922	.731	1.376	1.046		.571	.702
	A	.541	.677	.771	.913		.410	.546

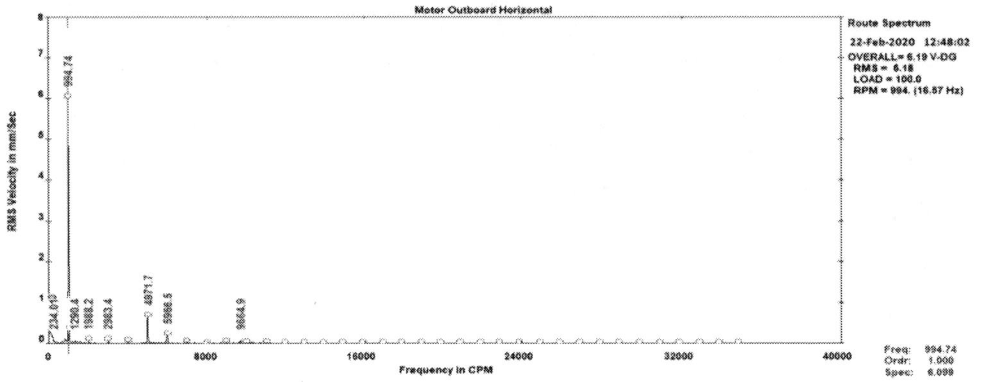

Figure 2: Vibration spectrum of ID Fan-Inboard/ Motor outboard (before balancing)

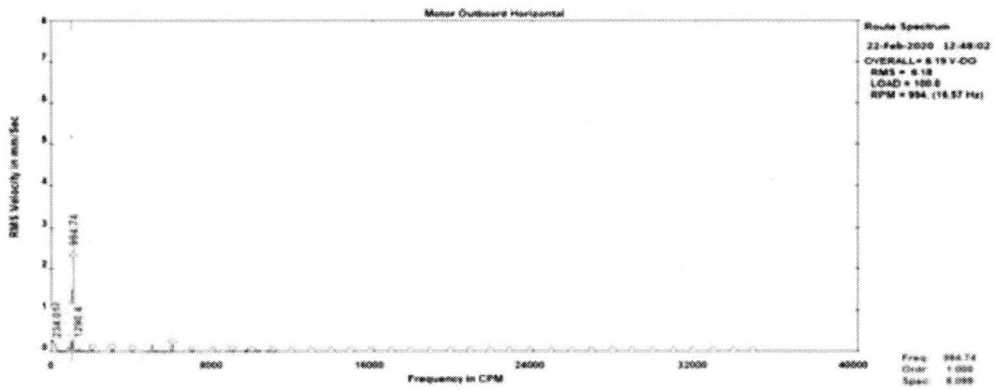

Figure 3: Vibration spectrum of ID Fan-Inboard/ Motor outboard (after balancing)

3.3 Observations and Analysis

i. After evaluating the vibration (rms) readings (from Table 2), it was observed that (on 22nd Feb.) the motor drive end & non-drive end vibrations amplitude in the horizontal direction were comparatively higher than the vibration values of other directions and very much nearer to the critical limit. That indicates that there is some problem in rotating masses of ID fan.

ii. It is observed from the spectrum (Figure 2) thatthe vibration amplitude at rotational frequency/speed (i. e. 994 cpm) is quite high. It indicates there is unbalance of rotating mass.

iii. There are no peaks at non-synchronous frequency. So, there is no possibility of bearing defect.

As per recommendation, balancing of fan impeller has been done.It is observed that due to handling of hot gases/fumes, an impeller-blade is worn out, thus making the impeller unbalanced. After balancing of the impeller, the ID fan was put into operation. Subsequently, the vibration measurement in was taken & the RMS value was found to be normal as indicated in Table-2 on 28th Feb.The 1xRPM amplitude in the vibration spectrum taken after the correction(Shown in Figure 3) has also decreased considerably. In last 6 years, for the first-time balancing has been carried out for a FTP-ID Fan. All the remaining ID fans in smelter plant are normal.

IOP Conf. Series: Materials Science and Engineering 998 (2020) 012008 doi:10.1088/1757-899X/998/1/012008

4. Summary

Condition monitoring is used to find out the forthcoming defect (or abnormality) of the machine and machinery element. Among all techniques, vibration analysis is very strong and reliable technique for condition monitoring of rotating machineries like; ID fan in Aluminium smelter plant etc. Vibration measurement and analysis can be online or off line. If the measured values deviated from the usual trend by a certain extent, then the data and trend are studied thoroughly. According to the trend and symptoms, the appropriate measures are suggested for the suitable maintenance. The timely repair and maintenance increase the availability of plant machineries and equipments. Ultimately, the condition monitoring helps in improving the efficiency, reliability and safety of the plant.

References

[1] Mandin, Philippe,Whrich, Rolf, Roustan, H, (2009), Industrial Aluminium Production: the Hall-Heroult Process Modelling, *The Electrochemical Society Transactions*,Vol. 19 (26), pp. 1-10.

[2] Mohanty, A. R.(2018) *Machinery condition monitoring: Principles and practices*. CRC Press, London.

[3] Mohanty, J. K., Dash, P. R., & Pradhan, P. K.. (2019), FMECA analysis and condition monitoring of critical equipments in super thermal power plant. *International Journal of System Assurance Engineering and Management*, 1-17, (doi:10.1007/s13198-020-00945-4).

[4] Jardine, A. K., Lin, D., & Banjevic, D. (2006), A review on machinery diagnostics and prognostics implementing condition-based maintenance. *Mechanical systems and signal processing*, 20 (7), 1483-1510.

[5] Kim, H., Na, M. G., & Heo, G. (2014), Application of monitoring, diagnosis, and prognosis in thermal performance analysis for nuclear power plants. *Nuclear Engineering and Technology*, 46(6), 737-752.

[6] Mohanty A. R., Pradhan P. K., Mahalik N. P. and Dastidar S. G., (2012), Fault Detection in Centrifugal Pump Using Vibration and Motor Current Signature Analysis, *International Journal of Automation and Control*, Vol 6 (3/4), 261-276.

[7] Bagavathiappan, S., Lahiri, B. B., Saravanan, T., Philip, J., & Jayakumar, T, (2013), Infrared thermography for condition monitoring–A review. *Infrared Physics & Technology*, *60*, 35-55.

[8] Mohanty J.K., Adarsh A., Dash P.R., Parida K. and Pradhan P.K, (2019), Integrated Condition Monitoring of Large Captive Power Plants and Aluminum Smelters, *Sound & Vibration*, 53 (5), 223-235.

[9] Sugimura, J., Takesue, M., & Yamamoto, Y. (1998), Wear debris analysis for condition monitoring of gears. *Tribology Series*, 34, 431-442.

[10] Collacott, R. A. (1975), Condition monitoring by sound analysis. *Non-Destructive Testing*, *8* (5), 245-248.

[11] Pradhan, P. K., Roy, S. K., & Mohanty, A. R. (2019). Detection of Broken Impeller in Submersible Pump by Estimation of Rotational Frequency from Motor Current Signal. *Journal of Vibration Engineering & Technologies*, 1-8. (doi.org/10.1007/s42417-019-00165-6).

IOP Conf. Series: Materials Science and Engineering 998 (2020) 012009 doi:10.1088/1757-899X/998/1/012009

Design, analysis and optimization of 12 cylinder v-type diesel engine components

Kumar Aman[1], Ankit Chandanpat[2], Pawar Kiran[3], Shivashankar Angadi[4] and Nilesh Totla[5]

[1]Mechanical Engineering Department, MIT Academy of Engineering, Pune, India

[2]Mechanical Engineering Department, MIT Academy of Engineering, Pune, India

[3]Mechanical Engineering Department, MIT Academy of Engineering, Pune, India

[4]Mechanical Engineering Department, MIT Academy of Engineering, Pune, India

[5]Mechanical Engineering Department, MIT Academy of Engineering, Pune, India

E-mail: amkumar@mitaoe.ac.in

Abstract: The designing, material selection and stress analysis are important processes in mechanical engineering design. The below brief study deals with the afore-mentioned processes conducted on the piston, connecting rod and crankshaft of a 12 cylinder V-type engine used for wood cutting application. The objective of this study was to optimize the weight of the given parts and increase their capacities to withstand different loads by means of iterative design changes as well as material selection for the three components. Pre-existing engine data and empirical relationships were used to generate a mathematical model for the design procedure and Ashby standards were followed for material selection for the above-mentioned parts. The Computer Aided Design (CAD) modelling and analysis of these parts was conducted in the Solidworks software.

Keywords: Optimization, piston, connecting rod, crankshaft, material selection.

1. Introduction

Three of the very important components of V12 engine are connecting rod, crankshaft and piston which are involved in most of the engine failure cases. Hence, we have taken these components as the focus of our study. Thus, we have made the following objectives to achieve optimized strengths for the three components.

The effect of reciprocating inertia forces and bending forces plays a vital role in the performance of a connecting rod. Connecting rod undergoes large repetitive loads of varying values and hence we need to increase its fatigue strength.

The key component of the engine that converts the reciprocating linear motion of the piston to rotational motion is the crankshaft. The cyclic changes in gas pressure inside the cylinder encourage the crankshaft torsional vibrations. In order to eliminate the leading cause responsible for part failure, it needs modification for withstanding the high torsional stress.

Engine performance is improved, if the weight of the components is reduced optimally. Weight can be reduced by changing the size, shape or material of the component. In this study, we have aimed to reduce the weight of the components through changes in their shapes and materials.

The above mentioned objectives were decided for a 12 cylinder V-type engine used in wood cutting application. The following table gives the engine specifications and operating parameters that are necessary for calculations and modifications. A V12 engine is a 12-cylinder piston engine where two banks of six cylinders are arranged in a V configuration around a common crankshaft. Each bank of a V12 engine essentially functions as a straight-six engine, which itself has perfect primary and secondary engine balance. The following table gives the engine specifications and operating parameters that are necessary for calculations and modifications.

Table 1. Engine specification and objective parameters

Given inputs:				
Sr. No.	Parameters	Symbol	Values	Units
1	Maximum Combustion pressure	p_{za}	11.307	MPa
2	RPM	n_N	2600	RPM
3	Max Speed in idling	$n_{id\,max}$	2700	RPM
4	Lambda	λ	0.27	
5	Mass of connecting rod assembly	$m_{c.r}$	3.39	kg
6	Length Of connecting rod	$l_{s.e}$	460	mm
7	Outer diameter of small end	d e	64	mm
8	Internal diameter of small end	d	50	mm
9	Youngs modulus of Con rod grade 40X steel	$E_{c.r.}$	220000	MPa
10	Inner radius of big end of connecting rod	r_1	43	mm
11	Big end length	l_c	33	mm
12	Mass of crankshaft side conrod	$m_{cr,c}$	2.458	kg
13	Mass of connecting rod assembly	m_{cr}	3.39	kg
14	Length of connecting rod shank	L_{cr}	222	mm
15	Cylinder Diameter	D	120	mm
16	Piston Stroke	S	120	mm
17	Piston area	F_p	113	cm^2
18	Maximum rated force for piston	N_{max}	0.00697	MN
19	Mass of piston assembly	m_p	2.94	kg
20	Piston height	H	120	mm
21	Piston crown wall thickness	s	12	mm
22	Thickness of the top ring land	hl	6	mm
23	Youngs modulus of piston pin	E	100000.00	MPa
24	Outer diameter of main journal	d_{mj}	90	mm
25	Length of journal	l_{mj}	37	mm
26	Crank pin inner diameter	d_{cpi}	30	mm
27	Length of crankpin	l_{cp}	68	mm
28	Crank radius	R	60	mm
29	Crank Pin Diameter	d_{cp}	80	mm

2. Literature review

The study includes the effects of the thermal conductivity of the material of piston, piston rings, and combustion chamber wall. Results show variation of temperature, stresses, and deformation at various points on the piston [1]. With this study we came to know the designing of strong cross-sectioned parts and optimizing the material thermal conductivity.

The mechanical properties of the Titanium composites were remarkably improved by adding a small amount of carbon black at 0:070:16 mass%. The increases in the yield stress of the extruded sponge and fine Titanium matrix composite were 70.0 MPa and 291 MPa, while the tensile strength increases were 67 MPa and 231 MPa,

IOP Conf. Series: Materials Science and Engineering 998 (2020) 012009 doi:10.1088/1757-899X/998/1/012009

respectively, compared to those of extruded pure Ti with no reinforcement [2]. The TMC material has excellent mechanical properties at the same time being lightweight. Hence it has been used as a candidate for the piston.

Results show that the mechanical properties such as tensile strength of the heat-treated Al-Zn-Mg-Cu (AA7068) alloy at 210° C aging temperature has increased by 155% as compared to as cast alloy. The compressive strength and hardness of the alloy at 210° C aging temperature shows improvement of 41% and 54% respectively as compared to as cast alloy [3]. Since connecting rod undergoes significant amount of compressive forces, thus we have included Al-7068 (or AA7068) as a strong candidate material for it.

To design a new cross section of connecting rod for optimizing its weight, three cross sections: T Section, C section and hollow C section were designed and analyzed. An optimized C section was obtained with overall 14% weight reduction using response surface optimization methodology [4]. The method used in this study was understood and similar iteration were applied in the project.

Different charts compare materials on various properties such as strength, ductility, stiffness, density, thermal conductivity and cost[5]. Hence, this information about the selection of material and selection of property chart using the performance index formula has been studied from this data book and the knowledge has been applied for material selection in this project.

The author has performed the calculation of bending stress, torsional stress, stress concentration factor, fatigue strength, acceptability criteria for shrink-fits and semi-built crankshaft. These methods were applicable for both main propulsion and auxiliary purpose and being designed for continuous operation at the specified (nominal) rating [6]. Hence, through this we take away the formulas required for the calculation of crankshaft.

The author describes the process of selection of material for connecting rod where the constraints are to make light weight and yet strong enough to carry the peak load. Candidate materials has been selected using CES EduPack section chart and alloy steel comes out to be their final one [7]. Hence with this information we carry out the material selection of the connecting rod.

The webpage basically discuss the methods which has to be used in order to perform the research work [8]. This section explains us what and how do we perform the research in such a way that readers could easily evaluate the reliability and validity of the work.

The book aim to impart the advanced skills of design for analysis of mechanical components [9]. We took the generic method of designing and selection of materials for the selected engine parts.

The author in the book explained the materials to be used in industrial applications [10]. We understand the methods involved in the material selection such as digital logic method and learned how to find the performance index value.

3. Methodology

The research that has been performed regarding 12 cylinders V-type engine calculation, material selection, design and analysis comes under the quantitative method. Meanwhile during calculation of the parameters, most of the data was received from the project sponsorship company whereas some data such as maximum combustion pressure, air-fuel equivalence ratio, fuel-air equivalence ratio, etc. was more generic.

Data collection has been done primarily from the respective source [6,7,9]. The inputs and objectives of the project are largely based on customer specifications. While as a project-based learning initiative our team has done research at the ARAI library, Pune and as a result we understood the design methodologies of various engine parts and the generic procedure of material selection in automobile engine.

Before the calculation of the engine, the data sheet was thoroughly observed and the useful and necessary data was identified. Data related to piston, connecting rod and crankshaft was preferred over that of the carburetor and other engine components. Using empirical relations, a mathematical model was constructed for calculating various dimensions of the three components. When it comes to material selection process the performance index has been characterized differently according to the load applied and shape of the parts used in it. Iterative structural and thermal analysis were conducted to verify the results of the design calculations and material selection.

The tools and techniques that were used for design calculations, material selection, and analysis are: We used the MS-excel software for performing various calculations, the Granta software was used for material screening, the digital logic method was used for determining various weighing factors in the material selection process and the CAD designing and analysis was done in Solidworks.

4. Material selection

Material and process selection of the following three parts (piston, connecting rod and crankshaft) has been performed by using the digital logic method which is one of the type of weighted property method [10]. This method is generally used where numerous material properties are given and the relative importance is not very much clear. Hence, determination of their weighting factors are intuitive in nature. When it comes to the selection of required materials the method is purely based on its performance index (C) where, [C=Young's modulus/Density]. The C values is different for different shapes and the type of load applied on it. For example, E/ρ=C for (stiff ties, centrifugal loading), E^(1/2)/ρ=C for (stiff beam and columns) and , E^(1/3)/ρ=C for (stiff plates) etc. A logarithmic chart is designed to display these groups or "material indices" from which the material families which satisfy the performance index's criteria can be selected. The chart used here for material selection of connecting rod is young's modulus v/s density with equation $E^{1/2}/\rho = C$ (minimum weight design of stiff beam). The chart does not give us the final material, but are valuable in guiding us quickly and efficiently to a subset of materials worth considering as shown in figure 2. It is also useful in making sure that we do not overlook a promising candidate material [5]. The materials selected from this chart using the Granta software are listed below. Similar procedures were followed for the other two components.

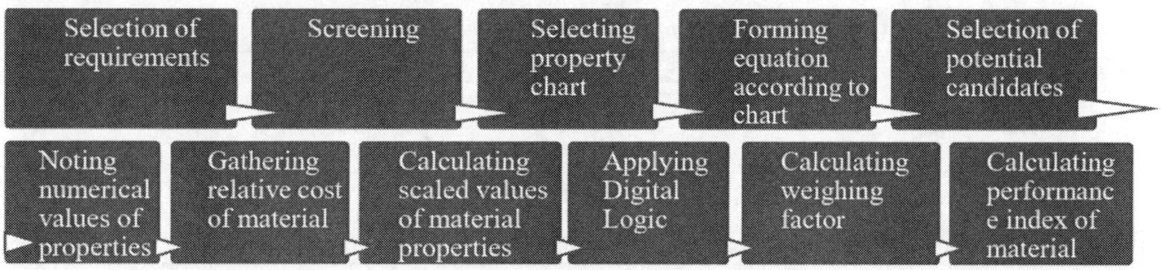

Figure 1. Material selection procedure

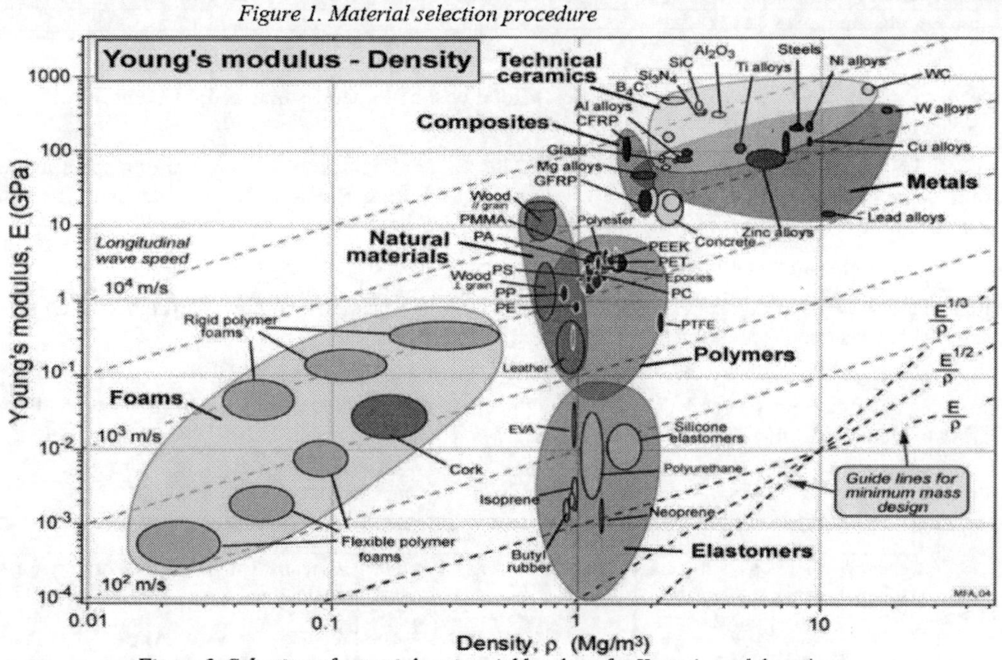

Figure 2. Selection of materials using Ashby chart for Young's modulus v/s

The above figure shown is an example of material selection of light and stuff components

4.1 Piston

- Screening:-{Function: To take gas force; Objective: To reduce the weight and increase the strength; Variable: Density; Constraints: Force, cross sectional area, length}

IOP Conf. Series: Materials Science and Engineering 998 (2020) 012009 doi:10.1088/1757-899X/998/1/012009

Table 2. Properties required for piston material selection

Material	Properties						
	Frictional coefficient	Wear rate	Thermal capacity [J/K]	Thermal conductivity [W/m-K]	Specific gravity	Yield strength [MPa]	Tensile strength [MPa]
Grey cast iron	0.41	2.36	0.46	80	7.2	460	455
Ti-alloy	0.34	246.3	0.58	17.58	4.42	700	1014
Ni-alloy	0.39	5.32	0.41	12.6	8.4	851	1100
Al-alloy	0.36	2.89	0.714	190	2.73	480	510
Aluminium matrix composite	0.35	3.25	0.98	155	2.7	276	310
Aluminium matrix composite	0.44	2.91	0.92	180	2.8	425	485
Titanium matrix composite	0.31	8.19	0.51	17.85	4.68	700	1029

Table 3. Performance index and final material for piston

Material	Properties		
	Relative cost	Performance index	Final material
Grey cast iron	1	51.04	51.04
Ti-alloy	20	77.	1552.30
Ni-alloy	17	69.21	1176.635
Al-alloy	10	48.08	480.84
Aluminium matrix composite	2.8	42.99	120.39
Aluminium matrix composite	2.6	52.46	136.40
Titanium matrix composite	**20.5**	**58.06**	**1190.23**

Relative cost = (Total cost of low carbon iron steel)/(total cost of a specific material). The more is the relative cost lesser will be the overall cost.

Final material = (Relative cost*Performance index). This value indicates the final rating of a specific material. Higher the value of final material results in greater chance of getting selected. But importance of the value of relative cost is more than that of final material.

Scaled value = (Minimum value in the list*100)/(Numerical value of property).

Performance index = $\sum$(Scaled property)i*(weightage factor), {i= property number starting from top left corner}. Weightage factor has been calculated by using the digital logic method [10].

4.2 Connecting rod

- Screening: - {Function: To carry the material; Objective: To reduce the weight and increase the fatigue cycle; Variable: Density; Constraints: Shape and force}

Table 4. Properties required for connecting rod material selection

Material	Properties							
	Density [kg/m^3]	Yield strength [MPa]	Ultimate tensile strength [MPa]	Brinell hardness	Fatigue strength [MPa]	Fracture Toughness [MPa-m$^{1/2}$]	Working stress [MPa]	FOS
C-70	7850	420	640	190	270	28.82	106.667	6
Al-360	2680	260	310	75	82	17.3	51.667	6
T-2024	2780	324	469	120	138	23.3	78.167	6
Al-7068	2850	683	710	190	220	35	118.333	6

IOP Conf. Series: Materials Science and Engineering 998 (2020) 012009 doi:10.1088/1757-899X/998/1/012009

Table 5. Performance index and final material for connecting rod

Material	Properties		
	Relative cost	Performance index	Final material
C-70	1.5	152.00	228.00
Al-360	1.2	65.08	78.10
T-2024	6.3	90.29	568.87
Al-7068	**7**	**140.16**	**981.15**

4.3 Crankshaft

Table 6. Properties required for crankshaft material selection

Material	Properties			
	Density [kg/m^3]	Youngs Modulus [MPa]	Yield Strength [MPa]	Thermal Expansion Co-efficient [K^{-1}]
Forged steel	7700	190	241	0.000015
42CrMo4 Steel	7850	210	750	0.000012
Tungsten	19250	411	1510	0.0000046

Table 7. Performance index and final material for crankshaft

Material	Relative Cost	Performance Index	Final Material
Forged steel	**11.9**	**50.260**	**598.102**
42CrMo4 Steel	6.5	54.317	353.060
Tungsten	2.1	84	176.4

5. Calculations & design:

The design of connecting rod, piston and crankshaft is based on evaluation of safety against fatigue in highly stressed areas.

A. CONNECTING ROD

1. BIG END:

i. The max. inertial force = -0.031405 MN
ii. The bending stress for cap and shell= -164.83 MPa

2. SMALL END

- Design of I-I section

i. Maximum stress in pulsating cycle= 30.4 N
ii. Mean Stress= 15.19 MPa;

3. CONNECTING ROD SHANK

i. Max stresses caused by compression forces = 79.447 MPa
ii. Min. stress caused by a tension force = 72.369 MPa

B. PISTON

- Compression stress at section X-X:

i. Max compression force = 0.127769 MN;
ii. Compression stress = 37.81141 MPa;

- Calculations for PISTON RINGS:

i. Bending stress of the ring in operation = 233.609 MPa
ii. Bending stress in slipping the ring over a piston = 298.2715 MPa

- Calculations for PISTON PIN

i. Tangential shear stress= 109.4785 MPa
ii. Maximum Ovalization increase in the pin diameter = 4.18E-05 m

C. CRANKSHAFT

- Calculations for CRANK WEB

IOP Conf. Series: Materials Science and Engineering 998 (2020) 012009 doi:10.1088/1757-899X/998/1/012009

i. Maximum twisting moments = 220185 N-mm
ii. Maximum tangential stress = 8580544 Pa
iii. Maximum normal stress = 50743628 Pa

- Calculations for CRANK PIN
  i. Moment resisting to twisting the crankpin = 98542.93 mm^3
- Calculations for MAIN JOURNAL
  i. Main journal moment resisting to torsion = 143138.8 mm^3
  ii. Maximum tangential stress = 2.07E+01 MPa
- Calculations for unit area pressure on CRANKPIN
  i. Max pressure on crankpin = 41.03448 MPa
  ii. Max pressure on main journal = 23.95062 MPa

DESIGN CHANGE IN CONNECTING ROD :

The different design iterations for the connecting rod are as shown below in the figure. The figure (a) is the original design that on which shape iterations were conducted by keeping the applied material same throughout it. The first iteration, as shown in figure (b) has square-shaped slots made in the shank. The second iteration (figure (c)), shows the connecting rod with a longitudinal slot in its shank.

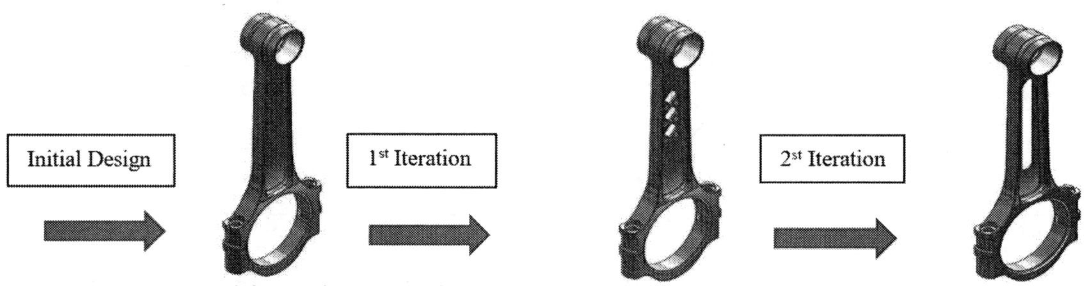

Figure 3. (a)Original design, (b)Design with square slots, (c)Design with longitudinal slot along the shank

6. Analysis

The materials selected for specific components after the material selection procedure needs to be verified by application of forces on them and analyzing whether they can withstand the operating conditions. Hence, the CAD models of the three components were generated. The static analysis of the components gave results for four parameters: Von Mises Stress, Strain, Displacement, and Factor of Safety. Material iterations and their analysis were carried out on the piston, the connecting rod and the crankshaft CAD parts to verify the results of the material selection process. Different shape iterations were conducted for the connecting rod shank, and then its static analysis was conducted.

6.1 Piston

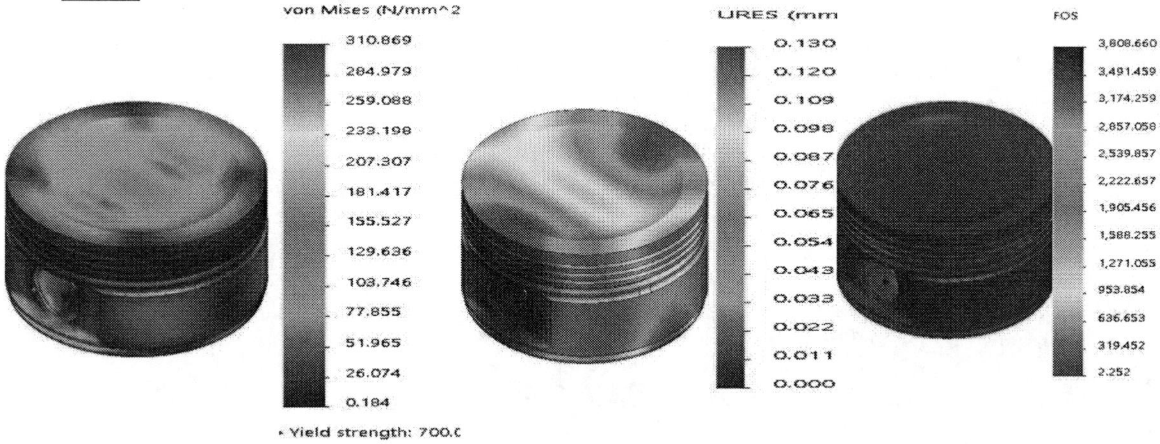

IOP Conf. Series: Materials Science and Engineering 998 (2020) 012009 doi:10.1088/1757-899X/998/1/012009

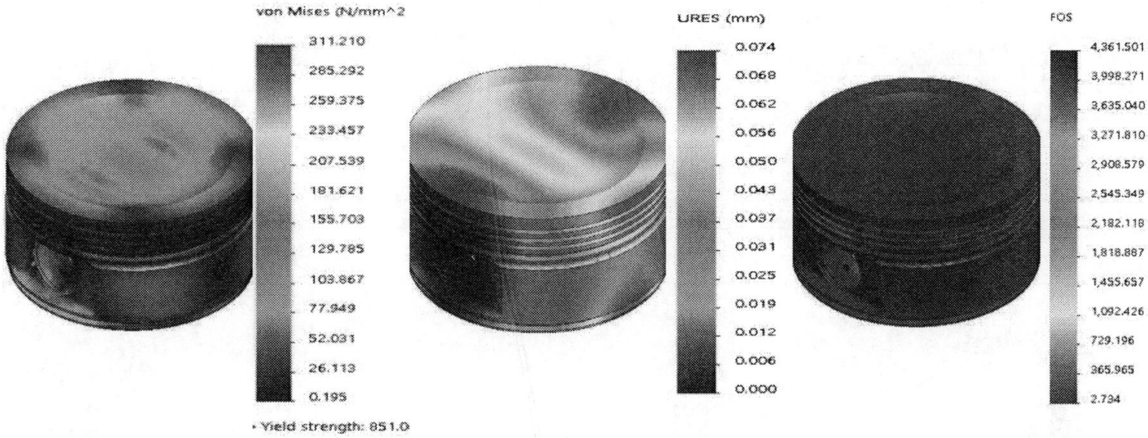

Figure 4. Analysis of Ti-matrix composite piston and Ni-alloy piston:(a) von Mises stress, (b)Displacement, (c)FOS for piston

Table 8. Result table for piston

Sr. No.	Parameter	Type/ Value			
1.	Mesh type	Solid Mesh (Standard mesh)			
2.	Load applied	Normal force and pressure			
3.	Fixture type	Fixed geometry			
		Previous (Ni alloy)		**Optimized (Ti-matrix composite)**	
		Min	**Max**	**Min**	**Max**
4.	Stress (von Mises Stress)	0.245 [MPa]	311.095 [MPa]	0.184[MPa]	310.869 [MPa]
5.	Displacement	0 [mm]	0.073 [mm]	0.000 [mm]	0.130[mm]
6.	FOS	2.735	3,477.899	2.252	3,808.660

6.2 Connecting rod

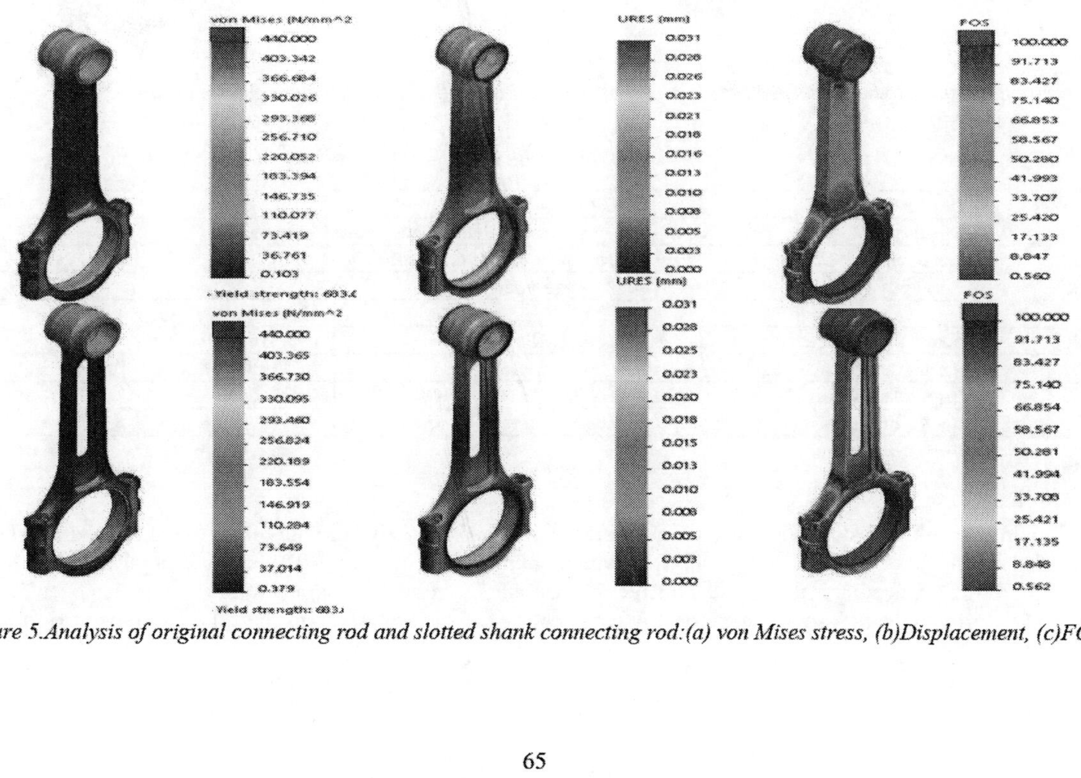

Figure 5.Analysis of original connecting rod and slotted shank connecting rod:(a) von Mises stress, (b)Displacement, (c)FOS

Table 9. Analysis result table for connecting rod

Sr. No.	Parameter	Type/ Value			
1.	Mesh type	Solid Mesh (Blended curvature-based mesh)			
2.	Load applied	Normal force and pressure			
3.	Fixture type	Fixed geometry			
		Without slotted shank, Al-7068		**With slotted shank, Al-7068**	
		Min	**Max**	**Min**	**Max**
4.	Stress	0.103 [MPa]	1,219.8 [MPa]	0.379 [MPa]	1215.7 [MPa]
5.	Displacement	0.000 [mm]	0.031 [mm]	0.000 [mm]	0.031 [mm]
6.	Factor of Safety	0.560	6,620.746	0.562	1803.243

6.3 Crankshaft

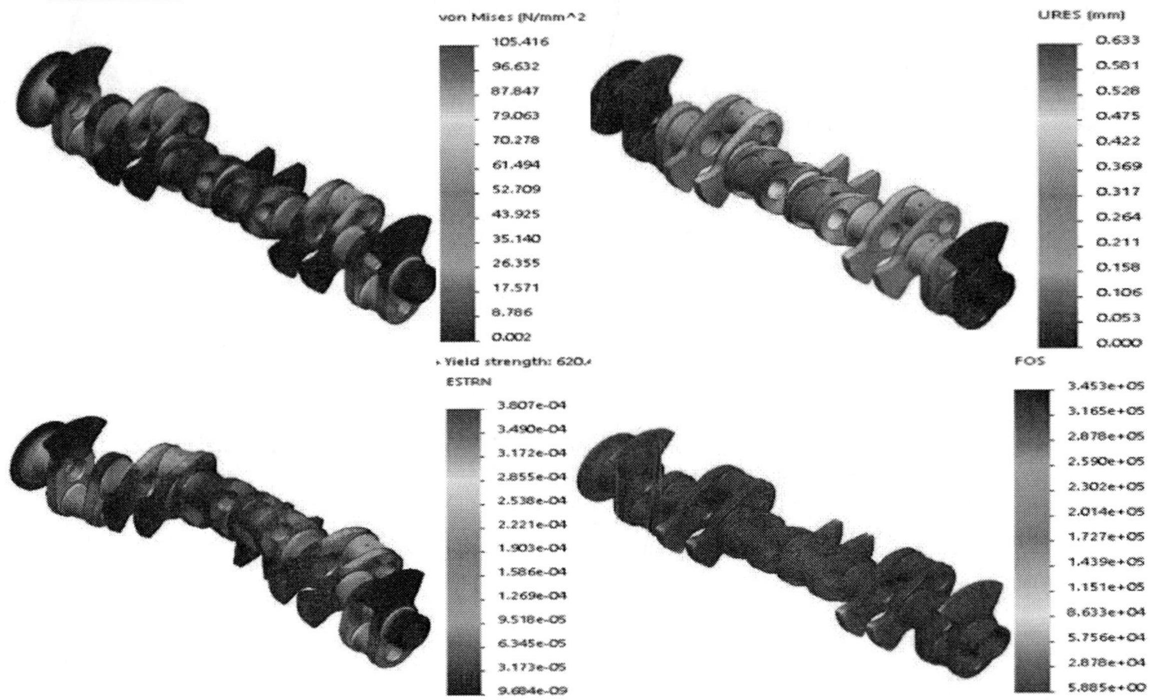

Figure 6. Analysis of "forged steel" crankshaft:(a) von Mises stress, (b) Displacement, (c) Strain, (d) FOS

Table 10. Analysis result table for crankshaft

Sr. No.	Parameter	Type/ Value	
1.	Mesh type	Solid Mesh (Blended curvature-based mesh)	
2.	Load applied	Normal force and pressure	
3.	Fixture type	Fixed geometry	
		Min	**Max**
4.	Stress (Von Mises Stress)	0.002 [MPa]	105.416 [MPa]
5.	Resultant displacement	0.000mm	0.633mm
6.	Equivalent strain	9.684e^-09	3.807e6-04
7.	Factor of Safety	5.855	3.453e^05

7. Results

The yield strength of piston is reduced as more importance has been given to the relative cost of the material and also the weight has been reduced by 45.01% which satisfies our objective. The yield strength and the fatigue strength of the connecting rod is constant as the material has been kept the same but the weight has been reduced by 6.25% through the design optimizations. After selecting the material of crankshaft the modulus of rigidity has

IOP Conf. Series: Materials Science and Engineering 998 (2020) 012009 doi:10.1088/1757-899X/998/1/012009

been increased and at the same time the crankshaft's weight has been reduced by 1.8% since the forged steel is having large shear modulus and less density as compared to that of 42CrMo4 steel.

Table 11. Result table

Component	Parameter	Previous design		Optimized design	
		Values	Material	Values	Material
Piston	Mass	3.31 [kg]	Nickel alloy	1.82 [kg]	Ti composite Matrix
	Yield strength	851 [MPa]		700 [MPa]	
Connecting rod	Mass	416 [g]	Al-7068	390 [g]	Al-7068
	Yield strength	100 [MPa]		100 [MPa]	
	Fatigue strength	220 [MPa]		220 [MPa]	
Crankshaft	Mass	50.18 [kg]	42CrMo4 steel	49.28 [kg]	Forged steel
	Modulus of rigidity	75 [GPa]		79 [GPa]	

8. Conclusion

The pre-existing engine data was used to create a mathematical formulation in MS excel and empirical relationships were used to calculate different parameters for the three components i.e., piston, connecting rod and crankshaft. Literature related to the objectives of the project was studied and relevant information was collected. By using the weighted property method and Ashby chart, the materials for the three components were selected. An optimum design has been achieved through iterations in the design of the connecting rod resulting in its reduced weight. By using a material with greater shear modulus for the crankshaft, its torsional strength has been increased. The material change for piston from nickel alloy to Ti-composite matrix has resulted in a drastic reduction in its weight and cost as well since the relative cost of Ti-composite is maximum amongst all . These changes have reduced the weights of the target components, while at the same time keeping their respective strengths at optimal levels. Hence, these changes will work well when incorporated into the design of the whole engine and thus a whole comprehensive approach to combine these parts into the engine is required.

9. Acknowledgment

The work was supported by the company Mechatol Engineering Solution. The method of project-based learning is being used and the project plan has been stated solely by the company. Hence the activities which have been implemented are based upon the data and guidance provided by them.

10. References

[1] Subodh Kumar Sharma, 2015, P. K. Saini, N. K. Samria, Experimental Thermal Analysis of Diesel Engine Piston and Cylinder Wall, Hindawi Journal of Engineering, Volume 2015.

[2] Thotsaphon Threrujirapapong, Katsuyoshi Kondoh, Hisashi Imai, Junko Umeda and Bunshi Fugetsu, 2009, Mechanical Properties of a Titanium Matrix Composite Reinforced with Low Cost Carbon Black via Powder Metallurgy Processing, Materials Transactions, Vol. 50, No. 12.

[3] Akhilesh Soni, R K Mandloi, Mar-2018, Mechanical and Tribological Behaviour of Artificially Aged (T6) Al-Zn-Mg-Cu Alloy, International Journal of Innovative Technology and Exploring Engineering (IJITEE), Volume-9 Issue-3, January 2020.

[4] Priyanka Inderchandji Mutha. To Design a New Cross Section for Connecting Rod with a Target of 10% Weight Reduction, Volume 05.

[5] CES EduPack software, 2010, Granta Design Limited, Cambridge, UK.

[6] DNV.GL CLASS GUIDELINE, 2015, Calculation of crankshafts for reciprocating internal combustion engines, DNVGL-CG-0037, Norway, Edition November 2015.

[7] Somnath Chattopadhyay, 2010, Selection of material, Shape and Manufacturing Process for a Connecting Rod, American Society for Engineering Education, AC 2010-926

[8] Information on https://www.scribbr.com/dissertation/methodology/

[9] Kalaikathir Achchagam, 2015, Design Data: Data Book Of Engineers, 2nd edition, Coimbatore.

[10] Myer Kutz, 2002, Handbook of materials selection, John willy & Sons, New York.

Linear and non-linear optical properties of reactive magnetron DC sputtered ZnTe thin films for opto-electronic devices

B Hymavathi[1]*, B Rajesh Kumar[2] and T Subba Rao[3]

[1] Department of Physics, ANITS, Visakhapatnam - 531 162, A.P, India.
[2] Department of Physics, GITAM (Deemed to be University), Visakhapatnam - 530 045, A.P, India.
[3] Department of Physics, S.K. University, Anantapuramu - 515 003, A.P, India.

Email: *brhymavathi@gmail.com

Abstract. Zinc Telluride thin films were prepared by reactive magnetron direct current sputtering method and the temperature of the glass substrates changed from 150 to 350 °C. The band gap of ZnTe films increased from 2.32 to 2.42eV. The normal dispersion of refractive index (n) for ZnTe films is described using the model of Wemple-Di Domenico (WDD) single-oscillator. The oscillator energy, dispersive energy and static refractive index (n_o) of ZnTe films were reported. The Verdet coefficient (V) is estimated from the normal dispersion studies. The first-order (I^{st}) and third-order (IIIrd) non-linear optical susceptibilities ($\chi^{(1)}$, $\chi^{(3)}$), refractive index (n_2) of ZnTe films were calculated.

1. Introduction

ZnTe is a direct band gap semiconductor with 2.20-2.32 eV at room temperature (RT) and exhibits p-type semiconducting behaviour. It is a cheap effective material with a large absorption-coefficient value and exhibiting photovoltaic properties [1, 2]. Recently, this compound found applications in photovoltaic cells, LED, manufacturing of opto-electronic and energy efficient devices [3, 4]. ZnTe compound exhibits cubic structure with a lattice parameter (a) of 6.103 Å [5]. The detailed data on optical constants, dispersion parameters and non-linear properties of these films have less reports in the literature review. The non-linear properties had importance in integrated optical devices (switches, filter, modulators, etc). The main objective of this research study is to analyze the effect of substrate temperature on the linear and non-linear optical parameters of ZnTe thin films.

2. Experimental details

ZnTe films were prepared by using reactive magnetron DC sputtering process with the metallic targets of Zinc and Tellurium. ZnTe films were deposited at room temperature (RT) and changing the temperature of glass substrates from 150 to 350 °C. The targets used for sputtering had a dimension of 2-inch diameter and 4 mm thickness of Zinc (with purity 99.99%) and Tellurium (with purity 99.99%)

IOP Conf. Series: Materials Science and Engineering 998 (2020) 012010 doi:10.1088/1757-899X/998/1/012010

had been used in this work. The sputtering gas of argon was released into the chamber by mass flow controllers (MFC). The base pressure in the vacuum chamber was 2.4 x 10^{-6} Torr, the substrate and target is at a distance of 60 mm. After pre-sputtering with argon gas for 10-15 min, the working pressure of 3 mTorr was maintained during the deposition. The power during sputtering was set to 110 W for Zinc and 85 W for Tellurium. The thickness of the films measured by surface profilometer (Talysurf) was in the range of 320 - 360 nm. X-ray diffraction (XRD) profiles of ZnTe thin films were measured with X-ray diffractometer having radiation of CuK_{α}. The X-ray tube is maintained with current 25 mA, voltage of 30 KV and the scan speed of 0.02(2θ) per sec. Optical absorption spectra of ZnTe films was measured at the wavelength of 300 - 2500 nm using UV-Vis-NIR spectrophotometer (Model: JASCO V-670, Japan).

3. Results and discussion

Figure 1(a) illustrates the absorption spectra of ZnTe films prepared at various substrate temperatures. The shifting of absorption edge towards lower energies of photon for the thermal treated layers can be due to change in the grain size. Figure 1(b) shows the extrapolation of $(\alpha h\nu)^2$ versus hν of ZnTe films. The intercept corresponding to straight line on hν axis represents the band gap of ZnTe films. The optical band gap (E_g) of ZnTe films increased from 2.32 to 2.42 eV with an increment of substrate temperature. The obtained values of E_g for the films were found in good agreement with the previous reported works [6-8].

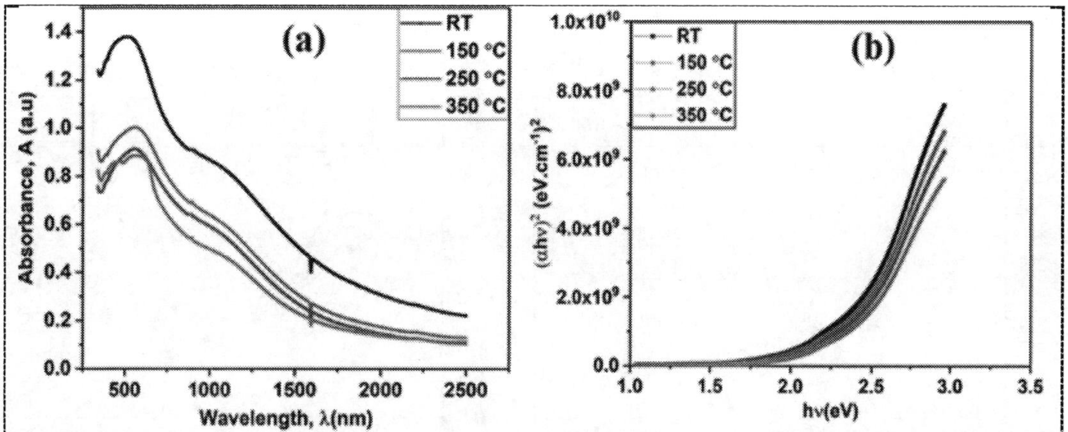

Figure 1. (a) Optical absorbance spectra of ZnTe films and (b) Tauc's plot for ZnTe films prepared at various substrate temperatures.

The refractive index (n) of ZnTe films were estimated from the equations provided in our research works [9]. The change in 'n' of the films with wavelength (λ) is shown in Fig. 2(a), the decreasing trend in refractive index value with an increase of wavelength (λ) and saturated at longer wavelengths. Such behaviour shows ZnTe thin films were free from structural defects. At λ = 550 nm, the 'n' varies from 2.57 to 2.36 with an increment in substrate temperature which may be because of change in the packing density. In other words, the variation in the refractive index is due to the stoichiometric in the ZnTe films and uniformity, which makes shifts the band gap of the ZnTe films.

The optical dispersion study of refractive index(n) for ZnTe thin films has been determined from the model related to single oscillator reported by Wemple and DiDomenico (WDD). From this model the oscillator energy (E_o) and dispersion energy (E_d) are determined and the refractive index (n) of ZnTe films were determined from the Wemple-DiDomenico relationship [10]

$$n^2 = 1 + \left[\frac{E_o E_d}{E_o^2 - (h\nu)^2} \right] \quad (1)$$

From the plot of $(n^2-1)^{-1}$ versus $(h\nu)^2$ and a straight-line fit is shown in Fig. 2(b), the single oscillator (SO) parameters E_o and E_d are found from the slope of $(E_oE_d)^{-1}$ and intercept (E_d/E_o). The values of E_o and E_d are shown in Table 1, these values were decreased with the increase of substrate temperature because of ordered structure and reconnection of molecular and atomic liaisons. The plot of $(n^2-1)^{-1}$ against $(h\nu)^2$ can be used to find the n_o from the equation

$$n_o = \left[1 + \left(\frac{E_d}{E_o}\right)\right]^{\frac{1}{2}} \tag{2}$$

Extrapolate the WDD relation to the zero value of hν. The estimated values of n_o for ZnTe films were reported in Table 1. The optical moments M_{-1} and M_{-3} were determined using the equations [11]

$$E_o^2 = \frac{M_{-1}}{M_{-3}} \tag{3}$$

$$E_d^2 = \frac{M_{-1}^3}{M_{-3}} \tag{4}$$

These optical moments measure the average bond strength and the dielectric response of the material. The optical moments M_{-1} and M_{-3} of ZnTe thin films were found from the estimated values of E_o and E_d are reported in Table 1.

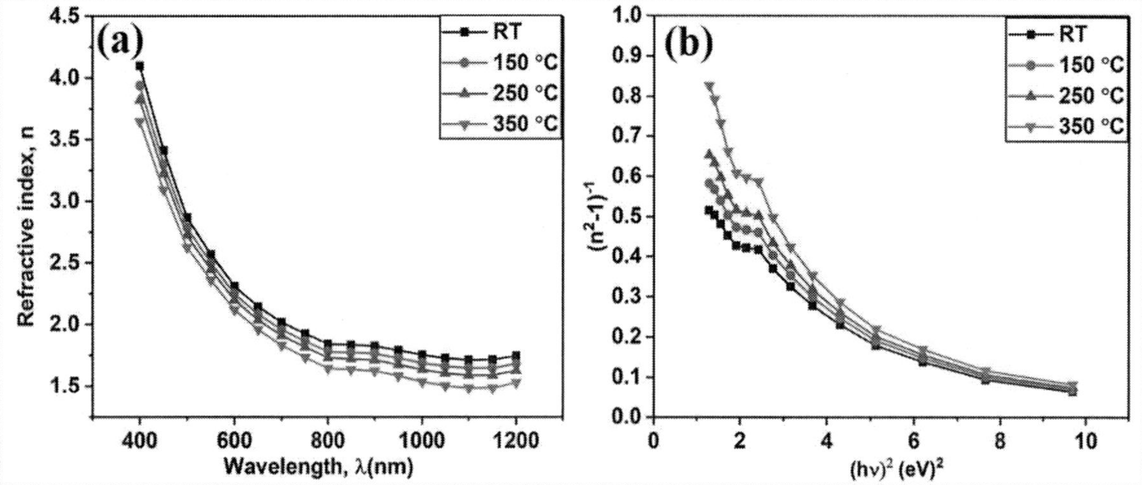

Figure 2. (a) Refractive index (n) of ZnTe films with the variation of wavelength (b) Variation of $(n^2-1)^{-1}$ vs $(h\nu)^2$ for ZnTe films.

The optical properties of ZnTe films can be understood by optical dispersion (OD) parameters. Figure 3(a)-(c) shows the dispersive refractive index, group velocity (U_g) and phase velocity (V_p) of ZnTe films with the wavelength. The group velocity, U_g is determined from the relation [12]

$$U_g = \frac{d\omega}{dk} = \frac{c}{\left[n - \lambda\left(\frac{dn}{d\lambda}\right)\right]} \tag{5}$$

where c = light velocity in the presence of vacuum, k = wave number, and ω = angular frequency. The wavelength dependence of the group and phase velocity for ZnTe films is shown in Fig. 3(b) and 3(c). The results show that the group and phase velocity for ZnTe thin films improves with an increment in the wavelength.

Refractive index is related to Verdet coefficient (V) with the following relation [13]

$$V(\lambda) = \mu_o r\left(\frac{e}{2mc}\right)\lambda\left(\frac{dn}{d\lambda}\right) \qquad (6)$$

where μ_o is the vacuum permeability, r = 0.28 (magneto-optical anomaly factor) for the substances with the covalent bond characteristics and nearer to one for ionic type bonding, e = electron charge, m = electron mass, c = light velocity and $dn/d\lambda$ is the optical dispersion of the material. The wavelength dependence of Verdet co-efficient for ZnTe films prepared by varying substrate temperatures are illustrated in Fig. 3(d). It is observed the positive (+) and negative (-) values of Verdet coefficient which depends on the negative values of E_g.

.

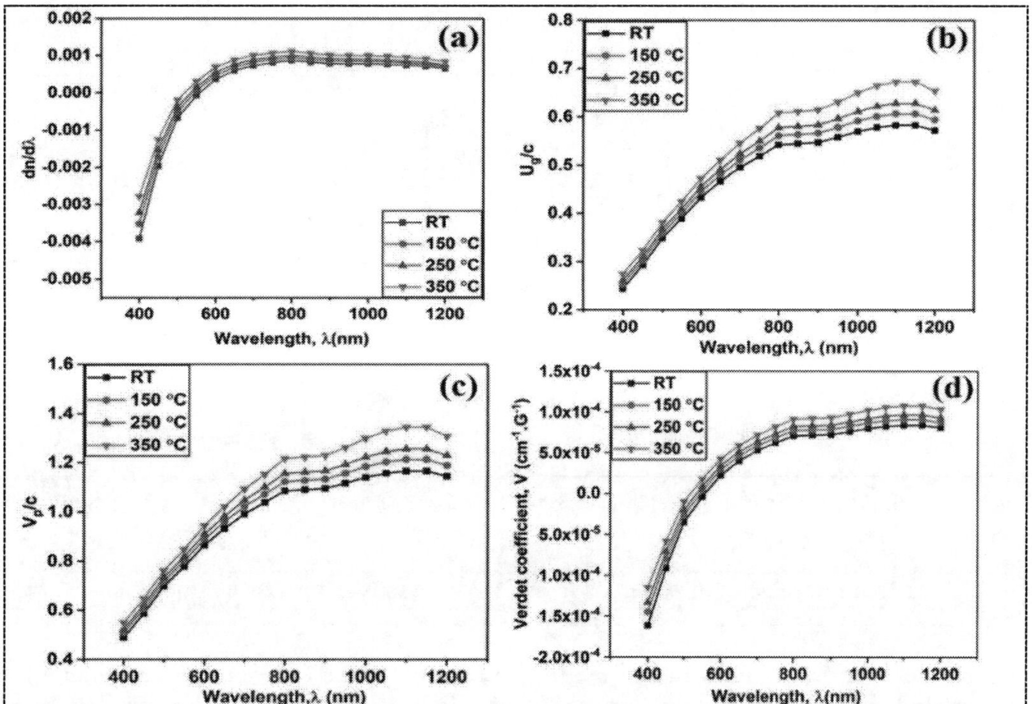

Figure 3. Plots of (a) optical dispersion, $dn/d\lambda$; (b) group velocity, U_g/c ; (c) phase velocity, v_p/c ; (d) Variation of Verdet coefficient with wavelength for ZnTe thin films.

The complex dielectric constant provides the information on electronic structure of the material and helps in the design of highly efficient opto-electronic devices. The dependent of 'n' on the real and imaginary dielectric constants (ε_1 and ε_2) are explained by the Spitzer-Fan model is [14]:

$$\varepsilon_1 = n^2 - k^2 = \varepsilon_\infty - \left[\frac{e^2}{\pi c^2}\right]\left[\frac{N}{m^*}\right]\lambda^2 \qquad (7)$$

$$\varepsilon_2 = 2nk = \left[\frac{\varepsilon_\infty \omega_p^{\,2}}{8\pi^2 c^3 \tau}\right]\lambda^3 \qquad (8)$$

where ε_∞ = dielectric constant with high frequency, e = electronic charge, c = light speed in vacuum, N/m^* = ratio of optical carrier concentration to the effective mass of electron, ω_p = plasma frequency and τ = optical relaxation time. The plot of real dielectric constant (ε_1) versus λ^2 for zinc telluride films prepared at various substrate temperatures are illustrated in Fig. 4(a). It is noticed the linear dependent of ε_1 on λ^2 with higher wavelengths, the ε_∞ and N/m^* values were obtained from extrapolating λ^2 to zero and the slope. Figure 4(b) shows the change in imaginary dielectric constant (ε_2) with λ^3 for ZnTe films, the ε_2 is lower than the value of ε_1 which indicates the light energy loss is minimum. The variation of ε_1

and ε_2 with the substrate temperature is attributed due to the defects such as stresses, voids, inhomogeneity, discontinuities, grain boundaries etc. The decrease in ε_1 and ε_2 for ZnTe films with a variation of substrate temperature (ST) from 150 to 350 °C is because of loss of energy which occurs due to the absorption of energy by free carriers present in the medium.

The plasma frequency (ω_p) is estimated from the relation

$$\omega_p^2 = \left[\frac{4\pi Ne^2}{\varepsilon_\infty m^*}\right] \quad (9)$$

From the ω_p, the relaxation time τ is determined and noted in Table 1. By the observation of both ε_∞ and N/m* values increased with the increase of ST from 150 to 350 °C

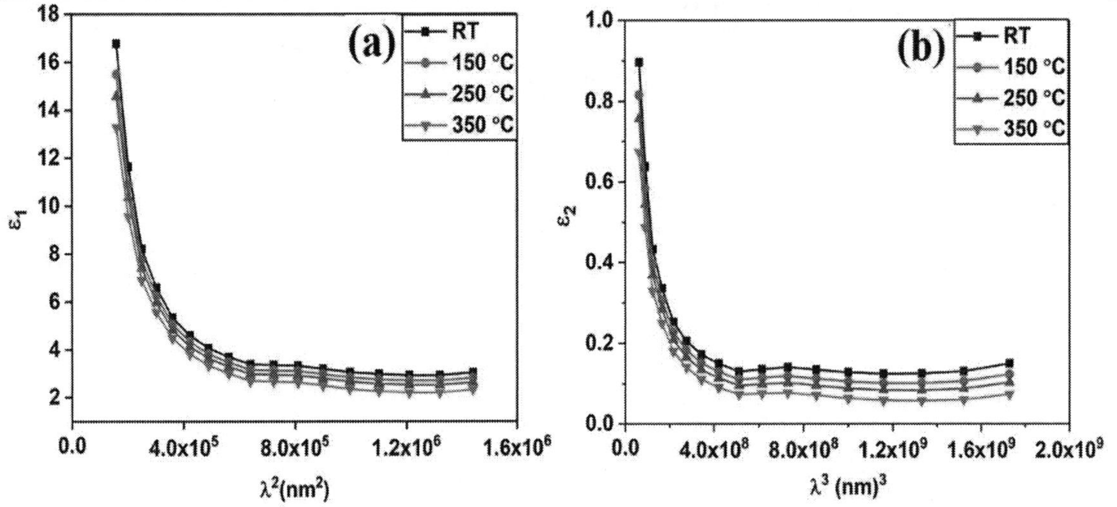

Figure 4. (a) Plot of ε_1 versus λ^2 and (b) ε_2 versus λ^3 for ZnTe thin films.

The non-linear refractive index (n_2) of the films is derived from Tichy and Ticha relation [15, 16]. From the Miller's generalized rule, the static refractive indices (n_o) of ZnTe films were deduced from Wemple-DiDomenico model

$$n_2 = \left[\frac{12\pi\chi^{(3)}}{n_o}\right] \quad (10)$$

where $\chi^{(3)}$ is the IIIrd non-linear susceptibility, which is given as

$$\chi^{(3)} = A\,(\chi^{(1)})^4 \quad (11)$$

where

$$\chi^{(1)} = \left[\frac{E_d}{4\pi E_o}\right] \quad (12)$$

where A is 1.7×10^{-10} esu, $\chi^{(3)}$ is given by the relation

$$\chi^{(3)} = \left[\frac{A}{(4\pi)^4}\right](n_o^2 - 1)^4 \quad (13)$$

The values of n_2, $\chi^{(1)}$, and $\chi^{(3)}$ are given in Table 4. It is noticed that these values decreased with an increment in the substrate temperature from 150 to 350 °C. Thus, ZnTe films is suitable to change refractive index and oscillator energy parameters by varying substrate temperature. Therefore, the ZnTe films were found in designing nonlinear optical device applications/optoelectronic devices.

Table. 1 Optical dispersion and non-linear optical properties of ZnTe films.

Parameter	Substrate temperature			
	RT	150°C	250°C	350°C
E_g (eV)	2.32	2.34	2.40	2.42
E_o (eV)	2.01	2.08	2.19	2.21
E_d (eV)	5.51	4.62	3.79	3.05
M_{-1}	2.73	2.22	1.73	1.45
M_{-3} (eV^{-2})	0.67	0.51	0.36	0.33
n_o	1.93	1.79	1.65	1.56
ε_∞	3.37	3.22	2.73	2.46
N/m^*(10^{49} $cm^{-3}g^{-1}$)	0.988	0.985	0.992	1.04
ω_p (10^{14} Hz)	1.02	0.97	0.92	0.87
τ (10^{14} s)	0.97	1.03	1.08	1.15
$\chi^{(1)}$ esu	0.22	0.17	0.14	0.11
$\chi^{(3)}$ (10^{-13} esu)	3.88	1.66	0.61	0.31
n_2 (10^{-12} esu)	2.32	2.34	2.40	2.42

Conclusions

ZnTe films were grown from magnetron DC sputtering by changing substrate temperature from 150 to 350 °C. The dispersive study of ZnTe films were analysed by the Wemple-Di Domenico single-oscillator (SO) model. It was examined that the decrease of oscillator energy E_o and dispersion energy E_d with an increment in substrate temperature. The Verdet coefficient for the ZnTe films was deduced from the refractive index that exhibits a negative range depending on optical band gap. In addition, non-linear properties such as $\chi^{(1)}$, $\chi^{(3)}$, and n_2 are also reported. We conclude that ZnTe films will be useful for futuristic optical devices and opto-electronic devices.

References

[1] Akkad F El and Abdulraheem Y 2013 J. Appl. Phys. **114** 183501
[2] Kshirsagar SD, Ghanashyam Krishna M and Tewari SP 2013 Mater. Sci. Semicond. Process. **16** 1002
[3] Devaraj Soundararajan, Young Joo Lim, Myong-Pyo Chun and Ki Hyeon Kim 2013 Electron. Mater. Lett. **9** 177
[4] Tanvir Hussain, Al-Kuhaili MF, Durrani SMA and Qayyum HA 2018 Ceram. Int. **44** 10130
[5] Hossain MS, Islam R, Shahjahan M and Khan KA 2008 J Mater Sci: Mater Electron **19** 1114
[6] Essam R.Shaaban, Ishu Kansal, Mohamed SH and Joes M.F.Ferreira 2009 Physica B Condens. Matter. **404** 3571
[7] López JD, Tirado-Mejía L, Ariza-Calderón H, Riascos H, de Anda F and Mosquera E 2020 Mater. Lett. **268** 127562
[8] Mahalingam T, John VS, Rajendran S and Sebastian PJ 2002 Semicond. Sci. Technol. **17** 465
[9] Rajesh Kumar B and Subba Rao T 2013 Appl. Surf. Sci. **265** 169
[10] Wemple SH and Didomenico Jr., M 1971 Phys. Rev. B **3** 1338
[11] Wemple SH and Didomenico Jr., M 1973 Phys. Rev. B **7** 3767
[12] Abdelkader D, Chaffar Akkari F, Khemiri N, Miloua R, Antoni F, Gallas B and Kanzari M 2018 Physica. B: Condens. Matter. **546** 33
[13] Venkata Veera Prasad M, Thyagarajan K and Rajesh Kumar B 2019 J. Nanosci. Tech. **5** 688
[14] Spitzer WG and Fan YH 1957 Phys. Rev. **106** 882
[15] Hymavathi B, Rajesh Kumar B and Subba Rao T 2017 Opt. Quant. Electron. **49** 86
[16] Ticha H and Tichy L 2004 J. Optoelectron. Adv. Mater. **4** 381

IOP Conf. Series: Materials Science and Engineering 998 (2020) 012011 doi:10.1088/1757-899X/998/1/012011

Structural and optical properties of amorphous TiO_2 films deposited by reactive DC magnetron sputtering

M Lavanya[1], S Sunita Ratnam[2], B Rajesh Kumar[3*] and T Subba Rao[4]

[1,4] Department of Physics, S.K. University, Anantapuramu – 515 003, A.P, India.
[2] Department of Physics, Malla Reddy Engineering College for Women (Autonomous), Secunderabad -500 100, Telangana, India.
[3] Department of Physics, GITAM, Visakhapatnam – 530 045, A.P, India.

Email: *rajindphy@gmail.com

Abstract. Titanium dioxide films were grown by magnetron DC sputtering technique and post-annealed in vacuum from 150 to 350 °C. X-ray diffraction profiles of TiO_2 films exhibits amorphous nature. The optical absorbance spectra show the absorption edge tends to shift towards higher wavelength and band gap (E_g) of TiO_2 thin films decreased from 3.44 to 3.35 eV with the increment of annealing temperature from 150 to 350°C. The decrease in band gap of the films leads to an increase in the size of the grain with an increment of annealing temperature.

1. Introduction

TiO_2(Titanium dioxide) had unique physical-chemical properties with a high value of refractive index, dielectric constant, electrical resistivity with above 10^8 Ω.cm which makes suitable for different applications [1,2]. TiO_2 films is used as filters in optics, highly efficient dielectrics, and solar cells, self-cleaning glasses etc. [3-6]. TiO_2 can be in amorphous, rutile, anatase, and brookite crystalline phases [7].

There are various physical and chemical methods for the deposition of TiO_2 films [8-12]. Among them reactive magnetron DC sputtering method had good features such as low temperature of the substrate, film adhesive, uniformity, large deposition rate and large-area film deposition for practical applications [13]. The present work is devoted to the preparation of TiO_2 films by magnetron DC sputtering method and post-annealed the films from 150 to 350 °C. The main objective is to report the structural and optical properties of annealed TiO_2 films.

2. Experimental details

TiO_2 films were grown by reactive direct current (DC) magnetron sputtering process at room temperature (RT) and post-annealed from 150 to 350 °C. First, the substrates washed with deionized water followed by acetone and absolute ethanol for the removal of contamination on the surface. The target Ti (purity of 99.99%) of 2-inch diameter and 3 mm thickness were used in the present investigation; the targets were pre-sputtered for ten minutes to remove an oxide layer on the surface of the target. The working pressure maintained in the vacuum chamber was 5.20×10^{-3} Pa. The substrate and Titanium target are fixed at 65 mm. The argon (99.99%) and oxygen (99.99%) flow rate was maintained at 32 and 2 sccm. The DC bias voltage was fixed at 480 V for Ti target and maintained for 2 hours for the deposition of TiO_2 film. The structural information of TiO_2 films were determined by X-ray analysis using radiation of CuK_α with λ=1.54 Å. The XRD profiles were taken with a voltage of 42 kV, current 36 mA and 2θ (=10 - 80°). The surface morphological of TiO_2 films were studied by FESEM (ULTRA-55, KARL ZIESS). The quantitative elemental analysis of TiO_2 films were estimated by EDS (Make: Oxford Instruments) combined with FESEM. The measured thickness of the films by surface profilometer is found to be 260-320 nm. The optical absorbance spectra of TiO_2 films were recorded from the wavelength of 300 to 1800 nm using UV-Vis-NIR spectrophotometer.

3. Results and discussion

X-ray diffraction (XRD) profiles of TiO_2 films annealed at various temperatures are illustrated in Fig. 1. The amorphous structure of TiO_2 films is due to less surface mobility of deposited particles [14-16]. TiO_2 thin films can be crystallized at temperature ≥450 °C, indicating the onset temperature of crystallization will be occurred at 450 °C.

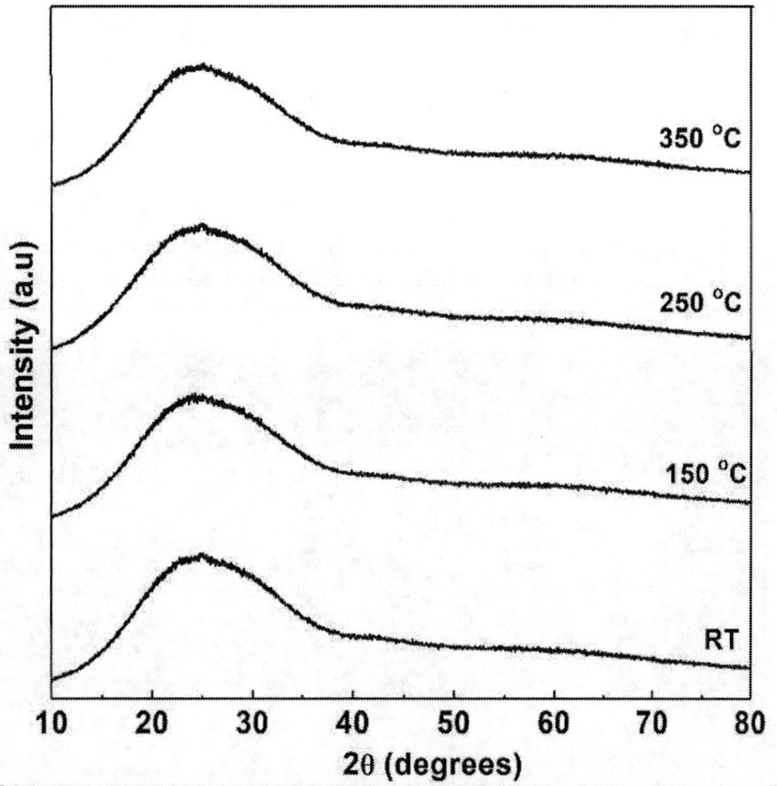

Figure 1. XRD profiles of annealed TiO_2 films.

Figure 2(a) - (d) illustrates the FESEM images of annealed TiO_2 films at various temperatures. The surface of the films exhibits dense and smooth morphology which confirms amorphous nature of the films with the presence of clusters. The high annealing temperature leads to gain required energy for grain boundaries, reduces the internal stress and forms large grain with the clusters in the films. Figure 3 shows the EDS spectra of TiO_2 thin film annealed at 350 °C. The identified peaks are labelled according to atomic species and show the strongest peaks belonging to Ti and O elements. In the EDS spectra, the titanium peaks were observed at 0.42 and 4.50 keV and the oxygen peak is observed at 0.52 keV.

IOP Conf. Series: Materials Science and Engineering 998 (2020) 012011 doi:10.1088/1757-899X/998/1/012011

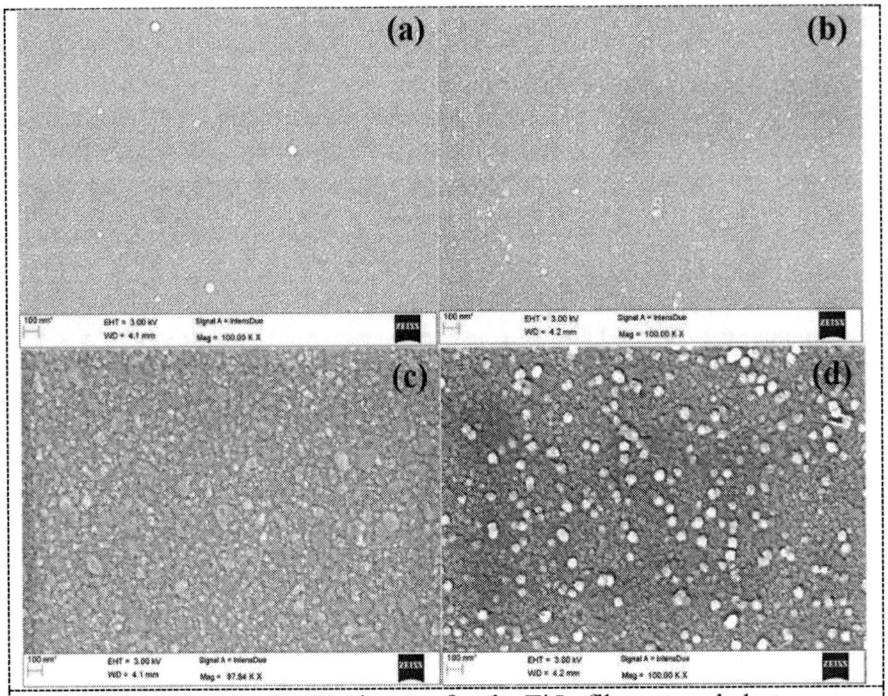

Figure 2. FESEM images for the TiO_2 films annealed at (a) RT, (b) 150, (c) 250, and (d) 350 °C

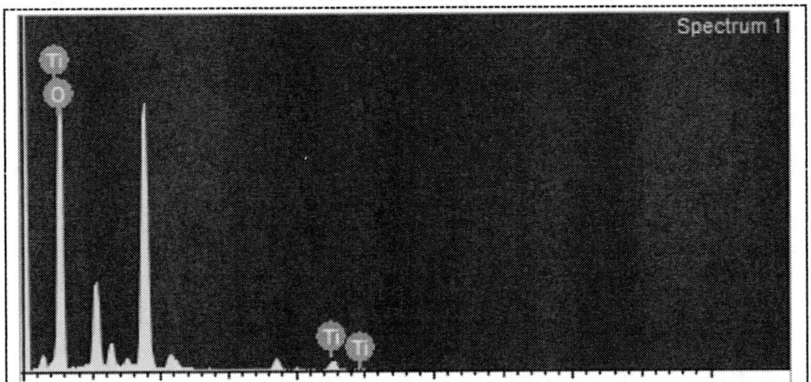

Figure 3. EDS spectra for TiO_2 film annealed at 350 °C

Figure 4 shows the optical absorbance spectra for the films of TiO_2 annealed at various temperatures. The absorbance edge shifts at higher wavelength regions indicates the decrement in the band gap (E_g). The increment in the grain size leads to decrease the defect concentration in the films as well as reduces the lattice strain. The E_g of TiO_2 films were determined using the relation of Tauc's as illustrated in Fig. 5. The E_g of TiO_2 films prepared at RT, post-annealed at 150, 250 and 350 °C are found to be 3.44, 3.42, 3.39 and 3.35 eV. The decrement in the E_g of the prepared films with annealing process leads to increase the levels of VB (valence band) and CB (conduction band) [18]. The results were in good correlation with the earlier literature works [19, 20]. The tuning of band gap of TiO_2 films is beneficial to design a window material for the manufacturing of solar cell and optical designing applications.

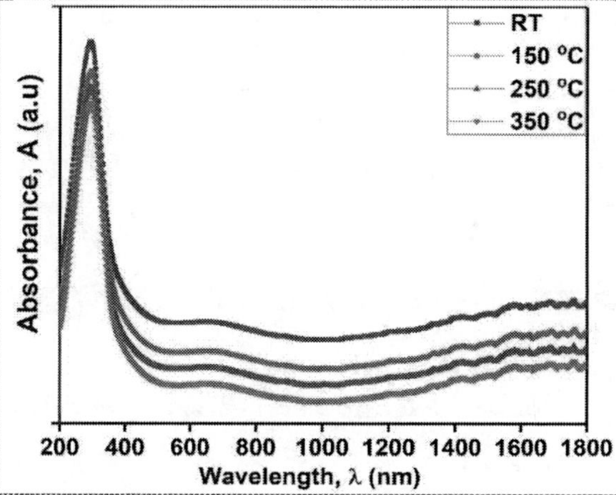

Figure 4. Optical absorbance of TiO_2 films.

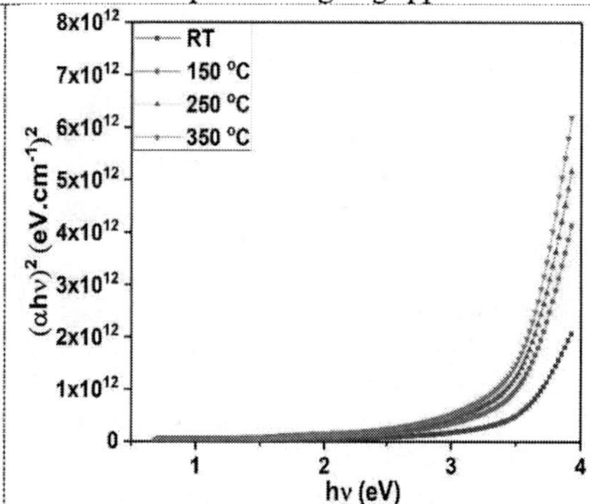

Figure 5. Tauc's plot for TiO_2 films.

To calculate the particle size the effective mass approximation (EMA) method is implemented and estimated by

$$E_g = E_{gb} + \left(\frac{h^2 \pi^2}{2m^* R^2} \right)$$

where E_{gn} and E_{gb} represents nanocrystalline and bulk TiO_2 (E_{gb} = 3.2 eV) band gap, m* = effective electron mass and R = particle radius [21]. The size of the particle resulted from the absorption spectra for TiO_2 films prepared at RT, post-annealed at 150, 250 and 350 °C are found to be 16, 18, 21 and 23 nm respectively.

Summary

TiO_2 films were grown by magnetron DC sputtering process and post-annealed the films from 150 to 350 °C. XRD studies reveal that the TiO_2 films shows the amorphous nature. Surface morphology of TiO_2 films shows a homogenous and dense distribution of grains composed of small spherical particles with increasing grain size. The optical absorbance edge of TiO_2 films shifts to higher wavelengths and the band gap of TiO_2 films decreased from 3.44 to 3.35 eV. The increment in grain size of the films with annealing temperature weakens the quantum size effects. The estimated size of the particles determined from the band gap shift confirms the nano-structured TiO_2 films. The amorphous nature of TiO_2 films shows the tunable optical band gap with the effect of annealing temperature and it is found to be suitable for its applications in optical coatings especially in the visible range.

Acknowledgments

The author acknowledges to APPCB (Andhra Pradesh Pollution Control Board) for financial support (Ref. No. APPCB/RFS/UH3/2016-23/01/2017).

References

[1] Ben Jemaa I, Chaabouni F and Abaab M *Phys. Status Solidi A* 2017 **214** 1600426
[2] Sung YM and Kim HJ *Thin Solid Films* 2007 **515** 4996
[3] Okada M, Tazawa M, Jin P, Yamada Y and Yoshimura K *Vacuum* 2006 **80** 732
[4] Fujishima A, Rao TN and Tryk DA J. Photochem. Photobiol. C *Photochem. Rev.* 2000 **11**
[5] Sun H, Wang C, Pang S, Li X,Tao Y, Tang YH and Liu M *J. Non-Cryst. Solids* 2008 **354** 1440
[6] Haseeb ASMA, Hasan MM and Masjuki MM *Surf. Coat. Technol.* 2010 **205** 338

IOP Conf. Series: Materials Science and Engineering 998 (2020) 012011 doi:10.1088/1757-899X/998/1/012011

[7] Yongfeng Ju, Mahua Wang,Yunlong Wang, Shihu Wang and Chengfang Fu *Adv. Condens. Matter. Phys.* 2013 **2013** 365475/1-5

[8] Oja I, Mer A, Krunks M, Solterbeck CH and Es-Souni M *Solid-state Phenom.* 2004 **99-100** 259

[9] Wang Z, Helmersson U and Käll PO *Thin Solid Films* 2002 **405** 50

[10] Yamamoto S, Sumita T, Miyashita A and Naramoto H *Thin Solid Films* 2001 **401** 88

[11] Horikawa T and Asano H *Int. J Mod. Phys.* B 2011 **25** 4171

[12] Ben Jemaa I, Chaabouni F and Abaab M *Phys. Stat. Sol. A* 2017 **214** 1600426

[13] Won Gyu Lee, Seong Ihl Woo, Jong Choul Kim, Soo Han Choi and Kye Hwan Oh *Thin Solid Films* 1994 **237** 105

[14] Löbl P, Huppertz M and Mergel D *Thin Solid Films* 1994 **251** 72

[15] Meng L-J, Andritschky M and dos Santos MP *Thin Solid Films* 1993 **223** 242

[16] Berg S and Nyberg T *Thin Solid Films* 2005 **476** 215

[17] Sankar S and Gopchandran K *Cryst. Res. Technol.* 2009 **44** 989

[18] Kadhim A. Aadim, Khalid H. Abass and Qunoot M. Hadi *Int. Lett. Chem. Phys Astron* 2015 **56** 63

[19] Muaz AKM, Hashim U, Fatimah Ibrahim, Thong KL, Mas S. Mohktar and Wei-Wen Liu *Microsyst. Technol.* 2016 **22** 871-881

[20] Devjyoti Lilhare, Sandhya Pillai and Ayush Khare *J Electron. Mater.* 2018 **47** 6532

IOP Conf. Series: Materials Science and Engineering 998 (2020) 012012 doi:10.1088/1757-899X/998/1/012012

Effect of rotational speed on structural and optical properties of spin coated SnO thin films

L Priyanka[1], B Hymavathi[2] and B Rajesh Kumar[3*]

[1,3] Department of Physics, GITAM (Deemed to be University), Visakhapatnam-530045, A.P, India.
[2] Department of Physics, ANITS (Autonomous), Visakhapatnam-531162, A.P, India.

Email: *rajindphy@gmail.com

Abstract. Tin monoxide films were synthesized by sol-gel process followed by spin coating at the spinning speeds from 1000 to 3000 rpm. X-ray diffraction profiles of the films exhibits amorphous nature. Surface morphological and elemental confirmation of SnO films were studied by scanning electron microscope and energy dispersive of X-ray spectroscopy. The optical transmittance of the films had an increment from 60 to 72% with the increase of spinning speed from 1000 to 3000 rpm. The band gap of SnO thin films varied from 2.96 to 3.52 eV with an increment in rotational speed due to Burstein-Moss (B-M) effect. The estimated particle size from the optical absorbance spectra using effective mass approximation method is decreased from 23.6 nm to 11.6 nm with an increment in the spin speed from 1000 to 3000 rpm.

1. Introduction

Tin (II) monoxide (SnO) had a potential significance because of p-type semiconducting nature and stabilization in structural and optoelectronic properties. It had a wide band gap ranging 2.5-3.4eV, which will be for most opto-electronic devices like diodes, transistors, complementary circuits, and gas sensitive materials etc. [1-3]. Though there are different process for the synthesis of SnO films [4-6]. The simple method is sol-gel spin coating with cost effective, and ability for uniformity films with good adherent, with a high surface area, small pore size, microstructural control of particles and uniformly distributed particles [7]. The main objective of this work is to study the structural, optical properties of SnO films grown by sol-gel spin coating (SGSC) process with varying rotational speeds from 1000 to 3000 rpm.

2. Experimental details

Tin oxide (SnO) precursor solution is prepared by using tin (II) chloride dihydrate ($SnCl_2.2H_2O$), 2-methoxyethanol (2-ME) as solvent, and monoethanolamine (MEA) as stabilizer to from a solution. The solution is stirred for 2 hours at the temperature 80 °C. After aging the synthesized solution for 48 hours, the resulting solution was spin-coated on glass substrates by varying rotational speed from 1000 to 3000 rpm for 45 s. The films were dried in an oven at 150°C for 15 min. This process will be continuously repeated upto 3 to 5 times in order to obtain a desired thickness of the films. The crystallinity and structure of SnO thin films were obtained from XRD data (Model: Bruker D-8 diffractometer) with the radiation of CuK_α having a wavelength λ=0.154 nm. Surface morphological and elemental analysis of SnO films were examined by FESEM combined with EDS. The absorption and transmission spectra of thin films were measured from the wavelength of 300 to 2500 nm by UV-Vis-NIR spectrophotometer (Model: Hitachi U-2900).

3. Results and discussion

XRD profiles of SnO films prepared by varying spinning speed from 1000 to 3000 rpm is illustrated in Fig. 1. From XRD, no extra peaks correspond to tin oxide is noticed indicating the amorphous nature. A broad hump in the spectra is because of amorphous nature. Similar behaviour in SnO films has been noticed in the previous research works [8, 9].

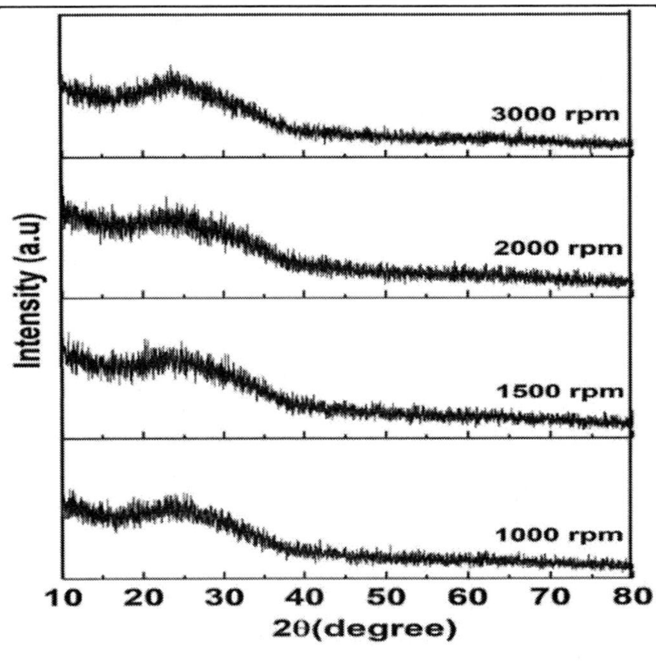

Figure 1. XRD profiles of SnO films

Gravimetric weight difference method is used to determine the film thickness by the equation [10]

$$t = \frac{m}{A.\rho_b} \quad (1)$$

where, 'm'=film mass, A is the substrate area and ρ_b=density of the bulk material. In the thickness calculation, the density of the bulk material (SnO) was taken as 6.40 gm cm^{-3} [11]. The thickness of the films prepared at 1000, 1500, 2000 and 3000 rpm were found to be 230, 210, 180 and 165 nm respectively. With the increase of rotational speed (rpm), the film thickness decreased because of increase in the centrifugal force of the solution acting on the glass substrate.

FESEM images of SnO films prepared at various rotational speeds are shown in Fig. 2(a)-(d). The surface morphology looks like featureless smooth surface, with lesser grain boundaries with the increment of rotational speed from 1000 to 3000 rpm, which can be attributed due to the amorphous state of the films. The surface is disordered with an irregular grain could lead to amorphous structure as seen in XRD studies. Generally, the SnO films prepared <400 °C forms amorphous nature with short-range and long-range disorder. Figure 3 shows the EDS of SnO films prepared at 2000 rpm. The strong peaks confirm the evidence of Sn and O elements in the energy dispersive spectrum. No other impurity peaks of the elements were present in the EDS spectra, which shows that the synthesized SnO thin films were quite pure. The atomic wt% of the elements Sn and O is 43.3 and 56.7 respectively. In the EDS spectra, the element tin and oxygen peak are observed at 3.44 keV and 0.52 keV respectively.

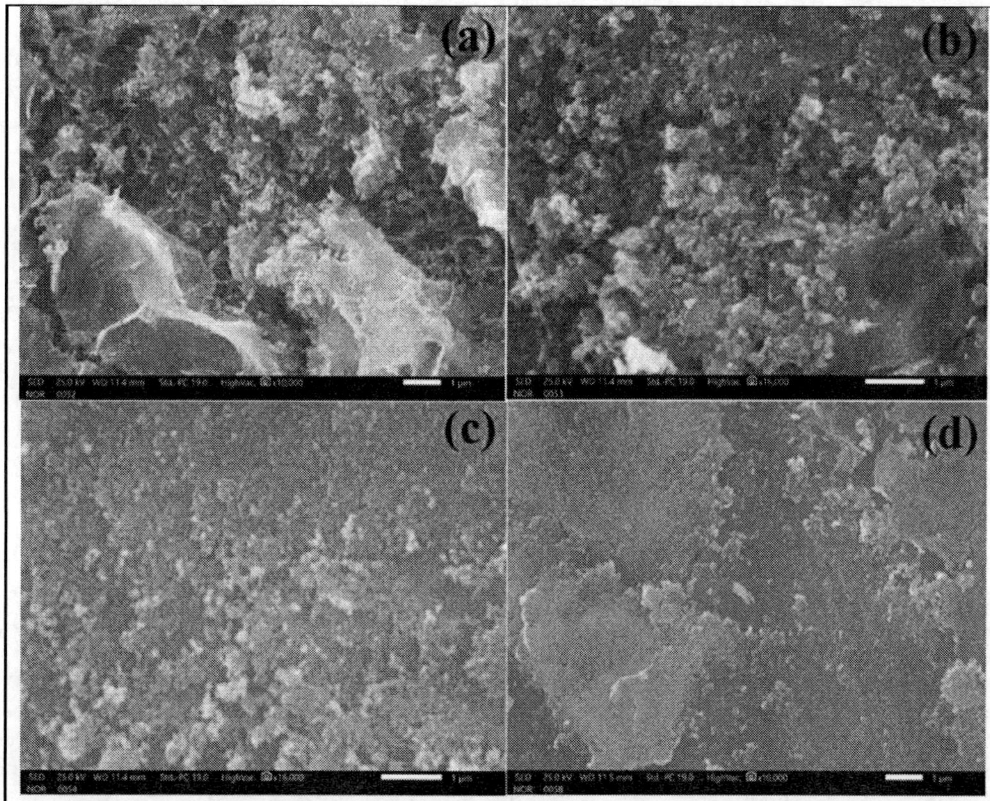

Figure 2. FESEM images of SnO films prepared at (a) 1000; b) 1500; (c) 2000; and (d) 3000 rpm.

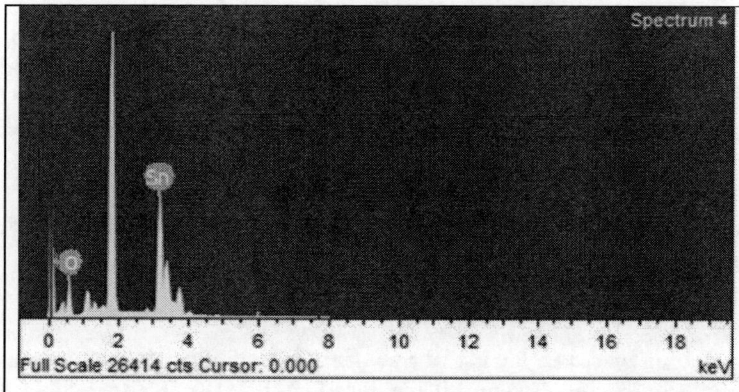

Figure 3. EDS of SnO thin film prepared at 2000 rpm

Figure 4 illustrates the absorbance spectra of SnO films grown by changing rotational speeds from 1000 to 3000 rpm. The spectra show low absorbance in the visible and near infrared regions, while the optical absorbance is higher ultra violet region. In the IR region, the absorbance decreases and the films

IOP Conf. Series: Materials Science and Engineering 998 (2020) 012012 doi:10.1088/1757-899X/998/1/012012

became transparent. The decrement in the absorbance of the films with an increment in the wavelength is due to high energy absorption. In absorbance spectra, the absorption edge shifts towards the shorter region of wavelength with an increasing in the spin speed from 1000 to 3000 rpm. The average optical transmittance of SnO thin films is illustrated in Fig. 5, which is between 60-72%. The increment in the transmittance with an increment of rotational speed because of reduction in the film thickness and reduction of light scattering. The transmittance spectra of about 70% was obtained for 3000 rpm at the visible wavelength region. The increase in transmittance explains the reduction of light scattering because of lower film thickness.

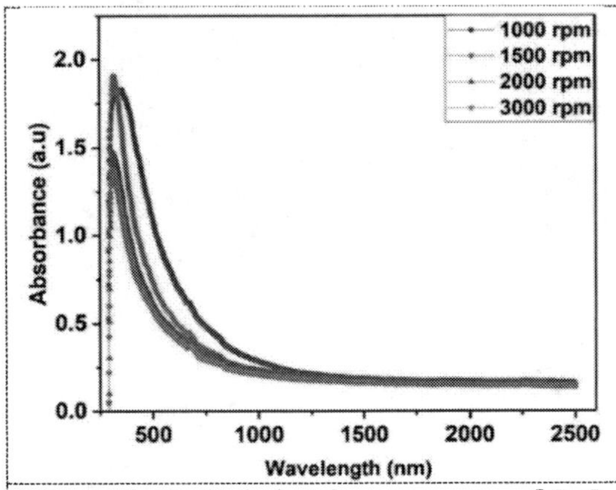

Figure 4. Optical absorbance spectra of SnO films

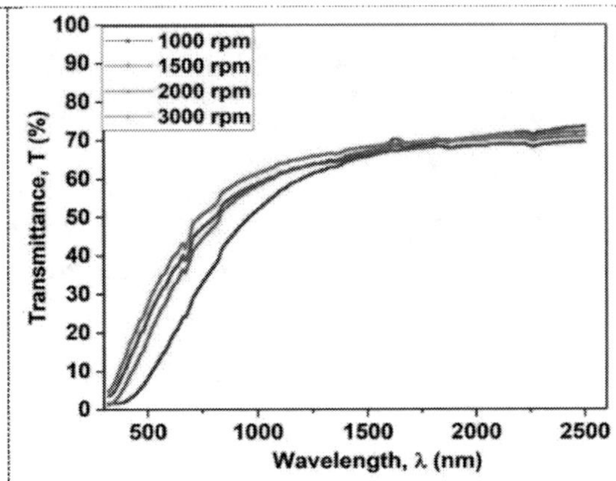

Figure 5. Transmission spectra of SnO thin films

The absorption coefficient (α) of the films were calculated from the relation [12]

$$\alpha = 2.303\left(\frac{A}{t}\right) \quad (2)$$

where A = optical absorbance and t = film thickness. The films exhibit $\alpha > 10^4\ cm^{-1}$ which is consistent with the optical absorption behaviour indicating the visible light produce electron-hole pairs which can be used as absorber layer for solar cells.

The absorption coefficient (α) and optical band gap (E_g) is related with the relation [13]

$$(\alpha h\nu)^n = B\left(h\nu - E_g\right) \quad (3)$$

where B =constant, hν = photon energy and E_g = optical band gap. For n = 2, 2/3,1/2 and 1/3 the transition corresponds to direct allowed, direct forbidden, indirect allowed and indirect forbidden [8, 9]. Figure 5 exhibits the Tauc's plot of $(\alpha h\nu)^2$ versus hν for SnO thin films exhibits a sharp absorption edge indicating the high quality of the film. The band gap (E_g) of the SnO films are varied from 2.36 to 3.08 eV with an increment in rotational speed due to the effect of Burstein-Moss (B-M).

The related equation for energy band gap widening ΔE_g and optical carrier concentration(n_{opt}) is [14]

$$\Delta E_g = \left(\frac{h^2}{8m^*}\right)\cdot\left(\frac{3n_{opt}}{\pi}\right)^{\frac{1}{2}} \quad (4)$$

The band gap widening increases the n_{opt} of SnO films. The optical carrier concentration of SnO thin films is obtained from the equation (4) and it increased from 1.18×10^{19} to $1.53\times10^{20}\ cm^{-3}$. The n_{opt} of SnO thin films increased with an increment of rotational speeds from 1000 to 3000 rpm. The increase of n_{opt} may also be attributed to decrement rate release of free carriers from trap centres in thicker films. Since the thicker films are less defective, the scattering of carriers by defects will be also less.

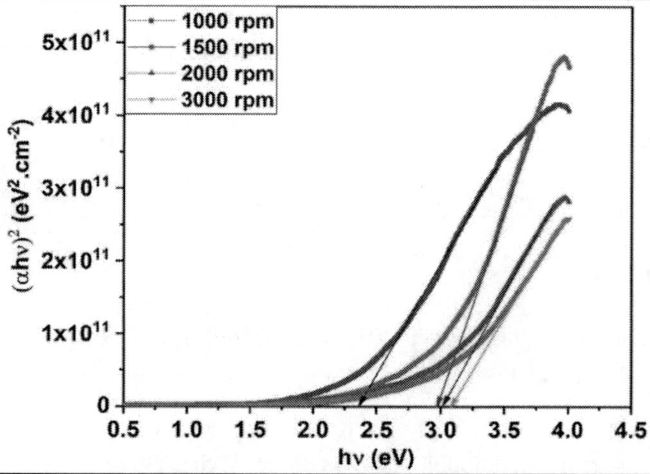

Figure 6. Plotting of $(\alpha h\nu)^2$ versus hν for SnO films

The particle size is calculated from effective mass approximation method using the relation [15]

$$E_g = E_{gb} + \left(\frac{h^2\pi^2}{2m^*R^2}\right) \quad (5)$$

where E_g and E_{gb} are the band gap values of SnO and bulk SnO semiconductor (E_{gb} = 2.4 eV), m* (=0.27m_e) = effective mass of the electron and R is the particle radius. The particle size obtained from the absorbance spectra decreased from 23.6 to 11.6 nm with an increase of rotational/spinning speed from 1000 to 3000 rpm. The optical parameters of SnO films prepared at different rotational speeds are reported in Table 1.

Table 1. Optical properties of SnO thin films.

S.No	Rotational speed (rpm)	Optical band gap, E_g (eV)	Particle radius, R (nm)	Optical carrier concentration, n_{opt} (cm^{-3})
1	1000	2.36	23.6	1.18 x 10^{19}
2	1500	2.97	13	1.12 x 10^{20}
3	2000	3.02	12.3	1.30 x 10^{20}
4	3000	3.08	11.6	1.53 x 10^{20}

Conclusions

SnO films were synthesized by sol-gel spin coating (SGSC) process at the spin speeds 1000, 1500, 2000 and 3000 rpm. The decrement of film thickness from 230 to 165 nm with an increment of rotational speed from 1000 to 3000 rpm. XRD patterns of SnO films confirms the amorphous nature. The surface morphology of SnO films shows very smooth, homogeneous and irregular grains. The average transmittance of SnO films decreases with an increment of rotations speed due to the decrease of film thickness. The band gap of SnO films increased from 2.36 to 3.08 eV due to Burstein-Moss (BM) with an increase of rotational speed. The optical carrier concentration of SnO thin films increased from 1.18×10^{19} to 1.53×10^{20} cm^{-3} with an increase of spinning speed from 1000 to 3000 rpm. From these results we conclude that SnO films can be utilized for novel opto-electronic devices.

References

[1] Jesus A. Caraveo-Frescas, Pradipta K. Nayak, Hala A. Al-Jawhari, Danilo B. Granator, Udo Schwingenschlogl and Husam N. Alshareef 2013 *ACS Nano* **7** 5160
[2] In-Tak Cho, Myeonghun U, Sang-Hun Song, Jong-Ho Lee and Hyuck-In Kwon 2014

Semicond. Sci. Technol. **29** 045001
[3] Kachirayil J.Saji, Venkata Subbaiah Y, Kun Tian and Ashutosh Tiwari 2016 *Thin Solid Films* **605** 193
[4] Ahmed Yousef Mohamed, Seung Jun Lee, Younjin Jang, Jun Shik Kim, Cheol Seong Hwang and Deok-Yong Cho 2020 *J Phys. Condens. Matter.* **32** 065502
[5] Ebitha Eqbal and E.I.Anila 2018 *Physica B Condens Matter.* **528** 60
[6] Sima Rezalou, Tuba Öznülüer and Ümit Demir 2018 *Appl. Surf. Sci.* **448** 510
[7] Aegerter MA, Almeida R, Soutar A, Tadanaga K, Yang H and Watanabe T 2008 *J. Sol. Gel. Sci. Tech.* **47** 203
[8] Po-Ching Hsu, Shiao-Po Tsai, Ching-Hsiang Chang,Chao-Jui Hsu, Wei-Chung Chen, Hsing-Hung Hsieh and Chung-Chih Wu 2015 *Thin Solid Films* **585** 50
[9] Zhong-guo Li, Ling-yan Liang, Hong-tao Cao and Ying-lin 2017 *J Phys Conf Ser.* **844** 012017
[10] Sakhare RD, Khuspe GD, Navale ST, Mulik RN, Chougule MA, Pawar RC, Lee CS, Shashwati Sen and Patil VB 2013 *J Alloy Compd.* **563** 300
[11] Matthias Batzill and Ulrike Diebold 2005 *Prog. Surf. Sci.* **79** 47
[12] Ibrahim R. Agool, Evan T. Salem and Marwa A. Hasan 2011 *Int. J Mod. Phys. B* **25** 1081.
[13] Rajesh Kumar B and Subba Rao T 2013 *Appl. Surf. Sci.* **265** 169
[14] Rajesh Kumar B and Subba Rao T 2015 *Materials Today: Proceedings* **2** 1502
[15] Devjyoti Lilhare, Sandhya Pillai and Ayush Khare 2018 *J Electron. Mater.* **47** 6532

A Critical Review on Machining Of Titanium and Its Alloy under Cryogenic Cooling Environment

Rajeev Sharma[1], Binit Kumar Jha[1], Vipin Pahuja[2]
[1] Department of School of manufacturing Skills, Bhartiya Skill Development University Jaipur 302037, India
[2] Department of HVAC, Bhartiya Skill Development University Jaipur, 302037, India

E-mail: Rajeev.sharma@ruj-bsdu.in

Abstract. Titanium and its composite has superb properties, for example, high clear cut quality, consumption obstruction, light weight which makes is generally fitting for application in biomedical, aviation and vehicle area and so forth Titanium composite has helpless machinability on the grounds that the low warm conductivity & high pressure so extortionate tool wear during machining of titanium combination and lead to higher machining cost. In this circumstance, cryogenic machine set-up is used for development of machinability and lower instrument wear rate. This exploration article accommodating for give great rundown on machining of titanium amalgams with cryogenic cooling climate. Additionally, these audit paper can be strong to survey, comprehend the machinability of titanium combinations and determination of the machining cycle.
Keywords: Titanium alloy; Tool wear; Cryogenic cooling; Machinability.

1. Introduction

Titanium alloy has, superb properties e.g. strength retention, corrosion resistance at high temperature due to these properties it is mostly applicable in biomedical, aerospace, petroleum and automotive industries. Fig. 1. Represents the early milestones in cryogenic technology.

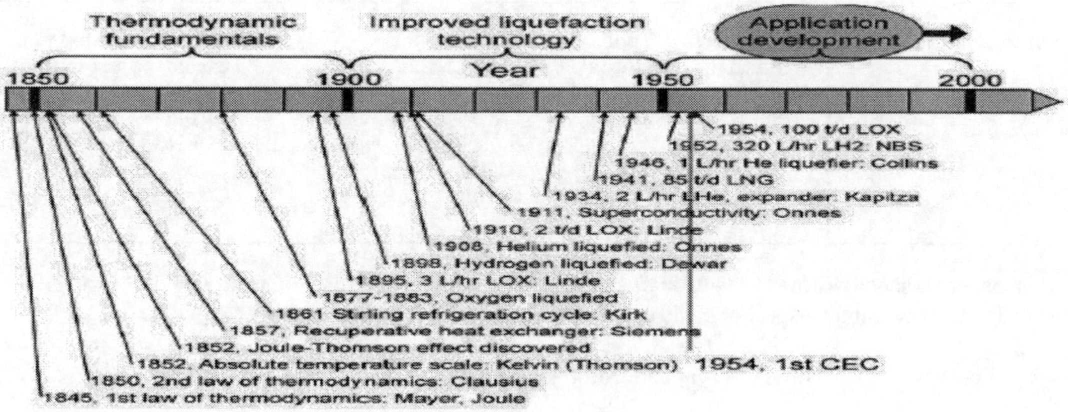

Fig. 1. History of cryogenic machining process [6]

IOP Conf. Series: Materials Science and Engineering 998 (2020) 012013 doi:10.1088/1757-899X/998/1/012013

Titanium alloy hard to cut materials on the grounds that contrasted with aluminum, carbon, hardened steel and so forth cutting velocity of these material, is exceptionally restricted, for example, if speed of titanium compound are between 60 m/min to 100m/min then tungsten carbide and HSS instrument is utilized separately. Cryogenic and MQL machining is utilized for improve machinability of difficult to cut material. Therefore, MQL and cryogenic cooling machining are considered as feasible answer for machining of various to cut materials [1].

In 1953 initially imagined machining by cryogenic cooling CO2 [2]. In 1961W.S.Holi expressed that when machining of titanium compound utilizing cryogenic CO2 as a, coolant at that point device life is additionally increment [3]. Airplane producing by Grumman revealed about expands MRR (Metal expulsion rate) in machining of titanium with LN2 and CO2 [4]. Tested works completed on machining of titanium amalgam, tempered steel and carbon with cooling climate and locate that distinctive conduct for various metals and having impact on device life, SR and flank wear [5]. Various sorts of cryogenic coolant are accessible for machining yet generally CO2 and LN2 is utilized as a coolant. To recognize fluid LN2and CO2, we need to investigate further the system making the low temperatures happen. Fig. 2 shows schematic phase diagrams for nitrogen gasses & carbon dioxide.

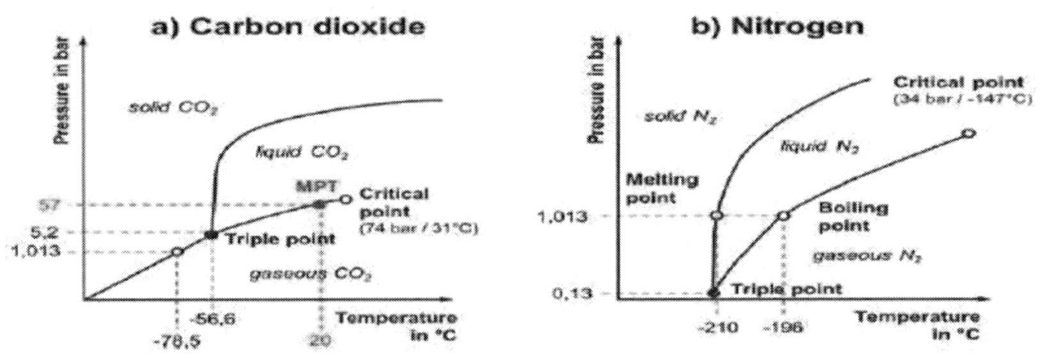

Fig. 2. Schematic phase diagrams for CO2 and nitrogen. [6]

Machining measure assume significant function in assembling businesses. Numerous individuals were researched on the diverse machining cycle and cooling climate to upgrade the machining execution of titanium and its combination. For maximize the tool life, different types of coated tool is used e.g. PVD, coated carbide, CVD coated carbide, CBN and PCBN. These all types of coated tool are providing longer tool life.

Table 1.

Work Piece	Cooling Environment	Response Variables	Tool
Titanium & its alloy	Cryogenics cooling (Internal, external, hybrid & indirect cooling)	Tool life(TL), Ra, Cutting force, Chip morphology (CM) Cutting temperature	CVD coated carbide tool PVD coated carbide tool PCBN CBN Ceramic

In this paper a short writing survey on machining of titanium and its composite under various cooling climate.Table.1 present the aspect of study covered in this review paper.

2. Process on machining of cryogenic cooling system

In this machining cycle cryogenic fluid (LN2 and CO2) is provided by various way and machine arrangement to cutting area. These cryogenic fluids are put away fit as a fiddle barrel shaped and circular with pressure control and vaporizer. During the time spent splashing the cryogenic cooling, the weight in the tank itself powers the coolant to the cutting zone and no supplemental energy is required.

The inside cooling is given through an extraordinarily planned device where the nitrogen or CO2 enters the cutting zone in the most proximate vicinity to the apparatus chip-work piece interface surface. The outer cooling can either be given utilizing by a spout which is arranged to point the coolant where required or through a cap-like repository which can be used in turning activities where the supply is put over the addition. Fig. 3 show machining set-up of cryogenic innovation.

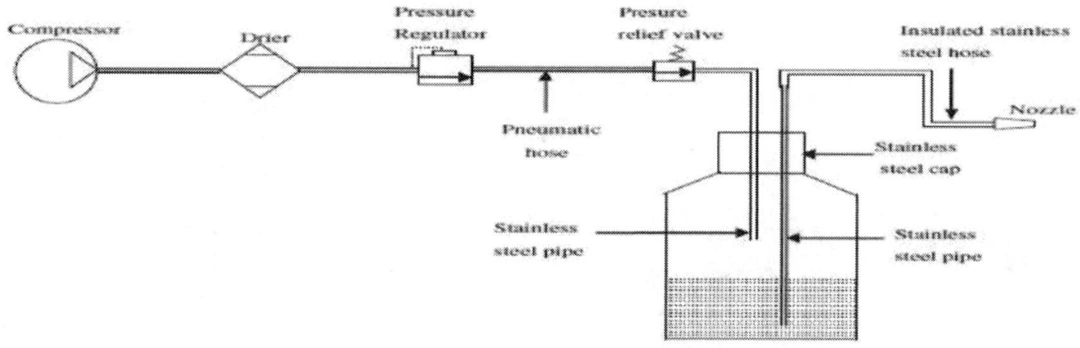

Fig. 3. Schematic diagram of cryogenic cooling setup [8]

3. Literature survey

L.Shakeel Ahmed et.al [09] in this research paper experiment work was carried out on titanium ASTM B265 grade utilized as work piece with PVD coated insert. In the experiment input process parameter e.g. feed rate & cutting speed with constant depth of cut. After this experiment work ascertain that 6-65% minimization in cutting temperature when LN2 is utilized, for LN2 coolant condition higher thrust force were recorded and higher surface roughness (Ra) but hole quality is not auspicious in LN2 coolant supply. **K.A.Venugopal, S.Paul et.al [10]** in this paper trial examination on Ti-6A-4A by turning activity with microcrystalline uncoated carbide embeds. Investigation work was completed under wet, dry and cryogenic cooling condition in the cutting speed scope of 70-100m/min. Also, found that apparatus life is augmented by utilized fluid nitrogen as a coolant through diminishing in grip disintegration device wear through controlled machining temperature and machining parameters. **Shane Y.Hong et.al [11]** has been developed an incipient cooling approach. The cooling method / approach implemented a minimum quantity of LN2 injected by micro nozzle setup between the tool rake and the chip breaker. In this approach, not wasted of coolant in dispensable areas. This types of cooling approach is better than other cooling approach because in this approach improvement of implement life. **Kyung-Hee park et.al [12]** in this paper have been analyzed the machining performance with different cutting approach such as cryogenic, flood cooling and MQL on end milling machine. Developed a specially liquid nitrogen spray system for evaluating cutting force and tool wear of machining condition. After this investigation found that improve tool life by the combination of MQL and internal cryogenic cooling environment compared to other conventional cooling method. Cutting force also reduced. **Mohd Azlan Suhaimi et.al [13]** in this paper, for improve machinability and tool life the novel cryogenic (indirect cryogenic) method is developed. In this type of mechanism novel cryogenic cooling approach compared with other cooling method and LN2 is utilized as a coolant medium. For novel cryogenic machining, a specially designed tooling unit that able to supply the liquid nitrogen. After this investigation found that novel cryogenic cooling method improve machinability of Ti-6A-4V, improve tool life by 90% and reduced cutting force by 54% compared to conventional flood cooling method. Fig. 4. show cutting force underdifferent machining conditions. And Fig.5. show flank wear under different cooling cooling conditions.

Alborz shokrani et.al [14] in this paper built up another cooling approach, for example, half and half cryogenic cooling MQL cooling/grease contrasted & flood cooling, MQL and ordinary cryogenic cooling. The examination work was completed on Ti-6Al-4V with covered strong carbide device in end processing machine set-up and discovered another mixture cryogenic MQL approach expanded instrument life of multiple times and half expanded profitability contrasted and other cooling method. **Upendra kumar et.al [15]** in this paper built up another cooling approach, for example, half and half cryogenic cooling MQL cooling/grease contrasted and MQL, flood cooling and ordinary cryogenic cooling. The examination work was completed on Ti-6Al-4V with covered strong carbide device in end processing machine set-up and discovered another mixture cryogenic MQL approach expanded instrument life of multiple times and half expanded profitability contrasted and other cooling method. **Damir et.al [16]** in this paper research work carried out on

Ti-6Al-4V with wet cooling and cryogenic cooling approach in turning operation. After this investigation found that cryogenic cooling approach significantly improved tool life and surface roughness compared to other cooling conditions.

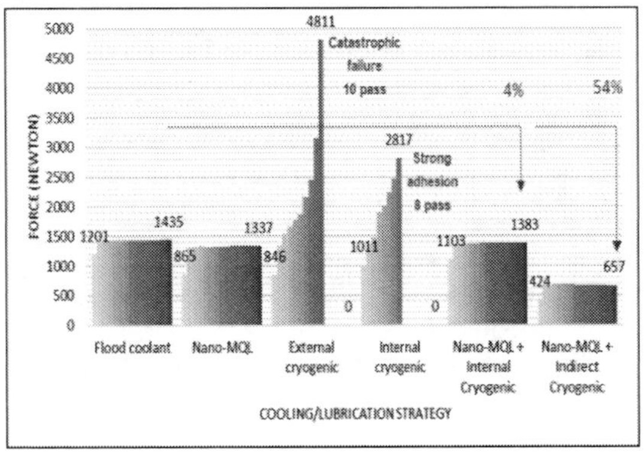

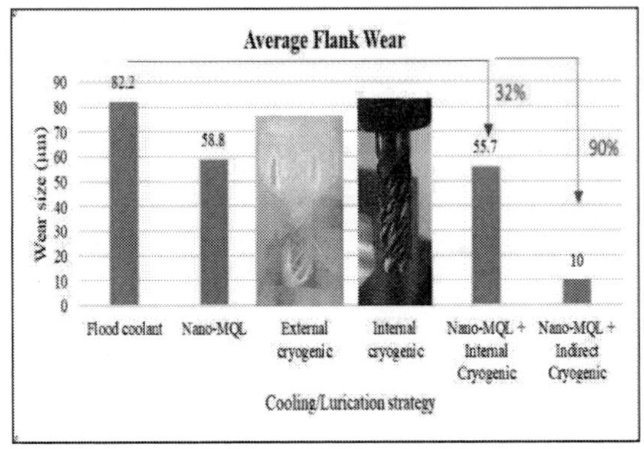

Fig.4. Showing force for all lubrication methods [13]

Fig. 5. Showing flank wear for all lubricant methods [13]

Vinothkumarsivalingam et.al [17] in this paper experimental work was carried out on Ti-6Al-4V alloy with PVD NBN coated tungsten carbide tool under different condition such as untreated condition and cryogenic treated condition (i.e. 24h and 48h). After this experimental investigation found out that machinability and tool life increased by utilized cryogenic treated condition. **Sanjeev kumar et.al [18]** experimental work was carried out on Ti-5Al-2.5sn alloy by using Electro discharge machining under different process parameter. And find out that current is most significant parameter which is affected on TWR compared to other process parameter. Table 2. Shows experiment result.

Table 2.

Source of variation	DF	Seq.SS	Adj.MS	F	P	Remark
CT of electrode	1	11.147	11.147	15.19	0.005	Significant
Peak current (Ip)	2	402.958	201.479	274.54	0.000	Most Significant
Pulse-on-time (Ton)	2	9.168	4.584	6.25	0.023	Significant
Pulse-off-time (Toff)	2	4.099	2.049	2.79	0.120	Not Significant
Flushing pressure	2	7.995	3.998	5.45	0.032	Significant
Residual error	8	5.871	0.734			
Total	17	441.238				

Renu.K.Shastri et.al [19] has been investigation on cryogenic treated copper and brass tool which were used for machining of Ti-6Al-4V. The experimental work was carried out on these parameters such as voltage (V), pulse on time(Ton), pulse off time(Toff) and current(I) with response viz. TWR, radial overcut and metal removal rate(MRR). After this investigation observed that radial over cut and tool wear rate reduced by using cryogenic treated copper and brass tool. Fig.6. represent graph between MRR with input variables. Fig.7. similarly shown variation of TWR with input variables.

Y.Sun,B.Hung et.al [20] main focus of this paper is that improvement of surface roughness and hardness by utilizing cryogenic machining compared to the flood cooling and dry machining. after this investigation found that cryogenic machining amend surface roughness by 35% and 6.6% respectively. Additionally, the hardness is increased 33.6% and 14.7% by utilizing cryogenic machining compared to other machining.

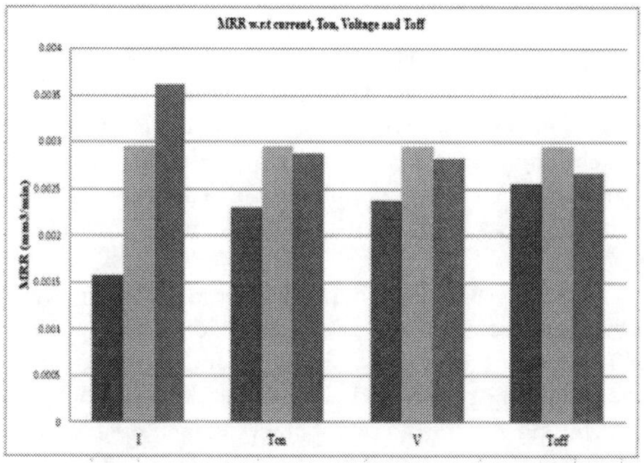

Fig. 6. Showing MRR together I, Ton, Toff and V for Cryogenic method treated copper electrode [19]

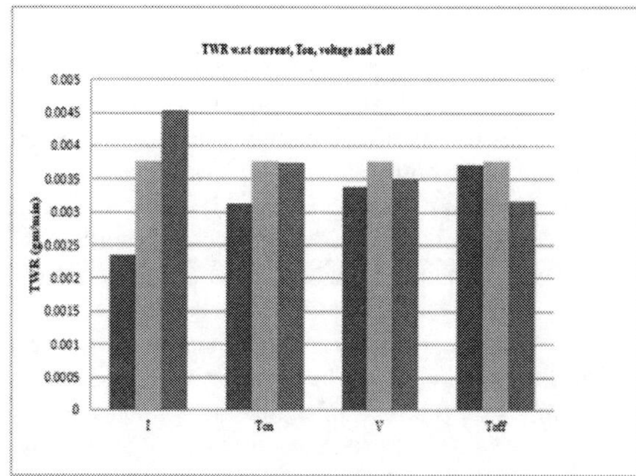

Fig. 7.Showing of TWR with Ton, I, Toff and V [19]

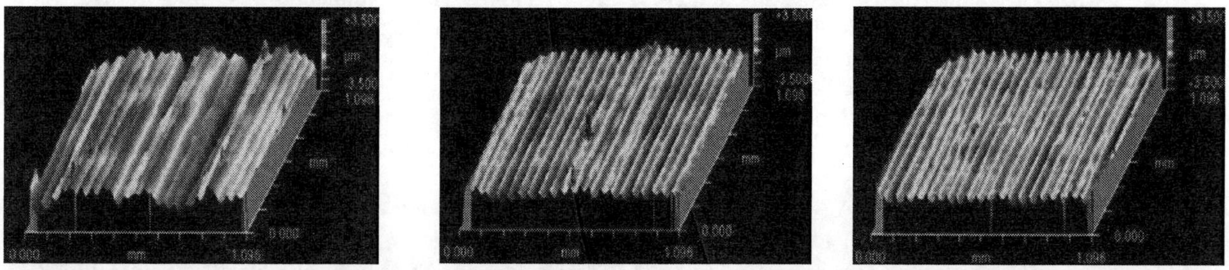

(a) (b) (c)

Fig. 8 Ra of machined surface of of Ti-6Al-7Nb alloy: (a) Showing Surface of dry machining; (b) Showing Surface of flood- cooled machining; (c) Showing Surface of cryogenic machining system [20]

Shoujin sun et.al [21] the paper presents a effect of cryogenic compressed air on cutting force and tool life during machining of Ti-6Al-4V. After this investigation found that tool life is increased compared to dry machining process. Fig. 9 shows snapshots of infra-red image.

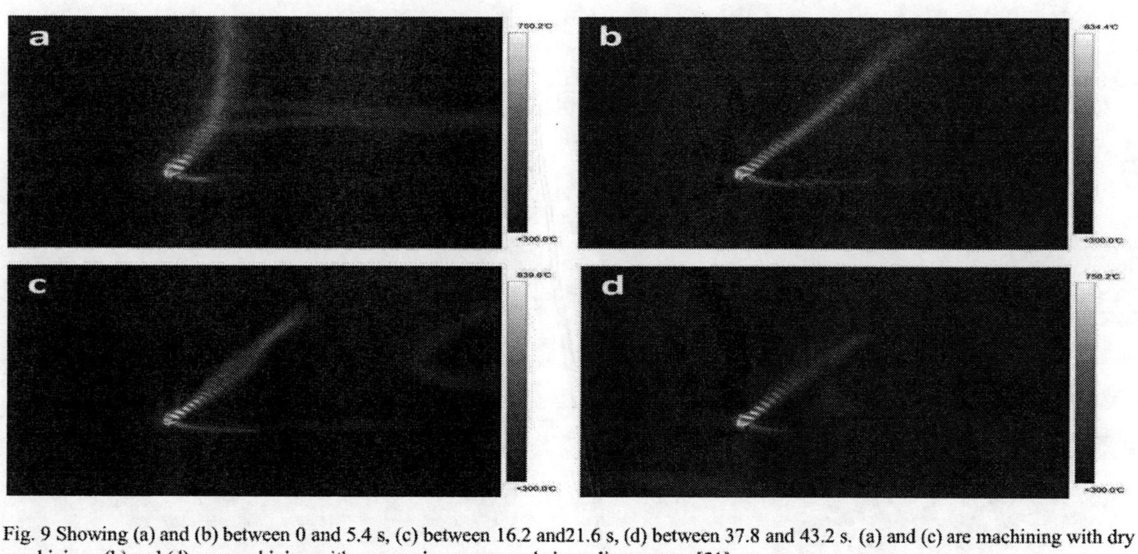

Fig. 9 Showing (a) and (b) between 0 and 5.4 s, (c) between 16.2 and21.6 s, (d) between 37.8 and 43.2 s. (a) and (c) are machining with dry machining , (b) and (d) are machining with cryogenic compressed air cooling system [21]

IOP Conf. Series: Materials Science and Engineering 998 (2020) 012013 doi:10.1088/1757-899X/998/1/012013

Stano imbrogno et.al [22] for forecast of cutting power, temperature and machining-instigated miniature basic adjustments during machining of Ti-6Al-4V under dry with cryogenic cooling condition built up a 3-D limited component model. The mathematical model is tentatively adjusted and approved. **Y.Sun et.al [23]** in this paper trial examination was done on Ti-5553 under cryogenic machining contrasted and the flood cooling, MQL for investigation of cutting power, device wear and surface harshness. After this examination discover that slicing power decline up to 30% by used cryogenic machining and MQL give higher surface harshness. Additionally, created limited component model of reproduce cutting power.

M.Dhananchezian et.al [24] in this examination article, cryogenic machining contrasted and the wet machining for investigation of cutting power, surface harshness, device wear and cutting temperature of Ti-6Al-4V on turning activity. After this examination results was that cutting temperature, surface unpleasantness, cutting power and flank wear diminished 66%, 36%, 35-42%, 27-35% separately by used cryogenic machining contrasted with other machining process. **Alborz shokrani et.al [25]** concentrated on impact of cryogenic machining on surface of Ti-6Al-4V work piece by used end processing machine. The outcome was that cryogenic machining improve surface of Ti-6Al-4V compound in end processing machine. A.Bordin et.al [26] this paper presents the component of hardware wear when turning of Ti6Al4V under dry and cryogenic machining. This examination was acting in EBM and utilizing distinctive investigation procedures. SEM (Scanning electron microscopy) is actualized for measure flank wear at commotion area and Energy dispersive X-Ray spectroscopy utilized for investigated the work piece material followed on the bleeding edge and rake face of cathode. After this examination discover that cryogenic machining decreased instrument wear contrasted with dry machining process. **Ampra aramcharoen [27]** in this paper analyzed the effect of cryogenic machining on chip formation and tool wear rate in turning operation for Titanium alloy. After this investigation results show that tool wear resistance is increasing by utilized cryogenic machining. Also, the thickness of secondary deformation zone and different chip formation if machining by oil-based coolant.

Christian machai et.al [28] in this exploration article, carbon dioxide is utilized as a coolant for turning activity of Ti-10V-2fe-3A. The coolant carbon dioxide is given in a pressurized tank and flexibly to apparatus tip by openings in the instrument holders clasping jaw. The coolant carbon dioxide contrasted and the flood emulsion cooling. The outcome shows that apparatus life was expanded and instrument wear rate is diminishing. Fig.10. speak to trial set-okay with machining. **S.M.Yuan et.al [29]** in this paper, experimental work was carried out under different coolant conditions such as wet, dry, MQL and MQL with cooling air for study cutting force, tool wear, chip morphology and surface roughness on milling machine of Ti-6Al-4V alloy with uncoated carbide tool. After this experimental investigation result was that MQL with cooling air is most significant parameter compared to other tested parameters. According to experimental observation MQL with cooling air -15° provide most significant result compared to 0°,-30°,- 45°. Fig. 12 shows the experimental set-up of cryogenic machining. **Simranpreet Singh Gill et.al [30]** in this paper concentrated on device life diminishing by using profound cryogenic treated addition contrasted with untreated supplement shows the ruinous impact of profound cryogenic temperature (— 196 1C) on TiAlN covered supplements. This is additionally upheld by VDI-3198 space. Test.

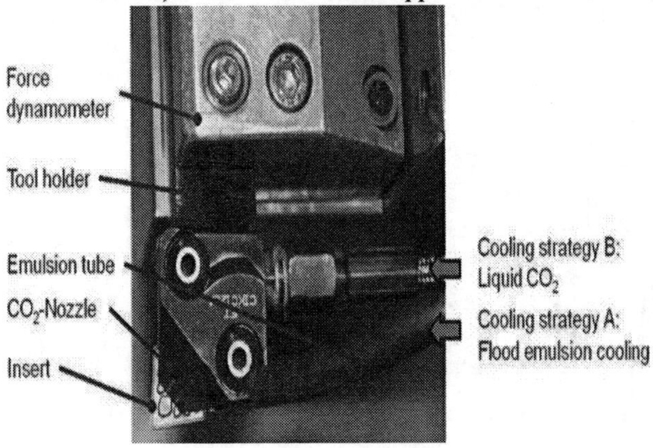

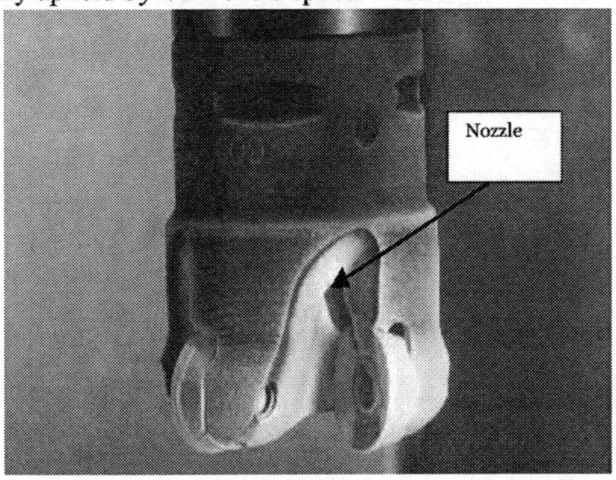

Fig. 10. Experimental set-up [28]. Fig. 11. Shape of tool for cryogenic machining[33]

M.J.Bermingham et.al [31] this paper zeroed in on chip morphology and device life during machining on Ti-6Al-4V composite. After this exploratory examination found that high weight water based emulsion offer somewhat better device life contrasted with cryogenic coolant. Along these lines, the best boundary is the coolant spout position in machining, set-up. **K.A.Venugopal et.al [32]** in the current examination concentrated on device life and apparatus wear under various cooling machining, for example, dry, wet, cryogenic machining on Ti-6Al-4V compound. After this exploration found that cryogenic machining is most noteworthy machining contrasted with wet & dry machining. Likewise, the SEM method is utilized for disciple affidavit of titanium chip material on the addition under all machining climate. **M.Ibrahim sadik et.al [33]** in this paper fluid carbon dioxide (Co2) is utilized e.g. coolant in face processing activity of Ti-6Al-4V as an instrument with PVD, covered supplement and furthermore investigation of coolant stream rate on apparatus life. Apparatus life is improve with higher coolant stream rate, the impact being more grounded in cryogenic contrasted with wet processing because of the way that the cryogenic coolant defers the wear advancement. **Table 3.** Represent conclusion of all literature survey in tabular form. Table 3 following.

Table 3. Summary of literature survey [9-33]

S.No.	Author Name	Material	Machining Type	Tested parameters	Response
1	L.Shakeel Ahmed et.al	W/P-Titanium ASTM B 265, Grade=2 Tool- PVD coated carbide insert	CNC vertical machining centre	Coolant- Wet & LN2 Variables- Feed rate (f), Cutting speed(vc , Depth of cut(d) = constant	Cutting Temp., Ra, Hole quality,
2	K.A.Venugopal et.al	W/P-Ti-6A-4V Tool- Uncoated carbide insert	Centre lathe	Wet, Dry & LN2 cooling under cutting range 70 to 100 m/min.	Tool Wear.
3	Shane Y.Hong et.al	W/P-Ti-6A4V Tool- ANSI C2-C3	CNC Turning centre	Dry cooling, cooling tool back by Liquid nitrogen, pre-Cooling w/p through LN2 process.	Tool life.
4	Kyung-Hee park et.al	Tool-Coated tool(AlCrN) W/P-Ti-6A-4V	Mori Seikri NVD- 400-DCG-HSC 3-Axis vertical milling centre	Flood cooling, nano+MQL, Nano-MQL+ internal spray, external internal spray.	Tool wear, Cutting force.
5	Mohd Azlan Suhaimi et.al	Tool-(AlCrN) W/p-Ti-6A-4V	3-Axis vertical milling centre	Cooling- Indirect, internal and External spray system. Variables-Cutting speed, Feed rate, Depth rate, Width of cut, Cutting length.	Tool life (TL), Cutting force.
6	Alborz shokrami et.al	Tool-coated solid carbide W/P.Ti-6Al4V	End milling machine	Flood cooling, MQL, Cryogenic cooling, Hybrid	Tool life, Tool Wear,(Ra).

				Cryogenic MQL.	
7	Upendra kumar et.al	Tool-uncoated tungsten carbide W/p- Ti-6Al-4V	CNC lathe machine	Cryogenic treated and untreated of uncoated WC insert	Cutting force, chip morphology, Ra
8	A.Damir et.al	Tool-CVD coated insert W/P-Ti-6Al-4V	CNC turning centre	Flood cooling, cryogenic cooling.	Tool life, Surface roughness.
9	Renu.K.Shastri et.al	Tool- brass&copper W/P-Ti-6Al-4V	EDM setup	Cryogenic treated and untreated tool under I, Ton, V, Ton off parameters.	MRR, TWR, Radial overcut.
10	Y.Sun et.al	Tool-TiAlN coating CNMG 431RP W/P-Ti-6Al-7NB	HAAS TL-LC CNC lathe	Dry, flood cooled and cryogenic machining with constant input parameters.	Surface roughness, hardness.
11	Vinothkumar sivalingam et.al	Tool- PVD coated tungsten carbide W/P- Ti-6Al-4V	End milling machine.	Cryogenic treated (i.e. 24h & 48h) and untreated tool.	Machinability, tool life, cutting force.
12	Sanjeev kumar et.al	Tool-copper tungsten carbide W/P- Ti-5Al-2.5sn	CNC EDM machine	Cryogenic treatment of electrode, peck current, Ton, pulse off, flushing Pressure.	Tool wear rate (TWR).
13	Stano Imbrogno et.al	Tool-Uncoated tungsten carbide W/P. Ti-6Al-4V	CNC lathe machine.	Validate experimental and theoretical result under dry and cryogenic Environment.	Cutting force, machining Temperature.
14	Y.Sun et.al	Tool- TiCN coated W/P-Ti-6Al-4V	CNC lathe machine.	Compared cryogenic machining with MQL and flood cooling under input Process parameters.	Cutting force, SR, Tool wear.
15	Alborz Shokrani et.al	Tool- TiN-TiAlN coated solid carbide W/P-Ti-6Al-4V	CNC end milling.	Dry, flood and cryogenic machining compared for Surface integrity.	Surface integrity.
16	A.Bordin et.al	Tool- TiAlN coating deposed by PVD. W/P. Ti-6Al-4V	Electron Beam Machining.	Compared dry and Cryogenic machining.	Surface roughness and tool wear.
17	Ampra Aramcharoen et.al	Tool- coated carbide inserts W/P- Ti-6Al-4V	CNC lathe Machine.	Compared dry and cryogenic machining with constant input Variables.	Tool wear, chip formation.
18	Christian machai et. al	Tool-cemented carbide W/P-Ti-10V-2f- 3Al	CNC turning lathe Machine.	Coolant-(Co2 & emulsion) Depth of cut, cutting speed, cutting length, feed Rate.	Tool life.
19	S.M.Yuan et.al	Tool- uncoated cemented carbide inserts W/P. Ti-6Al-4V	CNC milling Machine.	Dry, wet, MQL and MQL with cooling air under constant input variable.	Tool wear, cutting Force, Ra, chip formation.
20	M.Dhananchezian	Tool- PVD TiAlN	CNC turning	Wet and cryogenic	Cutting

	et.al	coated tungsten carbide tool W/P- Ti-6Al-4V	centre and inserts modified by EDM.	Machining.	Temperature, cutting force, Ra, tool wear.
21	Simpranpreet singh gill et.al	Tool- TiAlN coated tungsten carbide inserts W/P-Ti-6Al-4V	CNC turning centre.	Shallow cryogenically treated, Deep cryogenically treated, untreated.	Tool life.
22	M.J.Bermingham et.al	Tool-MF1 WC inserts. W/P- Ti-6Al-4V	CNC lathe Machine.	LN2, Dry and emulsion cooling environment with constant cutting parameter and constant nozzle Position.	Tool wear, Chip morphology.
23	K.A.Venugopal et.al	Tool- uncoated carbide inserts. W/P-Ti-6Al-4V	11KW centre lathe.	Dry, wet and cryogenic machining with constant cutting Parameters.	Tool life.
24	M.Ibrahim sadik et.al	Tool- PVD inserts W/P- Ti-6Al-4V	Hermle C40U Dynamic centre.	Different flow rate in cryogenic and wet machining.	Tool life, tool wear Rate.

4. Conclusion

These articles present a basic survey on machining of Titanium amalgam under cryogenic climate. These cryogenic machining compared with the different cooling environment for analysis the tool life, TW, Ra, CF, cutting temperature and chip morphology of work specimen. Today's increasing demand for efficient machining process and sustainable environment so many people interesting in new cooling approach that most effective on response. So, in this review paper present different cooling conditions and different types of tool.

- The primary circumstance of hardware disappointment is attachment dispersion and all instrument endured disappointment mode.
- High weight emulsion and fluid nitrogen are successfully machining contrasted with dry machining. However, high weight emulsions give longer device life contrasted with fluid nitrogen.
- MQL cooling condition show better surface harshness contrasted with dry machining. However, best cooling condition was accomplished under half breed cooling climate.
- As per grater durability and instrument wear opposition CVD covered device give better outcome contrasted with PVD covered device.

References

[1] Yakup Yildiz, Muammer Nalbant, A review of cryogenic cooling in machining processes, International Journal of Machine Tool & Manufacturing 48(2008) 947-964.

[2] V.C. Venkatesh, S.Izman, T.C.Yap, P.V.Brerern, N.S.M. EI-Tayeb, Precision cryogenic drilling, turning and grinding of Ti-64 alloys , Int. Jr. of precision tech., Vol.1.NOS.3/4, 287-301.

[3] Hollis, W.S, The application and effect of controlled atmospheres in the machining of metals, Int. Jr. Of machine tool design and research (1961), 59-78.

[4] Ghosh, Ranajit, Technology Assessment on Current Advanced Research Projects in Cryogenic Machining , the association for manufacturing technology, (2016), thesis.

[5] Kramer, Alxander, Lung, Dieter & Klocke, High Performance Cutting of Aerospace Materials, 4th manufacturing engineering society international conference (2011), 1-4.

[6] Timmerhaus, Klaus D.Och reed, P.Richard, Cryogenic Engineering , 5th years of progress, Springer science + Business medical LLC (2007), ISBN-10;0-387-33324-X.

[7] Wit Grzesik, Advanced machining process of metallic materials (2017). 2nd edition.

IOP Conf. Series: Materials Science and Engineering 998 (2020) 012013 doi:10.1088/1757-899X/998/1/012013

[8] Alborz Shokrami, Vimal Dhokia, S.T.Newman, Investigation of Cutting Parameters in Sustainable Cryogenic End Milling, 2nd International Conference on Sustainable Design and Manufacturing,(2017),UK, 1-13.
[9] L.Shakeel ahmed, V.Govindaraju & M.Pradeep kumar, Experimental Investigations on Cryogenic Cooling in the Drilling of Titanium Alloy, Materials and manufacturing process, (2015), 1-5.
[10] K.A.Venugopal, S.Paul, A.B.Chattopadhyay, Tool wear in cryogenic turning of Ti-6Al-4V alloy, Cryogenics 47 (2007), 12-18.
[11] Shane Y.Hong, Irel markus, Woo-Cheol Jeong, New cooling approach and tool life improvement in cryogenic machining of titanium alloy Ti-6Al-4V, International journal of machine tools & manufacturing 41 (2001), 2245-2260.
[12] Kyung-Hee Park, Gi-Dong Yang, M.A.Suhaimi, Dong Yoon Lee, Taegon Kim, Dong-Won Kim & Seok-Woo Lee, The effect of cryogenic cooling and minimum quantity lubrication on end milling of titanium alloy Ti-6Al-4V, Journal of Mechanical Science and Technology 29 (12) (2015), 5121-5126.
[13] Mohd Azlan Suhaimi, Gi-Dong Yang, Kyung-Hee Park, Mohd Juzaili Hisam, Safian Sharif & Dong-Won Kim, Effect of cryogenic titanium alloy based on indirect, internal and external spray system, Procedia Manufacturing 17 (2018) 158-165.
[14] Alborz Shokrami, Ihsan Al-Samarrai, Stephen T. Newman, Hybrid cryogenic MQL for improving tool life in machining of Ti- 6Al-4V titanium alloy, Journal of Manufacturing, 43(2019) 229-243.
[15] Upendra kumar, P.senthil, A comparative machinability study on titanium alloy Ti-6Al-4V during dry turning by cryogenic treated and untreated condition of uncoated WC inserts, Journal of Materials Today: Proceedings (2019).
[16] A.Damir, A.Sadek, H.Attia, Characterization of machinability and environmental impact of cryogenic turning of Ti-6Al-4V, Procedia CIRP 69 (2018) 893-898.
[17] Vinoth kumar Sivalingam, Jie Sun, Bin Yang, Kai Liu, Ramesh Raju, Machining performance and tool wear analysis on cryogenic treated insert during end milling of Ti-6Al-4V alloy, Journal of Manufacturing Process 36 (2018) 188-196.
[18] Sanjeev Kumar, Ajay Batish, Rupinder Singh, Anirban Bhattacharya, Effect of cryogenically treated copper-tungstenelectrodeontoolwear rateduringelectro-dischargemachiningofTi-5Al-2.5Snalloy, Wear 386-387 (2017) 223-229.
[19] Renu Kumar Shastri, Chinmaya P. Mohanty, Prashant S. Jadhav, Study of the Effect of Cryogenically treated tools during Rapid Drill Electro-discharge machining of Ti-6Al-4V alloy, Materials Today: Proceedings 18 (2019) 3122-3127.
[20] Y.Sun, B.Huang, D.A. Puleo, J.Schoop, I.S.Jawahir, Improved Surface Integrity from Cryogenic Machining of Ti-6Al-7Nb Alloy for Biomedical Applications, 3rd CIRP Conference on Surface Integrity (CIRP CSI), Procedia CIRP 45 (2016) 63-66.
[21] Shoujin Suna, Milan Brandt, Suresh Palanisamy, Matthew S. Dargusch, Effect of cryogenic compressed air on the evolution of cutting forceand tool wear during machining of Ti–6Al–4V alloy, Journal of Materials Processing Technology 221 (2015) 243–254.
[22] Stano Imbrogno, Stefano Sartori, Alberto Bordin, Stefania Bruschi, Domenico Umbrello, Machining simulation of Ti6Al4V under dry and cryogenic conditions, Procedia CIRP 58 (2017) 475-480.
[23] Y.Sun, B. Huang, D.A.Puleo, I.S.Jawahir, Enhanced Machinability of Ti-5553 Alloy from Cryogenic Machining: Comparison with MQL and Flood-cooled Machining and Modeling, Procedia CIRP (2015) 477-482.
[24] M.Dhananchezian, M.Pradeep Kumar, Cryogenic turning of the Ti–6Al–4V alloy with modified cutting tool inserts, Journal of Cryogenics (2011) 34-40.
[25] Alborz Shokrami, Vimal Dhokia, Stephen T.Newman, Investigation of the effects of cryogenic machining on surface integrity in CNC end milling of Ti–6Al–4V titanium alloy, Journal of Manufacturing Process 21 (2016) 172-179.
[26] A.Bordin, S.Brauschi, A.Ghiotti, P.F.Bariani, Analysis of tool wear in cryogenic machining of additive manufactured Ti6Al4V alloy, Wear 328-329 (2015), 89-99.
[27] Ampara Aramcharoen, Influence of cryogenic cooling on tool wear and chip formation in turning of titanium alloy, Procedia CIRP 46 (2016) 83-86.
[28] Christian Machai, Dirk Biermann, Machining of -titanium-alloy Ti–10V–2Fe–3Al under cryogenic conditions: Cooling with carbon dioxide snow, Journal of Materials Processing Technology 211 (2011), 1175-1183.
[29] S.M.Yuan, L.T.Yan, W.D.Liu, Q. Liu, Effects of cooling air temperature on cryogenic machining of Ti–6Al–4V alloy, Journal of Materials Processing Technology 211 (2011) 356-362.
[30] Simranpreet Singh Gill, Jagder Singh, Harpreet Singh, Rupinder Singh, Investigation on wear behaviour of cryogenicallytreatedTiAlNcoated tungsten carbideinsertsinturning, International Journal of Machine Tools & Manufacturing 51 (2011), 25-33.
[31] M.J.Bermingham, S.Palanigamy, D. Kent, M.S. Darguseh, A comparison of cryogenic and high pressure emulsion cooling technologies on tool life and chip morphology in Ti–6Al–4V cutting, Journal of Materials Processing Technology 212 (2012), 752- 765.
[32] K.A.Venugopal, S.Paul, A.B.Chattopadhyay, Growth of tool wear in turning of Ti-6Al-4V alloy under Cryogenic cooling, Wear 262 (2007), 1071-1078.
[33] M.Ibrahim Sadik, Simon Isakson, Amir Malakizadi, Lars Nyborg, Influence of coolant flow rate on tool life and wear development in cryogenic and wet milling of Ti-6Al-4V, Procedia CIRP 46 (2016), 91-94.

IOP Conf. Series: Materials Science and Engineering 998 (2020) 012014 doi:10.1088/1757-899X/998/1/012014

Prediction of solid rocket motor performance characteristics using computational fluid dynamics and validation with experimental data

Venkata Naga MohanManchiraju*[$a], Vijay Kumar Dwivedi[$b]
*Scientist, Advanced Systems Laboratory, Dr.APJ Abdul Kalam Missile Complex, DRDO, Ministry of Defence, Government of India, Hyderabad-500058,
[$]Department of Mechanical Engg., IET, GLA University, Mathura, U.P., India
Email: [a]mvnmohan33@gmail.com, [b]vijay.dwivedi@gla.ac.in

Abstract. Solid Propulsion is widely used in Missile and Space applications. Accurate prediction of Specific impulse (I_{sp}) is important for design of Solid rocket motor. These predictions can be done using empirical relations. Obtaining these empirical relations require large amount of data. The renowned aerospace organizations use their own code for prediction. These codes are not open source and also developing a code requires big team effort with challenges of verification and validations. The present work is aimed at developing numerical model for prediction of Specific Impulse. The model is validated against the experimental data of static test of solid rocket motor taken from literature.

Key words: Computational Fluid Dynamics, Specific Impulse, Solid Propulsion

1.NOMENCLATURE

1.1 Symbols

F = Thrust in N
$\dot{m}$ = mass flow rate in kg/s
A= Area in m^2
V = Velocity in m/s
P = pressure in Pa
ρ = density in kg/m^3
I_{sp}= Vacuum Specific Impulse
g_0= Acceleration due to gravity in m/s^2

1.2 Subscripts

1 nozzle inlet or chamber
2 nozzle exit
3 atmospheric or ambient

2. INTRODUCTION

Propulsion is the act of changing the motion of a body. The propulsive force of a solid rocket motor is obtained by ejecting propellant at high velocity. Accurate prediction of propulsion parameters is important for design of Solid rocket motor. CFD is a useful tool for the prediction of propulsion

IOP Conf. Series: Materials Science and Engineering 998 (2020) 012014 doi:10.1088/1757-899X/998/1/012014

parameters. Many researchers have made use of CFD for their research work to study, design, analyze or predict solid propulsion elements and parameters.

CFD has a very important role in rocket propulsion. Chongankim*et al.*[1] explained the role of CFD in development of rocket propulsion. Nozzle is a component that increases the performance of air breathing and non-air breathing engines. G Srinivas*et al.* [2] analyzed the performance of convergent nozzles. It has a good comparison with experimental results. To achieve dependable CFD predictions, it is important that the numerical model be based on first principle. The models based on correlation may not be fully reliable. Jiri blazek*et al.* [3] described a flow solver useful to simulate rocket motors. The reliability of code is of utmost importance for analysis of flow in nozzles. Bogdon-AlexandruBelega*et al.* [4] used fluent 6.3 for analysis of convergent divergent nozzle with GAMBIT 2.4 software for numerical modeling and generation of mesh. Nathan Spotts*et al.* [5] used Metacomp CFD++ software to study compressible flow through convergent conical nozzles. The results have good comparison with experimental data.

Laura *et al.* [6] used Hopsan a multi domain software to analyse the functioning of sounding rocket. The results obtained are comparable with real performance. The simulations were further useful for improvement of performance and increase in altitude a rocket can reach. Supersonic exhaust diffusers were also analyzed using CFD. M Srinivasa Rao *et al.* [7] performed numerical simulations for various rocket chamber pressures. The results have good comparison with experimental data.SukantaRoga*et al.* [8] analyzed the scramjet combustor using CFD. He identified important parameters for optimization of injection system. K. Schomberg*et al.* [9] use CFD for design of high-area-ratio nozzle contours using circular arcs. The design offers an improvement in thrust coefficient and reduction in average length. S. Saha*et al.* [10] analysed combustion instability using CFD in solid rocket motors. The analysis was validated against motor test data available in literature.

Specific Impulse is one of the important characteristic of rocket propulsion. It is a measure of fuel efficiency of the rocket, that is, the thrust imparted to the rocket per kilogram of the propellant expelled. If two different rocket motors have twodifferent values of specific impulse, then the motor with higher value of specific impulse is treated more efficient. This is because the motor will produce more thrust for the same amount of propellant. It gives us an easy way to size a motor during preliminary analysis. The rocket weight will define the required value of thrust. Dividing the thrust required by the specific impulse will tell us how much weight flow of propellants our motor must produce. This information determines the physical size of the motor. Accurate prediction of Specific impulse (I_{sp}) is important for design of Solid rocket motor.

3. OBJECTIVE

- To perform computational study on the rocket motor nozzle using AnsysFluent 14.0 for understanding the flow losses and compare results with published data
- To understand the flow losses involved in solid rocket propulsion and predict the vacuum specific impulse.

4. METHODOLOGY FOR CFD PREDICTIONS FOR ROCKET PROPULSION

The work flow for the CFD predictions for rocket propulsion is given below:

4.1Selectionof CFD Software

In the present paper, the computational study is done on the rocket motor nozzle using ANSYS Fluent 14.0 for understanding the flow losses involved in nozzle and predicting the vacuum specific impulse I_{sp} and thrust. The code solves the following Navier-Stokes equations through finite volume method.

IOP Conf. Series: Materials Science and Engineering 998 (2020) 012014 doi:10.1088/1757-899X/998/1/012014

4.2 Input Parameters

The rocket motor used for testing had an overall length of 720 mm and outer diameter of 198 mm. The throat diameter is 41 mm and nozzle exit diameter is 317.7 mm. The motor had a propellant weight of 23.345 kg and operational pressure of 7.34 MPa.

4.3 Pre-Processing

4.3.1Geometric Modelling. Using the coordinates as in ref 5 a 2D axi-symmetric geometric modelling is done using AutoCAD software. The Model is given in 'Figure 1'.

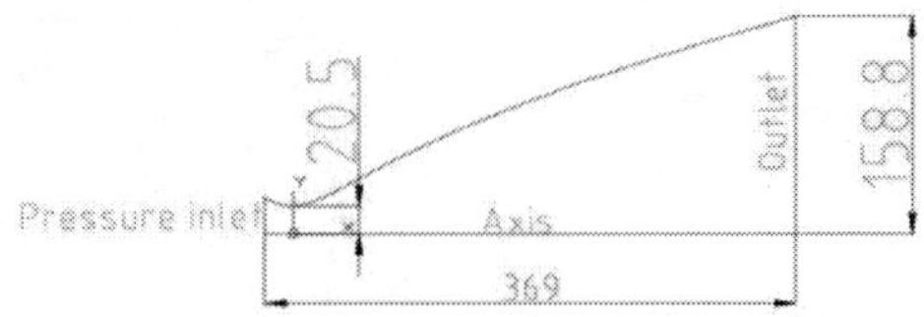

Figure 1.2axi-symmetric Geometric Model

4.3.2Meshing and Validation. Four different mesh sizes were taken for the setting up the model. All the meshes were Cartesian meshes as this kind of mesh is best suited for the problem. The elements chosen were quads. The four different quad sizes selected are 1mm, 0.5 mm, 0.3 mm and 0.1 mm. The meshes 0.3 mm and 0.1 mm are close to the actual tested mass flow rate within 2 percent. But the number of elements of mesh 0.1 mm is around 91,000. The initial solution took about 3 days to solve in a desktop. Due to computational constrain mesh of 0.3 mm is used for further calculation and prediction. The solutions arrived at are mesh independent.

4.4 CFD Setup

4.4.1 Boundary Conditions and Gas Properties. The CFD setup containing Boundary conditions and Gas properties is tabulated in Table1

Table 1CFD setup

General	Solver Type: Density Based 2D Space: Axis Symmetric Model Ideal Gas equation
Models	Energy Equation: on Viscous Model: k-omega model
Boundary Conditions	Inlet Pressure: 7.34 MPa Inlet Temperature: 3410 °K Outlet pressure: 0.001 atm Nozzle wall: No slip and adiabatic heat flux.

4.4.2 Run forChosen No of Iterations and Validation. For running the iterations the residuals for each flow equation was set at 10-10. Initially 10 iterations were run for initialization under hybrid initialization. Then multiples of 500 iterations were run till the residual condition was met. It was noticed that there was no further change after 3700 iterations. The residuals were also found within acceptable limits. The residual plots is shown below in 'Figure 2'.

IOP Conf. Series: Materials Science and Engineering 998 (2020) 012014 doi:10.1088/1757-899X/998/1/012014

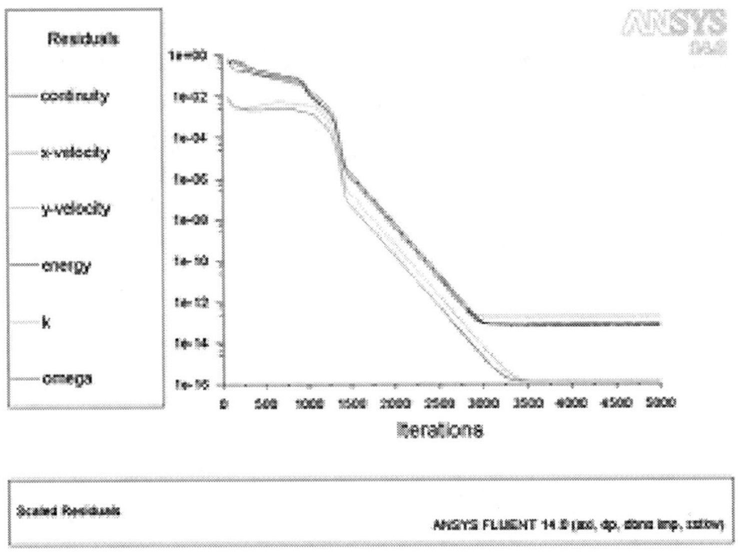

Figure2. Validation graph

4.5 Post Processing

4.5.1 Contour Plots

4.5.1.1 Pressure Contour Plot. The pressure contour plot is shown in 'Figure 3'the boundary condition defined at chamber inlet is 7.34 Mpa. The CFD predicted the outlet pressure at nozzle exit as 3.39kPa. This is because of gas expansion in the nozzle. This is a shock free expansion in the nozzle. The expansion waves are noticed at the start of nozzle divergent contour

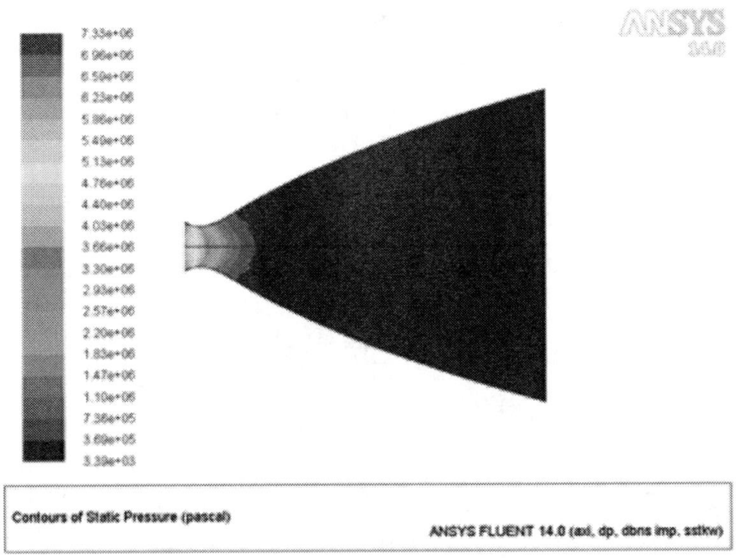

Figure 3. Pressure Contour Plot

4.5.1.2 Temperature Contour Plot. The boundary condition of gas temperature defined at chamber inlet is 3410°K. The gas temperature dropped from 3410°K to 938°K at nozzle exit. This is because of gas expansion in the nozzle. The kinetic energy required to expand into empty space comes from heat energy that gives the gas temperature. So the temperature is dropped. The 'Figure 4' shows variation in temperature across length of nozzle.

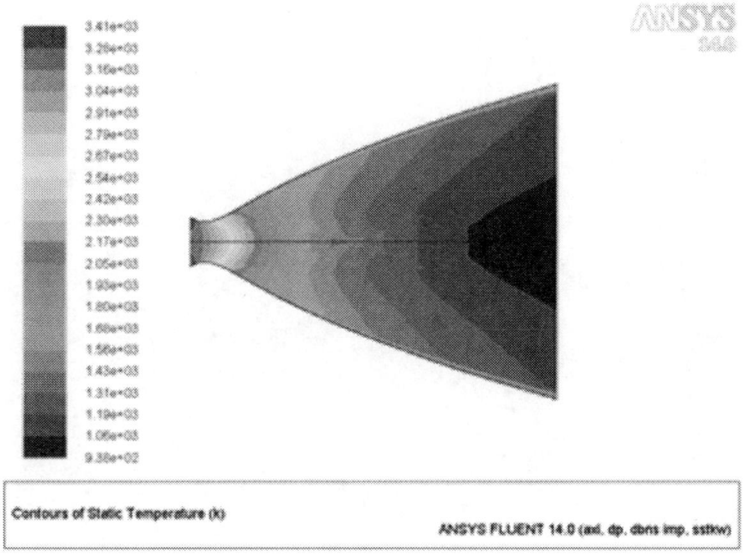

Figure 4. Temperature Contour Plot

4.5.1.3 Velocity Contour Plot. The gas velocity increasing form stagnant condition to 3070 m/s at the nozzle exit as solved by CFD. As the temperature is decreasing in nozzle, this energy is utilized by the gas to gain kinetic energy, thus increasing the velocity. The thrust developed by the nozzle increases with increase in exit velocity. The velocity contour plot is shown in 'Figure 5'.

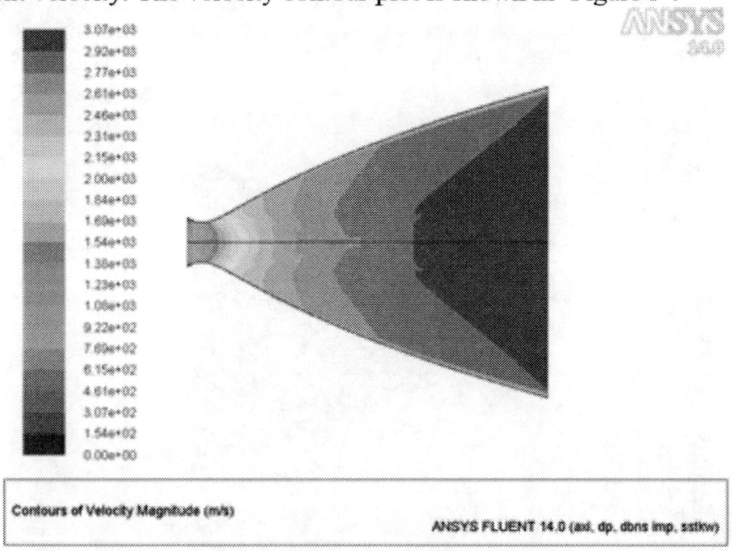

Figure 5. Velocity Contour Plot

4.5.1.4 Mach No Contour Plot. The Mach No, a ratio of gas velocity to sound velocity, is increasing from stagnant condition to Mach 5 at the nozzle exit as solved by CFD. Mach No is increasing due to increase in gas velocity in the nozzle as well as reduction in sound velocity. The flow is sub sonic in the convergent section, sonic at the throat and Supersonic in the Divergent section of the nozzle. The thrust developed by the nozzle increases with increase in exit velocity. The velocity contour plot is shown in 'Figure 6'.

IOP Conf. Series: Materials Science and Engineering 998 (2020) 012014 doi:10.1088/1757-899X/998/1/012014

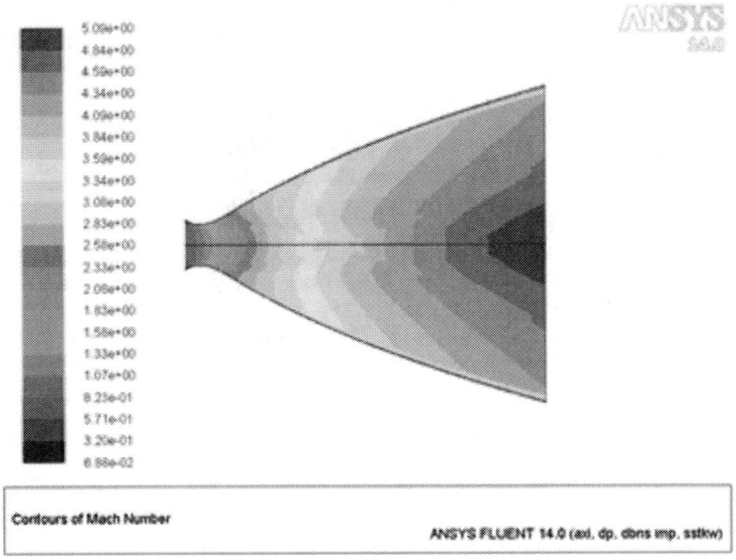

Figure 6. Mach No Contour Plot

4.5.1.5 Density Contour Plot. The gas density at the chamber was 6.67 kg/m^3. It has decreased to 0.012 kg/m^3 at the nozzle exit as solved by CFD. The mass flow rate is a function of Density, Area and Gas velocity. The mass flow rate is constant along the length of the nozzle. As the area of nozzle and gas velocity are increasing, the density of gas is decreasing. The density contour plot is shown in'Figure 7'.

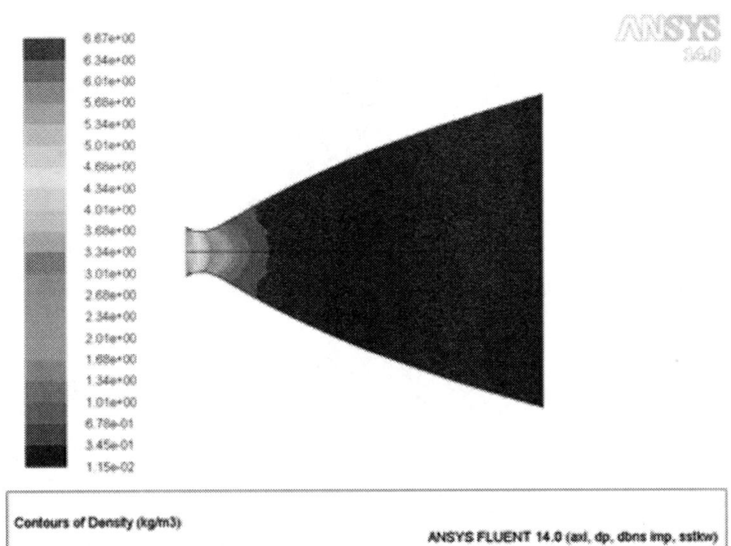

a

Figure 7. Density Contour Plot

5.0 RESULTS AND DISCUSSION

5.1 Calculation of I_{sp}

The Mathematical calculations to determine the Isp and Thrust are given below:

Thrust = $F = \dot{m}v_2 + (p_2 - p_3)A_2$
Where
F = Thrust in N
$\dot{m}$ = mass flow rate in kg/s = 5.96 kg/s
A_2 = Area at nozzle exit = 0.07926 m^2
V_2 = Axial Velocity at nozzle exit = 2859.5 m/s
P2 = Static Pressure at nozzle exit = 9833.24 Pascal
P3 = Ambient Static Pressure (Vacuum) = 101 Pascal
From this equation the thrust obtained from the CFD solution is 17814 N
The equation for $I_{sp}$is:

$$I_{sp} = \frac{F}{\dot{m}g_0}$$

Where
I_{sp}= Vacuum Specific Impulse
F = Thrust in N = 17814 N
$\dot{m}$ = mass flow rate in kg/s = 5.96 kg/s
g_0= Acceleration due to gravity = 9.81 m/s^2

The vacuum I_{sp}calculated using the results of CFD is 305.1 s

5.2 Comparisons of propulsion characteristics CFD vis-à-vis Experimental data

Table 2 Comparison of Propulsion characteristics

Parameter	From CFD code	Experimental data	Accuracy
I_{sp}	305.1 s	296.7 s	2.9%

The propulsion characteristics predicted through CFD is compared with the experimental data. The comparison table is given in Table 2. The I_{sp}predicted using model developed in fluent is 305.1s and matches very closely to experimental data. The results are within 3% accuracy. The pressure, velocity, temperature, Mach and density plots as discussed above follow the trend expected in an isentropic nozzle expansion. Thus it can be safely said the model developed is validated and the results can be used for design purpose.

6.0 CONCLUSIONS

CFD is a useful methodology in predicting the rocket motor / nozzle performance. The current project reveals that the accuracy of the prediction is better than 3%, which makes it reliable for the designers, scientists and engineer community. The CFD results can be used to accurately predict other parameters like temperature, pressure, velocity and density at any point in the nozzle. This data will help designers for selection of appropriate materials fornozzle fabrication. Flow features like expansion waves and oblique shocks occurring inside the nozzle can be captured accurately. This will help the designer to design the divergent contour free of any flow irregularities and flow separation. There is a future scope of improving the accuracy by using 3D simulations, which can be taken for further tasks up on availability of computational hardware.

7.0 REFERENCES

[1] Chongankim 2016 Roles of CFD simulations in developing rocket propulsion systems *The 2016 structures congress (structures 16)*
[2] G Srinivas*et al* 2017 *IOP Conf. Series: Materials Science and Engineering***197**, 01208
[3] J Blazek 2003 Flow Simulation in Solid Rocket Motors Using Advanced CFD, *39th AIAA/ASME/SAE/ASEE Joint Propulsion Conference and ExhibitAIAA 2003-5111*
[4] Bogdan-alexandrubelega andTrungducnguyen 2015 Analysis of flow in convergent-divergent rocket engine nozzle using computational fluid dynamics,*International conference of scientific paper, AFASES 2015*
[5] NSpotts andSGuzikandXGao2013 A CFD Analysis of Compressible Flow through Convergent-Conical Nozzles *49th AIAA/ASME/SAE/ASEE Joint Propulsion ConferenceAIAA 2013-3734*
[6] Laura, Navarrete-Martin and PetterKrus 2017 Sounding Rockets: analysis, simulation and optimization of a solid propellant motor using Hopsan*6th CEAS air & space conference aerospace Europe 2017, CEAS 2017*
[7] M. Srinivasa Rao, AfrozJaved and DebasisChakraborty 2017 Numerical Characterization of Supersonic Exhaust Diffusers,*Defence Science Journal* **67**(2) pp. 219-223
[8] SukantaRoga 2019 *J. Phys.: Conf. Ser.* **1276** 012041
[9] K. Schomberg and J. Olsen 2016 Design of High-Area-Ratio Nozzle Contours Using Circular Arcs *Journal of propulsion and power.* **32**(1), pp.188-195
[10] S. Saha and D. Chakraborty 2016 Computational Fluid Dynamics Simulation of Combustion Instability in Solid Rocket Motor: Implementation of Pressure Coupled Response Function *Defence Science Journal.* **66**(3) pp. 216-221
[11] Sutton G P 2001 Rocket Propulsion elements (New York: John Wiley & Sons)

ACKNOWLEDGEMENTS

While performing this work I came across many people, whose contributions helped my field of research in various ways. I express my gratitude to all of them. I thank Advisory Group for Aerospace Research and Development that published Technical report 230. The experimental data was taken from this technical report. I am thankful to Dr. MRM BABU,Director, Advanced systems Laboratory, Defence Research and Development Organization, Hyderabad for all the support extended for the current work. My specific thanks are due to Mr.Mrinmoy Biswas, Scientist, Advanced Systems Laboratory, and Hyderabad for extending his valuable support and resources in executing the current work. I thank GLA University for extending all the cooperation complete this work.

Investigation of Rupture in SS304 Steel Sheet using Macroscopic Criteria

Appurva Jain[1], Abhishek Mishra[1], and Vikrant Tiwari[2]

[1] Department of Mechanical Engineering, National Institute of Technology Delhi, Sector A7, Narela, Delhi 110040, India
[2] Department of applied mechanics, Indian Institute of Technology Delhi, New Delhi-110016, India

abhishekmishra@nitdelhi.ac.in

Abstract. Macroscopic ductile rupture prediction in materials always has been great interest in the investigation of rupture. Numerical simulations with finite element codes allow investigating rupture prediction in real scale model. This paper is focused on discussion of procedure for prediction of the rupture in tensile test for a typical SS 304 Stainless Steel sheet. For this procedure, experimentation of tensile test with the conventional Universal Testing Machine (UTM) and Digital Correlation Devices (DIC) are needed to be used. The data obtained from experimentation is used for determining Young's modulus (E), $R_{p0.2\%}$ and R_m These data further used for simulation upto rupture. The output result obtained from simulation is used in macroscopic ductile fracture criteria which gives damage value to predict onset rupture.

1. Introduction

Materials may remain un-ruptured in plastic deformation up to finite strain, but it rupture takes place when it reaches to critical strain value of 1. The material rupture phenomena are represented by macroscopic rupture criteria. Rupture in material is defined as break or sudden burst or initiation of crack in material [1].

The very first criterion was given by freudenthal [2], it was about generalized plastic or energy work. Parameters obtained from the calculation of stress and strain histories of both experimental and simulated data, are used for different ductile fracture criteria [2-5]. In this paper, experimental and simulation procedure for prediction of rupture is discussed. For this, example of a typical SS304 Stainless Steel sheet is considered. The author Cockroft-Latham [7] proposed criteria based on effect of highest tensile stress, this criterion was modified by Brozzo [4] to introduce the effect of hydrostatic stress in an explicit from.

Further for prediction of rupture, different macroscopic ductile fracture criteria's can be applied to tensile test of SS 304 Stainless Steel sheet. To obtain stress and strain histories simulation procedure is also discussed in this paper. These values are mostly implemented in ABAQUS/Standard software for simulation. The objective of this paper is to discuss the methodology for prediction of rupture using different macroscopic fracture criteria.

2. Experiments

IOP Conf. Series: Materials Science and Engineering 998 (2020) 012015 doi:10.1088/1757-899X/998/1/012015

Uniaxial tensile test upto fracture is discussed in this paper for study of mechanical behaviour in tension. From uniaxial tensile test, behaviour of Nominal and Cauchy stress can be understood. The Nominal stress start to reduce when necking starts in the middle section in tensile test and Cauchy stress can calculate only upto commencement of necking where nominal stress is highest. A typical curve of Nominal and Cauchy stress curves vs. logarithmic strain curve of SS304 steel sheet in 0° rolling direction is shown in Figure 2.

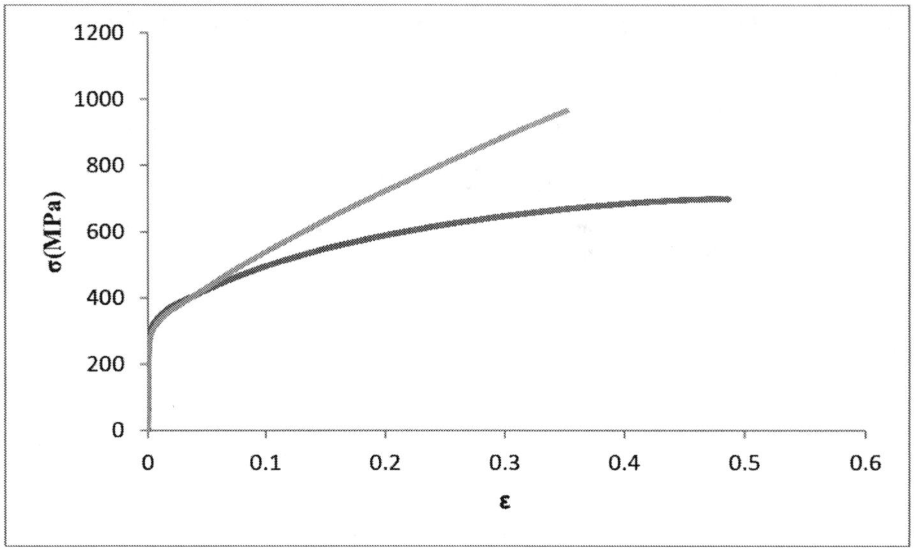

Figure1. Cauchy and Nominal stress vs. Strain diagram at 0° to the rolling direction.

3. Strain at rupture measured from Digital Image Correlation (DIC) device

Digital image correlation (DIC) system camera are used for measurement of normal Strain tensor components. The test should be considered as long as the test is similar for both the normal strain determined after the extensometer and the comparing DIC data are equivalent. The outcomes got from the DIC Device (normal values) were subsequent worked to calculate the stress vs. strain curves. A frame of typical SS304steel sheet is indicated in Figure 2: length is shown on y-axis, width is shown on x-axis and thickness is shown on z-axis of typical SS304 steel sheet tensile sample. Dispersion of the strain ε_{yy} at the step just prior to rupture is shown in Figure 2.

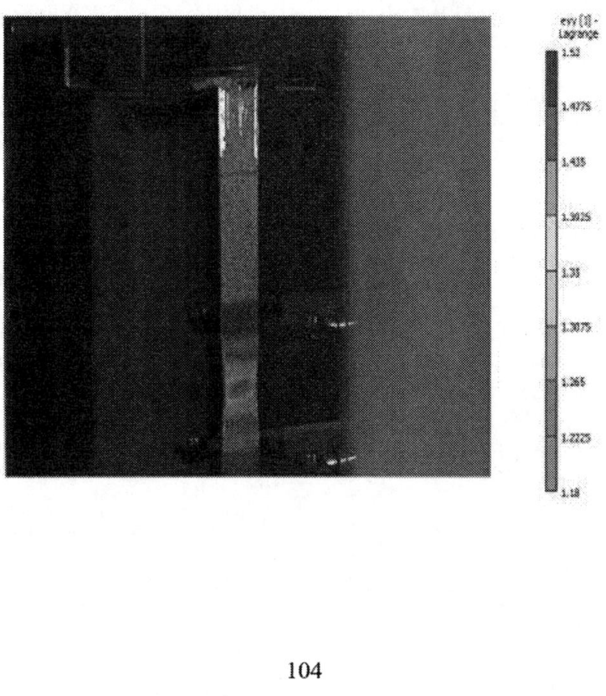

Figure 2. A typical image of SS304 steel sheet dispersion of the strain ε_{yy} in 0° to the rolling direction measured with DIC Device just prior to rupture in tension

4. Young's modulus (E), $R_{p0.2\%}$ and R_m

For determination of Young's modulus (E), Yield point ($R_{p0.2\%}$) and Maximum ultimate stress (R_m), extensometer data of a typical SS304 Stainless Steel sheet is used. Young's modulus (E) was attained using measurement data of extensometer within the elastic limit. The yield point $R_{p0.2\%}$ was attained using 0.2% parallel line to the elastic limit, R_m is the maximum ultimate stress value which is attained by curve Nominal stress vs. Logarithmic strain diagram.Values attained of Young's modulus (E), $R_{p0.2\%}$ and R_m in 0° to the rolling direction of a typical SS304 steel sheet is shown in Table 1.

Table 1 Measurement of Young's modulus (E), $R_{p0.2\%}$ and R_m in 0° to the rolling direction.

Exp. No.	**E (GPa)**	**$R_{p0.2\%}$ (MPa)**	**R_m(MPa)**
sample, 0° / RD 1	173632.26	280	725

5. Simulation procedure

Uniaxial tensile experiment is used to know material properties. Mostly Abaqus Standard/Explicit module is used for simulation purpose. Material properties are used in elastic-plastic condition and boundary conditions are applied accordingly for simulation of tensile test. Output of stress strain histories is obtained by simulation which is used for calculating damage value using different ductile fracture criteria. These damage values are calculated using trapezoidal rule from numerical integration method. Damage value shows the rupture initiation. Both experimental and simulation damage values are compared for onset prediction of rupture. Flow chart of simulation procedure is shown in Figure 3.

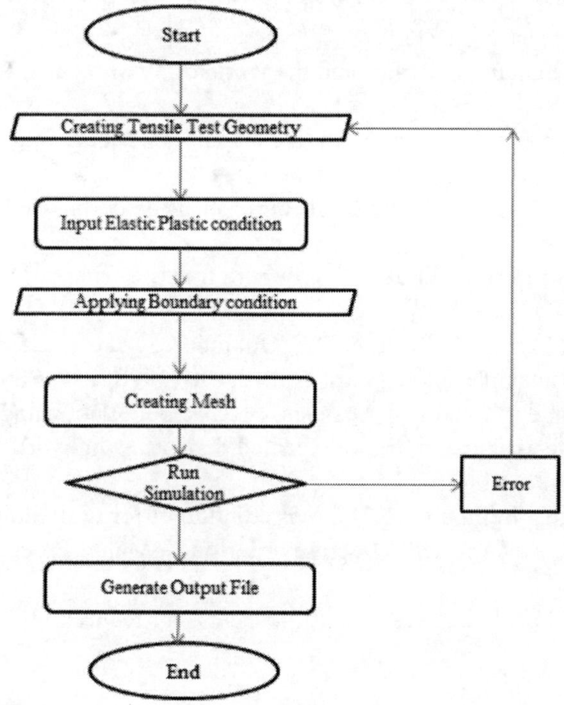

Figure 3. Flow chart of simulation of Tensile Test procedure.

6. Conclusion

Tensile test of SS304 Stainless steel sheet is discussed for simulation, in order to describe the feature of equivalent strain at rupture, corresponding to local strain values which are measured by digital image correlation. Influence of material anisotropy is neglected in this work. Damage value obtained from macroscopic ductile fracture criteria using finite element simulation depicts onset rupture.

Acknowledgments
The authors are gratefully acknowledge Mechanical Engineering Department, National Institute of Technology Delhi (India) for computing facilities and applied mechanics department of Indian Institute of Technology Delhi (India) for providing experimentation facilities to carry out this research work.

References

[1] M. Ayada. T. Higashino, K. Mori, in: K. Lange (Ed.), Advanced Technology of Plasticity, vol. 1, Springer, Berlin, 1987, pp. 553-558.
[2] P. Brozzo, B. deLuka, R. Rendina, A new method for the prediction of formability in metal sheets, in: Proceedings of the Seventh Biennial Conference on Sheet Metal Forming and Formability, International Deep Drawing Research Group, 1972.
[3] S.I. Oh, C.C. Chen, S. Kobayashi, Ductile fracture in axisymmetric extrusion and drawing, J. Eng. Ind. Trans. ASME 101 (1979) 36– 44.
[4] S. Storen, J.R. Rice, Localized necking in thin sheets, J. Mech. Phys. Solids 23 (1975) 421–441.
[5] A.L. Gurson, Continuum theory of ductile rupture by void nucleation and growth. Part I. Yield criteria and flow rules for porous ductile materials, J. Eng. Mater. Technol. Trans. ASME 99 (1977) 2–15.
[6] S.K. Esche, G.L. Kinzel, T. Altan, Review of failure analysis in sheet metal forming simulation, in: Proceedings of the Numisheet'96, 1996.
[7] M.G. Cockcroft, D.J. Latham, Ductility and the workability of metals, J. Inst. Met. 96 (1968) 33–39.
[8] M. Oyane, Criteria of ductile fracture strain, Bull. Jpn. Soc. Mech. Eng. (JSMN) 15 (1972) 1507–1513.
[9] A.M. Freudenthal, The Inelastic Behavior of Engineering Materials and Structures. John Wiley & Sons, New York, 1950.
[10] F. Ozturk, Analysis of forming limits using ductile fracture criteria, Doctoral Thesis, Rensselaer Polytechnic Institute, 2002.
[11] D.M. Norris Jr., J.E. Reaugh, B. Moran, D.F. Quinones, A plasticstrain, mean-stress criterion for ductile fracture, J. Eng. Mater. Technol. Trans. ASME 100 (1978) 279–286.
[12] R.L. Burden, J.D. Faires, Numerical Analysis, Brooks/Cole Publishing Company, 1997.
[13] A.G. Atkins, Possible explanation for unexpected departures in hydrostatic tension–fracture strain relations, Met. Sci. 15 (1981) 81– 83.
[14] H. Takuda, K. Mori, H. Fujimoto, N. Hatta, Prediction of forming limit in deep drawing of Fe/Al laminated composite sheets using ductile fracture criterion, J. Mater. Process.Technol. 60 (1996) 291–296.
[15] S. Storen, J.R. Rice, Localized necking in thin sheets, J. Mech. Phys. Solids 23 (1975) 421–441.

Priority Based Train Scheduling Method Using ACO in Indian Railway Perspective

Abhishek Mishra [1*], Neeraj Kumar [2] and Seema Kharb [3]

[1] Mechanical Engineering Department, National Institute of Technology, Delhi India
[2] Mechanical Engineering Department, National Institute of Technology, Delhi India
[3] Computer Science and Engineering Department, National Institute of Technology, Delhi India

corresponding author's e-mail: abhishekmishra@nitdelhi.ac.in

Abstract. The aim of this paper is to solve a train scheduling problem (TSP) with different train priorities and travelling at different speeds in the Indian railways (IR) perspective. Initially, a mathematical model for a TSP is developed and then the Ant Colony Optimization (ACO) technique is applied to resolve the problem. A real-world train timetable problem is presented to make a comparison between the manual solutions method used by the Indian Railway (IR) and the solution obtained from ACO and CPLEX solver to check for quality and accuracy in terms of total travel time. The result shows that there is a 6.827% improvement in travel time using ACO over manual solution. A collision free train schedule is determined based on the obtained train sequence.

Keywords: Indian Railways Train Scheduling, ACO, Information Flow System in Indian Railways

1. Introduction

Indian Railway operates in passenger and freight transportation as a major contributor. Approximately 8116 million passengers have travelled through the Indian railway in recent years contributing to a turnover of 462.80 billion INR. in the year 2018-19[1]. The rank of IR infrastructure i.e. the number of trains and length of tracks is 4th in the world. Unpredicted events like technical breakdown, equipment failures, sudden increase in passenger demand, track accidents, changed weather conditions causing disruption to segments track, varying speed trains running on tracks, different train types, train bypassing, siding, etc. causes the lateness of the train, triggering a knock-on effect in entirely railroad lines & connections, and hence generating the need of continuous scheduling. In this era of automation and IoT scheduling problem can be remarkably improved by using computational techniques. Further, the Indian Railway system has a heterogeneous nature (i.e. different types of train run on the same track). On the basis of speed, trains[2], [3] are classified as Mail, Passenger, Express, and freight, etc. As the whole territory of IR divide into control areas (zones) and these control areas further divided into division or control sections. A railway line connecting these different zones that have different control sections, containing single multiple tracks on which all types of trains run between the stations. Due to the limited capacity of tracks between the stations, it is important to have a fast and corrective way of scheduling trains. Initially scheduling in IR is done according to the First Come First Serve policy. Shortest Job First method was also implemented with different train priorities. For the last five decades, researchers have examined various parts of the train planning issue & TSP. For solving train scheduling problem many techniques are used such as Mathematical programming, discrete event simulation, and metaheuristic approaches, etc. A simulated annealing (SA) technique has been used by [4] for TSP. Van Wezel et al.[5], [6] implemented a Genetic algorithm (GA) to develop train timetables and TSP. Gorman [7] developed a hybrid method, consisting of GA and tabu search for solving the problems of routing and TSP. A neural network has been used [8] for routing in a railway system. The key to this article is (a) framework for real-time operational practices in Indian Railways for TSP (b) ACO based approach for solving train scheduling problem in IR (c) validation of model on a real railway network. The rest of the paper is organized as follows. Section 2 describes the basic concepts related to TSP. In Section 3, we give a detailed description mathematical model. Section 4 is devoted to the comparative study of computational results of different techniques implemented on the real IR network. Section 5 concludes and provides future directions.

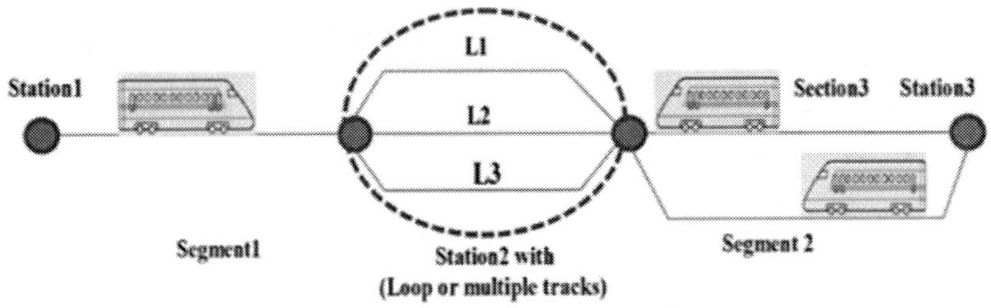

Figure 1. Represents a graphical view of an IR line network.

2. Problem Definition

Planning is an ever-interesting research topic, whether we are talking about manufacturing, goods or services. The TSP is one of the challenging problems in the railway system. At present most of the scheduling is done manually by human operators (called dispatchers) in IR. Due to the limited computational capability of a humankind, it is difficult to combat the complex nature of the scheduling problem and provide optimal solutions quickly. On the basis of infrastructure details, a network can be categories in to three parts. The first one is microscopic having a high level of detail, whereas mesoscopic representation has moderate. This paper considers the macroscopic (low level) representation of the IR network. A network contains nodes (station) & lines (segments) joining nodes. Generally, several tracks (loops) are established at almost every station where overtaking or crossings can be done in an effective manner. A segment can permit one-way traffic movement or two ways. consist of three stations joined with segments. Generally, In IR every station contains multiple tracks (loops) as shown in station2. In this work smaller details of infrastructure consisting of signaling distance, interlocking etc. are not considered. The railway planning operation depends upon the following inputs: commodity and demand data, number (no.) of trains, halt, and speed of trains, priority etc.[9] .In Indian railways priority of a train generally is based on their speed. Usually, a cargo train had a lower priority than a passenger train. During the scheduling process in IR, the higher priority train gets more importance in comparison with lower priority trains in term of resource usage. In the case of a single-track railway line, a segment can be only occupied by one train at a time. Simply, the no. of trains occupying tracks at a station should be equal or less than the no. of existing tracks at that station.

3. Mathematical Model

Railway scheduling is a very complex problem, it is necessary to state entirely the model's restrictions, assumptions and inputs. The model in this paper studies a single railway line TSP on which trains are going in both directions. This model is similar to that of Higgins [10] with slight modifications for achieving the infrastructure requirements. In this model, it has been assumed that the trains will be dispatched from the first and last stations only. The situation of crossing and overtaking of trains can

IOP Conf. Series: Materials Science and Engineering 998 (2020) 012016 doi:10.1088/1757-899X/998/1/012016

only be performed at the stations or sidings. On the basis of directions, the train is named as UP and DOWN. UP train movement is form right to left while down train traverse from left to right. Further it has been assumed that each train will fit on any track. Below the definitions and symbols used in this paper to design and implement the model are described.

Parameters and Decision variable

n	No. of trains
m	No. of stations
B	A sufficiently large number
$TD_{i,k}$	Actual departure time of train i from station k
$X_{i,j,k}$	Binary decision variables
$HT_{i,k}$	Halt time of train i in station k
$TT_{i,k}$	Travel time of UP train i between stations k = 1…..m-1 and k + 1
$TT_{j,k}$	Travel time of DOWN train j between stations k = 2….m and k – 1
$AA_{i,D}$	Actual arrival time at station
$AA_{i,O}$	Actual Departure time from origin station
OPT_{gap}	Optimality gap
$Saving_{TT}$	Total Travel time saving

Objective function

The objective function for the problem of interest (TSP).

$$\text{Min } Z = \sum_{i=\text{train}} w_i \left(AA_{i,D} - AA_{i,O}\right) \quad [1]$$

As an objective function paper focuses on minimizes the weighted sum of travel time of the UP and DOWN trains. At the end stations, halt time is zero, so this will give the actual arrival times of trains at their destinations. The flow diagram of infrastructure topology is represented in figure 1. And the timetables are described with the help of a graphical method termed as Train Graph.This train graph represents the time on one axis taken by the trains to travel from one station to another and on the second axis specify the number of stations and distance between the stations. The time distance graph is reversely represented by many countries i.e. on the minor axis travel time of train and major axis number of stations .

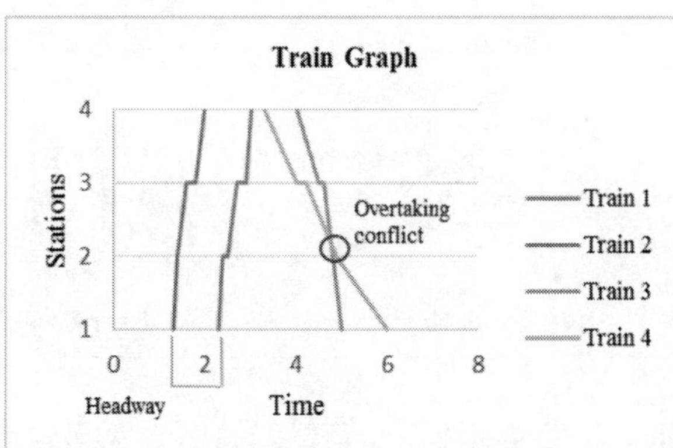

Figure 2. Train overtaking conflict b/w Train 3 &Train 4

In figure 2 train graph represents a schedule with four trains and four stations including headway (separation time between two trains i.e. Train1 & Train 2) and overtaking conflict.

4. Computational Results

The results obtained from ACO are compared with those of the exact optimization method, manual method for TSP. In the present scenario, IR has been using a manual solution technique to produce a conflict-free train schedule. The mathematical model transforms into program code in MATLAB and CPLEX solver which is specially oriented to handle matrices. CPLEX solver is a well-suited tool for solving mixed-integer linear programming-based problems and is inbuilt with branch & bound method. For obtaining an optimal solution the search tree is considered. The nodes indicate the conflicts and branches solutions & order of search indicated with the help of numbers. As this problem is NP-hard signifies that it increases exponentially with the no. of conflicts. Two problems are studied, first one is the example problem and the second is Indian line railway network problem. The data are first described then, the computational results are discussed. Every train has provided departure time from its starting station. A train transverse on rail segments with specific speed. The Halt time for every train at each station is deterministic. This gives an initial schedule specifying train departure time, stoppage time with conflict.

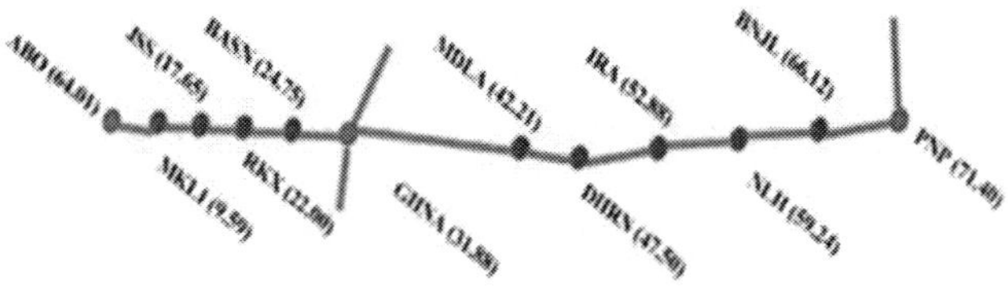

Figure 3. Schematic illustration of a test network

The objective value (z) obtained from the above schedule becomes the upper bound value. Similarly, in MATLAB, the ideal solution is generated with conflict known as conflict detection block. After implemented all the constraints with the objective function the optimal value is obtained using CPLEX solver and conflict free schedule is created.

Table 1. Train data for IR line

Train / Line data	Example problem data	Real network data
Length	297 Km	71.40 Km
Line (Single Line)	A-B	ROK-PNP
Max. Permissible Speed	60 Km/hr.	50 Km/hr.
Total Station	5	12
Total Train	3 (2 in UP & 1 in DOWN direction)	12 (M/Ex. 2 in UP & in 2 DOWN direction) (4 Passenger in UP & 4 in DOWN direction)
Priority Weight	1 for each train	1 for Passenger & 2 for Ex. Trains

The saving in total travel time is given as the difference between the travel times obtained from ACO and the manual method. All these methods are implemented on a real railway network at which12 trains are running with different priorities. A simple example of a problem is also illustrated. Results Show a 3.070%, 6.827% improvement in ACO solution as compare to the manual solution.

$$\%\text{improvement} = \frac{(\text{Solution from manual method} - \text{Solution from ACO})}{\text{Solution from manual method}} \times 100$$

5. Conclusion

In this paper, a metaheuristic technique (ACO) was implemented for the Single Line TSP. The comparison results show that ACO solutions similar to the optimal solutions obtained by CPLEX for example problem, but correspondingly leave behind the manual method used by IR for scheduling of trains. For the real network problem, we found that the % improvement is better in CPLEX as compare to both ACO and manual method. The obtained results show significantly better performance with regard to the solution quality, computational time.

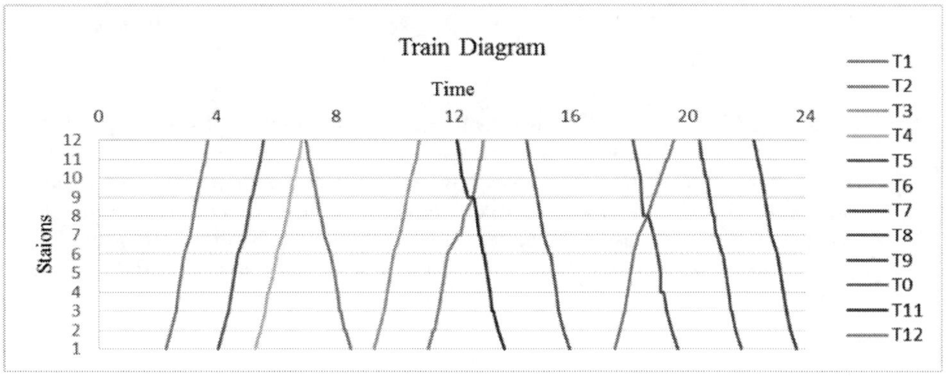

Figure 4. Train schedule diagram

For future direction, there is a significant need for extension of metaheuristic techniques for dynamic train scheduling problems with dynamic variables, with unexpected events (passenger demands and link travel times). For future works, it is proposed to develop software to represent the unquantifiable constraints that dispatchers face in the scheduling of rail traffic; i.e. (train length, capacity constraints).

References

[1] Indian Railways, 2017 In India Railways Facts & Figures, pp.1-48.
[2] V.N. Mathur, 2004 Operating Manual for Indian Railway, In Evolution 1–171.
[3] N. Rangaraj, 2008 A note on section scheduling on the Indian Railways, Industrial Engineering and Operations Research, IIT Bombay.
[4] C.L. Huntley, D.E.Brown, D.E. Sappington, and B.P. Markowicz, 1995 Freight routing and scheduling at CSX transportation, Interfaces 25, no.3, pp.58-71.
[5] V.Wezel, C.Michiel, J.N. Kok, J.V.D.Berg, and W.V.Kampen,1994 Genetic improvement of railway timetables, In International Conference on Parallel Problem Solving from Nature,Springer, Berlin, Heidelberg, pp. 566-575.
[6] K. Nachtigall, and S. Voget, 1996 A genetic algorithm approach to periodic railway synchronization. Computers & Operations Research, 23(5), pp.453-463.
[7] M.F. Gorman, 1998 An application of genetic and tabu searches to the freight railroad operating plan problem. Annals of operations research, 78, pp.51-69.
[8] D.R. Martinelli, and H. Teng, 1996 Optimization of railway operations using neural networks,Transportation Research Part C: Emerging Technologies 4, 1, pp.3-49.
[9] N.Kumar, V.Jangid, P.Kesarawani, A. Mishra, 2017 Network Design and Scheduling of Single-Track Trains: A Review , In NACRAME ,IEEE,pp.90–95.
[10] A.Higgins, E.Kozan and L. Ferreira,1996 Optimal scheduling of trains on a single line track,Transportation research part B:Methodological 30, 2, pp.147-161.

IOP Conf. Series: Materials Science and Engineering 998 (2020) 012017 doi:10.1088/1757-899X/998/1/012017

Design of framework of Hybrid RMS for Ball Manufacturing Industry

J Patel[1], N Sanghvi[1], A Khatri[2], D Vora[1]

[1] UG Student, Department of Mechanical Engineering, DJSCE, Vile Parle, Mumbai - 400056, India

[2] UG Student, Department of Production Engineering, FCRCE, Bandra, Mumbai - 400050, India

E-mail: jash.ravariya98@gmail.com, s.neel1998@gmail.com, aayushikhatri1@gmail.com, dakshvora1902@gmail.com

Abstract. This paper discusses the possibility of implementation of a reconfigurable manufacturing system to aid a manufacturing organization to manufacture tennis and rubber balls simultaneously or exclusively with minimum changeover time. The tennis balls and rubber balls come under a single part family, that is, Ball. They constitute 8 types of products in total. The proposed manufacturing system consists of three individual and parallel manufacturing lines that are capable of manufacturing both tennis and rubber balls. As per the change in demand, the line can be reconfigured to manufacture any product within the part family.

1. Introduction

In recent times, the focus on product design and manufacturing with respect to time and money has increased the importance of development of a high ranged system to maximize desired outputs. A lead in developing a system for a variety of products will help in better market exposure and eventually success in sales.

The manufacturing systems are classified as Dedicated Manufacturing System (DMS), Adjustable Manufacturing System (AMS), Flexible Manufacturing System (FMS) and Reconfigurable Manufacturing System (RMS). G. Zhang et al [1] has compared the cost and performance of DMS, AMS, FMS, and RMS analytically. A DMS is used to produce a single product for mass volumes. It consists of special-purpose machines with a fixed control system following a fixed path. An AMS is used to produce a range of products for mass volumes that are almost similar and can be produced with minor adjustments in the system. It consists of standardized machines with limited adjustable control system following a fixed path. An FMS is used to produce small batches of a large group of products. It consists of CNC machining centres with a programmable control following a flexible path. An RMS is used for producing mass customizable products. It is a platform that can be deployed to manufacture a family of products with minimum change over time, with no sacrifice in quality, even though production quantity for a given specific product is not high [2]. A reconfigurable manufacturing system is specifically designed to handle the demands of a highly volatile global market where product demand forecasting is a complex problem. [3]

RMS was developed in recent times and is confused with FMS by a lot of individuals. The goal of FMS is to maximize the variety of the generated products, while, RMS targets to give a quick response to the customer and the prevailing market. FMS offers general flexibility while RMS offers customized flexibility. Meaning that, its flexibility is limited only to the development of a part family. RMS benefits in fast scalability to the target quantity and convertibility at a reasonable cost to the manufacturers. Y. Koren et al [4] defined 6 core characteristics for an RMS as Modularity, Integrability, Customization, Scalability, Convertibility, and Diagnosability. M. Nakao et al [5] used an RMS in Mold design and manufacturing industry to reduce lead-time for less

IOP Conf. Series: Materials Science and Engineering 998 (2020) 012017 doi:10.1088/1757-899X/998/1/012017

quantity but more variety in production. Deif et al [6] demonstrated a systematic approach for designing an RMS and controlling the design process by development of an open mixed architecture for PCB assembly line.

A study indicates that the valuation of the global sporting goods market was 60900 mn USD in 2018 and will rise to 83300 mn USD in 2025. It can be said that this industry is of a substantial size, and optimization in this area can save tremendous amounts of energy, time, and financial resources. It is often observed that the manufacturing processes used to produce these articles often consist of significant input by a human directly without the use of any machine. And these inputs impart quality, accuracy, and finesse to the product. For example, two oblong pieces of fabric are carefully stuck to the rubber surface to form the outer covering of the tennis ball. This step has to be done with care so that the uniformity is seen in all balls, and the worker is able to do this very fast, considering the rate at which production is done. The key attribute of an RMS is combined functionality and scalability, it is capable of responsiveness to intrinsic (machine failure) and external (market changes) unpredictability. This is desirable for the sports goods manufacturing industry where the demand for a particular product in a product family may keep in varying with uncertainty.

In this paper, the authors have made an attempt to develop the framework for a Reconfigurable Manufacturing System that can be applied to a factory setup where tennis balls and rubber balls are being manufactured simultaneously.

2. Existing manufacturing system

The authors did a field visit to a manufacturing organization that manufactured tennis and rubber balls and studied the process for tennis ball and rubber ball manufacturing. As shown in Figure 1, Ball as a part family has two variants – Tennis and Rubber balls. The sub-variants for Tennis balls are Hard and Soft Tennis ball with two color variants each – Yellow and Maroon. The sub-variants for Rubber balls are basically color variants – Red, Blue, Orange and, Purple. The study allowed to understand the process and pick out some common processes and process equipment.

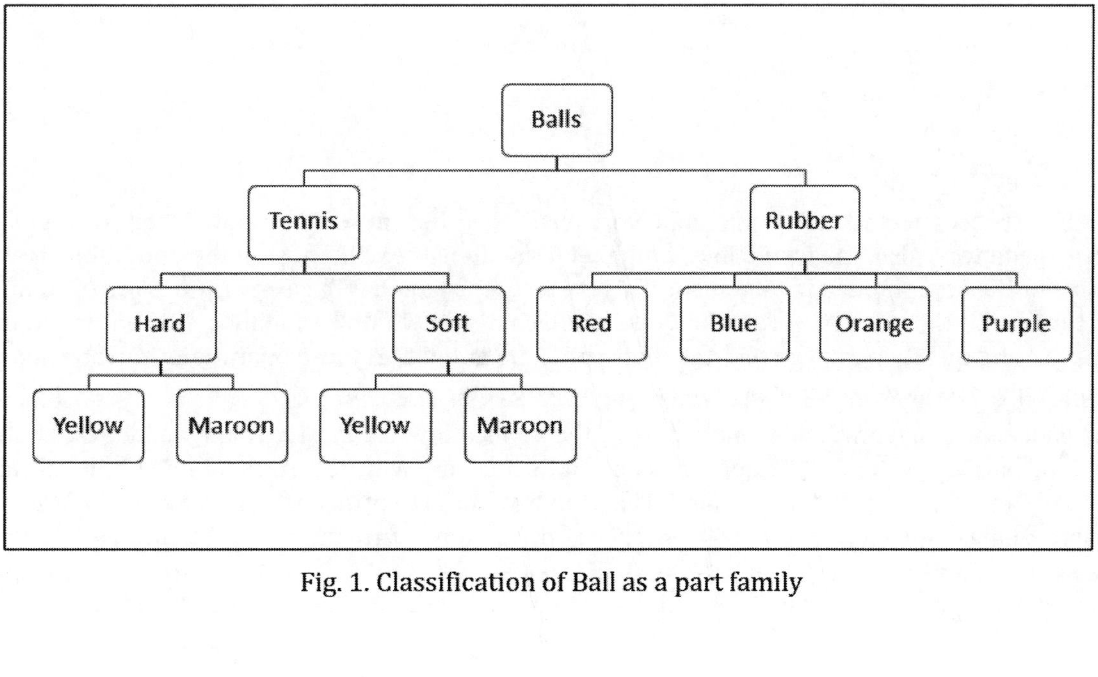

Fig. 1. Classification of Ball as a part family

IOP Conf. Series: Materials Science and Engineering 998 (2020) 012017 doi:10.1088/1757-899X/998/1/012017

The process involved in hard and soft tennis ball manufacturing comprises of the following steps:

- Preparation of standard rubber base consisting of latex and necessary additives.
- Liquidizing the above-prepared base to the right consistency by using thermal methods.
- Injection molding using a liquid base to prepare ball halves of different wall thickness for hard and soft variants.
- Gluing and joining of halves by using thermally activated adhesive.
- Cutting of fabric to obtain pieces of required shape.
- Application of thermally activated adhesive to the faces and sides of fabric.
- Joining of fabric to prepare a hollow rubber sphere by activating adhesive by heat curing.
- Labelling of ball with company logo.
- Quality testing
- Individual packaging
- Box packaging
- Warehousing

The process involved in rubber ball manufacturing of any given colour comprises of the following steps:

- Preparation of standard rubber base consisting of latex and necessary additives.
- Liquidizing the above-prepared base to the right consistency by using thermal methods.
- Preparation of irregular shaped rubber lumps with imparted coloring and heat-activated chemical additives.
- Enclosing of rubber lumps into spherical metal enclosure.
- Heat treatment of specimen in the enclosure to obtain uniformly expanded specimen in form of hollow ball.
- Cooling of prepared ball.
- Quality testing
- Individual packaging
- Box packaging
- Warehousing

Both the processes and their equipment were placed in different sections and spaced at irregularly large distances. Also, the whole line is highly labor-intensive with most of the operations having some or the other human involvement. This is due to the fact that these operations are of complex nature and if these tasks were to be done by machines, it would be highly inefficient from time, energy, and quality perspectives. And it also must be noted that these operations must be done at a high speed, which a trained person is capable of doing.

When it comes to tennis ball manufacturing, the application of adhesive to the molded ball halves is done entirely by hand. All operations except heat curing with regard to joining of the fabric to the rubber base are also done by hand. These operations are cutting of fabric, gluing of fabric, and the pressing of the fabric onto the rubber. The application of the company logo onto the prepared ball surface is also done partially by hand. In the case of rubber ball manufacturing, the insertion of prepared lumps into the metal enclosure is done by hand. All the inter-operational transport in the line and the addition of chemical compounds into the machine with specific proportions for chemical processes are also done manually. The quality testing, box packaging and warehousing processes are done manually as well

3. Proposed manufacturing system

As shown in Figure 2, except the fabric cutting, gluing, and heat curing machinery, the remaining machines are utilized by manufacturing processes of both the balls either directly or with a product family specific attachment. The previous sentence actually states a feature of the system that gives it a minimum changeover time when there is a change in the type of product that is being manufactured. This is the major feature of any Reconfigurable manufacturing system. To be able to manufacture a variety of products of a product family with minimum changeover time with no loss in quality and no substantial decrease in the rate of production.

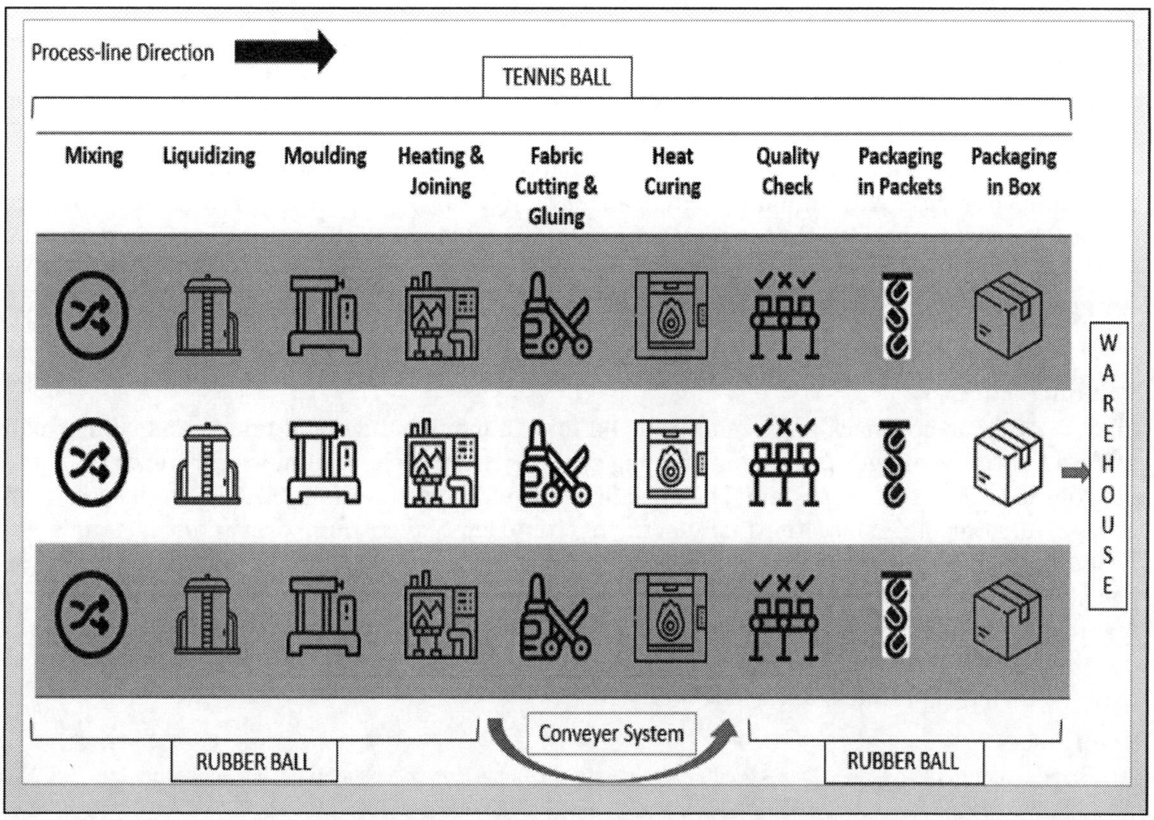

Fig. 2. Proposed manufacturing system

When rubber balls are manufactured, a conveyor system can be used to transport the balls across the fabric gluing and curing platforms. The proposed system actually consists of three parallel manufacturing lines. Each line is capable of manufacturing, both rubber and tennis balls.

The idea is to use the central line to manufacture tennis or rubber as dictated by the demand. The mixing and liquidizing machinery can be used to prepare the rubber base for tennis or rubber balls. The molding machinery can be used to prepare halves of hard or soft tennis balls based on the mold used, or can be used to make rubber ball lumps with chemical additives. This can be done by using adjustable attachments. Adhesive will be applied to ball halves and the heating - joining arrangement can be used to activate the adhesive to prepare tennis ball rubber base or it can be used to expand the specimen to form a rubber ball inside the metal enclosure. In the case of tennis balls, the fabric gluing and curing machinery is used to make the outer fabric cover of the ball and make a complete tennis ball. After this, the quality testing phase is similar for the entire family of products. The packaging and warehousing method and machinery are also the same. The human involvement in the existing system remains the same but it takes place in process-line and the process is not halted for the same.

In this RM system, the mechanisms involved are the same mechanisms involved in the manufacture of tennis and rubber balls. There is no overall specific mechanism involved. This system focuses more on the simple modification that can be done to the manufacturing system as a whole in response to the change in demand. Hence, this reconfigurable manufacturing system can be effectively used to manufacture tennis and rubber balls simultaneously in a ratio that is decided by the instantaneous market demand.

4. Results

It can be very well said that the proposed implementation of a reconfigurable manufacturing system for manufacturing the concerned family of products will benefit the manufacturing organization tremendously. Unprecedented fluctuations in the demand will be handled with confidence and ease with minimum changeover time with no depreciation in finished product quality. It must be noticed that the proposed RMS is a hybrid RMS, with humans participating directly as a system element. The participant is skilled in manufacturing operations required for the manufacturing of various products.

5. Conclusion

In this paper, an approach has been proposed for the manufacturing of tennis and rubber balls using a hybrid reconfigurable manufacturing system (HRMS) where human workforce is a direct part of the manufacturing system. The specific feature that is paramount in the proposed system is high functionality, since the manufacturing route employed for manufacturing tennis and rubber balls has little differences. Changeover effort is lower compared to completely machine-based RMS since a part of the manufacturing system is humans which are well versed with the manual procedures involved in the manufacturing process. The work done in this paper can be developed further to implement such kinds of reconfigurable manufacturing systems for manufacturing of products in various industries

References

[1] Zhang G, Liu R, Gong L and Huang Q 2006 An analytical comparison on cost and performance among DMS, AMS, FMS and RMS *Dashchenko A (eds) Reconfigurable Manufacturing Systems and Transformable Factories. Springer, Berlin, Heidelberg.*

[2] Koren Y 2020 The emergence of reconfigurable manufacturing systems (RMSs) *Benyoucef L (eds) Reconfigurable Manufacturing Systems: From Design to Implementation. Springer Series in Advanced Manufacturing. Springer, Cham.*

[3] Telsang M 2018 Industrial engineering and production management **IIIrd Edition** S. Chand Publishing, India pp.1082-1100.

[4] Koren Y 2006 General RMS characteristics comparison with dedicated and flexible systems *Dashchenko A (eds) Reconfigurable Manufacturing Systems and Transformable Factories. Springer, Berlin, Heidelberg.*

[5] Nakao M 2006 Example of reconfigurable manufacturing system in mold design and manufacturing *Dashchenko A (eds) Reconfigurable Manufacturing Systems and Transformable Factories. Springer, Berlin, Heidelberg.*

[6] Deif A, ElMaraghy W 2006 A systematic design approach for reconfigurable manufacturing systems *ElMaraghy H, ElMaraghy W (eds) Advances in Design. Springer Series in Advanced Manufacturing. Springer, London.*

Heat Transfer Characteristics of Water Cooled Minichannel Heat Sink Using Different Fluid Flow Geometries

Jugal Shrinivas Makam[1] and Nilesh Balkishanji Totla[2]

[1]Student, MIT Academy of Engineering, Pune, India.
[2]Sr Assistant Professor, MIT Academy of Engineering, Pune, India.

[1]Makam.jugal412@gmail.com, [2]nbtotla@mech.maepune.ac.in

Abstract: Efficient heat transfer for compact electronic components has become a major need in these days. For this purpose, we require a minichannel heat sink with a liquid cooling system that makes these requirements as possible. In this work, both analytical and CFD investigations have been carried out to calculate the heat transfer performance of a minichannel heat sink consisting of cross sections with three different shapes of a rectangular, trapezoidal and hexagonal in which working fluid is made to flow. The simulation is conducted with water as the working fluid for three different shapes with Reynolds number ranging from 200 to 1000 and analyzed numerically for the single-phase laminar flow with the boundary condition of constant heat flux. The channel dimensions and inlet velocity effects on the Nusselt number, pressure drop, and heat transfer coefficient are studied. The 3D steady-state, laminar flow, and governing equations of heat transfer are solved using the finite volume method by commercial ICEM-Fluent software.
Keywords: Computational fluid dynamics, Heat transfer, Minichannel heat sink, ICEM-Fluent CFD.

1. Introduction

In the present years of modernization and due to the technology advancement, the need of liquid cooling in various compact electronic devices and other electrical devices has been a major area of concern. The heat generated by electronic circuits must be dissipated out of the system so we require different cooling systems for these devices to work properly and effectively. Since the last few years minichannel cooling system has been a subject of research. The amount of work on the effect of heat transfer on flow geometries in mini/microchannels is fast and growing. Many researchers have emerged with a different cross section of the mini/microchnnel but for sure there is an effect of the configuration of these channels on heat transfer coefficient, Nusselt number, and Reynold number. The main advantage of using mini/microchannel is to achieve a high heat dissipation rate in a smaller space. It has been proved that single-phase, laminar flow, and forced convection in minichannels is one of the effective cooling ways in a variety of applications.

The microchannel heat sink was first developed and proposed by Tuckerman and Pease [1] in 1981. For VLSI circuits, an investigation has been done by authors with a high performance heat sink. Designed and tested water cooled based silicon minichannel heat sink used for integrated circuits. They also stated that reducing the dimension to the micro scale of a liquid cooling heat sink is helping in enhancing the performance of heat transfer. The experimental investigation and numerical analysis are presented by Lee et al. [2] to explore the validity of the classical correlations for analyzing the thermal behavior in single phase rectangular microchannels that are based on conventional sized channels. The microchannel is made up of copper and de-ionized water was used for the experiment and the Reynolds number was ranging from 300 to 3500. Numerical predictions those obtained were based upon the continuum and classical approach. Those predictions were found in an appropriate agreement with the data showing an average deviation of 5%. Qu and Mudawar [3] an experimental and numerical investigation is carried out for characteristics of heat transfer and pressure drop of oxygen free copper made heat sink and deionized water as the working fluid. The microchannel was studied for a rectangular array with a hydraulic diameter of 231μm and 713μm. The numerical investigation and effect of parameters of geometries on water flow and characteristics of heat transfer are studied with three different geometry shapes (Rectangle, Trapezoidal, and Triangle) and heat flux is added to the top plate of the aluminum microchannel. The results show that with an increase in heat flux pressure drop decreases from the top plate of the heat sink for the same Reynolds number. Further, they studied different flow paths such as step, curvature, and zigzag, etc. Results stated that for microchannel with

IOP Conf. Series: Materials Science and Engineering 998 (2020) 012018 doi:10.1088/1757-899X/998/1/012018

equal cross section, zigzag microchannel was having a high coefficient of heat transfer among all channel shapes [4, 5].

There was a study of comparison between circular and square microchannel with ammonia as a coolant for optimization of the thermal performance by using a genetic algorithm. Ghazali-Mohd et al. [6] reported that for circular channels thermal resistance is lower by 21% and 35% at the lowest and highest pumping power than square microchannel. Circular geometry performed better thermally and hydrodynamically. Chai and Tassou [7] investigated the effect of cross-section geometries to predict the heat transfer and pressure drop characteristics of supercritical Co_2 working fluid in minichannels. They presented six geometries with a cross-section of a square, semicircle, circle, equilateral triangle, ellipse, and rectangle with the same hydraulic diameter of 1.22mm. The numerical results show that the highest heat transfer coefficient in circle and ellipse as compared to other geometries. Thermal performance and hydraulic parameters were studied by Ghasemi et al. [8] with the experimental investigation. The heat sink is fabricated with aluminum material with four minichannels of circular cross sections with three different values of hydraulic diameter. The lower thermal resistance was shown by 4mm hydraulic diameter than 6mm and 8mm hydraulic diameter. Kewalramani et al. [9] reported an empirical correlation for laminar flow with forced convection in the trapezoidal microchannel. The experimental study and numerical model were developed for trapezoidal geometries with side angles of 30^0 and 60^0 with an aspect ratio in the range of 0.1 to 10 to carry out the poiseuille number and Nusselt number. Su et al. [10] studied, the numerical investigation of the elliptical minichannel heat sink and purposed a general correlation for the first time for the apparent friction coefficients and entrance length for the elliptical channel. They have examined the Reynolds number effects and aspect ratio on friction coefficient, entrance length, and Nusselt number in detail. The Reynolds number ranging from 14-450 for minichannel for enhancement of heat transfer and pressure drop was experimentally investigated by Moghanlou et al [11]. There are accuracies of 17%, 8%, and 14% were observed for thermal resistance, Nusselt number, and friction factor respectively. The experimentation is done on a square configuration with a heater fixed in its bottom.

From the literature review, it can be concluded that many kinds of researches have been focused on changing the cross-sectional shapes of fluid flow geometry and its configuration of minichannels. Furthermore, a small amount of work has been focused on investigating the heat transfer characteristics and the effect of geometrical parameters on the cooling performance, particularly for trapezoidal and hexagonal shapes. In this study, both analytical and CFD investigation is to be carried out to calculate the heat transfer performance of minichannel heat sink with water as working fluid for three different geometrical configurations (Rectangular, Trapezoidal, and Hexagonal) & compare the heat transfer performance of three different cross-sectional shapes and the decision is made for optimum geometry which a gives a better cooling performance.

2. Physical Model

The rectangular CAD model of the minichannel is shown in Figure 1. All minichannel heat sinks having a base size of 26.9 mm x 50 mm and a total depth of 6 mm. The dimensions of the rectangular shape are taken from Ho et al. [12]. The desired model of the heat sink consists of rectangular, trapezoidal, and hexagonal cross-sections in which working fluid is made to flow which is shown in Figure 2. There is a total of 8 parallel channels in each heat sink and the top surface is assumed to be covered with insulating material to prevent splashing.

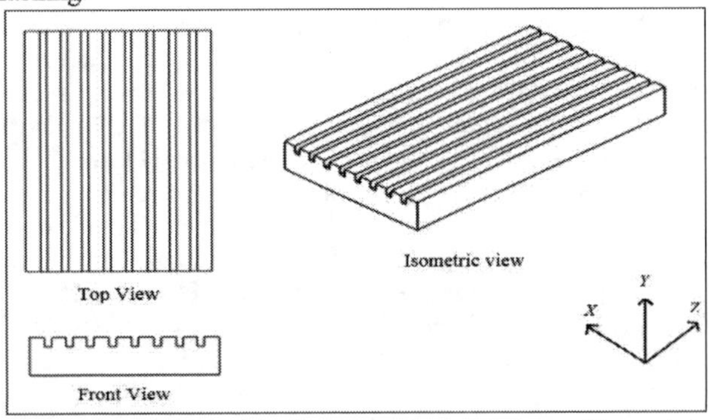

Figure 1. CAD model of rectangular minichannel heat sink

Dimensions of trapezoidal and hexagonal shapes are calculated, such that all geometries are designed by keeping hydraulic diameter and surface area as constant. In this study, all factors like fluid type, flow type, and governing equation did not change and only geometries of fluid flow were changed. The small hydraulic diameter of a minichannel can lead to appreciable heat dissipation, which is desirable in electronic systems. Therefore, a strong understanding of the heat transfer rate, Reynolds number, heat transfer coefficient, Nusselt number, flow rate, and velocity change has more importance in the design of minichannel. The detailed dimensions of the models are given in table 1.

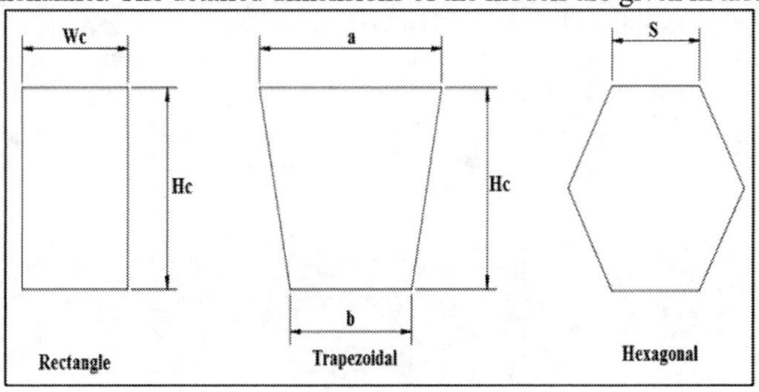

Figure 2. Different types of geometrical

Table 1. Dimensions of fluid flow geometries

Names of geometry	Dimensions
Rectangular	W_c=1.0mm, H_c=1.5mm, D_h=1.2mm
Trapezoidal	a=1.4mm, b=0.8mm, H_c=1.5mm, D_h=1.2mm
Hexagonal	S=0.7mm, D_h=1.2mm

3. The Computational Domain

In the present study, a 3D convective heat transfer simulation of the minichannel heat sink was carried out. The heat sink was made up of copper material and water as the working fluid was considered in the simulation. The following assumptions were made while modeling minichannel heat sink for analyzing cooling performance.

(1) Both heat transfer and fluid flow are in steady state and three dimensional.

(2) Fluid is incompressible, single phase, and the flow is fully developed laminar flow.

(3) The fluid thermo-physical properties are constant, negligible viscous dissipation, and negligible radioactive.

(4) The effect of body forces are neglected.

(5) All minichannel are supposed to be equal in heat transfer so one channel can be chosen for numerical simulation, the computational domain is used which is shown in Fig. 3. This problem is converted into symmetric boundary conditions it includes fluid flow channel height, width, and solid wall thickness which is equal to half of the thickness between two channels.

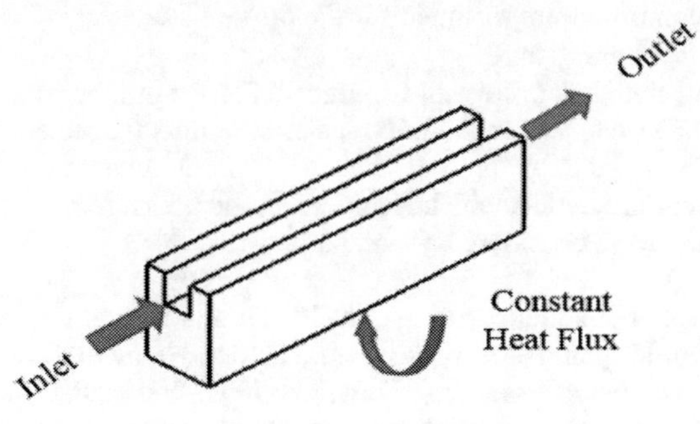

Figure 3. Computational domain of single channel

4. Grid Independence Study & Boundary Conditions

After generating the computational model, the meshing of the model is done on the ICEM CFD module in ANSYS software. The meshed model of rectangular, trapezoidal, and hexagonal minichannel heat sinks are shown in below Figure. 4. The fluid domain and solid domain has meshed with the same mesh density. The discretization has been done with Tetra/Mixed elements using the GSF method. Global scale factor (GSF) is one of the meshing types and is defined as the ratio of maximum length to the minimum length of an element in geometry. The grid independence test was conducted by adopting six different global scale factors of 3, 3.25, 3.5, 3.75, 4, and 4.25 to maintain the accuracy of the calculations and it is observed that the last two mesh models are given less pressure loss difference with a variation of 2%. Thus, the global scale factor (GSF) 4 mesh is adopted for this simulation. The same grid sensitivity approach was used for all cases in the present study. If we increased mesh density the simulation show less than 0.1% variation in pressure drop.

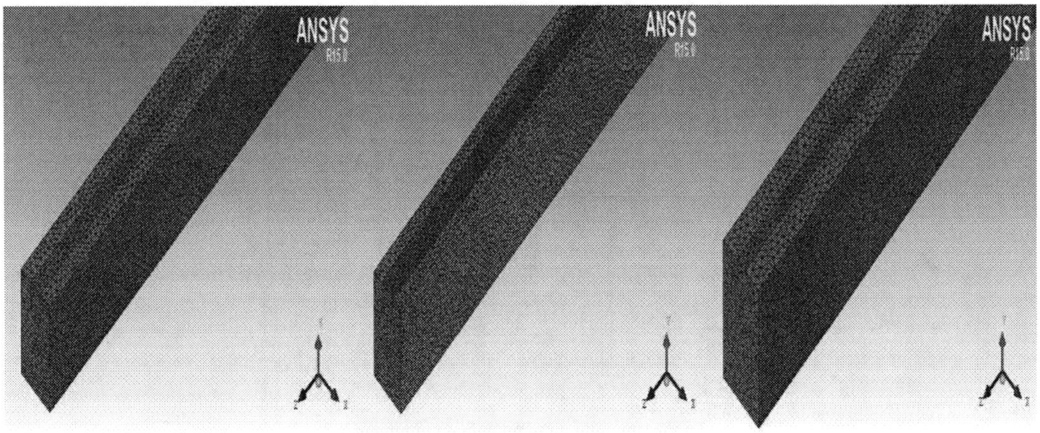

Figure 4. Meshed Geometries

Table 2. Meshing Parameters

Types of minichannel	Number of elements	Number of nodes
Rectangular minichannel	275353	46524
Trapezoidal minichannel	275486	46513
Hexagonal minichannel	269255	45491

The heat transfer performance of rectangular minichannel is compared with the trapezoidal and hexagonal minichannel heat sink, which has the same hydraulic diameter. The discretization of the domain and the governing equations are calculated by finite volume method and by using the SIMPLE algorithm these governing equations are solved with pressure-velocity coupling. The second-order upwind scheme is applied for all the variables and a pressure-based solver is used for numerical simulation.

The following boundary conditions are assumed for the present study:
1. The inlet temperature of water is 300K.
2. The inlet velocity is calculated from Reynolds number. For Re number 200, 400, 600, 800, 1000 the velocities are 0.167m/s, 0.334m/s, 0.501m/s, 0.664m/s & 0.884m/s respectively.
3. The constant heat flux is added to the bottom surface of the minichannel which is 32000W/m^2.
4. The outlets are set as pressure outlets which is at zero gauge pressure.
5. The remaining side walls were considered as adiabatic walls.

Various physical models are applied for carrying out CFD simulations of minichannel for three different cross-sections. As the Reynolds number of working fluid is found to be in the laminar zone. So, a viscous laminar flow modeling scheme is used for carrying out CFD simulation. In, order to study the temperature distribution, the energy model is turned on. Due to which temperature of the working fluid is specified.

5. Data Reduction from CFD Results

The water entered from the inlet of minichannel and convective heat transfer takes place in the system in which working fluid is absorbed heat from the heat sink and water leave through the outlet of the heat sink. So, heat is removed by circulating water through the heat sink. After boundary condition is applied to the computational model and the results obtained from that is taken to the CFD post software for the post-processing process. In CFD post software, the temperature and pressure variations are obtained by plotting straight line axis in both solid and fluid domain and data were collected. The line was created at mid of fluid domain and on the solid base of the heat sink.

1. The average temperature of fluid is calculated as,

$$(T_f)_{avg}=(T_{fi}+T_{fo})/2 \ \text{K}$$

2. The temperature between solid base average temperature and fluid average temperature is calculated as,

$$\Delta T= [(T_w)_{avg} - (T_f)_{avg}] \ \text{K}$$

3. Mass flow rate of working fluid is calculated as,

$$\dot{m}= \rho A_{cs}V \ \text{m/s}$$

4. Heat transfer from inlet to outlet is calculated as,

$$Q= \dot{m}c_p(\Delta T)_f \ \text{Watt}$$

Where, $(\Delta T)_f = T_{fi} - T_{fo}$ K

5. Heat gained by water through convection is same as that of heat transfer through convection heat transfer.

$$Q= \dot{m}c_p(\Delta T)_f=hA_s\Delta T \ \text{Watt}$$

6. The average heat transfer coefficient is calculated as,

$$h= Q/A_s\Delta T \ \text{W/m}^2\text{K}$$

7. Nusselt number is obtained by below equation,

$$Nu=hD_h/K_w$$

8. Reynolds number is calculated as,

$$Re= \rho VD_h/\mu$$

9. The hydraulic diameter is calculated from below equation,

$$D_h= 4A_{cs}/P$$

6. Results & Discussions

In this study, the phenomena of liquid cooling in the minichannels based heat sink have been studied. For this purpose, a copper heat sink has been selected having three different cross-sectional shapes and water has been taken as a coolant for the cooling system. The main objective of this study is to analyze and evaluate the heat transfer parameters like Nusselt number, heat transfer coefficient, and pressure drop of the heat sinks and to compare the heat transfer performance of the heat sinks with each other.

6.1 Validation of numerical approach

Table 3. Validation Results

Re	CFD Results ΔP (Pa)	Analytical Results ΔP (Pa)	Difference (%)
200	184.19	170.82	7.25
400	372.96	341.91	8.32
600	598.24	512.87	14.27
800	805.87	683.83	15.10
1000	1052.89	854.68	18.82

The numerical results are validated through the analytical solution method. Co-relations given by Khandlikar [13] are used to predict pressure drop (ΔP) in fully developed laminar flow across a straight rectangular channel which is shown in table 3. From the above table, it is shown that CFD results and analytical results are matching with some eligible error difference and there is a total average deviation of 12.75% in the analytical result and the numerical result of a pressure drop.

Figure. 5. represents the comparative analysis on the performance of minichannel based heat sinks and variation in the Nusselt number employing at different Reynolds numbers. It can be observed that as the Reynolds number increases Nusselt number also increases because the thermal boundary layer thickness decreases with an increase in fluid velocity. It is seen that the Nusselt number is higher for the rectangular shape among trapezoidal and hexagonal. The Nusselt Number results obtained for rectangular parallel minichannel is matched with the results obtained by Ho et al. [12] with a maximum error of 2%. Figure. 6. Represents the comparative analysis of the three different minichannel heat sink and it is clear that the value of the heat transfer coefficient goes on increasing with an increase in the value of Reynolds Number as the inlet velocity of the working fluid increases. It is seen from the figure that the value of the heat transfer coefficient for the rectangular shape is higher as compared to the remaining two shapes. The hexagonal shape showed the lower values of the coefficient of heat transfer.

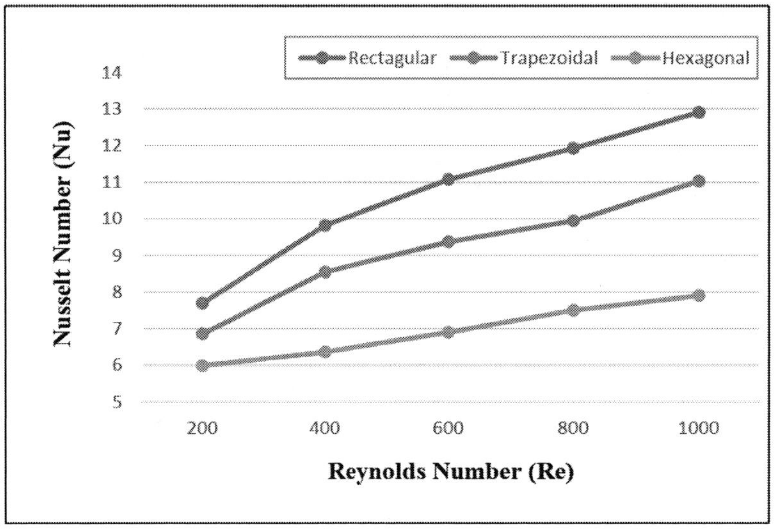

Figure 5. Nusselt number variation with respect to Reynolds number

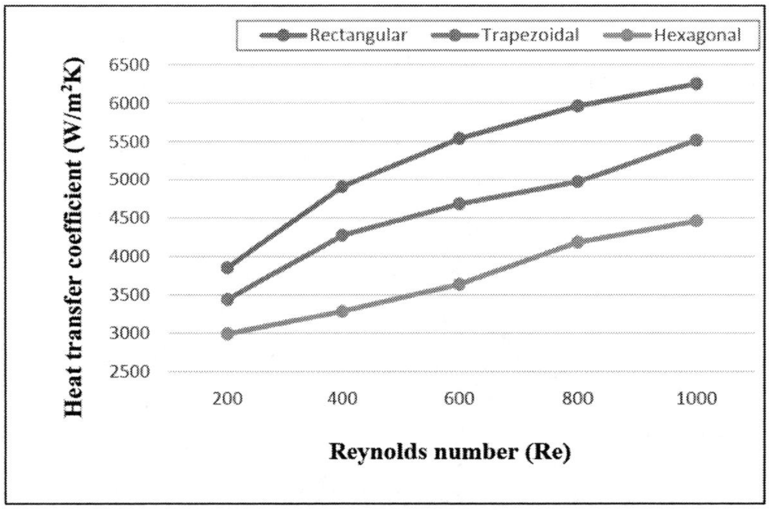

Figure 6. Variation of heat transfer of coefficient with respect to Reynolds number

IOP Conf. Series: Materials Science and Engineering 998 (2020) 012018 doi:10.1088/1757-899X/998/1/012018

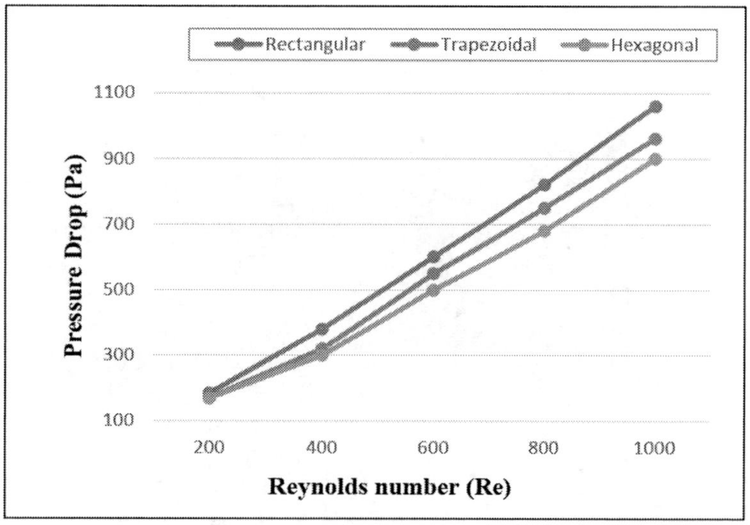

Figure 7. Variation of Pressure drop with respect to Reynolds number

Figure. 7. Shows the variation of pressure drop at Reynolds number 200-1000. From these figures, it is clear that the value of pressure drop goes on increasing with the increase in the value of the Reynolds number. While the hexagonal shape has the lowest pressure drop and the rectangular shape have the highest pressure drop compared to the trapezoidal shape. It should be noted that the fluid path in the numerical simulation, the pipes, and fittings effects between pressure transducer which are present in the experimental analysis are not included in the simulation.

6.2 Temperature Variations

CFD simulated temperature results of minichannel of three different cross-sectional shapes will give an idea about the heat transfer phenomenon in detail. From temperature contours, the temperature distribution of working fluid is shown, from inlet to outlet. It also illustrates how working fluid is picking up, and thereby help in cooling of the minichannel. It has also shown that for a working fluid and at the same flow rate, minichannel of rectangular-shaped has shown better heat transfer as compared to among other shapes. As more heat is picked up by the working fluid, which is apparent from the greater rise in temperature as compared to the inlet. All the below represented figures are taken when the Reynolds no. is 200.

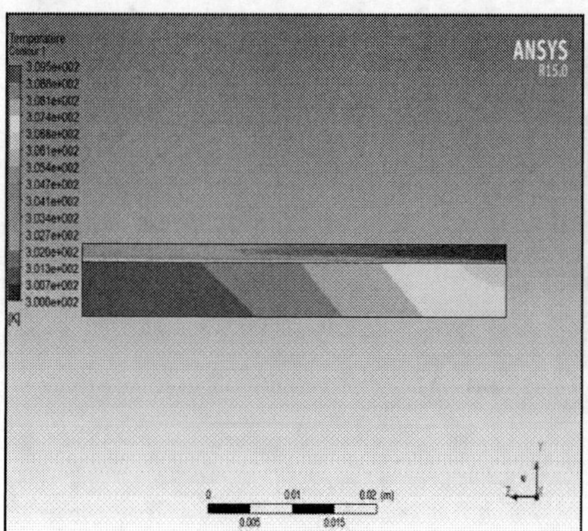

Figure 8. Temperature contour of rectangular minichannel heat sink

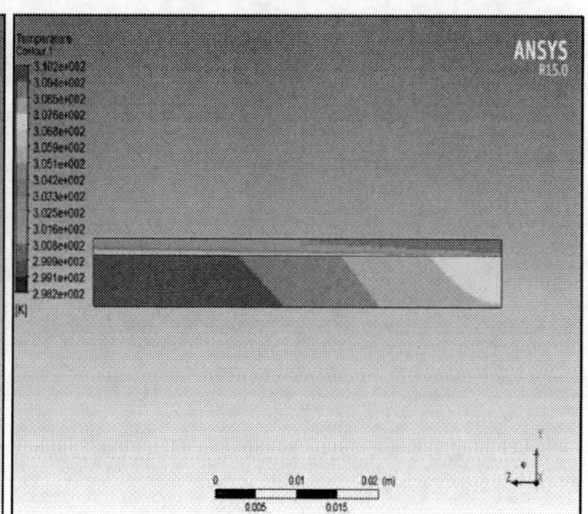

Figure 9. Temperature contour of trapezoidal minichannel heat sink

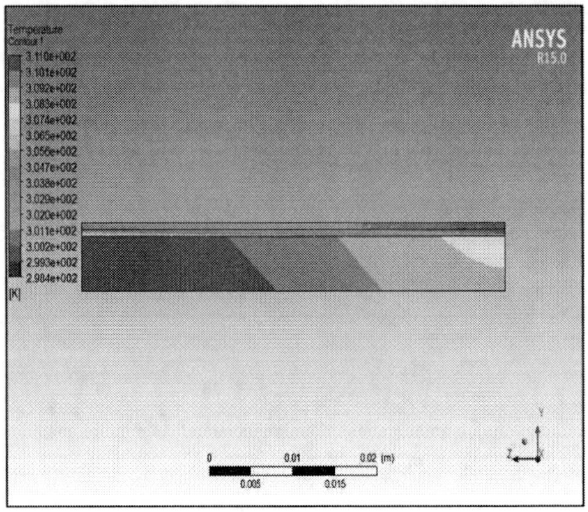

Figure 10. Temperature contour of hexagonal minichannel heat sink

6.3. *Pressure Variations*

The pressure variations of three geometrical configurations of minichannel will give an idea about the fluid flow structure in detail. Below are the pressure contours of three geometries which are showing the pressure distribution of heat sink from inlet to outlet along the length. All the below represented figures are taken when the Reynolds no. is 200

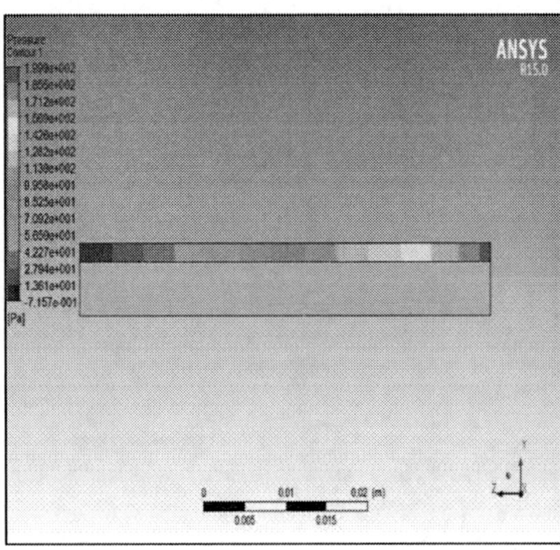

Figure 11. Pressure contour rectangular minichannel heat sink

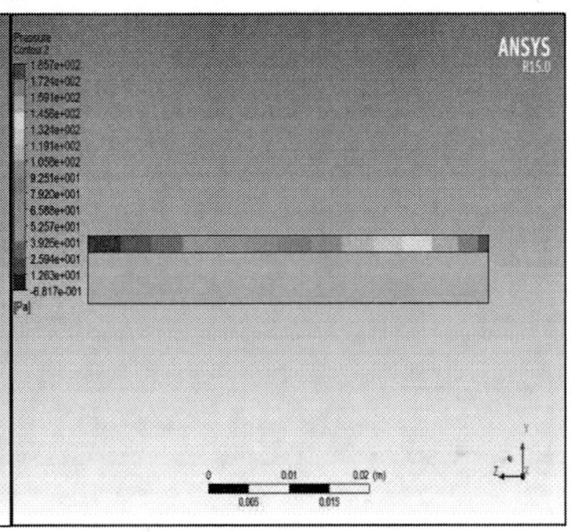

Figure 12. Pressure contour of trapezoidal minichannel heat sink

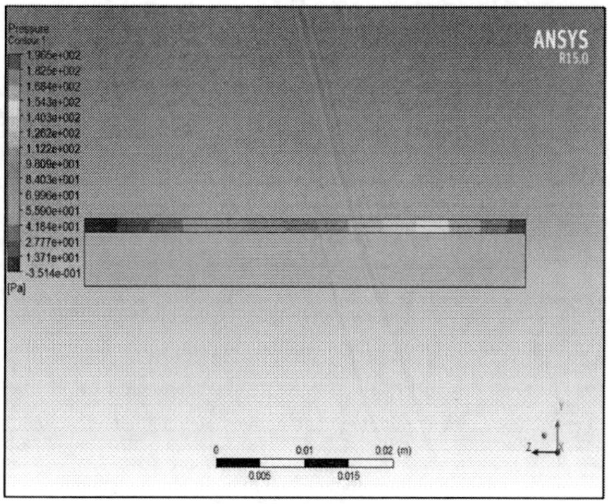

Figure 13. Pressure contour of hexagonal minichannel heat sink

7. Conclusion

In the present work, ICEM-FLUENT CFD (ver. 15.0) software is used for calculating heat transfer performance of minichannel heat sink for different flow rates or Reynolds number.

1) Rectangular shaped cross-section of the minichannel has shown better cooling performance as compared to the trapezoidal and hexagonal cross-section of minichannel with the same working fluid and at the same flow rate.

2) The thermal performance of rectangular shape is a 14% and 32.12% increase in heat transfer rate as compared to the trapezoidal and hexagonal shapes respectively.

3) The average heat transfer coefficient of rectangular, trapezoidal, and hexagonal shapes are 5301.83 W/m^2K, 4577.32 W/m^2K, and 3711.82 W/m^2K respectively. This shows that the heat transfer coefficient is higher for rectangular shape and there is a 13.66% and 29.98% increase in heat transfer coefficient as compared to trapezoidal and hexagonal shapes respectively.

4) An increment in pressure drop is seen, with a corresponding increase in the Reynolds number. Higher values of pressure drop are seen in rectangular minichannel and least in hexagonal minichnnel. While the trapezoidal minichannel is in between rectangular and hexagonal minichannel.

8. References

[1] D.B. Tuckerman, R.F. Pease, *High performance heat sinking for VLSI,* IEEE Electron. Devices Lett. (1981) EDL **2** pp.126–129.

[2] P.S Lee, S.V Garimella, D. Liu, *Investigation of heat transfer in rectangular microchannels*, International journal of heat and mass transfer, **48** (2005) 1688–1704.

[3] W. Qu, I. Mudawar, *Experimental and numerical study of pressure drop and heat transfer in a single-phase micro-channel heat sink*, International Journal of Heat and Mass Transfer **45** (2002) 2549–256.

[4] P. Gunnasegaran, H.A. Mohammed, N.H Shauib, R.Saidur, *The effects of geometrical parameters on heat transfer characteristics of microchannels heat sink with different shapes*, International communications in heat and mass transfer **37** (2010) 1078-1086.

[5] H.A. Mohammed, P. Gunnasegaran, N.H Shauib, *Influence of channel shape on the thermal and hydraulic performance of microchannel heat sink*, International communications in heat and mass transfer 38 (2011) 474-480.

[6] N. Ghazali-Mohd, O. J. Taek, N. B. Chien, C. Kwang, R. Ahmad, *Compariosn of the optimized thermal performance of square and circular ammonia-cooled microchannel heat sink with genetic algorithm*, Energy conversion and management (2015).

[7] L. Chai, S.A. Tassou, *Effect of cross-section geometry on the thermohydraulic characteristics of supercritical* Co_2 *in minichannel*, Energy Procedia **161** (2019) 446–453.

[8] S.E. Ghasemi, A.A. Ranjbar, M.J. Hosseini, *Experimental and numerical investigation of circular minichannel heat sinks with various hydraulic diameter for electronic cooling applications*, Microelectronics Reliability **73** (2017) 97-105.
[9] G.V. Kewalramani, G. Hedau, S.K. Saha, A. Agrawal, *Empirical correlation of laminar forced convective flow in trapezoidal microchannel based on experimental and 3D numerical study*, International Journal of thermal science **142** (2019) 422-433.
[10] L. Su, Z. Duan, B. He, H. Ma, G. Ding, *Laminar Flow and Heat transfer in the entrance region of elliptical minichannels*, International Journal of Heat and Transfer **145** (2019) 118717.
[11] F.S. Moghanlou, A.S. Khorrami, E. Esmaeilzadeh, H. Aminfar, *Experimental study on electrohydrodynamically induced heat transfer enhancement in a minichannel*, Experimental Thermal and Fluid Science **59** (2014) 24–31.
[12] C.J. Ho, P.C. Chang, W.M. Yan, P. Amani, *Thermal and hydrodynamic characteristics of divergent rectangular minichannel heat sinks*, International Journal of heat and mass transfer **122** (2018) 264-274.
[13] S.G. Khandlikar, *Single-Phase Liquid Flow in Minichannels and Microchannels*, Elsevier publication, 2014.

NOMENCALTURE

A_{cs}	Channel flow area, m^2
A_s	Channel surface area, m^2
a	Channel width of trapezoidal, mm
b	Bottom channel width of trapezoidal, mm
S	Hexagonal side length, mm
C_p	Specific heat at constant pressure, J/Kg K
D_h	Hydraulic diameter, mm
h	Convective heat transfer coefficient, W/m^2 K
K	Thermal conductivity of water, W/m.K
V	Fluid velocity, m/s
L_c	Length of channel, mm
P	Channel perimeter, mm
$\dot{m}$	Mass flow rate of cooling water, kg/s
N	Number of parallel minichannels
q''	Heat flux, W/m^2
$(T_f)_{avg}$	Average Mean temperature of fluid, K
$(T_w)_{avg}$	Temperature of solid wall, K
T_{fi}	Inlet temperature, K
T_{fo}	Outlet temperature, K
Nu	Nusselt number
Re	Reynolds number
Q	Heat rate, W

Greek symbol

ρ	Density of water, kg/m^3
μ	Dynamic viscosity, N s/m^2
Δp	Pressure drop, Pa

Dynamic Analysis of Fixed-Fixed Beam Under Mechanical Shock Loads

G.Sharma [1], Abhishek Mishra [1*] and P. Selvaraj [2]

[1] Department of Mechanical Engineering, National Institute of Technology Delhi, Sector A-7, Narela, Delhi – 110040, India
[2] Laboratory for Electro-Optics Systems, Mechanical Systems Division, Indian Space Research Organization, Bengaluru – 560058, India

corresponding author's e-mail: abhishekmishra@nitdelhi.ac.in

Abstract. Aerospace missions are very expensive and any carelessness may lead to mission failure. So, each component that should be installed in a launch vehicle and spacecraft essentially be designed and developed to ensure its reliability. In this paper, a beam made up of Al 6061 T6 material is modeled in the ANSYS 18.0 software and subject to a similar condition to the component of the launcher vehicle. Transient acceleration load having magnitude approximately equal to the amplitude of acceleration developed during pyrotechnic events is applied in X, Y and Z direction of the beam and response of the beam at different locations of the beam is calculated and based on the response acceleration time history a Shock Response Spectrum (SRS) is plotted for locations of interest on the beam. Furthermore, Response Spectrum Analysis is performed on the beam based on the SRS of pyroshock event, provided by the Laboratory for Electro-Optics Systems (LEOS), Indian Space Research Organization (ISRO).

Keywords: Beam, Response Spectrum Analysis, Modal Analysis, Electro-Optics Systems

1. Introduction

In spacecraft and launch vehicles, different components are mounted and subjected to mechanical shock from multiple sources. The components mounted in the spacecraft should be sufficiently durable to bear a series of shocks that are noticed during their service life. Typically, these shocks are a result of several events, such as satellite launch, phase separation, and nut pyrotechnic separation. Pyrotechnic devices commonly used in space missions generate high frequency and high amplitude shock energy as a by-product. The shock produced by these pyrotechnic devices is referred to as pyroshock having a frequency range between 100Hz and 1,000,000Hz. Structural reaction to this type of shock can lead to electronic components malfunctioning, cracks, and wear in a mounted component. So, it becomes very important for the structural response to such load conditions to be studied. Response spectrum analysis is readily used in the industry to record structure response to transient loading. However, there are other methods available to evaluate the structural response to such complex loading conditions as Transient Structural Full Method and Transient Structural Modal Superposition. In this paper, we're concerned with shocks, so it's called the Shock Response Spectrum. Developing SRS is itself a tedious process, using Fast Fourier Transform (FFT) to convert data from time to frequency domain. In this work, we first performed the modal analysis by measuring the natural modal frequencies and mode shapes considering the factor of mass participation. After that, Response Spectrum Analysis (RSA) is performed to predict a structure's dynamic response with the help of a commonly used tool called shock response spectrum (SRS). The transient structural analysis is used to determine the structure's reaction when exposed to time-dependent loads. This analysis is used to evaluate the structure's response to the time-varying load. The paper is structured as follows. The next section shows the related work to this problem. Section 3 analyses our problem formulation and methodology. The paper ends with our Section 4 conclusion and future scope.

2. Related Works

Weihua et al. [1] structurally represented flexible rockets as linear beams. Pourtakdoust et al. [2] investigated flexible missile bending due to thrust. Virgil et al. [3] developed a technique focused on fundamental beam structural analysis to estimate the load-bearing weight of the fuselage and propellant tanks that are integral to the fuselage. Xuesong et al. [4] explored various advantages of alloys and composites used in aerospace, such as Ti-based alloys. Rambabu et al. [5] analysed historical advances in aluminium alloys used in aerospace. Jayaraman et al. [6] conducted a response spectrum analysis on a printed circuit board to predict its response when subjected to a classic 100 G half-sine shock pulse over 11 milliseconds. The Glass encapsulated magnetic reed switches were examined by Mihai et al. [7] by exposing them in the pyroshock setting. The response analysis was based on the transient structural analysis used to extract the reed switch loads. Experimental data compare with the FEM findings. Kenneth [10] in this work performed shock analysis of IBM Model 0663 3.5" fixed disk drive to evaluate the shock response of the drive at different shock inputs so that the design of packaging which will carry the drive during transportation can be improved so that it can bear such type of shock loads. David [11] performed a shock analysis on a blade of mixed flow fan by modelling it as a cantilever beam using the finite element method and investigated the mechanical effects by varying the width in the acceleration input. As per MIL STD 810 half-sine acceleration is given as input as a representation of ballistic shock, gun firing shock and road shock in the lateral direction to perform transient analysis in FEA software. Suresh et al. [12] performed a shock test on a printed circuit board in a wireless telephone handset to design a good suspension. Three suspensions were subjected to a spectrum of shock pulse and the shock response spectrum is plotted which showed a resonant response of the PCB. Output results also illustrate the effectiveness of using high damping in material used in suspension in reducing the peak acceleration response. Hongda et al. [13] developed a finite element model for simulating the whole separation process of pyrotechnic separation nuts and the shock generated due to the separation is analysed. They found out that pyrotechnic shock and the strain energy shock are the main sources of shock in the separation process. Pre stressing of nuts strengthens the severity of the shocks.

3. Problem Definition and Methodology

In aerospace applications, it is well known that the components of aerospace vehicles are subjected to pyrotechnic events due to which pyroshocks are developed and they can be very fatal for an aerospace mission as the pyroshock can cause temporary or permanent damage to the critical electronic components mounted on the aerospace vehicle which may lead to failure of a mission. In the paper, a beam having fixed-fixed boundary conditions have been modelled in FEA software ANSYS 18.0 in

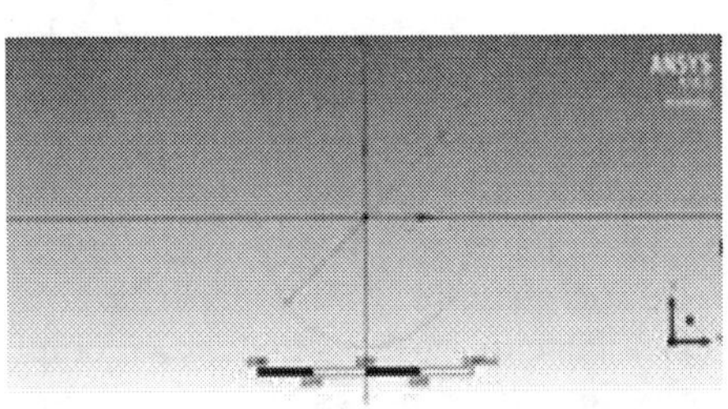

Figure 1. Cross-sectional sketch of the beam.

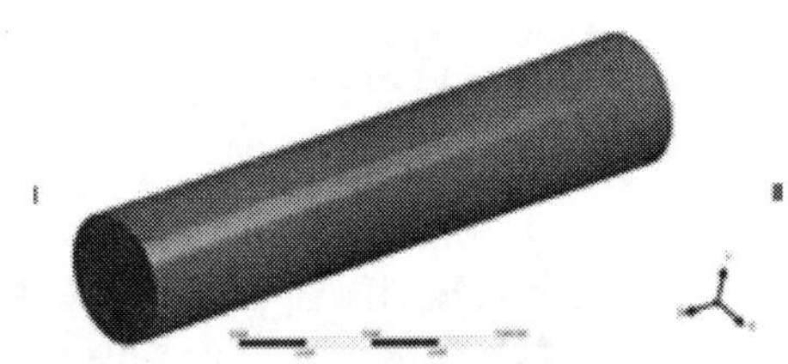

Figure 2. Extruded model of the beam.

order to perform analysis and predict the response of the beam when it is subjected to mechanical shock load as shown in figure 1 & figure 2. The finite element method is nothing but a step-by-step procedure that includes (a) Pre-processing: defining a physical model as simple forms of selection known as finite elements. (B) A solution of governing equations describing problem physics in a matrix format. (C) Post-processing: Analysis and discussion of findings. Finite element software packages are ABAQUS, ANSYS, MSC NASTRAN, etc. Aluminium alloy was considered in the presented work Al 6061 T6 [8] because it is commonly used in aerospace structures and geometric modeling is performed in the design modeling tab as shown in Figure 1 & Figure 2. The discretization of the beam into finite elements is achieved using ANSYS 18.0 mesh module. Both beam ends are considered fixed in research. Classical half-sine pulse shock load of magnitude 1000G is applied throughout 11 milliseconds in the X, Y, and Z direction of the beam. As shown in Figure 3 three Step Approach to analyse the fixed-fixed beam under mechanical shock loads is used. In this paper, we're concerned with shocks, so it's called the Shock Response Spectrum. Developing SRS is itself a tedious process, using Fast Fourier Transform (FFT) to convert data from time to frequency domain. In this work, we first performed the modal analysis by measuring the natural modal frequencies and mode shapes considering the factor of mass participation. After that, Response Spectrum Analysis (RSA) is performed to predict a structure's dynamic response with the help of a commonly used tool called shock response spectrum (SRS). The transient structural analysis is used to determine the structure's reaction when exposed to time-dependent loads. This analysis is used to evaluate the structure's response to the time-varying load.

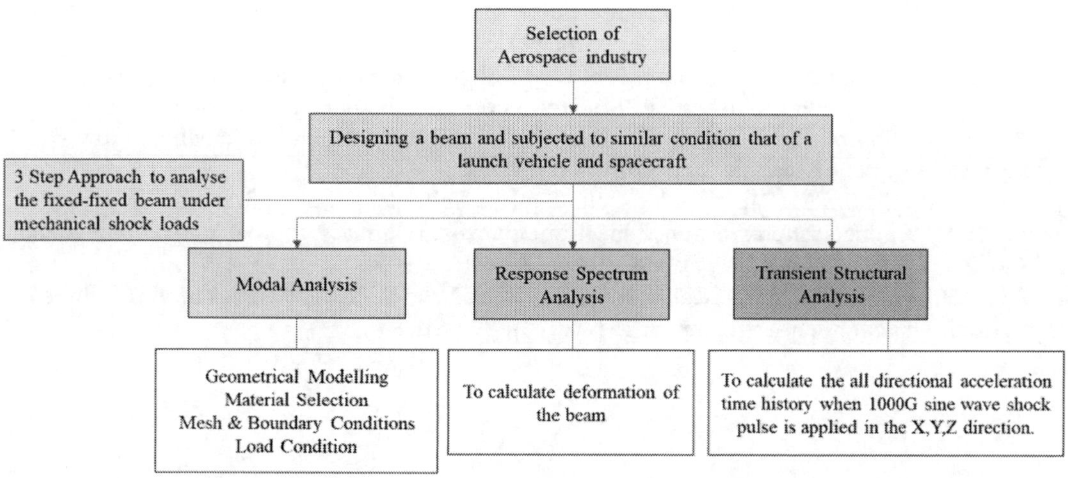

Figure 3. Flow chart of the proposed methodology

4. Conclusions

RSA was performed on a beam structure under two load conditions in the presented work. First, the classic half-sine shock pulse of magnitude 1000 G applied to the beam in all three directions, and secondly pyroshock response spectrum provided by the ISRO. Transient structural analysis findings show that components in Z direction loading condition cannot withstand high acceleration compared to when a load is applied in X and Y directions. A more realistic launcher vehicle model can also be implemented in simulation, and transient structural analysis can be performed by taking the pyroshock event's real time-acceleration history data. A dedicated software programme that can perform transient structural analysis, modal analysis, and response spectrum analysis can also expand the present work.

References

[1] Su, Weihua, et al. "Dynamic beam solutions for real-time simulation and control development of flexible rockets." Journal of Spacecraft and Rockets 54.2 (2017): 403-416.

IOP Conf. Series: Materials Science and Engineering 998 (2020) 012019 doi:10.1088/1757-899X/998/1/012019

[2] Pourtakdoust, S. H., and N. Assadian. "Investigation of thrust effect on the vibrational characteristics of flexible guided missiles." Journal of Sound and Vibration 272.1-2 (2004) : 287-299.
[3] Hutchinson, Virgil, and John Olds. "Estimation of launch vehicle propellant tank structural weight using simplified beam approximation." 40th AIAA/ASME/SAE/ASEE Joint Propulsion Conference and Exhibit. 2004.
[4] Zhang, Xuesong, Yongjun Chen, and Junling Hu. "Recent advances in the development of aerospace materials." Progress in Aerospace Sciences 97 (2018): 22-34.
[5] Rambabu, P. P. N. K. V., et al. "Aluminium alloys for aerospace applications." Aerospace Materials and Material Technologies. Springer, Singapore, 2017. 29-52.
[6] Jayaraman, S., et al. "Response spectrum analysis of printed circuit boards subjected to shock loads." Procedia Engineering144 (2016): 1469-1476.
[7] Vladimirescu, Mihai, Andre Zybura, and Regina Kwiatkowski. "3D finite element analysis of the glass encapsulated magnetic reed switches under pyroshock environment." 2010 Proceedings of the 56th IEEE Holm Conference on Electrical Contacts. IEEE, 2010.
[8] "ASM Material Data Sheet", asm.matweb.com, http://asm.matweb.com/search/SpecificMaterial.asp?bassnum=MA6061t6.
[9] Irvine, By Tom, and Random Vibration. DB/Octave Calculations for Log-Log Plots. Vol. 2, no. 2, 2009, pp. 5–7. and Operations Research, IIT Bombay.
[10] Neuburg, Kenneth. "A Study in the application of shock response, spectrum analysis to disk drive shipping and handling shock tests." (1995).
[11] Campos, David Jonathan González. A Study of Shock Analysis Using the Finite Element Method Verified with Euler-bernoulli Beam Theory; Mechanical Effects Due to Pulse Width Variation of Shock Inputs; and Evaluation of Shock Response of a Mixed Flow Fan. Diss. California Polytechnic State University, 2014.
[12] Goyal, Suresh, Jim M. Papadopoulos, and Paul A. Sullivan. "Shock protection of portable electronic products: shock response spectrum, damage boundary approach, and 56 beyond." Shock and Vibration 4.3 (1997): 169-191.
[13] Zhao, Hongda, et al. "Numerical study on separation shock characteristics of pyrotechnic separation nuts." Acta Astronautica 151 (2018): 893-903.

IOP Conf. Series: Materials Science and Engineering 998 (2020) 012020 doi:10.1088/1757-899X/998/1/012020

Prediction of Lateral-Vibration Responses of A Full Size Cantilever Beam by Using Those of Its Scale Models

S. Naresh Kumar[1,a], **S. Sairam**[2,b*], **Shakti P Jena**[3,c]

[1,3]Department of Mechanical Engineering, Vardhaman College of Engineering, Hyderabad, Telangana, India.

[2]PG Scholar, Department of Mechanical Engineering, Vardhaman College of Engineering, Hyderabad, Telangana, India.

E-mail: [a]`2naresh@gmail.com`, [*b]`sairamsundaragiri51@gmail.com`, [c]`shaktipjena@gmail.com`

Abstract. The main objective of this article is to study and calculate lateral vibration responses of the cantilever beam by using a scale model and the associated scaling laws. Lateral vibration responses of a cantilever beam are analyzed theoretically and to validate results using a method of finite element to calibrate vibration response of a similar beam. We get the results and then we had compared both scale model and full-size cantilever beam model. Initially derive the solution of free response and steady-state response for the full and scaled model. Later, the scaling-laws for all the applicable parameters concerning the cantilever beam in the size of full and its scale beam of free vibration are get from the Euler's equation of motion of the last two systems. Next, by dimensional analysis find the scaling factors for applicable dimensions and properties of required two systems. And in Finite element method introduce property matrices and equations of motion of cantilever beam system and also systematic procedure to get element property matrices of the beam is shown. In the end it is solved by Method of Eigen value problem with the help of Mat lab software. Lastly the impact of specimen dimensions on the similarity basis, natural frequencies of cantilever beam specimen is investigated.

1. Introduction

The aim of this project is to introduce the problems of measuring instruments due to its cost and access where we used to vibrating bodies of huge in size. In fields of structural and mechanical-engineering application, the lateral-vibration responses of cantilever beams, for instance those of supporting beams, etc. are the essential data for engineers. Researchers have considered the vibration responses in machine members. Firoozian and Zhu [1] have applied the transfer-matrix method to estimate the lateral vibration responses of the rotor bearing model. Besides, Wu and Yang [2], Qing and Cheng [3] and Al-Bedoor [4, 5] have perform on the connected rotor-bearing system to calculate torsional-lateral vibration responses using methods of transfer matrix, finite element, respectively. Rezaeepazhand et al. [6, 7] and Simitses and Rezaeepazhand [8] have examine the buckling and free vibration responses of full size laminated shell and develop scale down models. Wu [9] has applied dynamic loads on a scale flat plate and compare with those of flat plate in full dimensions to predict dynamic responses by using respected scaling-laws. Further, Wu [10] has applied the lateral vibration responses of a scale rotor bearing arrangement

IOP Conf. Series: Materials Science and Engineering 998 (2020) 012020 doi:10.1088/1757-899X/998/1/012020

Properties	**Fullsize beam**	**Scalesize beam**
Mass(m)(Kg)	2	2×0.5^3
Damping(c)(N-s/m)	4	4×0.5^2
Stiffness(K)	8	8×0.5
Force(F)	$3 \times \sin(5t)$	$3 \times 0.5^2 \sin(5 \times 0.5^{-1} \times t)$
Frequency(ω)	5	5×0.5^{-1}
Natural Frequency(ω_n)	2	2×0.5^{-1}

Table 1: Physical parameters of full size and scale size vibration system

and to predict those of the rotor-bearing at the place of application (full size) due to free and forced vibrations by using corresponding scaling laws.

Initially, the article delivers the scaling-factors for all appropriate parameters concerned to cantilever beam. For validation, the lateral-vibration responses of a full size cantilever beam and its scale beam specimen are calculated theoretically and numerically and achieve valuable information on results. And the formulation of a cantilever beam in the finite element method of the equation of motion is solved by the help of mat lab Eigen value problem.

2. Steady & free vibration

The properties of full scale and scale size beam is shown in the given table.

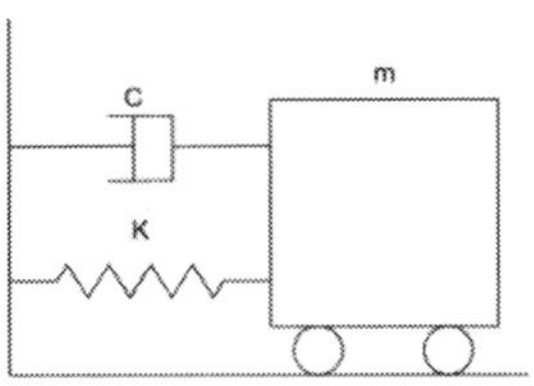

Figure 1: Full size

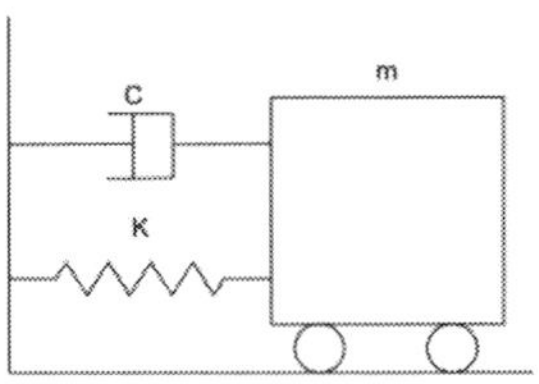

Figure 2: Scaled size.

2.1. Free vibration solution for scale model

The equation of motion is given by

$$m\ddot{x} + c\dot{x} + kx = 0 \tag{1}$$

For steady - state responce

$$x(t) = X \times e^{\xi \omega_n t} \times \sin(\sqrt{1-\xi^2} \times \omega_n \times t + \phi) \tag{2}$$

c_c is the critical damping of the system, ϕ is the phase angle, damping ratio

$$\xi = \frac{c}{c_c} = \frac{c}{2 \times \sqrt{k \times m}} \tag{3}$$

$x(0) = 0$; $\dot{x}(0) = 0$, by sloving eq(2) and eq(4), we get $\tan(\phi) = 1.732$

$$X = \frac{1}{\sin(60)} = 1.1547, \phi = 60^0 \tag{4}$$

$$x(t) = 1.1547 \times \sin(1.732 \times t + 60^0) \times e^{-t} \tag{5}$$

2.2. Free vibration solution for full model

$$x(t) = X \times e^{-\xi\omega_n t} \times \sin(\sqrt{1-\xi^2} \times \omega_n \times t + \phi) \tag{6}$$

$$\xi = \frac{c}{c_c} = \frac{c}{2 \times \sqrt{k \times m}} = 0.5 \tag{7}$$

$x(0) = 0.5$; $\dot{x}(0) = 0$,
by sloving eq(6) and eq(7), we get $\tan(\phi) = 1.732$

$$x(t) = 0.57735 \times \sin(3.4641 \times t + 60^0) \times e^{-2t} \tag{8}$$

From Equations (5), (8), frequency is doubled and amplitude is halved.

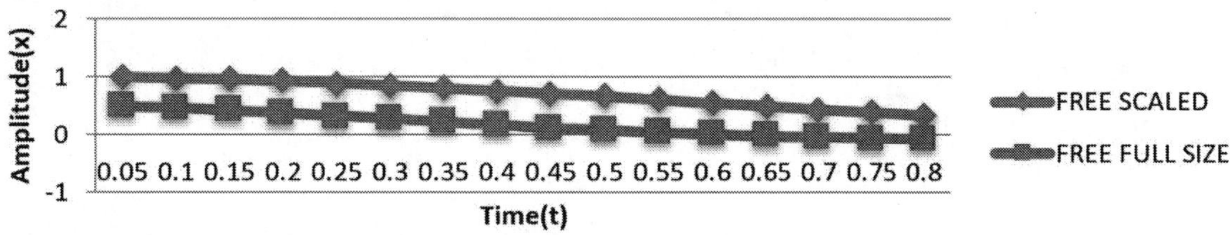

Figure 3: Amplitude v/s Time graph for full and scale model in free vibration

2.3. Forced vibration for scale model

For steady state response,

$$x(t) = \frac{F_e}{\sqrt{((k - m\omega^2)^2) + (c\omega)^2}} \times \sin(\omega t - \phi) \tag{9}$$

$$\phi = \tan^{-1}\left(\frac{c \times \omega}{k - m(\omega)^2}\right) = \tan^{-1}\left(\frac{4 \times 5}{8 - 2(5)^2}\right) = -25.4633 \tag{10}$$

$$x(t) = 0.064490037 \times \sin(5t + 25.4633) \tag{11}$$

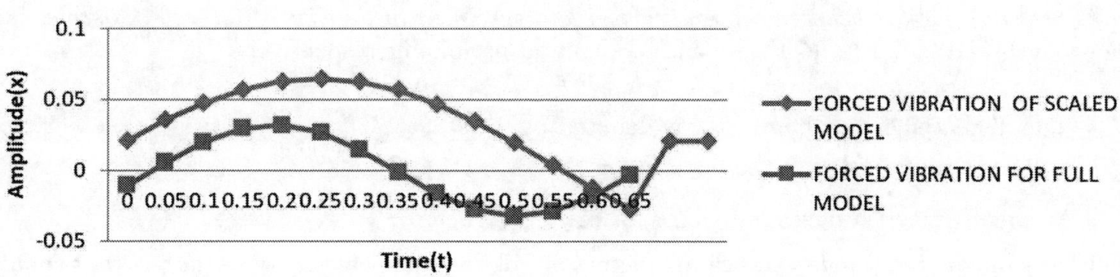

Figure 4: Amplitude v/s Time graph for full and scale model in forced vibration

2.4. Forced vibration for full model

$$x(t) = 0.03224 \times \sin(10t + 25.4633) \tag{12}$$

3. Determination of scaling-laws & scaling-factors for physical dimensions of a cantilever beam

The scaling study is to get terminology of the scaling-laws and scaling-factors for the dynamic or static-similitude between the *full size* beam specimen and its *scale-size* beam specimen.

3.1. Scaling laws between the cantilever beam in full size and its scale size beam

For a free-vibrating system, based on Euler-theory, we substituted following parameters in the equation of motion.
Let m(x) = distributed mass/unit length
L=length of beam-specimen
y=deflection of beam
ω=angular velocity of the beam-specimen
E=Young's modulus of the beam-specimen
I= moment of inertia of the beam-specimen
λ=scaling variable of the beam
Dynamic load per unit length

$$\frac{d^2}{dx^2}EI(X)\frac{d^2Y(x)}{dx^2} \tag{13}$$

Centrifugal force per unit length = $m(x)Y(x)\omega^2$

$$\frac{d^2}{dx^2}EI(X)\frac{d^2Y(x)}{dx^2} = m(x)Y(x)\omega^2 \tag{14}$$

After solving, we get

$$EI(X)\frac{d^4Y(x)}{dx^4} = \omega^2 m(x)Y(x), \lambda = \frac{d_i^s}{d_i^f}(d = m, y, I, \omega, \eta) \tag{15}$$

$$\left(\frac{\lambda_\eta^3}{\lambda_\eta}\lambda_\eta\left(\frac{1}{\lambda_\eta}\right)^2\right)m(x)Y(x)\omega^2 = \left(1.\lambda_\eta^4.\frac{\lambda_\eta}{\lambda_\eta^4}\right).EI\frac{d^4y}{dx^4}.\left(\frac{\left(\frac{\lambda_\eta^3}{\lambda_\eta}.\lambda_\eta.\left(\frac{1}{\lambda_\eta}\right)^2\right)}{\left(1.\lambda_\eta^4.\frac{\lambda_\eta}{\lambda_\eta^4}\right)}\right) = 1 \tag{16}$$

the above equation is called the *scalling laws* and it is the requirement for the similitude between the cantilever beam specimen in the size of full and its scaled specimen.

3.2. Scaling - factors between the cantilever beam specimen in the size of full and its scale - size

For length and breadth, depth use as fundamental parameter respectively

$$\lambda_l = \lambda_\eta, \lambda_b = \lambda_\eta, \lambda_h = \lambda_\eta$$

- The displacement scaling factor is denoted by λ_η
- A mass scaling factor of the beam $\lambda_m = \lambda_\eta^3$
- Moment of inertia of scaling factor for cantilever beam $I = \frac{bh^3}{12} = \frac{\lambda_\eta.\lambda_\eta^3}{12} = \frac{\lambda_\eta^4}{12}$

For the same material following scaling factors of full and scale cantilever beam does not change.

- The ratio of Young's Modulus of scale and full-size cantilever beam is equal to 1
- The ratio of the density of scale and full-size cantilever beam is equal to 1
- Stiffness scaling factor of the cantilever beam $\lambda_k = \lambda_m \times \lambda_\omega^2, \lambda_k = \lambda_\eta^3 \times \frac{1}{\lambda_\eta^2} = \lambda_\eta$

IOP Conf. Series: Materials Science and Engineering 998 (2020) 012020 doi:10.1088/1757-899X/998/1/012020

4. Free vibration of a cantilever beam Mathematical Analysis

If a cantilever beam specimen subjected to free vibration, then the specimen arrangement is taken as a continuous system. In this case the mass of a beam is considered as a well propagated with respect to stiffness of the beam specimen. The free vibration equation can be taken from earlier equation (13). Where, E modulus of rigidity of beam, I- moment of inertia of the rectangular cross sectional beam specimen, Y(x) is deflection, ω -circular natural frequency, m(x)-mass per unit length, $m(x) = \rho A(x)$, ρ - density of specimen, x- linear distance calculated from the fixed end.

We have following boundary conditions of a cantilever beam specimen (Fig. 5)

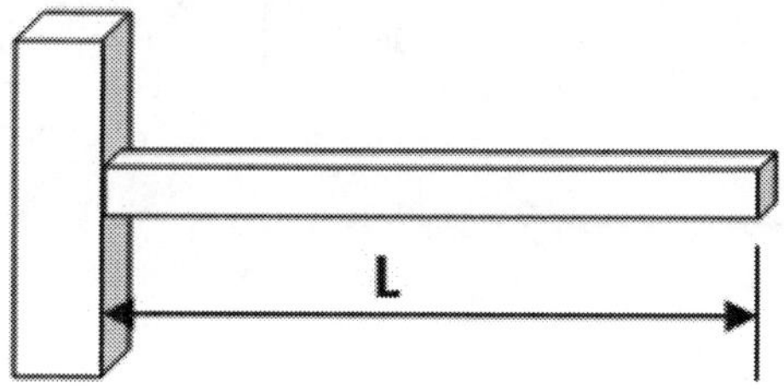

Figure 5: A cantilever beam,ss

$$@x = 0 \Rightarrow Y(x) = 0, \frac{dY(x)}{dx} = 0\&@x = L \Rightarrow \frac{d^2Y(x)}{dx^2} = 0, \frac{d^3Y(x)}{dx^3} = 0 \tag{17}$$

For a homogeneous cross sectional beam specimen subjected to natural vibration from equation (13), Can form following equation

$$\frac{d^4Y(x)}{dx^4} - \beta^4 Y(x) = 0 \qquad \beta^4 = \frac{\omega^2 m}{EI} \tag{18}$$

The general equation of mode shapes for a structural beam specimen is written as a closed form of the natural frequency ω_{nf} , after applying boundary conditions to above equation, final form be written as below.

$$f(x) = A_n\left((\sin\beta_n L - sinh\beta_n L)(\sin\beta_n x - \sinh\beta_n x) + (\cos\beta_n L - \cosh\beta_n L)(\cos\beta_n x - \cosh\beta_n x)\right) \tag{19}$$

$$\beta_n L = n\pi \qquad \omega_{nf} = \alpha_n^2\sqrt{\frac{EI}{mL^4}} \tag{20}$$

where $\alpha_n = 1.875, 4.694, 7.895$ & $n = 1, 2, 3, ...$

IOP Conf. Series: Materials Science and Engineering 998 (2020) 012020 doi:10.1088/1757-899X/998/1/012020

Material	Density	E
Steel Beam	7850 Kg/m^3	2.1e11

Table 2: Material properties of steel

Length(L)	Width (b)	height(d)
0.45 m	0.02 m	0.003 m

Table 3: Geometry of the beam

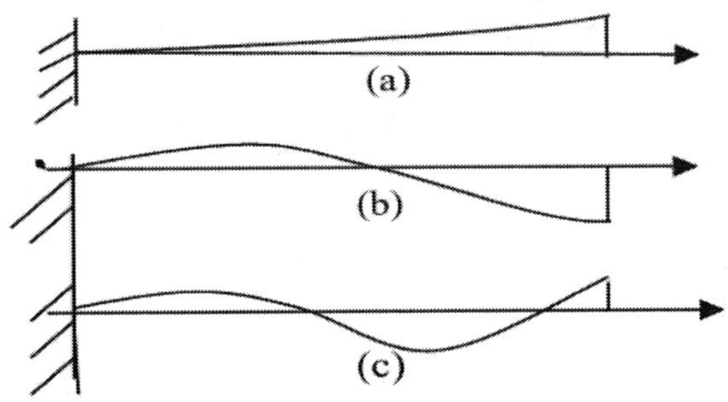

Figure 6: First three un-damped natural frequencies, mode shapes of cantilever beam.

$$\begin{aligned} \omega_{nf1} &= 1.875^2\sqrt{\frac{EI}{mL^4}} \\ \omega_{nf2} &= 4.697^2\sqrt{\frac{EI}{mL^4}} \\ \omega_{nf3} &= 7.855^2\sqrt{\frac{EI}{mL^4}} \end{aligned} \quad (21)$$

The relation between natural frequency, circular natural frequency is written as

$$f = \frac{\omega_{nf}}{2\pi} \;\&\; I = \frac{bd^3}{12}$$

In the above formula, I referred as moment of inertia of the rectangular specimen.

Example: find out the un-damped natural frequency of a cantilever beam with the above dimensions. Based on scaling ratio for scale cantilever beam of length, breadth and depth dimensions are halved & following results can be obtained.

Frequency	**Full Model**	**Scale Model**	$\frac{scale}{full}$
First Natural Frequency	12.37	24.74	2.00
Second Natural Frequency	77.53	155.06	2.00
Third Natural Frequency	434.22	868.44	2.00

Table 4: Theoretical results on scale and full size cantilever beam

5. Finite Element Formuation

Finite element method is a numerical method for solving differential equations and also estimate the second natural frequency of a cantilever beam by the method of finite element formulation. The basic method is explained here.

For a cross sectional uniform beam, the stiffness matrix and the consistence mass matrix are given below.

$$[K]^e = \frac{EI}{l^3}\begin{bmatrix} 12 & 6l & -12 & 6l \\ 6l & 4l^2 & -6l & 2l^2 \\ -12 & -6l & 12 & -6l \\ 6l & 2l^2 & -6l & 4l^2 \end{bmatrix} \; and \; [M]^e = \frac{\rho Al}{420}\begin{bmatrix} 156 & 22l & 54 & -13l \\ 22l & 4l^2 & 13l & -3l^2 \\ 54 & 13l & 156 & -22l \\ -13l & -3l^2 & -22l & 4l^2 \end{bmatrix} \quad (22)$$

IOP Conf. Series: Materials Science and Engineering 998 (2020) 012020 doi:10.1088/1757-899X/998/1/012020

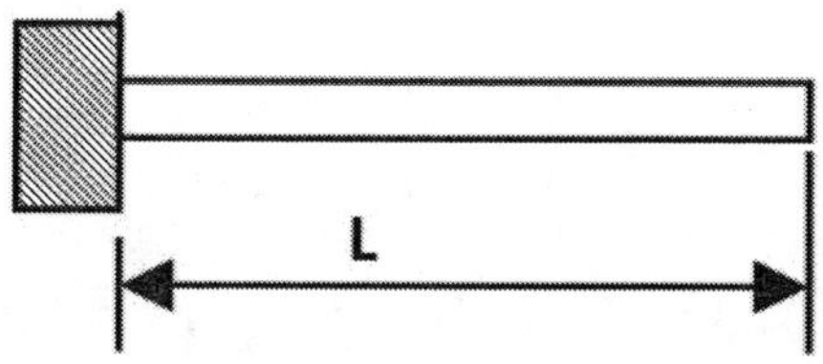

Figure 7: A cantilever beam

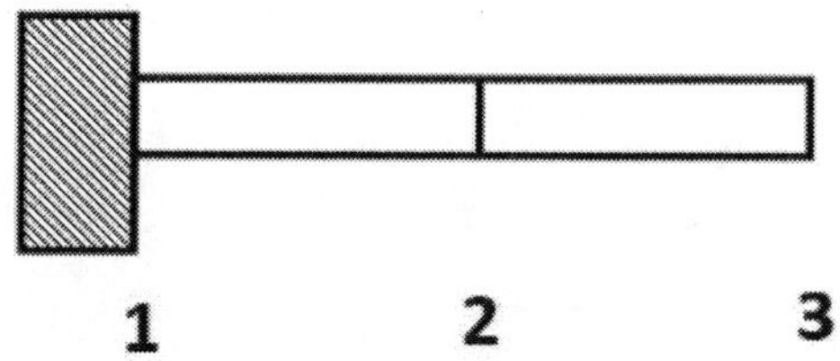

Figure 8: Discretization of beam into 2 elements

Step 1: Discrete the experimental beam arrangement into three small parts as shown in Fig.8. In Fig. 8, numbers 1, 2, 3 refer to the nodes.
Step 2: Arrange the elemental equations. For the first part, from equation (22), we have

$$\frac{\rho Al}{420}\begin{bmatrix} 156 & 22l & 54 & -13l \\ 22l & 4l^2 & 13l & -3l^2 \\ 54 & 13l & 156 & -22l \\ -13l & -3l^2 & -22l & 4l^2 \end{bmatrix}\ddot{x} + \frac{EI}{l^3}\begin{bmatrix} 12 & 6l & -12 & 6l \\ 6l & 4l^2 & -6l & 2l^2 \\ -12 & -6l & 12 & -6l \\ 6l & 2l^2 & -6l & 4l^2 \end{bmatrix}x = F \tag{23}$$

Here X refer to the vector node variables (i.e. at the first and second nodes). In this scenario node variables are slope and displacement, the vector F be a symbol of force.
Again we get the second step that also has the same elemental equation as equation (23).
Step 3: combine above first and second element matrices to form a global matrix. The dimensions of the beam are multiplied with a scaling factor 0.5.
Step 4: The final step is to find out the solution by using following equations

$$\begin{bmatrix} 0.0066 & 0 & 0.0011 & 0 & 0 & 0 \\ 0 & 0 & 0 & 0\ 0\ 0 & & \\ 0.001 & 0 & 0.006 & 0 & 0.0011 & 0 \\ 0 & 0 & 0 & 0\ 0\ 0 & & \\ 0 & 0 & 0.0011 & 0 & 0.003 & 0 \\ 0 & 0 & 0 & 0\ 0\ 0 & & \end{bmatrix}\ddot{x} + \begin{bmatrix} 3.36 & 0 & -1.68 & 0.063 & 0 & 0 \\ 0 & 0.0063 & -0.063 & 0.0016 & 0 & 0 \\ -1.68 & -0.063 & 3.36 & 0 & -1.68 & 0.063 \\ 0.063 & 0.0016 & 0 & 0.0063 & 1.68 & -0.06 \\ 0 & 0 & -1.68 & -0.063 & 1.68 & -0.06 \\ 0 & 0 & 0.063 & 0.0016 & -0.063 & 0.003 \end{bmatrix}x = F \tag{24}$$

And for free vibration

$$\ddot{x} = -\omega_{nf}^2 x$$

Hence, above equations can be written as

$$([M]\omega_{nf}^2 + [K])x = 0 \tag{25}$$

6. Eigen Value Problem of Cantilever Beam Solve by Mat Lab Program for Scale Model

Here the dimensions of the cantilever beam are replaced with half of the dimensions of full size cantilever beam, then we get following results as seen in the MATLAB software .
End Result

Frequency	**Full Model**	**Scale Model**	$\frac{scale}{full}$
First Natural Frequency	12.40	24.755	2.00
Second Natural Frequency	77.80	155.1375	2.00
Third Natural Frequency	219.90	434.4331	2.00

Table 5: shows the results of Scale and Full Size Cantilever Beam by Using FEA

The above end results obtained at the end of Eigen value problem in Mat Lab Program by the finite element analysis coincide with results arriving from theoretical analysis.

7. Conclusion

- In this paper, one of the most popular technique scale-down model tests is used to analyse the theoretical results of a large size structural beam specimens.
- Numerical results in a scale model are calculated through the mathematical equations by presenting the scaling laws and the corresponding scaling factors, and these results are utilized to predict the responses of free lateral vibrations of a cantilever beam in size of full.
- To validate the theoretical results and developed mathematical analysis of cantilever beam to find out natural frequencies, the effect of various dimensions of component member on the similarity of a free lateral vibrations and the influence of external loads are induced which creates unbalance force on the free end of cantilever beam have been studied.

References

[1] Aleyaasin M and Ebrahimi M, Whalley R 2000 *Vibration analysis of distributed lumped rotor systems*,Comput. Methods Appl. Mech. Eng. 189545–558.
[2] Firoozian R, Zhu H1991 *A hybrid method for the vibration analysis of rotor-bearing systems*, Proc. Inst. Mech. Eng. C 205 131–137.
[3] Qian Y, Swanson S RandNuismer R J,Bucinell R B1990 *An experimental study of scaling rules for impact damage in fibre composites*, J. Compos. Mater. 24559–570.
[4] W.E. Baker and Westine P S, *Dodge F T 1973Similarity Methods in Engineering Dynamics: Theory and Practice of Scale Modelling*, Hayden Book Company, Inc., Rochelle Park, NJ.
[5] Vassalos D, 1999 *Physical modelling and similitude of marine structures*, Ocean Eng. 26 111–123.
[6] RezaeepazhandJ, Simitses G J and Starnes J H, 1995 *Use of scaled-down models for predicting vibration response of laminated plates*, Compos. Struct. 30 419–426.
[7] RezaeepazhandJ , Simitses G J , Starnes J H, 1996 *Design of scaled-down models for predicting shell vibration response*, J. Sound Vibr. 195 (2) 301–311.
[8] Simitses G J, Rezaeepazhand J, 1993 *Structural similitude for laminated structures*, Compos. Eng. 3751–765.
[9] Wu J J, 2003 *The complete-similitude scale models for predicting the vibration characteristics of the elastically restrained flat plates subjected to dynamic loads*, J. Sound Vibr. 268 (5) 1041–1053.
[10] Wu J J, 2007 *Prediction of lateral vibration characteristics of a full-size rotor-bearing system by using those of its scale models*, Finite element in analysis and design 43 803-816.

Experimental Investigations of Canola Oil Lubrication with Nano-Crystalline MoS_2 Additives in CNC End Milling of Aluminium Alloy

Suresh Kumar Mandru[1], P. Ramamurthy Raju[2] and CH. Ratnam[3]
[1]Research scholar, Department of Mechanical Engineering, A.U College of Engineering, Andhra University, Andhrapradesh, India.
[2]Professor, Department of Mechanical Engineering, S.R.K.R Engineering College, Andhrapradesh, India
[3]Professor, Department of Mechanical Engineering, A.U College of Engineering, Andhra University, Andhrapradesh, India

Email: msuresh.auce@gmail.com

Abstract. Since the past few decades, manufacturing operations including the milling of metals use various types of coolants and lubricants. These are composed of chemicals and other pollutant factors, which are very dangerous and hazardous to the nature and human. And also the increasing demands of high productivity for the machining requirements leaves no choice but to use high cutting velocity along with higher feed rates. These renewed machining conditions leads to the generation of high cutting temperatures, which reduces tool life with poor product quality. Hence there is a need to find the alternative coolants and lubricants, which reduces the above factors in addition to machining cost. In general lubrication is attributed for the solving of heat generation related issues at the cutting zone. Due to some inherent properties like cooling, lubrication and functions like flushing of chips, the cutting fluids leads to better performances during milling operations. As the conventional way of using fluids are less efficient during high production machining, it is observed that nano fluids have better properties in terms of transfer of heat from the cutting zone. Economic aspects are considered during the use of nano fluids by applying the phenomenon of Minimum quantity lubrication (MQL). The work is an attempt in the direction of using nano fluids as alternative lubricants in milling operations by the study of tool temperature and quality of surface finish.
Keywords: Milling, Solid lubricant, MQL, Canola oil, MoS_2, Nano lubrication.

1. INTRODUCTION

Radical technological changes are observed in the use of lubricants in the solid form in various machining operations, which can be attributed to the advancement in modern tribology. Many solid lubricants have been identified that have the capacity to sustain its inherent property of lubricity with the change in temperature at regular intervals. Some of the powders used are classified as lamellar solids which are being used as lubricants. They are graphite, molybdenum disulfide, tungsten disulfide and calcium fluoride [1, 2]. Experimentations have shown that solid lubricant could effectively control the machining zone temperatures by intensively removing heat during the machining process. Experimentations in order to study the end effect of the solid lubricant used were carried out by analyzing the surface finish and chip thickness after machining [3]. It is observed that heat generated during machining is dissipated when cutting fluids are introduced into the cutting zone. The conventional cutting fluids which create some problems like environmental pollution, water pollution, and keeping in mind the economic aspects [4], substitutions are being found out that ns normally meet the functional requirements by the coolants with some other means. It is found that a few other approaches are by using cryogenic coolants and lubricants which are bio-degradable [5, 6]. Keeping in mind the environmental aspects, use of environment friendly lubricants are the best alternatives. These include vegetable oils like rapeseed, corn or soybean oil, esters, etc. Considering the economic aspects, the concept of MQL is an aspect which researchers have pondered upon keeping in mind the parameters like cutting temperature, tool wear, surface roughness and dimensional deviations. Experimentally it is found that there is a decrease in the temperature at the chip–tool interface by 10% at various cutting conditions [7].

Sharif et al. [8] investigated the use of various cutting fluids (fatty alcohol, palm oil, palm oil with additive A and palm oil with additive B) as lubricants during the end milling operation of hardened stainless steel (AISI 420) using coated carbide tool materials (TiA1N and A1TiN). Using various input parameters like cutting speed, feed rate, axial depth of cut and radial depth of cut along with the end mill diameter, he has experimentally studied its effect on Cutting forces, tool life and surface roughness. Observations revealed that the progress in the tool wear is gradual during the use of palm oil and fatty alcohol, but rapidly progressed when dry and flood type of lubrication is used. It is also observed that the flank wear progressed rapidly with machining time for dry and flood cutting, while the tool wear increased very slowly when palm oil and fatty alcohol coolant is used. It is also observed that the surface properties were not good initially with the use of palm oil and fatty alcohol which happened to improve in the later stages.

End milling process is widely being used in various industries like automobile, ship building. aerospace, textile and other manufacturing industries due to its varied applications. The present work encompasses the use of MoS_2 particles in vegetable oils in order to check its machining performance with a special focus on the heat generated at the primary zone. MoS_2 is characterized by its high temperature thermo physical property. Experiments are carried out on CNC Milling machine by using MoS_2 as solid lubricant mixed in vegetable oil.

2. MATERIALS AND METHODS

Machining experiments are conducted on *7011H ALUMINIUM* steel with the application of nano crystalline MoS_2 powder as a lubricant. Nano crystalline MoS_2 powder with size ranges of 30-35 nm, 65-70 nm and 85-90 nm are used. CNC Milling machine is used for the purpose of machining. The experiments are conducted using 6mm diameter HSS material Type-A, Type-N, Flat ball end mill bit. The input parameters which are varied are speed, depth of cut, feed and particle size of the nano crystalline powder. Full Factorial Design of experiments are conducted to analyze the output parameters- Temperature and Surface Roughness.

3. EXPERIMENTATION

Initial experimentations involved the finding of minimum quantity lubrication, and later experimentations were conducted. The temperature was measured during the machining process using a Infra-red thermometer. Surface roughness is measured off line after experimentation using a portable stylus-type TALYSURFF. The evaluation length of 2.5mm is used to measure response Ra value in µm. The complete methodology adopted in carrying out the experimental work is illustrated in Fig. 3.1. An indigenously developed liquid particulate dispensing system is attached to the machine for dispersing the lubricant mixture and to take care of agglomeration of the MoS_2 powder mixed in the carrying unit. A CNC Milling Machine is used for the experimentation. Specifications of the Milling Machine are as follows:

(i)	Make:	CNC MILL HYTECH
(ii)	Rapid feed rate in each axes:	600mm/min
(iii)	Minimum increment:	0.005mm
(iv)	Milling head : Power - 2HP Spindle motor - DC Motor, 2HP	Spindle speed - 200-3000rpm

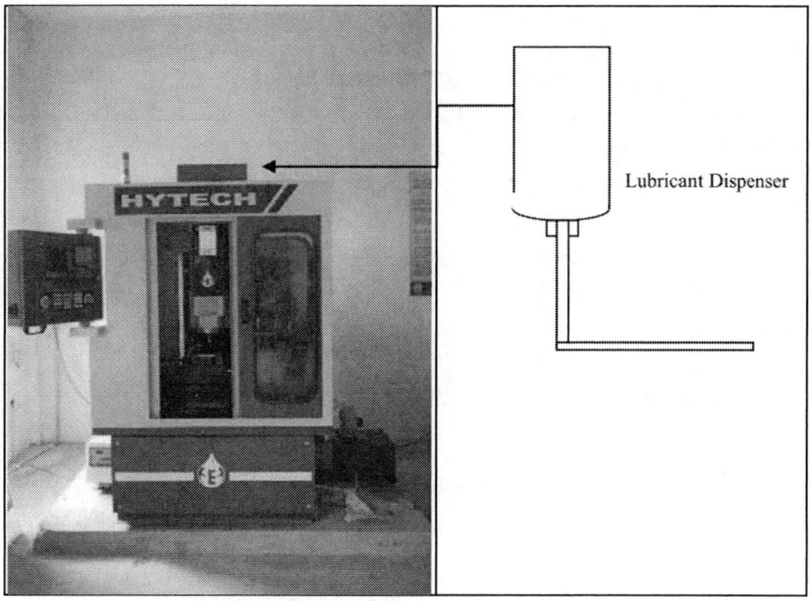

Fig. 1: Experimental setup

The behavior of MoS_2 powder in the nano-crystalline form as a lubricant which is mixed in a carrying medium Canola is studied. The size of the nano-crystalline MoS_2 powder considered are 30-35 nm, 65-70 nm and 85-90 nm. Preliminary tests are conducted to find out the optimal weight percentage of the nano-crystalline MoS_2 powder to be mixed in the carrying medium. It is found that a 0.5 wt. % is optimal for the purpose of experimentation. The evaluating parameters of temperature and the surface roughness are considered while machining Aluminium. The temperatures are measured while the experimentation is being carried out and the surface roughness is found out off-line at the end of each machining process.

4. RESULTS & DISCUSSIONS

To compare the performance of the measured parameters, a base line is followed in which the experimental values are plotted against the cutting speed for all the sizes of the nano-crystalline MoS_2 powder. Machining is performed for 6 minutes for each sample. The flow rate of the lubricant is maintained at 5 ml./min. The values plotted are the average of the values obtained while machining for 6 minutes. Fig. 2 represents the values with respect to Tool temperature.

The affect of temperature due to the variation of crystalline size of the lubricant used is represented in Fig. 2.The temperature is represented on the y-axis and the change in the cutting speed along the x-axis.

Fig. 2 (a) illustrates the change in Tool temperature values at a depth of cut of 0.50 mm and feed rate of 50 mm/min. while Fig. 2 (b) and Fig. 2 (c) represents the Tool temperature at a feed rate of 70 mm/min and 90 mm/min respectively. It is observed that the Tool temperature at 0.50 mm depth of cut and if the feed rate is changed from 50 to 90 mm/rev., it increases from 38 °C to 74.5 °C for a crystalline size of 85-90 nm when the speed is increased from 1000 to 2000 rpm. This clearly shows that the Tool temperature changes with the increase in the cutting velocity and change in the depth of cut.

`At higher cutting speeds and when the depth of cut is increased, there is a drastic increase in the Tool temperatures. The temperatures increases to a maximum of 175 °C for a crystalline size of 85-90 nm and 210 °C for a crystalline size of 30-35 nm. This clearly shows that with the increase in the size of nano-crystalline MoS_2, there is a decrease in the Tool temperature. While comparing the average increase in the Tool temperature with respect to the crystalline size of the MoS_2 particles, the increase is 1.09 times of the average temperature when using 30-35 nm crystalline particles than by using 85-90 nm nano-crystalline particles. The temperature developed is the maximum when the smallest crystalline particles (30-35 nm) are used. The nano sized particles that enter into the contact interface between the tool and the work piece gets agglomerated. At higher temperatures, Canola oil loses its viscosity and only MoS_2 particles remain in the contact interface. Since, MoS_2 powder has high friction coefficient than that of Canola oil, the friction between the chip-tool interface increases thus increasing the temperature.

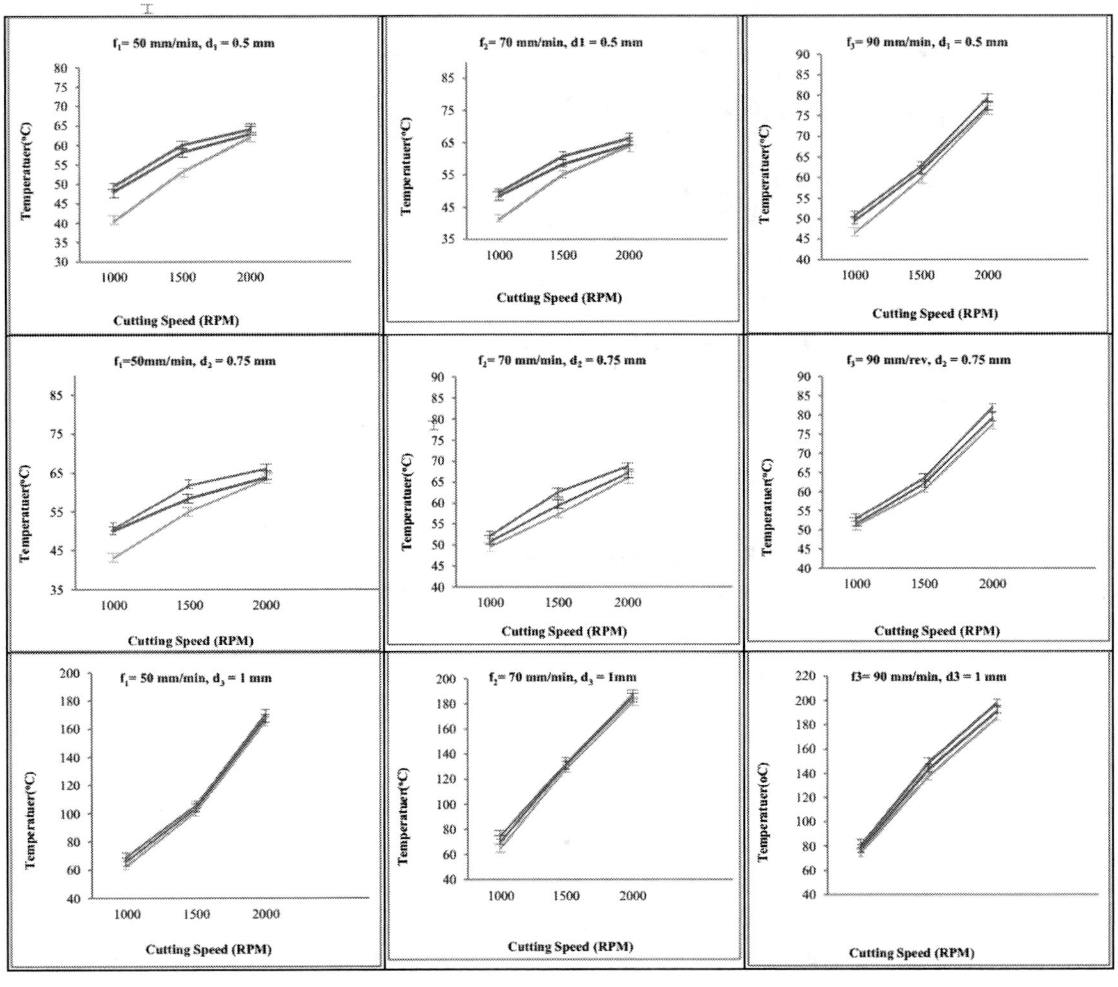

Fig. 2: Graphs related to Temperature

Agglomeration of nano particles at the interfacing zone of the tool and work piece increases the tool temperature, thus increasing the temperature at the cutting zone. This shows that with an increase in cutting speed, the temperature increase is drastic. The effect of cutting speed on Tool temperature is predominant when compared to the other parameters like depth of cut and feed rate. At higher cutting speeds and at higher depth of cuts, more amount of material comes in contact with the tool thus leading to stagnation of the tiny nano particles between the asperities. This in turn impairs the heat transfer mechanism from the cutting zone leading to increase in temperature. The increase in temperature in turn decreases the viscosity of the carrying medium thus leading to further increase in temperature at the cutting zone.

IOP Conf. Series: Materials Science and Engineering **998** (2020) 012021 doi:10.1088/1757-899X/998/1/012021

Fig.3 depicts the values of the Surface roughness which are experimentally found out. The graphs represent the values at a particular cutting speed with various crystalline sizes of the solid lubricant. It is to be noted that these values are obtained after the machining process is carried out. When considered at a particular speed, the quality of the surface is better when 85-90 nm sized crystalline MoS_2 powder is used when being compared to smaller crystalline sizes of the solid lubricant. It is observed from the experimental values that the quality of surface is poor when smaller crystalline nano sized solid lubricant is used and better when large sized nano particles are used.

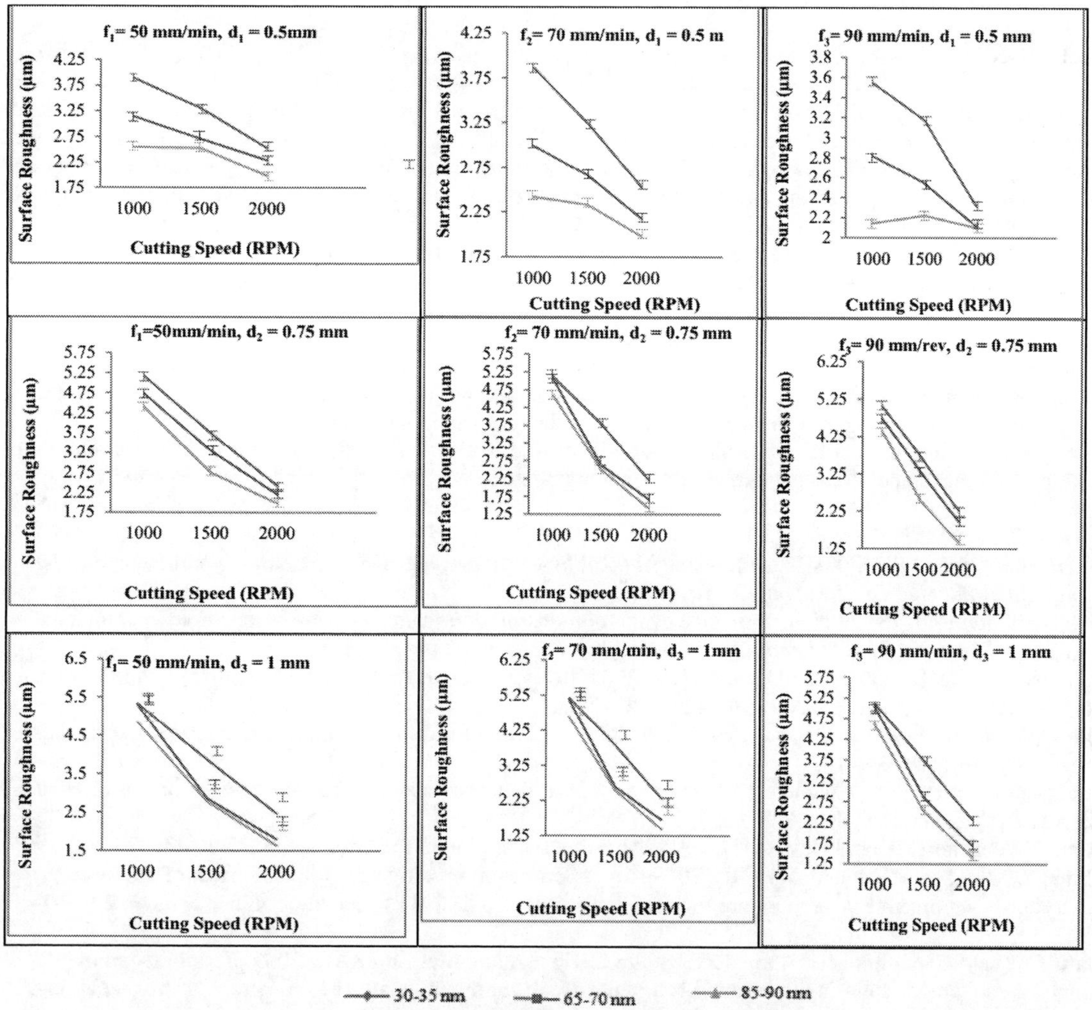

Fig. 3: Graphs related to Surface Roughness

It is observed that built up edges are formed when the cutting speeds are lower which is the reason for poor surface quality. These built up edges disappear when the cutting speeds are increased. This phenomenon is visible in the experimentation process where it is found out that the surface quality improves when the cutting speeds are increased. There is a proven work where the quality of surface finish depends on the size of the particle when the particle size used ranges from micron level to the nano level. It is observed during the experimentation that surface finish is better during the use of larger particle size in the nano regime which is contrary to the experimental findings when micron sized particles are used.

The surface quality deteriorates more with the increase in the feed rate which is a dominating factor when compared to other cutting parameters considered for the experimentation. Experimentally it is a proven fact that the quality of surface depends on the feed. It is also proven that the quality of surface, which is a function of feed at a particular nose radius, which changes with the square of the feed rate value. At higher cutting speeds, the built up edges disappear and there is a direct contact between the tool and the work piece with the lubricating layer between the machining parts. The nano particles act as good thermal conductors which dissipates heat from the cutting zone. As the built up edges disappear, the surface quality improves due to the fact that the nano particles act as lubricants between the machining components.

CONCLUSIONS

The present experimental investigations considering the input parameters like feed rate, cutting speed, depth of cut and size of the nano-level MoS_2 powder. It is experimentally proved that the said input parameters greatly influence the output factors like the tool temperature and the quality of the machined surface. The results obtained experimentally can be used for developing analytical models which can be used to predict the surface quality and the tool temperature. The data obtained gives a scope for analyzing the effect of nano-level variation in the particle size of MoS_2 powder while machining Al.

i. It is observed that the tool temperature is depends on the size of the particle. As the particle size decreases, the tool temperature increases. It implies that more heat is generated at the tool chip interface as a result of greater rubbing action at the interface.
ii. The other observation is with respect to the surface quality. The experimental investigation revealed an inverse phenomenon with respect to the change in the particle size at the nano level of MoS_2 powder during the machining process of Al. This phenomenon can be because of the increase in the cutting forces, higher tool temperature and greater physical interaction at the tool – chip interface.

It can be concluded that the particle size plays a dominant role in the nano regime with respect to the quality of surface finish and the temperatures generated in the machining zone.

REFERENCES

[1] Rao DN and Krishna PV, 2008, The influence of solid lubricant particle size on machining parameters in turning. *Int J Machine Tools Manuf*; **48**: 107–111.
[2] Krishna VP and Rao DN, 2008 Performance evaluation of solid lubricants in terms of machining parameters in turning. *Int J Machine Tools Manuf*; **48**: 1131–1137.
[3] Mukhopadhyay D, Banerjee S and Reddy SKN, 2007, Investigation to study the applicability of solid lubricant in turning AISI 1040 steel. *Trans ASME*, **129**: 520–526.
[4] Cakir O, Yardimeden A and Ozben T, 2015, Selection of cutting fluids in machining processes. *J Achievements Mater Manuf Eng* , **25**: 99–102
[5] Wang B and Tao D, 2005, Characteristic study of biodegradable soybean oil. *In: Proceedings of world tribology congress III*, Washington, DC, **12**(16):367–371.
[6] Inasaki I, Toenshoff HK and Howes TD, 1993, Abrasive machining in the future, *Ann CIRP*, **42**: 723–732.
[7] Rahman MM, Khan MMA and Dhar NR, 2009, An experimental investigation into the effect of minimum Quality lubricant on cutting temperature for machine ability of AISI 9310 steel alloy. *Eur J Sci Res*; **29**: 502–508.
[8] Sharif S, Yusof NM, Idris MH, Ahmad ZB, Sudin I, Ripin A and Mat Zin MAH, 2009, Feasibility study of using vegetable oil as a cutting lubricant through the use of minimum quantity lubrication during machining, Research vote No 78055.

IOP Conf. Series: Materials Science and Engineering 998 (2020) 012022 doi:10.1088/1757-899X/998/1/012022

Experimental investigation of free connective heat transfer augmentation using transformer oil-Al_2O_3 nanofluid

S Ravi Babu[1]* A Saras Chandra[1] P Ramesh Babu[2]
[1] Department of Power Engineering, GMR Institute of Technology, Rajam, India.
[2] Department of Mechanical Engineering, SITE, Tadepalligudem, India.

E-mail: ravibabu.s@gmrit.edu.in

Abstract. There is a definite need to improve the heat transfer capability of the transformer oil as it plays a vital role in the life span of transformers. As the thermal conductivity of the transformer oil is very less compared to the conventional liquids and those liquids cannot be used as dielectric liquids as they do not possess electrical insulation properties. It is very much essential to improve the thermal conductivity of the transformer oil in any manner in order to augment the "heat transfer or thermal management performance" of the transformer oil. Dispersion of metallic or metallic oxide nanoparticles into the base fluid in a systematic way is a proven method to enhance the thermal properties for various practical heat transfer applications like forced convection, boiling, refrigeration, solar and electronic cooling. In this study, alumina nanoparticles are dispersed in transformer oil to synthesize stable, homogenous transformer oil – Al_2O_3 nanofluid and to investigate the free convection "coefficient of heat transfer performance" along a vertical slender cylinder submerged in the liquid. Experiments are conducted using the transformer oil - Al_2O_3 nanofluid at different particle loadings or fractions starting from 0.05 % to 0.6 %. It is learnt that the highest "coefficient of heat transfer performance" occurs at 0.1 Vol% among all the volume fractions. It has been acknowledged that the "coefficient of heat transfer" is got better up to 0.1 % particle loading and then decreased with additional accumulation of nanoparticles.

1. Introduction

Transformer oil is a protecting oil and it possesses great electrical shielding insulating properties, cools transformers and stable even at significantly at higher fluid temperatures. So, it must possess "high dielectric strength", "chemical stability", "thermal conductivity k", "viscosity μ" and very much importantly it must keep its thermodynamic and electrical properties for longer spans of time and at higher temperatures [1, 2]. In general, for practical transformer applications purified form of mineral oil will be used as a transformer oil. In practical, due to higher fluid temperatures, after consuming it for some time the transformer oil loses its properties [3, 9]. If the transformer oil loses its properties, thermal management or heat transfer from transformer will not happened effectively and this leads to generate excessive heat inside transformer and eventually causes the failure of it. For several years ago, researchers tried to mix micro sized particles and observed more settling and then they tried mixing nano sized particles in carrier fluid in order to improve their properties in addition to retain them stable for several days at constant heat flux without settling. Then after researchers have tried to synthesize transformer oil-based nanofluids (TONF) by dispersing the nano sized particles in the transformer oil [9].

2. Literature Review

Asefi et.al, [2] synthesized Transformer oil-diamond nanofluids and observed fluid properties for various concentrations of nanoparticles and observed that at 0.15 vol%, 38% enhancement in breakdown strength and 73% increase in thermal conductivity. Hwang et.al.,[3] equipped TONF nanofluids and compared the electrical breakdown strength with pure oil and observed the enhancement. Lv et.al.,[4], Rafiq et.al.,[7] reviewed various studies based on transformer oil TONF nanofluids and electrical insulation properties. Wang et.al.,[5] prepared three different types of TONF nanofluids and compared the effect of nanoparticle concentration on electrical properties. Du et.al,[6] prepared transformer oil-boron nitride and observed electrical properties and thermal properties and observed enhancement. Dong et.al.,[8] analyzed the "breakdown or dielectric strength of transformer oil-SiO_2 nanofluids" and witnessed the improvement in augmentation of "breakdown or dielectric strength" up to 0.06 vol% and later it decreases. In author's previous works [9] transformer oil based Al_2O_3 nanofluids is synthesized and electrical insulation properties are measured using TAGUCHI design of experiments for finding optimum parameters and observed that at 0.1 vol% enhancement in breakdown strength is more also heat transfer performance is analyzed experimentally using water based Al_2O_3 nanofluid. From the literature, it is observed that very few works are done on the "natural convection heat transfer performance" using transformer oil based Al_2O_3 nanofluid. In the present work, transformer oil based Al_2O_3 nanofluid is prepared and experiments are conducted to determine the "natural convective thermal or heat transfer performance" as it is very much importance in case of transformer applications [10].

3. Synthesis and Characterization of Nanofluids

Nanofluids are prepared using two step method [11] by taking transformer oil as base fluid and Al_2O_3 nanoparticles are dispersed in the required proportion. The quantity of Al_2O_3 nanoparticles to be added can be determined from the eq., (1) for a particular volume fraction. Initially after adding nanoparticles it is stirred for 1 hour using magnetic stirrer and then it is sonicated using ultrasonic sonicator for 3 hours with a model like 15 minutes sonication and then 15 minutes break alternatively to avoid the excess temperature in the nanofluid. Thermo-physical properties of transformer oil and Al_2O_3 nanoparticles at room temperature is specified in table 1 and 2 respectively.

$$\text{Volume fraction} = \frac{v_{np}}{v_{np}+v_{bf}} = \frac{(m_{np}/\rho_{np})}{(m_{np}/\rho_{np})+(m_{bf}/\rho_{bf})} \tag{1}$$

Table 1. Thermo-physical properties of Transformer oil [11]

Base fluid	Density ρ (Kg/m³)	Thermal conductivity k (W/mK)	Specific heat Cp (J/kgK)	Thermal expansion coefficient β (K^{-1})	Dynamic viscosity μ (Ns/m^2)
Transformer oil	873	0.11	1982	0.78×10^{-3}	13.44×10^{-3}

Table 2. Thermo-physical properties of Al_2O_3 nanoparticles at room temp. [11]

Thermal conductivity k (W/mK)	Density ρ (kg/m^3)	Specific heat Cp (J/kgK)	Thermal expansion coefficient β (K^{-1})
38	3970	765	$0.85\ 10^{-5}$

4. Experimental Setup

Experimental setup is equipped to conduct the experiments to determine the heat transfer or heat transfer coefficient for various test fluids. In earlier experiments, authors conducted experiments using water, water + Ethylene glycol (3:1 ratio) as base fluids and as nanofluids with the addition of Al_2O_3 nanoparticles in the required quantity based on the volume fraction to determine the heat transfer performance [10]. Photograph and schematic diagram of the experimental arrangement are shown in Fig.1. Physical dimensions of the set-up are also given in Fig.1. Surface area of the vertical cylinder is 0.0099 m^2 and l/d ratio is 19.68. In order to get the steady state conditions a shell with cooling water circulation arrangement. K-type thermocouples are used in order to measure the temperature on the surface of vertical cylinder. Before conducting experiments all thermocouples are calibrated and fans are switched off to ensure that there is no movement of fluid due to external agencies. Power supply is given to the heater, which is present exactly inside the vertical cylinder without any gap in between them. MgO is filled in between heater and cylinder in order to ensure that there is no heat loss between them as it is having high thermal conductivity. At a particular heat input, flow rate of cooling water is adjusted such that the heat is received by the test fluid is taken away by cooling circulated water in order to get the steady state condition. Temperature readings are noted down after steady state conditions prevailed. The cold water temperature is maintained constant at 25°C.

For the heat inputs or heat fluxes given to the cylinder from 3030 W/m^2 to 5050 W/m^2, the "flow rate of cooling circulated water is wide-ranging from 1.43 lpm to 2.23 lpm. At conditions of steady state, the amount of heat input given to the cooling water circulation will carry the metallic cylinder away and thereby it is assumed that steady state condition is attained" [10]. When steady state conditions are prevailed, temperature readings of the surface of vertical cylinder are noted down. Experimental setup is validated by taking water as test fluid, determined the average Nusselt number, and compared with McAdam's equation [10].

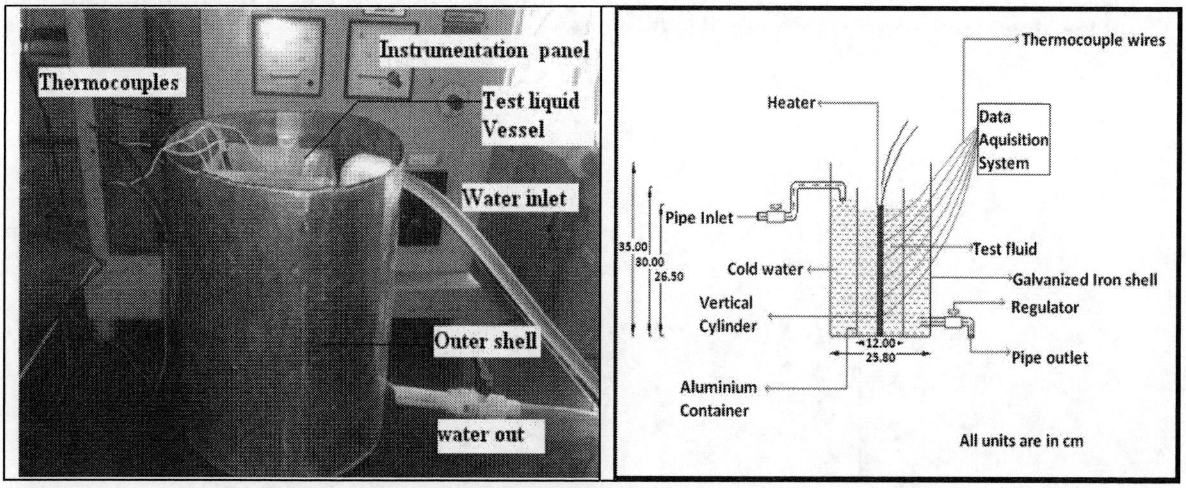

Fig.1 Photograph and Schematic diagram of experimental setup [10]

5. Results and Discussion

Experiments are carried out at various heat input given to the heater ranging from 30 W to 50 W. Surface temperatures are noted down after steady state conditions prevailed. Initially experiments are performed transformer only as a test fluid for different heat inputs and then with transformer oil – Al_2O_3 nanofluid at various concentrations of nanoparticles and at various heat inputs. Variation in surface temperatures of cylinder along axial directions is noted and presented in Fig.2.

It is observed that in both cases the trend is similar but in case of nanofluid, surface temperature is less compared to pure transformer oil. Test fluid temperature is varied with heat input and temperature difference of cylinder and test fluid at middle of cylinder is shown in fig.3. Reason for this could be reduction in temperature is due to more heat from the cylinder is carried by the nanofluid. Using eq (2) "local heat transfer coefficient" for thermal management performance is calculated for both base fluid and transformer oil – Al_2O_3 nanofluid and presented in Fig.4. Where ΔT_x is the temperature difference between a point on cylinder and a point in the fluid at same level outside the boundary layer formation.

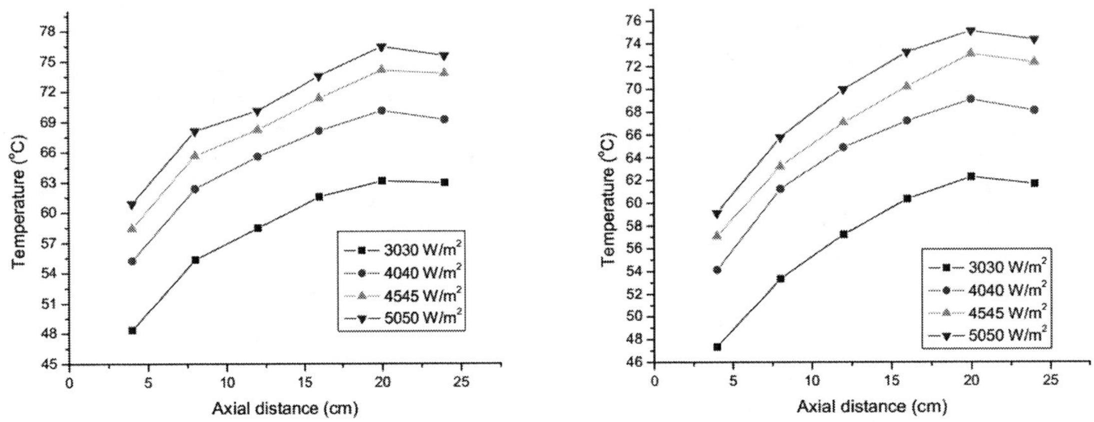

(a) (b)

Fig.2.Variation in "surface temperatures of vertical cylinder" along its length for (a) transformer oil (b) transformer oil – alumina nanofluid at 0.1 vol%.

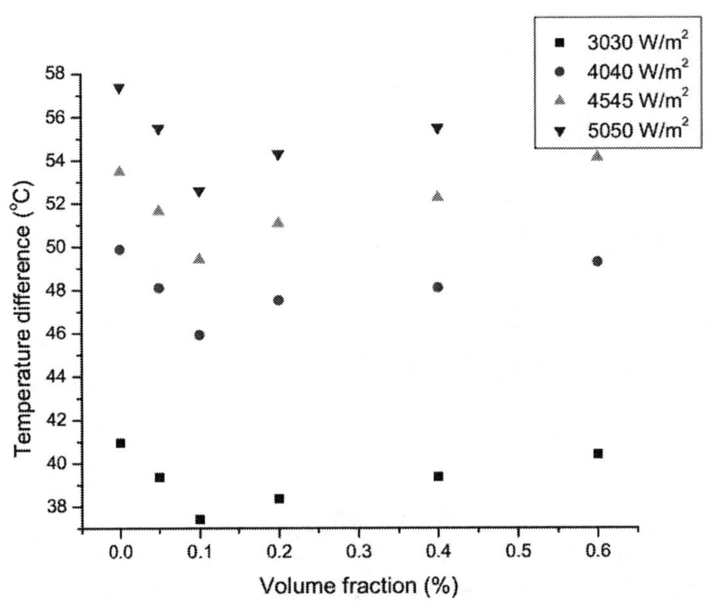

Fig.3 Temperature difference of cylinder and test fluid at middle of cylinder

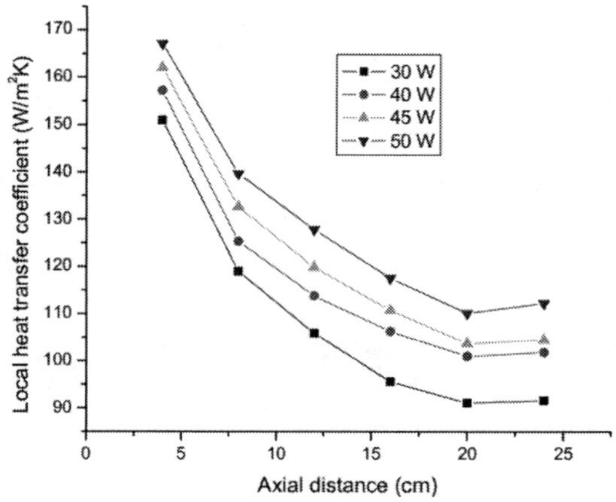

(a)

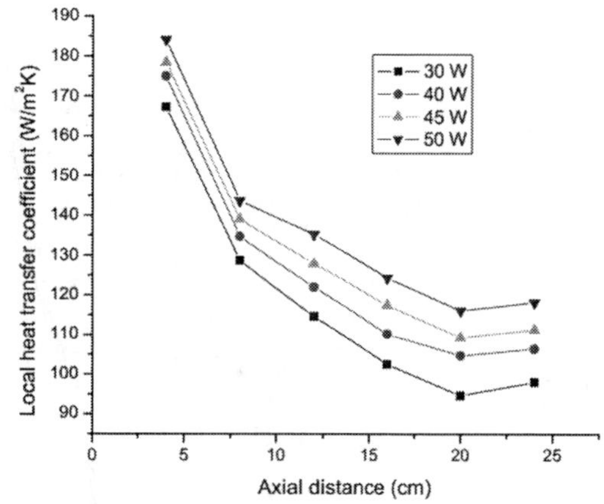

(b)

Fig.4 Local heat transfer coefficient for (a) transformer oil (b) transformer oil - Al_2O_3 as medium as medium

$$h_x = \frac{Q}{A(\Delta T_x)} \quad (2)$$

It is also identified that at the lower part of the cylinder "coefficient of local heat transfer performance" is high and low at the higher cylindrical parts. It is happened because of, at the bottom portion of cylinder the thickness of the boundary layer is very small and more thickness at the higher parts of the cylinder [10]. All thermos-physical properties are measured at average temperature of surface temperature and fluid temperature i.e., film temperature. At this film temperature and at various particle loadings of nanoparticles thermal conductivity k, viscosity μ, density ρ are measured. Volumetric expansion coefficient β is determined from the model taken from eq. (3) [1].

$$\beta_{nf} = (1-\varphi)\,\beta_f + \varphi\,\beta_{np} \quad (3)$$

Average "coefficient of local heat transfer performance" of transformer oil based alumina nanofluids is determined and its variation with alumina nanoparticle loading for various heat inputs is depicted in fig.5. It has been acknowledged that "average heat transfer performance coefficient" is amplified up to 0.1 vol% concentration of nanoparticles and then with additional accumulation of nanoparticles starts diminishing. The maximum improvement observed in the natural convective "coefficient of local heat transfer performance" is 8.7 % as "coefficient of heat transfer" is augmented from 127 W/m^{2}K to 138 W/m^{2}K at 0.1 % volume of particle loading for transformer oil - alumina nanofluids. The enrichment in the thermal performance of the transformer oil based alumina nanofluid occurs because, thermal conductivity dominates at lower concentrations than viscosity of nanofluid and at higher concentrations, and viscosity plays significant role rather than thermal conductivity [10].

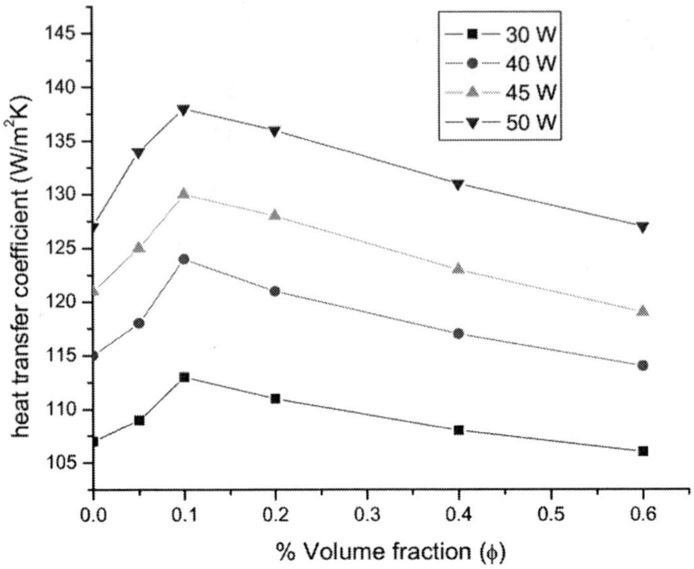

Fig.5 Variation of "average heat transfer coefficient" with particle loading of alumina nanoparticles for various heat inputs.

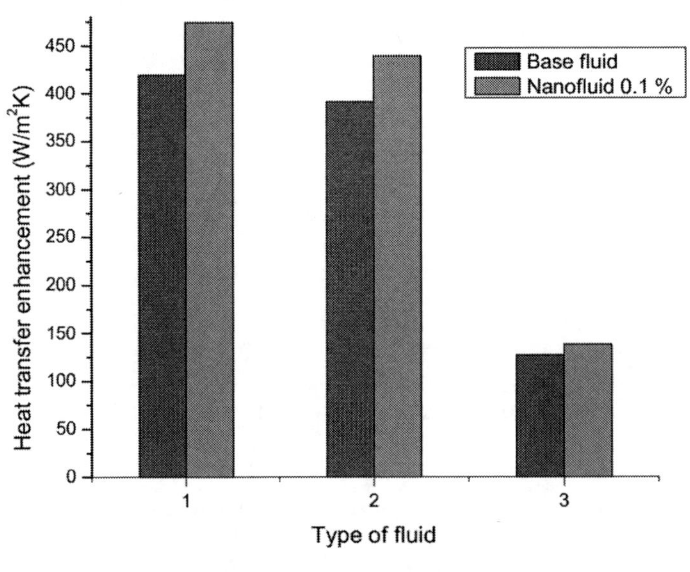

Fig. 6 Heat transfer enhancement for (1) water (2) water + EG (3) Transfer oil based Al_2O_3 nanofluids

But the enhancement in heat transfer coefficient is less compared to water- Al_2O_3 nanofluid because of high viscosity of the transformer oil. Due to this high viscosity, nanoparticles do not move fast in case of transformer oil unlike water. Heat transfer augmentation is compared for water, water + EG,

Transformer oil based Al_2O_3 nanofluids at 0.1 % volume fraction and presented in fig. 6. In case of water- Al_2O_3 nanofluid the augmentation is 13.7%, water +Ethylene glycol (3:1 ratio) is 11.5% [10], in transformer oil - Al_2O_3 nanofluid it is observed as 8.7%.

6. Conclusions

In this work, transformer oil based Al_2O_3 nanofluids are synthesized using two step method for conducting the experiments for determining the performance of thermal management of the prepared nanofluids. Alumina nano sized particles are dispersed in transformer oil to synthesize stable, homogenous transformer oil – Al_2O_3 nanofluid and to investigate the free convection "heat transfer performance" in terms of "heat transfer coefficient" of a vertical cylinder submerged in the test liquid. Experiments are conducted using the transformer oil - Al_2O_3 nanofluid at various particle loadings of nanoparticles starting from 0.05 % to 0.6 %. It is perceived that the highest "coefficient of heat transfer" occurs at 0.1% volume particle loading among other volume fractions. It is also witnessed that the "heat transfer coefficient" is amplified up to 0.1 % volume concentration and then decreased with extra accumulation of alumina nano sized particles. Performance of water, water + EG, transformer oil based Al_2O_3 nanofluids is compared and presented.

References

[1] Agwu Nnanna A G 2007 *ASME J. heat transfer* **129** 697
[2] Asefi M et al 2016 *IJE Transactions A: Basics* **29** 10 1332
[3] Hwang J G et al 2010 *J. Appl. Phy.* **107** 014310 1
[4] Lv Y Z et al 2014 *IEEE Electrical Insulation magazine* **30** 5 23
[5] Wang Qi et al 2016 *J. Nanotechnology* 5802753 1
[6] Du B X et al 2015 *IEEE Transactions on Dielectrics and Electrical Insulation* **22** 2 851
[7] Rafiq M et al 2016 *J. Nanomaterials* 8371560 1-23
[8] Dong M et al 2017 *AIP Advances* **7** 025307 1-8
[9] Babu S R and Rao G S 2018 *AIP. Conf. Proc.* 020075 1-7
[10] Babu S R and Rao G S 2018 *J. Therm. Sci. Engg. Appl.* **10** 031005 1-7
[11] American Society of Heating, Refrigeration and Air-Conditioning Engineers (ASHRAE) Handbook, Fundamentals, Inc, Atlanta, 2015.

Dynamic Analysis of High Speed Spindle

Prasad V V S H[1,2] and Kamala V[3]

[1] Associate Professor, Department of Mechanical Engineering, Institute Of Aeronautical Engineering, Hyderabad, India – 500043

[2] Corresponding author email address: visuvaranasi@gmail.com

[3] Retd. Deputy General Manager, Bharat Heavy Electricals Limited- Corporate R&D Division, Hyderabad, India – 500037

Abstract. Assessment of the thermo-mechanical behaviour of machine-tool spindles is a requisite for the reliable operation of high-speed CNC machine tools. The thermal behaviour and applied loads affect the performance of these high-speed spindles. The primary source of heat generation in the spindle is due to friction torque in angular contact ball bearings, cutting forces, centrifugal force, gyratory effects and bearing preload. The spindle shaft is treated as beam element with one translational and one rotational degree-of-freedom at each node. Angular contact ball bearings are modelled to predict heat generation due to load, viscous, and spin-related effects. Mutual interlinkage between the thermal and structural behaviour of both spindle shaft and bearings is modelled using the effects caused due to the thermal expansion and rate of heat generation. The bearing preload is pushed up due to centrifugal force and temperature rise due to high spindle speed. Components matrices are assembled to configure a finite element model for the thermo-mechanical analysis of spindle-bearing systems using Euler-Bernoulli beam theory. Further this work presents ANSYS simulations on CNC spindle system with angular contact ball bearings using 103 Cr1steel material and hybrid type to predict thermal stress. Static and dynamic analysis is carried out on the spindle with and without thermal load and cutting load to predict the displacement and slope.

Keywords: CNC spindle, thermo-mechanical, finite element model, gyratory effects, Hybrid bearing.

1. Introduction

High speed machining (HSS) is a promising advanced manufacturing technology for increasing productivity and reducing production costs dramatically. High speed machine tool and its spindle rigidity are crucial for high speed machining which is of considerable significance in the fields of research and development.

The spindle is the principal mechanical rotating component in milling /turning centers. The machine productivity and finish quality of the workpiece depends upon the static and dynamic stiffness of the spindle. The constitutional properties of the spindle are based on the geometry of the shaft, bearings, motor, type of tool holder, and the design configuration of the spindle assembly. The bearing arrangements are governed by type of cutting operation, applied cutting load, spindle stiffness and limiting speed of the bearings. The main focus of HSS design is to attain long-term precision and high machining productivity for minimizing the operating costs of the machine. In high speed machine spindle assembly's bearings plays a major role when it comes to high speeds. Generally angular contact ball bearings are used in spindle assemblies as they can take both axial loads and radial loads. Hence a study is made in this paper on the behaviour of these angular contact ball bearings at different speeds and also its effects on the spindle and machining. Forces acting on the high speed spindle during HSM are shown in the Figure 1.

The machine tool spindle is subjected to combined torsion, bending, axial, and impact loads during metal removal work cycles producing reaction forces at bearing supports and deflection at the nose end. In order to provide structural stability, the spindle design and bearing selection are critical to producing high-quality components and also for reliable operation. Today CNC machines are used for machining different work materials and operations at high cutting speeds and feed rates to achieve enhanced performance. At this very high spindle speeds, a large amount of heat generates due to the cutting force and viscous spin torque, which leads to excessive thermal preload on the bearings inducing stresses and deformation.

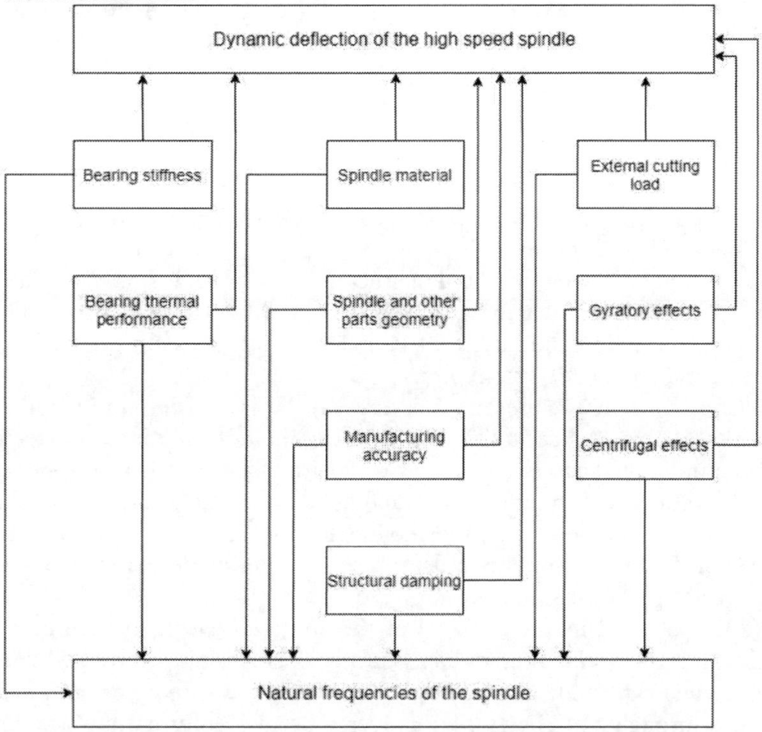

Figure 1: Factors influencing spindle bearing system (SBS)

2. Nomenclature

l	Beam elemental length	H_b	Heat generated in bearings (W)
E	Modulus of elasticity of Spindle material (N/mm^2)	n	rotational speed in RPM
		M_f	Total frictional torque (N-mm)
F	Cutting force (N)	M_1	Frictional torque because of applied load (N-mm)
K	Stiffness matrix (N/mm)		
C_s	Static equivalent load rating (from SKF table)	M_2	Viscous torque owing to lubricant(N-mm)
d_m	Mean diameter of bearing (mm)	M_s	Spin torque (N-mm)
F_r	Radial force acting on bearing (N)	f_1	Factor based on bearing design and

IOP Conf. Series: Materials Science and Engineering **998** (2020) 012023 doi:10.1088/1757-899X/998/1/012023

F_a	Axial force acting on bearing (N)		relative load
X_0, Y_0	Bearing coefficient values in a tandem arrangement (from SKF table)	Z	Factor for angular contact ball bearing with 25° attitude angle (from SKF table)
F_{th}	Thermal preload	Y	the factor for angular contact ball bearing with 25° attitude angle (from SKF table)
ϵ_t	Thermal displacement (μm)		
K_r	Bearing elastic constant	F_β	Dynamic equivalent load (N)
I	Mass moment of inertia (mm^4)	F_s	Static equivalent load (N)
n_1	Constant based on the type of bearing, 1.5 for angular contact ball bearings	v_0	Kinematic Viscosity (centi-Poise)
		θ	Bearing attitude angle
ϵ	Elliptical area integration	F_0	Factor-based on type of bearing
M	Mass matrix	μ	Coefficient of friction
d	Global displacement vector	a	Major axis of the elliptical contact zone in bearing
HSS	High speed spindle		
E	Young's Modulus	HSM	High speed machining
v	Poisson's Ratio		

3. LITERATURE SURVEY:

Abele et al.(2010) presented in the review article, a hollow spindle shaft is preferred to meet the objectives of the machine tool spindle and mounted on angular contact ball bearing in headstock housing. The angular contact ball bearings possess both axial and radial stiffness at high external load at high rotational speeds. Bearing models are used for obtaining preload, and load due to thermal expansion during the running of the spindle.

Lin et al.(2013) presented review on the dynamic analysis of high speed spindle and discussed the governing equation of motion which includes gyratory and centrifugal force and its effects on the spindle performance. The hollow spindle shaft is analyzed by modeling in the FEA environment by using element models namely beam/ brick/ pipe.

Arotaritei and Constantin(2016) presented review on the modelling and simulation of high speed spindle and dynamics of spindle bearing system to study the deformations and temperature profile in the assembly. The spindle analysis is conducted using Euler-Bernoulli beam theory with gyroscopic and centrifugal effects along with spindle inertia and bending moment in the governing equation.

Long and Su(2017) conducted virtual orthogonal simulation for optimization in the ANSYS software for a CNC machine tool spindle to determine the spindle deflections using input variables as size of the spindle ,bearing stiffness and cutting force.

Alfares et al.(2019) studied thermal effects of the spindle bearing system of a grinding machine wheel head mounted in the angular contact ball bearings. The simple steady and transient heat transfer thermal resistance network discretization is implemented for the spindle bearing system to predict the thermal deformations through MATLAB ODE solver. Thermal deformations are directly proportional to spindle speed and influencing the thermal preload in the dynamic model.

Liu et al.(2019) modelled the thermo –structural integration of the spindle bearing system by thermal contact resistance network. The heat generation in the bearings is predicted by combined viscous friction and equivalent external load due to thermal deformations with external load as well as interference contact at solid joints. Li and Shin(2004) presented the bearing arrangement has a major influence on spindle stiffness. The tandem arrangement of bearing results in a higher stiffness and natural frequency of the system compared to DF arrangement (X configuration).

Zahedi and Movahhedy(2012) developed a comprehensive model for the thermomechanical performance of high-speed spindle units considering bearings, shaft, headstock housing. The temperature variation along the length of the spindle is plotted at 10000 RPM. The highest temperature rise is 60°C in the bearing at 400 mm and 40°C on the housing. The bearing stiffness reduces as the spindle speed increases.

Uhlmann and Hu(2012) performed a Transient heat transfer analysis is carried at an ambient temperature of 19°C to predict the deformations w.r.t tool canter point in VMC at a spindle speed of 12000 RPM for a time period of 480 minutes. The drift of tool canter points is 2 μm, 10 μm and 50 μm in X, Y and Z axes respectively from the simulations. Experimental temperature measurements are carried out under similar operating conditions using thermal imaging and thermocouples. The experimental axes deformations are 5 μm, 10 μm and 50 μm in X, Y, and Z axes, respectively.

Sathiya Moorthy and Prabhu Raja(2014) presented a model for the heat generation in the bearings for low as well as for higher rotational speeds. The proposed model is validated with three sets of experimental results at various speeds. The analytical model has upper deviation of 20% when the rotational speed is less than 2000 RPM, while the deviation is reduced to 5% in the rotational speed range of 2000-11000 RPM. .

Lu et al.(2007b) studied the dynamic characteristics of the high-speed integral spindle in FEM using ANSYS software. The integral spindle is subjected to static and dynamic loads, including centrifugal force and gyroscopic moments at high rotational speeds. Using the standard bearing equations, the axial and radial stiffness is calculated for hybrid angular contact bearings with specification HCB 7011 and HCB7008CDGA/HCP4 in tandem arrangement using spacer and locknuts for initial bearing preloading. The bearings are life time grease lubricated. Experimental temperature measurements are verified with the ANSYS model at 2000, 4000, and 6000 RPM spindle speeds. Further, the thermal displacement at the spindle nose is 3.51, 11.11, and 11.02 μm in the X, Y and Z directions respectively.

Lu et al.(2007a) conducted the experimental work on the high speed spindle to measure the temperature raise at the front, and rear bearing which is found to be 60°C and 67°C respectively when the spindle speed varies from 5000-20000 RPM. Hence, the external cooling of the bearing is done.

Chen and Hsu(2003) implemented an auto regression dynamic thermal error model in place of the static regression model for prediction of thermal deformation of the rotating spindle. The new thermal model proposed is found to be accurate compared to the temperature-based model and validated experimentally.

Lin et al.(2003) presented their work on high-speed rotational effects and softening of the spindle due to centrifugal force. Theoretical and experimental studies are carried on high-speed spindle including gyroscopic, centrifugal and cutting loads. Angular contact ball bearings with hybrid material are used for high-speed spindles. The increased radial stiffness compensates for the spindle softening effect.

Choi and Lee(1997) investigated the thermal behavior of the spindle bearing system with a bevel gear located on the bearing span. Since the drive is taking place form the front bearing zone, the temperature rise is higher than rear bearing for 30 minutes operation. The experimental work on the B7020C and B7016C bearings is carried out to measure the temperatures using J type thermocouples up to spindle speeds of 3500 RPM. ANSYS simulation and experimental results are compared for validation. The heat generated for B7020C at 3500 RPM is 216 W for duration of 1200 s with a temperature rise of 46°C.

4. MODELLING OF HIGH-SPEED SPINDLE

In the present work, a CNC vertical machining Centre (VMC) model spark-E (ACE Micromatic) is taken for the study of the thermo-mechanical functioning of the high-speed spindle bearing system under static and dynamic conditions. The spindle is mounted in the headstock housing supported by DB (O configuration) bearing arrangement. The abutment dimensions and bearing specifications are shown in Table 1 and Figure 2 respectively. The SOLIDWORKS software version, 2016 (Dassault Systems(2016)), is used for the CAD drawings, presented in Figure 2and ANSYS work bench version 16.0 for analysis. The maximum operating spindle speed is up to 6000 RPM. The bearings are preloaded by mechanical locknut through cylindrical spacers under medium preload conditions.

The spindle material is EN36C which is case-carburized and hardened steel up to $58^{\pm 2}$ HRC. The standard bearing materials are as per SAE 103Cr1 series, which are used up to 6000 RPM. Hybrid bearings with ceramic balls and steel rings are used for higher spindle speeds beyond 6000 RPM for the better thermomechanical performance of the spindle bearing system.

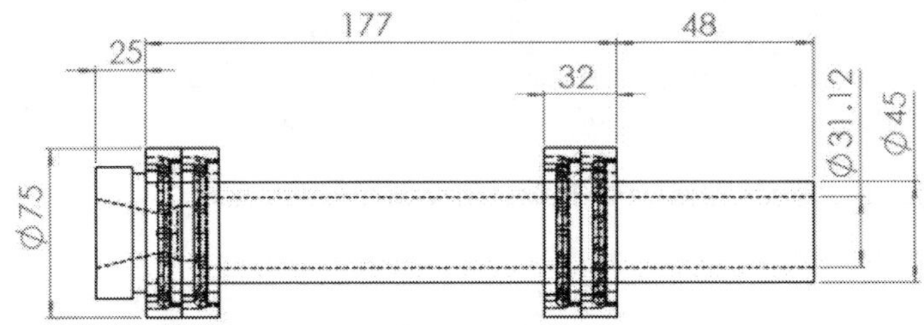

Figure 2: Geometrical dimensions of the Spindle

5. THERMAL MODELLING

The structural and thermal processes in a bearing are combined. This integration demonstrates itself in the thermal deformation of structural components and the change of thermal properties due to heat transfer through spindle assembly mechanical parts. The temperature at which a bearing operates based on factors, such as operating speeds, applied load, bearing arrangement, lubricant properties, housing design, and working environment. All these factors affect either heat generation or heat transfer. The thermal analysis includes heat generation and heat transfer through different modes within the bearing and spindle.

Table 1: Abutment dimensions and load ratings Schaeffler group(2008a)

Bearing Designation	Boundary dimensions (mm)					Abutment Dimensions (mm)			Basic load rating (kN)		Preload (N)	Number of balls	Ball diameter(mm)
									Dynamic	Static			
B7009ETP4S	45	75	16	1	1	53.6	64.2	0.6	26.5	20.0	202	21	7.144
HCB7009ETP4S	45	75	16	1	1	53.6	64.2	0.6	18.0	14.0	90	21	7.144

Table 2: Material property CMTI Machine tool design handbook(2015)

Component	Material	ρ (Kg/m^3)	K (W/m K)	E (GPa)	ν	α ($\times 10^{-6}$)	σ_Y (MPa)
Spindle	21Cr1Mo28	7870	45	210	0.3	11.9	770
Bearing race	SAE103Cr1	7860	46	208	0.3	11.9	2500
Steel bearing ball	SAE103Cr1	7860	46	208	0.3	11.9	2500
Ceramic bearing ball	Si_3N_4	3200	32	314	0.27	3.2	700
Free cutting steel	14Mn1S14	7750	45	210	0.3	11.9	490

The heat generated in the steel bearing is estimated due to the effects of cutting load, viscous friction and spin friction.

5.1. Heat generation due to load:

The primary heat generation of the system is mainly caused by two sources, i.e. cutting load and friction between balls and races of bearings. The heat generated in bearings is the primary cause of temperature change. The heat generated in the bearing is computed as Harris and Kotzalas(1996)

$$H_b = 1.047X10^{-4}\ n\ M_f \quad (1)$$

WhereH_b, n and M_f are heat generated, rotational speed and total frictional torque respectively
The total frictional torque M consists of three components, torque M_1 which is due to cutting force and the other torque M_2 due to viscosity of the lubricant, and M_s is the spin-torque, $M_f = M_1 + M_2 + M_s$.

5.2. Frictional torque due to cutting force (M_1):

The torque due to cutting force is calculated asHarris and Kotzalas(1996)

$$M_1 = f_l \times F_\beta \times d_m \quad (2)$$

Where f1 denotes the factor based upon bearing design and corresponding load. For angular contact ball bearings is given asSchaeffler group(2008b)

$$f_1 = z(\frac{F_S}{C_S})^Y \quad (3)$$

The static equivalent load, F_s is given as (Schaeffler group(2008b))

$$F_s = X_oF_r + Y_oF_a \quad (4)$$

For angular contact ball bearings, the dynamically equivalent load F_β depends upon the magnitude and orientation of the applied load, and is given as Schaeffler group(2008b)

$$F_\beta = \max\ (0.9F_a\cot\theta - 0.1F_r) \quad (5)$$

Solving the equation (2) by taking pitch diameter of bearing d_m = 60 mm, the torque developed due to applied load is given as M_1 = 7.4 N mm.

5.3. Viscous Friction Torque (M_2):

For bearings, the viscous friction torque is expressed empirically as, Harris and Kotzalas(1996)

$$M_2 = 10^{-7} f_o\ d_m^3\ (v_o n)^{2/3} \quad (6)$$

The value of v_o =20 is taken at the uniform temperature of 40°C.
f_o is the factor depending upon the type of bearing and method of lubrication.f_o = 2 (adapted from Harris and Kotzalas(1996)), M_2 = 48.02 N mm, Total M=55.42 N mm.

The heat generated in the bearing is computed as H_b = 12.128 W at 2000 RPM. Similarly, the heat generated at various speeds is calculated for steel and hybrid bearings, as shown in Figure 3. The heat generated in the hybrid bearing is lower due to parameters such as Fa ,F_s, ,$f_1$and f_b in comparison with steel bearing. The heat generated in hybrid bearing is 10.18 w at 2000RPM.

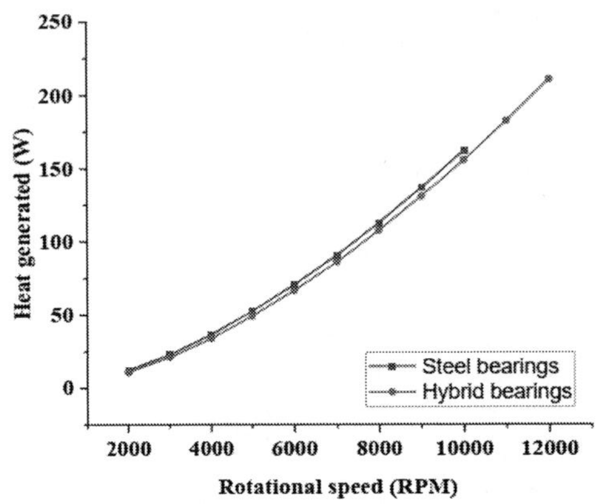

Figure 3: Heat generation in steel and hybrid bearings

5.4. Heat generation due to spin:

$$M_s = \frac{3\mu a\varepsilon F_p}{8} = 35.43 \times 10^{-3} \text{ (N-mm)} \tag{7}$$

Where μ, a, ε, and F are the friction factor, the major axis of the elliptical contact region, the elliptical integral of 2nd kind over the contact zone and preload respectively as presented by
$\mu = 0.01$, $a=3$, $\varepsilon = 0.7$, and $F_p = 300$ N.
The thermal preload (F_{th}) on the bearings due to the thermal deformations is given as Takabi and Khonsari(2013)

$$F_{th} = K_r \varepsilon_t^{n_1} \tag{8}$$

Where K_r, ε_t and n_1 denote the radial elastic constant (adapted from Harris and Kotzalas(1996)), thermal displacement (based on contact stress theory) of the bearing, and constant based on the type of bearing respectively.
From the thermal model of the bearing, the cutting load-induced friction torque is lower, whereas the viscous friction torque is higher at the same spindle speed. Hence, as the spindle speed increases, the heat generation increases, this causes the thermal deformation of the bearings thus leading to an increase in the thermal preload. The temperature increases in the bearing by thermal conductive heat transfer network and temperature is controlled by grease packed in bearings(Jorgensen and Shin(1997)). The thermal preload is added to the initial preload (through lock nut), and the spindle speed is 6000 RPM as the FEM boundary conditions.

6. Finite element formulation of the spindle:

The spindle is discretized as one-dimensional beam elements by applying Euler –Bernoulli beam theory. The spindle is idealized as a five node beam element having two degrees of freedom (one translational and one rotational) at each node, as shown in Figure 4 and Figure 6. The FEM dynamic model for the spindle bearing system constitutes discretized shaft elements with coupling effects. The model is similar to be a rotor-dynamic system. Both static and dynamic analysis is performed by imposing appropriate loading and boundary conditions. The cutting load is applied on the node 1and thermal load is applied at nodes 2, 3 which are bearing supports.

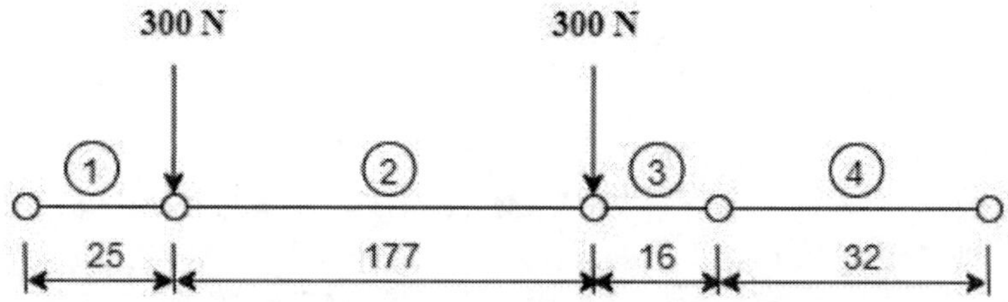

Figure 4: Spindle discretization - FEM model

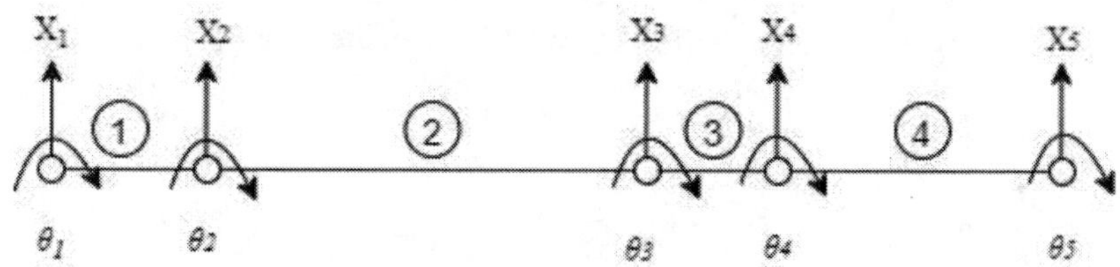

Figure 5: Representation of Finite element model

$$\text{Global Nodal displacement vector},[d] = [x_1\theta_1 x_2\theta_2 x_3\theta_3 x_4\theta_4 x_5\theta_5]^T \tag{9}$$

Where x_i and θ_i are respective node deflection and slope.

The governing equation of motion for the spindle deformations under static and dynamic conditions is given by,

$$[M][\ddot{d}] + \omega\,[G]\,[\dot{d}] + ([K] - \omega^2[M_\omega])[d] = [F(t)] \tag{10}$$

Where, [M], [G], [K] and [F] denote assembled mass, skew symmetric gyroscopic, centrifugal, stiffness and force matrices respectively The term.$\omega^2 M_\omega d$ represent the softening effect of centrifugal force which reduces the stiffness of the spindle bearing system and the term ω [G] $[\dot{d}]$ represent gyroscopic moment The matrices are calculated based on the geometry, speed and mass of the spindle shaft.

The stiffness matrix for i_{th} element of the beam is given as,

$$[K_i] = \frac{EI}{l^3}\begin{bmatrix} 12 & 6l & -12 & 6l \\ 6l & 4l^2 & -6l & 2l^2 \\ -12 & -6l & 12 & -6l \\ 6l & 2l^2 & -6l & -4l^2 \end{bmatrix} \tag{11}$$

Global stiffness matrix $[K] = [K_1] + [K_2] + [K_3] + [K_4]$

0.00076	0.0096	−0.00079	0.0096	0	0	0	0	0	0
0.0096	0.16	−0.0096	0.08	0	0	0	0	0	0
−0.00076	−0.0096	0.00077	−0.0094	−0.000002	0.00019	0	0	0	0
0.0096	0.08	−0.0094	0.182	−0.00019	0.0112	0	0	0	0
0	0	−0.000002	−0.00019	0.0029	0.0232	−0.0029	0.023	0	0
0	0	0.000191	00.112	0.0232	0.272	−0.023	0.125	0	0
0	0	0	0	−0.0029	−0.023	0.0039	0.0066	−0.00102	0.0164
0	0	0	0	0.023	0.125	0.0066	0.61	−0.0164	0.175
0	0	0	0	0	0	−0.00102	−0.0164	0.00102	−0.0164
0	0	0	0	0	0	0.0164	0.175	−0.0164	0.351

Elemental mass matrix, $[M_i] = \frac{\rho A_i l_i}{2}\begin{bmatrix}1 & 0 & 0 & 0\\ 0 & 0 & 0 & 0\\ 0 & 0 & 1 & 0\\ 0 & 0 & 0 & 0\end{bmatrix}$ (12)

Global mass matrix [M] = $[M_1] + [M_2] + [M_3] + [M_4]$ (Jorgensen and Shin(1998))
Elemental gyroscopic matrix [G]

$$[G_i] = \frac{\rho I_x}{30l(l+\omega)^2}\begin{bmatrix} -36 & 3l(1-5\omega) & 36 & 3l(l-\omega)\\ -3l(1-5\omega) & l^2(4+5\omega+10\omega^2) & -36 & -l^2(1+5\omega-5\omega^2)\\ 36 & -3l(1-5\omega) & -36 & -3l(1-5\omega)\\ -3l(1-5\omega) & -l^2(1+5\omega-5\omega^2) & 3l(1-5\omega) & l^2(4+5\omega+10\omega^2)\end{bmatrix} \quad (13)$$

Global gyroscopic matrix [G] = $[G_1] + [G_2] + [G_3] + [G_4]$

Elemental centrifugal matrix $[K_{c,i}] = m\,\omega^2\begin{bmatrix}1 & 0 & 0 & 0\\ 0 & 0 & 0 & 0\\ 0 & 0 & 1 & 0\\ 0 & 0 & 0 & 0\end{bmatrix}$ (14)

Global centrifugal matrix $[K_c] = [K_{c,1}] + [K_{c,2}] + [K_{c,3}] + [K_{c,4}]$

The centrifugal force, gyratory moments, thermal, and cutting loads are considered in the dynamic analysis. During HSM, the bearing ball experiences gyroscopic moment due to an increase in friction by the sliding motion. The gyratory moment, G on each bearing ball is calculated as Harris and Kotzalas(1996).

$$G = J\,\omega_o\,\omega_b\sin\beta \quad (15)$$

Where, mass moment of inertia of each ball, $J = \frac{1}{60}\times\rho\pi D^5$

ω_o, ω_b, n_c, n_b denote the orbital speed of the ball, the rotational speed of bearing element, cage speed (RPM), and ball speed (RPM), respectively.

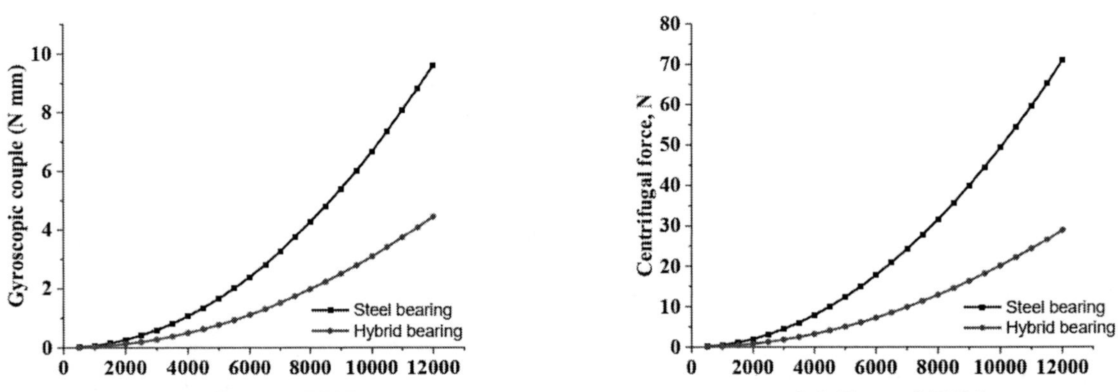

(a) (b)

Figure 6: Gyroscopic couple and Centrifugal effects on high-speed spindle

The centrifugal force (F) at the bearings calculated as $F = mR\omega^2$ (16)

Where m, ω, and R denote the mass, spindle rotational speed, and pitch circle radius of the bearing, respectively.

Table 3: Gyratory and Centrifugal force

	Gyratory moment (Nmm)		Centrifugal force (N)	
Speed in RPM	Steel bearing	Hybrid bearing	Steel bearing	Hybrid bearing
6000	2.401	1.113	17.760	7.231
12000	9.604	4.451	71.043	28.923

The gyroscopic moment and centrifugal force are function of rotational speed. At 6000 RPM, the gyroscopic moment is 2.401 N-mm and has little effect on the spindle bearing dynamics. But at higher spindle speeds, the frequency value beyond 1st order is getting lowered. The effect of centrifugal force is to produce a spin softening effect and reduces the bearing stiffness. Further, the reduced bearing stiffness lowers the natural frequency by 5% up to 12000 RPM, as presented by Wang et al.(2011)

Table 4: Tangential cutting force and bearing temperatures during high-speed machining

S.No	Cutting velocity, m/min	Spindle speed, RPM	DOC, mm	Feed $\frac{mm}{min}$	MRR cc/min	Tangential cutting force (N)	Power at cutter (kW)	Power at motor (kW)	Bearing temperature (°C)
1	500	1989.4	2	994.7	119.36	310	2.6	3.06	26
2	800	3183	1.5	1591	143.24	232.5	3.1	3.65	28.5
3	1000	3978	1	1989	119.37	155	2.6	3.06	31.2
4	1250	4973.6	0.9	2486	134.29	139.4	2.9	3.41	33.8
5	1500	5968	0.75	2984	134.29	116.3	2.9	3.41	36.9

DOC and MRR represent the depth of cut and material removal rates, respectively.

During machining of free cutting steel 14Mn1S14, the tangential cutting force decreases as the cutting velocity increases. However, the material removal rate increases with cutting velocity. The spindle power utilized also increases with the material removal rate. . Experimentally, the temperature observed is 36.9 °C using an infrared temperature gun on the on the spindle inside at front bearing zone. From Figure 7, the bearing temperature steadily increases, although the tangential force is decreasing. This experimental result indicates that at higher cutting velocities, the viscous friction in the bearing is predominant to the applied tangential force in the increase of bearing temperature. Since feed rate is increasing with increase in cutting velocity and MRR is increasing (Altintas and Cao(2005))

Tangential Cutting Force, $P_Z = \frac{6120\ P_w}{V_c}$ (CMTI Machine tool design handbook(2015)).

As the spindle power is calculated based on the cutting parameters, the observations from the Table 3 are the depth of cut is reduced for chatter free cutting. Hence the tangential cutting force is decreasing.

$$\textit{Cutting velocity,} V_C = \frac{\pi D n}{1000} \text{ m/min}$$, n is the spindle rotational speed (RPM).

D is the diameter of the work material and cutting tool for turning and milling respectively

$$f_r = Z \times f_Z \text{ mm/rev}, f_m = f_r \times n \text{ mm/min}$$

f_m , Z, f_z is the feed per minute, number of teeth, and feed in mm/tooth respectively.

Power at tool tip, $P_w = UK_nK_rQ$ (CMTI Machine tool design handbook(2015))

Material removal rate , $Q = f_m d V_c$ (CMTI Machine tool design handbook(2015))

Material removal rate (Q), unit power (U), Correction factors for flank wear (K_n) and rake angle (K_r).

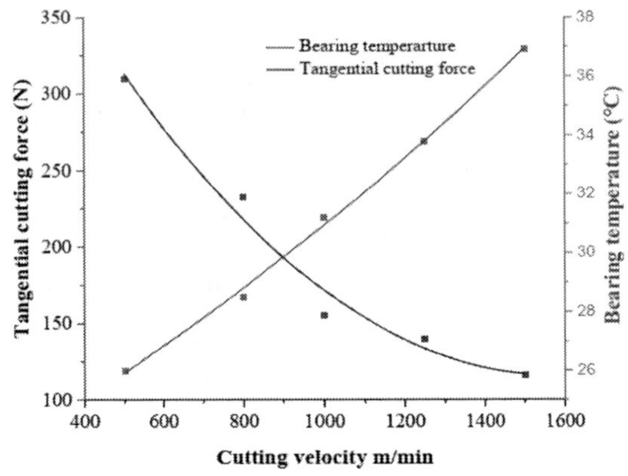

Figure 7: Effect of cutting speed on tangential cutting force and bearing temperature

6.1. For static analysis:

An analysis is carried out by mechanical lock nut fixed bearing preload and interference effects at the support locations along with a static load of 300 N at node 1 (cutting load). The beam deflections and slopes are computed from static deflection equation of FEM. (17).

$$[\mathrm{d}] = [K]^{-1}\{[\mathrm{F}] + [F_c]\} \quad (17)$$

$$= 10^{-5}[3.022 \quad 58.9 \quad 1305 \quad 44.18 \quad 49.61 \quad -10.32 \quad -30.23 \quad 1.085 \quad -24.97 \quad 5301]^T$$

6.2. For dynamic analysis:

Initially, the analysis is carried out with beam inertia, gyroscopic, centrifugal, and cutting load effects as the inputs. The structural stiffness matrix is coupled with a centrifugal matrix. Hence,

$$[K]_{global} = [K] + [K]_c$$

The beam deflections and slopes are computed from (2).

$$[\mathrm{d}] = [Z]^{-1}\{[\mathrm{F}] + [F_c]\} \quad (18)$$

Where, $Z = -\omega^2 Sin(\omega t)[M] + \omega Cos(\omega t)[G] + Sin(\omega t)[K]_{global}$

$$= 10^{-10}[-1836 \quad -51.62 \quad -9392.1 \quad 2506.8 \quad -4310 \quad 4.219 \quad -8164 \quad -2.324 \quad 9422 \quad 18.41]^T$$

6.3. For dynamic with thermal load:

A similar analysis is performed by incorporating the thermal load and the beam deflections and slopes are computed using (19).

$$[\mathrm{d}] = [Z]^{-1}\{[\mathrm{F}] + [F_c] + [F_{th}]\} \quad (19)$$

$$= 10^{-10}[-4642 \quad -236 \quad -2.49\times10^6 \quad 2506.3 \quad -1.55\times10^5 \quad 35.01 \quad -2\times10^5 \quad 6.33 \quad -9399 \quad 13.49]^T$$

The spindle deflections and slopes with static and dynamic (with and without thermal load) analysis are calculated using equations (21), (22), and (23), the results are presented in Table 5 and Table 6 along with trend graphs shown in Figure 9 (a), (b), (c) and (d).

7. Analysis of the high speed spindle

The stress induced, displacement and strain are extracted using ANSYS finite element analysis at 2000, 3000, 4000, 5000 and 6000 RPM for thermo - Structural behavior by applying fixed bearing boundary conditions with and without external cutting load. Variation in the displacement, strain and stress w.r.t spindle rotational speeds for steel bearing and hybrid bearing are shown in Figure 10 (a) (b) (c) and (d). The maximum Von-Mises stress and deformation in steel bearing are 439.87 MPa, and 0.71 μm whereas the same in hybrid bearing are 460.67 and 0.69 respectively (Lu et al.(2007b) [18]. The stress in the hybrid bearing is higher than steel bearing due to low thermal conductivity and deformation is lower than steel bearing due low coefficient of expansion. Therefore hybrid bearings are preferred for high speed spindles and also lower centrifugal force and Gyratory moment due to lower density as shown in the Table 2.

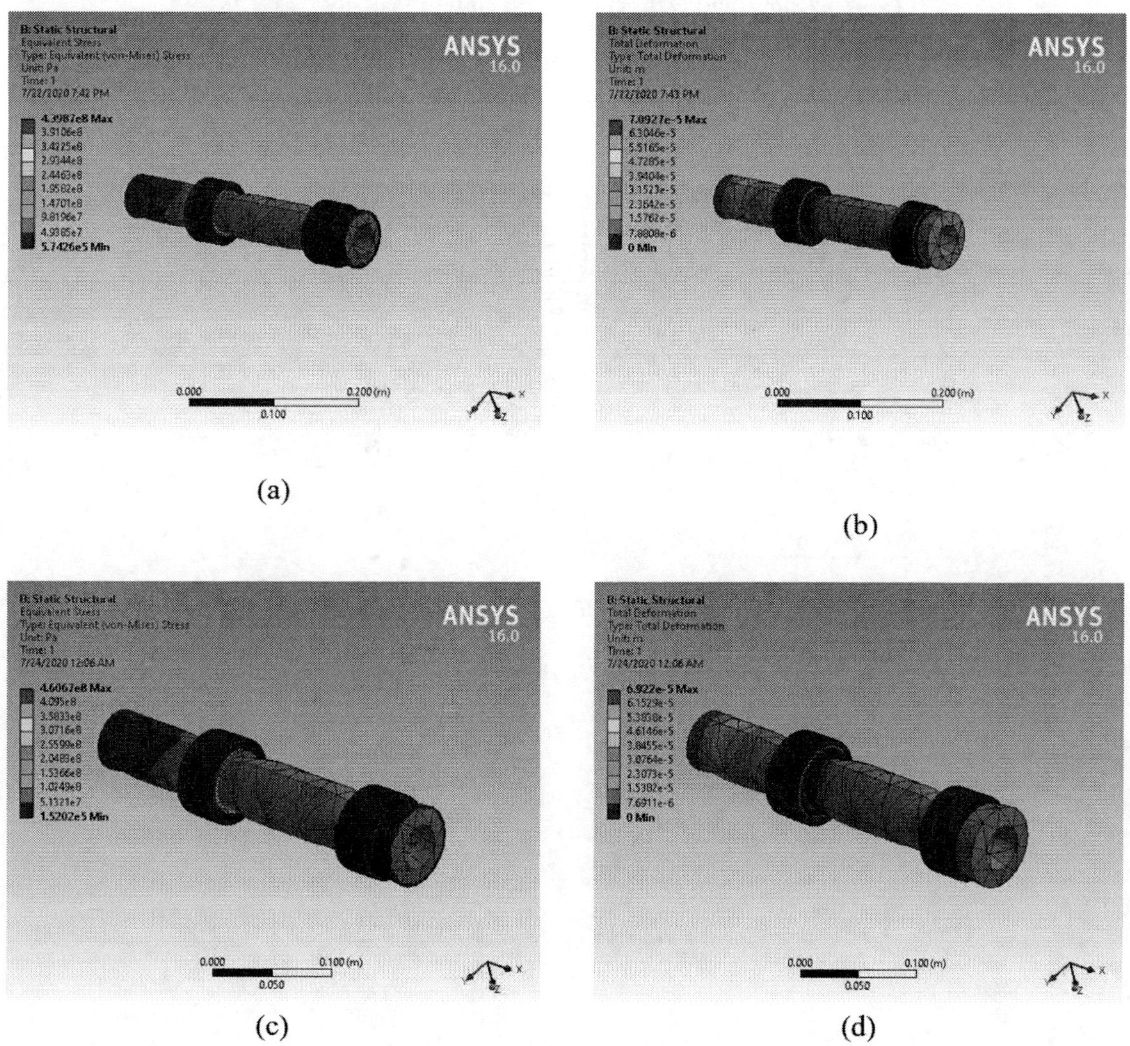

Figure 8: Analysis results of spindle with (a), (b) steel and (c), (d) hybrid bearings

Table 5: Slope and Deflections without cutting load at 6000 RPM

Node number	Deflections					
	Static		Dynamic		Dynamic with thermal load	
	Translation (x, μm)	Slope, rad ($\theta \times 10^{-6}$)	Translation (x, μm)	Slope, rad ($\theta \times 10^{-6}$)	Translation (x, μm)	Slope, rad ($\theta \times 10^{-6}$)
1	0.0133	258	$-1.818\text{x}10^{-3}$	$-2.278\text{x}10^{-3}$	$-4.635\text{x}10^{-2}$	-2132
	0.056	211	$-0.955\text{x}10^{-2}$	$-2.018\text{x}10^{-4}$	-0.241	$0.927\text{x}10^{-3}$
3	0.2066	-62.5	$-2.627\text{x}10^{-4}$	$4.17\text{x}10^{-4}$	$-1.761\text{x}10^{-4}$	$4.727\text{x}10^{-3}$
4	-0.272	5.26	$-6.648\text{x}10^{-5}$	$4.43\text{x}10^{-4}$	0.12	$3.24\text{x}10^{-3}$
5	-0.0836	0.65	$-2.461\text{x}10^{-5}$	$1.793\text{x}10^{-3}$	0.1946	$-4.727\text{x}10^{-3}$

Table 6: Slope and Deflections with cutting load at 6000 RPM

Node number	Deflections					
	Static		Dynamic		Dynamic with thermal load	
	Translation (x, μm)	Slope, rad ($\theta \times 10^{-6}$)	Translation (x, μm)	Slope, rad ($\theta \times 10^{-6}$)	Translation (x, μm)	Slope, rad ($\theta \times 10^{-6}$)
1	0.0302	589	$-1.836 \text{ x } 10^{-3}$	$-5.1623 \text{ x } 10^{-3}$	$-4.642 \text{ x } 10^{-3}$	$-2.36 \text{ x } 10^{-3}$
2	13.05	441.8	$-9.39 \text{ x } 10^{-3}$	0.2506	-0.249	0.2506
3	0.4961	-103.2	$-4.31 \text{ x } 10^{-4}$	$4.219 \text{ x } 10^{-6}$	$-1.559 \text{ x } 10^{-2}$	$3.501 \text{ x } 10^{-3}$
4	-0.302	10.85	$-8.164 \text{ x } 10^{-4}$	$2.324 \text{ x } 10^{-4}$	$-2.002 \text{ x } 10^{-2}$	6.336×10^{-4}
5	-0.2497	5301	$-9.422 \text{ x } 10^{-4}$	$1.841 \text{ x } 10^{-3}$	-9.399×10^{-3}	$1.349 \text{ x } 10^{-3}$

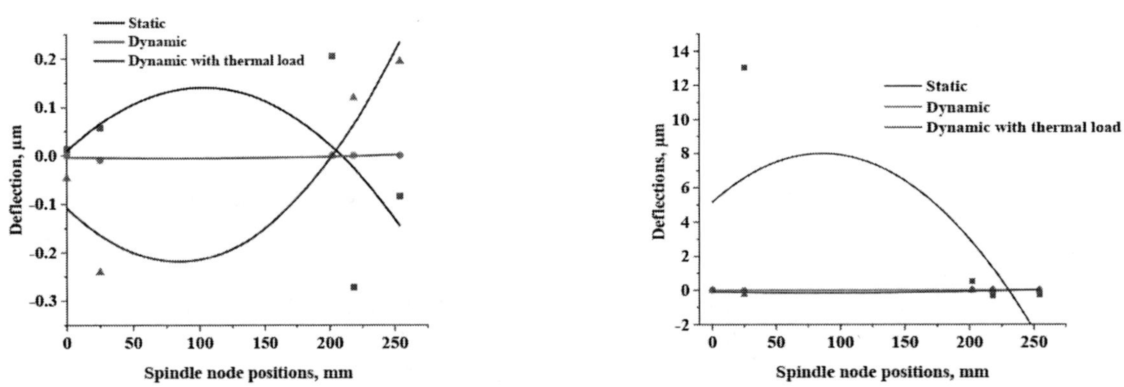

(a) Without cutting load

(b) With cutting load

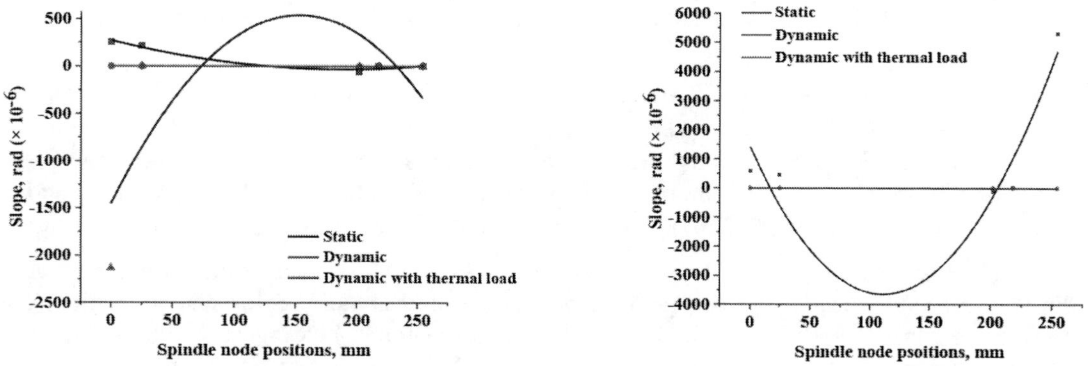

(c) Without cutting load (d) With cutting load

Figure 9: Nodal deflections and slope in spindle

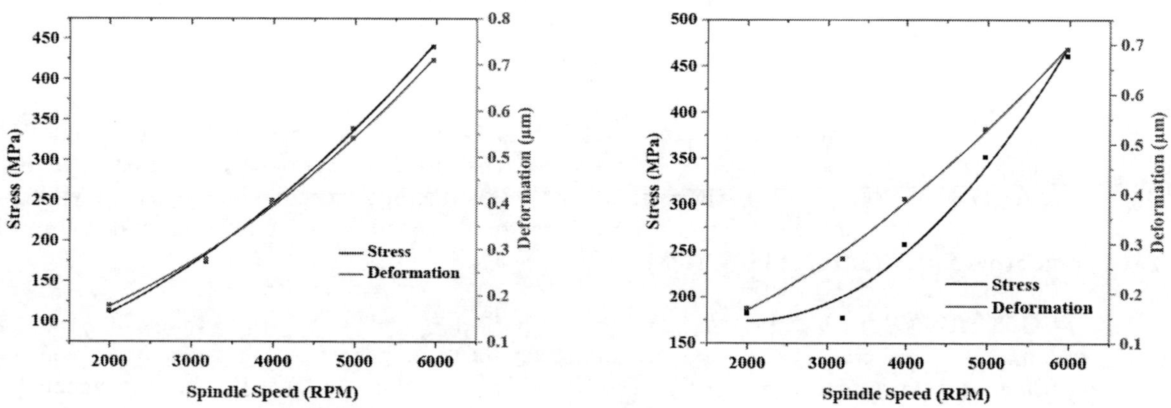

(a) (b)

Figure 10: Stress and displacement w.r.t spindle speed

Table 7: ANSYS simulation results at 6000 RPM

Type of bearing	Stress induced (MPa)	Deformation (μm)
Steel	439.87	0.71
Hybrid	460.67	0.69

8. RESULTS AND DISCUSSION

1. The angular contact ball bearing with lifetime Li-based grease lubrication in a tandem arrangement is used. The heat generated at 6000 RPM is 70.66 W, and the bearing temperature is 41.7 °C at an ambient temperature of 28°C.Analytical and experimental studies are performed using the same bearing configuration and specifications. The variation in temperature of 4.8 °C is due to conductive and convective heat transfer in the experimental work. . The bearings are provided with grease lubrication up to 6000 RPM and oil jacket cooling beyond 6000 RPM to ensure reliable thermal bearing performance during high-speed machining (Lu et al.(2007a)) .The experimental

value is considered for estimating the thermal preload on the bearing and is computed 39.99N from the equation (12). The critical speed of the spindle is increased with increased thermal preload up to a certain extent beyond the bearing failure takes place.

2. The coupling of gyratory, centrifugal, and cutting load has significant effect on the spindle dynamic deflections From the Figure 6(a) and (b) and Table 3, it is observed that the gyroscopic moment and centrifugal force in hybrid bearing are 46.36% and 40.72% of steel bearing respectively at all spindle speeds and effect are minimized by using hybrid bearing. A design guideline is to use hybrid bearings beyond 6000 RPM to minimize the coupling effects as presented in the table 3.

3. It is observed from Table 2, that the tangential cutting force decreases with increased speed. Further, it also depends on other cutting parameters and spindle-motor characteristics (power – torque-speed). The correlation of two variables (cutting force, bearing temperature) w r t cutting velocity is shown in figure 7. The spindle attains its full power utilization with a tangential cutting load of 116.3 N during machining, as shown in Table 4.

4. Table 5 represent the FEM solutions of spindle deflections and slope at nodal points with and without the application of cutting loads in the force vector, respectively. The nodal displacements and slopes are shown in Figure 9 (a), (b), (c), and (d). The maximum spindle deflection by FEM solution is 13.5 μm, whereas; the simulation result is 0.71 μm as shown in the Figure 8. The results are within the limit as per the CNC machine tool geometrical testing standards. O Kilicay(1983) determined the deflection of the spindle using the beam theory for a simply supported with overhanging load condition.

5. The ANSYS simulations for steel and hybrid bearings are performed up to 6000 RPM to determine the Von-mises stress and displacement. The results are shown and plotted in figures 8 (a) (b) (c) and (d), 10(a) and (b). The results indicate that the high-speed spindle is stable with 2.5 μm deflection by attaining equilibrium state due to speed-related effects compared with the results of Uhlmann and Hu(2012) at 12000RPM.

9. CONCLUSION

A comprehensive process is presented to predict the heat generation in the angular contact ball bearings at different operating speeds. The bearing oil cooling is required to be implemented in the high-speed spindle for reliable thermal performance during HSM. The high-speed spindle dynamic analysis is more complicated due to speed-dependent non-linear and time-varying effects when compared to the low-speed spindle. FEM solutions are implemented using the Euler- Bernoulli beam with coupling effects to obtain the nodal deflections and slopes of HSS up to 6000RPM.

10. ACKNOWLEDGMENTS:

The authors would like to thank the Principal of Institute of Aeronautical Engineering for his encouragement, support and granting permission to publish this work. The authors also express their sincere thanks to the authorities of Osmania University, Hyderabad for their support and permission to present this work

11. REFERENCES

[1] Abele, E., Altintas, Y., Brecher, C. (2010). Machine tool spindle units. CIRP Ann. - Manuf. Technol. 59, 781–802. https://doi.org/10.1016/j.cirp.2010.05.002

[2] Alfares, M., Saleem, O., Majeed, M. (2019). Analytical study of thermal variation impact on dynamics of a spindle bearing system 0, 1–28. https://doi.org/10.1177/1464419319841687

[3] Altintas, Y., Cao, Y. (2005). Virtual Design and Optimization of Machine Tool Spindles. CIRP Ann. 54, 379–382. https://doi.org/10.1016/S0007-8506(07)60127-9

[4] Arotaritei, D., Constantin, G. (2016). Modeling and Simulation of High Speed Spindle, Current

Problems and Optimizations – A Survey.
[5] Chen, J.-S., Hsu, W.-Y. (2003). Characterizations and models for the thermal growth of a motorized high speed spindle. Int. J. Mach. Tools Manuf. 43, 1163–1170. https://doi.org/10.1016/S0890-6955(03)00103-2
[6] Choi, J.K., Lee, D.G. (1997). Characteristics of a spindle bearing system with a gear located on the bearing span. Int. J. Mach. Tools Manuf. 37, 171–181. https://doi.org/10.1016/S0890-6955(96)00017-X
[7] CMTI Machine tool design handbook (2015). CMTI Machine tool design handbook, 39th ed. McGraw Hill Education Pvt Ltd(India)., New Delhi.
[8] Dassault Systems (2016). SOLIDWORKS.
[9] Harris, T.A., Kotzalas, M.N. (1996). Rolling bearing analysis, fifth. ed. John Wiley, New York.
[10] Jorgensen, B.R., Shin, Y.C. (1998). Dynamics of Spindle-Bearing Systems at High Speeds Including Cutting Load Effects. J. Manuf. Sci. Eng. 120, 387–394. https://doi.org/10.1115/1.2830138
[11] Jorgensen, B.R., Shin, Y.C. (1997). Dynamics of Machine Tool Spindle/Bearing Systems Under Thermal Growth. J. Tribol. 119, 875–882. https://doi.org/10.1115/1.2833899
[12] Li, H., Shin, Y.C. (2004). Analysis of bearing configuration effects on high speed spindles using an integrated dynamic thermo-mechanical spindle model. Int. J. Mach. Tools Manuf. 44, 347–364. https://doi.org/10.1016/j.ijmachtools.2003.10.011
[13] Lin, C.W., Lin, Y.K., Chu, C.H. (2013). Dynamic models and design of spindle-bearing systems of machine tools: A review. Int. J. Precis. Eng. Manuf. 14, 513–521. https://doi.org/10.1007/s12541-013-0070-6
[14] Lin, C.W., Tu, J.F., Kamman, J. (2003). An integrated thermo-mechanical-dynamic model to characterize motorized machine tool spindles during very high speed rotation. Int. J. Mach. Tools Manuf. 43, 1035–1050. https://doi.org/10.1016/S0890-6955(03)00091-9
[15] Liu, J., Ma, C., Wang, Shilong, Wang, Sibao, Yang, B., Shi, H. (2019). Thermal-structure interaction characteristics of a high-speed spindle- bearing system. Int. J. Mach. Tools Manuf. 137, 42–57. https://doi.org/10.1016/j.ijmachtools.2018.10.004
[16] Long, X., Su, D. (2017). Research on the Optimization Design of Spindle Load of CNC Machine Tools 59, 67–72. https://doi.org/10.3303/CET1759012
[17] Lu, Y., Yao, Y.X., Hong, R.H. (2007a). Finite Element Analysis of Thermal Characteristics of High-Speed Motorized Spindle. Appl. Mech. Mater. 10–12, 258–262. https://doi.org/10.4028/www.scientific.net/AMM.10-12.258
[18] Lu, Y., Yao, Y.X., Xie, W.Z. (2007b). Finite Element Analysis of Dynamic Characteristics of High-Speed Motorized Spindle. Appl. Mech. Mater. 10–12, 900–904. https://doi.org/10.4028/www.scientific.net/AMM.10-12.900
[19] O Kilicay (1983). Determination of bearing deflections. IMechE 197, 277–281.
[20] Sathiya Moorthy, R., Prabhu Raja, V. (2014). An Improved Analytical Model for Prediction of Heat Generation in Angular Contact Ball Bearing. Arab. J. Sci. Eng. 39, 8111–8119. https://doi.org/10.1007/s13369-014-1351-9
[21] Schaeffler group, I. (2008a). Super precision bearings catalogue, FAG.
[22] Schaeffler group, I. (2008b). FAG super precision bearings AC41 130/7 A. catalogue.
[23] Takabi, J., Khonsari, M.M. (2013). Experimental testing and thermal analysis of ball bearings. Tribol. Int. 60, 93–103. https://doi.org/10.1016/j.triboint.2012.10.009
[24] Uhlmann, E., Hu, J. (2012). Thermal modelling of a high speed motor spindle, in: Procedia CIRP. pp. 313–318. https://doi.org/10.1016/j.procir.2012.04.056
[25] Wang, B., Sun, W., Xu, K.P., Wen, B.C. (2011). Speed Effects on Inherent Rotating Frequency of Motorized Spindle System. Adv. Eng. Forum 2–3, 900–905. https://doi.org/10.4028/www.scientific.net/aef.2-3.900
[26] Zahedi, A., Movahhedy, M.R. (2012). Thermo-mechanical modeling of high speed spindles. Sci. Iran. 19, 282–293. https://doi.org/10.1016/j.scient.2012.01.004

IOP Conf. Series: Materials Science and Engineering 998 (2020) 012024 doi:10.1088/1757-899X/998/1/012024

Effect of Geocell on the behaviour of Soil

Dharmesh Lal[1], K. Laxmi Prasanna[2], K. Rohith Rao[3], Prithhe Kamalekkar[4], J. Bajrang[5]

[1]Assistant Professor, Department of Civil, Vardhaman College of Engineering, Hyderabad, India

[2,3,4,5] B.Tech Student, Department of Civil, Vardhaman College of Engineering, Hyderabad, India

E-mail:dharmeshlal34@gmail.com

Abstract. As the use of plastic has rapidly increased in past few years, attempts are made to use it as reinforcement material to escalate the soil properties. For the present study, plastic bottles are cut and made into geocell to use it as the reinforcement to the soil. Use of plastic as reinforcement has great advantages, as it is low cost and easily available. Results of plate load tests on soil and plastic bottle reinforced soil are presented and compared. Finally, it is observed that plastic bottle reinforced soil provides improved performance when compared to that of unreinforced one. An improvement of 1.86 times the plain sand is attained when plastic bottle reinforcement was used.

1. Introduction

Reinforced earth is relatively new construction technique, which has only been used commercially for past 40 years or so for Bearing capacity Improvement. Engineers effectively use soil as a reliable construction material in civil engineering application by introduction of the sand reinforcing techniques

In this present study, an attempt is made to stabilize natural sand with the use of plastic geocell. The term geocell has two parts, first is "geo" which means earth or soil and second is "cell" which means cellular type of shape for infill material such as soil. Geocell confinement effectively increases stiffness and strength of soil, while reducing vertical settlement and lateral spreading of stress. Use of geosynthetics as reinforcement for improving the performance of shallow foundations has been studied by engineers over the past two decades. But most developed and developing countries all over the world have huge resources of waste materials. They should be banned for the better good of the environment for a sustainable world.

The current work deals with the plate load test data of plain sand and plastic bottle geocell reinforced sand. The results obtained for the unreinforced and reinforced sand with plastic geocell are compared. It was observed that reinforced sand has more bearing capacity than unreinforced sand. The confinement effect of geocell on reinforced sand beds were studied by [1]. The occurance of an apparent cohesion was observed by [2-5]. Effect of plastic as soil reinforcements was studied by [4]. The tensile strength was the sole reason for the additional strength. Similar remarks were made by [6,7]. The studies revealed an escalation is soil strength with the addition of plastic.

2. Test materials

IOP Conf. Series: Materials Science and Engineering 998 (2020) 012024 doi:10.1088/1757-899X/998/1/012024

2.1 Sand

Sand was collected from a nearby river source. The size distribution and properties of the collected sand is shown in Fig.1 and Table 1 respectively. The properties of soil were calculated according to the standard Indian codes.

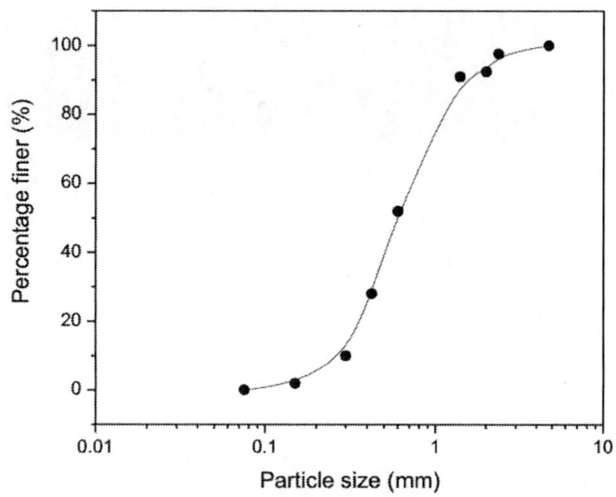

Figure 1 Grain size distribution of sand

[redacted]*Properties of soil*

Property value	**Value**
Specific gravity	2.37
Effective size (D_{10})	0.30
Uniformity coefficient (C_u)	2
Coefficient of curvature (C_c)	1.02
Dry density (Max) kN/m^3	17.8
Dry density (Min) kN/m^3	15.63
Angle of friction from direct shear test	25°

2.2 Plastic bottles

Plastic bottles used in this study is locally available in the stores (Bisleri bottle).

Figure 2 Plastic bottle geocell

The bottles were cut and made into geocells (Figure 2). Plastic bottle dimensions are :- Height – 8cm; Length:- 20cm; Width :- 20cm. A similar process was done by [8,9].

3 Test set up and procedure

Tests were conducted on a 900mm x 900 mm x 900mm test tank (Figure 3). For model circular footing (Figure 4), the dimensions were 150 mm diameter and 25 m height.

Figure 3 Plate load tank

Figure 4 Circular footing

The layout the test tank and reinforcement is shown in the figure 5. The parameters u, b &B are explained in the figure. The depth and width are standardized by dividing with the foundation width for wider applicability [10, 11].

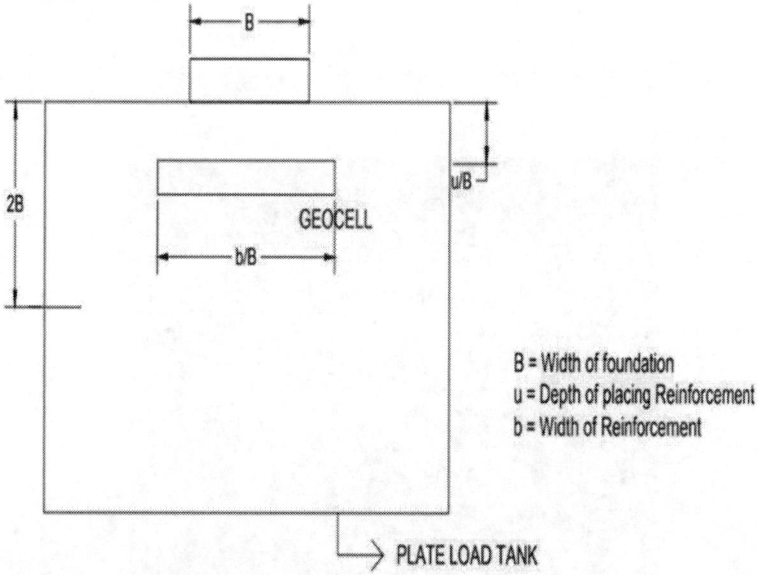

Figure 5 Layout of Reinforcement

Relative density of 55% was adopted because medium dense is the most common condition in the field. Weight of the sand required to achieve this condition was found out mathematically. Sand was emptied in the test tank and vibrated with the electric vibrator as shown in the figure 6 to get the required relative density. Plastic bottle geocell was then placed in the test tank after proper levelling. After proper placement of geocell reinforcement as shown in figure 7, sand was put up to the footing.

IOP Conf. Series: Materials Science and Engineering 998 (2020) 012024 doi:10.1088/1757-899X/998/1/012024

Figure 6 Vibrator

Figure 7 Placement of geocell reinforcement

Circular footing was placed centrally after filling the sand till the top such that distribution of load is homogeneous throughout. Above the footing, hydraulic jack was carefully placed as shown in figure 8. The load values are noted for regular interval of time and bearing capacity is calculated by plotting a graph between Settlement and bearing pressure.

Figure 8 Hydraulic jack on the footing

4 Results and Discussion

4.1 Comparison between reinforced and unreinforced soil

Figure 9 shows the Results of plate load tests on plain soil and plastic bottle reinforced soil placed at different levels. Plastic geocell escalates the performance of sand through the reinforcement effect provided. It shows that optimum depth of placement is at 0.3 since the maximum settlement occurs at the top but further at u/B = 0.1 the reinforcement gets damaged due to foundation. Therefore, overburden sand is required, which prevents contact with footing and the reinforcement.

IOP Conf. Series: Materials Science and Engineering 998 (2020) 012024 doi:10.1088/1757-899X/998/1/012024

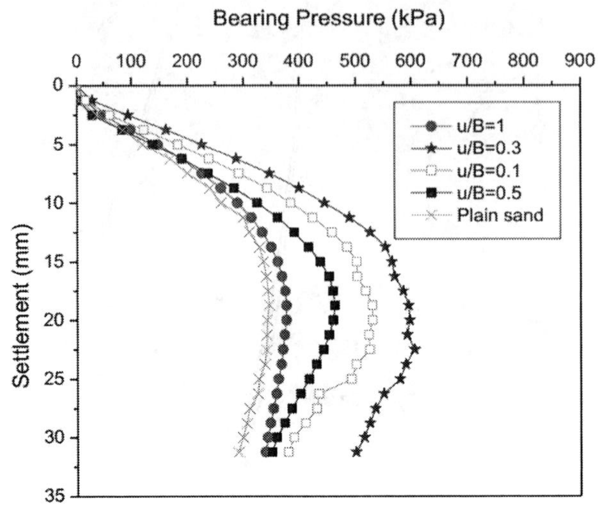

Figure 9 Comparison between reinforced and unreinforced sand

As soil generally fails with lateral spreading, however the reinforcement prevents this spreading and hence enhances the bearing capacity of soil. Figure 11 shows that improvement factor increases up to u/B=0.3, and thereafter, a decrease was indicated. Table 2 shows the list of bearing capacities of both unreinforced and reinforced sand

[redacted]*Bearing capacities of reinforced and unreinforced sand*

REINFORCEMENT DEPTH	**BEARING CAPACITY (kPa)**
Plain sand	320
0.1	530
0.3	597.64
0.5	463.52
1	376.47

5 Improvement factor

Improvement factor can be calculated by eq (1). Table 3 shows the list of improvement factors of reinforced sand and Figure 11 shows the graph between placement of geocell vs Improvement factor. This factor gives an idea of how much enhancement has taken place in soil.

Improvement Factor = Bearing capacity of plastic bottle reinforced/ Bearing capacity of unreinforced eq (1)

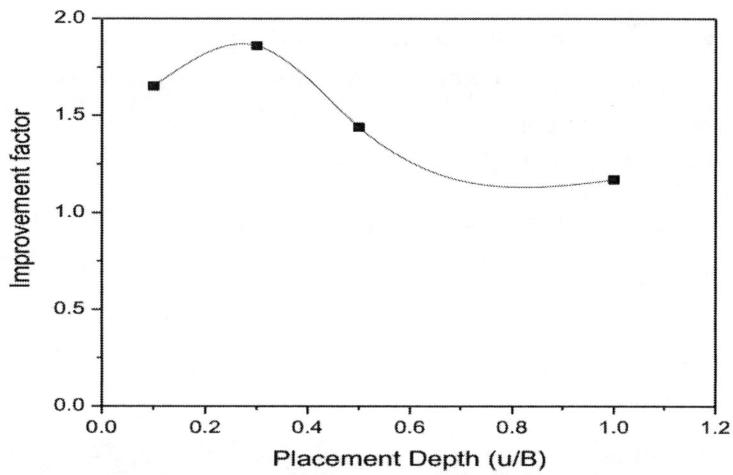

Figure 10 Placement of geocell vs Improvement factor

6. Conclusion

In the present study, plastic bottles were cut and made into geocells. The major observations that can be seen from the study are:

1. Placing of plastic geocell have increased the soil strength
2. It was found that optimum depth was at u/B = 0.3 i.e. at 4.5cm from the surface of the sand.
3. An improvement of 1.86 times the plain sand is attained when plastic bottle reinforcement was used.

Acknowledgement

The authors would like to thank Dr. HariprasadChennarapu for his invaluable support and Mahindra Ecole Centrale for giving permission to conduct the experimental study in the college.

References

[1] AdemIsik, AyhanGurbuz, 2019. Pullout behavior of geocell reinforcement in the cohesionless soils. Geotextiles and Geomembranes, 48 (1), 71-81.

[2] GhotbiSiabil M, S N MoghaddasTafreshi, A R Dawson, 2020. Response of pavement foundations incorporating both geocellsnd expanded polystyrene geofoam, Geotextiles and Geomembranes, 48 (1), 1-23.

[3] KoushikHalder, Debarghya Chakraborty, 2020. Influence of soil spatial variability on the response of strip footing on the geocell-reinforced slope, Geotechnics, 122, 103533

[4] Dash, S.K., Rajagopal, K., Krishnaswamy, N.R., 2004. Performance of different geosynthetic reinforcement materials in sand foundations. Geosynth. Int. 11 (1), 35–42.

[5] Mehrjardi G T ,Behrad R, MoghaddasTafreshi S N, 2019. Scale effect on the behaviour of geocell-reinforced soil, Geotexlites and Geomembranes, 47 (2), 154-163.

[6] Lal D, Sankar N, Chandrakaran S 2017. Effect of reinforcement form on the behaviour of coir geotextile reinforced sand beds. Soils and Foundations, 57(2), 227–236.

[7] Lal D, Sankar N, Chandrakaran S 2018. Effect of reinforcement form on the behaviour of coir geotextile-reinforced sand through laboratory triaxial compression tests, International Journal of Geotechnical Engineering, 12 (3), 309-315.

[8] Lal D, Sankar N, Chandrakaran S 201. Behaviour of square footing on sand reinforced with coir geocell. Arabian Journal of Geosciences, 10(15), 345.

[9] Lal D, Sankar N, Chandrakaran S 2017. Surface heave behaviour of coir geotextile reinforced sand beds. J InstEng India SerA, 98(1–2), 121–125.

[10] Lal D, Sankar N, Chandrakaran S 2017. Performance of shallow foundations resting on coir geotextile reinforced sand bed. Soil Mechanics and FoundationEngineering, 54(1), 60–64.

[11] GholamhoseinTavakoliMehrjardi, FaribaMotarjemi, 2018. Interfacial properties of geocell-reinforced granular soils, Geotextiles and Geomembranes, 46 (4), 384-395.

[12] Khing K H , Das B M, Puri V K, Cook E, Yen S C, 1993. The bearing capacity of a strip foundation on geogrid reinforced sand, Geotext. Geomembr,12, 351–361.

[13] Krishnaswamy N R, Rajagopal K, Latha G M, 2000. Model studies on geocell supported embankments constructed over soft clay foundation. Geotech. Test. J, ASTM 23, 45–54.

[14] Latha G M, Murthy V S, 2007. Effects of reinforcement form on the behavior of geosynthetic reinforced sand., Geotext. Geomembr. 25 (1), 23–32.

[15] Latha, G.M., Somwanshi, A., 2009a. Bearing capacity of square footings on geosynthetic reinforced sand. Geotext. Geomembr. 27 (4), 281–294.

[16] Latha G M, Somwanshi A, 2009. Effect of reinforcement form on the bearing capacity of square footings on sand. Geotext. Geomembr, 27 (6), 409–422.

[17] Sanat k. Pokharel, Robert L. Parsons (2018), "Experimental evaluation of geocell-reinforced bases under repeated loading", j.ijprt.2017.03.007, Volume 11, Issue 2, pages 114-127.

[18] Benson C H and KhireM , 1994. Reinforcing Sand with Strips of Reclaimed High-Density Polyethylene, Journal of Geotechnical Engineering, 121,838-855.

[19] Bueno, Benedito de Souza, 1997. The Mechanical Response of Reinforced Soils Using Short Randomly Distributed Plastic Strips, Recent Developments in Soil and Pavement Mechanics, Almeida (ed.) Balkema, Rotterdam, 401-407.

[20] Anas Ashraf, A Sunil, J Dhanya, M Joseph, M Verghese, M Veena, 2011. Soil Stabilization Using Raw Plastic Bottles,Proceedings of Indian Geotechnical Conference December, 15-17, Kochi (paper no. H-304).

[21] Akinmusuru J O, Akinbolade J A, 1981. Stability of loaded footings on reinforced soil, J. Geotech. Eng. Div, ASCE, 107, 819–827.

[22] Dash, S K, Krishnaswamy NR, Rajagopal K, 2001. Strip footing on geocell reinforced sand beds with additional planar reinforcement, Geotext. Geomembr, 19, 529–538.

[23] Sitharam TG and Sireesh S, 2004. Model studies of embedded circular footing on geogrid-reinforced sand beds, Ground Improvement, 28, No.2, 69-75.

A STUDY ON PROPERTIES OF RECYCLED COARSE AGGREGATE AND IT'S CONCRETE

Harish B A[1*], Dr. N Venkata Ramana[2], Dr. K Gnaneswar[3]

[1] *Assistant Professor, Department of Civil Engineering, GMIT, Davanagere. 577006, India.*
[2] *Associate Professor, Department of Studies in Civil Engineering, UBDTCE, Davanagere. 577004, India.*
[3] *Deputy Executive Engineer, Water Resources Department Nandyal, Kurnool (Dist.) Andhra Pradesh, India.*
[*] E-mail: harishba129@gmail.com Contact Number: +91-8884764227.

ABSTRACT: This paper presents recycled coarse aggregate and its concrete properties. Recycled coarse aggregate exhibits lower values for loose and dry-rodded unit weight and high water absorption values than the natural coarse aggregate. Natural aggregate concrete exhibits higher mechanical strength than recycled coarse aggregate concrete. Results pronounced that the percentage increment of recycled aggregate content is inversely proportional to the strength and concluded that optimum dose of recycled aggregate is 50% and proposed a model to estimate compressive, split tensile and flexural strengths of recycled coarse aggregate concrete. The maximum replacement of RCA in concrete is 50%, and the proposed models show excellent compatibility with experimental results.

Keywords: NCA; RCA; Physical Properties; Mechanical Properties; Regression Analysis:

1. BACKGROUND

Processing of used material and using processed material in the new product is called recycling. Advancement of technology in the construction industry pushing towards sustainability. To employ recycled aggregate in a reinforced concrete structure, properties like compression, crushing resistance, the ratio of water absorption, and modulus of elasticity of concrete are to be known to understand mechanical behavior and reliability. Reused coarse total cement is another solid, which is like characteristic total cement. However, the main distinction is that the total utilized is shown up from the wrecked solid waste. Till today the examination of reused coarse total cement is, for the most part, done in nations such as the US, Japan, Europe, etc., in India; its exploration is at the starting stage. Propelled nations are utilizing reused coarse total cement for the most part for non-basic components because the exploration is likewise constrained to essential mechanical and physical properties, for more prominent uses and cook ecological characteristics in creating nations and created nations, it is necessary to research the possibility of delivering distinctive reused coarse total cement auxiliary components with locally accessible waste cement.

2. RELATED WORKS

In recycled aggregate concrete, amount of followed mortar in the first total is relative to the quality of the first concrete, low-quality cement has less developed mortar, but high-quality cement has more followed mortar [1], the undesirable surface of aggregate add to influence the holding capacity among aggregate and solid paste in concrete [2]. Recycled aggregate concrete exhibits lesser density than conventional cement concrete; its specific gravity is lower than natural aggregate [3], [4]. Water assimilation proportion is very much higher in recycled aggregate [5], [6], From the above, it is clear that functionality of solid waste total is low and could be clarified the higher water ingestion of solid waste total hence the compressive quality of concrete made with reused total is lesser than the regular concrete [7], splitting tensile strength also decreased as recycled aggregate content increased [8]. There was no incredible contrast between the flexural quality of reused coarse total concrete and common sand with ordinary concrete [9].

From the above literature study, it is clear that in India, its research is at the initial stage, hence planned to conduct tests on natural coarse aggregate, recycled coarse aggregate, natural aggregate concrete (NAC), and recycled coarse aggregate concrete (RCAC) and to compare test results. Objectives of the proposed research are,

Viability RCAC for construction works, to examine RCAC behavior in compression, split tension and flexure to evaluate the optimal quantity of RCA in RCAC and to correlate experimental results with theory.

3. EXPERIMENTAL PROGRAM

The exploratory program was isolated into four different stages, the main stage related to the evaluation of used material's physical properties. The second stage compared to the blend plan of cement dependent on the first stage results and determination of ideal w/c proportion. Third stage connected with the throwing of cubes, cylinders, and beams that were utilized in the examination with the solid dependent on the second stage blend extents. The fourth stage related to the testing of cubes, cylinders, and beams to arrive mechanical properties of the solid utilized. An experimental program was planned and is presented in Table 1. Cubes, cylinders for compression, cylinders for split tension, and beams for flexure are cast and tested.

Table 1: Experimental program

Sl.No.	Mix	No. of specimens			
		Cubes	Cylinders	Beams	*1.Cube:* 150x150x150mm [Compression Test]
1.	RCAC0	3	6	3	*2.Cyliner:* 150x300mm
2.	RCAC50	3	6	3	[Compression & Split Tesile Test]
3.	RCAC100	3	6	3	*3.Beam:* 150x150x700mm
	Total	9	18	9	[Flexure Test]

- RCAC indicates recycled coarse aggregate concrete, suffixed 0, 50 and 100 indicate replacement % of natural coarse aggregate (NCA) by recycled coarse aggregate (RCA)

4. MATERIALS: Materials that are used in the research area,

4.1 Cement

Ultra Tech, 43-grade cement, conforming to IS: 8112-1989 [21] standard was used in research. Tests conducted are fineness, consistency of cement, setting times, soundness, specific gravity, and compressive strength test of cement. The test results and their limiting values, as mentioned in IS: 8112-1989 [21] is reported in Table 2. From the test results, it is noticed that the obtained results are within the IS specified limits. It inferences that the cement is in good condition, and it may be used for present work.

Table 2: Cement test results

SL.No	Test	Test value	IS:8112-1989 Specified limits
1.	Specific gravity	3.15	3.0 to 3.15
2.	Fineness of cement	5.0%	Shouldn't be greater than10%
3.	Normal consistency	30%	Shouldn't be less than 27%
4.	Initial setting time	35 minutes	Shouldn't be less than 30 minutes
5.	Final setting time	295 minutes	Shouldn't be greater than 600 minutes
6.	Soundness	6 mm	Maximum of 10 mm
7.	Compressive Strength (MPa)		
	3 Days	28.23	Minimum 16 MPa
	7 Days	37.23	Minimum 22 MPa
	28 Days	46.93	Minimum 43 MPa

4.2 Fine Aggregate

Sand from Bhadra River was used as fine aggregate, confirming to zone III of Table 4 of IS: 383-1970 [16]. The test results of fine aggregate are tabulated in Table 3 and Table 4.

Table 3: Test results of fine aggregate

SL.No	Test	Value
1.	Specific gravity	2.6
2.	Water absorption	1%
3.	Fineness modulus	2.5

Table 4: Sieve analysis of fine aggregate

Sieve size (mm)	Wt. of sample retained (g)	Cumulative wt. of sample retained	Cumulative % wt. retained	Percentage finer N=100-C	IS:383-1970 Table-4
4.75	49	49	4.9	95.1	90-100
2.36	14	63	6.3	93.7	75-100
1.18	96	159	15.9	84.1	55-90
0.60	236	395	39.5	60.5	35-59
0.30	459	854	85.4	14.6	8-30
0.15	132	986	98.6	1.4	0-10
Pan	14	1000	100	0	-
		TOTAL =	250.6		
Fineness Modulus 2.5					
Zone III					

4.3 Natural coarse aggregate (NCA)

Crushed granite of 20mm size is used as NCA. Tests are performed as per IS: 383-1970 [16], and results are tabulated in Table 5 and Table 6.

Table 5: Test results of NCA

Test	Test values
Specific gravity	2.48
Fines modulus	7.65
Water absorption	0.199%
Bulk Density (loose)	1447.48 Kg/m^3
Bulk Density (rodded)	1529.92 Kg/m^3
Impact value	29.76%
Flakiness index	14.16%
Elongation index	29.16%
Percentage of voids	38%
Void ratio	15.00

Table 6: Sieve analysis of NCA

IS Sieve size (mm)	Wt. of aggregates retained (gm)	Cumulative wt. retained (gm)	Cumulative % of wt. retained	% finer (100-c)
80	0	0	0	100
40	0	0	0	100
20	1301	1301	26.02	73.98
10	2913	4214	84.28	15.72
4.75	671	4885	97.70	2.30
Pan	15	5000	100	0
			TOTAL = 292.00	
Fineness modulus = 7.92				

4.4 Recycled Coarse Aggregate (RCA)

It was acquired from demolished building waste. The concrete blocks were crushed in machine crusher and received aggregates of size 12.5mm and 20mm. Tests are similar to NCA, and results are presented in Table 7 and Table 8.

Table 7: Test results of RCA

Tests	Tests value
Specific gravity	2.23
Fines modulus	8.05
Water absorption	4.38%
Bulk Density (loose)	1200 Kg/m^3
Bulk Density (rodded)	1305.83 Kg/m^3
Impact value	42.65%
Flakiness index	3.88%
Elongation index	24.12%
Percentage of voids	42%
Void ratio	15.76

Table 8: Sieve analysis of RCA

IS Sieve size (mm)	Wt. of aggregate retained (gm)	Cumulative Wt. retained in (gm)	Cumulative % Wt. retained	% finer (100-c)
80	0	0	0	100
40	0	0	0	100
20	1017	1017	20.34	79.66
10	2761	3778	75.56	24.44
4.75	1165	4943	98.86	1.14
Pan	57	5000	100	0
			TOTAL = 305.20	
Fineness modulus = 8.05				

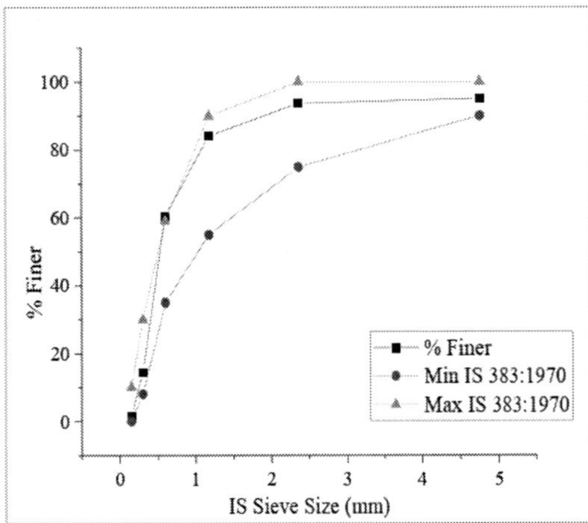

Figure 1: Grading curve for fine aggregate

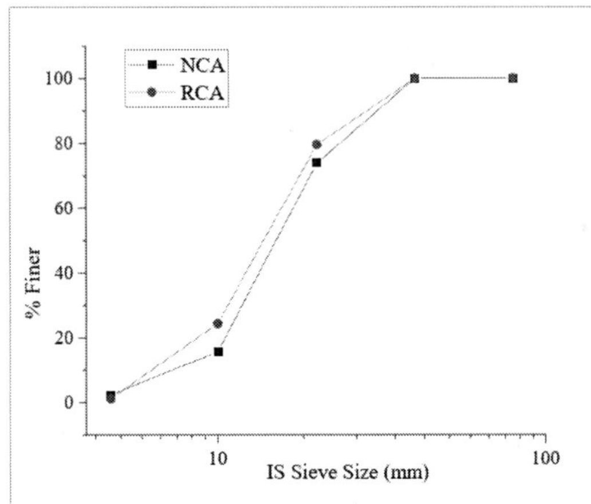

Figure 2: Grading curve for coarse aggregate

4.5 Water: Drinking water is used for mixing and curing of concrete specimens.

Table 9: Physical properties of coarse aggregate

Property	RCA0	RCA50	RCA100
Specific gravity	2.48	2.39	2.23
Fineness modulus	7.65	7.80	8.05
Water absorption (%)	0.199	2.26	4.38
Loose Bulk density (Kg/m^3)	1447.48	1336.80	1200
Rodded Bulk density (Kg/m^3)	1529.92	1431.86	1305.83
Impact value (%)	29.76	39.32	42.65
Flakiness index (%)	14.16	8.18	3.88
Elongation index (%)	29.16	26.52	24.12
Percentage of voids (%)	38	40	42
Void ratio	15.00	15.39	15.76

5. METHODOLOGY: Mix design

To achieve the desired performance of concrete, material selection and mix design are important. Grade of concrete chosen is M20, mix design procedure adopted is per IS: 10262-2009 [12], mix proportions are tabulated in table 10.

Table 10: M20 Grade concrete mix proportions as per IS: 10262-2009

W / C Ratio	Ingredients (kg/m^3)				Mix Proportion
	Water	Cement	Fine aggregate	Coarse aggregate	
0.50	197	394	634.31	1076.12	1:1.61:2.73

6. CASTING OF CONCRETE SPECIMENS

Ingredients except water are mixed in dry condition, for this blended water is added as per the proportion and mixed thoroughly by hand mixing. For casting, all moulds are cleaned, oiled, and secured tightly to achieve proper dimension. The mixed material is placed in the mould and is placed on a vibrating table and vibrated as per the requirement. These specimens are allowed for twenty-four hours, and after that demoulded and during demolding, care should be taken to see that no edges are harmed, and these specimens are cured in water pond for the required number of days.

7. TESTS ON FRESH CONCRETE

7.1 Slump Cone Test: It is performed per IS: 1199-1959 [14] to check the workability of freshly made concrete.

7.2 Compaction Factor Test: It is also used to examine newly formed concrete; it is carried per IS: 5515-1983 [19].

8. TESTS ON HARDEN CONCRETE

Cubes, cylinders, and beams are tested in the compression testing machine per IS: 516-1959 [18] and IS 5816-1999 [20].

Figure 3: Cube Compressive strength test

Figure 4: Cylinder Compressive strength test

Figure 5: Split tensile strength test

Figure 6: Flexural strength test

8.1 SUMMARY

M20grade concrete mix is designed per IS: 10262-2009 [12]. The distinct description of the mix proportion has been implemented. Compaction factor, slump values were found. Eventually, for right mix percentage cubes, cylinders and beams were cast and cured.

9. EXPERIMENTAL RESULTS

Experimental results of concrete at fresh and hardened state are tabulated in table 11 to table 17.

Table 11: Slump and Compaction factor values

Mix	Slump Value (mm)	Compaction Factor
RCAC0	100	0.91
RCAC50	20	0.90
RCAC100	10	0.89

Table 12: 28 day's Cube Compressive strength of concrete (f_{ck})

Mix	Average Cube Compressive Strength (MPa)	% decrement when compared to RCAC0
RCAC0	39.70	-----
RCAC50	36.00	-9.31
RCAC100	22.96	-42.16

Table 13: 28 day's Cylinder Compressive strength of concrete (f_c)

Mix	Average Cylinder Compressive Strength (MPa)	% decrease when compared to RCAC0
RCAC0	14.71	-----
RCAC50	13.39	-8.97
RCAC100	13.01	-11.55

Table 14: Recommended empirical relationships between elastic modulus and compressive strength (E_c)

Mix	As per IS Code $E = 5000\sqrt{f_{ck}}$ (MPa)	As per EC-02 $E = 9500(\sqrt{fc} + 8)*0.33$ (MPa)
RCAC0	31503.96	37103.85
RCAC50	30000.00	36551.70
RCAC100	23958.29	36387.74

Table 15: 28 day's Split Tensile strength of concrete (f_{sp})

Mix	Average Split Tensile Strength (MPa)	% Decrease when compared to RCAC0	As per ACI Code $f_{sp} = 0.49\sqrt{f_{ck}}$	As per GB Code $f_{sp} = 0.19f_{ck}$^0.75	A.R. Santhakumar (1/10th of the f_{ck})
RCAC0	2.92	-----	3.08	3.00	3.97
RCAC50	2.56	-12.32	2.94	2.79	3.60
RCAC100	2.42	-17.12	2.34	1.99	2.29

Table 16: 28 day's Flexural strength of concrete (f_r)

Mix	Average Flexural Strength (MPa)	% Decrement when compared to RCAC0	As per IS Code $f_t = 0.7\sqrt{f_{ck}}$	As per ACI Code $f_t = 0.54\sqrt{f_{ck}}$	As per CEB-FIB Code $f_t = 0.81\sqrt{f_{ck}}$	As per EC-02 Code $f_t = 0.3f_{ck}$^0.67
RCAC0	11.18	-----	4.41	3.40	5.10	3.53
RCAC50	8.96	-19.85	4.20	3.24	4.86	3.31
RCAC100	8.81	-21.19	3.35	2.58	3.88	2.44

Table 17: Density of concrete (γ)

Mix	Average Density (kg/m^3)	% decrease when compared to RCAC0
RCAC0	2401.77	-----
RCAC50	2329.77	-2.99
RCAC100	2312.88	-3.70

10. DISCUSSION

Cube Compressive strengths (f_{ck}) of RCAC are in the range of 36.00 to 22.96 MPa, whereas RAC0, i.e., NCAC, is having a cube CS of 39.70 MPa. By replacement of NCA with RCA in concrete, CS decreased but not less than target strength 20 MPa. Hence it can be concluded that RCAC is not inferior. RCAC50 CS was close to target mean strength.

Using BIS and EC-02 code estimated modulus of elasticity in terms of compression and are presented in table12.

Cylinder Compressive strengths (f_c) of RCAC are in the range of 13.39 to 13.01 MPa, whereas RAC0, i.e., NCAC, is having a cube CCS of 14.71 MPa. By replacement of NCA with RCA in concrete, CCS decreased.

Split tensile strengths (f_{sp}) of RCAC are in the range of 2.56 to 2.42 MPa, whereas RCAC0 or NCAC is having an STS of 2.92 MPa. The habitually tensile strength of concrete is estimated using compressive strength. ACI, Chinese codes, and equations suggested by A R Santhakuma [10] are used to determine STS in terms of compression. STS of concrete is relatively lower than corresponding compressive strength, and RCA use in concrete further decreased its STS.

Flexural strengths (f_r) of RCAC are in the range of 8.96 to 8.81 MPa, but NCAC is having a flexure strength of 11.18 MPa. With the replacement of NCA with RCA in concrete, FS decreased but not less than the target strength. IS code, ACI code, CEB-FIB code, and EC-02 code is used to estimate FS in terms of compression.

The Density of RCAC is in the range of 2329.77 to 2312.88 Kg/m^3, whereas RAC0, i.e., NCAC is having a density of 2401.77 Kg/m^3. By replacement of NCA with RCA in concrete, Density decreased.

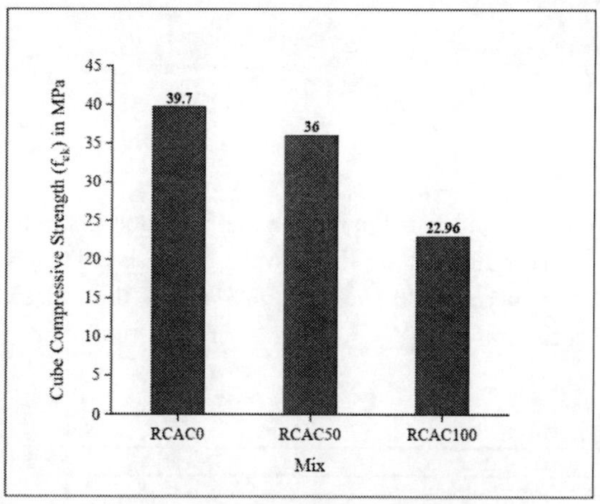

Cube Compressive strengths (f_{ck})

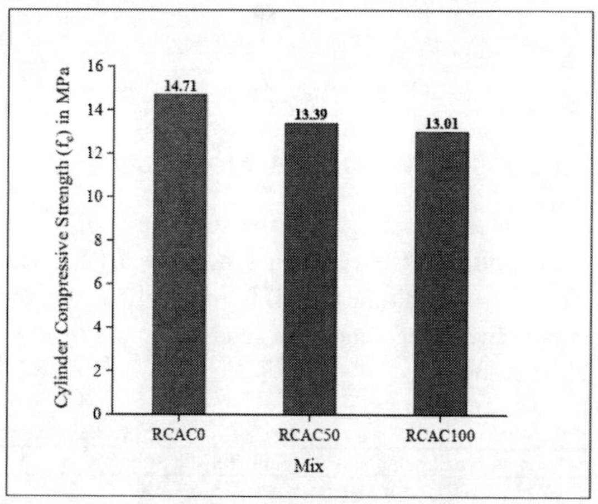

Cylinder Compressive strengths (f_c)

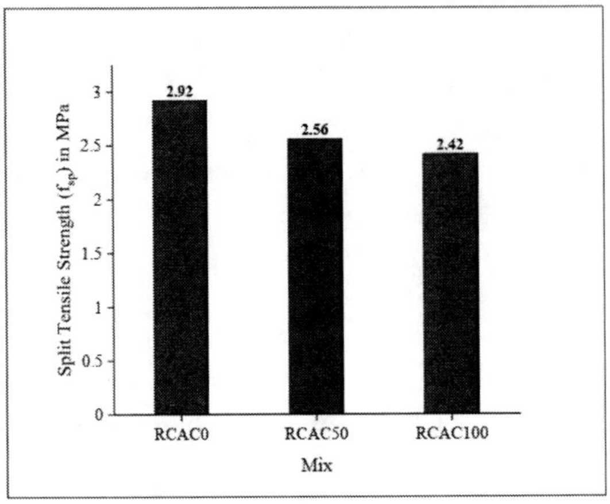

Split tensile strengths (f_{sp})

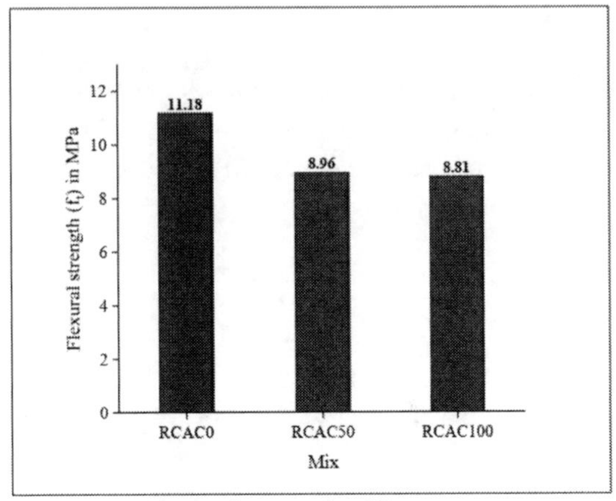

Flexural strengths (f_r)

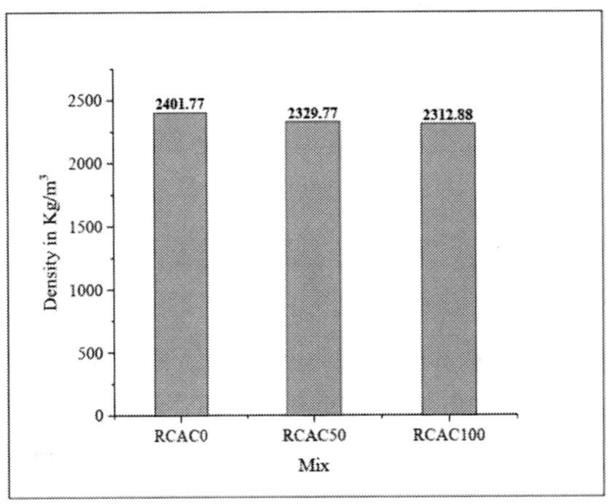

Density (γ)

11. REGRESSION ANALYSIS: Method of Least Squares

Regression analysis, a statistical approach, is used and proposed a model to estimate Cube CS, STS, FS, and Density of RCAC. The method of least squares says that a line can be fitted to the given data set such that the errors are minimized. This implies that one can determine α and β such that the sum of the squared distances between the data points and the line $Y = \alpha + \beta X$ is minimized. ($Y = \alpha + \beta X$ where α is the intercept and β is the slope).

Table 18: Linear Regression Details for RCAC

Cube Compressive Strength	Split Tensile Strength	Flexural strength	Density
$Y = \alpha + \beta X$	$Y = \alpha + \beta X$	$Y = \alpha + \beta X$	$Y = \alpha + \beta X$
$\alpha = 41.25$	$\alpha = 2.88$	$\alpha = 10.83$	$\alpha = 2392.59$
$\beta = -0.1674$	$\beta = -0.005$	$\beta = -0.0237$	$\beta = -0.8889$
$R^2 = 0.905$	$R^2 = 0.939$	$R^2 = 0.797$	$R^2 = 0.883$
$r = \sqrt{R^2} = 0.95$	$r = \sqrt{R^2} = 0.96$	$r = \sqrt{R^2} = 0.89$	$r = \sqrt{R^2} = 0.94$
X = replacement of NCA with RCA			

Table 19: Comparison of Experimental Results and Regression analysis results

Mix	Average CS (MPa)			Average STS (MPa)			Average FS (MPa)			Average Density (kg/m^3)		
	EXP	RM	EXP/RM	EXP	RM	EXP/RM	EXP	RM	EXP/RM	EXP	RM	EXP/RM
RCAC0	39.70	41.26	0.96	2.92	2.88	1.01	11.18	10.83	1.03	2401.77	2392.58	1.00
RCAC50	36.00	32.91	1.09	2.56	2.63	0.97	8.96	9.65	0.92	2329.77	2348.14	0.99
RCAC100	22.96	24.56	0.93	2.42	2.38	1.01	8.81	8.46	1.04	2312.88	2303.69	1.00

12. CONCLUSIONS

From experimental results and discussion on a recycled coarse aggregate and its concrete, the following conclusion is drawn.

RCA obtained after crushing building demolition waste can be used for structural concrete works as its flakiness and elongation are superior to NCA, and grading limits are per BIS (USL: 383-1970). Despite lower specific gravity, lower density, and higher water absorption, higher impact value, concrete made with RCA are in no way inferior to conventional concrete. Loss of workability with time increased with an increase in RCAC. With the increment in RCA in concrete, strengths are reduced compared to NCA concrete. The maximum replacement of RCA in concrete is 50%, and the proposed models show excellent compatibility with experimental results.

REFERENCES

[1] Hasaba S, M. Kawamura, K. Torik, K. Takemoto, "Drying shrinkage and durability of the concrete made of recycled concrete aggregate," *Transactions of the Japan Concrete Institute* **3** (1981), pp. 55 – 60.

[2] Tasong, W. A., Lynsdale, C. J. and Cripps, J. C., "Aggregate-Cement Paste Interface: Influence of Aggregate Physical Properties," *Cement and Concrete Research*, 1998; Vol. **28**, No. 10, pp. 1453 – 1465.

[3] Gonzalez-Fonteboa B, Martinez-Ablle F, "Concretes with aggregates from demolition waste and silica fume. Materials and mechanical properties", *Building and Environment*, **43** (2008), pp. 429 – 437.

[4] Prasad M.L.V and Rathish Kumar P, "Strength studies on glass fiber reinforced recycled aggregate concrete," *Asian Journal of Civil Engineering (Building and Housing)*, Vol.**8**, No.6 (2007), pp. 677- 690.

[5] Topcu I.B., "Using waste concrete as aggregate," *Cement and Concrete Research*, **25** (1995), pp. 1385-1390.

[6] Topcu I.B., Sengel S., "Properties of concretes produced with waste concrete aggregate," *Cement and Concrete Research*, **34** (2004), pp. 1307 – 1312.

[7] Nixon PJ., "Recycled concrete as an aggregate for concrete – a review," *Material and Structures*, RILEM 1978, **11**(65), pp. 371 – 378.

[8] Cong Kou S., Poon C.S., Chan D., "Influence of fly ash as cement addition on the hardened properties of recycled aggregate concrete," *Materials and Structures* (RILEM), 2007.

[9] Ravindrarajah R.S, C.T. Tam, "Properties of concrete made with crushed concrete as coarse aggregate", *Magazine of Concrete Research*, **37** (1985), pp.29 – 38.

[10] A.R. Santhakumar, (Second Edition), "Concrete Technology," OXFORD Higher Education, New Delhi.

[11] Shetty.M.S, (2005), "Concrete Technology Theory and Practice," S.Chand & Company Ltd., New Delhi.

[12] Indian Standard Code (IS: 10262-2009), "Concrete mix proportioning-guidelines," New Delhi.

[13] Indian Standard Code (IS: 1199-1959) "Methods of sampling and analysis of concrete," New Delhi.

[14] Indian Standard Code (IS: 1199-1959), "Methods of sampling and analysis of concrete," New Delhi.

[15] Indian Standard Code (IS: 2386-1997), "Method of test for aggregates for concrete," New Delhi.

[16] Indian Standard Code (IS: 383-1970), "Specifications of coarse and fine aggregates from natural sources for concrete," New Delhi.

[17] Indian Standard Code (IS: 456-2000), "Plain and reinforced concrete- Code of practice," New Delhi.

[18] Indian Standard Code (IS: 516-1959), "Methods of tests for strength of concrete," New Delhi.

[19] Indian Standard Code (IS: 5515-1983), "Specification for compacting factor apparatus," New Delhi.

[20] Indian Standard Code (IS: 5816-1999), "Splitting tensile strength of concrete- Method of test," New Delhi.

[21] Indian Standard Code (IS: 8112-1989), "Specifications for 43 grade ordinary Portland cement", New Delhi.

[22] ACI: 221R-96-(1997). "Guide for Use of Normal Weight and Heavyweight Aggregates in Concrete, ACI Manual of Concrete Practice, Part I: Materials and General Properties of Concrete," American Concrete Institute, Farmington Hills, MI.

[23] British Standards Institution (BS: 8110-97), "Code of practice for the structural use of concrete-Part-1,", London.

[24] GB (Chinese Standard) (2002) Code for design of concrete structures GB: 50010-2002, Chinese Building Press, Beijing, China.

[25] Euro code 2, "Design of concrete structures- Part 1-1: General rules and rules for buildings", CEN, EN 1992-1-1, Brussels.

IOP Conf. Series: Materials Science and Engineering 998 (2020) 012026 doi:10.1088/1757-899X/998/1/012026

A Review on Identification and Elimination of Ovality in Circular Holes During Manufacturing of Hollow Cylindrical Components

SHRINATH MANE[1] and JAMIL DABIR[2]

[1,2]Mechanical Engineering Department, Walchand College of Engineering, Sangli, India
Email: shrinathmane97@gmail.com

Abstract. This review paper aims toward the investigation of manufacturing process of hollow cylindrical component to identify the defects causing ovality in circular holes. The manufacturing refers to the processes required to convert raw material into the final product which may include casting, machining, welding, finishing, heat treatment processes and ends with final inspection to check for defects. Mostly the circular holes of hollow cylindrical components show ovality after manufacturing, exceeding tolerance values and such defect leads to poor fittings. After careful study of the variety of research papers related to ovality, it was found that the ovality may occur due to component characteristics or due to defects in manufacturing process of the component. The results of these research papers concluded that machining parameters, unbalanced workpiece, clamping pressure, clamping geometry and clamping technique influence the dimensional deviations and it can be prevented by improving the production with corrective techniques.
Keywords. Defects, Ovality, Manufacturing, Component characteristics, Production

1. Introduction

Manufacturing can be defined as the process of converting the raw materials into useful products and this is achieved by following some steps such as design of product, raw material selection, and machining. There are various conventional and non-conventional manufacturing processes to manufacture a product, each process has its advantages and disadvantages. The manufacturing processes have been evolved from traditional approaches to the recent advancements due to the requirement of high quality finished good.

In this paper, study of manufacturing a cylindrical component is investigated to identify the defects that can cause ovality in the component. The machining parameters as well as geometry of the component are also investigated to find the source of error causing the ovality. The ovality is usually expressed as deviation from a circular periphery i.e. total difference at the cross-sections between the individual maximum and minimum diameters. Ovality can be expressed as shown in Equation 1.

$$e = 2 \times \frac{b-a}{b+a} \tag{1}$$

Where, e is ovality, b and a are the major axis and the minor axis of elliptical hole respectively. The holes are mostly drilled in the workpieces or present during the casting and further finished by the boring operation. Hence, the workpieces need to be clamped while machining to avoid slipping. During machining of these holes in the workpiece, elastic residual stresses are developed which gets relaxed after removal of clamps and causes deformation. In industry, holding multi-rod as well as gear systems in such hole results in failure due to poor fitting.

2. Literature Review

Kulkarni and Bagale (2017) studied the processes involved in manufacturing of components and how parameters of those processes affect the final product. They identified defects leading to outside

IOP Conf. Series: Materials Science and Engineering 998 (2020) 012026 doi:10.1088/1757-899X/998/1/012026

or inside diameter undersize, oversize and oval shapes during CNC machining and are presented in Table 1.

Table 1 Problem, cause and solution when inaccuracy occurs.

Problem	Cause	Solution
Concentricity or offset, Parallelism error	Centerline or Angular misalignment of chuck	Check the alignment of chuck, tailstock and tool before start of production properly.
Roundness deviation	Non-uniform geometry of workpiece	Ensure proper clamping by checking workpiece alignment with the rotating or cutting axis
Tool or workpiece chatter	Vibration of machine	Proper tool setup and insert installation reduces tool chatter and controlling the feed rate reduces the workpiece chatter.
Incorrect program	Syntax or motion or setup mistakes	Proper code writing and verification eliminates this defect.

Han et al. (2010) focused their study on the ovality occurred in round castings in continuous castings because of the mechanical as well as thermal strain. They selected diameter of strand and hot pressure of molten metal as variables. The calculation analysis concluded that, the large mechanical deformations are generated because of the excessive hot pressure and the groove design of drive roller along with range of round section size should be considered while determining the optimum value of hot pressure.

Ghetiya and Panchal (2014) worked towards reducing the problem of ovality in turning of a clutch component. The principal parameter causing the ovality has been detected within machining variables. They used the RSM and investigated the radiography, work piece balancing, and used finite element analysis for determining the centrifugal force in the component.

- ❖ Ovality depends on the depth of cut as it increases, workpiece experiences more force.
- ❖ High feed rate leads to high heat generation, eventually plastic deformation starts taking place with higher stress which decreases the ovality.
- ❖ Cutting speed and feed rate causes less ovality compared to depth of cut. With increasing cutting speed till 0.150mm/min, ovality decreases but further increasing the cutting speed, increases ovality.

Pandya et al. (2015) analyzed the effect clamping force, hole eccentricity, hole diameter, and angular position of hole on the ovality of holes occurring after the drilling process. They created a simulation plan of L25 orthogonal array by using Taguchi method with 2 mesh settings. To identify very small deformations of cylinders, it is important to consider a fine mesh.

After analyzing the simulation results, they concluded:

- ❖ To reduce oval shape, the hole to be machined should be aligned with clamps at 45°.
- ❖ Ovality does not affected by hole diameter as it is a dimensionless entity.
- ❖ The displacement of hole center is function of clamping force and it should be as minimum as possible while ensuring firm contact.

Nowag et al. (2007) investigated the influence of machining parameters (feed rate, cutting speed, depth of cut and nose radius) on distortion of bearing rings made of 100Cr6 (SAE 52 100). Along with that they analyzed the two clamping techniques and Figure 1 shows both the clamping techniques i.e. external clamping and internal clamping. After turning, the geometry and residual stresses along the circumference of ring is measured.

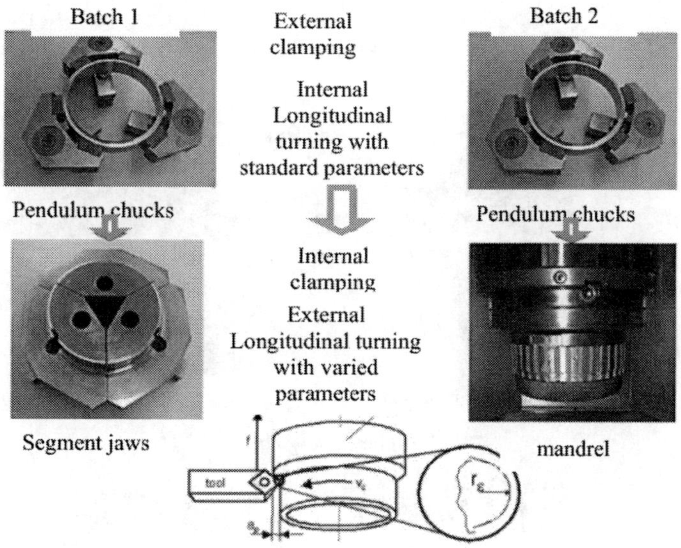

Figure 1. Machining and clamping of rings.

The results analysis concluded:

- ❖ With the changing the clamping strategy, residual stress distribution around the ring's circumference can be altered. The ideal condition of clamping the ring was from inside the diameter with the mandrel.
- ❖ With rising feed and nose radius, the surface residual stresses increase and there is no significant impact of speed, feed, depth of cut and nose radius on ring distortion, when using perfect clamping system i.e. internal clamping.

Gorog and Gorogova (2018) studied the impact of clamping forces on roundness of turned pipes made of steel S355J2H (STN 41 1503). The machining condition used were cutting speed- 180m/min, depth of cut- 1mm and feed rate- 0.2mm. The machined pipes were measured on Rondcom 16A roundness machine.

This paper concluded:

- ❖ Wall thickness of the pipe affects the roundness. With smaller wall thickness of pipe, the roundness is higher.
- ❖ Roundness depends on the clamping forces or tightening torque. The roundness increases with increasing clamping forces.

Gorog et al. (2012) focused their study on deviations generated during the turning process. The clamping pressure was assumed to influence the deviations. The material was a hollow cylinder with the outer diameter of 60mm made of low carbon steel. In order to measure the deviation, they used ZVL MK 300C device with three different filters.

After the analysis, they concluded that:

- ❖ The clamping pressure induces elastic deformations, causing the roundness deviation in machined hollow cylinder and therefore the shape of the part before and after machining is not equal.
- ❖ The roundness deviation proportionally depends on set out of the part from the chuck as well as on clamping pressure.
- ❖ To avoid deformations and obtain lower roundness deviation a different clamping system e.g. collet should be used.

Shao et al. (2013) proposed an algorithm which can predict the machining error generally developed due to release of the clamping force. The Figure 2 shows the oval holes formed due to clamping pressure.

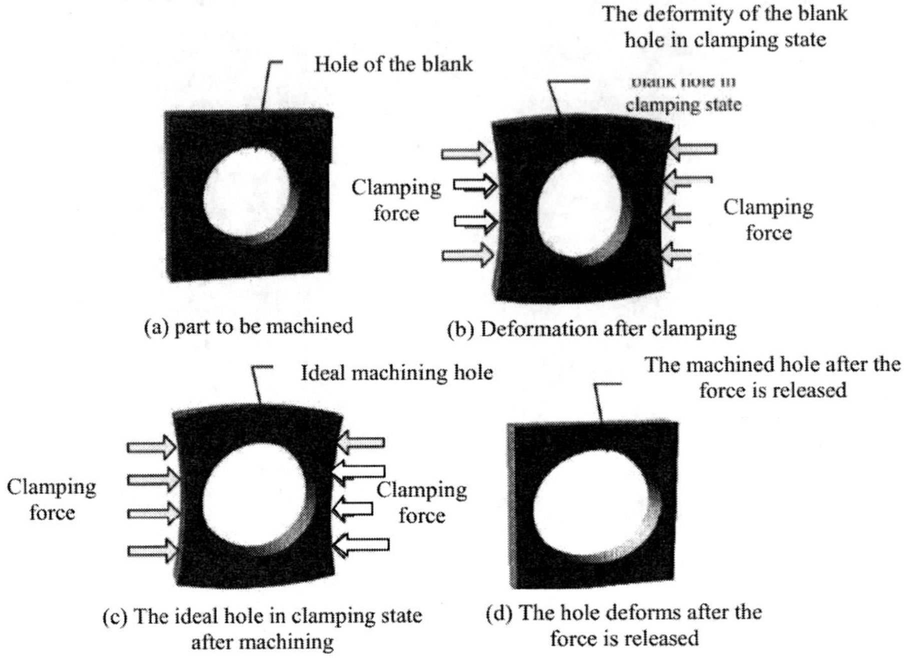

Figure 2. Machining error caused by clamping force.

After analysis, the proposed algorithm was found to be effective because, the deformation due to clamping force with FEM can be predicted, calculates the spring back deformation using grid mapping method, and obtains the machining error by error analysis and also the experimental values follow the simulation value.

They concluded:

- ❖ The error due to factors which are cutting speed, vibration, cutting heat, and tool wear is higher as compared to clamping released error.
- ❖ The clamping error can be minimized by improving the manufacturing planning and setup.

Winzenz and Dyler (2001) focused their study on effects of rotating unbalanced workpieces on the final turned product and its precision. The analytical as well as experimental results shows that in order to have tight concentricity, roundness, or surface finish tolerances of eccentric parts or workpieces, it is necessary to balance the chuck and workpiece assembly. The workpiece gets unbalanced due to casting weight variation or due to machining parts with interrupted cuts.

Zohoor and Yousefi (2019) studied the surface roughness and dimensional accuracy of hardened steel machined using CBN. The results showed that surface roughness changes with changing feed rate and dimensional accuracy gets affected by cutting depth and spindle speed. Nose radius does not affect dimensional accuracy. Dimensional accuracy also a function of the tool wear and vibration as they have major impact on it.

Yang and Hu (2016) investigated the relation between machining parameters and ovality in hollow shafts in Cross Wedge Rolling (CWR) process by examining the effect of circular as well as helical contact area between workpiece and tool with the help of finite element method. They concluded that ovality of hollow shafts can be minimized by decreasing stretching angle and rising the forming angle, also the ovality can be predicted by defining the coefficient which will make selecting process parameters simple.

IOP Conf. Series: Materials Science and Engineering 998 (2020) 012026 doi:10.1088/1757-899X/998/1/012026

Shendage and Borkar (2019) focused their study on reducing the ovality of KTM bike SMC (Starter motor cover) parts by optimizing the process parameters. The machining processes were facing, turning, step turning, grooving followed by boring. The analysis was done by using Taguchi method, ANOVA and RSM techniques. The conclusion was that ovality increases by increasing cutting speed, increasing depth of cut and decreasing feed rate and the defect arouse due to lower wall thickness and high heat generation. The optimized process parameters were feed rate (0.08mm/rev), depth of cut (0.3mm) and cutting speed (420rpm).

Ji et al. (2016) established a numerical simulation model to predict the ovality of hollow engine valve without mandrel in CWR process. After analyzing the experimental and numeric results, it was concluded that ovality was best with forming angle 30°-34°, stretching angle 5°-7°, mold void width 20mm-30mm and area reduction 65%-70%.

Ramesh et al. (2013) studied the effect of machining parameters on ovality of drilling holes on non-laminated Glass Reinforced Plastic (GFRP) composites. Taguchi's orthogonal array and ANOVA were used to study the correlation between ovality and parameters. The results identified the optimal parameter values which were 1000 rpm cutting speed, 10mm diameter twist drill and 0.15mm/rev. The ovality varies with drill rod geometry and the type of laminated composites.

Shastry et al. (2014) investigated the effect of process parameters on ovality of carbon-carbon composite plates which were used in space crafts. The HSS and TiN coated carbide material tools were used while drilling on CNC drilling machine. The comprehensive analysis concluded that better hole quality was achieved with TiN coated carbide tools and the optimal parameters were point angle 118o, cutting speed 2000 rpm, feed rate 100mm/min.

More et al. (2015) focused their study on optimization of process parameters while drilling GFRP composites using tungsten carbide tool. Taguchi method as well as L9 orthogonal were used for this designing the experiments and the experimental results concluded that feed rate is most influential factor in inducing ovality.

Shrivanshu et al. (2013) performed experiments on flow forming machine with AA6082 alloy flow formed tube considering axil feed of the roller, roller radius and speed of the mandrel as machining parameters. The analysis found that ovality is influenced by roller feed and controlling it can reduce this inaccuracy.

Nichit et al. (2019) eliminated the ovality problem arose during manufacturing of valve seat by clamping the workpiece in machine chuck by using fixture. Computer Aided fixture design, Feature Based fixture design, Flexible design fixture and the Computer aided mass balancing method is used while designing the fixture.

Hoffman et al. (2009) investigated the roundness deviation caused due to residual stresses arose during manufacturing of bearing rings. The experiments were carried with changing clamping conditions i.e. clamping the bearing from inner and outer diameter with simultaneous quenching during turning operation. After analysis, it was concluded that residual stresses induced after unclamping the bearing rings i.e. stress relief which causes plastic deformation and eventually roundness deviation.

IOP Conf. Series: Materials Science and Engineering 998 (2020) 012026 doi:10.1088/1757-899X/998/1/012026

Abe et al. (2016) studied the effects of process conditions on ovality caused during cold pilgering of tubes made of stainless steel and zircaloy-4. The experiments conducted with feed rate (1, 1.5 ,2 and 2.5 mm/stroke), stroke speed (140, 180 and 200strokes/min) and turn angle (40°, 60° and 80°). The results concluded that with increasing feed rate the residual stresses developed in tubes and hence spring back effect takes place which lead to ovality. Also, the turn angle has significant impact on ovality.

Kopac et al. (1999) worked for improving dimensional accuracy and reducing production cost during finishing of cold formed parts. The analysis concluded that the tolerances should be met by using near net shape machining technology and advanced tools.

Dijkman et al. (2007) provided an experimental setup for controlling the turning operation online i.e. during process. The setup includes ultrasonic measurement device, global control unit and fast tool unit. The tests of this setup showed that the clamping force of chuck can be compensated equally to eliminate ring wall thickness and diameter deviations.

Cicek et al. (2012) investigated the influence of deep cryogenic treatment along with drilling parameters while drilling of AISI 316 austenitic SS by M35 HSS twist drills. Taguchi technique was used for evaluating optimal control parameters and obtained minimum geometric deviations with 14m/min cutting speed and 0.08 mm/rev feed rate and they also found that changing cryogenic temperatures affects roundness deviation.

Deng et al. (2005) investigated the hole roundness in BTA deep hole drilling process by applying Taguchi method. The analysis concluded that the roundness deviation of holes can be minimized by optimizing control parameters which were tool diameter, shaft length, feed rate, rotational speed and day of the week.

3. Conclusions

This review paper identifies the reason behind occurrence of ovality in cylindrical components. It also puts forward the powerful tools and techniques to eliminate ovality problem.

1. In casting process, the mechanical deformations are dependent on the diameter of strand and hot pressure of molten metal. The thermal and/or mechanical strains can introduce ovality at circular sections. This can be reduced by using optimum value of hot pressure and diameter of strands.
2. In machining processes like turning, drilling and boring, the machining parameters such as cutting speed, depth of cut and feed rate plays an important role in causing inaccuracies in the component. The faulty selection of values can cause vibrations, large heat generation, chatter which can give rise to deformations and product rejection problems. Hence, optimization of machining parameters is important to control output parameters.
3. The ovality can also be introduced in the component due to elastic residual stresses. The elastic residual stresses get retained while drilling a hole in a component in clamping condition and when clamps are removed the geometry of the hole changes.
4. Sometimes the geometry of component is not symmetric which results in unbalancing of workpiece. This causes difficulty in clamping of component as well as unbalanced rotation of workpiece. In such case increase in clamping pressure may ensure sufficient clamping but majorly impacts on the shape change of the component. The fixture design is important in this case.

References

[1] Kulkarni C and Bagale G 2017 Defect identification in CNC manufactured products and their control measure, International Journal of Innovative Research in Science, *Engineering and Technology* **6**, pp 13930-13936.

[2] Han J, Liu K and Zhang J 2011 Ovality Defect Control for Round Casting based on Mechanical Deformation Analysis, *Advanced Materials Research* **154**, pp 151-157.
[3] Ghetiya N and Panchal S 2014 Some Studies on Reduction in Turned Component -Case Study, Nirma Univeristy *Journal of Engineering and Technology* **3**, 39-46.
[4] Apte P, Shah V, Pandya P and Ahmed P 2015 Minimization of Ovality of Circular Holes in Drilling Process, *Applied Mechanics and Materials* **800**, pp 393-396.
[5] Nowag L, Solter J, Walter A and Brinksmeier E, Effect of Machining Parameters and Clamping Technique on Residual Stresses and Distortion of Bearing Rings, *Mat. -wiss. u. Werkstofftech* **37**, pp 45-51.
[6] Görög A., Görögová I., Research of the Influence of Clamping Forces on the Roundness Deviations of the Pipes Turned Surface, Research Papers Faculty of Materials Science and Technology in Trnava Slovak University of Technology in Bratislava,26, 47-54, 2018.
[7] Maračeková M, Zvončan M, and Görög A 2012 Effect of clamping pressure on parts inaccuracy in turning, *Technical Gazette* **19**, pp 509-512.
[8] Shao X, Liu S, Zhang L and Lin Z 2013 Simulation of Workpiece Deformation caused by Releasing the Clamping Force, *Transactions of the Canadian Society for Mechanical Engineering* **37**, pp 703-712.
[9] Winzenz W and Dyer S 2001 Effects of Rotating Unbalance on Turning Precision: Analytical and Experimental Investigations and Real-Time Compensation, *American Society for Precision Engineering* **16**, pp 1-6.
[10] Yousefi S and Zohoor M 2019 Effect of Cutting Parameters on the Dimensional Accuracy and Surface Finish in the Hard Turning of MDN250 Steel with Cubic Boron Nitride Tool, For Developing A Knowledge Base Expert System, *International Journal of Mechanical and Materials Engineering* **14**, pp 1-13.
[11] Yang C and Hu Z 2016 Research on the Ovality of Hollow Shafts in Cross Wedge Rolling with Mandrel, *International Journal of Advanced Manufacturing Technology* **83**, 67-76.
[12] Shendage D and Borkar B 2019 Experimental Effect of Various Machining Parameters on Machining of Cylindrical SMC Part, *International Journal of Advance Research and Innovation Ideas in Education* **5**, 481-492.
[13] Ji H, Liu J, Wang B, Lin J and Tang X 2016 The Process Parameters Effect of Ovality in Cross Wedge Rolling for Hollow Valve without Mandril, *MATEC web of conference* **80**, pp 1-6.
[14] Ramesh B, Elayaperumal A, Balaji A and Rakesh N 2013 Optimization of Ovality on Drilling Glass Fiber Reinforced Plastic Composites with Coated Tungsten Carbide Tool, International Journal of Innovative Research in Science, *Engineering and Technology* **2**, pp 2801-2819.
[15] Shastry K, Rao V, Kumar M and Velayudham A 2014 Experimental Analysis of Hole Quality in Drilling of Carbon-Carbon Composites, *Applied Mechanics and Materials* **592**, pp 294-301.
[16] More U 2015 Influence of Process Parameters on Ovality in Drilling of GFRP Composites using Tungsten Carbide Tool, *International Journal of Advances in Computing and Management* **4**, pp 1-5.
[17] Shrivanshu M, Komaraiah M and Rao K 2013 Application of Response Surface Methodology to Predict Ovality of AA6082 Flow Formed Tubes, *International Journal of Manufacturing, Materials, and Mechanical Engineering* **3**, 52-65.
[18] Nichit V, Khandarkar S, Karwa N, Sonawane V and Patil K 2019 Ovality Correction in Manufacturing of the Valve-Seat, *International Journal for Science and Advance Research in Technology* **5**, pp 1158-1166.
[19] Hoffman F and Surm H 2009 Influence of clamping conditions on distortion during heating of bearing rings, *Mat. -wiss. u. Werkstofftech* **40**, pp 396-401.
[20] Abe H, Iwamoto T, Yamamoto Y, Nishida S and Komatsu R 2016 Dimensional Accuracy of Tubes in Cold Pilgering, *Journal of Materials Processing Technology* **231**, pp 277-287.
[21] Kopac J and Sokovic M 1999 Dimensional Accuracy and Cost Optimisation in the Finish Machining of Cold Formed Parts, *Journal of Materials Processing Technology* **92**, pp 335-343.

IOP Conf. Series: Materials Science and Engineering 998 (2020) 012026 doi:10.1088/1757-899X/998/1/012026

[22] Dijkman M, Stobener D, Allers S, Kuhfuss B and Goch G 2007 In-Process Control of Cutting Depth during Turning, *SICE annual conference*, pp 2984-2989.
[23] Cicek A, Kivak T and Samtas G 2012 Application of Taguchi Method for Surface Roughness and Roundness Error in Drilling of AISI 316 Stainless Steel, *Journal of Mechanical Engineering* **58**, pp 165-174.
[24] Deng C snd Chin J 2005 Hole Roundness in Deep-Hole Drilling as Analysed by Taguchi Methods, *International Journal of Advanced Manufacturing Technology* **25**, pp 420-426.

IOP Conf. Series: Materials Science and Engineering 998 (2020) 012027 doi:10.1088/1757-899X/998/1/012027

Lead Time Reduction for managing Engineering Changes

S. Kolapkar[1], R. S. Bharsakade[2] and A. U. Rajurkar[2]

[1] John Deere India Pvt. Ltd, Pune, Maharashtra, INDIA

[2] Department of Industrial & Production Engineering, Vishwakarma Institute of Technology, Pune, Maharashtra, INDIA

E-mail: kolapkarswapnil@johndeere.com

Abstract. Engineering changes need to be done in a product because of design improvement, product and process optimization, government regulations, etc. Implementation of those changes in the production shop floor involves multiple stakeholders and functions within and outside the company. Engineering change management (ECM) is the process of communication between all stakeholders to implement and document the engineering change on the shop floor. This involvement of multiple stakeholders and responsibilities increases ECM lead time and creates a challenge for change implementation on the shop floor. This paper presents such a case of engineering changes for factory shop floor implementation. Causes like delay in decisions behind an increase in lead time have been investigated. Process automation solutions that can reduce the ECM lead time drastically have been proposed and implemented. The developed solution for tool selection and layout problems achieves almost a 66% reduction in lead time and helps the factory for process standardization and resource optimization.

Keywords: Engineering change Management, lead time, Process standardization

1. Introduction

Engineering change management (ECM) is related to are changes made in forms, fits, functions, materials, dimensions, etc., of products or Processes. ECs usually induce a series of downstream changes. Multiple stakeholders are, therefore, involved in change management [1]. The involvement of various stakeholders and responsibilities increases ECM lead time and creates a challenge for change implementation on the shop floor. Multiple communications between stakeholders add waiting time in ECM process. Most of the research conducted so far is on ECM requirements and the effect of engineering changes for different product types. However, research needs to be done on the micro-level processes of ECM and interrelation between ECM stakeholder's activities.

In this study, we used lean methodology to solve this problem here as we know lean tools are the best way to reduce non value added activities and improve efficiency. In this study, we first mapped the current state for the ECM process. After current state mapping, we analyzed the process in detail to investigate the root causes for ECM higher lead time .as a result of analysis we could find that multiple communications with stakeholder for Torque assignment is the primary root cause contributing highly to increase ECM lead time. We developed a process automation application that can suggest preferences for tool assignment based on the current tooling data. With the help of this application, decision making related to tooling assignment is done with a single communication. ECM lead time reduced by reducing the waiting time and process standardization.

IOP Conf. Series: Materials Science and Engineering 998 (2020) 012027 doi:10.1088/1757-899X/998/1/012027

2. Engineering Change Management

Engineering change management (ECM) is an organized way to document changes in the product through effective planning from its identification to implementation. Before starting work for solutions on this problem, a detailed literature review has been done. But there are very few case studies or ECM related papers available. Research has been done mainly on the comparison between EMC paperwork web solutions have also tried [1] for ECM smooth flow. It was mostly related to the reduction of ECM throughput time by web base solution.

There have been several industrial case studies on the ECM subject. Saeed et al. [2] have conducted an intensive case study for studying the effect of sound process knowledge on the Engineering changes. In the case of our problem statement, despite strong Process knowledge, we need to work on EC due to engineering changes and continuous improvement activities.so for us, EC reduction is not an option. Hegde et al. [3] have carried out a field investigation on the impact of Engineering Change Orders (ECOs) on the completion time of jobs in repetitive manufacturing environments. In this study, the effect of engineering changes on part supplier and their quality issues and effect on the overall impact on Manufacturing Schedules. This study is talking about external factors that may not always be in our control. Balcerak and Dale [4] have examined ECM in an automotive manufacturer and made several specific suggestions on EC classification and prioritization, and selectivity analysis. Balcerak and Dale put the focus on the difference between purpose and reason for the engineering change. Based on this literature review, we defined the purpose of our project is to improve efficiency by reducing decision-making time. Harhalakis [5] has carried out industrial case studies on engineering change management for made to- order products. He studied the effect of many Engineering changes on the bill of material and material requirement planning and came up with a system that integrates these Processes. Watts [6] has told his experience in re-engineering the EC control process in an electronic product manufacturer. Perhaps, Dale's early work in 1982 was among one of the best papers that provided a comprehensive framework and yet specific proper practice guidelines [7]. Based on his company case studies, Harhalakis [5] also outlined an ECM system. These contributions not only identified ECM activities but also described responsibilities. The literature has highlighted a clear message that EC is a severe issue that cannot be underestimated, and ECM is of significant concern to most companies that design and manufacture products. This has been confirmed in a survey conducted by the authors within the UK manufacturing industries [8]. In addition, the limited number of references has also highlighted that ECM is, in general, under-researched, and ECM deserves much more attention compared with its extent of severity, as highlighted in the literature.

3. Objectives of the study

Considering all the literature research, so far, research has happened on the ECM process, its components, Style of ECM execution (Paper-based or web-based). The main objectives of this study:

i) To identify and remove internal non-value-added activities carried out in the process.
ii) To study inter-relation between various ECM activities and the stakeholder involved.
iii) Develop effective communication strategy among various stakeholder, and avoid decision waiting time between multiple stakeholders to reduce ECM lead time

3.1 Problem Statement -

A manufacturing factory has some process change activities going on because of changes by product engineering. Those changes must be absorbed on the shop floor to start the production. To check its feasibility and system integration, those changes are updated in the system before it goes for implementation on the shop floor. This task of updating changes and its documentation in the system has been assigned to a manufacturing service provider. The service company works remotely and has only virtual access to the factory shop floor. Therefore, at present service company has to rely on the factory for most of the inputs and decision making.

Since the service company person is not able to make all decisions, he must wait for the response from the factory. In most of the cases, response time is up to 3 days. Hence, till the time the current ECM is waiting for a response, one has to work on multiple ECMs in parallel. Various touch points for single ECM lead to an increase in the execution time of ECM by 15%, depending on the complexity of ECM. As in this problem, we are dealing with waste like waiting time for decision and multiple touch points for an ECM, and we can use lean methodology.

4. Problem Priority Matrix

After analysis of the complete Engineering change management data from history, it has been observed that some steps in change management are heavily contributing to an increase in change management lead time increase. Execution time vs Lead time comparison shows that for ECMs, execution time is 0.5 to 2 hrs whereas the lead time for these ECMs is up to 72 hrs (Shown in fig.1).
Lead time = time elapsed between receiving requests from factory for Engineering changes implementation and the change implementation and documentation completed in the system by the service company.
Execution time = Actual time spent by service company resource for implementation and documentation of the change.

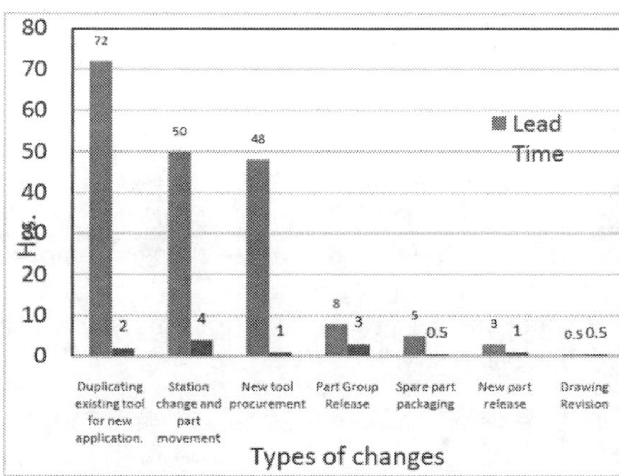

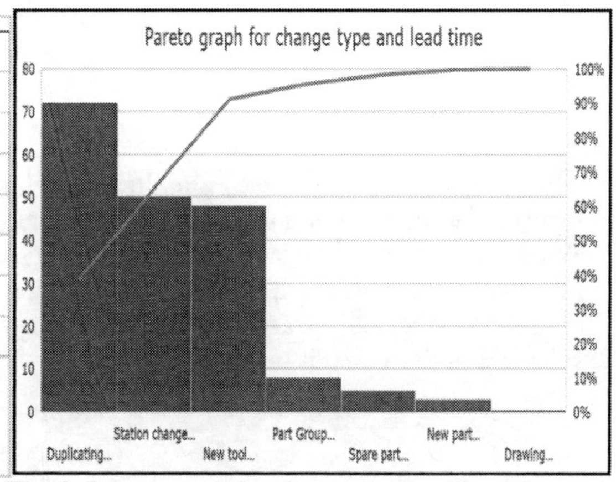

Fig.1. Execution & Lead Times for types of changes Fig. 2. Pareto graph for change type and lead time

Lead time = Execution time + time for decision making on changes + waiting time for inputs based on a decision made

From problem priority matrix and Pareto analysis for various types of changes handled by a service company and execution time required for these changes, below are the changes are contributing for more lead time for changes implementation because of waiting and decision-making time (As per Fig. 2) -

- Duplication of existing tooling's for a new application.
- Assembly part station changes
- New tool procurement for assembly stations

Therefore, based on the Pareto chart, changes that are related to new assembly tooling and torque controls, existing tooling duplications are contributing heavily to an increase in change implementation lead time. So, only these types of changes have been considered as a scope for this project.

5. Current state mapping

Current state Mapping shows the current Process flow. It shows the current relationship between multiple stakeholders involved in Engineering change management activities.

5.1 Change management Process flow chart

Change management process flow chart shows typical 4 phases of change management like input, Analysis, Execution, and Delivery to customers.

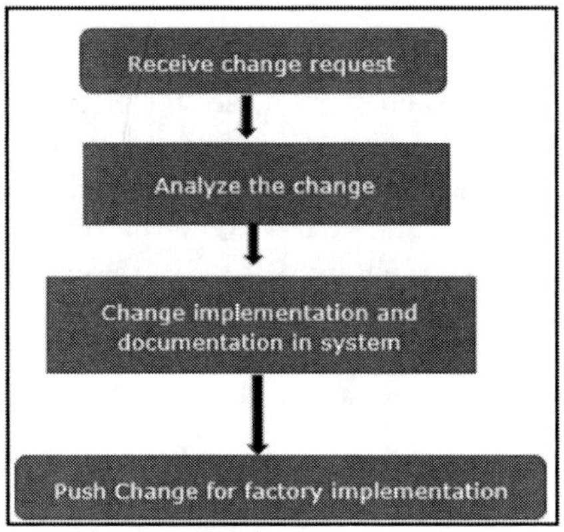

Fig. 3. ECM flow process chart

5.2 SIPOC

In Order to understand the complicated relationship between stakeholders and process, we have mapped the Engineering Change Management process under SIPOC. (Fig.4)

Wherein the supplier is a factory that provides inputs like design and Process changes happened. In the process, we have listed down the detailed process of tooling assignment ECMs, which gives a comprehensive understanding of process flow. Output in this process is the changes made in the system and delivered it back to the customer, which is Factory ME.

SIPOC DIAGRAM

SUPPLIERS	INPUT	P	PROCESS	OUTPUT	CUSTOMER
S: who is providing input to a process	I: resource provided by supplier for incorporation to process	P	steps taken to convert input to output	O: resource resulting from process	C: receiver of newly created output
Factory ME team	Process changes	1	Study the change	Complete the ECM with correct tool selection & assignment	Factoy ME team
		2	Find out new/changed part & sub-parts.		
Product Engineering team	Parts Changes in Product	3	Study assembly instruction for the options for details of torque values on assembly		
		4	Locate the work-center where new/changed parts get assembled		CUSTOMER REQUIREMENTS
		5	Compare new/changed part by ECM with other parts		
		6	If we find same/similar part then no new tool is needed		
		7	If new/changed part is not matching to other then we need new tool for the same		
		8	Identify new tooling requirements		
		9	Tools selection & assignment with discussion with ME team in Factory		
		10	Complete tool assignment		
		11	Review the ECM		
		12	Push the ECM in Next stage		

Fig. 4. SIPOC

5.3 Swim Lane diagram

Based on the detailed Process flow from SIPOC Diagram, we mapped the flow or Steps of the transactions that happen between stakeholders like engineering, Unit ME, and service company.

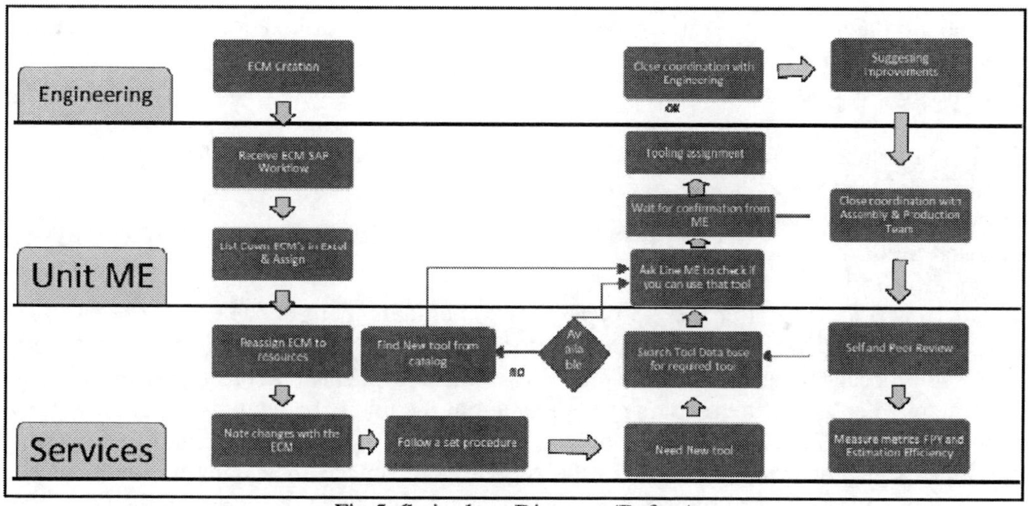

Fig.5. Swim lane Diagram (Before)

6. Analysis

6.1 Fishbone Diagram

In Current state mapping (Fig. 5) we could understand that multiple transactional activities between Factory ME and Services for information and decision making are causing higher ECM lead time.

To understand root causes for this multiple communication for information's and decisions we plotted Fishbone diagram.

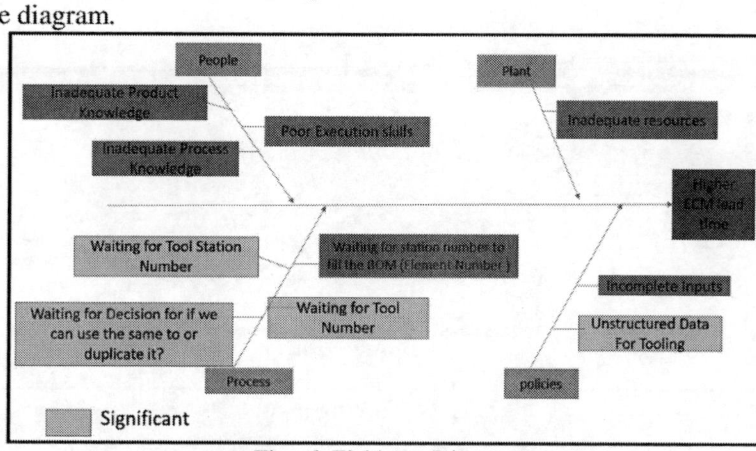

Fig. 6. Fishbone Diagram

6.2 Why-Why Analysis

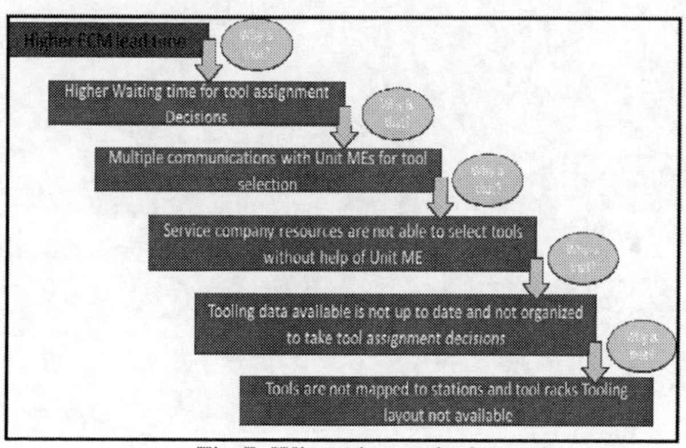

Fig.7. Why-why analysis

From fishbone diagram (Fig. 6) it can be seen that waiting for tool station number, waiting for the decision for if we can use the same tool or duplicate it, waiting for tool number are the reasons from method category and unstructured data tooling from material category leading to higher ECM lead time

Tools are not mapped to the station, and tool racks layout not available is the significant root cause for higher ECM lead time.

7. Proposed Solution

7.1 Tooling Layout

There are 10 ME looking after these 3 departments the main changes have been observed is related to standardization of data storage related to tooling especially not standard tooling used on the assembly line because standard tools which are connected to manufacturing execution system can easily be tracked or assigned based on system data, but the main challenge is about custom tool built for specific product order use. These tools are not mapped to stations rack, and there is no custom tooling racks standard layout as such, so the service company team assigns the tools based on old references, which leads to quality issues. (Need to create tooling layout and map the racks to stations with the alignment of factory Engineer)

So, we developed the tooling layout and mapped all stations to particular tool racks (see Fig.8). Then with alignment with factory ME rules have been set for the use of tools from specific shelves. These rules have been used to set logic for tool assignment automation.

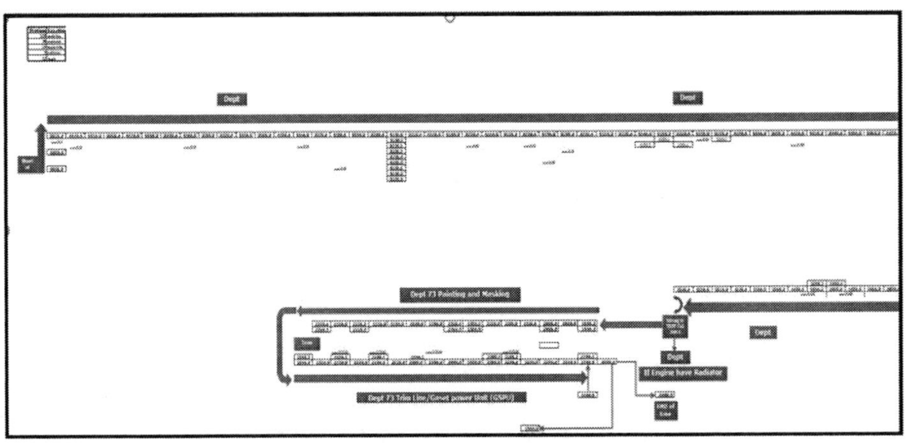

Fig. 8. Tooling layout

7.2 Tool selection Automation

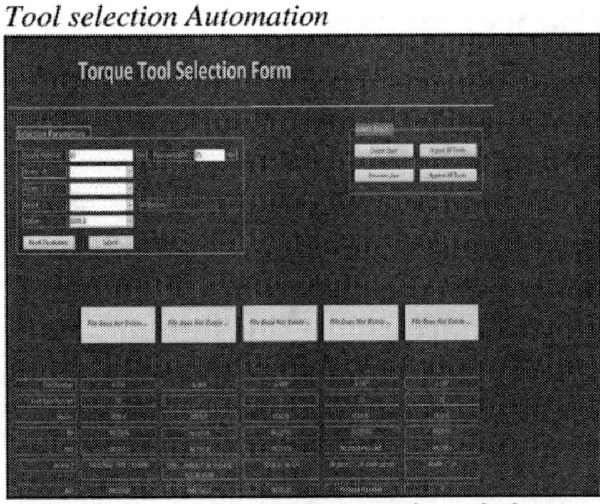

Fig. 9. Tool selection Automation window

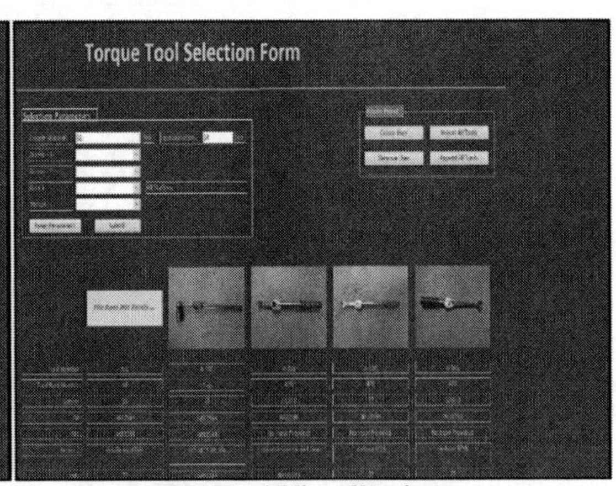

Fig.10. Tool Visualization

Another problem observed while analysis of the process is that there is no standard procedure available to track tooling information. So, each ME keep that information in a different format as per different departments requirement. (Need to create a location where all ME will update the tooling information at the standard format). After making tooling layout, tool assignment rules, data tracking standardization, we developed tool assignment applications (Fig. 9 and 10). It gives preferences for tool assignment based on hardware attributes, assembly station, and torque specification. It also suggests preferences for tools duplication and torque standardization across multiple stations.

We added all images of tools and labels for visualization and easy identification of the tool. We defined the nomenclature for the tools. So, it can be easily identified and have the track for the stations and tool rack.
3D models for these tools are not available in PDM as these are a non-standard and custom tool; adding images and labels helps service company person to validate tools reachability and accessibility considering other parts to be installed on a particular station. (Fig 11 and 12 shows the difference in tool socket length)

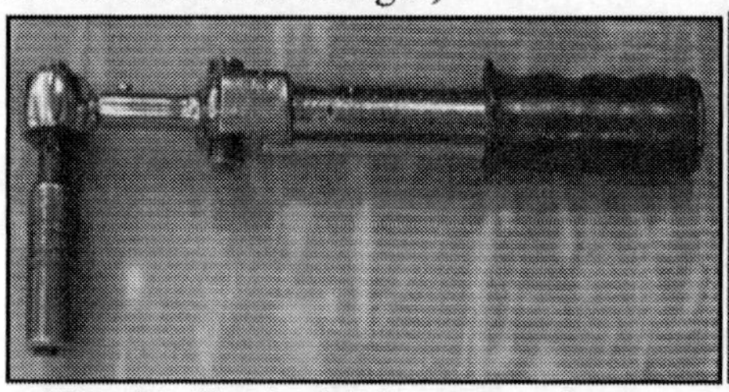
Fig. 11. Tool with longer socket

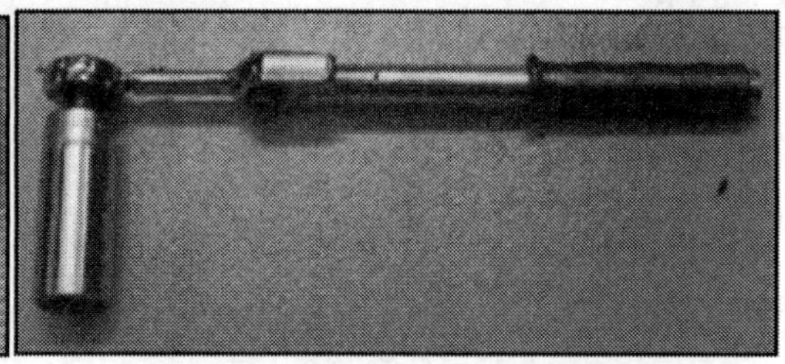
Fig. 12. Tool with short socket

8. Implementation

This solution has been successfully implemented for both manufacturing factory for all tool tracking standardization such as tool data tracking, tool racks mapping, tool usage rules and service for tool selection automation.

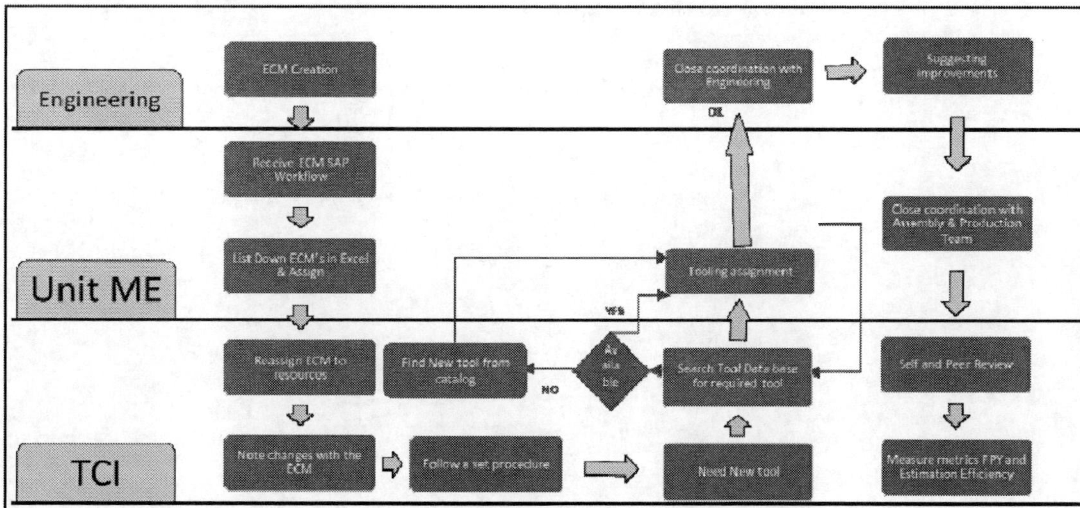

Fig.13. Swim lane Diagram (After)

Success for this project can be seen from the above swim lane diagram after implantation of this project. Tooling assignment multiple communication and waiting for decision has been completely removed with the new process. New tooling addition and old tool deletion form has been provided in application to be able to update the tooling database in standard way. Limited user access has been provided so as to keep data up to date and controlled.

9. Results

Lead time for tooling assignment ECM has been reduced from 3 days to 1 day (66% reduction). That too, this 1 day is for complete looping of ECM. The number of tooling assignment related communication has been reduced to 1 from 3 earlier. Tooling assignment time has been reduced to on an average 4.34 min from 12.7 min earlier (65% reduction).

10. Conclusions

In this project, we have successfully used lean tools for service companies and reduced the service lead time by reducing the waiting time for decision making. This project has reduced the multiple communications between manufacturing factory and service industry. Hence, there is reduced the dependency on manufacturing factory to get all inputs. Also, numerous touch points per ECM has been minimized, which at the end resulted in overall execution time for change implementation in the system. Also, the factory is getting a lot of suggestions torque standardization by using this application which in turn going to reduce the process complexity online and reduce the number of tooling count in factory.

Acknowledgments

Authors express their sincere gratitude towards the management of Vishwakarma Institute of Technology Pune and John Deere India Pvt Ltd, Pune, for their resource courtesy.

References

[1] G Q. Huang, W.Y. Yee, K. L. Mak 2001 *Robotics and Computer Integrated Manufacturing* **17** 255:267
[2] Saeed I, Bowen D M, Sohoni V S.1993 *IEEE Transactions On Engineering Management*, Vol.**40**, no.1
[3] Hegde G G, Kekre S H, Su H.1992. *Elsevier Internutional Journul of Production Economics*, **28**
[4] Balcerak K J, Dale B G. 1992 *Computer-Integrated Manuf Systems*;**5** (2) 125 32
[5] Harhalakis G.1986; *Elsevier Science Publishers* B.V **4** 19 36
[6] Watts F.1984 *Prod Invent Manage* J;**25** (Part 4) 55 62
[7] Dale B G.1982 *Engineering Management international Elsevier* **1** 201-208
[8] Huang G Q, Mak K L.1997 *Proceedings of International Conference on Managing Enterprises*, Loughborough University

IOP Conf. Series: Materials Science and Engineering **998** (2020) 012028 doi:10.1088/1757-899X/998/1/012028

Effect of Spindle Speed, Feed Rate and Cooling Medium on the Burr Structure of Aluminium through Milling

Aman Sharma[1] and Vijay Kumar Dwivedi[1]

[1]IET Department of Mechanical Engineering
GLA University Mathura 281406

Email: meamansharma02@gmail.com

Abstract. The burr is formed during milling, when the tool exits the work piece in any machining operation. The burr formation on a milling machine with the use of end mill cutter with different speed and different cooling mediums like air, water, and cutting fluid are discusses. The tool used for the machining operation is an end mill cutter tool. This work is taken up to examine the effect of different speeds and different cooling mediums on burr formation in the milling operation of aluminium in a dry and wet atmosphere. Results are found on the exit edge of the work piece, found during the milling operation. This paper manages with the aspects that influence the formation of burr and type of burr formed while doing experiment. The parameters considered are, the maximum length of stroke, drive speed (step cone pulley), range of speed arrangement groove and the selection of a cooling medium. The drive speed of the milling machine was measuring by Tachometer in all drive stages. The pictures of produced burrs were taken with the support of a digital microscope and digital camera.

1. Introduction

Burr control and evacuation have been among the most relevant economic considerations in machining processes and so have been the objective of research in metalworking activities over the past five decades [1]. Milling machines are mechanical devices which by method for chip evacuation give work pieces the ideal shape, measurement and surface accomplishment. They produce, for the most part, level surfaces, shoulders, grooves, and comparative shapes. It is, not withstanding, additionally possible to make circularly curved surfaces. The cutting is affected with single-edged tools in the interfered with cut with a working stroke and an arrival stroke. Milling machine is one of the utmost adaptable and predictable machine tools with a varied variety for the capability of metal cutting. Most of the complex operations such as straddle milling, gang milling, indexing and making of gears and slots etc. can be passed out on a different type of milling machine as per the requirement of operation. This preparation module is proposed to give you a decent valuation for the kind of milling machines and the different sorts of processing forms. Accentuation is set on its mechanical applications, activities, and the determination of suitable cutting instruments. Since molding can be comprehended as turning of workpieces with boundlessly huge distance across, there are numerous comparable perspectives among shaping and turning and shaper and turning devices. The primary reason for milling is much of the time to expel generally enormous measures of material as chips. This is applied, for instance, to evacuate the aluminium scale and to get flat and adjusted surfaces or to cut long or substantial parts (for machining plate edges in addition to other things). There are different types of operation which have been performed on milling machine; they are horizontal surfaces, upright surfaces, sharp surfaces, asymmetrical sides, cutting holes, channels and keyway, and gear cutting. The size of the milling machine is determined by the dimensions of table, speed of spindle and the feed rate, ram travel, ram rotation angle, spindle rotation angle are available. In this investigation the formation of the burr on a milling with different speed measures and different cooling mediums like air, water and cutting fluid. The added, non-productive deburring process is employed for the subtraction of burr in practice. The tool used for the machining operation is an end mill cutter tool, and the various experimental outcomes have been performed on the formation of burr for its amount. This work is taken up to examine the effect of different speeds and different cooling mediums on burr formation of aluminium in a dry and wet atmosphere, and validating the experimental results. Numerous researchers have been worked on the formation of burr to minimize the deburring method usually on the development and deburring of burr. He finds that the price of process of deburring had been about 30% of the total cost of production of precise component.

In direction of drilling holes though, the specific material component of the stacks places very different demands with favor to cutting edge design and optimal process parameters. To one side from that, the requirements of quality are very high which is concerning the consistency for the assembly of joints. The results of errors on drilling parameters of surface roughness and roundness were inspected in drilling of material Al 6061 alloy using with the help of twist drill of high-speed steel (HSS) by Reddy Sreenivasulu and Srinivasa Rao [2].

Eynian presented a connection between the cutting forces during the machining and the wear condition of twist drills and the relation between the forces during the cutting process as well as between the formation of burr and the quality of surface for titanium alloys. As per a result of his study, Eynian found that the normal force on the tool and the feed force on the tool improved to approximately four times of the amount with growing of wear. Torque, still, rise only slightly nearby 70 % with the growing wear [3]. Bajpai et al. inspected that formation of burr of titanium alloy "Ti6Al4V" in high speed micro-milling. They determined that most critical and larger type of burrs is side exit burr in up milling sideways. The height of burr decreases when the feed and speed of spindle increase. With the increment in depth of cut the stature of burr rises. The width of burr is only affects the depth of cut. The width of burr rises with the increasing in depth of cut. The proportions of burr does not change knowingly by the tool diameter and the number of flutes.[4]

Most of the work was done on the mechanism of formation of burr via the finite element method, and the simulation is to validate the mechanism [5-6]. Saha and Das investigational found out that there is a considerable effect of exit edge beveling on the formation of burr in milling [7,8]. Saha and Das and Das et al. [12-13] performed an experimental examination and analysis of stress via the finite element method to figure out the influence on orthogonal machining operation for the exit edge bevel angle on the burr and formation of foot through machining process [9,10].

According to [11], the machining of different ductile materials has a tendency to form bigger burrs, predominantly at a complex level of feed rate and cutting speed of tool. According to Ko and Lee [12], the properties of their material had added an outcome on the burr size while drilling than the feed rate. The investigative model proposed in [13] that were accomplished to forecast the type and size of ductile materials while machining operations were performing on milling, drilling, grinding machine of exit burrs.

Form the literature review it is originate that the groove and the development of burr are generally exaggerated by different parameters like depth of cut spindle speed and feed rate and the ram travel, Different coolants are used (air or water and a mixture of cutting oil and water), the material of the workpiece used and the tool used for the machining operation. All the said literature points visibly, which supports the declaration of influencing the aspects for the formation of burr in the process of milling. In these experimentations, there are three input parameters that well-thought-out the ram travel, speed of spindle type of tool angle of tool, angle of tool and feed rate and the selection of cooling medium. All of these parameters are self-determining to each other and distinct cases were occupied in all these steps of examination. By the use of tachometer drive speed of milling machine is being measured in each of the cases, "while a channelized path was created for the cooling medium with the help of Drip-Case attached to the water bottle and the end mill cutter tool was taken for making the grooves".

To conclude that the pictures of different produced burrs on different speeds and different cooling mediums were taken with the help of digital camera and digital microscope. "This experiment will support in choosing the suitable, material and tool, cooling medium for a particular milling machining process and upcoming work on other materials possibly will be done by varying the different limitations revealed above as per the necessity".

There are different types of burrs on varying different speed on the vertical milling machine and different cooling medium. All of them lies below the group of horizontally or vertical burrs. They consist of unbroken burr. This kind of burr is insignificant and consistently spread around the groove. The following figures illustrate these figures in their particular order.

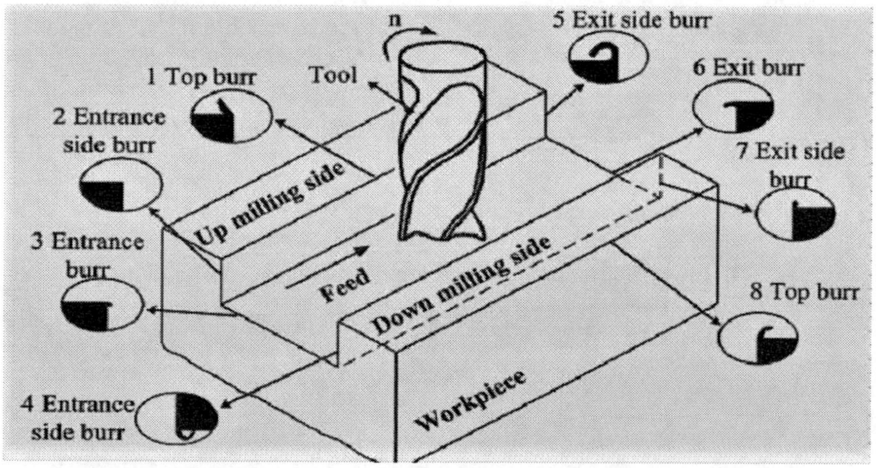

Figure 1. *Different types of burrs form on milling machine* [14]

2. Experimental Methodology

This investigation, is carried out on the vertical milling machine and the parameters and working consideration of vertical milling machine are in the given Table.1.Differentparameters like depth of cut and different spindle speed in Table 2, spindle feed rate, rotation of spindle angle, travel length of ram speed of milling and three cooling medium viz, air, water and the mixture of cutting oil and water have been used separately in a different way to understand the shape and size of burr which is formed on milling machine . 16mm end mill cutter tool is used for milling operation as shown in Fig 3. Different speed rates has been calculated and explained in the Table 3 and the operation on a milling machine is going thru with different speeds (rpm), and different cooling mediums, the machining operation is done with the use of end mill cutter tool freely. To ensure that the non-stop flow of water during the groove process, with the use of a drip case, has remained fixed to the bottle and the adjustment of knob for managing the movement (flow) of the cooling medium, and the burrs formed during without use cooling medium.

2.1 Working Considerations

Table 1 shows that the type of machine tool and tool used for machining, geometry of tool and its angles and type of material for machining operation and the physical and mechanical properties for material which is aluminium, and the dimensions for the specimen and the machining conditions like type of machine, speed drive, depth of cut and environment for operation like air (dry), cutting oil + water (wet).

Table 1. Working Considerations

Type of Machine Tool	Vertical milling machine
Tool Used	High-speed steel (HSS) end mill cutter
Geometry of tool	Cutter dia 1 /32, number of grooves 4, helical angle 35°, diameter 16mm, rake angle 7°, gash angle 40°, radial primarily clearance angle 8° radial secondary clearance angle 15°
Material for workpiece and its properties	Aluminium, Density 2720 kg/m^3, Melting point 650.5°C, Tensile Strength (50-600) MPa, Compressive Strength (100-600) MPa
Dimensions of workpiece	10 mm thickness, 90mm length, 75mm width
Machining Conditions	Type of machining: Orthogonal shaping Speed of spindle 260, 340, 550 (rpm) Depth of cut (t): 4 mm, width of cut: 16 mm Atmosphere: air (dry), cutting oil+ water (wet)

Table 2 shows the different speeds of spindle of vertical milling machine.

Table 2. Showing different speeds taken for experiment

No.	Speed of spindle in (rpm)
1.	260
2.	340
3.	550

Different speeds of spindle of milling machine drive speed and different feed rates, time, number of rotations and depth of cut were occupied and revealed in the Table 3 for the investigation, and obtained dissimilar results. Technique for the accomplishment the different speeds of spindle of the vertical milling machine. The speed of the spindle of vertical machine is set on a specific spindle and the speed was record by means of the "tachometer" by engaging over the groove. Different speeds of spindle (rpm) and feed rates in (mm) have been recorded.

Table 3. For measuring the different speeds and different rates

S.No	Rpm	Time (sec)	Feed rate (mm)	Depth of cut (mm)
1	260	124	0.037	4
2	340	62	0.075	4
3	550	40	0.204	4

3. Types of cutting tools used for this experiment

In this experiment of finding the shape of burr, an end mill cutter tool has been used at each of the speeds as mentioned above as without a cooling medium and with the use of air-cooled medium and water-cooled medium and the mixture of cutting oil and Water, Fig 2 show the cutting edges of end mill cutter with 4 flutes and Fig 3 shows the end-mill cutter.

IOP Conf. Series: Materials Science and Engineering **998** (2020) 012028 doi:10.1088/1757-899X/998/1/012028

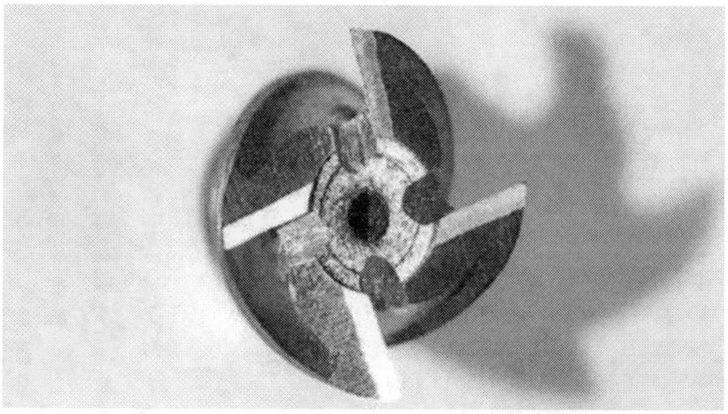

Figure 2.. Cutting edges of end mill cutter (4 Flutes)

Figure 3. End mill cutter

4. Results and Discussions

Here are the different forms of burr that can be shaped from machining process: "Poisson burr, rollover burr, and breakout burr". The burrs which have been formed during the machining operation on vertical milling machine with the use of end mill cutter are the rollover burrs and the breakout burrs. One of the most common type of burr is rollover burr. The three cases of burr formation on aluminium workpiece with different speeds that are mention in Fig. 4. Slot 1 in Fig. 5 is done with cutting oil and slot 2 is done without cutting oil respectively.

In this we achieve that the formation of burr though machining on milling machine on different parameters like speed of spindle feed rate and depth of cut with different medium like air and cutting oil with water. While using cutting oil the formation of burr is less and the surface finish is good as compared to that of air medium.

4.1 Findings

The different processes parameters which effect the nature of burr formed like, depth of cut, type of tool and angle of the tool, speed of spindle, feed rate, and the proper proportion for the combination of the cutting fluid.

In situation of without using the cooling medium, the creation of burr is high when the speed is high while at the low speed the formation of burrs are uniform, rollover burrs are commonly forms in this. In case of using a cooling medium. The formation of burr is less because of cutting fluid and good surface finish, breakout burr is found in this practical investigation.

IOP Conf. Series: Materials Science and Engineering **998** (2020) 012028 doi:10.1088/1757-899X/998/1/012028

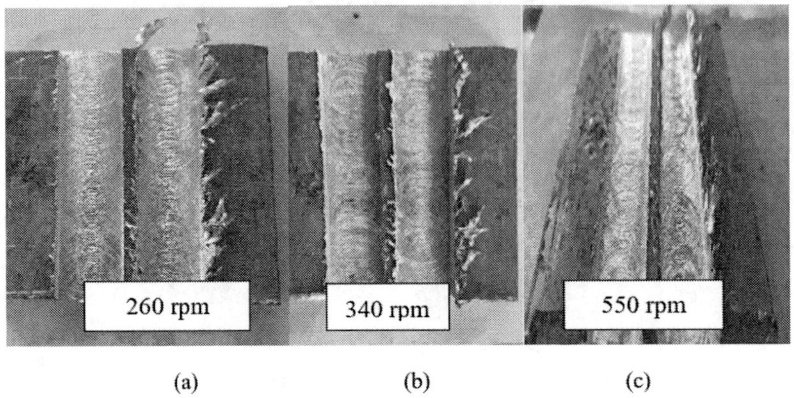

Figure 4. Formation of burrs on different speeds with cutting oil or without cutting oil.

From the above Fig 4 shows the three cases of burr formation on aluminium work piece with different speeds that are mention in Table 2 respectively.

Figure 5. Machining without cutting fluid or with cutting fluid

Fig 5 shows that the two conditions of burr formation on milling with the use of end-mill cutter the slot one show the formation of burr without using cutting fluid and the slot 2 with the use of cutting fluid we see that the in case 1 the burr are in large shaped and pointed structure burr like rollover burr and breakout burr. The depth of cut is 4mm. Fig 6 shows the manual deburring.

Figure 6. Manual De-burring

5. Conclusions

One and only of the most significant and valuable machining activities is milling establishes to 30-40% amongst all the forms of machining like "flat surfaces, vertical surfaces, precise surfaces, irregular surfaces, slot cutting, grooves and keyways, or gear cutting". The arrangement of burrs is one of the most open and unwanted things that proceeds while performing machining tasks. The burr which is answerable for diminishing the at long last fabricated part and furthermore makes its quality terrible. In this way, it is attractive to evacuate the burrs formerly the last completing activity. Accordingly, to identify the type burr shaped and the parameters that influence the superiority of burr shows countless reputation.

6. References

[1] Stuhrmann, J.HerausforderungenbeimBohren von CFK und CFK-Mischverbundenbeim Airbus A350XWB, 6. IfW-Tagung. Bearbeitung von Verbundwerkstoffen – SpanendeBearbeitung von CFK, 20.10.2016, Stuttgart.

[2] Reddy Sreenivasulu & Srinivasa Rao Ch, Application of Gray relational analysis for surface roughness and roundness error in drilling of Al6061 alloy, International J of Lean Thinking, 3 (2) (2012), 67 – 78.

[3] Eynian, M.; Das, K.; Wretland, A. (2017): Effect of tool wear on quality in drilling of titanium alloy Ti6Al4V, Part I: Cutting Forces, Burr Formation, Surface Quality and Defects. In: High Speed Machining 3 (1), 1–10.

[4] Bajpai V, Kushwaha AK, Singh RK. Burr formation and surface quality in high speed micro milling of titanium alloy (Ti6Al4V). Proceedings ASME Int Manufacturing Sci Eng. Conf, Wisconsin, USA 2013.

[5] Saha, P. P., and Das, S., 2011, "Burr Minimization in Face Milling: An Edge Beveling Approach", Proc. Institution of Mechanical Engineers, Part B: Journal of Engineering Manufacture.

[6] Saha, P. P., and Das, S., 2011, "An Experimental Approach to Investigate the Effect of Process Condition on Exit Burr Formation in Shaping an Aluminum Alloy", Manufacturing Technology .

[7] Saha, P. P., and Das, S., 2010, "A Study on the Effect of Process Parameters and Exit Edge Beveling on Foot and Burr Formation During Machining of Medium Carbon Steels", Proc. National Conf. on Recent Advances in Manufacturing Technology and Management, Jadavpur, Kolkata, pp.30-35

IOP Conf. Series: Materials Science and Engineering **998** (2020) 012028 doi:10.1088/1757-899X/998/1/012028

[8] Das, S., Saha, P. P., and Das, A., 2011, "An Experimental Study on Burr Minimization in Shaping of En25 Steels", Proc. 5th Int. Conf. on Advances in Mechanical Engineering, Surat, Gujarat, India, pp.701- 705.

[9] Das, S., Saha, P. P., and Das, A., 2011, "An Experimental Study on Burr Minimization in Shaping of En25 Steels", Proc. 5th Int. Conf. on Advances in Mechanical Engineering, Surat, Gujarat, India, pp.701- 705.

[10] Das, A., Saha, P.P., and Das, S., 2011, "Investigation on Burr Formation in Shaping of En25 Steels: Through Experimental and Stress Analysis Approach",

[11] Kim J, Dornfeld DA. Development of an analytical model for drilling burr formation in ductile materials. Journal of Engineering materials and technology. 2002;124(2):192-8.

[12] Ko S-L, Lee J-K. Analysis of burr formation in drilling with a new-concept drill. Journal of materials processing technology. 2001;113(1):392-8.

[13] Segonds S, Masounave J, Songmene V, Bès C. A simple analytical model for burr type prediction in the drilling of ductile materials. Journal of Materials Processing Technology.2012; 213(6): 971–977.

[14] Burr Controlling in Micro Milling with Supporting MaterialMethodZhaojun Kou, Yi Wan*, Yukui Cai, Xichang Liang and Zhanqiang Liu1 Key Laboratory of High Efficiency and Clean Manufacturing, Shandong University,CHINAProcediaManufacturingVolume 1, 2015, Pages 501–511.

EFFECT OF TWO STAGES ARTIFICIAL AGEING ON MICROSTRUCTURE AND MECHANICAL BEHAVIOR OF LM 25 ALUMINIUM ALLOY

BHASKARA RAJU VVS[1], SAMBI REDDY SEVANAM[2], ANANDA BABU VARADALA[3], SURESH CHITTURI[4*]
[1,3,4]*Mechanical Engineering Department, Vignan's Institute of Engineering for Women, Visakhapatnam, India.*
[2]*Retired NSTL Scientist D, Visakhapatnam, India*
Email: chitturisuresh007@gmail.com

Abstract. In the present investigation the influence of two stages artificial ageing, 1st stage at 160^{0}C followed by 2nd stage at 140^{0}C on microstructure and mechanical properties of LM 25 alloy was studied. It is observed that the microstructure changed under the influence of two stages ageing. It is also observed that the decrease in ageing temperature in the second stage, increases mechanical properties of the alloy. The tensile strength of the alloy in single stage ageing at 160^{0}C is studied and compared with two stages ageing.

Keywords: Artificial ageing; Microstructure; Tensile strength; Hardness.

1. Introduction

Aluminium and its alloys exhibit multiple desirable properties like high resistance to corrosion, light weight and good strength, which made it possible for its extensive use in very broad range of applications from regular house hold to advanced applications in defence. Aluminum alloys are primarily classified as wrought and cast alloys depending upon the ability of fabrication. The wrought and cast alloys are labelled with four-digit numerals from 1xxx to 9xxx groups but in cast alloys a decimal is introduced before the fourth digit as 1xx.x to 9xx.x. Mechanical properties of aluminum alloys possessing high solid solubility, which are called heat-treatable alloys are enhanced by thermal treatment, quenching and age hardening, whereas other Aluminum alloys which are non-heat treatable are subjected to Strain hardening or Cold working. Aluminum alloys with silicon as a main alloying element form the most significant part of critical shape castings, especially in under water applications because of its high strength to weight ratio, good processability, weldability and corrosion resistance. Apart from casting process, microstructure of cast Al-Si is influenced by solidification rate, dendrite arm space. The micro structural characteristics such as morphology of silicon and intermetallic compounds depend upon the presence of alloying element. Factors like solutionizing time & temperature, quenching temperature & cooling rate, artificial ageing delay period are strategic heat treatment methods which are essential for control of mechanical and microstructure properties .Of late a variety of studies have been carried to study the affect of age hardening on Al-Si alloys with microstructure and mechanical properties as main objective. In age hardening of these alloys, the alloy was heated to solutionizing temperature and then water quench followed by a single stage artificial ageing, to enhance the mechanical properties. Elagin et al. [1] experimentally studied methods for producing high strength and high temperature structural aluminum alloys, observed improvement in strength with the incorporation of transition metals and earth race metals in to their composition.

Zakharov et al. [2] examined and developed high strength corrosion resistance weldable aluminum alloy by using Al-Zn-Mgs system with the additives of Cu, Ti, Zr and Sn. The incorporation of Cu element enriches corrosion resistance and the addition of Sn, Zr, Ti enriches the weldable characteristics of the aluminum alloy. Zhang et al. [3] studied precipitation of excess silicon in the cast alloy under different heat treatment process. They do not observe silicon precipitate at initial stage of ageing with high quench rate and at low quench rate the silicon precipitate forms during ageing. ESKIN et al. [4] studied precipitation kinetics through TEM and observed two types of interfaces in Al-Mg-Si alloys. Matsuda et al. [5] studied β" crystal structure in Al–1.0mass%Mg_2Si–0.4mass%Si alloy and observed needle shape β" with their chemical composition as Mg:Al:Si =1:3:6. Cavazos et al. [6] studied precipitation in heat treatable aluminum cooled at different rates. They performed experiments at various age temperatures 130°C, 180^{0}C, 230^{0}C and observed the precipitation behaviour and concluded that reduction in hardness if the sample cooled below 10^{0}C/s. Naga Raju et al. [7] studied AA 2219 aluminum alloy altered with Sc, Mg and Zr additives and observed these modified compositions have equiaxed grains and high temperature withstand precipitations. Lynch et al. [8] investigated micro analysis of precipitates in aluminum alloys, observed maximum hardness at early stage of precipitation. Wahi et al. [9] studied metastable Phase β" in Al-Si-Mg alloy and observed different ageing behaviour in the two alloys based on the present of Cu in their composition. Matsuda et al. [10] studied precipitates morphology in Al-1wt%Mg_2Si alloy and classified orientation relationships of precipitates and matrix. They also observed that precipitates are elongated in prescribed direction in the matrix. Gupta et al. [11] analysed hardening of 6061/10% SiC composite and observed increased activation energies of β" precipitations with elevation of solution temperature from 510 to 600^{0}C. Their micro structural analysis revealed increase in grain size at high solution temperature. Takeda et al. [12] examined of Al–Mg–Si ternary alloys and concluded β" metastable precipitates plays a vital role in enhancing the components hardness.

The current research is an attempt to investigate the combined effect of 1st stage ageing at 160^{0}C for one hour followed by 2nd stage ageing at 140^{0}C for different holding periods, on tensile properties and microstructure of LM 25 (Ternary alloy of Al-Si-Mg). The binary alloys of Al-Si are not responding to heat treatment. But by the addition of Mg to this binary Al-Si alloy, respond to heat treatment by which the mechanical properties are enhanced. This alloy was chosen as it is used for many of the torpedo components, especially for shells.

2. Experimental Procedure

2.1 Casting test bars

The test bars used for experiment are prepared by melting commercially available LM 25 alloy ingots in an electrical muffle furnace. The constitution of the alloy is presented in Table 1. Degassing of the alloy has been carried by adding 0.05 wt. % degassing agent in which 2/3 sodium fluoride and 1/3 sodium chloride are present. The molten alloy was poured in a metallic mould having proper riser and runner mechanism. About 15 sets, each set with 3 samples are cast and used for tensile testing.

Table 1. Chemical constitution of LM alloy

Element	Si	Mg	Fe	Mn	Cu	Zn	Ti	Sn	Al
Wt.%	7.47	0.74	0.043	0.82	0.028	0.028	0.004	0.009	90.906

2.2 Age hardening of cast bars and fabrication of test samples

The cast bars used for tensile testing and metallographic examination are solutionized at 520^{0}C for 6 hrs and water quenched at 20^{0}C. The bars cooled to room temperature are artificially aged in two stages. The 1st stage ageing is carried out at 160^{0}C for 1hr and then cooled to 140^{0}C in the furnace where the 2nd stage ageing is done. The 2nd stage ageing is performed for the different periods from 4 to 40 hrs. After age hardening, the tensile test bars have been fabricated as per ASTM E8M specification. The metallographic and hardness samples are made of diameter 10 mm and height of 20 mm. The heat treatment cycle for two stage artificial age hardening is as shown in figure 1.

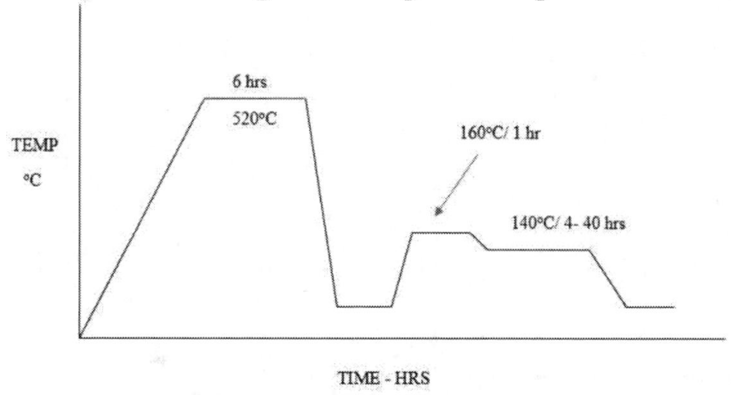

Figure 1. Two stage artificial ageing cycle

2.3 Evaluation of microstructures
The samples with diameter of 10 mm and height 20 mm cut from the cast bars for micro structural studies are heat treated along with tensile and hardness samples. The samples are polished with a standard metallographic technique. Refined samples are etched by use of Keller's reagent. Olympus make inverted optical microscope is used for micro structural examination.

2.4 Evaluation of mechanical properties
The samples for hardness testing are polished with 2/0 grade emery paper and hardness tests are carried out by using Reicherter make computerized Universal hardness testing machine in Vickers scale. The hardness is evaluated at various places over the entire cross-section of the test specimen using a load 5 kgf and the average hardness is taken. The tensile tests are conducted on specimens prepared as per ASTM E8M specification at a constant strain rate of 2mm/min using FIE make UNITEK 95100 computerized tensile testing machine of 10 tons capacity. The ductility and UTS are calculated in terms of % of elongation.

3. Results and discussion
3.1 Influence of two stages ageing on microstructure
The microstructure of LM25 as cast condition is shown in figure 2. It indicates that the structure is coarse eutectic with primary Si dendrites which may be due to slow cooling during solidification. The microstructure in solutionized condition is shown in figure 3. It was observed from the structure that a small fraction of un dissolved Si is present in the super saturated solid solution of alpha aluminum. The microlevel crystalline structure of alloy in the 1st stage ageing at 160°C for one hour is shown in figure 4. It can be seen from the structure that high density of second phase particles are nucleated at grains and grain boundary regions. It is obvious to say that the nucleation and growth rate of precipitates increase with increase in ageing temperatures. The microstructures of the alloy in 2nd stage ageing at 140°C with different holding time are shown in figures from 4 to 7. The microstructure of the alloy in 1st 4 hours ageing time shows a mixture of coarse and fine precipitates. This may be revealed that the precipitates, which are already nucleated in the 1st stage, are coarsen during furnace cooling to 140°C and the fine precipitates are the freshly nucleated particles at 140°C. From the microstructures of the alloy in 2nd stage ageing at different holding times, it was observed that the structure consists of different forms and sizes of precipitates. First the Guinier Preston (GP) zones form. These are the clusters of solute atoms which are only few atoms thick. Then intermediate precipitates ß'' and ß' formed which are much larger than GP zones and have definite composition and crystal structures slightly different to equilibrium precipitate Mg_2Si. The sequence of precipitates is GP zones ß'' - ß' - Mg_2Si.

Figure 2. As cast – Microstructure at 200X

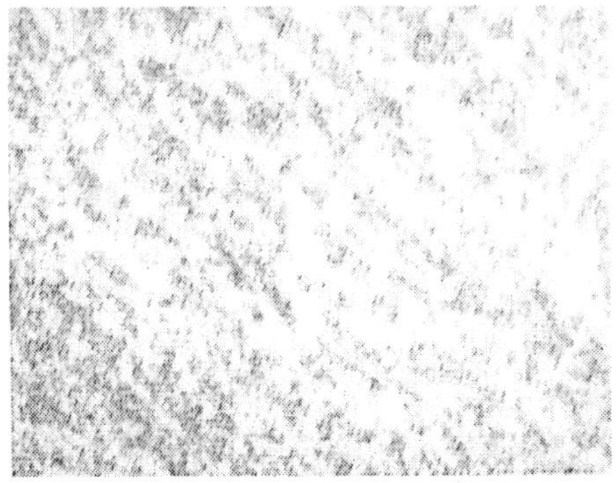

Figure 3. Solutionized – Microstructure at 200X

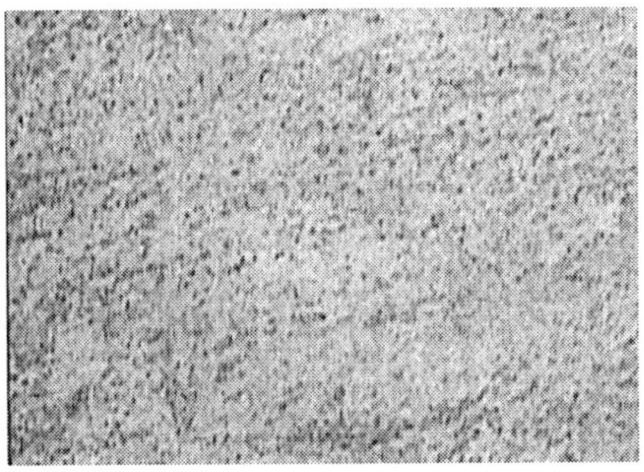

Figure 4. 1st stage aged at 160⁰C for 1 hr - Microstructure at 200X

Figure 5. 2nd stage aged at 140⁰C for 12 hrs – Microstructure at 200X

Figure 6. 2nd stage aged at 140⁰C for 24 hrs – Microstructure at 200X

Figure 7. 2nd stage aged at 140⁰C for 40 hrs – Microstructure at 200X

3.2 Influence of two stages ageing on mechanical properties

The mechanical properties of the alloy as cast conditions are shown in table 2. In this condition, the properties are low due to slow cooling during solidification and the presence of Si dendrites in the microstructure. The mechanical properties in solutionized condition are shown in Table 2. The results

IOP Conf. Series: Materials Science and Engineering 998 (2020) 012029 doi:10.1088/1757-899X/998/1/012029

indicated that improvement in the properties than in cast condition. This can be stated that in solutionized condition the solid solution strengthening was occur due to formation of supersaturated solid solution, by quenching the alloy from solutionizing temperature. Long holding time at terminal solid solution temperature results homogenization of alloying elements, might be another reason for improvement in tensile properties than in cast properties. The mechanical properties of the alloy in 1st stage ageing at 160°C for one hour are shown in table 2. Further improvement in strength and hardness were found in this condition. This can be reviewed that the high temperature ageing accelerates the nucleation and growth rate of precipitates. The mechanical properties of the alloy in 2nd stage ageing at 140°C for different holding times are shown in table 2. It is reported that strength and hardness increase with ageing time and reaches to peak values at 16 hrs of holding time. With further increase in ageing time, the strength and hardness decreased. The increase in hardness and strength to peak values with increase in ageing time can be obtained due to the growth of precipitates to a critical size and due to strong coherency of these precipitates with parent matrix. With further increase in ageing time tends to decrease in strength. This phenomenon is called over ageing. During coarsening, the size of the particles increased the number of particles decreased and the interparticle distance decreased. This leads to less resistance to dislocation motion in the matrix and makes the alloy soft. The variation of mechanical properties with ageing time in 2nd stage ageing under various holding times is shown figure 8 & 9.

Table 2. Hardness, UTS & elongation results

Condition of the material		Hardness VHN	UTS (MPa)	Elongation (%)
In cast condition		98	161	1.21
In solutionized condition		89	195	1.75
1st stage aged condition, aged at 160°C for 1 hour		106	216.5	1.21
2nd stage aged condition, aged at 140°C for different holding times	4hrs	105	207	1.92
	8hrs	111	241	1.86
	12hrs	117	260	1.73
	16hrs	128	270	1.14
	20hrs	120	235	2.95
	24hrs	110	226	3.50
	32hrs	112	235	3.64
	40hrs	97	161	3.64

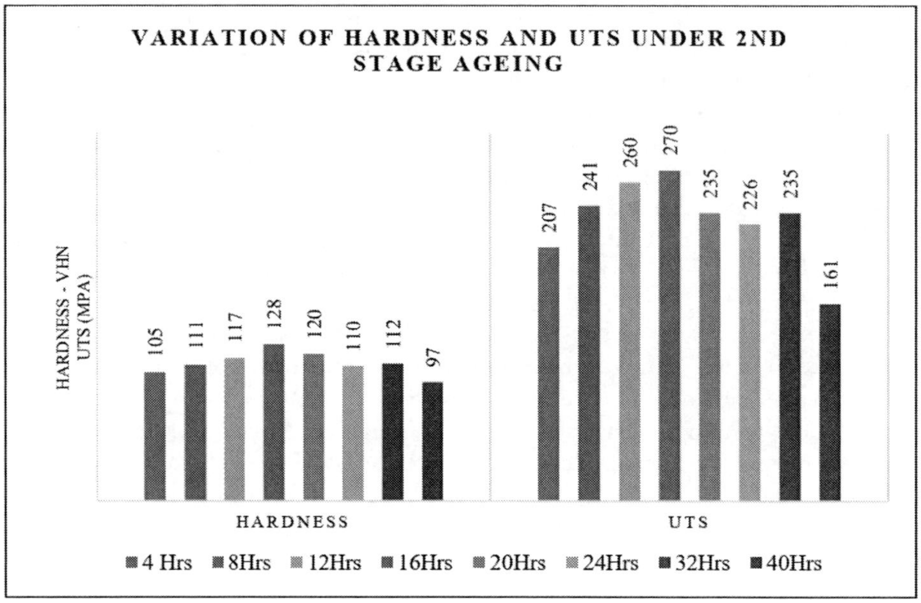

Figure 8. Comparison of Hardness, UTS and under 2nd stage ageing

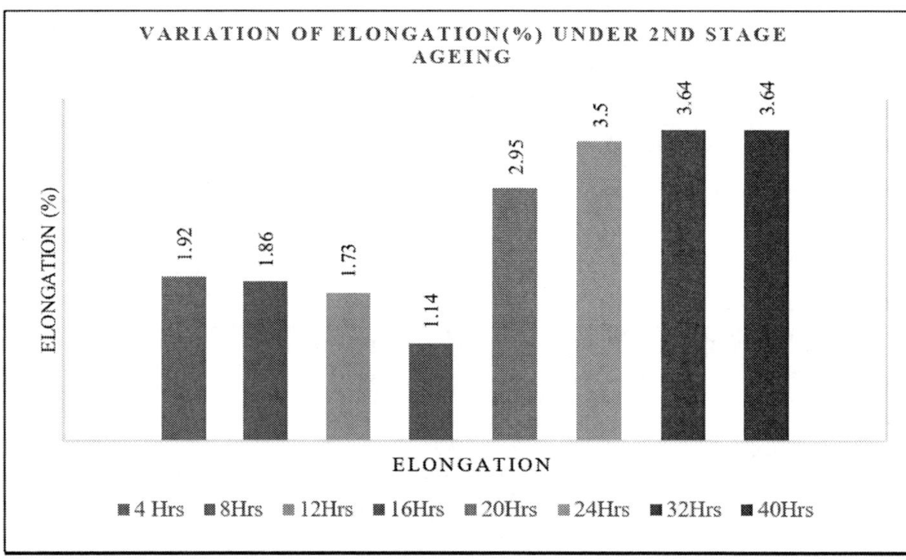

Figure 9. Comparision of Elongation (%) under 2nd stage ageing

The tensile strength in single stage ageing at 160^{0}C for different holding times is shown in table 3. It was observed that enhancement in strength with ageing time and reaches to peak value at 4 hours ageing time and then decreased with further ageing time. The peak strength with short duration might be due to accelerated nucleation and growth rate of precipitates with increase in ageing temperature.

Table 3. UTS under 1stageing at various holding times

Hoding time (hrs)	1	2	4	6	8	10
UTS (MPa)	216	238	255	250	242	214

4.Conclusions

The microstructure of LM25 in as cast condition shows coarse eutectic with primary Si dendrites due to slow cooling during solidification. The microstructure of LM 25 in as solutionized condition shows small fraction of un dissolved Si in the super saturated solid solution of alpha aluminum. The microstructure of the alloy in 1st 4 hours ageing time shows a mixture of coarse and fine precipitates. This may be revealed that the precipitates, which are already nucleated in the 1st stage, are coarsen during furnace cooling to 140^{0}C and the fine precipitates are the freshly nucleated particles at 140^{0}C. The microstructures of the alloy in 2nd stage ageing at different holding times, it was observed that the structure consists of different forms and sizes of precipitates. First the Guinier Preston (GP) zones form. These are the clusters of solute atoms which are only few atoms thick. Then intermediate precipitates ß'' and ß' formed which are much larger than GP zones and have definite composition and crystal structures slightly different to equilibrium precipitate Mg_2Si. The sequence of precipitates is GP zones ß''- ß' - Mg_2Si. In two stage ageing, 5-8% improvement in tensile strength was observed than in single stage ageing. In two stage ageing, peak values of strength and hardness were achieved with long ageing periods.

References

[1] Elagin V I 2007 Ways of developing high-strength and high-temperature structural aluminum alloys in the 21st century *Met. Sci. Heat Treat.* **49** 427–34
[2] Zakharov V V., Rostova T D and Fisenko I A 2005 High-strength weldable corrosion-resistant aluminum alloy for bearing building structures *Met. Sci. Heat Treat.* **47** 377–82
[3] Zhang D L 1996 Precipitation of excess silicon during heat treatment of cast Al-7wt%Si-0.4wt%Mg alloy *Mater. Sci. Forum* **217–222** 771–6
[4] Eskin D G, Massardier V and Merle P 1999 Study of high-temperature precipitation in Al-Mg-Si alloys with an excess of silicon *J. Mater. Sci.* **34** 811–20
[5] Matsuda K, Naoi T, Fujii K, Uetani Y, Sato T, Kamio A and Ikeno S 1999 Crystal structure of the β″ phase in an Al-1.0mass%Mg2Si-0.4mass%Si alloy *Mater. Sci. Eng. A* **262** 232–7
[6] Cavazos J L and Colás R 2001 Precipitation in a heat-treatable aluminum alloy cooled at different rates *Mater. Charact.* **47** 175–9
[7] Naga Raju P, Srinivasa Rao K, Reddy G M, Kamaraj M and Prasad Rao K 2007 Microstructure and high temperature stability of age hardenable AA2219 aluminium alloy modified by Sc, Mg and Zr additions *Mater. Sci. Eng. A* **464** 192–201
[8] Lynch J P, Brown L M and Jacobs M H 1982 Microanalysis of age-hardening precipitates in aluminium alloys *Acta Metall.* **30** 1389–95
[9] Wahi R P and Von Heimendahl M 1974 On the Occurrence of the Metastable Phase β″ in an Al-Si-Mg Alloy *Phys. Status Solidi* **24** 607–12
[10] Matsuda K, Tada S and Ikeno S 1993 The morphology of precipitates in an Al-1 wt% Mg2Si alloy *J. Electron Microsc. (Tokyo).* **42** 1–6
[11] Lu L, Gupta M and Lai M O 1998 Analysis of precipitacion hardening of A6061/10% SiC composite by diferential scanning calorimetry *J. Therm. Anal.* **54** 825-35
[12] Takeda M, Ohkubo F, Shirai T and Fukui K 1998 Stability of metastable phases and microstructures in the ageing process of Al-Mg-Si ternary alloys *J. Mater. Sci.* **33** 2385–90

IOP Conf. Series: Materials Science and Engineering 998 (2020) 012030 doi:10.1088/1757-899X/998/1/012030

CFD Simulation and Analytical Validation of Rectangular and Square microchannel heat sink with liquid water as cooling medium

Saurabh Narayan Pawar[a], Nilesh Balkishanji Totla[b]

[1]Student, MIT Academy of Engineering, Alandi, Pune-412105, India
[2]Senior assistant professor, MIT Academy of Engineering, Alandi, Pune-412105, India
[a]snpawar@mitaoe.ac.in, [b]nbtotla@mech.maepune.ac.in

Abstract: In present work, both analytical and CFD investigation was performed to check the microchannel heat sink cooling performance, made up of copper with two different microchannel cross sections with shape of rectangle and square. CFD analysis is conducted with water as working fluid for both of the cross sections and Reynolds number ranging from 600 to 1000. A copper microchannel heat sink with bottom size of 25.4mm×25.4mm and height of 2.384mm is selected and CFD analysis was carried out and at bottom constant heat flux boundary condition is assumed. For comparison between two cross sections, the hydraulic diameter used is same for both. Rectangular cross section microchannel showed better cooling performance as compared to square cross section microchannel. Both analytical and CFD simulation results of pressure drop in rectangular microchannel heat sink are in good agreement with a difference of 18.9%.

Keywords: CFD, Heat sink, Microchannel, Rectangular, Square

NOMENCLATURE

L	Microchannel heat sink length, mm	$\dot{m}$	Mass flow rate, kg/s
W	Microchannel heat sink width, mm	C_p	Specific heat at constant pressure, J/kgK
H	Microchannel heat sink height, mm	V	Velocity, m/s
D_h	Hydraulic diameter, mm	Q	Heat rate, W
H_r	Height of rectangular microchannel, mm	Nu	Nusselt number
W_r	Width of rectangular microchannel, mm	Re	Reynolds number
H_S	Height of square microchannel, mm	A_s	Surface area, m^2
W_s	Width of square microchannel, mm	k	Thermal conductivity, W/mK
$T_{f,i}$	fluid temperature at inlet, K	h	Coefficient of Heat transfer, W/ m^2K
$T_{f,o}$	fluid temperature at outlet, K	N	No of microchannels
$(\Delta T)_f$	Fluid temperature difference, K		GREEK SYMBOLS
$T_{f,avg}$	Average fluid temperature, K	ρ	Density, Kg/m^3
$T_{w.avg}$	Average wall temperature, K	μ	Dynamic viscosity, Pa-s
(ΔT)	Average wall and average fluid temperature difference, K		
A_{cs}	Microchannel cross sectional area, m^2		

INTRODUCTION

The importance of minichannels and microchannels is ever increasing because of compact size and effective as well as efficient heat transfer performance in compact electronic circuits. Because of technological advancement, dependency on computers is increased hence use of smaller electronics circuits is essential in computers. Along with usage of circuits, problem of heat dissipation has also been raised and which has been successfully solved by using minichannels and microchannels. The amount of literature of the use of different geometries and different fluids to get better heat transfer effects in minichannels and microchannels has been increasing because it removes good heat amount from small area. For better cooling performance, the working fluid, geometry, material of the heat sink must be so chosen that better results are achieved. Many researchers have conducted their study in this field and have given good results about microchannels and minichannels. Rajesh Sharma, Amoljit Singh Gill and Vikas Dhawan [1] performed heat transfer study in single phase through microchannels. For microchannel heat sink copper material was used and water as cooling agent was used. They stated that classical approach and experimental results of microchannel heat transfer are in good agreement. They observed that with increase in velocity pressure drop increases. CFD numerical study was conducted by Oana Giurgiu, Angela Pleşa and Lavinia Socaciu [2] with two different minichannel models which were included in plate heat exchanger structure. The geometric characteristics influence comparative study of two plates on heat transfer intensification was conducted. Minichannels which were analysed had inclination angles of 30° and 60°. they stated that plate heat exchanger minichannel model with 60° inclination angle provide best heat transfer. Manoj Rao and Sameer Khandekar [3] presented experimental and computational study related to thermo-hydrodynamics of single phase liquid flow developing simultaneously. Seven parallel semi-circular channels array made by copper substrate was used. Experimental laminar and turbulent flow Po and Nu numbers matched with computational model value. Experimental Reynolds number was in range of 300-3200 and deionized, distilled and degassed water was used in this study. Dr. B. S. Gawali, V. B. Swami and S. D. Thakre [4] investigated rectangular cross section straight microchannel heat sink with theoretical and experimental approach. The distribution of temperature, pressure drop, coefficient of heat transfer were some of the selected criteria for study. Working fluid used was water and Reynolds number was in range of 200 to 700. They found that with increase in microchannel length, water temperature increases.

O. Abouali and N. Baghernezhad [5] investigated numerically rectangular and arc shaped two types of groove which are fabricated in microchannel surface. Microchannel heat sink having single phase laminar flow was studied numerically for heat transfer and pressure drop characteristic analysis. Result showed that higher heat removal flux is shown by arc groove as compared to rectangular groove. Results also showed that heat removal flux is higher for grooved microchannel with lower mass flow rate and higher wall thickness. Yang Liu, Jing Cui, WeiZhong Li and Ning Zhang [6] presented forced convection heat transfer numerical study in microchannel with different microstructure. They stated that with inlet Reynolds number heat transfer performance improves. They stated that highest heat transfer performance is of v-shaped grooved microstructure. Single phase liquid flow in microchannels and minichannels was studied by Satish G. Kandlikar [7]. Author addressed the fundamental issues in micro scale liquid flow and gave the relation to calculate pressure drop for single phase liquid flow. Experimental study was conducted for rectangular microchannels with single phase flow By Poh-Seng Lee, Suresh V. Garimella and Dong Liu [8]. Microchannels considered was having width in the range from 194 μm to 534 μm. Test piece contained ten parallel microchannels made up of copper. Reynolds number in range from 300 to 3500 was considered and Deionized water was used. Average deviation of 5% was seen between experimental data and numerical predictions obtained on basis of classical, continuum approach. By applying condition of uniform wall temperature circumferentially and uniform wall Heat flux axially to rectangular cross section microchannels, convective heat transfer of laminar flow in entrance region was studied by Poh-Seng Lee and S V. Garimella [9]. Microchannels with different aspect ratio was chosen. Aspect ratio was defined as channel height divided by channel width. Paisarn Naphon and Osod Khonseur [10] conducted experiments to study heat transfer characteristics and pressure drop in microchannel heat sink with constant heat flux consideration. Reynolds number was selected in range from 200-1000 and heat flux in range from 1.80-5.40 KW/m^2. it is stated that heat transfer performance enhancement and variation in pressure drop are affected by shape and size of microchannel surface roughness irregularities.

P. Promvonge, S. Sripattanapipat, S. Tamna, S. Kwankaomeng and C. Thianpong [11] numerically studied laminar periodic flow in square channel with three dimensional isothermal wall and 45° inclined baffles fitted on one channel wall. Reynolds number in range from 100 to 1200 was used in this study. The enhancement factor was found to be higher for 45° baffle than 90° baffle for all baffle height and Reynolds number. G. Gamrat, M. Favre-Marinet and S. Le Person [12] used two different approaches to study roughness effect on heat transfer in microchannel with laminar flow. 3D numerical simulation and 1D RLM models were used. Both the models agreed to show that with relative roughness Poiseuille number and Nusselt number increases. Xiaojin Wei and Yogendra Joshi [13] stated that microchannel stack require less pumping power than single microchannel for removing

IOP Conf. Series: Materials Science and Engineering 998 (2020) 012030 doi:10.1088/1757-899X/998/1/012030

certain rate of heat at smaller flow velocity at inlet because of large area of heat transfer. Authors designed simple thermal resistance network to study stacked microchannel heat sink and overall thermal performance. Tomasz A. Kowalewski, Jacek Szumbarski and Slawomir Blonski [14] investigated channels with wavy walls for instability of viscous incompressible flow with theoretical, numerical and experimental approach. It has been shown that correctly chosen waviness of wall will lead to flow destabilization at surprisingly low Reynolds number by linear stability analysis. Jing ZHOU, Zhiheng YAN and Qiang GAO [15] applied microchannel heat exchanger to heat pump and air conditioning system used residentially and commercially. Results obtained from experiment shown that by at most 4% heat pump system capacity can be increased with using microchannel heat exchanger. Also by at most 4% energy efficiency can be increased.

So, large amount of work is being done in the field of microchannels and minichannels with different working fluid and cross sections. But In the present study, comparison between rectangular and square cross section microchannel is studied and presented using CFD solution and analytical approach.

CAD MODEL

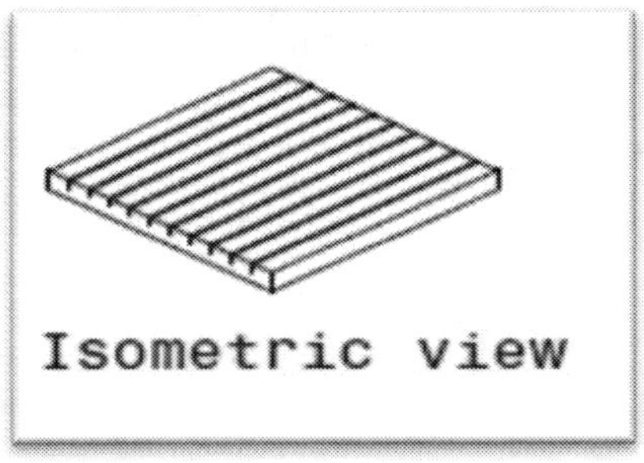

Figure 1: Isometric view of CAD model of rectangular microchannel heat sink

CFD which is Computational Fluid Dynamics, is a tool with which we can solve heat transfer and fluid flow related problems. The model of rectangular microchannel heat sink is drawn using CAD as shown in Fig 1. As given in [8], square copper block with dimensions of 25.4mm×25.4mm×70mm was used and machined to get heat sink. So in this present study, for CFD analysis purpose the dimensions of rectangular microchannel heat sink are taken to be L=25.4 mm, W=25.4mm, H=2.384mm.The dimensions of the rectangular microchannel heat sink are taken from [8]. The microchannel heat sink consists of N=10 parallel straight microchannels, same as that in [8], as shown in Fig 1. The fluid is allowed to flow through the microchannels for analysing the cooling performance as the base is given the constant heat flux. The working fluid used is water. For comparison purpose the CFD results of the rectangular microchannel was compared with square microchannel by keeping the hydraulic diameter of square microchannel same as that of the rectangular microchannel which is D_h=0.318 mm. The Reynolds number for current study was considered in range from 600 to 1000. The range of Reynolds number and working fluid was same for both rectangular and square microchannel. The number of microchannels was also same for both rectangular and square microchannel. The dimension of square and rectangular microchannel are shown in below table 1.

Table 1: Dimensions of rectangular and square microchannels

Dimensions of microchannels	height	width	D_h
Rectangular	H_r=0.884 mm	W_r=0.194 mm	0.318 mm
square	H_S=0.318 mm	W_S=0.318 mm	0.318 mm

IOP Conf. Series: Materials Science and Engineering 998 (2020) 012030 doi:10.1088/1757-899X/998/1/012030

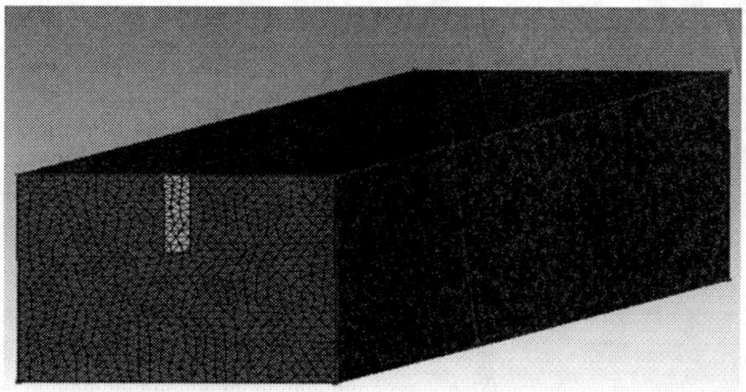

Figure 2: Meshed rectangular microchannel heat sink model

Fig 2. Represents rectangular microchannel heat sink after meshing, done using ICEM-CFD module. As shown in Fig 2. The central yellow zone is the microchannel with rectangular cross section. The liquid water flows through this zone from inlet to outlet and absorbs heat as constant heat flux was given to the base of the heat sink. Base is the bottom side in Fig 2.

For ease of CFD analysis, 1 microchannel was considered instead of 10 using symmetric boundary condition, as shown in Fig 2.

Boundary conditions and computational domain

For analysis purpose some of the assumptions are made for the present study:

1. The single phase laminar flow of working fluid is considered.
2. Steady state three dimensional fluid flow is considered.
3. All the properties of working fluid and material use as solid are considered to be constant.

The meshing of the model is done in ICEM-CFD module. For ease of analysis only one microchannel is considered out of ten and analysis is done. The mesh type is selected to be tetra/mixed and then using GSF meshing is done. GSF is global scale factor. It is defined as ratio of maximum length to minimum length of element in geometry. The grid independency was conducted by taking GSF value of 5 and 6 and it was found that for rectangular microchannel with Re=600, the pressure drop variation between GSF 5 and GSF 6 was found to be 6.01%, hence GSF 6 was chosen for study. The fluid and solid zones of the model were meshed with same mesh density. Output solver was considered to be Fluent_V6 and common structural solver ANSYS was selected.

The output of the ICEM-CFD module was then taken to the ANSYS Fluent software. This software solves the heat transfer and fluid flow related problems by using various standard equations of navier stokes equation. While solving, the energy equation was kept on and viscous laminar flow was selected. The material was selected to be copper as solid and liquid water as fluid. The various boundary condition are given as follows:

4. According to Reynolds number the inlet velocity was calculated for each flow. For Re number 600, 700, 800, 900 and 1000 the inlet velocities are 1.8924 m/s, 2.2078 m/s, 2.5232 m/s, 2.8386 m/s and 3.1540 m/s respectively.
5. The outlet was considered to be pressure outlet with guage pressure value considered to be 0 Pa.
6. At bottom of the channel constant heat flux of 450000 W/m^2 was provided.
7. The top surface of the microchannel was assumed to be insulated.

The ANSYS fluent output was taken to CFD post module where post processing was done.

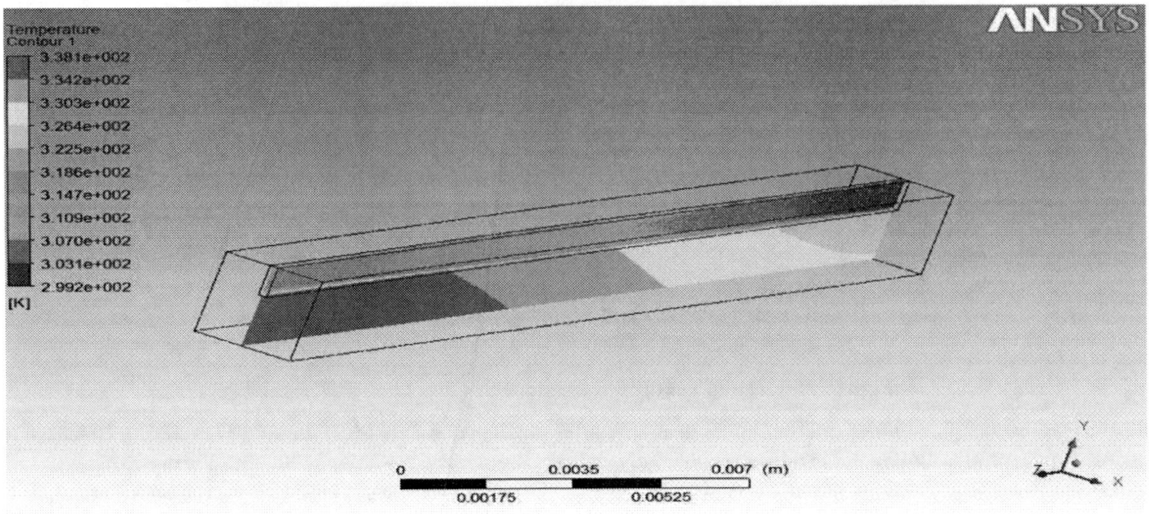

Figure 3: Temperature variation for rectangular microchannel heat sink for Reynolds number of 600

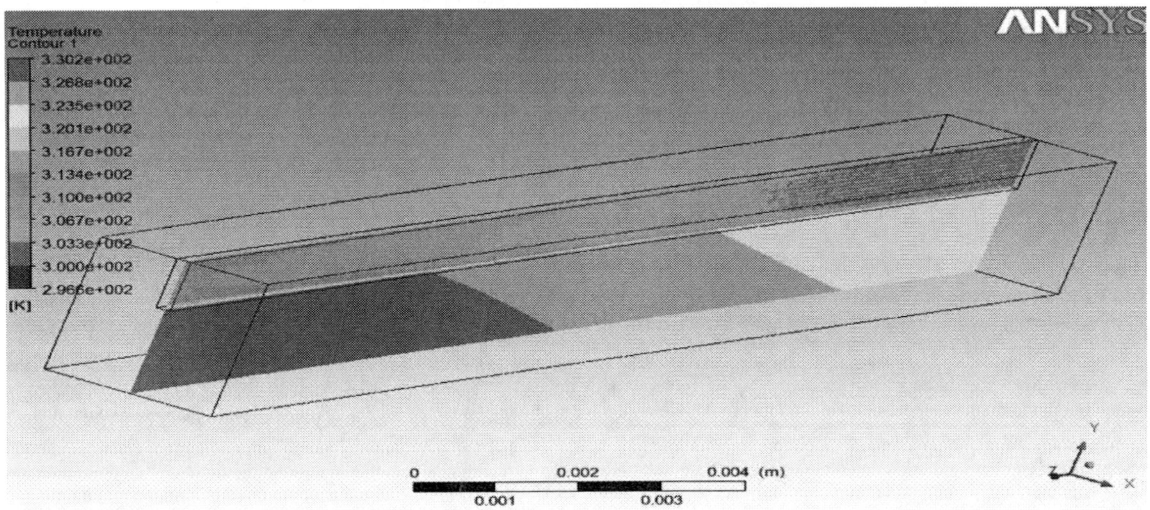

Figure 4: Temperature variation for rectangular microchannel heat sink for Reynolds number of 1000

The temperature variation for rectangular microchannel heat sink is shown in Fig 3. And Fig 4. For Reynolds number of 600 and 1000 respectively. For Re=600, the temperature range is 299.2 K to 338.1 K as shown in Fig 3. And that for Re=1000 is 296.6 K to 330.2 K as shown in Fig 4.

As per the colour contours for temperature shown in Fig 3. And Fig 4. It is cleared that temperature of the fluid increases from inlet to outlet as fluid absorbs heat as it passes through channel from inlet to outlet.

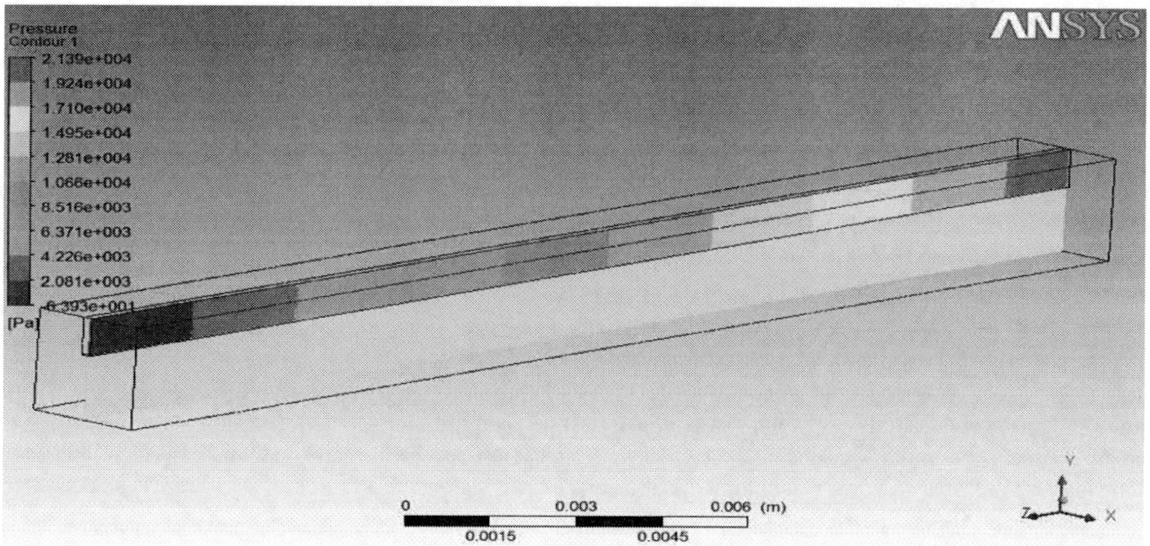

Figure 5: Pressure variation for rectangular microchannel heat sink for Reynolds number of 600

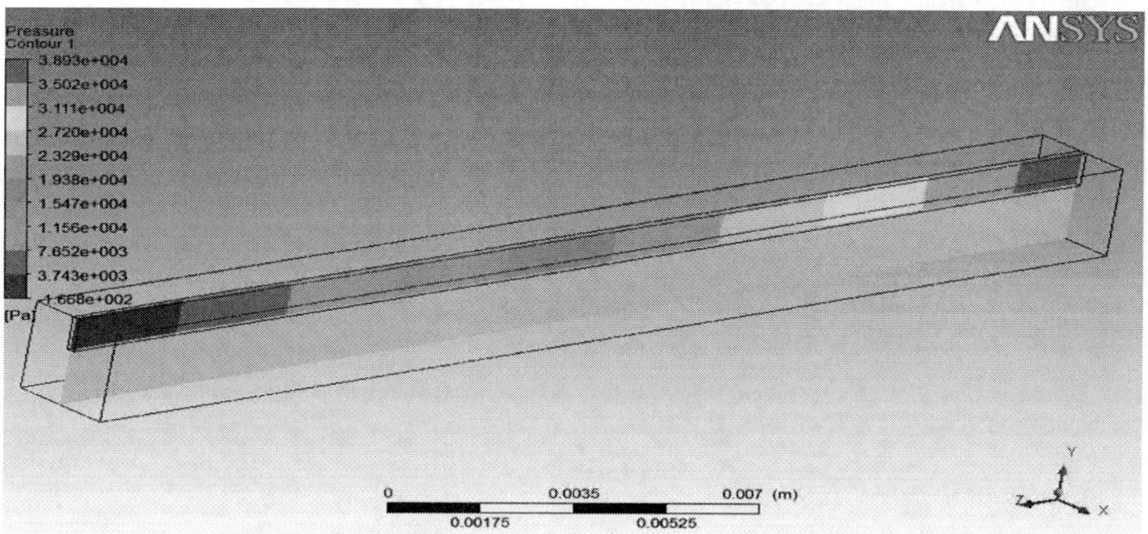

Figure 6: Pressure variation for rectangular microchannel heat sink for Reynolds number of 1000

Pressure variation for rectangular microchannel heat sink for Re=600 and Re=1000 is shown in Fig 5. And Fig 6. Respectively. From the pressure colour contour, it is observed that pressure drop of fluid takes place as fluid flows from inlet to outlet. This is because along flow direction loss of energy because of head losses takes place and that is why the pressure drop of fluid is sign of loss of energy.

CFD CALCULATION

IOP Conf. Series: Materials Science and Engineering 998 (2020) 012030 doi:10.1088/1757-899X/998/1/012030

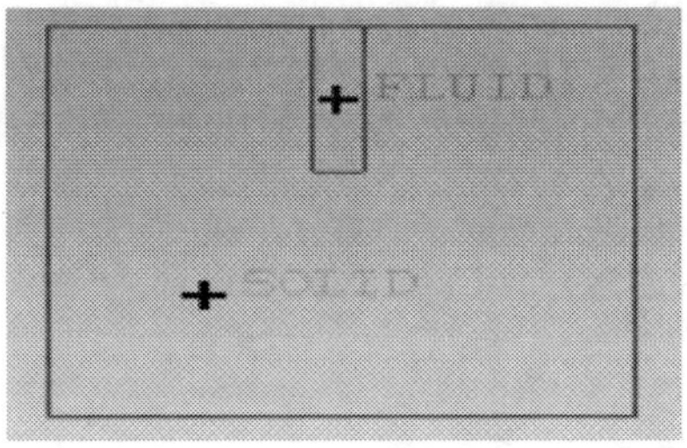

Figure 7: Front view of rectangular microchannel heat sink

In this study, convective heat transfer takes place and heat was passed to liquid water flowing through the microchannel which was being used as cooling medium. As shown in Fig 7. 2 zones were created in model, one was fluid zone and another solid zone. For analysis in CFD post, 4 lines were drawn out of which 3 are in fluid zone and one is in the solid zone. 3 lines were drawn in fluid zone from inlet to outlet exactly at a distance of half of the microchannel width from one microchannel wall, first at top, second at centre and third at base of fluid zone. One line in solid zone was drawn at bottom of solid zone exactly below the 3 lines in fluid zone. These lines were used for analysis in CFD post.

$$(\Delta T)_f = T_{f,o} - T_{f,i} \tag{1}$$

Then the average fluid temperature was calculated as,

$$T_{f,avg} = (T_{f,i} + T_{f,o})/2 \tag{2}$$

The temperature difference between fluid average and wall average temperature was calculated as,

$$(\Delta T) = T_{w.avg} - T_{f,avg} \tag{3}$$

Mass flow rate was calculated as,

$$\dot{m} = \rho A_{cs} V \tag{4}$$

Then heat transfer to fluid from inlet to outlet was calculated as,

$$Q = \dot{m} C_p (\Delta T)_f \tag{5}$$

Heat gained by fluid through convection is same as that of heat transfer through solid material through conduction which is,

$$Q = h A_s (\Delta T) \tag{6}$$

Then Nu number was calculated as,

$$Nu = (h D_h)/k \tag{7}$$

The Reynolds number was calculated as,

$$Re = (\rho V D_h)/\mu \tag{8}$$

Whereas the hydraulic diameter is calculated as,

$$D_h = (4A_{cs})/P \tag{9}$$

RESULT AND DISCUSSION

In present study, many factors like coefficient of heat transfer, pressure drop, Nusselt number are studied and comparison between rectangular and square microchannel heat sink is presented in the form of graph. Rectangular microchannel heat sink is showing good results as compared to square microchannel heat sink.

IOP Conf. Series: Materials Science and Engineering 998 (2020) 012030 doi:10.1088/1757-899X/998/1/012030

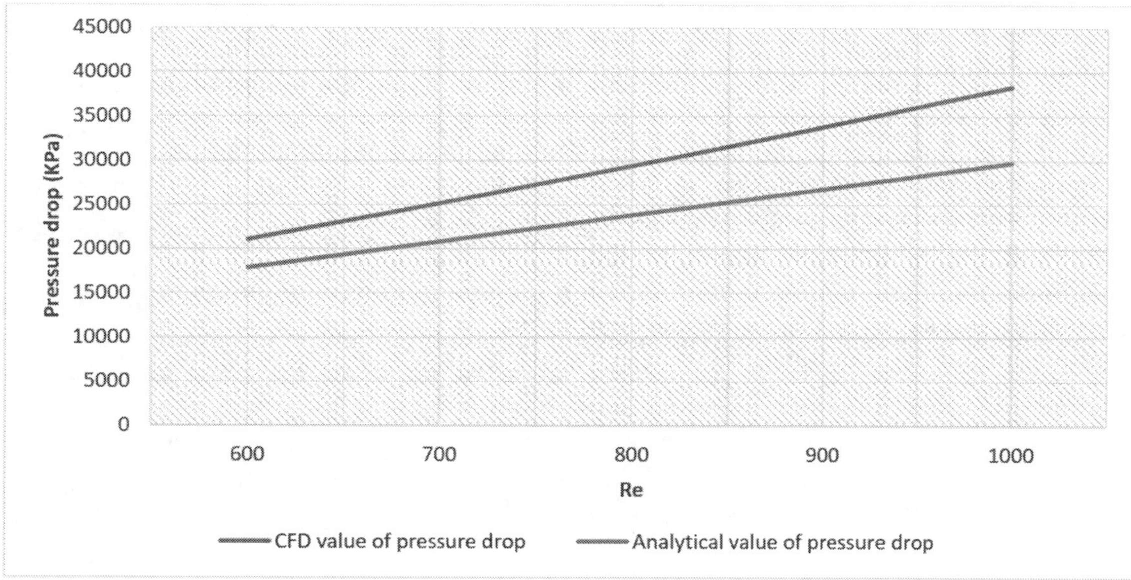

Figure 8: CFD and Analytical pressure drop value comparison of rectangular microchannel

As shown in Fig 8. The CFD value are on little higher side than that of analytical values of pressure drop for rectangular microchannel for each and every Reynolds number. The analytical value are calculated using formula in [7]. Pressure drop takes place along the flow direction as energy decreases and it is converted to head loss mainly due to friction. After comparison between values obtained, the percentage variation between CFD and analytical value for Reynolds number of 600, 700, 800, 900 and 1000 were observed to be 15.21%, 17.17%, 19.06%, 20.78% and 22.34% respectively. The average percentage variation after considering five percentage variation value was observed to be 18.9%.

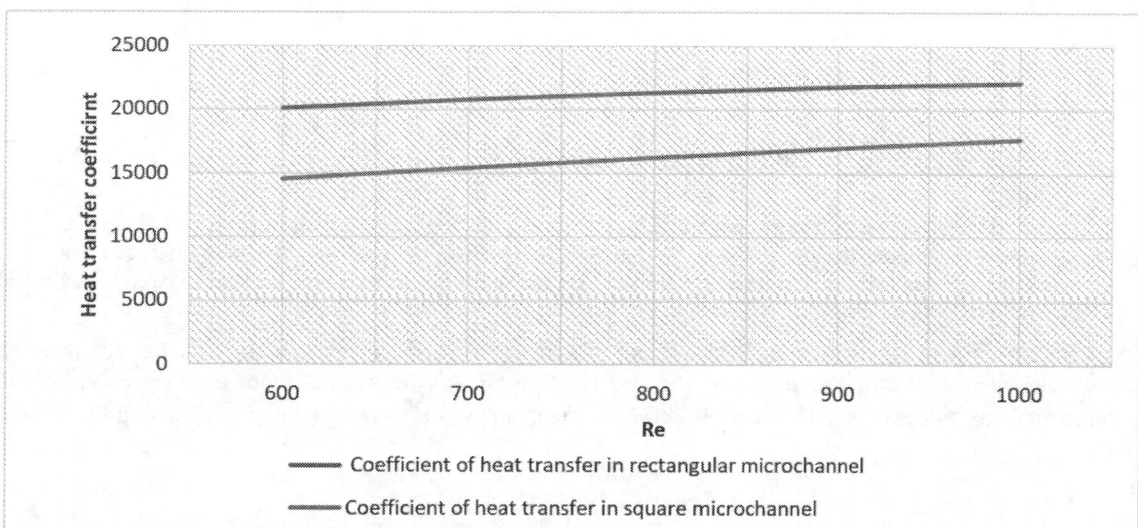

Figure 9: Rectangular and Square microchannel heat transfer coefficient value comparison

As shown in Fig 9. The rectangular microchannel is showing high coefficient of heat transfer value as compared to square microchannel for each and every Reynolds number. Coefficient of heat transfer is the main factor as it shows how much effectively heat is transferred between solid and adjacent flowing fluid. Hence we can say that rectangular microchannel is better as compared to square microchannel. The results obtained in [5] shows different trend and states that higher heat removal flux is shown by arc groove as compared to rectangular groove. Coefficient of heat transfer shows effectiveness of the system as it represents amount of heat removed effectively by cooling fluid from desired section.

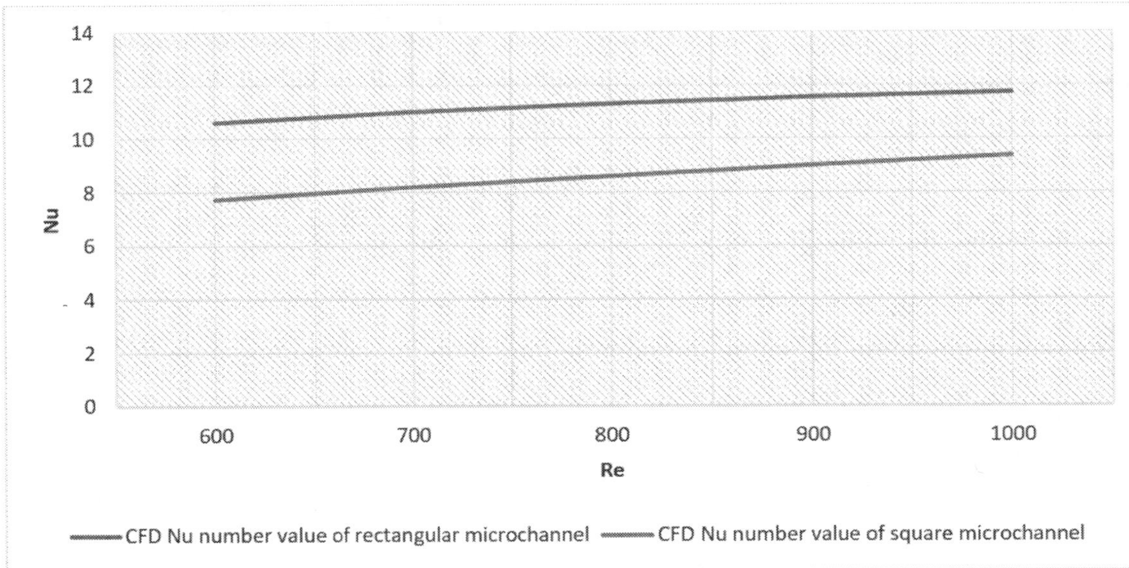

Figure 10: Rectangular and Square microchannel Nu number value comparison

As shown in Fig 10. Rectangular microchannel is showing higher Nusselt number values as compared to square microchannel for each and every Reynolds number. Nusselt number is directly proportional to coefficient of heat transfer hence the values are showing same trend as shown in Fig 9. Nusselt number is a dimensionless number used in heat transfer analysis.

Conclusion

As per the results obtained in each case, rectangular microchannel values are higher than that of square microchannel.

1. Heat transfer coefficient values were higher in each case for rectangular microchannel as compared to square microchannel so efficient heat transfer will take place in rectangular microchannel to the cooling fluid.
2. the rectangular microchannel heat sink is better to use for better cooling performance as Nusselt number values were higher in each case for rectangular microchannel as compared to square microchannel.
3. CFD pressure drop value of rectangular microchannel were higher in each case as compared to analytical values calculated using formula in [7].

So as per the results for better cooling performance it is better to use rectangular microchannel as compared to square microchannel. In the present study, the effects of piping and fitting and minor head losses present in experimental analysis are not considered. So these factors can have effect on the values obtained in the present study.

References:

1) Rajesh Sharma, Amoljit Singh Gill and Vikas Dhawan, "EXPERIMENTAL ANALYSIS OF HEAT TRANSFER AND FLUID FLOW IN MICRO-CHANNEL HEAT SINK", International Journal of Recent advances in Mechanical Engineering (IJMECH) Vol.5, No.3, August 2016.
2) Oana Giurgiu, Angela Pleşa, Lavinia Socaciu, "Plate heat exchangers - flow analysis through mini channels", Energy Procedia 85 (2016) 244 – 251
3) Manoj Rao, Sameer Khandekar, "Thermo-hydrodynamics of developing flows in a mini-channels array", 19th National and 8th ISHMT-ASME heat and mass transfer conference, Jan3-5, 2008, JNTU, Hyderabad, India.
4) Dr. B. S. Gawali , V. B. Swami , S. D. Thakre, "Theoretical and Experimental Investigation of Heat Transfer Characteristics through a Rectangular Microchannel Heat Sink", International Journal of Innovative Research in Science, Engineering and Technology, Vol. 3, Issue 8, August 2014
5) O. Abouali, N. Baghernezhad, "Numerical Investigation of Heat Transfer Enhancement in a Microchannel With Grooved Surfaces", Journal of Heat Transfer ,APRIL 2010, Vol. 132 / 041005-1
6) Yang Liu,Jing Cui,WeiZhong Li,Ning Zhang, "Effect of Surface Microstructure on Microchannel Heat Transfer Performance", Journal of Heat Transfer ,DECEMBER 2011, Vol. 133 / 124501-1
7) Satish G. Kandlikar, "Single-Phase Liquid Flow in Minichannels and Microchannels", Heat Transfer and Fluid Flow in Minichannels and Microchannels, 2014 Elsevier Ltd.
8) Poh-Seng Lee, Suresh V. Garimella , Dong Liu, "Investigation of heat transfer in rectangular microchannels", International Journal of Heat and Mass Transfer 48 (2005) 1688–1704
9) Poh-Seng Lee,S V. Garimella, "Thermally Developing Flow and Heat Transfer in Rectangular Microchannels of Different Aspect Ratios", International Journal of Heat and Mass Transfer, August 2005
10) Paisarn Naphon , Osod Khonseur, "Study on the convective heat transfer and pressure drop in the micro-channel heat sink", International Communications in Heat and Mass Transfer 36 (2009) 39–44
11) P. Promvonge , S. Sripattanapipat , S. Tamna , S. Kwankaomeng , C. Thianpong, "Numerical investigation of laminar heat transfer in a square channel with 45° inclined baffles", International Communications in Heat and Mass Transfer 37 (2010) 170–177
12) G. Gamrat , M. Favre-Marinet, S. Le Person, "Modelling of roughness effects on heat transfer in thermally fully-developed laminar flows through microchannels", International Journal of Thermal Sciences 48 (2009) 2203–2214
13) Xiaojin Wei and Yogendra Joshi, "Optimization Study of Stacked Micro-Channel Heat Sinks for Micro-Electronic Cooling", IEEE TRANSACTIONS ON COMPONENTS AND PACKAGING TECHNOLOGIES, VOL. 26, NO. 1, MARCH 2003
14) Tomasz A. Kowalewski, Jacek Szumbarski, Slawomir Blonski ,"LOW-REYNOLDS-NUMBER INSTABILITY OF THE LAMINAR FLOW BETWEEN WAVY WALLS", Proceedings of the Sixth International ASME Conference on Nanochannels, Microchannels and Minichannels ICNMM2008 June 23-25, 2008, Darmstadt, Germany
15) Jing ZHOU, Zhiheng YAN, Qiang GAO, " Development and Application of Micro Channel Heat Exchanger for Heat Pump", 12th IEA Heat Pump Conference (2017) P.3.1.5

IOP Conf. Series: Materials Science and Engineering **998** (2020) 012031 doi:10.1088/1757-899X/998/1/012031

Design and Development of an Indigenous Drum Dryer for Preheating of Plastic Scrap

Parvez Alam[1], Vijay Kumar Dwivedi[1], Anas Islam[1] and Rohit Sharma[1]

[1]IET Department of Mechanical Engineering
GLA University Mathura 281406

Email: alamparvez752@gmail.com

Abstract. In this experiment an indigenous plastic scrap preheating machine has been developed that preheats the plastic scrap very economically and may be used at small scale industries, as such types of industries largely uses sun drying method for heating the plastic scrap. The limitation of such method is that the time taken to heat the scrap is quite large and also there remains porosities inside the plastic scrap, these porosities create voids in these scraps that reduce the strength of the final manufactured plastic item in operations like injection mouldings.

1. Nomenclature

1.1 Symbols

I = Moment of inertia of the flywheel
ω = rotational velocity of the
flywheel k = inertial constant
M = mass of the wheel
R = radius of the fly wheel

2. Introduction

Plastic plays an important role in different sectors; the use of plastics varies from general day to day life to complex packing of items. The industries that use plastics are packaging, logistics, manufacturing etc. It is quite evident that plastic is also the main cause of environmental pollution. It is very dangerous to living being and affects the ecological system directly or indirectly. The consumption of plastics has several benefits such as in medical and packaging of surgical items. The quality of plastics is measured in microns, and they should be under 40 microns, as under 40 microns they can be decomposed into the soil after a specified amount of time. Plastics items that are over 40 microns e.g., toys, beverage bottles, etc are very common. Therefore they should be recycled in order to have a safe environment.

As plastics are very light in weight, resistant to moisture, economical and versatile therefore they are very vital and popular in today's era. The discovery of polymers has further broadened the utility of plastics. Due to their unique property of being easily molded into the desired shape and size it has a very large scope as far as its usage is concerned. The production of plastics is increased at an exponential rate in the few years thus became the first choice for the man made items as per the need. The main source of plastic production is petroleum, and it is non renewable energy source. The production of plastic is about 400 million tons per year and the recycled amount of plastic is very less i.e., around 9 % of its total production. In India on an average around 11 Kgs of plastic is consumed by one person, 60 % of the total is plastic is recycled and rest around 9400 tones is buried inside the

IOP Conf. Series: Materials Science and Engineering **998** (2020) 012031 doi:10.1088/1757-899X/998/1/012031

lands and oceans this results in creating the environment polluted and effects the water reserves negatively [1].

Plastics can be categorized into two categories i.e., thermo sets and thermoplastics and they are fabricated by adding and the condensation polymerization. The main advantage of thermoplastics over thermo sets is their ability of remolding as many times as required on the other hand the thermo sets are unable to remold and are permanent in shape. The manufacturing of thermoplastics is very easy and then precision is higher as compared to thermosets. The other qualities of thermoplastic are its light weight and the strength is also higher. Propylene, low density polythene (LDPE), High density polythene (HDPE).

In 1951 the first polymerization of polypropylene took place and it is later polymerized by German Italian scientists named Rana and Rehn [2]. The various natural methods by which the degradation of the plastic scarp could be done like Hydrolytic degradation, Photodegradation and Biodegradation, etc. Natural methods of degradation as mentioned above takes a very long duration to degrade the plastic waste which is near to 30-40 years. Hence, it is very important to adopt an alternative recycling process, as it will enable in restoring the natural resources. The other benefit of recycling plastic is to reduce the wastes that results from plastic scarp which pollutes the landfills and oceans. Management of plastic waste techniques are very necessary specially in those regions where unsegregated waste is very high. Government bodies that are responsible for controlling and managing the plastic pollution, adopted slogans such as R4 which means "Reduce, Reuse, Recycle and Recovery" and various public awareness programmes the purpose of such slogans is to motivate the public to minimize the waste. Plastic waste could be grouped by number of techniques; the foremost technique among all the methods is the floatation of plastic that depends upon its chemical and physical properties [3]. The most commonly used techniques for recycling the plastic are Mechanical methods [4], Pyrolysis, Solvent extraction [5], etc. The main sources of pollution in India are categorized as follows 64% is LDPE and HDPE wastes, 9.66% is PET, 11% is waste from PP and the remaining 9% is due to other sources of plastic wastes [6]. The data mentioned above clearly shows that HDPE and LDPE are the major pollution causing element followed by PP and PET.

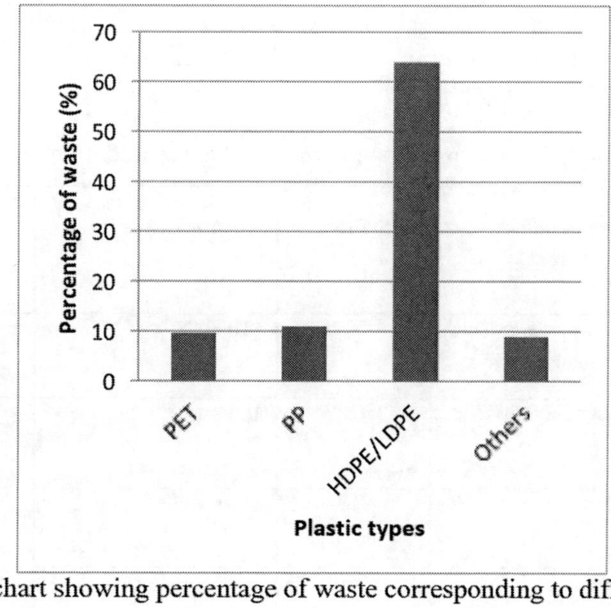

Figure 1. Bar chart showing percentage of waste corresponding to different plastic [6]

3. Motivation Behind

In this work an indigenous plastic scrap preheating machine has been developed that preheats the plastic scrap very economically and may be used at small scale industries, as such types of industries largely uses sun drying method for heating the plastic scrap. The limitation of such method is that the time taken to heat the scrap is quite large and also there remains porosities inside the plastic scrap, these porosities create voids in these scraps that reduce the strength of the final manufactured plastic item in operations like injection moldings.

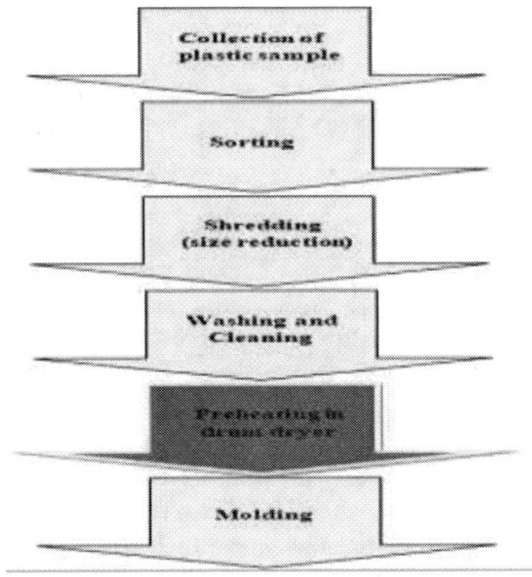

Figure 2. Stages of plastic recycling

4. Experimental Setup

The main motivation behind the design is to minimize the production of voids that are generated in the scrap due to ineffective procedure of heating the plastic scrap. As far as the small scale and cottage industries are concerned they do not adopt proper procedures for treating the plastic scraps due to the lack of resources and financial constraints (limitations) the mechanical properties of the final manufactured plastic items gets degraded.

Table 1. Pre-drying temperatures of plastic molding materials

Material name	Symbol	Pre-drying temperature (°C)	Drying time (Hrs)
Low-density polyethylene	LDPE	110~120	4~8
High-density polyethylene	HDPE	150~180	8~16
poly(ethylene terephthalate)	PET	120~165	4~8
Polypropylene	PP	140~175	6~12

IOP Conf. Series: Materials Science and Engineering **998** (2020) 012031 doi:10.1088/1757-899X/998/1/012031

In this work a machine has been designed and developed, for the purpose of optimal pre heating of the plastic scrap. Fig.3 Illustrating the Schematic diagram of Machine. It is V-belt drive system that is having a 12 inch pulley on shaft and motor is having a 3inch pulley system. The motor is having 60 number B-Type belt. Material of shaft is mild steel (MS) and shaft is double bearing having identification number as 6207 manufactured by SKF company.

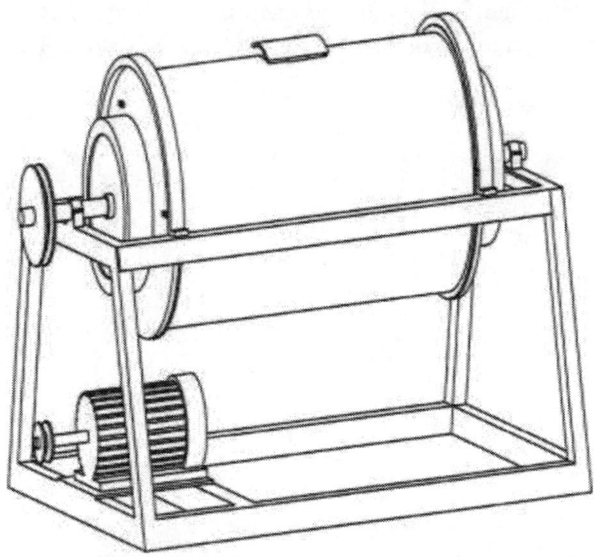

Figure 3. Schematic diagram of Plastic Pre heating machine

The main components of the machine are Bearings, Pulley, Motor, Blade, Frame (Casing). The workings of each part of the machine are as follows:

4.1 ***Drum***: Drum acts as a storing agent, the plastic scrap after cutting and washing transferred to the machine drum where the processing (preheating) of the scarp will take place on the rotation of the drum. Inside the drum the blades are fixed, the main aim of the blades attached is to collide with the plastic scraps so that its temperature gets increases, this increase in temperature is responsible for the pre heating of the scarp particles. A thermo couple is also attached to measure the temperature and after a certain (set temperature) range of the heat it directs the motor to stop, but due to the attachment of the fly wheels the drum keeps rotating even after the motor is stopped, thus the setting of temperature is kept below the desired temperature. Another advantage of the fly wheel is the low consumption of electricity.

The centrifugal force that is experienced by the drum can be found out by the equation as discussed below:

$$C.F = \boldsymbol{M} \times \boldsymbol{R} \times \omega^2 \quad (1)$$

Where, (**M**) is the mass of the drum and the moist scrap of plastic =80 Kgs;

(**R**) is the radius of the drum 3ft ~0.90m and

(**ω**) is the rotational velocity of the drum (1100 rpm) ≅ 110 rad/sec

Inserting the values in equation.1,

$$C.F = 80\times90\times110^2$$

$$\therefore \text{C.F} = 8\times 10^2 \text{ KN}$$

In the drum, blades are paced which are at an angle of 22.50 with X- axis inside the drum enables the fast heating of the scrap material. The thermocouple enables the operator to get the real time value of the temperature and they can set the cutoff temperature as many times as required according to the type of plastic or as per the need. The cut portion of the machine is as illustrated in the Fig.4.

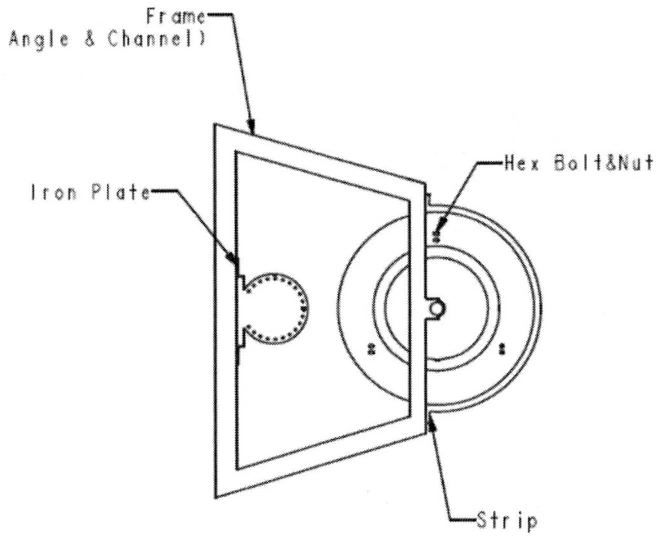

Figure 4. A cut section of the machine

4.2 ***Motor***: Motor is placed at the lower side (bed) of the casing or frame. The power of the motor is 5 H.P (horse power) and its rotational speed is 2200 rpm. The main purpose of the motor is to rotate (in circular motion) in order to rotate the pulley, which is further attached to the extended shaft of the drum. As the motor rotate it will rotate the drum in which the plastic scrap is already stored. This circular motion of the drum will change the state of the plastic scrap from rest to motion, thus the temperature of the plastic scrap rises and the pre heating of the scrap will come into effect. Various parts of the machine has been shown in Fig. 5.

Calculation for the energy consumed by the motor,

Energy consumed= 5 hp × 1 hour

5 H. P hour = 5 × 746 watts

= 3730 watts hour

= 3.7 units (approx)

Thus the motor will consume 3.7 units of electrical energy in one hour.

IOP Conf. Series: Materials Science and Engineering **998** (2020) 012031 doi:10.1088/1757-899X/998/1/012031

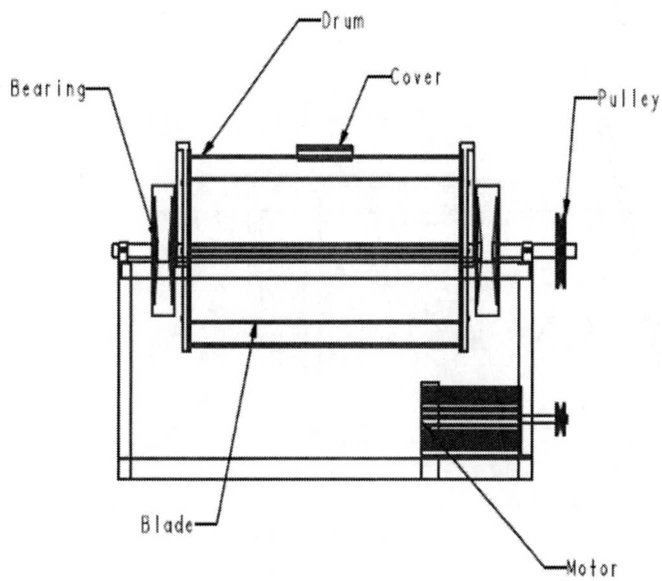

Figure 5. Showing the various parts of the Machine

4.3 ***Flywheel*:** Fly wheel is too attached to both sides of the drum with a view keeping in mind so that the drum keeps rotating even after given a small start, this saves energy. As fly wheel has the amazing ability of storing the rotational energy thus it will resist any change in speed due to its moment of Inertia enabling the drum attached to it to keep rotating. The rotational speed of fly wheel is 1100 rpm (110 rad/sec), Diameter of wheel is 2ft (0.61 m) and efficiency of the flywheel is around 80 %. Fig. 6 shows the other parts of the machine that works in line of action of the main parts as discussed in the above sections.

$$KE = \frac{1}{2} I \times \omega^2$$

$$= 1/2 \times 1.8 \times 1102$$

$$= 10.89 \text{ KJ}$$

Where, (**KE**) is the kinetic energy stored, (**I**) is the moment of inertia of the flywheel and, (**ω**) is the rotational velocity of the flywheel,

The moment of inertia of the fly wheel can be evaluated by the formula:

$$I = k \times M \times R^2$$

$$= 1/2 \times 40 \times 0.32$$

$$= 1.8 \text{ kg m}^2$$

k = inertial constant, $\frac{1}{2}$ (for solid disk),

M= mass of the wheel = 40 kg (approx.),

R = radius of the fly wheel

IOP Conf. Series: Materials Science and Engineering **998** (2020) 012031 doi:10.1088/1757-899X/998/1/012031

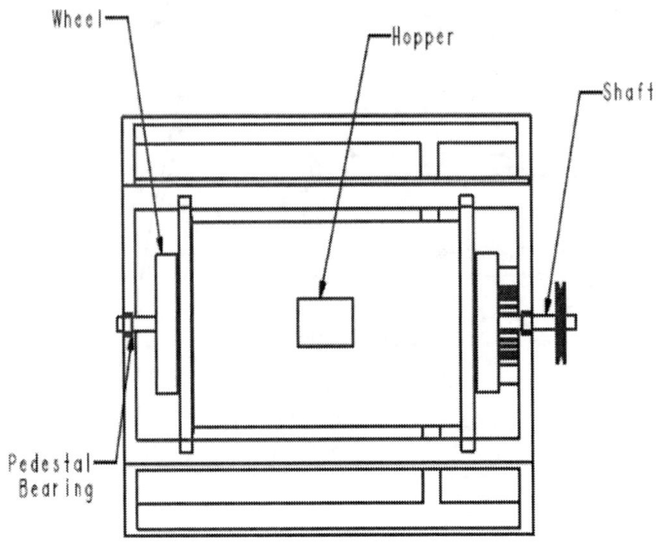

Figure 6. Showing the Hopper, shaft and pedestal bearing

Figure 7. (a) Shown the actual set up of machine while working (b) Tilting dryer for plastic scrap

5. Results and Discussions

The thermocouple enables the operator to get the real time value of the temperature and they can set the cutoff temperature as many times as required according to the type of plastic or as per the need. The heating process is also a step ahead than the conventional method this brings to the end of the problems of the creation of void and heats the scrap, so that the material gets heated without losing its strength that is nearly impossible in sun drying method, or use of hot blower method. The major advantages are as discussed below:

- Suited for fined grained as well as coarse grained material.
- Relatively less consumption of electricity.
- Sensitive to abrupt changes of the material size, moisture changes etc.
- Lesser price Controlled atmosphere inside the drum for de dusting of the particles
- Tolerant of operating faults
- Less wear and lesser part replacement requirement.

IOP Conf. Series: Materials Science and Engineering **998** (2020) 012031 doi:10.1088/1757-899X/998/1/012031

6. Conclusions

Due to its indigenous design the preheating of the plastic scrap has become economical thus can also be employed at small scale industries. The units of energy consumed per hour are 3.6 units which is quite less than conventional ones. Light weight because of the design of thin walls enables to easily carry it from one place to the other and its simple yet innovative design consideration makes the fabrication and manufacturing process easier and simpler.

The thermocouple which placed in drum is enables the operator to get the real time value of the temperature and they can set the cutoff temperature as many times as required according to the type of plastic or as per the need. The heating process is also a step ahead than the conventional method this brings to the end of the problems of the creation of void and heats the scrap, so that the material gets heated without losing its strength that is nearly impossible in sun drying method, or use of hot blower method. The major advantages includes it suited for fined grained as well as coarse grained material, Relatively less consumption of electricity, Sensitive to abrupt changes of the material size, moisture changes etc., controlled atmosphere inside the drum for de-dusting of the particles, Tolerant of operating faults, Less wear and lesser part replacement requirement.

7. References

[1] Plastic waste Management Swacha Bharat- http://164.100.228.143:8080/sbm/content/writereaddata/SBM%20Plastic%20Waste%20 Book.pdf.

[2] Everything You Need to Know About Polypropylene- https://www.creativemechanisms.com/blog/all-about-polypropylene-pp-plastic.

[3] Francis R. (2016) Recycling of Polymers: Methods, Characterization and Applications. Willey, ISBN: 978-3-527-33848-1.

[4] HuitingShent, R.J. Pugh, E. Forssberg(1999) A review of plastics waste recycling and the flotation of plastics. Resources, Conservation and Recycling, 25(2): 85-109. doi:10.1016/S0921-3449(98)00017-2

[5] Singh N, H. David (2017) Recycling of plastic solid waste: A state of art review and future application, Composites Part B 115: 409-422.doi.org/10.1016/j.compositesb.2016.09.013.

[6] Yi Bo, D. Xu, Hong Gang (2018) Solvent-based separation and recycling of waste plastics: A review, Chemosphere 209: 707–720.doi 10.1016/j.chemosphere.2018.06.095

[7] AchiliasD.S. (2007) Chemical recycling of plastic wastes made from polyethylene (LDPE and HDPE) and polypropylene (PP). J Hazard Mater 149(3): 536–542. doi:10.1016/j.jhazmat.2007.06.076

Mechanical Testing and Numerical Analysis of Flax/Glass Epoxy Hybrid Composite Material

S NARESH KUMAR[1], D VENKATESH[2], B. SUBBARATNAM[3] M SHEKAR[4]

[1,3]Department of Mechanical Engineering, Vardhaman College of Engineering, Hyderabad, India
[2]PG Scholar, Department of Mechanical Engineering, Vardhaman College of Engineering, Hyderabad, India
[4]Project Assistant, Instrumentation Group, CSIR-NGRI, Hyderabad, India

Corresponding author: venkateshd688@gmail.com

Abstract. The natural fibres, as reinforcing medium will tremendously use in polymer composite industries due to rapid development of research in this field to maintain the sustainable materials towards its application. In this paper, Bidirectional woven E-Glass and Flax Fabrics with Epoxy resin were used to fabricate the two different composites with (0^0, 45^0) orientation by hand layup method. Laminate-1 made of four flax/epoxy layers and four Glass/Epoxy layers consecutively arranged one after the other, Laminate-2 made of four Flax/Epoxy layers sandwiched between the two Glass/Epoxy layers top side and two Glass/Epoxy layers bottom side of the laminate. Tensile test and bending test results obtained as per ASTM standards. Finite element analysis was used in ABAQUS software for the simulation of important mechanical properties in tension and bending of the two laminates. Theoretical results obtained in FEA were compared with the experimental results of both laminates in terms of Force-displacement curve on the elastic domain. The error percentage of the maximum strength of two laminates were 4.8% and 8.04% respectively in the tensile test, similarly the maximum flexural strength of two laminates in order of 5.33% and 2.54% in bending test. Due to this experiment, it observed that the strength of the laminate is more in the hybrid composite than the composite made of purely natural fibres.

Keywords: Woven Flax, E-Glass, FEA model, Laminate, tensile test, bending test.

1. INTRODUCTION

The natural fibers (Jute, Hemp, Flax, Kenaf, and Sisal) have abundantly used in the process of fabrication of composites as reinforcement material for the last ten decades. The natural fibers were preferred than synthetic fibers (Example: E-Glass) because of increasing environmental concern and high need for eco-friendly material [1]. Sathish et al. [2] proved that laminate made of a combination of both natural fiber (jute) and synthetic fiber (E-Glass) with epoxy resin has more tensile strength & impact strength than laminate made only single type natural fiber. Ramesh et al. [3] showed that 0^0 flax fiber orientations have more value on important Mechanical properties than 90^0 orientations.

Camalia Cerbu et al. [4] paper reveals that the equivalent Young's modulus E_x in bending for two different hybrid composites (Flax/Epoxy and Glass/Epoxy) in weft direction was more than wrap direction because flax yarn use in wrap direction is not same with the yarn used in the weft direction. The mechanical properties of hybrid composites depend on the orientation of fibers, the method of fabrication, the addition of fillers, and some geometrical parameters of fibers [5-8].

Flax fibers were excellent material that can be hybridized easily with carbon fiber, E-glass synthetic fibers. Moreover, it has a better tensile strength (343 – 1035MPa) and cost effective fiber, so that these are preferable to fabricate composite laminates for various applications than other natural fibers like Hemp, Jute fibers [9]. Here, the bleached brown color fiber yarns used whose 200μm fiber diameter and chemical composition of 88 % cellulose, 0.6-4 % lignin with 3 to 4 % pectin.

In this paper, two types of laminates fabricated by using a hand layup method with different stacking sequences and tested experimentally in the laboratory. Numerical analysis performed in ABAQUS (student edition 2018 Dassault system) software and validated by comparing the results of mechanical properties such as maximum Tensile strength, maximum force, Young's modulus, Flexural strength, and impact strength, etc.

The main objective of this work is to determine important mechanical properties from the experimental & numerical method for two laminates, compare themselves as well as those with previously published papers, and observe the strength variation across the lamina stacking sequence with different plies orientation in bending test.

2. MATERIALS & METHOD

In this paper, Bidirectional plain-woven Flax and E-glass fabrics were taken whose size of yarn in the weft direction and wrap direction approximately same. The density of the E-Glass fabric is $\rho = 410$ g/m^2 while the density of the Flax fabric is $\rho = 250$ g/m^2. The Epoxy resin Araldite LY556 and Hardener Aradur HY951 were used as matrix material and both are applied in the ratio of 10:1, some of its mechanical characteristics without reinforcement are Tensile strength σt =75MPa, Flexural strength = 112MPa, Modulus of Elasticity E = 4.5GPa.

Two laminated composite sheets of size 330 x 300 mm^2 were fabricated by using Hand layup method in two different stacking sequences as shown in table 1.

Table 1. Laminate structure corresponding to the specimens tested

Composite material	Stacking sequence	Resin used	No of specimens	Thickness of specimens
Laminate-1	$[0^0G/0^0F/45^0G/45^0F]_2$	Epoxy	5	5.1mm
Laminate-2	$[(0/45)G/(0/45)F_4/(0/45)G]$	Epoxy	5	5.08mm

Four Glass/Epoxy and four Flax/Epoxy were consecutively arranged one after the other called Laminate-1 and a sandwich-type sheet consisting of two Glass/Epoxy layers as a base layer, four Flax/Epoxy layers as a center, and again two Glass/Epoxy layers as top layers called laminate-2. The fiber content was equal to 40 % wt. The fabricated sheets kept under the hydraulic press about 24 hours to remove excess resin while preparing laminates and give two weeks for curing at room temperature of 33±2^0, after that cut the laminates into specimens for tensile and bending as per ASTM standards.

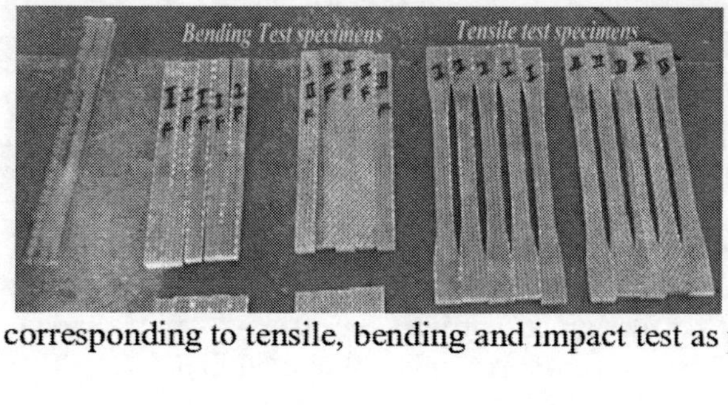

Figure 1. Specimens corresponding to tensile, bending and impact test as per ASTM standards.

IOP Conf. Series: Materials Science and Engineering 998 (2020) 012032 doi:10.1088/1757-899X/998/1/012032

2.1. *Experimental Work*

The Hybrid composites laminate-1 & laminate-2 after curing ready to tested for their mechanical properties. Here, the specimens cut as per ASTM standards in CNC machine with Router mill 4 x 18 x 50 x 2F tool and finished manually with fine Emery paper for accurate dimensions. The Universal Tensile Machine (UTM) was used for both the tensile test & bending test by changing specimen grips.

The tensile strength of a material is the maximum amount of tensile stress in the direction of force applied longitudinally that can happen before the fracture of a composite. Tensile specimens were testing on UTM as per ASTM D638 standards [10]. The size of each specimen was 165 x 19 mm^2 and the test speed for the crosshead displacement was 5mm/min and a gauge length of 55 mm as shown in figure 2.

A three-point bending test was a straightforward and easy testing method to evaluate material properties while the load was subjected to the transverse direction of the specific cross-section of the beam. It was conducted on UTM by fixing support anvil and specimen grips as shown in figure 3. This testing furnished the values for Flexural stress σ_f, flexural strain ε_f, maximum bending load and the modulus of elasticity in bending E_f. Specimens of both laminates were cut as per ASTM D790 [11] whose size of each specimen equal to 127 x 19 mm2 and specimens were placed onto two supports with a span to depth ratio of 16:1 and test speed equal to 5 mm/min.

Figure 2. Specimen was clamped on UTM

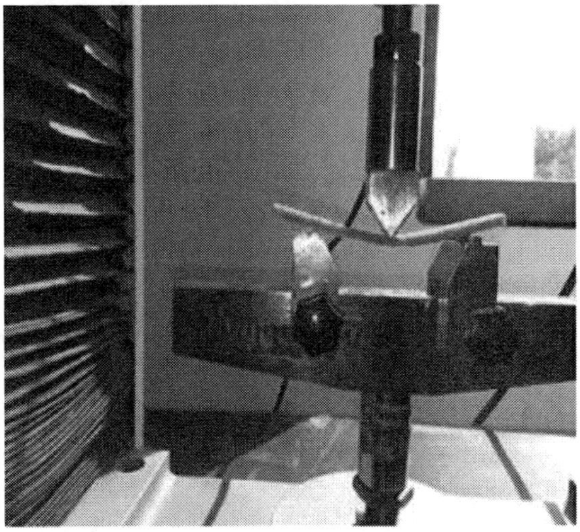

Figure 3. Specimen was simply supported on UTM

3. THEORETICAL APPROACH

Finite Element analysis (FEA) was done for both Tensile test & bending test in Abaqus software to obtain mechanical properties of two laminates. Specimen model was created in part module and simulation carried out by given input data [12] as shown in table.2

Table 2. Elastic constants & Poisson's ratios for Flax Fiber and Glass/Epoxy orthotropic materials.

Constants	E_1 (GPa)	E_2 (GPa)	E_3 (GPa)	G_{12} (GPa)	G_{13} (GPa)	G_{23} (GPa)	N_{12}	N_{13}	N_{23}
Flax fiber	52.04	10.61	11.49	3.17	2.92	2.75	0.473	0.161	0.279
Glass/Epoxy*	45	10	10	5	5	3.8462	0.3	0.3	0.4

*taken from ANSYS 16.0 material library.

A deformable 3D model created in dog bone shape as per ASTM D638 in tensile test. Composite Layup option makes facilitate to create required two laminates simply stacking sequence and assign each ply orientation as per design in software i.e. 0^0 and 45^0 as shown in figure (4a) & (4b) and stacking arrangement of two type laminates plies shown in figure 5.

Figure (4a). Axis 2 represents a longitudinal direction (say 0^0) of the fiber in woven fabric

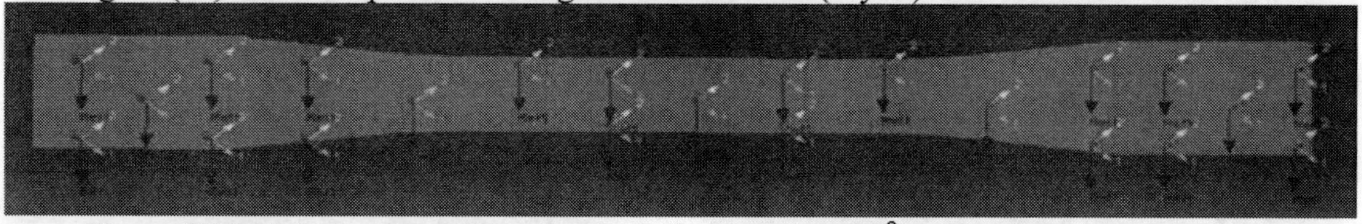

Figure (4b). Axis 2 represents a diagonal direction (say 45^0) of the fiber in woven fabric

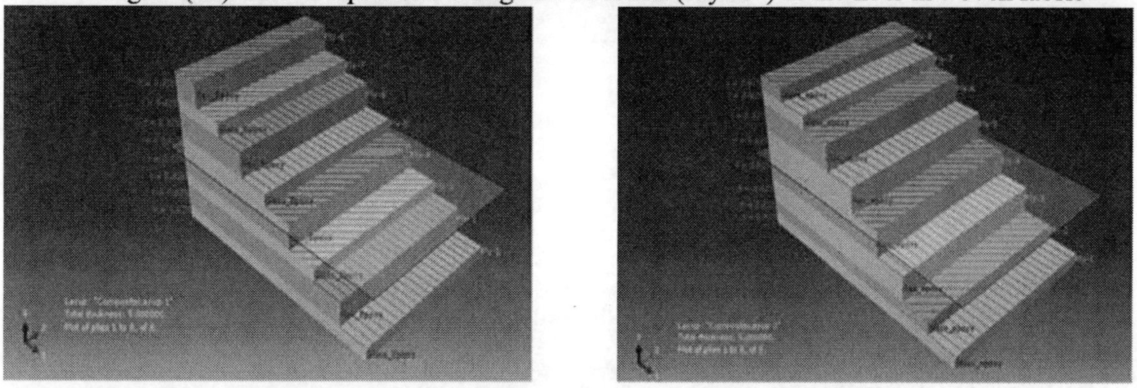

a. b.

Figure 5. Defining the layout layers (Stacking sequence) in case of two composite materials theoretically analyzed: (a) Laminate-1, (b) Laminate-2.

A parameterized model with the finite element of the two laminates were constructed in Abaqus software to perform the simulation of mechanical behavior such as Stresses, strain, and displacements in both tension test and bending test. Here, the results attained by FEA in the bending test would compare with experimental results for the elastic domain only. It is shown that the composite layup tool for shell-type parts has facilitated to create stacking sequence and ply orientation as per requirement. It was designed a shell whose dimensions are 127 x 20 mm in bending and 165 x 20 mm in the tensile test. The scheme of loading & mesh model in both bending and tensile test was as shown in figure 6 and figure 7.

In tensile test, Laminate-1 employed of maximum load of 4281 N with total elements about 672, Laminate-2 subjected to maximum load of 4030 N with 656 total elements and for bending test maximum force applied for elastic domain about 100N for both specimen composites. In this paper, the finite element model was defined with a 4-node doubly curved S4R element which was used for thin or thick shell element for both tests. Boundary conditions were applied according to loading for tensile and bending test separately.

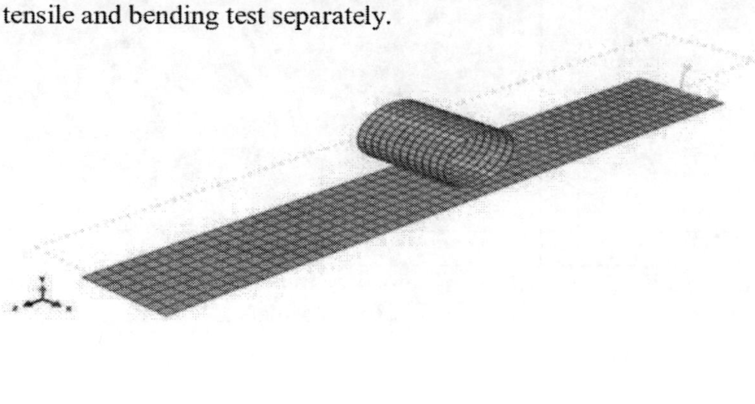

Figure 6. Model of the laminated composite material in bending with an applicator at the middle of the span

IOP Conf. Series: Materials Science and Engineering 998 (2020) 012032 doi:10.1088/1757-899X/998/1/012032

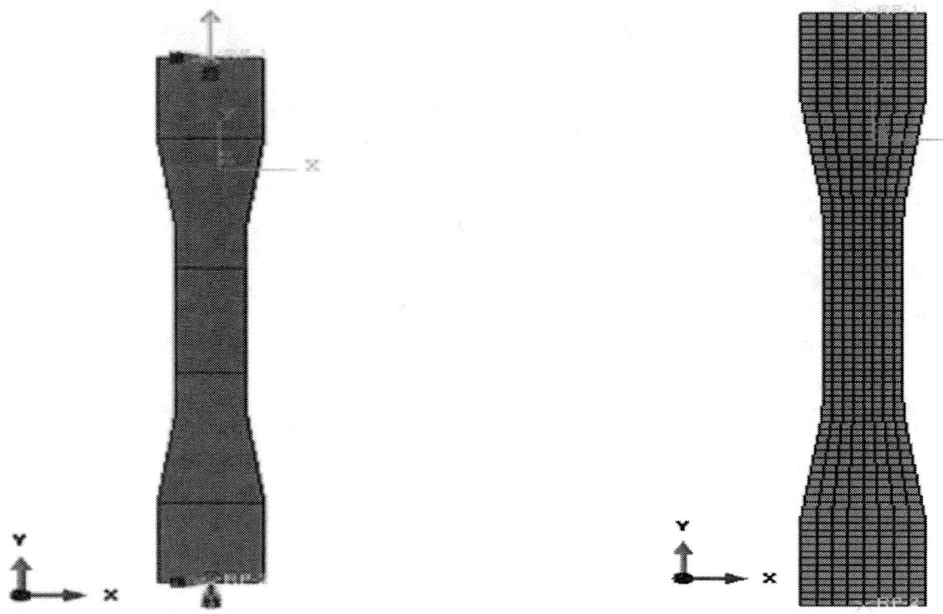

a. b.

Figure 7. Model of the laminated composite material in tensile test (a) scheme of loading (b) meshing with shell elements

4. THEORETICAL RESULTS

4.1. For Bending Test

The distribution of the normal stress generated σ_1=S_{11} across the thickness in bending through integration points for both laminates as shown in figure 8. Shell tool has defined that each layer of the specimen has three integration points, one is represented lower layer surface, middle point represented the center of the layer, and third layer represented a top surface of the layer. Abaqus automatically assign these integration points across the given thickness of shell to found the variation of mechanical property in each layer.

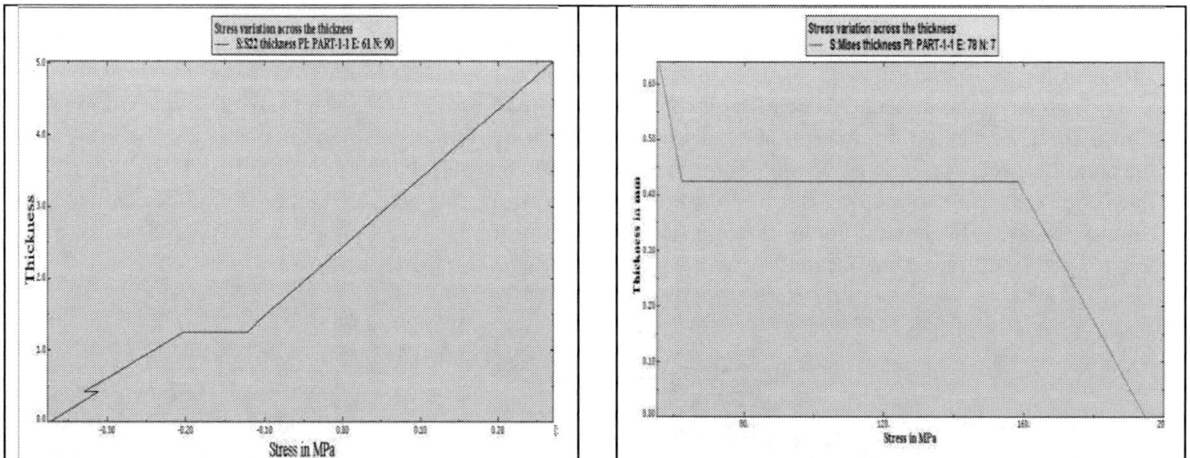

a. b.

Figure 8. Distribution of the normal stress σ_1=S_{11} across the thickness in bending in case of two composite structures analyzed: (a) Laminate-1 (b) Laminat-2

From above figure, it is noticed that the laminate-2 was subjected to maximum normal stress at bottom layer of the composite material and sudden decrease at the interaction of Flax/Epoxy and Glass/Epoxy layers. Laminate-1 has employed maximum normal stress at the top surface and minimum stress at lower layer as they were arranged alternately. The figure (9) & (10) shows the Distribution of vertical displacement U_2 and the distribution of stresses in the direction of loading in bending respectively after the simulation of specimens of both laminates.

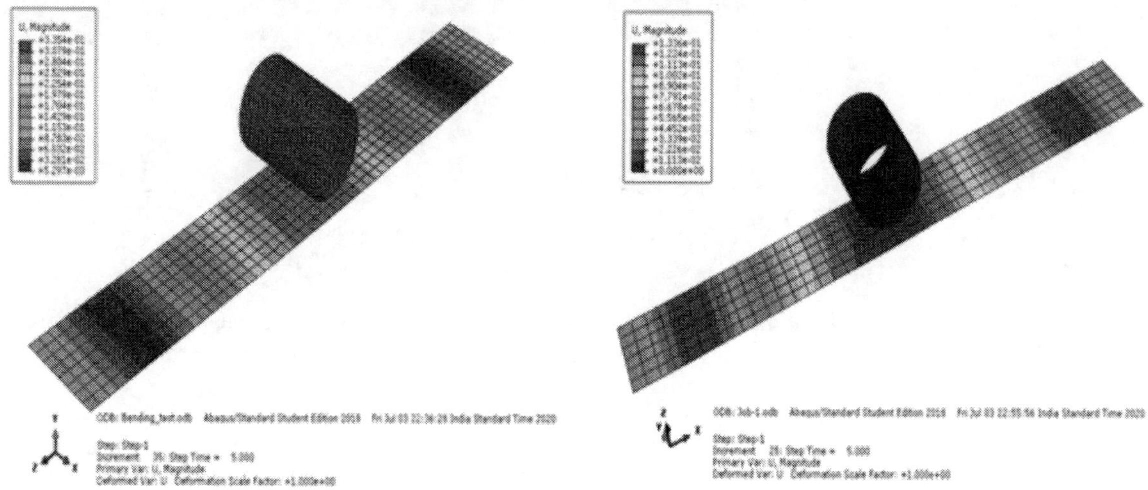

a. b.

Figure 9. Distribution of the vertical displacement U_2 in the direction of the axis OY in the case of (a) composite laminate-1, (b) composite laminate-2.

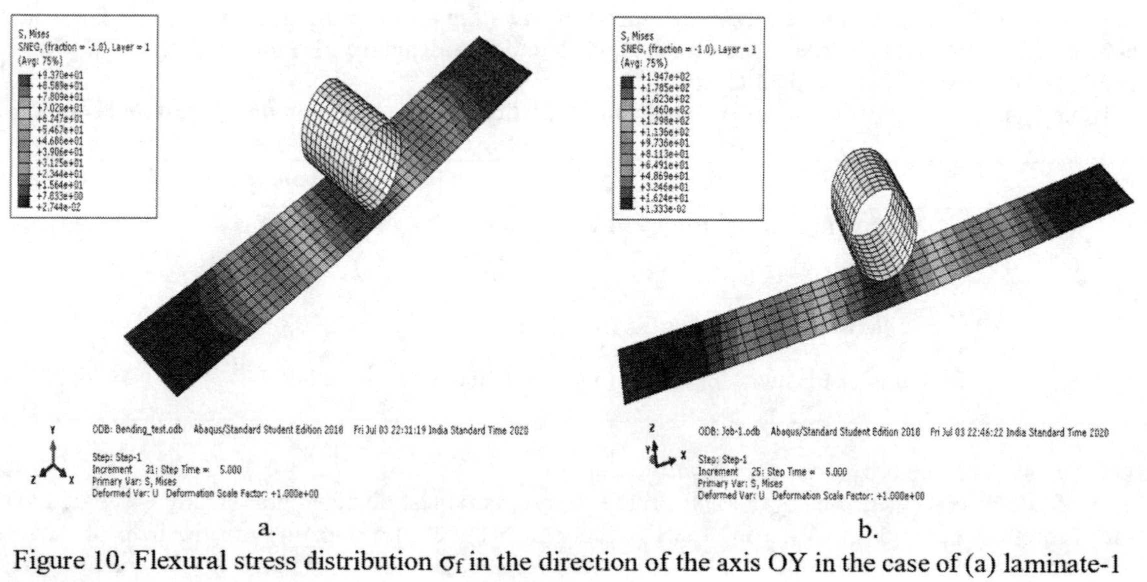

a. b.

Figure 10. Flexural stress distribution σ_f in the direction of the axis OY in the case of (a) laminate-1 and, (b) laminate-2.

4.2 For Tensile test

It is shown that the distribution of stresses in tension test in the direction of the axis OY of both laminate specimens as shown in figure 11.

IOP Conf. Series: Materials Science and Engineering 998 (2020) 012032 doi:10.1088/1757-899X/998/1/012032

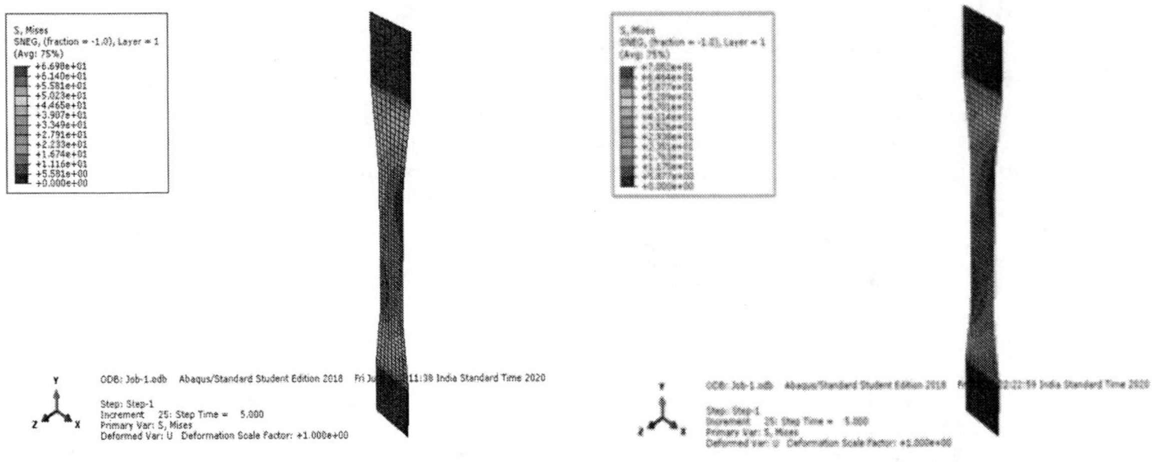

a. b.

Figure 11. Tensile stress distribution σ_{max} in the direction of the axis OY in the case of (a) laminate-1, (b) laminate-2.

4.3. *Experimental Results*

The average values of the mechanical properties of five specimens for each laminate were experimentally determined in table 3. In that, it is marked that the Laminate-2 shows greater values than Laminate-1 in the bending test because the Flax/Epoxy layers accumulated as reinforcement fully with Glass/Epoxy layers like a sandwich model whereas in tension testing the laminate-1 obtained somewhat better values than other laminate. It may also observe that the Flexural Modulus of Elasticity E_f in bending is 80 % higher in the case of laminate-2 than in the case of the laminate-1 whereas the modulus of Elasticity E_x in tensile testing is with 1.434 % more in case of laminate-1 than in case of laminate-2. Moreover, the maximum flexural strength in bending in the case of laminate-2 shows 92.2 % more than in the case of laminate-1 due to the advantage of reinforcing the Glass/Epoxy layers in both top and bottom shell layers.

Table 3. The average values of the mechanical properties measured in both tensile test and bending test in case of two composite materials tested

S.No	Type of the composite material	Laminate-1	Laminate-2
1.	Maximum Tensile strength σ_{max} , MPa	67.08	64.89
2.	Modulus of Elasticity E_x in Tensile test, MPa	753.99	743.33
3.	Maximum Flexural stress σ_f , MPa	98.698	189.742
4.	Modulus of Elasticity E_f in bending test, MPa	4549.742	8186.55

4.4. *Theoretical versus experimental results*

Here, force-displacement curves (F-U) from the experimental test obtained graphically compared with the theoretical curve drawn by using the FEA model in the elastic domain with the help of Abaqus software in the case of both laminates. In figure 12, the data obtained in the bending test experimentally has shown with different color lines for three specimens of both laminates while the theoretical curves (F-U) drawn by using red lines. Similarly, in figure 13, the data obtained in the tensile test has indicated with various color lines but the theoretical curves (F-U) drawn by using blue lines.it may be noticed that analysis model gets with available data in a beautiful manner in the case of both material.

IOP Conf. Series: Materials Science and Engineering 998 (2020) 012032 doi:10.1088/1757-899X/998/1/012032

It can be noticed that the FEA model force-displacement curve fits linearly with experimental data in good manner in case of both laminates. It is marked that the error percentage for both test recorded in table (4) that calculated by comparing the important mechanical properties of both laminates experimentally with the values evaluated theoretically from the FEA model.

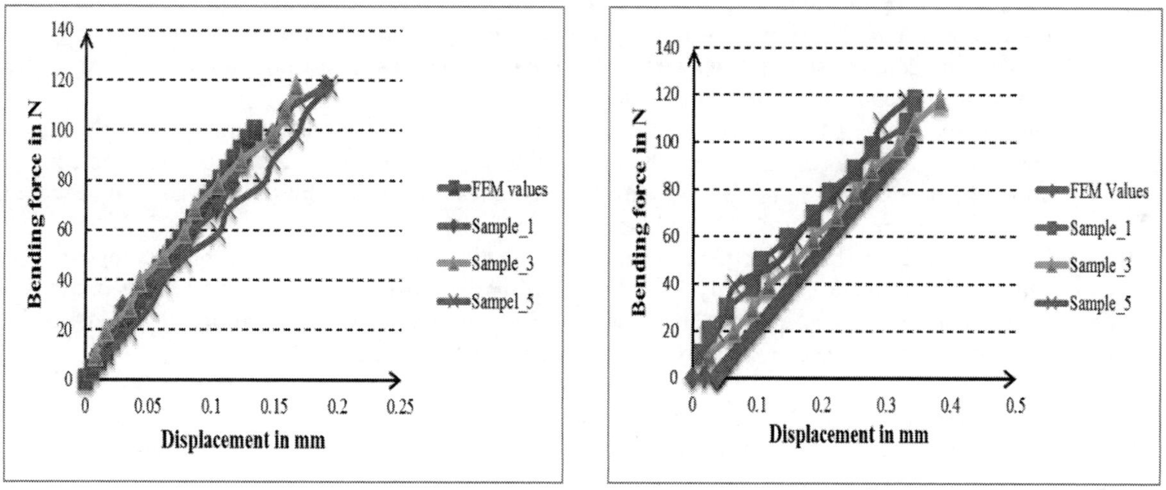

a. b.

Figure 12. Comparison between the results experimentally obtained in bending and ones obtained by FEA concerning the Force-Displacement curve (F-U) in the case of (a) Laminate-1, (b) Laminate-2.

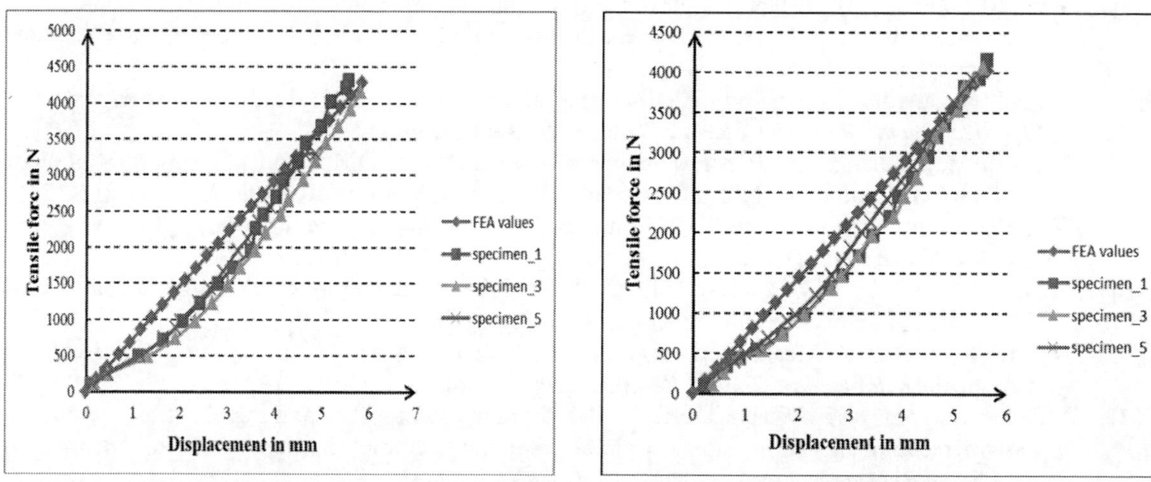

a. b.

Figure 13. Comparison between the results experimentally obtained in tensile test and ones obtained by FEA concerning the Force-Displacement curve (F-U) in the case of (a) Laminate-1, (b) Laminate-2.

Table 4. Comparison between the experimental and theoretical results obtained from FEA model.

Test method	Type of the composite material	Maximum strength in MPa		Error in (%)
		FEA model	experimentally	Exp. Versus FEA
Tensile Test	Laminate-1	64.0126	67.0811	4.793
	Laminate-2	70.5735	64.8976	-8.042
Bending Test	Laminate-1	93.7	98.698	5.334
	Laminate-2	194.7	189.742	-2.546

5. CONCLUSION

The main conclusions taken from experimental and numerical investigation of tensile and bending test on two laminated composites are as follows:

- Laminate-2 appeared in better results in terms of maximum flexural strength σ_f = 189.742 MPa in bending than the value σ_f = 98.698 MPa recorded in case of laminate-1.
- In tensile test both laminates showed almost same property maximum tensile strength that means changing the stacking sequence of same fibers in tensile test not enough to improve the mechanical properties of composite materials.
- Numerical analysis carried out in the tensile and bending test of two laminated composite materials and these results experimentally validated and therefore calculated error percentage about 4.8 % and 8.04 % respectively in the tensile test. In the bending test, it has found that an error percentage of about 5.3 % and 2.5 % respectively.

It has also noticed from this work that the mechanical characteristics of the hybrid composite are higher than composite made up of only Flax/Epoxy composite [4] material. These hybrid composite materials can be useful in the innovation of automotive and aerospace applications in the aspect of lightweight and cost effective eco-friendly manufacturing technology.

Acknowledgement
Authors thankfully acknowledge the department of Mechanical Engineering, Vardhaman College of Engineering for the encouragement and support, and to the Department of Mechanical Engineering Vignan Institute of Technology and Science Deshmukhi, for providing Composite Material laboratory.

References

[1] Yan L, Chouw N 2015 Effect of water, seawater and alkaline solution ageing on mechanical properties of flax fabric/epoxy composites used for civil engineering applications, *Construction and Building Materials* **99**, 118-127.
[2] Sathish Ganapathy, Bhoopathy T 2014 Experimental Testing on Hybrid Composite Materials, *Applied Mechanics and Materials* Vols. **592-594**, pp 339-343.
[3] Ramesh M, Sudharsan P 2018 Experimental Investigation of Mechanical Properties of Flax-Glass Fiber Reinforced Hybrid Composite using FEA, Volume **10**, Issue 3, Pages 747-757.
[4] Camelia Cerbu, Marius Botis 2017 Numerical modeling of the Flax/Glass/Epoxy Hybrid composite material in bending, *Procedia Engineering* **181** 308-3015.
[5] Wang H.W et al 2014 Analysis of Effect of Fiber Orientation on Young's Modulus for Unidirectional Fiber Reinforced Composites, *Composites Part B* **56** 733-739.
[6] Charlet K et al 2010 Mechanical Properties of Flax Fibers and of the derived Unidirectional Composites, *Journal of Composite Materials* Vol. **44.**
[7] Infanta Mary Priya I, Vinayagan BK 2018 Enhancement of bi-axial glass fibre reinforced polymer composite with graphene platelet nanopowder modifies epoxy resin, *Advances in Mechanical Engineering*, Vol. **10(8)** 1–10.
[8] Buket Okutan et al (2001) A study of the effects of various geometric parameters on the failure strength of pin-loaded woven-glass-fiber reinforced epoxy laminate, *Composite science and technology*, Vol **61** Pages 1491-1497.
[9] Djafari Petroudy 2017 Physical and mechanical properties of natural fibers, *Advanced High Strength Natural Fiber Composites in Construction*, 59–83.
[10] ASTM D638, standard test method for Tensile properties of unreinforced and electrical insulating materials, annual book of ASTM standards. philadelphia.
[11] ASTM D790-03, Standard test method for flexural properties of unreinforced and electrical Insulating Materials, Annual book of ASTM standards. Philadelphia.
[12] Rahman, Ziaur M 2017 Mechanical Performance of Natural Fiber Reinforced Hybrid Composite Materials Using Finite Element Method Based Micromechanics and Experiments

IOP Conf. Series: Materials Science and Engineering **998** (2020) 012033 doi:10.1088/1757-899X/998/1/012033

EXPERIMENTAL INVESTIGATION OF INCLINED CIRCULAR RING TURBULATORS ON HEAT TRANSFER AND FLUID FLOW CHARACTERISTICS.

Atul M. Suryavanshi[1], Avinash G. Kamble[2], Sandip S. Kore[3]
[1]Student, [2]Associate Professor at School of Mechanical Engineering, MIT Academy of Engineering, Alandi, Pune. [3]Associate Professor at Department of Mechanical Engineering, Vishwakarma Institute of Information Technology, Pune, Maharashtra, India, 412105.
Correspondence Author-atulsuryavanshi0712@gmail.com

Abstract

This paper demonstrates influence of various inclination of Circular Ring Turbulator (CRT) on thermal characteristics in circular type heat exchanger. "A wall linked circular inserts were analyzed having constant inner diameter to achieve uniform flow blockage area (FBA) of 40%". These CRTs located within the tube to prevailed different angular patterns (-10°, -15°, -20° and -25°) w. r. t. horizontal of tube. Influence of these geometries on fluid friction and heat performance were studied. Parameter constant throughout the investigation was diameter ratio and Pitch Diameter Ratio (PDR). Modified parameters throughout study were the angular inclinations of CRT and Reynolds number. Inner side of tube is preserved at constant heat flux having air as working medium at atmospheric temperature with variation of Reynolds number from 6,000 to 24,000 respectively.
Implementation of CRT as turbulator favorably improves rate of heat exchange within heated enclosure of tube to fluid domain throughout the test model .Out of all studied inserts, CRT having angular deflection of -25° gives peak value of heat transfer at extreme Reynolds number and at slightest Reynolds number CRT having angle inclination of -10° offers greater friction factor which is at 2.5 and 5.5 times than smooth tube. CRT with angular inclination of -25° at minimum Reynolds number offers extreme thermal performance factor is 1.7.

Keywords
Heat transfer enhancement; Circular ring turbulator (CRT); friction factor; Heat Exchanger; Angular deflection (-10°, -15°, -20° and -25°); Nusselt number; Reynolds number.

Introduction

Heat exchangers plays a vital role in different thermal power plant, process industries, condensers, evaporators, space vehicles and radiators in automobiles, electronic cooling units, nuclear reactors etc. The thermal characteristics of the heat transfer inventions depend on the impact of convective factor between heated surface and moving fluid. "The overall thermal resistance is proposed by the development of the boundary layer across the heated wall of the heat transferring device". In modern decades, most of investigators dealing with optimization techniques and overall stability of exchangers due to maximum material cost, which enhancing the transfer rate with less pumping energy [1]. Size as well as cost of heat exchanging device and pumping power may be shorten by raising the convective factor and shrinking pressure loss over the exchanger. "Heightening the turbulence by obstructing and blocking the fluid flow as well as swirl generation by suitable external geometry or surface modification causes increment in the rate of convective heat transfer". It advances not only severance of the boundary layer but also appropriate and swift domain mixing. Normally, improvement in heat exchange is carried by active and passive process. Among these, passive process is most preferred by researchers because it can't demand any external source of power, while active techniques require an additional source. In passive process, these geometries known as a turbulators through which working medium passing and fixed inner tube of exchanger. "In the process, along with enhancement in heat exchange, the pressure drop over the tube increases as a result of produced higher turbulence and secondary fluid flow". Hence efficient process of improvement in thermal stability of exchanger is the one improves the heat exchange expressively with less power [2].
Now a days, various process were designed and investigated solutions have been represented by Sakr [3] and Dewan [4]. Most of the investigators predicted that passive processes as emerging domain of research for thermal enhancement. Implicit of these processes of heat enhancements may be thoughtful in modified refrigeration methods by Jomde et al. 2018a; 2018b [5,6].This paper consists passive heat exchange improvement technique, most use passive methods and allied patterns from the survey data mentioned below.

Literature Review

IOP Conf. Series: Materials Science and Engineering **998** (2020) 012033 doi:10.1088/1757-899X/998/1/012033

Most of passive patterns were made for exchange of heat enhancement in working medium of circular tube which consist perforated twisted pattern [7], coil wire, circular rings [8-9], V-nozzle [10], mesh inserts, conical nozzle ring [11], conical ring, ribbed pipes, V-ring, brush and strip turbulators, baffles etc. Distinctive twisted patterns insert mostly beneficial over laminar region also geometry such as conical ring, CRTs, wire coil etc. are highly useful for turbulent model. Wire coil pattern doesn't mix the fluid domain satisfactorily; also unconducive for laminar zone which overcome efficiently with twisted type patterns. Because of this, twisted tapes are more useful in laminar region. Lot of researchers studied various geometries which generate the secondary flow and vortices by obstructing the flow effectively. "The inserts investigated were a V-nozzle, conical nozzles, conical ring, diamond-shaped turbulators [12], perforated ring [13], conical-nozzle with free-spacing entry, anchor-shaped inserts" [14], tube corrugation turbulators [15] baffle used geometry [16-17], wavy stripes [18], dimpled surface [19], converging-diverging ring, etc. An investigation by Eiamsaard was carried to predict thermal phenomenon throughout circular tube by twisted tape including conical-nozzles insert, shows better solution in overall performance coefficient. Heat flow improvement along with vibration produced in exchangers having conical-ring patterns, investigated by Yakut [20] with varying Reynolds number from 8000 to 18000. It concluded that, increment in Nusselt number possible by increasing Reynolds number at lowest pitch arrangement. Tandiroglu studied baffle inserts for convective heat exchange across turbulent model. Parameters varied of flow geometry were angular baffle orientation from 45°, 90°, 180° with PDR, empirical data was presented to calculate Nusselt number and fall in pressure [21]. Eiamsa-ard [22] concluded that, inserts created more frictional losses because of the flow blockage. An investigation by Rattanawong [23] stated that increment in the convective factor gained with propeller type swirl propagator effectively. An investigation carried on flow friction as well as heat exchange in round tube consisting porous type ring insert by Akansu [24]. The effect of inserts like inclined vortex rings (IVR) were studied with uniform heat flux in a round tube and its influence on heat transfer and fluid friction were investigated numerically by Promvonge [25].Effect of the IVRs in moving flow offers significant influence on thermal parameters and pressure loss compared to smooth tube. Promvonge [26] studied allied horseshoes baffles along with various angular inclinations in a round heat exchanger. "This type of turbulator was inserted significantly into the circular tube exchanger at different flow blockage ratios; width and PDR factor (25%, 50%, 100%, 200%)". The range of overall thermal performance coefficient was varied 1.34 to 1.92. The optimum scale observed at PDR of 25%, 50% having FBA of 25%, 100%. Chingtuaythong [27] investigated influence of V-type inserts on heat performance as well as friction factor under the uniform heat flux. In this work he presented the V-shaped ring inserts with four relative ring-pitches diameter ratio (0.5, 1.0, 1.5 and 2.0), three flow blockage ratios (0.1, 0.15 and 0.2), and angular orientation of 30^0 were fitted in tube. The overall factor was gained up to 1.36 to 1.63 across FBA 0.1 with relative diameter 1.0. V-ring achieves better thermal stability than other patterns such as twisted tapes, inclined rings and wire-coils. Singh [28] carried experiment with solid rings twisted tapes in a round heat exchanger to study heat transfer rate and fluid friction factor of moving flow along with uniform heat flux for the turbulent flow in which Reynolds number ranges from 6300 to 22,500. It shows at lowest pitch factor 1 with twist ratio 2 achieves excellent thermal performance. "Twisted ring (TRs) or Circular rings (CRs) inserts were comparatively tested for augmentation of friction factor and heat transfer in tube exchanger for Reynolds number 6000-20000". Most of CRs achieve highest Nusselt numbers with flow friction compare to TRs excluded slightest PDR and highest width ratio, studied by Thianpong [29]. Gear-ring turbulators (GRTs) investigated to demonstrate effects on friction phenomenon with heat exchange rate in round tube exchanger. Friction characteristics and heat transfer rate increase with decreasing free-space length ratio and tooth number. Study was carried with CRT along with PDR and different diameter factor by Nanan [30]. It showed a CRT having lowest diameter factor produces the largest pressure drop along with the higher Nusselt number value. As per the investigation at the extreme diameter ratio 0.7 and lowest pitch 6, higher thermal performance is achieved. Promvonge investigated effect of conical type inserts having numerous alignments (converge geometry [CG] converge-diverge geometry [CDG] and diverge geometry [DG]) to heat transfer and overall thermal performance in circular tube. "By using these alignments, DR array of conical rings achieves the higher results as far as the thermal parameters are concern" [31]. Effect to overall thermal performance as well as moving flow friction of V-nozzle and conical ring inserts was studied by Promvonge [32]. Sane presented modern variable as flow blockage area applied for circular inserts terminology. "Kore conduct investigation on circular rings of different inner diameters for the tests which creates 30%, 40%, and 50% flow blockage area [33]. These ring type turbulator were mounted one after another to create insert of converging, diverging and converging-diverging model". It showed that, CRT is efficient insert which improves transfer rate and fluid friction favorably with 40% blockage area. Also overall thermal performance decreased with higher Reynolds number.

Summary of study determine, turbulence magnitude may be predicted higher than smooth tube by obstructing flow beneficially by various arrangements in tube. Current manuscript comes with an idea of different angular inclinations of circular ring turbulators (CRTs). "Main focus of study is to demonstrate effect of different geometrical parameters (angular deviation w.r.t. horizontal) of CRT on rate of heat transfer and fluid friction". Each insert has a Circular ring with the same diameter ratio; PDR is 2 along with 40% FBA. The CRT of various angular inclinations (-10°, -15°, -20° and -25°) has been selected to check influence on thermal performance. The inserts were tested for turbulent model with Reynolds number from 6,000 to 24,000.

Model Description

Schematic configuration of simulation model is display in figure 1. Test section i.e. heat exchanger tube is open-type model having blower attachment of 2 HP capacity. To measure the flow rate, U-tube manometer is connected to orifice meter at the calming section. Calibration of orifice meter is done by using the anemometer and it has been fixed up according to ASME standard. According to the relation given [34] to smooth tube, distance of calming part is predicted at maximum Reynolds number. "It is taking 1.4 m which offers flow fully develops before entering the test section also uniform fluid entry inside the tube a bell mouth shape opening is given".

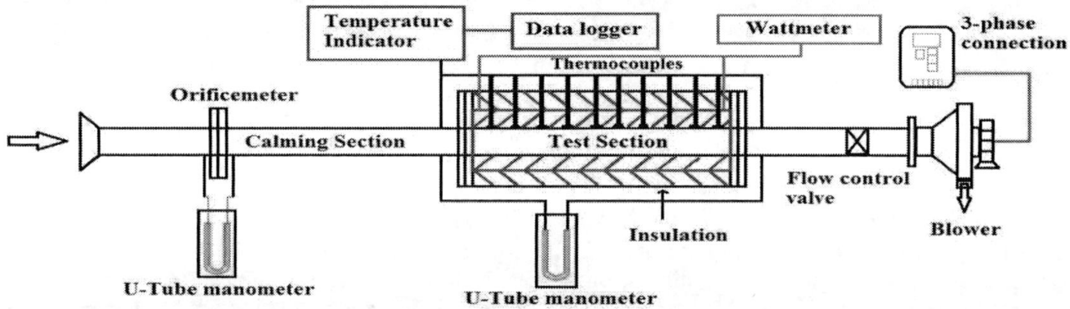

[35]

Figure 1.Illustration structure.

Testing model is a hollow circular tube consist length of 1 m with inside diameter 49 mm and thickness is provided 3 mm framed by galvanized iron. Selection of test section length is so to predict measurable pressure as well as temperature variation at moderate Reynolds number. Bunch of 16 K-type temperature indicator are mounted on tube wall to predict surface temperature (T_w) of test section. To permits accuracy in the range bound of ±0.1° thermocouples are calibrated accordingly. "Pressure indicator is fixed at the end of test section which is attached to micro-manometer to predict significant pressure drop along the test section, outside surface is enclosed with Nichrome winding wire heater having 1.2 Ω/m resistance and single phase variac transformer is adapted to maintain a tube wall condition at constant heat flux of 1000 W/m^2along with control input power".

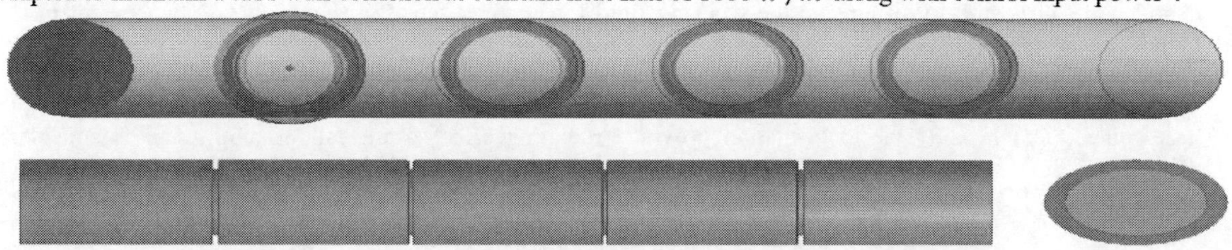

Figure 2.Typical circular ring turbulators.

Variables data	Specification
Dimension of Calm part (mm)	1400
Dimension of test model (mm)	1000
Inside dia. of tube (mm)	49
Outside dia. of tube (mm)	55
PDR	2.0
Re	6000–24000
Temperature of fluid (air)	At ambient condition
Angular deflection of CRTs	0°,-10°, -15°, -20°, -25°

Table1.Details of parameters

To minimize losses to surrounding from exchanger, asbestos insulation with glass wool layers of thickness 5 cm is provided. Two set of thermocouples used to distinguish working fluid temperature at inlet as well as outlet. While testing, fluid i.e. air is pass throughout model by blower. Flow control valve is attached to tube for adjusting flow inside the round test section and obtain required Reynolds number. The CRT made by mild steel having 3 mm thickness was fitted in the test section display in figure 2. Specification of dimensions and variables modify within experiment display in table 1.

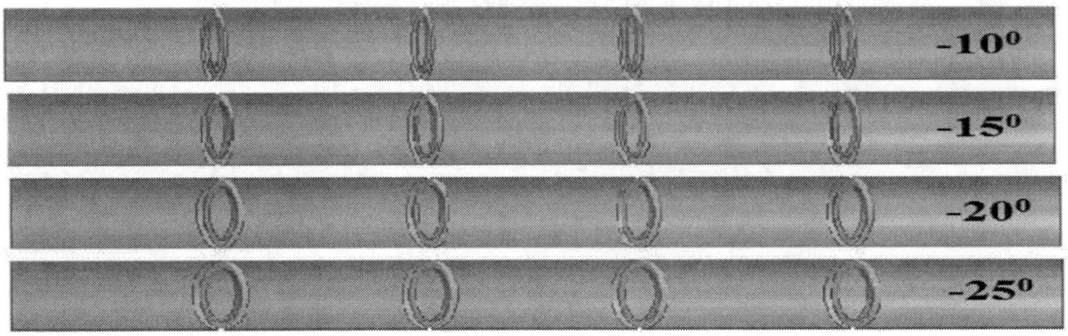

Figure 3.CRTs having angular deflection of -10°, -15°, -20° and -25° fitted in testing model.

Simulation models of various angular deflections -10°, -15°, -20° and -25° from the horizontal is shown in figure 3. McClintock and Kline adopted method for uncertainty inspection [36]. Uncertainty equation expressed as,

$$\mu_F = \left[\left(\frac{\partial F}{\partial a_1} \mu a_1 \right)^2 + \left(\frac{\partial F}{\partial a_2} \mu a_2 \right)^2 + \dots \dots + \left(\frac{\partial F}{\partial a_n} \mu a_n \right)^2 \right]^{\frac{1}{2}} \quad (1)$$

Such as F= $F\ (a_1, a_2, \dots, a_n)$ with a_n as parameter which influence on outcome of F.

"For parameter Reynolds number i.e. non-dimensional, friction factor and Nusselt number higher uncertainties were estimated to be 6.4%, 8.3%, and 7.2% respectively, for the average axial velocity predicted uncertainty was found to be less than ±7%; the uncertainty in pressure drop predicted has been found as ±6%, while the uncertainty in surface temperature was estimated about ±0.5%."

Data computation

Calculation of Parameters

At steady state condition all the data were measured during conducting tests. To achieve steady state phenomena boundary temperatures offers fixed data. Current simulation values was additionally utilize for prediction of the pressure drop, fluid friction, coefficient of heat exchange, Nusselt number, and thermal performance criteria for dissimilar geometry and motion condition. Reynolds number is given by $R_e = UD/v$ where D, U and v are the diameter of tube, velocity of air and the working fluid kinematic viscosity respectively. Enhancement in heat transfer indicates by Nusselt number which is dimensionless number expressed as $N_u = hD/k$ in particular operating criteria, where h is heat transfer coefficient and k is conductivity of material used.

"At steady state condition, the convective heat transfer ($Q_{convection}$) of tube from inner surface is equal to the heat gained (Q_a) by air. Due to insulation provided radiation effect from wall of tube to air is considered negligible".

$$Q_{convection} = Q_a \quad (2)$$

$$hA(T_w - T_b) = mC_p(T_o - T_i) \quad (3)$$

Where, $T_w = \frac{(T_1+T_2+\cdots+T_{16})}{16}$ And $T_b = \frac{(T_i+T_o)}{2}$

Parameter of heat transfer, h is predicted from enthalpy (3). Fluid friction, f is given by,

$$f = \frac{\Delta P}{\left(\frac{L}{D}\right)\left(\rho \frac{U^2}{2}\right)} \quad (4)$$

Prediction of the overall outcome of CRTs calculated within various conditions with modifies variables. Net overall ingredient (η) consists heat exchange along with moving flow factor synchronously having ratios of (Nu_t/Nu_p) to (f_t/f_p) expressed below.

$$\eta = \frac{(Nu_t/Nu_p)}{(f_t/f_p)^{\frac{1}{3}}} \quad (5)$$

Determination of results

Validation of testing model

Investigated data regarding friction factor as well as Nusselt number were predicted from the values obtained from round exchanger. Calculated data compared with outcome received from Blasius and Dittus-Boelter relation to validate experimental methodology adopted given in the literature. For fully developed turbulent fluid flow tube Dittus and Boelter correlation is expressed as

$$Nu = 0.023Re^{0.8}Pr^{0.4} \tag{6}$$

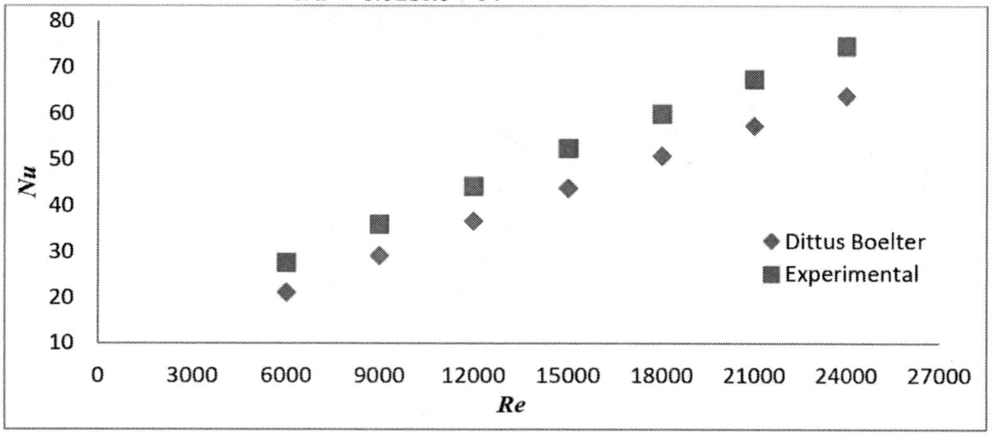

Figure 4.Validation of Nu for current model.

Blasius relation to evaluate fluid friction for the smooth tube in turbulent region, expressed as

$$f = \frac{0.316}{(Re)^{\frac{1}{4}}} \tag{7}$$

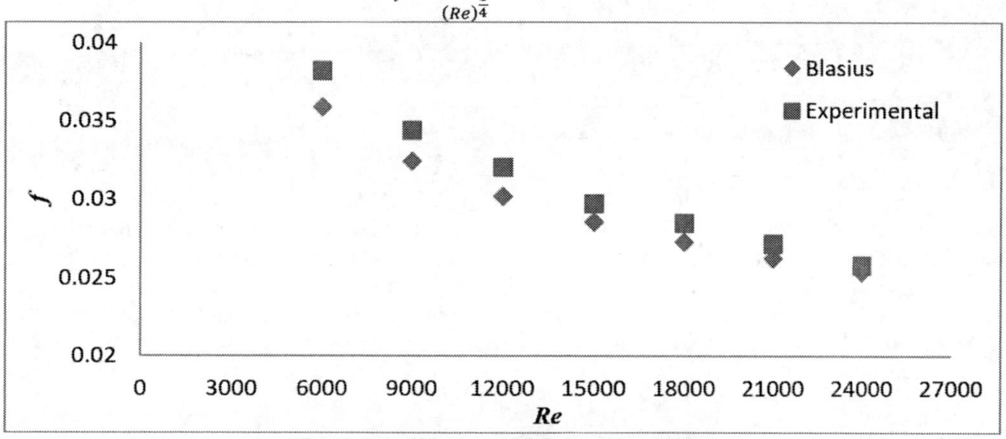

Figure 5.Validation of f for current model.

Image 4 shows that Nusselt number obtained from correlation expressed by Equation (6) are comparable with Nu determined from simulation values of circular exchanger with a highest deviation of 14%.

Friction factor value for smooth tube obtained from experiment was compared with those calculated from the Blasius correlation with highest deviation 8.4% which shown in Figure 5.

Numerical Study

Adoption of turbulence model

In practical engineering applications most commonly used turbulence models, consist the realizable k-ε, standard k- ω and SST k- ω model were helpful to predict three-dimensional turbulent flow, the steady-state and Thermal characteristics of model [37]. The simulation data obtained from each models was compared with testing result for smooth tube; observation made that calculation of Nu by realizable k-ε model is more correspondents to investigated values than other models.

"Comparatively, the highest variation for the average Nusselt number for numerical results and experimental data by using realizable k-ε model, the standard k-ω model and the SST k- ω model are 8.9%, 7.8% and 6.5%, respectively, concluding that the k-ε Model is more reliable than other models". Therefore, for further numerical investigations k-ε Model was applied in the present work shown in figure 6.

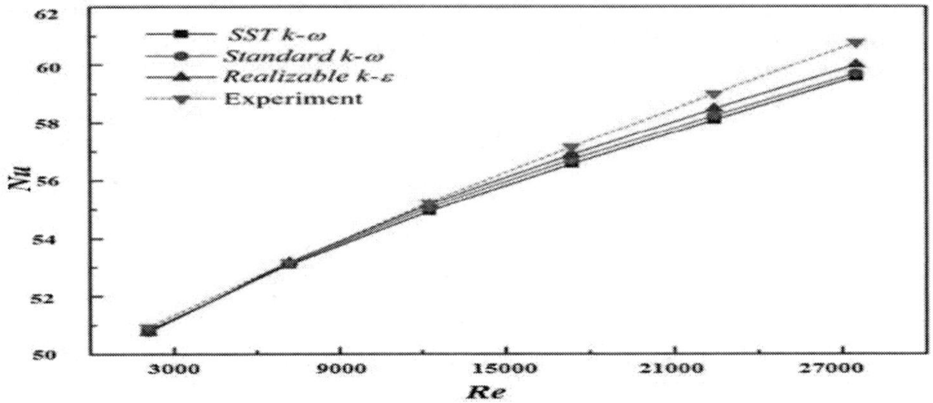

Figure 6 Selection of turbulence model for present smooth tube

Grid independence study

"The fluid domain was discretized by unstructured mesh generated with the commercial software ANSYS CFD 18.0 as shown in figure 7, to obtain more accurate simulation results it is ensure that y^+ remained less than 1, prism grids are orthogonal to the wall surfaces and with faces perpendicular to the boundary layer fluid flow direction, were extruded from the surface of the smooth tube".

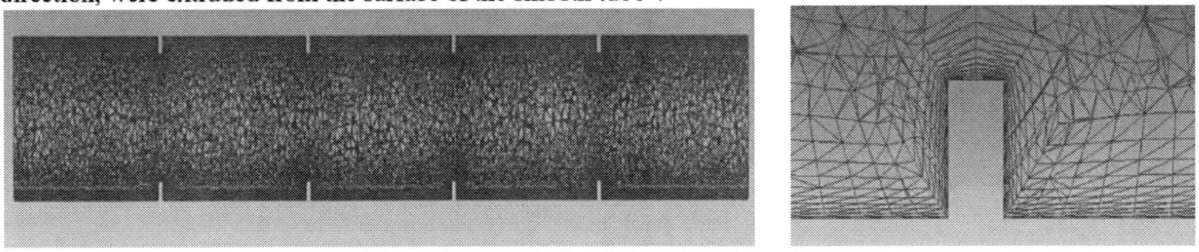

Figure 7 computational domain grid systems

"A hex core mesh, with hexahedron grids covering the majority of the volume, was created to obtain the characteristics of the fluid flow in the major region, to cover the areas inside the surface of the prisms and the hex core, tetrahedron grids used". Compute highest accuracy of simulations, six bunches of grid, with 286868, 377728, 506131, 842525, 996171 and 999845 count size, were offered in computation to check appropriate grid study at Re 24000. Comparative Nu as well as f data determined from grid is display in figure 8.

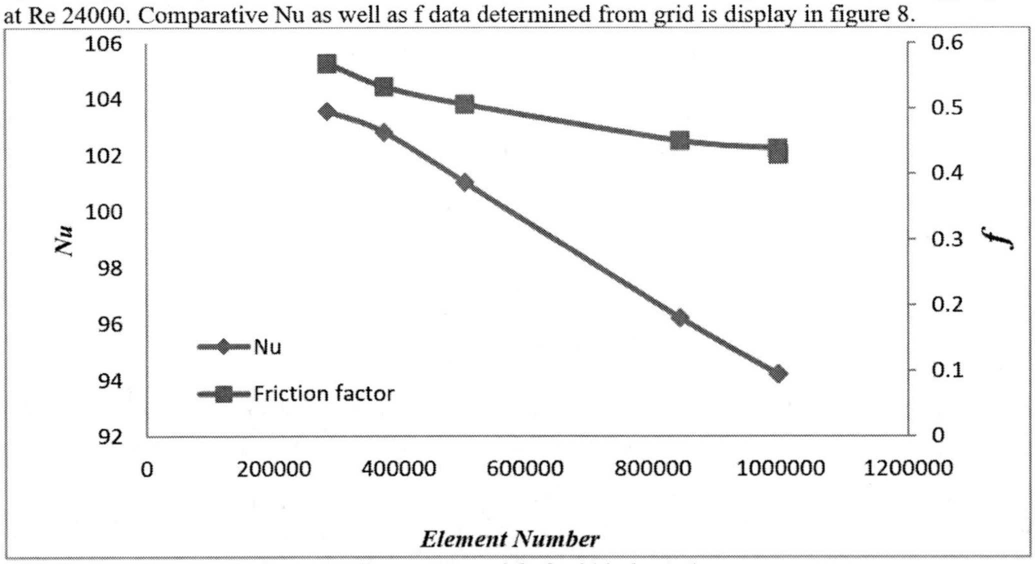

Figure 8 Effect on Nu and f of grid independence test

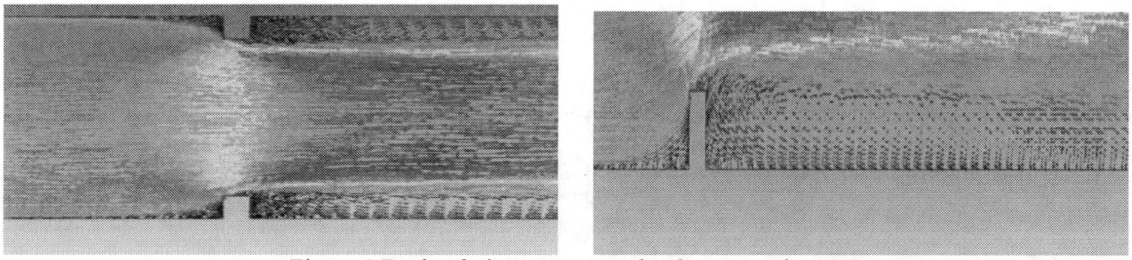

Figure 9 Recirculation over smooth tube across the CRT

"The results indicate that the deviation between the calculated values for grids comprising 996171 and 999845 elements is about 0.9% for Nusselt number and 1.2% for friction factor, indicating that the grid system with 996171 elements is adequately dense for the simulations shown in fig 9, hence grid system with 996171 elements was approved in the subsequent simulation reflecting a compromise of computational time and precision in solution".

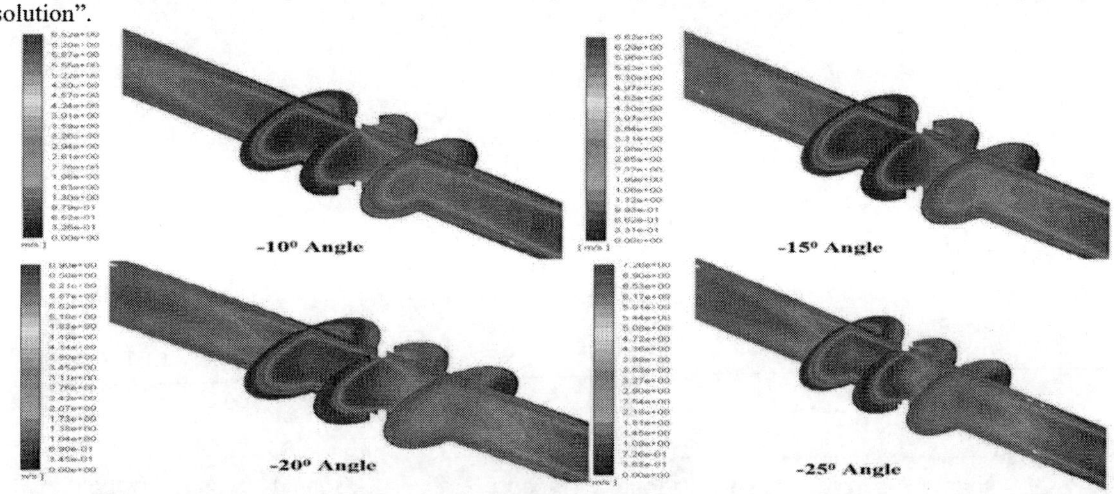

Figure 10a Velocity counters at various sectional planes of angular deflections -10°, -15°, -20° and -25° at Re 9000

In the present simulation work, a uniform heat flux of 1000 W/m^2 is specified as boundary condition on the tube wall. Before simulation inlet tube is extruded by 10 times diameter to achieve fully enlarge velocity outline. Pressure, boundary conditions concerned at outlet region. No-slip condition is offered on the surface of tube wall and turbulator surface.

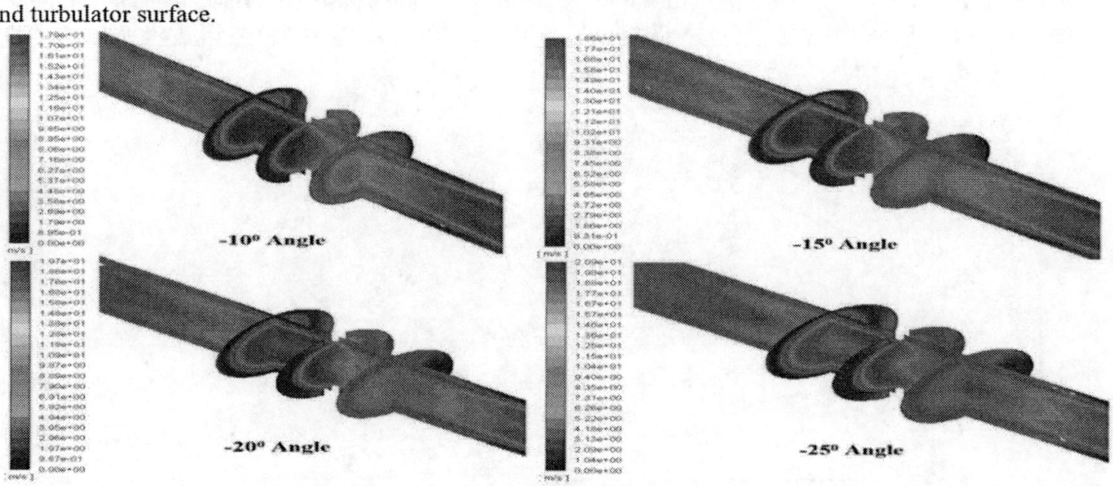

Figure 10b.Velocity counters at various sectional planes of angular deflections -10°, -15°, -20° and -25° at Re 24000

Governing factors such as energy, momentum and continuity equations are computed within steady state phenomena. "For momentum and convective diffusion evaluation, QUICK method (Quadratic upstream interpolation for convective kinetics differencing) scheme and SIMPLE (Semi-Implicit Method for Pressure-Linked Equations) method is used for velocity-pressure relation and for pressure variation". To capture accurate results near-wall, enhanced wall treatment model is used.

"A solution is supposed to be converged when residuals were less than 10^{-9} for the energy equation and less than 10^{-5} for other of the governing relation". Velocity variation contour at Reynolds number 9000 and 24000 displayed on three planes for various angular deflections of -10°, -15°, -20° and -25° shown in figure 10a and figure 10b. "In case of a CRTs, behind the ring the phenomenon of secondary reverse direction flow is higher which gives more effective separation of flow and tiny boundary thickness in that region, which promotes the strong wall shear stress on a tube".

Accomplishment of heat transmit

Figure 11[a, b] present influence of CRTs having angular deflections of 0°, -10°, -15°, -20° and -25° and Re from 6000 to 24000 on heat exchange (*Nu* and Nu_t/Nu_p) resp. Figure 11(a) gives diversity of Nu along with Re for tested turbulators.

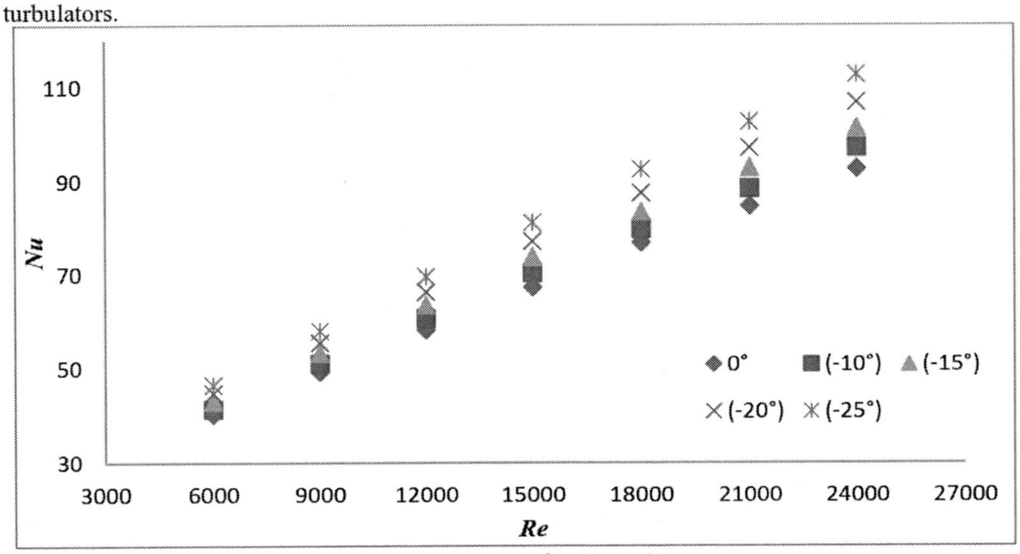

Figure 11a.Computing Re vs Nu

The main aim to predict; enhancement in heat transfer achieved, while testing with CRTs comparably higher 55% to 75% than the round plain tube. This is achieved because; it advances not only rupture of the thermal boundary layer but also appropriate and swift fluid mixing along with enhancement in heat exchange, produced higher turbulence and secondary fluid flow with eddy's as well as vortex structure across CRT shown in figure 9.

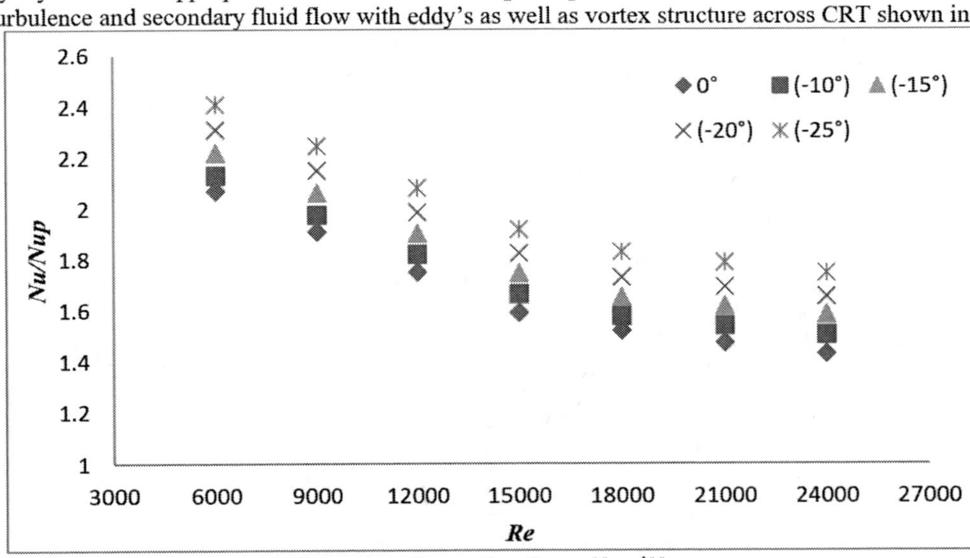

Figure 11b.Computing Re vs Nu_t/Nu_p

Figure 11(b) gives the ratio of the theoretical to plain tube Nu_t/Nu_p which reduces with raising Re in the range of 6000 to 24000 for all tested inserts. As results show, CRT having angular deflection of -10°, -15°, -20° and -25° improves rate of enhancement in heat transfer respectively around1.43-1.95, 1.58-2.06, 1.65–2.304 and 1.747–2.51 times than that of smooth tube exchanger.

It is observed that, obstruction across moving flow is higher in case of larger angular deflection i.e. -25⁰. "Because of additional blockage, extreme secondary flow is generated across the heated wall which leads to tiny thermal boundary thickness". Position at which CRT is in contact with a heated medium, which shows desolation of boundary thickness offers suitable circumstance to heat exchange. At, maximum turbulence magnitude with maximum Re shows convection transfer more favorably.

Characteristics of fluid friction

Effect of CRTs having angular deflection of 0°, -10°, -15°, -20° and -25° on fluid friction factor with Reynolds number vary from 6000 to 24000 shown in Figure 12 (a, b).

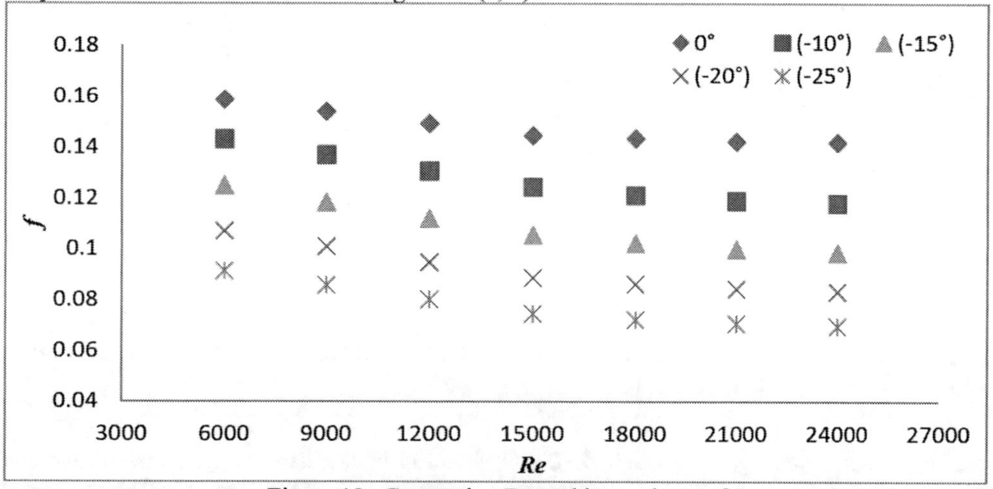

Figure 12a.Computing Reynolds number vs f

Figure 12[a] indicates, increment in Reynolds number causes slightly decrease in the friction factor for all tested inserts. Among the tested CRT with maximum angular deflection i.e. -25° indicates higher pressure drop for highest Reynolds number due to stronger secondary flow and higher turbulence intensity.

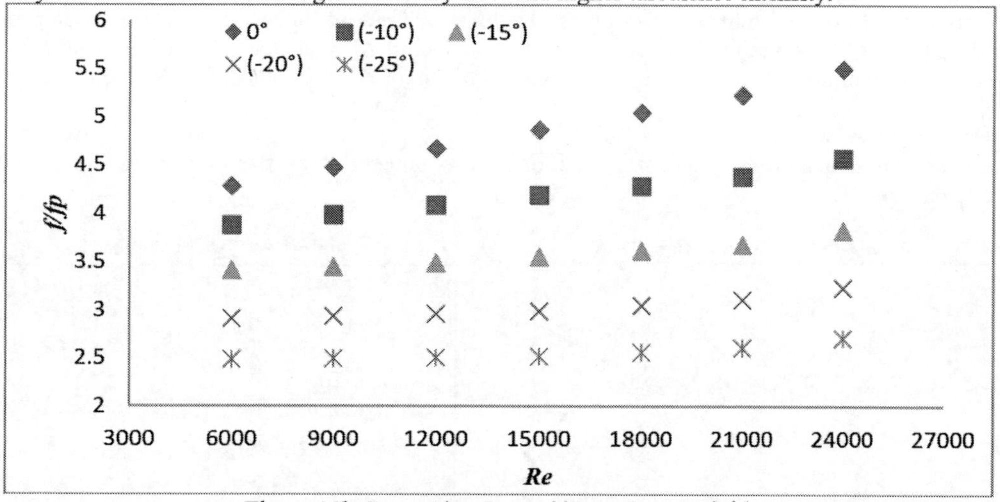

Figure 12b.Computing Reynolds number vs f_t/f_p

Figure 12(b) gives the ratio of the theoretical to plain tube friction factor (f_t/f_p) which increases with increases Re all tested inserts. This show at lower angular deflection value of friction factor is higher than other inserts. Also, the friction factor decreases as Reynolds number increase with higher CRT. The insert with angular geometry of -25° has the lowest friction factor ranging from 125%-210% which is greater than that of the smooth

IOP Conf. Series: Materials Science and Engineering **998** (2020) 012033 doi:10.1088/1757-899X/998/1/012033

round tube. In the range bound of 6000 to 24000 Reynolds number, CRT with angular deflection of 0° and -10° gives the maximum value of friction factor in the range of 165%-250% and 149%-230% respectively. The having angular deflection i.e. (-25°) and Re from 6000 to 24000 gives maximum pressure reduction 18.5 to 31.7 times additional than smooth tube.

Effect on overall achievement factor

To demonstrate overall achievement of CRTs having angular deviation of -10°, -15°, -20° and -25° for Reynolds number ranging 6000 to 24000, it's mandatory to measure overall performance factor (η).

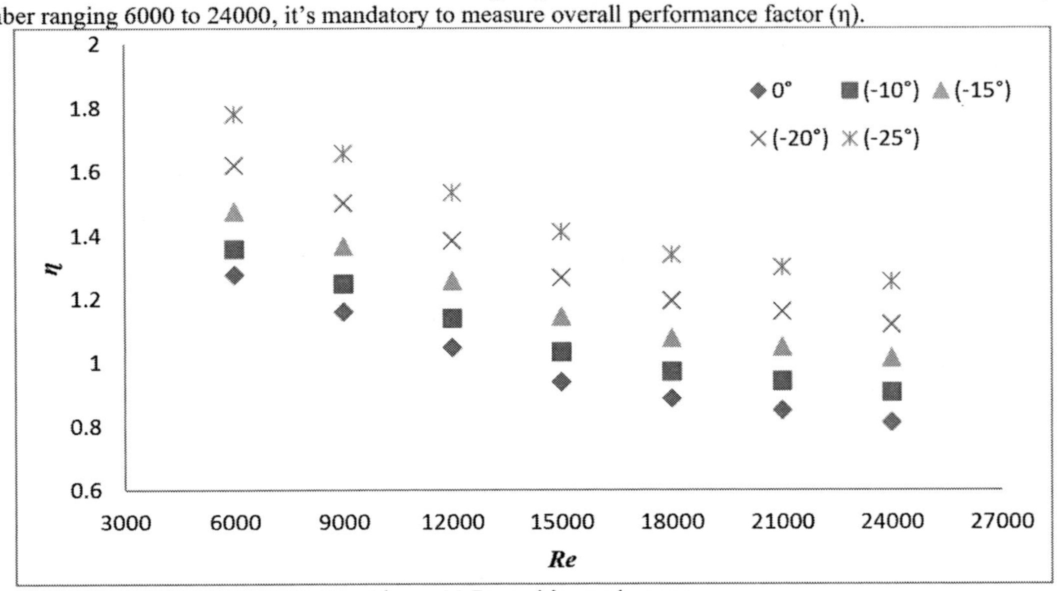

Figure 13 Reynolds number vs η

"Improvement in heat exchange as a result of flow separation is attendant by an increase in pressure drop. Overall performance factor includes heat transfer characteristics (Nu_t/Nu_p) and fluid friction performance (f_t/f_p) favorably". The significant of passive method adapted effectively if offers the η greater than 1. Influence of Re as well as CRT with inclination of -10°, -15°, -20° and -25° on η display in figure 13. It shows overall thermal characteristics for all CRTs under test decrease with increase in Reynolds number. This happens due to friction across fluid offers dominant effect on the heat transfer rate at higher Re. "CRT including maximum angular deviation of -25⁰ gives larger thermal performance factor. Also, decrease in Reynolds number tends to increases in η". Maximum η predicted is 1.77 for a CRT with angular deviation (-25⁰) at Reynolds number 6000.

Empirical correlations

Using present experimental data empirical correlations was obtained for Nu, f and η expressed by equation 8, 9 and 10 respectively.

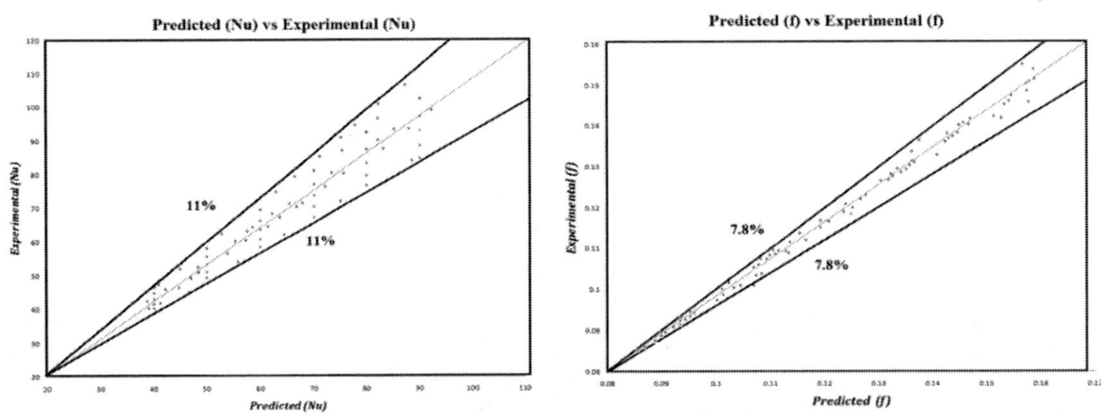

(a) (b)

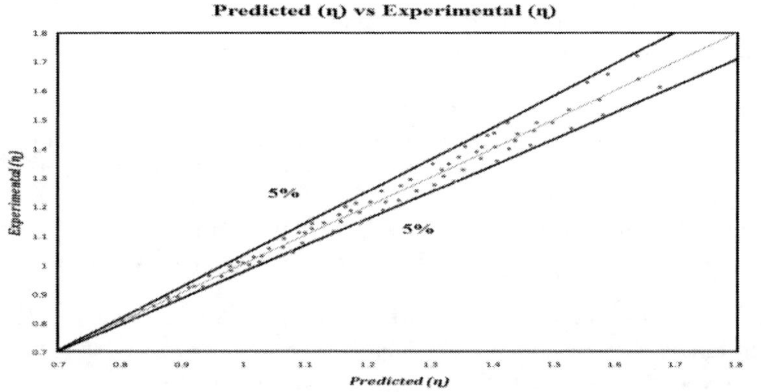

(c)

Figure 16.Comparison of experimental to predictive data (a) Nu (b) f (c) η

These are,

$$N_u = 0.1574Re^{0.662}Pr^{0.3} \quad (8)$$

$$f = 0.5632Re^{-0.209} \quad (9)$$

Figure 16 shows comparative investigation of predicted versus experimental sheet. "For Nusselt number, friction factor and Overall performance factor maximum fluctuation occurs in the predictive data is 11%, 7.8%, and 5% respectively".

$$\eta = (Nu_t/Nu_p)(f_t/f_p)^{-0.3602} \quad (10)$$

Installation and manufacturing of the CRT inserts used in the current work are very simple as well as cost-effective. Also comparison of wall shear stress for all tested inserts with smooth tube, indicates wall shear stress is high for higher Reynolds number and at maximum angular deviation (-25^0). At higher rate of change of shear wall stress we get higher heat enhancement at higher Reynolds

Illustration with literature

"Some of the numerically published data from the literature has been given in table 2 to compare the present investigation with equivalent methods". Installation and manufacturing of the CRT inserts used in the current work are very simple as well as cost-effective. Also comparison of wall shear stress for all tested inserts with smooth tube, indicates wall shear stress is high for higher Reynolds number and at maximum angular deviation (-25^0). At higher rate of change of shear wall stress we get higher heat enhancement at higher Reynolds.

Sr. No	Geometry	Nu_t/Nu_p	η
1	Conical Ring [38]	3.3	1.8
2	Circular Ring with twisted tape [39]	4.5	1.42
3	Conical ring with twisted tape [40]	3.67	1.95
4	Conical nozzle with snail entry [41]	4.1	1.10
5	Perforated conical ring [42]	3.8	54
6	Hexagonal ring turbulator [43]	3	1.34
7	**Present Work**	**2.5**	**1.72**

Table 2.Current work compare with other study

Conclusions

In current investigation work, a new modified geometry of turbulator i.e. Circular ring turbulator consisting negative angle deflection from the horizontal of tube was examined in rounded heat exchanger tube with working fluid air at ambient condition and influence on the rate of pressure drop, heat transfer and fluid friction with CRT including angular deviation of -10°, -15°, -20° and -25° were analyzed by changing the Re in the spam of 6000 to 24000. The observations defined on different parameters are summarized below.

(1) "CRT Insert improves the rate of heat transfer favorably as well as increases a pressure loss over a smooth tube within the tested range of all the parameters".

(2) "Heat transfer rate increases with increasing the angular deflection towards the inlet and increasing Reynolds number. Also, by increment in Reynolds number having higher angular deflection shows increment in pressure drop".
(3) "For all tested inserts with increase in angular deviation tends to slightly decreases friction factor with increasing Reynolds number. Friction factor (f_t/f_p) which increases with increases in Reynolds number in the range of 6000 to 24000 for all tested inserts. This demonstrates at lowest angular deflection geometry friction factor is higher than other tested inserts".
(4) "Overall performance factor increases with decrease in Reynolds number for all tested inserts. Also, for significant Reynolds number, overall performance factor decreases with increase in angular deflection across the flow".
(5) "Nusselt number (Nu_t/Nu_p) which decreases with increases in Reynolds number in the range of 6000 to 24000 for all tested inserts. Greater heat transfer improvement (Nu_t/Nu_p) 1.747–2.51 is obtained for insert angular deviation of (–25°). Also maximum thermal overall performance factor of 1.77 was achieved for a CRT with angular deviation (-25°) at Reynolds number 6000".

References

[1] Kumar A, Kumar M. and Chamoli S. 2016 Comparative study for thermal-hydraulic performance of circular tube with inserts. *Alexandria Eng. J.* vol. **55**, no. **1**, pp. 343–349. DOI: 10.1016/j.aej.2015.12.013.

[2] Agrawal P, Gautam A, Kunwar A, Kumar M and Chamoli S. 2018 Performance assessment of heat transfer and friction characteristics of a packed bed heat storage system embedded with internal grooved cylinders. *Sol. Energy*. vol. **161**, pp. 148–158. DOI: 10.1016/j.solener.2017.12.044.

[3] Liu S, and Sakr M. 2013 A Comprehensive Review on Passive Heat Transfer Enhancements in Pipe Exchangers. *Renewable and Sustainable Energy Reviews*, vol. **19**, pp. 64–81. doi:10.1016/j.rser.2012.11.021.

[4] Dewan A, Mahanta P, Raju S, and Kumar S. 2004 Review of Passive Heat Transfer Augmentation Techniques. *Proceedings of the Institution of Mechanical Engineers, Part A: Journal of Power and Energy*, vol. **218**, pp. 509–527. doi:10.1243/0957650042456953.

[5] Jomde A, Anderson A, Bhojwani V. and Deshmukh M. 2018a Performance Predictions and Parametric Analysis of a Valved Linear Compressor Using a Mathematical Model. *International Journal of Ambient Energy*, vol. **39**, pp. 456–461. doi:10.1080/01430750.2017.1318784

[6] Jomde A, Anderson A, Bhojwani V. and P. Khedkar. 2018b Modeling and Measurement of a Moving Coil Oil-Free Linear Compressor Performance for Refrigeration Application Using R134a. *International Journal of Refrigeration*, vol. **88**, pp. 182–194. doi:10.1016/j.ijrefrig.2018.01.002.

[7] Raj A, Suri S, Kumar A. and Maithani R. 2017 Experimental determination of enhancement of heat transfer in a multiple square perforated twisted tape inserts heat exchanger tube. *Exp. Heat Transfer*, vol **00**, pp. 1–21.

[8] Kongkaitpaiboon V, Nanan K. and Eiamsa-ard S. 2010 Experimental investigation of convective heat transfer and pressure loss in a round tube fitted with circular-ring turbulators. *Int. Commun. Heat Mass Transfer*, vol. **37**, no. **5**, pp.568–574. DOI:10.1016/j.icheatmasstransfer.2009.12.016.

[9] Kumar A, Chamoli S. and Kumar M. 2016 Experimental investigation on thermal performance and fluid flow characteristics in heat exchanger tube with solid hollow circular disk inserts. *Appl. Therm. Eng*, vol. **100**, pp. 227–236. DOI: 10.1016/j.applthermaleng.2016.01.081.

[10] Promvonge P. and Eiamsa-ard S. 2007 Heat transfer augmentation in a circular tube using V-nozzle turbulator inserts and snail entry. *Exp. Therm. Fluid. Sci*, vol. **32**, no. **1**, pp.332–340. DOI: 10.1016/j. expthermflusci.2007.04.010.

[11] Promvonge P. and Eiamsa-ard S. 2007 Heat transfer and turbulent flow friction in a circular tube fitted with conical-nozzle turbulators. *Int. Commun. Heat Mass Transfer*, vol. **34**, no. **1**, pp.72-82. DOI: 10.1016/j. icheatmasstransfer.2006.08.003.

[12] Promvonge P. and Eiamsa-ard S. 2010 Thermal characterization of turbulent tube flows over diamond-shaped elements in tandem. *Int. J. Therm. Sci.*, vol. **49**, no. **6**, pp.1051–1062. DOI: 10.1016/j.ijthermalsci.2009.12.003.

[13] Chamoli S, Yu P. and Kumar A. 2016 Multi-response optimization of geometric and flow parameters in a heat exchanger tube with perforated disk inserts by Taguchi grey relational analysis. *Appl. Therm. Eng*, vol. **103**, pp.1339–1350, DOI:10.1016/j.applthermaleng.2016.04.166.14.

[14] Chamoli S, Lu R, Xie J. and Yu P. 2018 Numerical study on flow structure and heat transfer in a circular tube integrated with novel anchor shaped inserts. *Appl. Therm. Eng.*, vol. **135**, pp. 304–324. DOI: 10.1016/j.applthermaleng.2018.02.052.

[15] Bhattacharyya S, Benim A. and Chattopadhyay H. 2018 Experimental investigation of heat transfer performance of corrugated tube with spring tape inserts. *Exp. Heat Transfer*, vol. **00**, pp. 1–15.

[16] Chamoli S. 2015 Preference selection index approach for optimization of V down perforated baffled roughened rectangular channel. *Energy*, vol. **93**, pp. 1418–1425. DOI: 10.1016/j.energy.2015.09.125.

[17] Chamoli S. 2015 Hybrid FAHP (fuzzy analytical hierarchy process) -FTOPSIS (fuzzy technique for order preference by similarity of an ideal solution) approach for performance evaluation of the V down perforated baffle roughened rectangular channel. *Energy*, vol. **84**, pp. 432–442. DOI: 10.1016/j.energy.2015.03.007.

[18] Saha S. and Saha K. 2013 Enhancement of heat transfer of laminar flow through a circular tube having integral helical rib roughness and fitted with wavy strip inserts. *Exp. Therm. Fluid. Sci.*, vol. **50**, pp. 107–113. DOI: 10.1016/j.expthermflusci.2013.05.010.

[19] Perwez A, Shende S. and Kumar R. 2018 Heat transfer and friction factor characteristic of spherical and inclined teardrop dimple channel subjected to forced convection. *Exp. Heat Transfer*, vol. **00**, pp. 1–20.

[20] Yakut K. and Sahin B. 2004 Flow-induced Vibration Analysis of Conical Rings Used for Heat Transfer Enhancement in Heat Exchangers. *Applied Energy*, vol. **78**, pp. 273–288. doi:10.1016/j.apenergy.2003.09.001.

[21] Tandiroglu A. 2006 Effect of Flow Geometry Parameters on Transient Heat Transfer for Turbulent Flow in a Circular Tube with Baffle Inserts. *International Journal of Heat and Mass Transfer,* vol. **49**, pp. 1559–1567. doi:10.1016/j.ijheatmasstransfer.2006.01.018.

[22] Eiamsa-ard S. and Promvonge P. 2010 Thermal Characterization of Turbulent Tube Flows Over Diamond-Shaped Elements in Tandem. *International Journal of Thermal Sciences,* vol. **49**, pp. 1051–1062. doi:10.1016/j.ijthermal sci.2009.12.003.

[23] Eiamsa-ard S, Rattanawong S. and Promvonge P. 2009 Turbulent Convection in Round Tube Equipped with Propeller Type Swirl Generators. *International Communications in Heat and Mass Transfer,* vol. **36**, pp. 357–364. doi:10.1016/j.icheatmasstransfer.2009.01.007.

[24] Akansu, S. 2006 Heat transfers and pressure drops for porous- ring turbulators in a circular pipe. *Applied Energy*, vol. **83**, pp. 280–298. doi:10.1016/j.apenergy.2005.02.003.

[25] Promvonge P, Koolnapadol N, Pimsarn M. and Thianpong C. 2014 Thermal Performance Enhancement in a Heat Exchanger Tube Fitted with Inclined Vortex Rings. *Applied Thermal Engineering*, vol. **62**, pp. 285–292. doi:10.1016/j.applthermaleng.2013.09.031.

[26] Promvonge P, Tamna S, Pimsarn M. and Thianpong C. 2015 Thermal Characterization in a Circular Tube Fitted with Inclined Horseshoe Baffles. *Applied Thermal Engineering, vol.* **75**, pp. 1147–1155. doi:10.1016/j.appltherma leng.2014.10.045

[27] Chingtuaythong W, Promvonge P, Thianpong C. and Pimsarn M. 2017 Heat Transfer Characterization in a Tubular Heat Exchanger with VShaped Rings. *Applied Thermal Engineering,* vol. **110**, pp. 1164–1171. doi:10.1016/ j.applthermaleng.2016.09.020.

[28] Singh V, Chamoli S, Kumar M. and Kumar A. 2016 Heat Transfer and Fluid Flow Characteristics of Heat Exchanger Tube with Multiple Twisted Tapes and Solid Rings Inserts. *Chemical Engineering and Processing: Process Intensification,* vol. **102**, pp. 156–168. doi:10.1016/j.cep.2016.01.013.

[29] Thianpong C, Yongsiri K, Nanan K. and Eiamsa-ard S. 2012 Thermal Performance Evaluation of Heat Exchangers Fitted with Twisted-Ring Turbulators. *International Communications in Heat and Mass Transfer*, vol. **39**, pp. 861–868. doi:10.1016/j.icheatmasstransfer.2012.04.004.

[30] Kongkaitpaiboon V, Nanan K. and Eiamsa-ard S. 2010 Experimental Investigation of Convective Heat Transfer and Pressure Loss in a Round Tube Fitted with Circular-Ring Turbulators. *International Communications in Heat and Mass Transfer,* vol. **37**, pp. 568–574. doi:10.1016/j. icheatmasstransfer. 2009.12.016.

[31] Promvonge P. 2008 Heat Transfer Behaviors in Round Tube with Conical Ring Inserts. *Energy Conversion and Management*, vol. **49**, pp. 8–15. doi:10.1016/j.enconman.2007.06.009.

[32] Eiamsa-ard S. and Promvonge P. 2006 Experimental Investigation of Heat Transfer and Friction Characteristics in a Circular Tube Fitted with V-Nozzle Turbulators. *International Communications in Heat and Mass Transfer*, vol. **33**, pp. 591–600. doi:10.1016/j.icheatmasstransfer.2006.02.022.

[33] Patil A. S. and Kore S. S. 2019 Experimental investigation on thermo-physical characteristics of fluid flow in circular tube enhanced with circular rings with different arrangements. *Int. J. Ambient Energy*, pp.1–10. DOI: 10.1080/01430750.2019.1608859.

[34] F. White 2010 Fluid Mechanics. *McGraw-Hill Education*, New York. 862.

[35] Patil A. S, Kore S. S. and Sane N. K. 2018 Experimental Investigation of the Effect of Flow Blockages on Heat Transfer and Fluid Friction in a Round Tube Using Wall-Attached Circular Rings. *Heat Tran. Res*, vol. **50,** no. **1**, pp.17–32. DOI: 10.1615/HeatTransRes.2018025424.

[36] Kline S. J. and McClintock F. A. 1953 Describing uncertainties in single-sample experiments. *Mech. Eng.*, vol. **75**, pp. 3–8.

[37] Meng J. 2003 Enhanced heat transfer technology of longitudinal vortices based on field-coordination principle and its application. *Power Engineering and Engineering Thermophysics*, Beijing, Tsinghua University.

[38] Promvonge P. 2008 Heat transfer behaviors in round tube with conical ring inserts. *Energy Convers. Manage*, vol. **49**, no. **1**, pp. 8–15. DOI: 10.1016/j.enconman.2007.06.009.

[39] Eiamsa-Ard S, Kongkaitpaiboon V. and Nanan K. 2013 Thermohydraulics of turbulent flow through heat exchanger tubes fitted with circular-rings and twisted tapes. *Appl. Therm. Eng.*, vol. **21**, no. **6**, pp. 585–593. DOI:10.1016/S1004-9541(13)60504-2.

[40] Promvonge, P. and Eiamsa-ard S. 2007 Heat Transfer Behaviors in a Tube with Combined Conical-Ring and Twisted-Tape Insert. *International Communications in Heat and Mass Transfer,* vol. **34**, pp. 849–859. doi:10.1016/j. icheat masstransfer.2007.03.019.

[41] Promvonge P. and Eiamsa-ard S. 2007 Heat transfer in a circular tube fitted with free-spacing snail entry and conical-nozzle turbulators. *Int. Commun. Heat Mass Transfer*, vol. **34**, no. **7**, pp. 838–848. DOI: 10.1016/j. icheatmasstransfer.2007.03.020.

[42] Sheikholeslami M, Gorji-Bandpy M. and Ganji D. 2015 Experimental study of the influence of perforated circular-ring on pressure loss and heat transfer enhancement using sensitivity analysis. *Appl. Therm. Eng*, vol. **91**, pp. 739–748. DOI: 10.1016/j.applthermaleng.2015.08.068.

[43] Patil A. S, Kore S. S. and Sane N. K. 2019 Thermal performance of tube exchanger enhanced with hexagonal ring turbulators. *Experimental Heat Transfer*, pp. 1–15. DOI: 10.1080/08916152.2019.1656302.

IOP Conf. Series: Materials Science and Engineering 998 (2020) 012034 doi:10.1088/1757-899X/998/1/012034

EXPERIMENTAL INVESTIGATION OF THERMAL PERFORMANCE OF MESH WICK HEAT PIPE USING NANOFLUID

AL RABBUL NIZAM[1], NAVEEN KUMAR GUPTA[1], PARVEZ ALAM[1]

[1]IET Department of Mechanical Engineering
GLA University Mathura 281406

Email: alrabbul.nizam@gmail.com

Abstract. In the present world with the advancement in technology and development of devices with high speeds and continuous runs, it has become the most serious challenge to effectively manage the thermal dissipation so that the devices are able to run smoothly and efficiently. With an effective thermal management system, the performance of these equipments or machineries can be improved to a great extent. This can be done by using heat pipes with nanofluids. The suspension of nanoparticles within the size range of 1-100 nm in the base fluids like water, ethylene glycol, etc. is called nanofluid. The aim of this research is to prepare Aluminium-water based nanofluids and to study its various properties. The aluminium nanofluid was synthesised using De-ionized water as the base fluid. The characterisation of the nanofluid was done by scanning electron microscopy (SEM) and transmission electron microscope (TEM) techniques.

1. Nomenclature

1.1 *Symbols*

k_{nf} = Thermal conductivity of nanofluid
k_b = Thermal conductivity of base fluid
k_{np} = Thermal conductivity of nanoparticles

2. Introduction

Heat pipe technology has become popular for heat dissipation due to its high efficiency. Heat pipe is a thermal transfer device which has capability to transfer heat and energy from one place to another. The rate of this thermal transfer can be 100 times quicker than those of available conventional methods, because of this characteristics they can be titled as thermal super conductors. In addition to the transferring of heat from one region to another, they can also be utilized to isothermalize the temperature distribution.

In a heat pipe, we see 3-sections viz. condenser section, evaporator section and adiabatic section. We allow the object to be cooled to meet with the evaporator section of heat pipe. In this section (evaporator section), after getting heat from the hot body, the working fluid is changed to vapour. The vapour that is formed in the evaporator now travels to cold region from the hot region, the reason for this is pressure difference. Now, at the next section, i.e. condenser section, vapour releases its latent heat and gets condensed to liquid. The working fluid is then returned to the evaporation section either by gravity or capillary action. As we are studying mesh-wick heat pipe, here the working fluid is returned with the help of capillary action.

The heat transfer of liquid can be improved by the inclusion of solid particles into it. After the addition of millimetre or micrometre particles to the liquids, the heat transfer performance did not enhance as

IOP Conf. Series: Materials Science and Engineering 998 (2020) 012034 doi:10.1088/1757-899X/998/1/012034

expected due to some constraints such as poor dispersibility, aggregation as well as adhering to inner walls of the setup. To overcome this hindrance, nanofluid was found favourable.

2.1 Applications of Heat Pipe

Aerospace – cooling satellite solar array, Electronics, Medical, Heat Exchangers

2.2 Advantages of Heat Pipe

Very high thermal transfer rate; No maintenance since it has got no moving parts; It can be utilized to isothermalize the space or component; With minimum heat loss, it can be adopted for transmitting heat over a long distance, hereby it allows the separation of heat source from the sink.

2.3 Objectives of The Research

The core purpose of this project is to study the thermal working of mesh wick heat pipe using aluminium oxide nanofluids with base fluid being de-ionized water.
The objectives of this study include:

- Characterisation of thermophysical properties of nanofluid
- Determining the augmentation in thermal working of heat pipe using nanofluid

3. Motivation For The Work

In the advancing world, the importance of phase change heat transfer is incredible to the technology. If particularly spoken, heat pipes have emerged as a crucial device that works on the concept of phase change heat transfer. It has a wide range of applications, which may include the energy conservation, space cooling or heating, electronic device/component cooling, etc. Earlier, these systems were equipped with the conventional cooling system, but the performance or capacity of those cooling systems was so low that the devices or systems could not function efficiently during high heat flux operations. As a result, the components failures were seen in a very short time. To get rid of this problem, the heat pipes are equipped to remove the heat efficiently whenever or wherever the high heat flux is developed

4. Literature Review

"Gaugler was the person who introduced the concept of heat pipe in 1942"[1]. He patented the Capillary-based heat pipes. The capillary design was anticipated as the channel for returning the working medium (liquid) from the condenser to the evaporator. According to him, the objective of this invention was "to cause absorption of heat". In 1963, George Grover individually fabricated capillary-based heat pipes and patented it to term it as "heat pipe". Several types of wick structures were proposed and many working fluids were used. Along with the conventional heat pipes increasing interests are also found now days in the area of flat rectangular /disk type /rotating/flexible/micro heat pipes
Nanofluid is a fluid which contains nanometre-sized (nm-sized) particles, called nanoparticles (typically of the order of 1-100 nm). Nanofluids with increased thermal conductivities are established convective heat transfer coefficients with respect to the base fluids. Previously, different-different base liquids, such as formamide, water etc. have been in use in many heat transfer applications. Many research works have been published by various researchers in this field. The published researches focus on nanofluids containing distinct nanoparticles with different size and volume concentration. Conventional fluids like metallic fluids, ethylene glycol, water etc. were extensively being used in the past as a working fluid [2]. Because of their suspension property, there were some limitations in enhancing the thermal properties.
Sukchana et al. [3]performed an experimental analysis of the effect of R- 134a and the adiabatic length of heat pipe on its heat transfer enhancement. Filling ratio viz. 10%, 15% and 20% and heat flux of the heat pipe from 1.97 to 9.87 kW/m^2 were varied. The results show that the adiabatic length as much as minimum produced the optimum performance for the filling ratio and heat flux of 15% and 5.92 kW/ m^2 respectively. Bing et al. fabricated and tested the different copper powder size for the micro heat pipe. The method of copper powder pouring and the heat transfer performance of heat pipe was initially analyzed. The copper powder size of 140 – 170 μm sintered at 900 - 950°C registered the best

IOP Conf. Series: Materials Science and Engineering 998 (2020) 012034 doi:10.1088/1757-899X/998/1/012034

performance. The authors also noted that the larger wick porosity and the heat transfer capabilities are the important parameters to improve thermal the performance of heat pipe.
Kim et al.[4] performed an experimental work on the effect of the shape of Al_2O_3 nanoparticles dispersed in acetone. The nanoparticles shapes are varied viz. sphere, brick and cylinder and the prepared nanoparticles are suspended in acetone using two-step method. The sphere shaped nanoparticles are enhances the thermal resistance about 33% compared with that pure acetone, whereas the brick and cylinder shaped produce only about 29% and 16% respectively.

5. Characterization

5.1 Method(s) of Nanofluid Preparation

Nanofluids are the mixture of solid particles (basically nano-sized) and the base fluid, but it is not like an ordinary mixture. Many parameters should be considered while the nanofluid preparations i.e., reduced agglomeration and increased stability, chemical aggregation of the fluid and positive/negative energy between the nanoparticles, etc. The nanofluid preparation is the most important step and it is majorly categorized into two groups i.e. Single step or 1-step method and 2-step method.

5.1.1 Single step method

In this method nanoparticles are produced by the evaporation or condensation process. The main drawback of this method is the nanoparticles and the working fluid could not be separated. Akoh et al.[5] developed the single step direct evaporation approach. In this method, nanoparticle size is reduced to its desired level in the base fluid. This method also had the limitation in the separation of dry nanoparticles from the nanofluid.

5.1.2 Two-step method

The 2-step method is commonly used for the preparation of nanofluids. The nanoparticles are separately distributed in the base fluid and then the solution is taken for the ultrasonication. Using the two-step method, Murshed et al.[6] experimentally produced the TiO2 nanoparticles suspended in water. The authors suggested that the two-step method is suitable for oxide nanoparticles compared with metallic nanoparticles produces less stability. Li et al. [7] utilized the two-step method to prepare the alumina oxide nanofluids. The prepared nanofluid was kept in the ultrasonic homogenizer to avoid the agglomeration of nanoparticles. Preparation of nanofluid by 2-step method is shown in Figure 1.

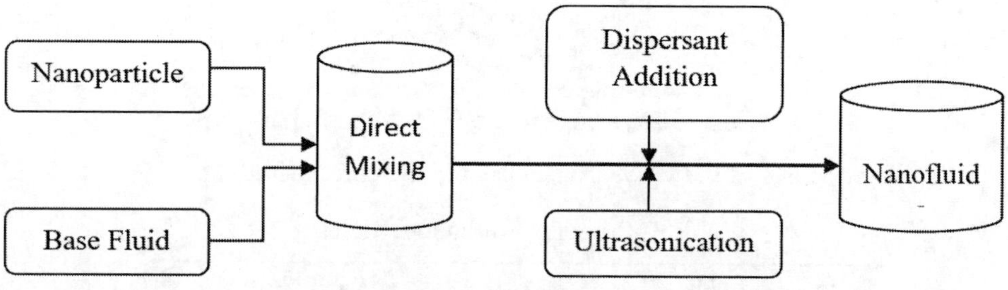

Figure 1. Nanofluid preparation by 2-step method

5.2 Equipments used for characterization of nanofluids

The equipments which were used for characterization of nanofluids are SEM, and TEM.

5.2.1 Scanning electron microscope (SEM)

- SEM uses focused beams of electrons to extract high resolution, 3-d images
- For the given sample, it provides information about its
  Topography morphology composition
- Modern SEM can provide a resolution of 1-20 nm

IOP Conf. Series: Materials Science and Engineering 998 (2020) 012034 doi:10.1088/1757-899X/998/1/012034

- SEM image of Al2O3 nanoparticles is shown in Figure 2

Figure 2. SEM image of Al2O3 nanoparticles

5.2.2 Transmission electron microscope (TEM)

- TEM in present days can be used for studying the defects, composition and layer growth in semiconductors. It can also be used to visualize the grain boundaries, etc. Hence, we can say that the TEM is a very strong and powerful tool for today's material science lab.
- The TEM works on the basic principles as light microscope works, the difference here is that it uses electrons replacing the light. As we have studied that the electrons wavelength is much smaller than that of light, hence the TEM images resolution obtained is many orders of magnitude better than the image obtained from a light microscope. Thus, we can say that the images obtained from the TEM can show the finest details of internal structure. Figure 3 shows the TEM image of Al_2O_3 nanoparticles

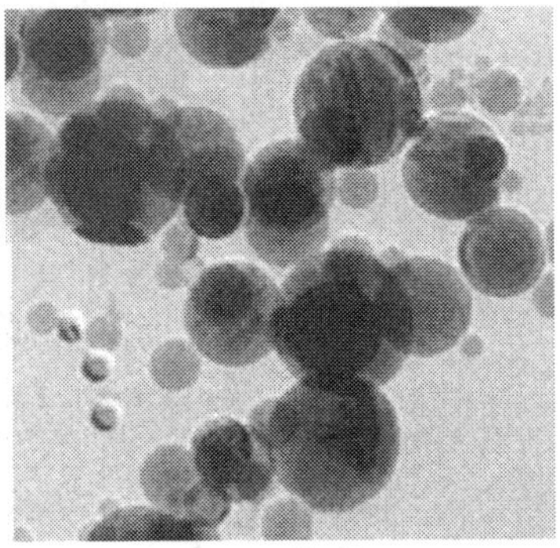

Figure 3. TEM image of Al_2O_3 nanoparticles

Table 1 describes the properties of the nanoparticles used in the current study.

Table 1. Properties of nanoparticles

Item	Description
Chemical Symbol	Al_2O_3
Density	3920 kg/m^3
Shape of particles	Spherical
Thermal conductivity	28-37 W/m K
Particle size	38-45 nm
Specific Heat Co-efficient	850 J/Kg K

5.3 Thermal Conductivity of Nanofluids

IOP Conf. Series: Materials Science and Engineering 998 (2020) 012034 doi:10.1088/1757-899X/998/1/012034

"For a nanofluid, the calculation of effective thermal conductivity can be done using the following equation [8]".

$$k_{nf} = \left[\frac{k_{np} + 2k_b + (k_{np} - k_b)\,\varphi}{k_{np} + 2k_b - (k_{np} - k_b)\,\varphi} \right] \qquad \text{(Eq. 1)}$$

where k_{np} and k_b are thermal conductivities of nanoparticles and the base fluid respectively and φ is the volume fraction of the nanofluid

Table 2. Thermal conductivity of some important solids and liquids

Component	Material	Thermal Conductivity (k) W/m-K
Metallic Solids	Aluminium	237
	Copper	401
	Silver	429
Non-metallic Solids	Alumina	40
	Silicon	148
	Carbon nanotubes	3000
	Diamond	3300
Metallic Liquids	Sodium @644K	0.613
Non-metallic Liquids	Engine Oil	
	Ethylene glycol	0.145
	Water	0.253

The thermal conductivity for the prepared nanofluids was measured at the room temperature for the chosen concentration range of volume and the obtained results have been plotted in the graph shown in Figure 4(a). These values are also plotted with some literature values in the Figure 4(b). In the present work, while analysing the thermal conductivity, there was no unusual improvement in the obtained results but some slight deviations were seen among the literature values. The reasons for these variations maybe – shape and size of nanoparticles, stability method, preparation method adopted for nanofluid preparation and the uncertainties associated with the instruments and experimentation

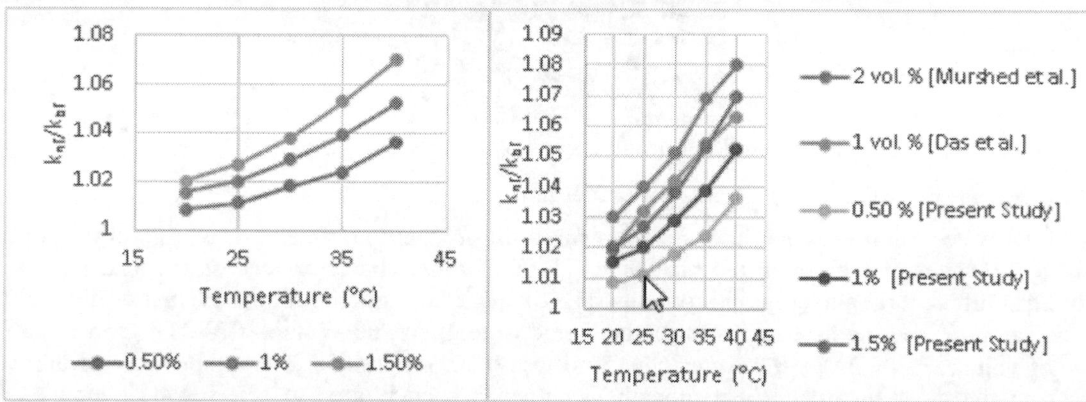

Figure 4. Thermal conductivity ratio of Al2O3 / DI water nanofluids (a) as a function of volume concentration and temperature (b) and its comparison with literature values

6. Experimentation

6.1 Design and specifications of heat pipe

Heat pipe material		: Copper
Wick material		: Stainless steel (SS304)
Wick type		: Screen wire mesh
Evaporator length	Le	: 150 mm

IOP Conf. Series: Materials Science and Engineering 998 (2020) 012034 doi:10.1088/1757-899X/998/1/012034

Adiabatic length	La	: 100 mm
Condenser length	Lc	: 200 mm
Total length of pipe	Lt	: 450 mm
Outer diameter of the pipe	do	: 24 mm
Inner diameter of the pipe	di	: 22.6 mm
Thickness of the pipe	t	: 1.4 mm
Wick permeability	k	: 3.672×10^{-10} m^2
Mesh size	n	: 50 in^{-1}
Wire diameter	d	: 0.041
No of wires/unit	N	: 2365 m^{-1}
No of layers		: 3
Wick porosity		: 0.64
Permeability	K	: $1.8\text{x}10^{-10}$
Aperture size per linear inch		: 17.18
Aspect Ratio (AR)		: 1

6.2 Experimental Setup

The setup used for this study is shown in Figure 5. It shows a photographic of the heat pipe setup.

Figure 5. Photographic view of experimental setup

7. Performance Evaluation And Enhancement

Al_2O_3 / DI water nanofluids that have been prepared till now, are now being used as working fluids, with no surfactants, in the modified heat pipe. The functional characteristics of the heat pipe were analysed at different tilt angles and heat inputs. The tilt angles considered for the evaluation are 0°, 30°, 45°, 60° and 90° and the heat inputs to the heat pipe were 40 W, 60 W and 80 W. The concentration range of volume % of Al2O3/DI water nanofluid was 0.50% to 1.50%. The temperature of the wall surface in the direction of the heat pipe the length has been plotted at selected heat inputs. The performance of heat pipe has been examined and the obtained values are discussed in detail.

Figure 6 (a –c) shows the variation of wall temperature the heat pipe for 0° tilt angle using DI water and Al_2O_3 / DI water nanofluids with the different volume concentration viz., 0.5%, 1% and 1.5% and different heat inputs i.e. 40 W, 60 W and 80 W. After observing the plots carefully, it was observed that the maximum wall temperature was seen for DI water and the minimum wall temperature was seen with 1.5% volume concentration. After increasing the volume concentration from 1% to 1.5%, it was observed that the temperature drop was less than that of 1%.

IOP Conf. Series: Materials Science and Engineering 998 (2020) 012034 doi:10.1088/1757-899X/998/1/012034

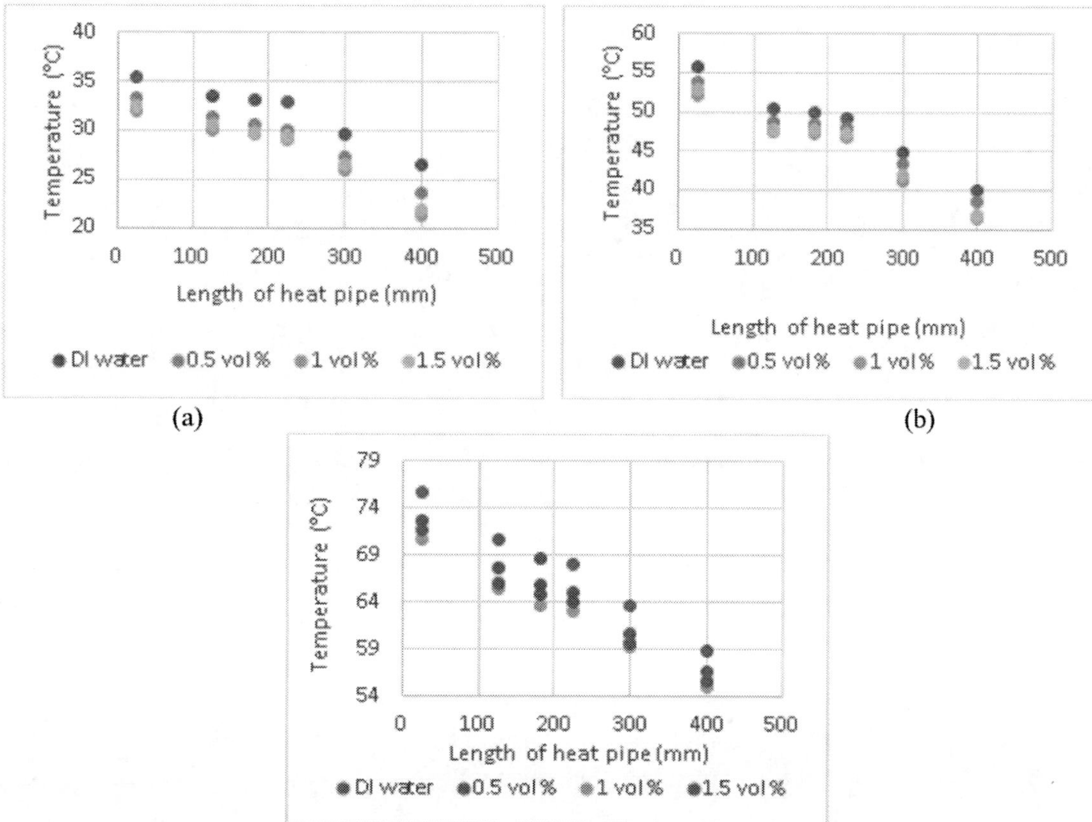

Figure 6. Wall temperature distributions along the length of the heat pipe for 0° tilt angle, a) 40 W, b) 60 W, and c) 80 W

It was observed that the maximum temperature was seen with DI water and the minimum temperature was seen with 1.5% volume concentration Al_2O_3/DI water nanofluid. Now, these fluids have been taken into consideration for analysis at different tilt angles i.e. 0°, 30°, 45°, 60° and 90°. The wall temperature of the heat pipe with DI water as the working fluid at different tilt angles has been shown in Figure 7. For all the heat inputs, the optimum tilt angle was found to be 30° as the lowest wall temperature was seen with 30° tilt angle while the maximum temperature was seen with 0° tilt angle.

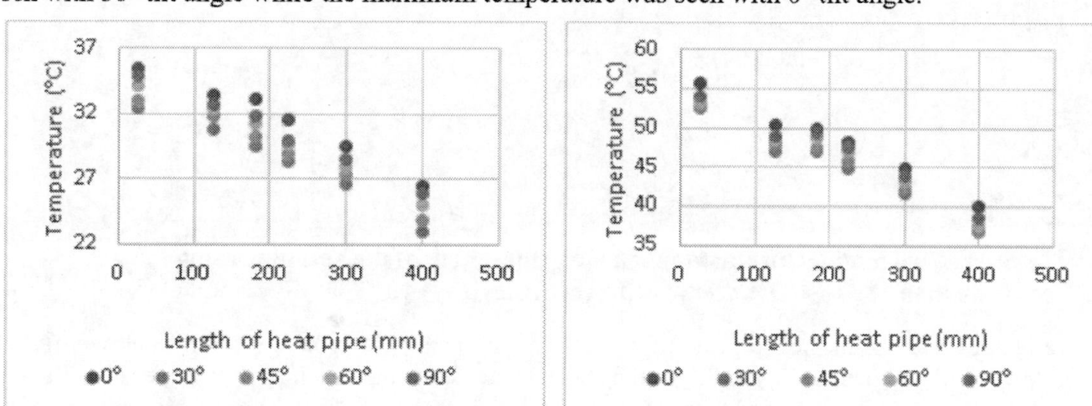

(a) (b)

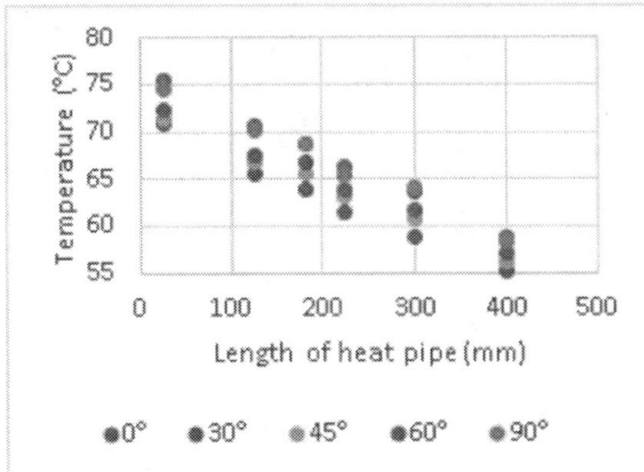

(c)

Figure 7. Wall temperature distribution along the length of the heat pipe using DI water at different tilt angle, a) 40 W, b) 60 W, and c) 80 W

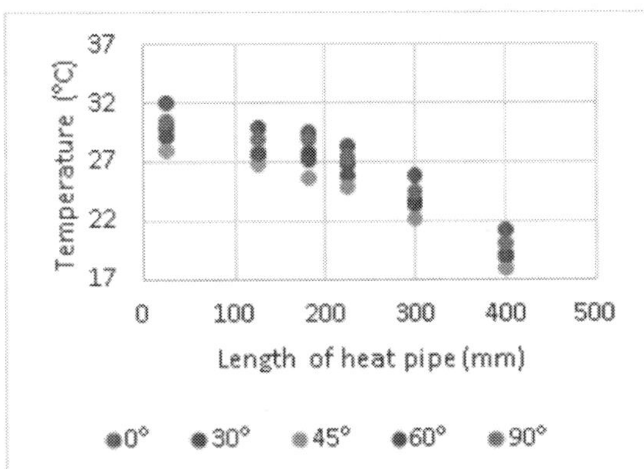

(a)

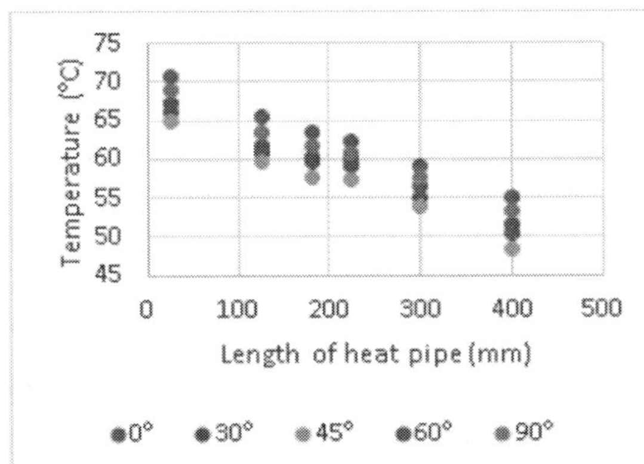

(b)

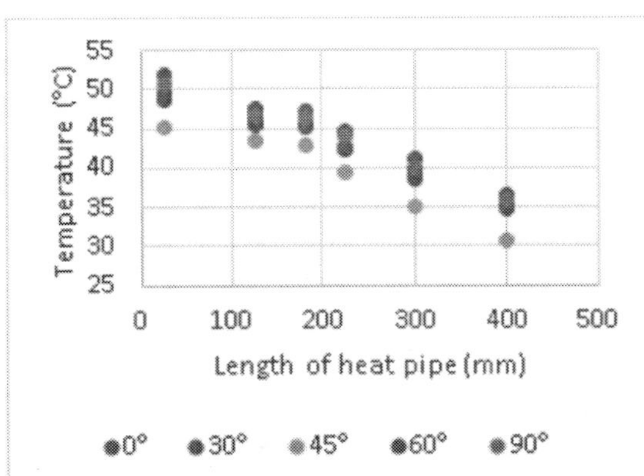

Figure 8. Wall temperature distribution along the length of the heat pipe using 1% vol. concentration of Al_2O_3/DI water at different tilt angle, a) 40 W, b) 60 W, and c) 80 W

The wall temperature of the heat pipe at different tilt angles using 1 vol. % Al_2O_3/DI water nanofluid as the working fluid is shown in Figure 8. For the chosen nanofluid and heat inputs, the optimum tilt angle was found to be 45° as the lowest wall temperature was seen at 30° tilt angle while the maximum temperature was seen at 0° tilt angle.

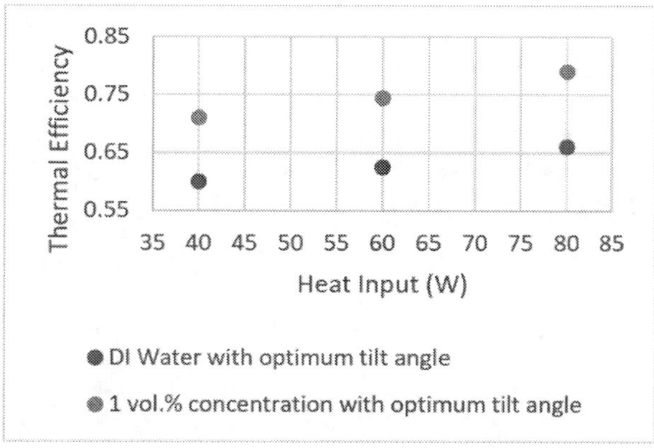

Figure 9. Thermal Efficiency of Heat pipe

Figure 9 shows the thermal efficiency of the heat pipe for different heat inputs. It was observed that there was a great improvement in the efficiency of the heat pipe when the working fluid was chosen as Al_2O_3 nanofluid. For the DI water alone, the efficiency was found in the range of 0.6 to 0.69, while for the nanofluid with 1 vol. % concentration this value was greatly improved to the range of 0.71 to 0.79.

8. Conclusions

- The thermal performance of the heat pipe was improved when the conventional working fluid was replaced with Al_2O_3 nanofluids.
- For DI water, the optimum tilt angle for the heat pipe was found to be 30°, while for 1 volume % Al_2O_3 it was found to be 45°.
- In accordance with the conclusion #2, it was observed that the optimum tilt angle for the heat pipe is not constant for different working fluids. It showed that the optimum tilt angle is also dependent on the working fluid.
- A good increase in efficiency of the modified heat pipe was observed when Al_2O_3/DI water nanofluid was used in place of DI water alone.

Future Scope

- The heat pipe performance can be investigated using different wick structure and the results can be used for the optimization of the system.
- Thermal performance of heat pipe can be studied by using the different volume concentrations of nanoparticles and by increasing or decreasing the nanoparticles size.
- For studying the thermal performance of heat pipe, different shapes of nanoparticles can also be used.
- The performance of heat pipes can be studied using different and the latest nanofluids available with better characteristics.

References

[1] P. D. Dunn and D. A. Reay, *The heat pipe*, 1973rd ed., vol. 4, no. 3. 1973.

[2] R. M. Mostafizur, A. R. Abdul Aziz, R. Saidur, M. H. U. Bhuiyan, and I. M. Mahbubul, "Effect of temperature and volume fraction on rheology of methanol based nanofluids," *Int. J. Heat Mass Transf.*, vol. 77, pp. 765–769, 2014, doi: 10.1016/j.ijheatmasstransfer.2014.05.055.

[3] T. Sukchana and C. Jaiboonma, "Effect of filling ratios and adiabatic length on thermal efficiency of long heat pipe filled with R-134a," *Energy Procedia*, vol. 34, pp. 298–306, 2013, doi: 10.1016/j.egypro.2013.06.758.

[4] H. J. Kim, S. H. Lee, S. Bin Kim, and S. P. Jang, "The effect of nanoparticle shape on the thermal resistance of a flat-plate heat pipe using acetone-based Al2O3 nanofluids," *Int. J. Heat Mass Transf.*, vol. 92, pp. 572–577, 2016, doi: 10.1016/j.ijheatmasstransfer.2015.09.013.

[5] S. Y. and A. T. Hiroshi Akoh, Yukihiro Tsukasaki, "Magnetic Properties of Ferromagnetic Ultrafine Particles Prepared By Vacuum," *J. Cryst. Growth*, vol. 45, pp. 495–500, 1978.

[6] S. M. S. Murshed, K. C. Leong, and C. Yang, "Enhanced thermal conductivity of TiO2 - Water based

nanofluids," *Int. J. Therm. Sci.*, vol. 44, no. 4, pp. 367–373, 2005, doi: 10.1016/j.ijthermalsci.2004.12.005.

[7] S. U. S. Choi, S. Li, and J. A. Eastman, "Measuring thermal conductivity of fluids containing oxide nanoparticles," *J. Heat Transfer*, vol. 121, no. 2, pp. 280–289, 1999, doi: 10.1115/1.2825978.

[8] Y. J. Lin and K. S. Hwang, "Effects of particle size and particle size distribution on heat dissipation of heat pipes with sintered porous wicks," *Metall. Mater. Trans. A Phys. Metall. Mater. Sci.*, vol. 40, no. 9, pp. 2071–2078, 2009, doi: 10.1007/s11661-009-9888-z.

IMPROVEMENT OF WELD JOINT STRENGTH BY APPLYING RANDOM VIBRATIONS ALONG WITH EXTERNAL MAGNETIC FIELD

SURESH CHITTURI[1], MOHANA KRISHNA CHOWDARY K[2], MUNMUN BHAUMIK[3], DHANUNJAY KUMAR AMMISETTI[4]

[1,3]*Mechanical Engineering Department, Vignan's Institute of Engineering for Women, Visakhapatnam, India*
[2]*Mechanical Engineering Department, Rajiv Gandhi University of Knowledge Technologies, Idupulapaya, India*
[4]*Mechanical Engineering Department, Lakireddy Bali Reddy college of Engineering, Mylavaram, India*
Email: chitturisuresh007@gmail.com, munmunbhaumik@gmail.com

Abstract. Welding is a metal joining process induces high residual stresses. These are strongly influencing the mechanical properties of weldment. In earlier days heat treatment and shot peening techniques were used to relieve these stresses. Due to the time consuming of these processes, in this research work we have applied random vibrations to relive the residual stresses to improve mechanical properties. Along with that, welding arc may get affected by magnetic field during welding. This leads to arc instability which is responsible for welding defects like lack of fusion, porosity. These reduce the quality and strength of weld. To overcome this, arc spattering with external magnetic field need to be reduced. In this research work, a setup has been designed for vibration assisted welding along with external magnetic field set up to improve the mechanical properties of Mild Steel weld joints by means of hardness and ultimate tensile strength. Welding had been performed with and without these setups. After performing welding work pieces have been tested both welding conditions and results have been compared.

Keywords: Hardness; Magnetic field; Mild Steel; Ultimate tensile strength; Vibration assisted welding

1. Introduction

Welding is a permanent joining process widely used in automobile, naval & construction applications. In all types of welding techniques shielded metal arc welding is most commonly used because of its advantages like cheap and easy. In shied metal arc welding (SMAW) flux coated consumable electrodes were used, during welding the flux melts and protect the weld pool from the atmosphere. In SMAW magnetic field is induced around the workpieces, electrodes & arc. The magnetic disturbance surrounding the arc may cause arc instability which is responsible for welding defects. This obviously reduces the quality and strength of weldment. The conventional shield metal arc welding induces residual stresses which influence the mechanical properties of the weldment. The following literature survey indicates the effect of vibrations and external magnetic field on mechanical properties of the weldments.

Many researchers have analyzed the effect of vibrations on microstructure and mechanical properties of welded joints and obtained a general conclusion that vibrations are able to change the microstructure thus improving the mechanical properties of welds. Singh et al. [1] investigated the influence of vibrations on mechanical properties of butt-welded joints using dynamic solidification technology. They observed improved yield strength and ductility under vibratory conditions of the welding. Pucko et al. [2] experimentally investigated the effect of vibration on weld metal hardness and toughness. They concluded that the improvement in hardness of the weld bead depends on type of welding process, welding parameters and filler materials used and reported maximum 80% improvement in fracture toughness during multi pass welding with vibration effect. Rao et al. [3] studied the influence of mechanical vibrations on the hardness of Al5052 weldments. They conducted experiments at various input voltages (70V, 160V,230V) of vibratory table. They observed maximum 12.36% improvement in hardness at 160V compared to without vibration. Aoki et al. [4] experimentally measured residual stresses by X-ray diffractometer and observed reduced tensile residual stresses under random vibrations compared to welding performed without vibration. Kalpana et al. [5] reviewed various techniques for improving the mechanical properties of the fusion welded joints and suggested vibratory assisted welding for improved properties compared to all other techniques. Balasubramanyam et al. [6] experimentally investigated the effect of vibrations on welding. Their results indicate 18.35% improvement in strength & 12.82% improvement in hardness compared to welding under without vibration. Jose et al. [7] studied vibration assisted welding process and their effect on quality of welds. They concluded that vibratory assisted welding helps in grain refinement and improvement in mechanical properties which in turn produce quality welds. Singh et al. [8] experimentally examined and observed 78% and 49% improvement in yield strength and elongation during welding under vibratory conditions. Sharma et al. [9] experimentally investigated the effect of magnetic field on weld quality and weld geometry. They observed that externally applied magnetic field deflects the arc and increases the bead width which in turn enriches the tensile strength. Pathak et al. [10] investigated the effect of welding current and torch angle on tensile strength of shield metal arc welded low carbon steel plates. They performed experiments with varying welding currents (95A-125A) and torch angles (65° to 85°)· They reported the improvement in tensile strength from 540Mpa to 588Mpa when the torch angle varies from 65° to 85° at 120A. Senapati et al.[11] studied the effects of external magnetic field on mechanical properties and observed 9.72%, 25.98% & 43.05% improvement in tensile strength, hardness and toughness respectively with vertically applied magnetic field. Singh et al. [12] used artificial neural networks to predict the weld bead geometry as function of depth of penetration and observed decreased depth of penetration with increased magnetic field. Queiroz et al.[13] studied the effects of external magnetic field on microstructure and observed decreased primary and secondary dendrite spacings.

From the above literature review it is evident that vibration assisted welding and external magnetic field improved the mechanical properties. Till now all the researchers studied the vibration assisted welding and external magnetic field individually. In the current research work an attempt has been taken to understand the combined effect of vibration assisted welding along with external magnetic field on mechanical properties (hardness and ultimate tensile strength) of the weldment.

2. Experimental Details

The welding performed by traditional SMAW process faces many welding defects like porosity, arc blow, undercut, lower penetration, and generation of residual stresses. In order to overcome these defects and relieve residual stresses, vibration assisted welding along with that external magnetic field set up is used. The vibrations are generated by using eccentricity mechanism and the external magnetic field is applied with the help of solenoid. The designed set up for vibrations and external magnetic field shown in the Figure 1 & Figure 2. In the current research work mild steel is used as a workpiece material. Welding was performed on the workpieces with varying the amplitude of vibrations by altering the speed of the motor from 721 rpm to 923 rpm & by altering the external magnetic field by varying currents from 0.2 amp to 0.3 amp. In this study initially the experiments were performed

IOP Conf. Series: Materials Science and Engineering **998** (2020) 012035 doi:10.1088/1757-899X/998/1/012035

without vibrations and external magnetic field. In second case the experiments were performed on specimens with vibrations (at 721 rpm, 823 rpm, 922 rpm) and without external magnetic field. In third case the experiments were performed on specimens with both vibrations (at 721 rpm, 823 rpm, 922 rpm) and external magnetic field (at 0.2 amp ,0.3 amp). After performing welding under three cases specimens are subjected to Brinell hardness test and the hardness number is calculated by using the equation 1. After measuring the Brinell hardness number the ultimate tensile strength is calculated by using the correlation shown in equation 2.

$$\text{Brinell hardness number (BHN)} = \frac{2p}{\pi D(\sqrt{D-(D^2-d^2)})} \quad (1)$$

P = Applied load; D=Diameter of indenter; d = Average diameter of indentation

$$\text{Ultimate tensile strength} = \text{Hardness} * k \quad (2)$$

Where k = 3.38 to 3.55

Figure 1. Vibration assisted welding set up

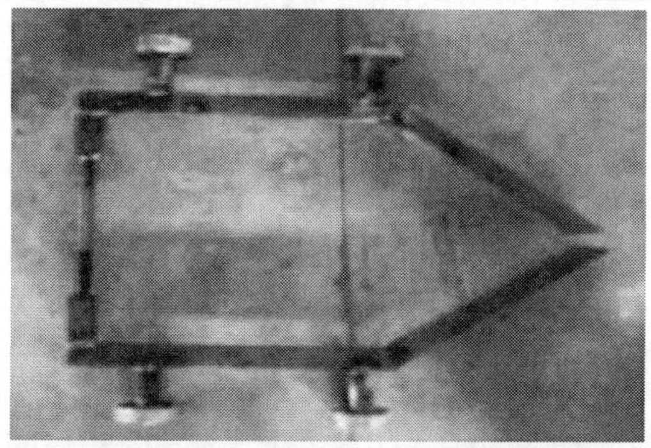

Figure 2. External Magnetic field set up

3. Results and discussion

3.1 Brinell hardness test

The Brinell hardness test is performed on the SMAW specimens under three cases such as without vibrations and external magnetic field, with vibrations, with both vibrations and external magnetic field. The hardness of the specimens is measured on the weld bead using the equation 1. Hardness is measured by taking the average of readings at three different places on the weld bead. The pictorial representation of the welded specimens under three cases shown in the Figure 3- Figure 5. After measuring the Brinell hardness number, ultimate tensile strength is calculated by using the following correlation shown in equation 2. Table 1 & Table 2 indicates the hardness and tensile strength data under three cases.

Table 1. Harness and tensile strength results under case1

Conditions on which welding performed	place on which hardness measured	Avg indentation Diameter (mm)	BHN	UTS (MPa)
Without vibration & magnetic field	Weld bead	4.6	170	574.6

Table 2. Hardness and tensile strength data under case 2 and case 3

Conditions on which welding performed	RPM	place on which hardness measured	Avg indentation Diameter (mm)	BHN	UTS (MPa)
With vibrations & without external magnetic field	721 rpm	Weld bead	4.2	207	699.66
	823 rpm	Weld bead	3.8	255	861.9
	922 rpm	Weld bead	3.4	321	1084.9
With both vibrations & external magnetic field at 0.2 amp	721 rpm	Weld bead	3.57	290	980.2
	823 rpm	Weld bead	3.5	302	1020.7
	922 rpm	Weld bead	3.27	348	1176.2
With both vibrations & external magnetic field at 0.3 amp	721 rpm	Weld bead	3.4	321	1084.9
	823 rpm	Weld bead	3.4	321	1084.9
	922 rpm	Weld bead	3.27	348	1176.2

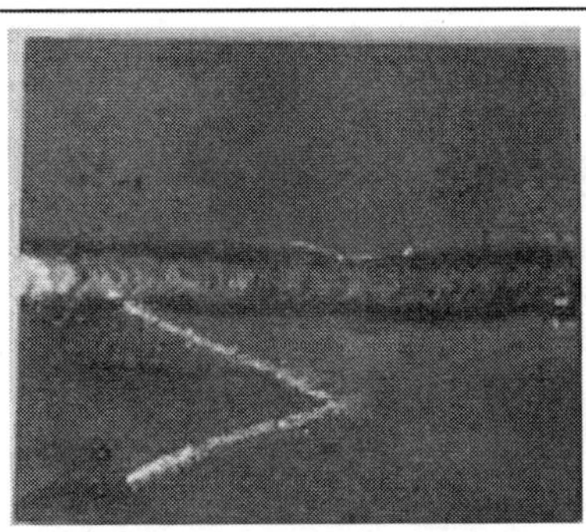

Figure 3. Without vibration and External magnetic field

Figure 4. With vibrations

Figure 5. With both vibration and External magnetic field

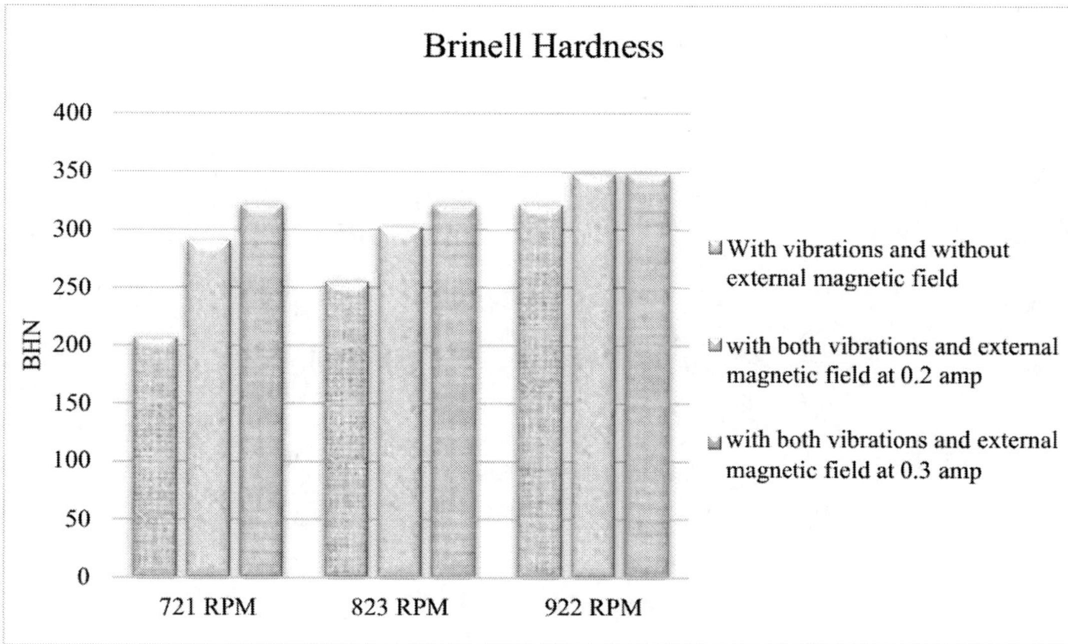

Figure 6. Comparison of Brinell Harness Number

Figure 6 indicates that the variation of hardness values with vibrations condition and with both vibrations and external magnetic field conditions at 0.2 amp and 0.3 amp current. Based on the results it is observed that the enrichment in hardness values with the amplitude of vibrations increase from 721rpm to 922 rpm. The results indicate that compared to conventional SMAW process vibration assisted welding and the combined vibration and magnetic field assisted welding shows improved results. The conventional SMAW induces residual stresses in the weldments, which are relieved by vibrations applied during welding and improve the mechanical properties. The applied random vibrations stir the molten weld metal pool and restricts the growth of the grains, due to that the grains do not achieve their original dendritic length. Such fragmented dendrites form new nucleation sites and produce new grains. In this way the number of grains increases continuously and reduce the size of the grains, which in turn avoid the porosity defects and voids between the grains and improve the hardness of the weldments [8]. The results also indicate the combined vibration and magnetic field assisted welding shows improved results compared vibration assisted welding. The externally applied magnetic field deflects the arc, increases the fusion zone and eliminates the arc blow defects which in turn enhance the weld quality as well as mechanical properties [11]. Vibration assisted welding shows improvement in hardness by 21.76%, 50% and maximum 88.82 % at 721 rpm, 823 rpm and 922 rpm respectively compared to conventional SMAW. Similarly, maximum 89.54% and 95.13% improvement in hardness observed for combined vibration and magnetic field assisted welding with 923 rpm at 0.2 amp and 0.3 amp respectively compared to conventional SMAW. The maximum 8.4 % improvement in hardness is reported with combined vibration and external magnetic field assisted welding compared to vibration assisted welding.

3.2 Ultimate Tensile Strength

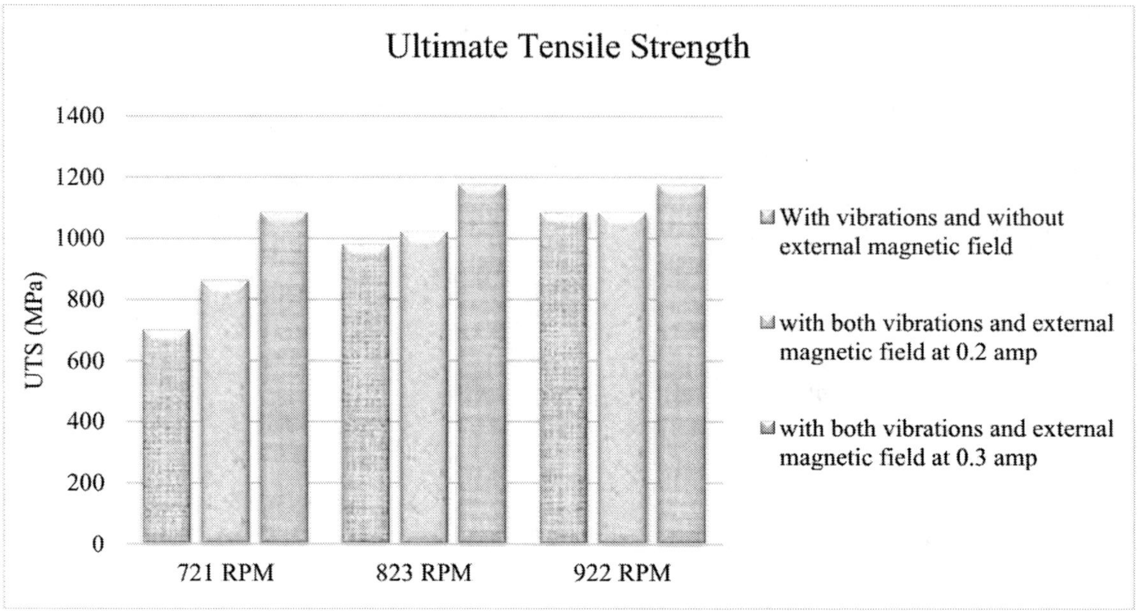

Fig 7. Comparison of Ultimate Tensile Strength

Figure 7 indicates the variation of ultimate tensile strength with vibration assisted welding and with combined vibration and magnetic field assisted welding processes. In this study the ultimate tensile strength is calculated by using the correlation which is function of Brinell hardness number shown in equation 2. It is observed that ultimate tensile strength increases when the amplitude of vibrations increase from 721 rpm to 922 rpm. The vibration assisted welding and combined vibration and magnetic field assisted welding show improved results compared to conventional SMAW process. Maximum 89.02% improvement in ultimate tensile strength was observed with vibration assisted welding compared to traditional SMAW process. Similarly, maximum 89.95% and 96.64 % improvement in ultimate tensile strength was observed with combined vibration and magnetic field assisted welding with 922 rpm at 0.2 amp and 0.3 amp respectively. The maximum 15.23 % improvement in ultimate tensile strength is reported with combined vibration and external magnetic field assisted welding compared to vibration assisted welding.

4. Conclusions

The present investigation aimed at improving the mechanical properties like hardness and strength of the welded joints by applying random vibrations along with external magnetic field. The analysis was carried out by fabricating the experimental setups, preparing the specimens and performing the hardness test on the specimens. The bar charts are drawn between the hardness number and rpm and also between the ultimate tensile strength and rpm. Finally, the results are compared for the three cases and observed the improvement in mechanical properties.

1. Hardness of the weld joint on weld bead zone is enriched with the increase of random vibrations. The maximum 88.82%, 95.13% improvement in hardness was observed with vibration assisted welding and combined vibration and magnetic field assisted welding. The maximum 8.4 % improvement in hardness was observed with combined vibration and magnetic field assisted welding compared to vibration assisted welding.

2. Ultimate tensile strength of the weld joint on weld bead zone is improved with the increase in random vibrations. The maximum 89.02%, 96.64% improvement in UTS was observed with vibration assisted welding and combined vibration and magnetic field assisted welding. The maximum 15.23 % improvement in UTS was observed with combined vibration and magnetic field assisted welding compared to vibration assisted welding.

References

[1] Sing, J, Kumar G and Garg N 2012 Influence of vibrations in arc welding over mechanical properties and microstructure of butt-welded-joints *Int. J. Sci. & Technol.* **2(1)** 1-6.

[2] Pucko B and Gliha V 2005 Effect of vibration on weld metal hardness and toughness *Sci. Technol. Weld. Join.* ***10*(3)** 335-338.

[3] Rao M V, Rao P S and Babu B S 2016 Investigate the influence of mechanical vibrations on the hardness of Al5052 weldments *Ind. J. Sci. Technol.* ***9*** 39.

[4] Aoki S, Nishimura T and Hiroi T 2005 Reduction method for residual stress of welded joint using random vibration *Nucl. Eng. Des.* ***235*(14)** 1441-1445.

[5] Kalpana J, Srinivasa Rao P and Govinda Rao P 2017 A review on techniques for improving the mechanical properties of fusion welded joints *Eng. Solid Mech.* **5(4)** 213–24.

[6] Balasubramanyam P N V, Uzwalkiran R, Kumar M N V R L and Ramgopal M 2016 Strength improvement of welded joint by using random vibrations *ARPN J. Eng. Appl. Sci.* **11**(18) 11068–73.

[7] Jose M J, Kumar S S and Sharma A 2016 Vibration assisted welding processes and their influence on quality of welds *Sci. Technol. Weld. Join.* **21(4)** 243–58.

[8] Singh P, Patel D and Prasad S B 2017 Investigation on the effect of vibrations on cooling behavior and mechanical properties of smaw butt welded joints *UPB Sci. Bull. Ser. D Mech. Eng.* **79(2)** 137–46.

[9] Sharma S S, Kumar M and Ojha P K Effect of Magnetic Field on Weld Quality and Weld Geometry 53–8.

[10] Pathak D, Singh R P, Gaur S and Balu V 2020 Experimental investigation of effects of welding current and electrode angle on tensile strength of shielded metal arc welded low carbon steel plates *Mater. Today Proc.* **26** 929–31.

[11] Senapati A and Mohanty S brata 2014 Effects of External Magnetic Field on Mechanical properties of a welded M.S metal through Metal Shield Arc Welding *Int. J. Eng. Trends Technol.* **10(6)** 297–303.

[12] Singh R P, Gupta R C and Sarkar S C 2012 The effect of process parameters on penetration in shielded metal arc welding under magnetic field using artificial neural networks *Int. J. Appl. Innovation Eng. Manage.* **1(4)** 12-17.

[13] Queiroz A V D, Femandes M, Silva L, Demarque R, Xavier C R and Castro J A D 2020 Effects of an external magnetic field on the microstructural and mechanical properties of the fusion zone in TIG welding. *Metals* **10(6)** 714.

IOP Conf. Series: Materials Science and Engineering 998 (2020) 012036 doi:10.1088/1757-899X/998/1/012036

Experimental Investigation and Micro Structural Characteristics of GTAW and FSW Welded Dissimilar Aluminum Alloys

Sairam Kotari[1, 2,a *],Eshwaraiah Punna[2,b] S M Gangadhar Reddy[1,c] S.Venukumar[1,d]

[2]Gitam Deemed to be University, Hyderabad, India

[1] Vardhaman college Of Engineering, Hyderabad, India

E-mail:[a]sairam.kotari87@gmail.com,[b]epunna@gitam.edu,[c]gangadhar.8011@gmail.com,[d]venukumar24@vardhaman.org

Abstract.
Current paper shares the study on optimization of parameters for GTAW and FSW Aluminum alloys. Base metals chosen were AA6061 and AA7075, filler metal used for TIG welding is AA4043. Welding Current, Welding speed and gas flow rate were taken as input parameters for GTAW and axial load, weld speed and tool rotation were choosen as variables for FSW. L9 orthogonal array is used for design of experiments. Analysis of variance is used for ranking the parameters. It was found that an average tensile strength of 127MPa is found for GTAW samples and 151 MPa for FSW samples. Thus an overall 19% increase in strength is achieved. Micro structural analysis is done to investigate the reason for variation of strength. It was found that the formation of Mg-Si precipitates has affected the increase in strength for FSW samples.

1 Introduction

Many methods which are available for joining aluminum are gas metal arc welding ,gas tungsten arc welding, joining by using adhesives, brazing ,soldering, riveting, friction stir welding, laser welding , ultrasonic welding [1-6]. When welding aluminum it typically creates a soft region in the base metal and fusion zone [7].Aluminum can also be brazed or soldered to many materials which include concrete, wood, ceramics. Manual aluminum brazing is difficult due to its variable colour change before melting. Aluminum oxide will prevent proper bonding between base metals. Many strong acids and fluxes can be used to weaker the oxidation. The alloys which are used for brazing must have low melting temperature that is below 660^0 C. Aluminum alloys with high magnesium can lower the melting temperature and can result for pure joint[8].
Because of higher thermal conductivity of Al alloys larger heat input is required to weld Al alloys. Because of difference in thermal conductivity Arc welding of different Al alloys (dissimilar) will give problems. High thermal conductivity will lead will make the heat to flow easily and low conductivity will lead to excess melting of material. So fusion welding of dissimilar Aluminum is interminable. Many researchers has studied the joining of dissimilar Al alloys by friction stir welding since the temperature achieved will be less than the melting point. Satisfactory results were obtained when harder material is placed on advancing position.(9-13)
AA7075 is a precipitation-hardened aluminum amalgam commonly used in aerospace applications owing to its high strength. In the traditional tungsten inert gas and laser welding processes, decorative structure establishes in the fusion zone which result in the radical decrease in the strength which is one of the important mechanical properties [14].The friction stir welding process is a solid-state welding process, so the solidification microstructure is not present in the welded metals and the presence of brittle inter dendritic and eutectic phases is avoided [15]. Generally, joints between dissimilar materials of 6061-T6 and 7075-T6 in aerospace structures are mostly prepared by riveting. This process causes stress concentration and due to which the final joints weight is increased, thereby limiting the application in the aerospace industry. Friction stir welding is used to increase the quality of the welding without the major loss of strength and corrosion. The process was used first in 1991 by TWI for joining of the aluminum amalgams.[16].
Geometrical parameters like the height of the pin and also the shape and the shoulder end details have its effect on both metal flow and the heat generation matured to frictional forces developed. With respect to this the super imposed on the rotating tool should be controlled throughout the process because the pressure developed on the tools shoulder end shows the amount of heat developed during the process. Dickerson and przydatek discovered that friction stir welded butt joints are defect free if all the precautions are followed accordingly and tuned with in a tolerance box for a particular alloy[17].Many analysts have come to know during their research that the relation between adjustable parameters, fatigue

IOP Conf. Series: Materials Science and Engineering 998 (2020) 012036 doi:10.1088/1757-899X/998/1/012036

characteristics, and mechanical and metallurgic properties of the weld in welding are equivalent to the aluminum amalgam[18-27].

2 Materials and Methods

There are different situations where different series of Aluminum are to be joined due to various service conditions. So proper joining process should be employed for better performance. Base metals chosen for present project were AA6061 and AA7075. Material composition and general properties of base elements were shown in Table 1 and Table 2.

Table 1: Chemical Composition Of Base and Filler metals

S. No	Chemical Composition	% Composition of base metal AA6061	% Composition of base metal AA7075	% Composition of filler metal AA4043
1	Silicon	0.7	0.4	4.5 – 6.0
2	Iron	0.6	0.5	0.80
3	Copper	0.3	1.5	0.30
4	Manganese	0.05	0.3	0.05
5	Magnesium	0.9	2.5	0.05
6	Zinc	0.2	5.6	0.10
7	Titanium	0.1	0.2	0.20
8	Beryllium	-	-	0.0003
9	Chromium	0.25	0.25	-
10	Aluminium	96.85	87.1-91.4	Rest
11	Other	0.05	0.15	-

Table 2: General properties of base metals.

S.No	Physical, Mechanical, Thermal Properties	AA 6061	AA7075
1	Density	2.7 g/cc	2.81 g/cc
2	Hardness, Vickers	107	175
3	Ultimate Tensile Strength	310 MPa	572 Mpa
4	Tensile Yield Strength	276 Mpa	503 Mpa
5	Melting Point	582 - 652 °C	477 - 635 °C
6	Specific Heat Capacity	0.896 J/g-°C	0.96 J/g-°C
7	Thermal Conductivity	167 W/m-K	130 W/m-K

Rolled AA6061 and AA7075 plates of 2mm thick were taken and they are sheared to a size of 75×75 mm each. A V Notch is grinded at one of the ends of the specimens as shown in Fig 1. Before welding Surfaces were marinated with NaOH and HNO3 and they are wire brushed and degreased. Both plates were clamped to a copper backing plate.Rather than performing full factorial experiment design which will increase the number of experiments and cost, in this paper Taguchi method is used which provide simple, efficient and systematic approach to optimize designs for performance quality and cost. L_9 orthogonal array is used to perform the experiments as shown in table 3. Both plates were TIG welded with AA4043 of 2mm diameter as filler rod. Welding Current, Welding Speed and Gas flow rate was taken as input parameters for TIG welding. After Tig welding another 9 set of specimens were taken for FSW. Tool rotation, Welding speed and axial load were taken as parameters. Tool material used was AISI H13 tool steel with taper pin. After welding the coupons were machined as per ASME standards by using electronic Elpuls 40A CNC wire cut EDM machine ,Tensile tests were conducted on Computerized Universal testing machine TUE-C-600. Micro hardness for the highest strength specimens were found by using Vickers hardness testing machine. As shown in Fig 2. Microscopic images were captured by etching the specimens by Kellers reagent. Further FESEM analysis was done in ARCI, Hyderabad on Gemini 500 (M/s Carl Zeiss) equipment.

IOP Conf. Series: Materials Science and Engineering 998 (2020) 012036 doi:10.1088/1757-899X/998/1/012036

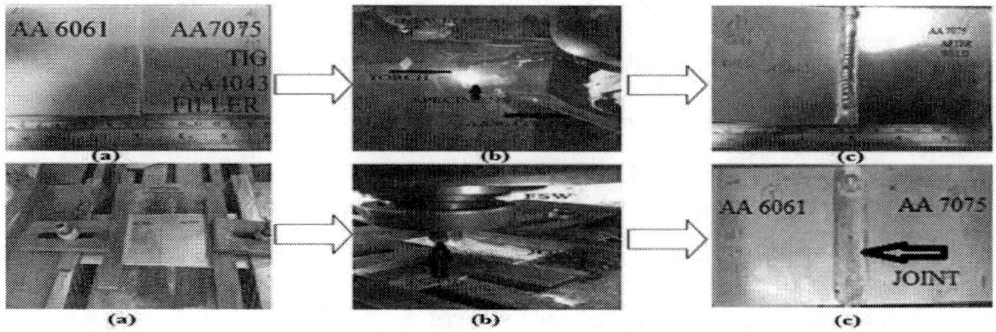

Fig 1: (a)Edge preparation (b)TIG,FSW processes (c)Welded samples

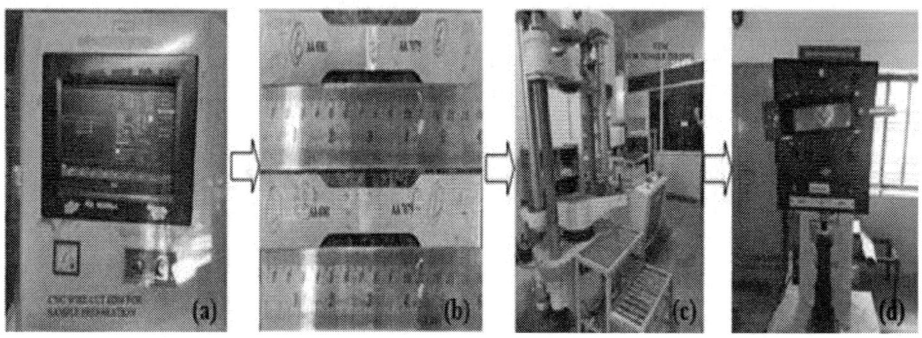

Fig 2: (a) CNC EDM Machine (b) Prepared specimens (c) UTM (d) Vickers Hardness testing

3 Results and Discussion

3.1 Tensile Test

The specimens for Tensile test were prepared as per ASTM E8 standard I section, on CNC Wire cut EDM machine. It is observed that the fracture is occurred at the joint. The tensile strengths for both TIG and FSW welded specimens were plotted in Fig 3.During the tensile test, all the examples constantly fizzled in the weld area. This shows the weld locale is nearly more vulnerable than different districts and thus the joint properties are constrained by weld district synthetic arrangement and microstructure. The transverse ductile properties of the welded joints recommend that the FSW joints show better tractable properties relative to the GTAW joints. It was found that in TIG welding at a current of 140 A, weld speed of 70mm/min and gas flow rate of 10 Lit/Min a maximum strength of 148MPa is achieved. The tensile strength of FSW welded sample at 1200 rpm tool rotation, 60mm/min welding speed and 5KN welding speed a maximum tensile strength of 180MPa is found to be maximum. Thus an 21.6 % increase of strength is observed.

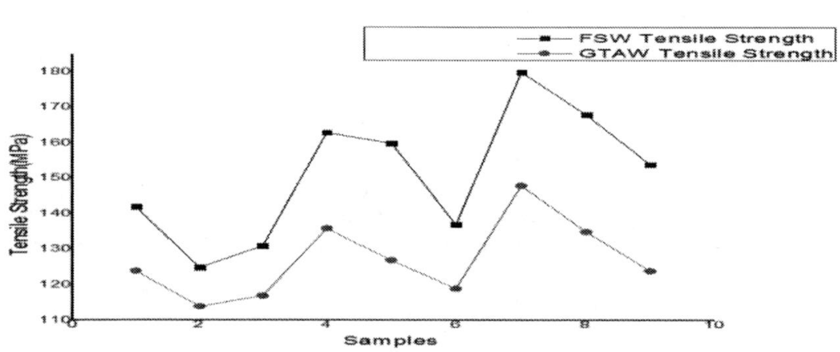

Fig 3: Tensile Strength of GTAW and FSW samples

3.2 Hardness Test

Tensile strength and Hardness are connected to each other since they depend upon the formation of metal. The more hardness the material possess the more brittle it becomes and eventually tensile strength gets decreased so in order to evaluate the behavior of metal both tensile and hardness of the joint is need to be analyzed. Hardness for the sample with high tensile strength is chose and values are determined. The graphs for hardness are shown in Fig 4 and Fig 5. In all coupons it was found that the hardness value seen decreasing from weld center.

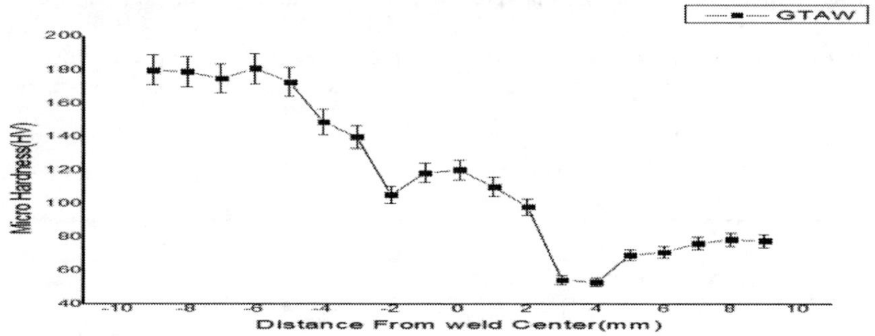

Fig 4: Micro Hardness for GTAW samples

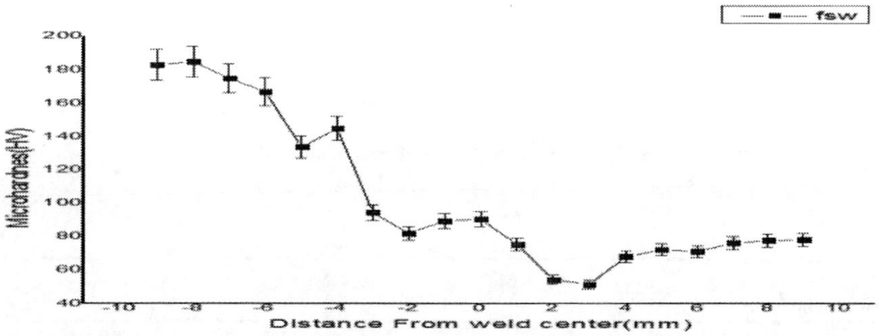

Fig 5: Micro Hardness for FSW welded samples

3.3 Analysis Of Variance (ANOVA)

Analysis of variance is used to determine the importance of one or more factors by comparing the response means at the various levels of factors. It works by comparing the variance between the group means and the variance within the groups. ANOVA table lists the sources of variation, their degrees of freedom, the total sum of squares and mean squares, F-statistics and P-Values. All these are used to determine whether the factors are significantly related to the response or not. Sum of squares between the factors and sum of squares within the factors (error) is used to calculate the total sum of squares.

Total sum of squares = Sum of squares between the factors + sum of squares within the factors

$$= \sum_{i=1}^{n} (y_i - \bar{y})^2 \quad \text{Eq-1}$$

Where y_i = value of tensile strength of a particular level.
$\bar{y}$ = mean of the tensile strength.

Mean of squares is calculated by dividing the sum of squares by the degrees of freedom. F-statistics is found by dividing the mean of squares by the error mean of squares. P-value is used to determine whether the factor is significant or not. If the P-Value is less than 0.05, then the particular factor is considered as significant.

$$\text{Percentage of contribution} = \frac{\text{sum of squares of individual factor}}{\text{Total sum of squares}} \quad \text{Eq-2}$$

IOP Conf. Series: Materials Science and Engineering 998 (2020) 012036 doi:10.1088/1757-899X/998/1/012036

Process parameters choosen were shown in Table 3. The Main effects of plots of S/N ratio for TIG and Friction stir welding are shown below. These graphs can be used to know the optimum input parameters for maximum tensile strength. In TIG welding at a welding current of 140Amps, welding speed of 70 mm/min, Gas flow rate 10 L/min a maximum tensile strength of 148Mpa is obtained. In Friction stir welding at a Tool rotation speed of 1200RPM, Welding speed of 60 mm/min and Axial force 5KN, a maximum tensile strength of 180Mpa is obtained. From Fig 8 it can be observed that in GTAW current ,speed and flow rate has taken first ,second and third preferences. Whereas in FSW Tool rotation, welding speed and load has taken first , second and third place respectively. The graphs are shown in Fig 7 and Fig 8. F values and P values are shown in Table 4 . The regression equations found from anova are shown in Eq 4 and Eq 5.

Table 3: Process Parameters for FSW and GTAW

FSW				GTAW			
Factor	Type	Levels	Values	Factor	Type	Levels	Values
Tool Rotation(Rpm)	Fixed	3	800, 1000, 1200	Welding Current(A)	Fixed	3	100, 120, 140
Welding Speed(Mm/Min)	Fixed	3	60, 70, 80	Welding Speed (Mm/Min)	Fixed	3	70, 80, 90
Axial Load(Kn)	Fixed	3	3, 4, 5	Gas Flow Rate (Lit/Min)	Fixed	3	8, 9, 10

Table 4: Analysis variation

FSW						GTAW					
Source	DF	Adj SS	Adj MS	F-Value	P-Value	Source	DF	Adj SS	Adj MS	F-Value	P-Value
Tool Rotation (Rpm)	2	1824.89	912.44	25.99	0.037	Welding Current (A)	2	450.89	225.44	15.98	0.059
Welding Speed (Mm/Min)	2	661.56	330.78	9.42	0.096	Welding Speed (Mm/Min)	2	398.22	199.11	14.11	0.066
Axial Load (Kn)	2	160.22	80.11	2.28	0.305	Gas Flow Rate (Lit/Min)	2	59.56	29.78	2.11	0.322
Error	2	70.22	35.11			Error	2	28.22	14.11		
Total	8	2716.89				Total	8	936.89			

IOP Conf. Series: Materials Science and Engineering 998 (2020) 012036 doi:10.1088/1757-899X/998/1/012036

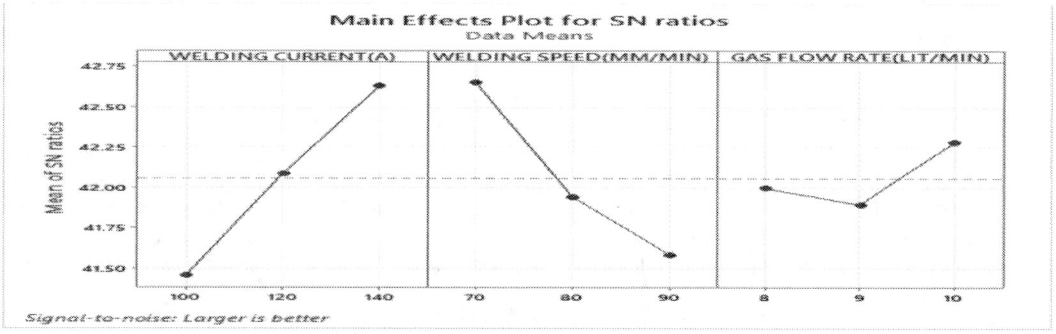

Fig 6: Main Effects Plot For SN ratios GTAW joints

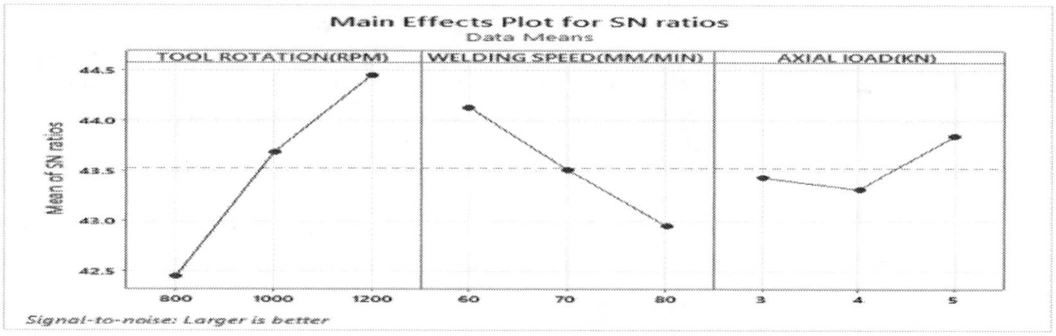

Fig 7 : Main Effects Plot For SN ratios FSW joints

The Regression equation For FSW is

Tensile Strength(Mpa) = 151.11 -18.44 Tool Rotation(Rpm)_800+ 2.22 Tool Rotation(Rpm)_1000 + 16.22 Tool Rotation(Rpm)_1200 + 10.56 Welding Speed(Mm/Min)_60 - 0.11 Welding Speed(Mm/Min)_70 - 10.44 Welding Speed(Mm/Min)_80 - 2.11 Axial Load(Kn)_3 - 3.78 Axial Load(Kn)_4 + 5.89 Axial Load(Kn)_5 Eq 4

The regression equation for GTAW is

Tensile Strength(Mpa) = 127.11 - 8.78 Welding Current(A)_100 + 0.22 Welding Current(A)_120 + 8.56 Welding Current(A)_140 + 8.89 Welding Speed(Mm/Min)_70 - 1.78 Welding Speed(Mm/Min)_80 - 7.11 Welding Speed(Mm/Min)_90 - 1.11 Gas Flow Rate(Lit/Min)_8 - 2.44 Gas Flow Rate(Lit/Min)_9 + 3.56 Gas Flow Rate(Lit/Min)_10 Eq 5

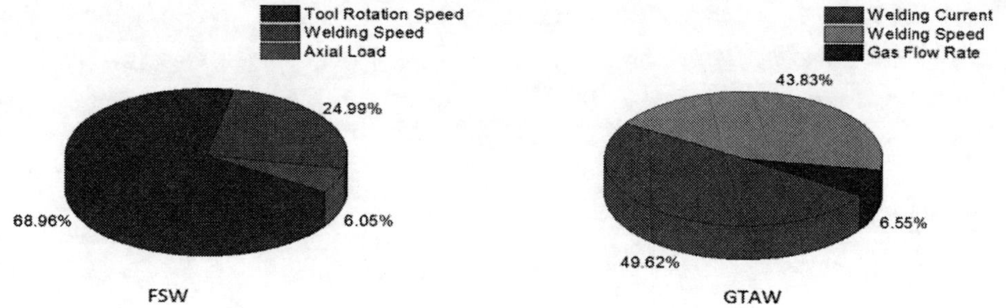

Fig 8: Percentage Contribution of Parameters

IOP Conf. Series: Materials Science and Engineering 998 (2020) 012036 doi:10.1088/1757-899X/998/1/012036

3.4 Microscopic Study

Usually base metals (BM) exhibit highest strength this is because of involvement of alloying partners like Mg and Si. These two mix together to form a much stronger precipitate Mg_2Si. Steady distribution of these precipitates all round the Al matrix gives high strength and hardness. In GTAW base metal gets diluted in weld zone , thus a huge amount of Si is readily available for precipitation ,but Mg because of this fusion gets evaporated so the formation of Mg_2Si is very low. Whereas in case of FSW since there is no inclusion of filler via fusion the base metal does not gets diluted or melted as a result the formation of Mg_2Si is high and high strength and hardness is achieved. The microstructures of base and welded specimens were shown in Fig 9.

One more comparison observed between GTAW and FSW samples were size of grains, the longitudinal grains observed in GTAW samples and fine grains were observed in FSW samples. we know that fine grains achieve high properties. In FSW the base metals are brought to plastic deformation because of higher rotation of pin, so the long grains are axed to short and coarse grains are achieved. Because of this the mechanical properties are found improved in FSW joint. When the microstructures are compared in FSW samples the original structure of the grains were completely refined and a new pattern of crystals were formed this process is known as Recrystallization. Even though there is no clear view of grain nucleation but new pattern of grain structure has evoked the study. This might one reason for the high mechanical properties of FSW Joints. FESEM images were shown in Fig 10.

Emission Scanning electron microscope (FESEM) is used for study of grain crystallization, the fractured samples were further studied and images were shown in Fig. 11. The difference in fractured specimens was clearly visible the formation of dimple and smooth zones were seen. An erratic fracture is seen in TIG welded samples. The mechanism of fracture is dimple blowout. The existence of voids and pores in TIG is main cause for fracture.

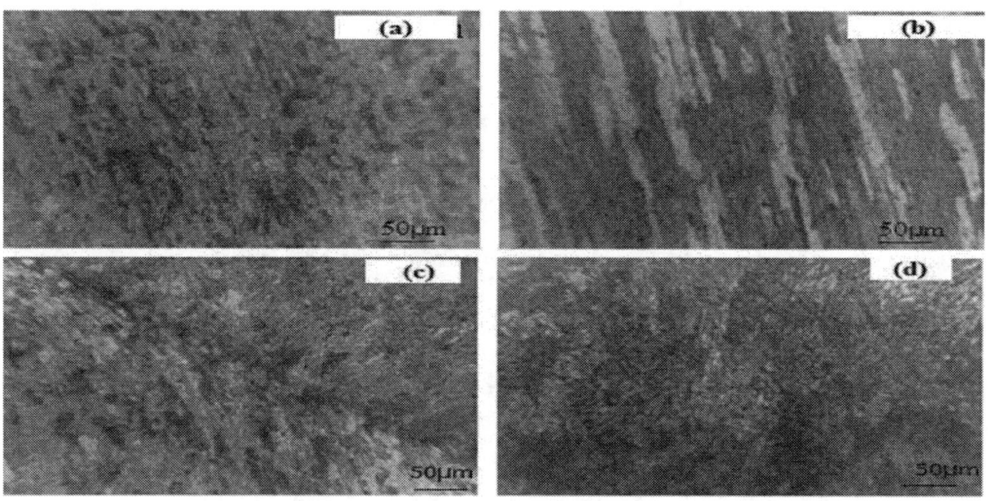

Fig 9: (a) AA 6061 Microstructure (b) AA 7075 Microstructure
(c)GTAW Microstructure (d) FSW Microstructure

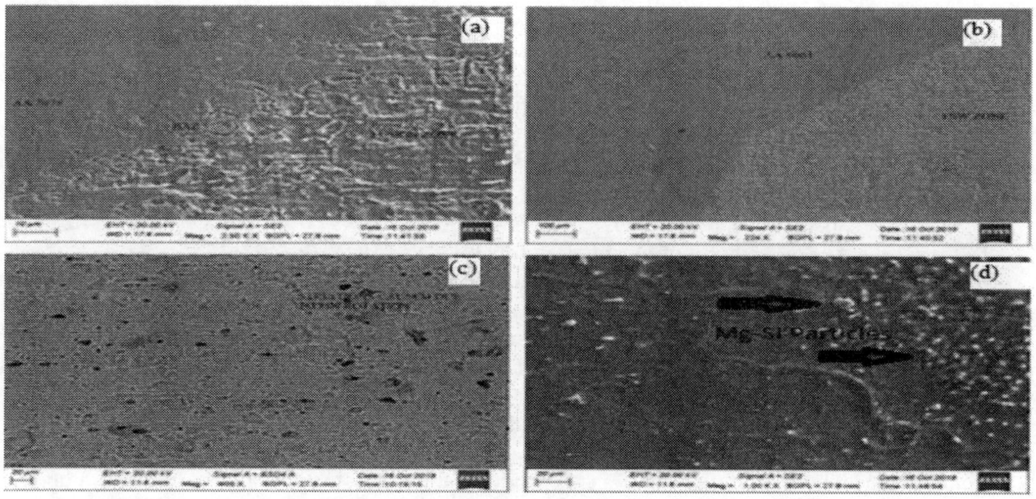

Fig 10: (a) FESEM of GTAW (b) FESEM of FSW (c) Striations in FSW (d) Intermetallics

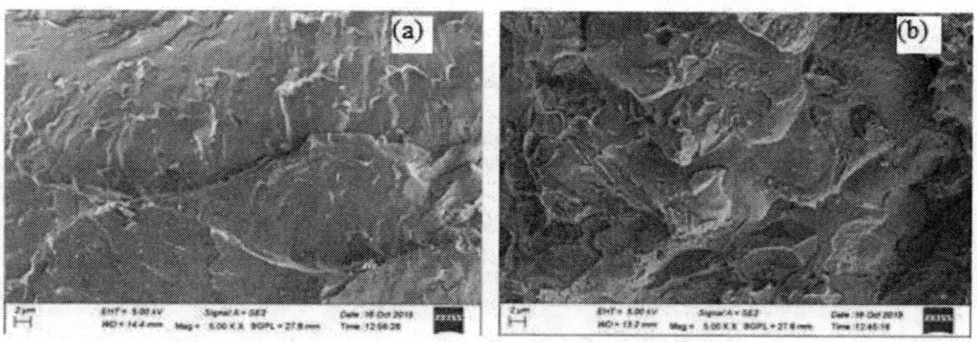

Fig 11: (a)Fracture in FSW (b)Fracture in GTAW

4 Conclusion

From the results and discussions made the following conclusions are drawn,

1. AA6061 and AA7075 Dissimilar Aluminum alloys were joined in the form of butt 2 mm thick sheets by GTAW and FSW. The joint was found successful.
2. An average tensile strength of 127 MPa is achieved by GTAW samples and 151 Mpa for FSW samples. An increase of 20% tensile strength is observed for FSW joints. Presence of magnesium has improved strength in FSW joints.
3. In TIG welding at a welding current of 140Amps, welding speed of 70 mm/min, Gas flow rate 10 L/min a maximum tensile strength of 148Mpa is obtained. In Friction stir welding at a Tool rotation speed of 1200RPM, Welding speed of 60 mm/min and Axial force 5KN, a maximum tensile strength of 180Mpa is obtained.
4. Involvement of β-Mg Si precipitates is main cause for improved mechanical properties of FSW joint.
5. Due to the stirring action longitudinal grains were axed to fine coarse grains in FSW weld area which improved the mechanical properties of joint.

References

1. M. J. Crooks, R. E. Miner, Journal of Metals, 1996 , 48, 13–15.
2. Y. Kurihara, Journal of Metals, 45 1993,11, 32–33.
3. G. P. Syukla, D. B. Goel, P. C. Pandey, All India Seminar on Aluminium, New Delhi, 1978.
4. Report of Investigation on the Technical Trend of Patent Applications in 2004: Automobile Weight reduction Technologies, Japan Patent Office, 2005.
5. Fundamentals of Aluminium Metallurgy, doi.org/10.1016/C2016-0-02117-4.
6. Bonenberger, R. Paul (2005). The First Snap-Fit Handbook. 6915Valley Avenue,Cincinnati, Ohio 45244-3029, USA: Hanser Gardner Publications, Inc. ISBN 1-56990-388-3.
7. Pocius, Alphonsus V. (2012). Adhesion and Adhesives Technology: An Introduction. 6915 Valley Avenue, Cincinnati, Ohio 45244-3029, USA: Hanser Publications. ISBN 978-3-446-43177-5.
8. A.Heidarzadeh, H.Khodaverdizadeh, A.Mahmoudi and E. Nazari. Tensile behavior of friction stir welded AA6061-T4 aluminum alloy joints. Journal OfMaterials&Design 2012; 37:166–173.
9. S.Sheikhi, and C.Bolfarini. Preliminary study on the microstructure and mechanical properties of dissimilarfriction stir welds in aircraft aluminium alloys 2024-T351 and 6056-T4 2007; 6:132–142.
10. P.Rao Microstructure and Mechanical Properties of Friction Stir Lap Welded Aluminum Alloy AA2014. Journalof Materials Science & Technology 2011; 28(5): 414–426.
11. M.Koilraj, V. Sundareswaran, S.Vijayan, andS. R. K. Rao, Friction stir welding of dissimilar aluminum alloys AA2219 to AA5083 – Optimization of process parameters using Taguchi technique. Materials and Design 2012;42:1–7.
12. P.Xue,D. R.Ni, D.Wang, B. L.Xiao, andZ. Y. Ma.Effect of friction stir welding parameters on themicrostructure and mechanical properties of the dissimilar Al – Cu joints 2011; 528:4683–4689.
13. S. Kotari et al. Mechanical and micro structural behaviour of flux coated GTAW and FSW joined AA6061 aluminium alloy.Materials Today: Proceedings 27 (2020) 1660–1667
14. R. G.Madhusudhana, P.Sammaiah andT. Mohandas. Influence of welding techniques on
15. microstructure and mechanical properties of AA6061(Al-Mg-Si) gas tungsten arc welds. In Proceedings ofNational Conference on Processing of metals, Miami, FL,2002.
16. J. Q.Su, T. W.Nelson, R.Mishra and M.Mahoney.Microstructural investigation of friction stir welded7050-T651 aluminium. ActaMater., 2003, 51, 713–729.
17. C. G.Mahoney and W. H.Bingel. Effects of friction stir welding on microstructure of 7075aluminum. Scripta Mater., 1997, 36, 69–75.
18. W. M.Thomas. Friction stir butt welding. US Patent no.5460, 317, USA, 1991.
19. H. J. Liu, H.Fujii, M.Maeda and K. Nogi.Tensileproperties and fracture locations of friction-stir-weldedjoints of 2017-T351 aluminum alloy. J. Mater. ProcessingTechnol, 2003, 142, 692–696.
20. C. G.Rhodes, M. W. Mahoney, W. H.Bingel,R. A.Spurling and C. C. Bampton.Effects of friction stirwelding on microstructure of 7075 aluminum. ScriptaMater., 1987, 36(1), 69–75.
21. D. J. Hattingha, C.Blignault, T. I.Niekerk andM. N.Jamesa. Characterization of the influences of FSW tool geometry on welding forces and weld tensilestrength using an instrumented tool. J. Mater. ProcessingTechnol., 2008, 203, 46–57.
22. A.Barcellona, G.Buffa, and L. Fratini, Processparameters analysis in friction stir welding of AA6082-T6sheets. Keynote paper of the VII ESAFORM Conference,Trondheim, Norway, 2004, pp. 371–374.
23. T. L. Dickerson, and J. Przydatek, Fatigue of friction stirwelds in aluminum alloys that contain root flaws. Int. J.Fatigue, 2003, 25, 1399–1409.
24. Y.Chen, H.LiuandFeng.J. Friction stir weldingcharacteristics of different heat-treated-state 2219 aluminum alloy plates. Mater. Sci. Engng A, 2006, 420, 21–25.
25. P.Cavaliere and F.Panella. Effect of tool position on the fatigue properties of dissimilar 2024–7075 sheetsjoined by friction stir welding. J. Mater. Processing Technol., 2008, 206, 249–255.
26. P.Cavaliere, De, A.Santis.F.Panella, and A.Squillace.Effect of welding parameters on mechanical and microstructural properties of dissimilar AA6082– AA2024 joints produced by friction stir welding. Mater. Des., 2008, 30, 609–616.
27. S. Kotari, E. Punna, S. M. Gangadhar et al., Dissimilar metals TIG welding-brazing of AZ31 magnesium alloy to 304 stainless steel, Materials Today: Proceedings, matpr.2020.05.553

Acoustic emission examination of high strength steel M250 pressure vessels for aerospace applications

Gowri Shankar Wuriti[*$a], Somnath Chattopadhyaya[$b], Tessy Thomas[#c]
*Scientist, Advanced Systems Laboratory, Dr.APJ Abdul Kalam Missile Complex, DRDO, Ministry of Defence, Government of India, Hyderabad500058,
#Director General, DRDO, Banglore, India,
$Department of Mechanical Engg., IIT, Dhanbad, India.
Email: [a]wuritigshankar@gmail.com, [b]somuismu@gmail.com, [c]tessy@gmail.com.

Abstract. This paper deals at experiences with the Acoustic Emission (AE) monitoring of critical pressure vessels during qualification proof pressure test. Conventional Non Destructive Evaluation (NDE) methods have limitations in evaluating the vessel. There is a high probability that they may not provide information required to evaluate the quality of vessel. In contrary, Acoustic emission technique has a high probability of detecting the likely defects in pressure vessels prior to the failure. This technique has to be applied under pressure load. These pressure loads create stresses in the vessel. During increasing pressure loads AE signals are obtained from discontinuities and other locations of stress concentrations. AE is very useful in estimating the soundness of critical pressure vessels used by aerospace industry that is made up of high strength Maraging steels. AE Testing and result analysis was carried out according to standard procedures. During the tests, AE signals were recorded and post analysis was done to understand the response from the material. The experimentation revealed that the amplitude of AE signals were less than the corresponding AE signals for critical crack opening, during pressurization cycle. Also, at proof pressure, a 2 minutes hold time was observed with no increase in amplitude of signals. AE activity can detect any defect growth during proof pressure test. Inspection was carried-out in accordance with a set of acceptance criteria based on available literature and specimen level tests.

Keywords: Acoustic emission; Pressure vessel (PV); Proof pressure testing

INTRODUCTION

Acoustic emission (AE) technique is a rapidly developing Nondestructive evaluation technique which is quite useful in the evaluation of structural integrity of pressure vessels during acceptance or pressure tests[1][2]. AE techniques play an effective role in proving such systems. The unique capability of the technique lies in its ability for the real time monitoring of active defects. AE monitoring during pressurization helps to identify growing defects. The initial pressurization cycle is characterized by considerable amount of emissions. If the subsequent pressure cycle is “quiet” then it is considered that the defects are sub - critical and does not lead to failure. The vessel may however still have large but “stable” defects [3][4]. Detailed analysis of the initial loading is carried out to ascertain the quality of the vessel. Acoustic Emissions are governed by the microstructure as well as the presence of inclusions and second phase particles.

The AE response is anticipated to be from active micro defects or localized stress raisers and weld mismatches that get corrected up possibly by micro yielding so long as the peak amplitude levels are nominal and the hit rates show a decline during 'pressure holds'. When micro cracks showing a sustained growth correspondingly AE shows continuous or intermittent emissions

IOP Conf. Series: Materials Science and Engineering **998** (2020) 012037 doi:10.1088/1757-899X/998/1/012037

and the severity that is indicated in terms of certain AE parameters such as peak amplitude, event duration, hit-rate and so on in real time[5][6].
In this study, the authors share their experience derived from the AE response during proof pressure testing on High strength steel Pressure vessels used in critical aerospace applications [7].

The chief Acoustic Emission sources in metallic components are from localized stress-raisers viz. cracks, weld mismatch, tailed porosity, weld defects etc. The other predominant source includes plastic deformation due to yielding.

The micro defect/cracks so available can now behave in three different fashions.

1. The defect can blunt itself and eventually AE from defect dies down during loading.
2. The defect has a proportional growth to applied load. Thereby Acoustic emission steadily increases with load applied
3. The defect grows faster during application of load i.e. AE increases rapidly with application of load.

The cases are respectively referred as sub-critical, critical and super-critical. In the critical and supercritical states, AE is characterized by emissions which have continuous higher amplitudes, counts and energy. Moreover severity is indicated by the respective rates [8].

In the Current study, acoustic emission parameter details have been captured and analyzed during pressure test of the vessel.

PROOF PRESSURE TEST OF PRESSURE VESSEL

Material	:	High strength M250 steel
Size	:	900 mm diameter, 2500mm length
Thickness	:	5.0 mm
Hydro test pressure	:	52 ksc

Proof pressure test was conducted on a typical pressure vessel. The schematic of the pressure vessel is given in 'Figure 1' that has four circumferential welds and 3 longitudinal seam welds. Welding is carried out by Gas tungsten arc welding process.

AE Instrumentation and Sequence of Pressurization

The number of sensors employed to carry out the testing is based on signal attenuation characteristics. A front end filter of 100 kHz to 400 kHz is applied and a pre amplification of 100 times is done with a pre amplifier. Prior to test, a pencil lead break method in accordance with ASTM E 976 is employed to ascertain sensitivity of AE sensor. This signal shall be comfortably picked up by the AE sensor layout. The pressure is raised in steps with each dwell period having a 'Hold time' of at least 1 minute. The vessel is pressurized up to the acceptance level – P_A (as per design) in steps. A repeat pressure cycle is resorted to a pressure value which is lower to the acceptance pressure say 0.9 P_A, so as to observe 'Kaiser Effect' (The *Kaiser Effect* is the non- existence of detectable acoustic emission until the previous maximum applied stress level surpassed). Comparison of relative emission density from welds over the entire vessel gives an indication of which areas might be having growing defects or have material problems. This makes follow-up NDT easy and efficient.

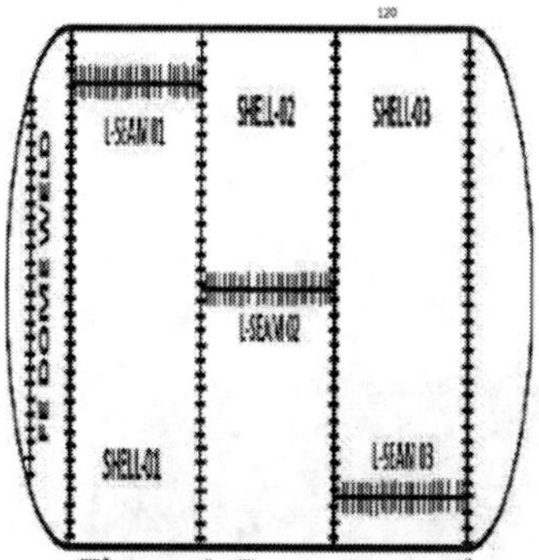

Figure 1. Pressure vessel

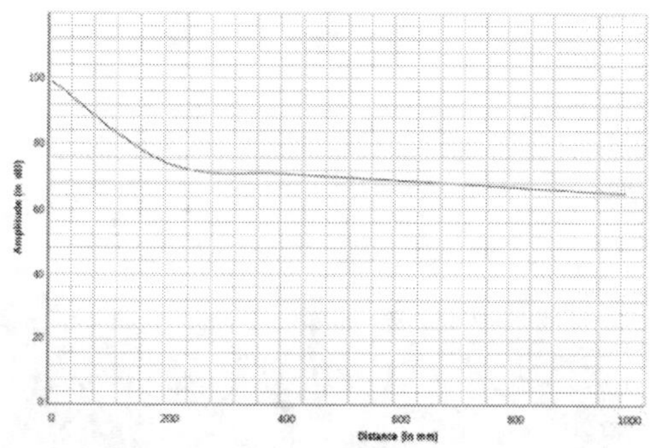

Figure 2. Attenuation study on Pressure vessel

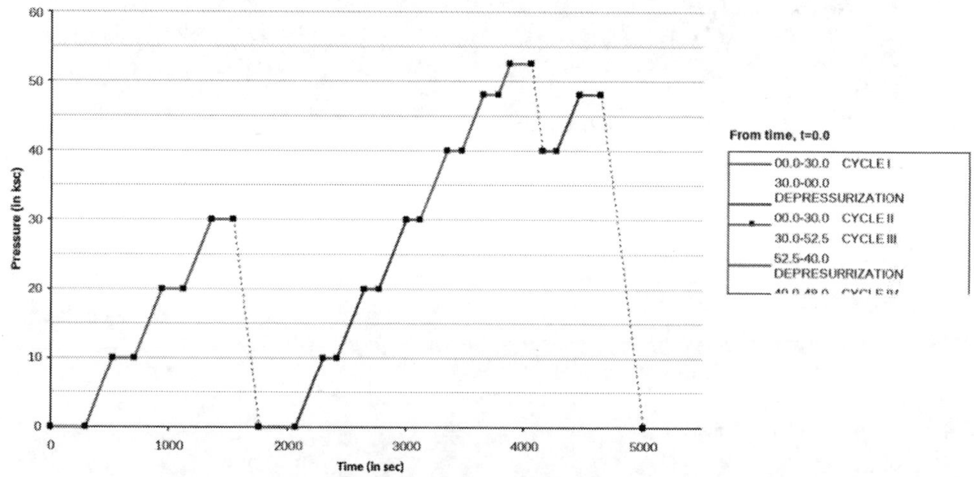

Figure 3. Pressurization Cycle

A typical pressure vessel with PA 52 ksc is tested and the results are presented. The first pressurisation appeared quite normal, with emission starting from low loads as expected as a result of stress relief. 'Figure 4' shows the history of amplitude and hits/counts for the entire vessel. There were few high amplitude events near weld zones of segment 2 during initial pressurisation. But subsequent radiographic inspection revealed no flaw. There were no significant hit or event during any of the hold period. "Hits" were evaluated based on number of counts it produced, total energy released during the process and the duration, it lasts. For e.g., a mechanical noise has a considerable amplitude but lasts for a longer duration[9].

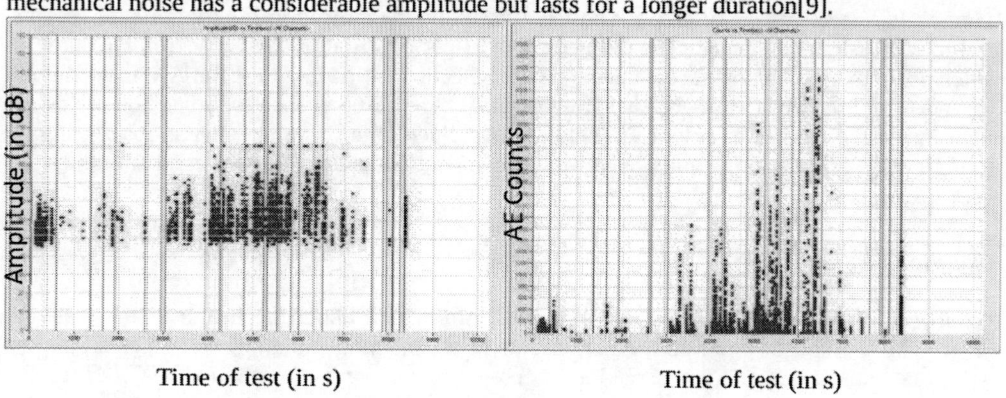

Figure 4. History of amplitudes and counts during test

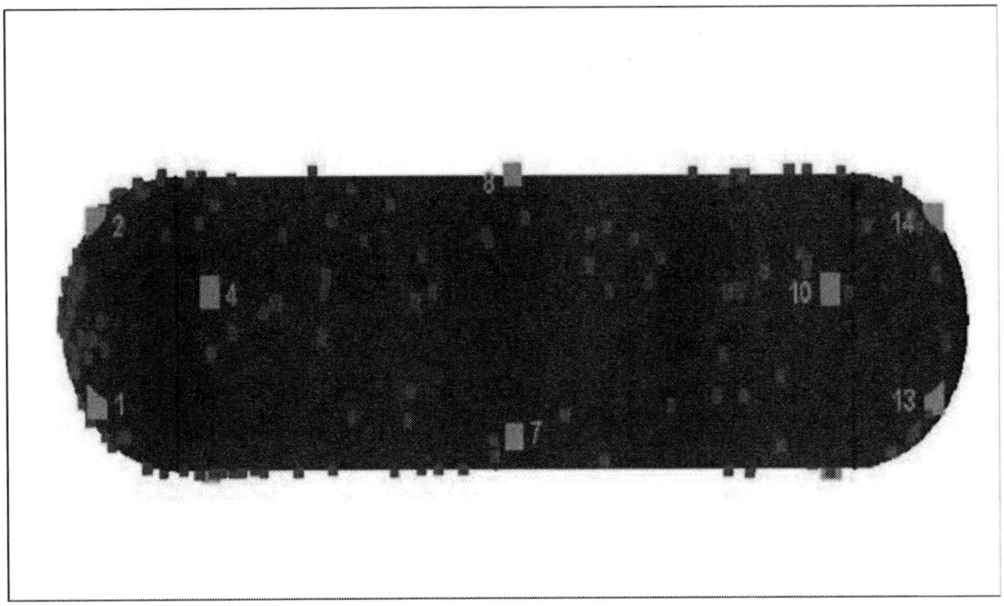

Figure 5. Location Plots on pressure vessel

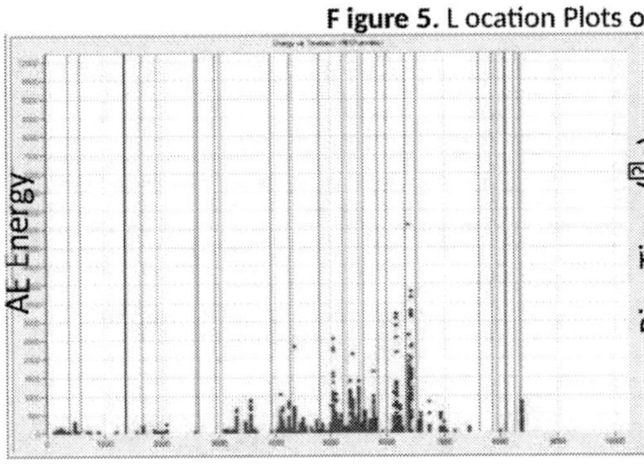

Time of test (in s)

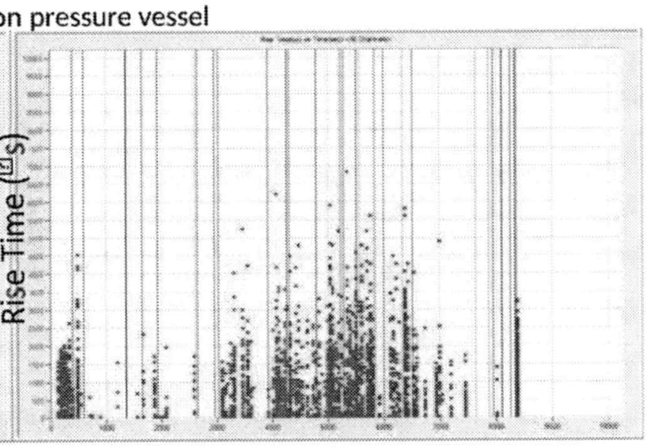

Time of test (in s)

Figure 6(a). Energy Graph

Figure 6(b). Rise Time Graph

'Figure 6(a)' is called energy graph. The felicity ratio is the indication of any pressure vessel's health. The Felicity Ratio (FR) is defined as ratio between the applied load (or pressure) at which acoustic emission reappears during next application of loading to the previous maximum applied load. Metallic vessels has a FR equal to 1. 'Figure 6(b)' indicates that, there was no rapid increase of counts in any of the hold periods. Such high count rates were observed only during cycle I and cycle III which were first pressurization cycles. There-pressurisation cycles were silent.

Acoustic Emission Criteria

Acoustic Emission examination provides useful information about the health of the material when loaded. Acoustic Emission tool is useful in ascertaining initiation of failure during an early state of testing. Any AE observation shall be correlated with conventional NDT test results viz.

Ultrasonic Inspection, Radiographic evaluation for better understanding of the source. The performed Acoustic emission test results shall be studied in accordance with an established acceptance criteria. Such criteria shall be derived based on series of sample testing. Apart from quantitative assessment based on acceptance criteria, qualitative assessment based on felicity ratio/Kaiser effect can also be carried out.

Acceptance criteria for Maraging steel 250 pressure vessels. In general, for metallic motors Kaiser Effect is to be observed. The felicity ratio for such motors is equal to 1. (FR=1). During first pressurisation cycles, there shall not be any rapid increase in AE events or AE counts. During the second pressurisation any AE event during any pressure hold is not acceptable.

STANDARDS AND GUIDELINES PERTAINING TO AE

Several standards have been published by standards publishing organizations like International organization for Standardisation (ISO), Comite Europeen de Normalisation (CEN), American Society for Testing and Materials (ASTM). Apart from these standards, the American Society for Mechanical Engineers (ASME), has issued, standard practices for AE testing metallic pressure vessels (Article 11 of ASME Section V).

Perspectives for future Standardization

AE testing is a capable of evaluating structural integrity assessment and health monitoring of structures. AE testing of civil structures such as bridges is useful in determining safety of those structures. However there is a need for detailed standards and acceptance criteria for various classes of test objects.

4.0 CONCLUSION

AE is useful tool during proof pressure testing. Emission from crack opening is difficult to differentiate with emission has stress relieve cycle. During second pressurization cycle no emission indicate that there are no incident defect growth or continues stress relief. “Abnormal” levels of emission from stress relief may also be the result of some manufacturing problems that leave high residual stresses in the welds [8]. During the second pressurization cycle, there might not be any indication of defects that have grown or yielded during the first proof pressure cycle and become stable. Hence analysis of first cycle results should also be done. After proof pressure test, inspection by other NDT techniques may be required to be performed is necessary if there are significant or alarming emissions. For acoustic emission to serve as a quantitative structural health monitoring system, reliable classification of emission source and severity must be known. These are obtained by specimen level tests. These tests reveal the “signatures” of the material at different states of stress.

5.0 REFERENCES

[1] S. Gholizadeh *et al* 2015 A review of the application of acoustic emission technique in engineering *Structural Engineering and Mechanics* **54** 6 1075-1095

[2] F.Witos et al 2019 Properties of Amplitude distributions of acoustic emission signals generated in pressure vessel during testing *Archives of acoustics* **44** 3 493–503

[3] N. Satyanarayana, S. V. Subba Rao, D. Jayaseelan and K. Viswanathan 1996 Acoustic Emission m onitoring of a Long Stored Rocket Motor Case *Trends in NDE Science & Technology Proceedings of the 14th World Conference on Non-Destructive Testing.* **4** 2525 – 2528.

[4] Phillip, Cole, Stephen, Gautrey 2004 Acoustic Emission Experience with AE Monitoring of New Vessels during Initial Proof Pressure Test *EWGAE 2004 DGZfP-Proceedings* . 75-81

[5] T. Chelladurai, et al 1996 Acoustic Emission Technique – an effective tool for the integrity evaluation of M250 Maraging steel aerospace pressure chambers *14th WCNDT*

[6] ASTM E1930 / ASTM E1930-17, Standard Test Method for Examination of Liquid-Filled Atmospheric and Low-Pressure Metal Storage Tanks Using Acoustic Emission, ASTM international, West Conshohocken PA, 2017

[7] Dennis A White 1988 Acoustic Emission Examination of Metallic Pressure Vessels during Pressure Testing,
ASME Boiler and Pressure Vessel Code Article 12 Subsection A Section V

[8] J. Brunner and J. Bohse 2002 Acoustic Emission standards and guidelines 2002: A comparative assessment and perspectives Internet Source: www . nd t .n e t - September 2002, Vol., No.91

[9] Ronnie K Miller 2005 Non-Destructive Testing Handbook Volume 5Acoustic Emission Testing, American Society for Non-Destructive Testing, pages 167-186.

ACKNOWLEDGEMENTS
Authors are thankful to DRDO & Director IIT Dhanbad for their support and grant of permission to publish the work.

OPTIMIZATION OF MODEL AND STRUCTURAL ANALYSIS HYPOID GEAR

D.V. Ramana Reddy[1,A], G. Revanth Kumar[2,B]

[1]Assistant Professor, Department of Mechanical Engineering, Vardhaman College of Engineering, Hyderabad, Telangana,

[2]PG Scholar, Department of Mechanical Engineering, Vardhaman College of Engineering, Hyderabad.

[a]Corresponding Author: ramana.mech@vardhaman.org

Abstract:
The purpose of this work is to analyze structural behaviour of differential hypoid gear. For the studying of the structural analysis of the gear is made up of the different composite materials & as well as alloy materials compared with conventional metallic materials. As, composite materials provide enough strength and emerged as a good alternative for replacing metallic gears which is Ni-Cr steel. The alloy materials considered here is Aluminium alloy, Grey cast iron, and the composite material is Epoxy E-Glass unidirectional and compare with existing gear material Ni-Cr steel. This work makes an attempt to replace the metallic gears of Ni-Cr steel with alloy & composite materials based on static structural analysis results. For the analysis purpose the solid model is created by using CATIAV5R20 design software and ANSYS 17.1 is used to determine the Total deformation, Von-Mises stress and the equivalent elastic strain for the different selected materials at three different speeds.

Key words: Hypoid gear, von mises stress, alloy steel, Epoxy E-Glass.

1. INTRODUCTION

Now a day's gears are used in many applications for transmission of motion from one shaft to another shaft. The gears drives are positives which are transmit exact velocity ratio, the gear drives are transmit motion when two shafts distances are less than compared to the belt, chains, rope drives. The gear drive is used when the shaft having long centre distances this time some idler gears are using between driver and driven gears for transmission of motion from one to another shaft. Thus providing the idler gears the cost also increased and maintained cast also high than compared to the other power transmission drives like belt, ropes and chain drives, for this reason the gear drives not recommend for transmission of motion from large centre distances from driving shaft to the driven shaft. Now a day's gears are various types are available in the market, based on the type of application, purpose and working condition the gear materials is selected, because the selection of the material is most important for manufacturing of the any component, because the tensile strength, young's modules, corrosion resistance, efficiency and life of the material etc. can be depended into the selection of the material.

This work makes an attempt to replace the conventional metallic gears of Ni-Cr steel gear material with the alloy & composite materials having application for high power transmission system & reducing of the stress same as a gearbox were in automobile industries. For this purpose 3-D solid model of the differential hypoid gear made in CATIA V5R20 design software and ANSYS 17.1 finite element method analysis software were used to carry out the static analysis in order to identify the behaviour of the conventional Ni-Cr steel gear material and identified replacements of different alloy & composite materials under the different loading criteria's. Then the simulation of the results determines the total deformation, Equivalent Von-Misses Stress, and Equivalent Strain of the different selected materials at three different rpms.

2. METHODOLOGY

1. To develop a solid model of Differential Hypoid gear by using CATIA V5R20 design software and data book.
2. Select the Conventional materials and alternative materials (alloys & composite).
3. Here we are comparing the Aluminium alloy Grey cast iron, and Epoxy E-Glass Unidirectional and compare with the Ni-Cr steel.
4. Apply the boundary conditions.
5. To calculate the Von Mises stresses strains and displacements by using ANSYS17.1 Analysis tool.
6. Analyze the static structural behaviour of various different selected materials under different torques.
7. Compare the analysis results of Von-misses stress, Equivalent elastic strain and Total deformation with the existing journal [6].

3. FEM ANALYSIS OF THE GEAR

In this project numerical method of the FEM based static structural analysis carried out by using ANSYS 17.1 analysis software. FEM (Finite element analysis) a computer-based analysis technique is used of the studying the behaviour of structures under the applied blundered conditions. The ANSYS 17.1 analysis software determine results like the total deformation, Equivalent Von – misses Stress, and Equivalent elastic strain under the applied boundary conditions. The 3D model for numerical analysis developed in CATIA V5R20 design and these models have been imported in ANSYS as IGES file format for further.

4. MATERIAL PROPERTIES

The various mechanical properties of the selected material were given in the table below.

Table 1. The various materials considered for analysis

Properties	Ni-Cr Steel	Aluminium Alloy	Grey cast iron	Epoxy E-Glass_UD
Model type	Linear elastic Isotropic	Linear elastic Isotropic	Linear elastic Isotropic	Linear elastic Isotropic
Failure Criteria	Max Von-Misses stress	Max Von-Misses stress	Max Von-Misses stress	Max Von-Misses stress
Yield Strength	172.339 MPa	280 MPa	138 MPa	350 MPa
Tensile Strength	413.613 MPa	310 MPa	240 MPa	650 MPa
Youngs Modules	180000 MPa	71000 MPa	110000 MPa	200000 MPa
Poisson's Ratio	0.28	0.33	0.29	0.22

IOP Conf. Series: Materials Science and Engineering 998 (2020) 012038 doi:10.1088/1757-899X/998/1/012038

5. TORQUE CONSIDERATION

For the analysis purpose of the differential hypoid gear the applied boundary conditions are friction less supports, that means all gears are rotated in applied rpms and three different rpms were uses for conducting of the static structural analysis. Based on the selection of the different materials and applied boundary conditions the results are generated. The various torques and rpms for the analysis is mention in the below table.

Table 2: different torque considered

Sr. no	TORQUE (N-M)	RPM
1	490	1800
2	390	3600
3	130	5400

6. STATIC STRUCTURAL ANALYSIS

This analysis, which is numerical method of the ANSYS software analysis, the static structural analysis which is a system based analysis, that means, resulting the static structural analysis is generated by computer based on the selection of various different materials an applied boundary conditions the computer solve the results with the help of preloaded formulas and theorems the results will be generated. The static structural analysis is mainly conducted to find the selected material for gear manufacture can withstand or not for the applied loads and generation stress, temperatures etc. during motion. To finding the best material for the manufacturing of the differential hypoid gear is mandatory.

In this work, the FEA tool were used to identify the structural behaviour of various alloy materials and composite material under certain boundary conditions there by determining total deformation, equivalent Von-mises stress and maximum strain for each alloy material and as well as composite material then the comparison was done with previous existing metallic gear material which is Ni-Cr steel. In this research work FEM structural analysis simulation results were shown the behaviour of Ni-Cr steel, Aluminium Alloy, Grey Cast Iron alloy materials and Epoxy E-Glass_UD composite material at the applied boundary conditions the results of the static structural analysis were shown below as:

For The Ni-Cr Steel Torque = 490 N-m

Figure 1: Von Mises Stress For Ni-Cr Steel = 118.05MPa

At Torque = 390 N-m

Figure3: Von Mises Stress For Ni-Cr Steel = 472.22MPa

IOP Conf. Series: Materials Science and Engineering 998 (2020) 012038 doi:10.1088/1757-899X/998/1/012038

Figure2: Total Deformation For Ni-Cr steel = 0.020104 mm

At Torque =130 N-m

Figure4: Total Deformation For Ni-Cr Steel = 0.080417 mm

For The Aluminium Alloy Torque = 490 N-m

Figure5: Von Mises Stress For Ni-Cr Steel = 1062.5MPa

Figure7: Von-Mises Stress For Aluminium Alloy = 40.628 MPa

Figure6: Total Deformation For Ni-Cr Steel = 0.18094 mm

For The Aluminium Alloy At Torque = 390 N-m

Figure 8: Total Defomation For Aluminium Alloy = 0.017663 mm

For The Aluminium Alloy At Torque = 130 N-m

Figure9: Von-Mises Stress For Aluminium Alloy = 162.51 MPa

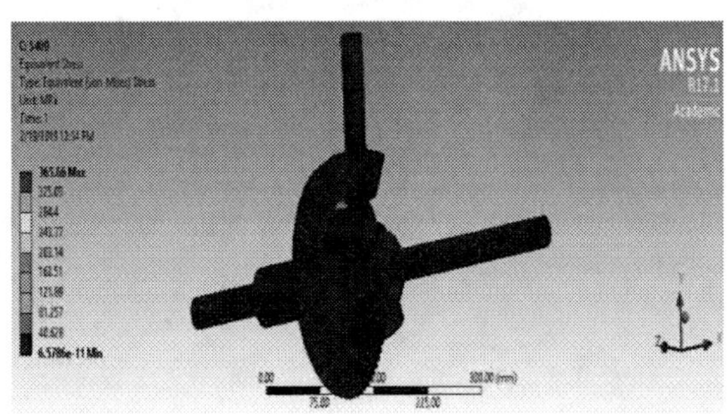

Figure11: Von Mises Stress For Aluminium Alloy = 365.66MPa

Figure10:Total Deformation For Aluminium Alloy =0.070653mm

Figure12:Total Deformation For Aluminium Alloy =0.15897mm

For The Epoxy E-Glass_UD Torque = 490 N-m

Figure13:Von-Mises Stress For Epoxy E-Glass_UD= 39.239 MPa

Figure14: Total Deformation For Epoxy E-Glass_UD = 0.067215mm

For The Epoxy E-Glass_UD Torque = 130 N-m

For The Epoxy E-Glass_UD Torque = 390 N-m

Figure 15: Von-Mises Stress For Epoxy E-Glass_UD= 156.96 MPa

Figure16:Total Deformation For Epoxy E-Glass_UD = 0.26886mm

For The Epoxy E-Glass_UD Torque = 130 N-m

IOP Conf. Series: Materials Science and Engineering 998 (2020) 012038 doi:10.1088/1757-899X/998/1/012038

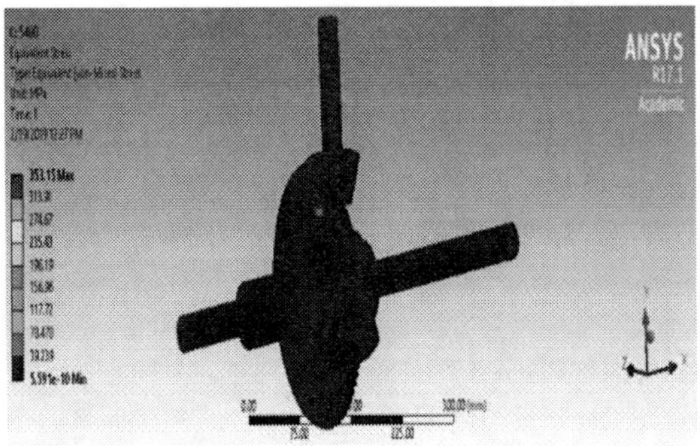

Figure17:Von-Mises Stress For Epoxy E-Glass_UD= 353.15 MPa

Figure18:Total Deformation For Epoxy E-Glass_UD = 0.60493mm

For The Grey cast Iron Torque = 490 N-m

@ Torque = 390 N-m

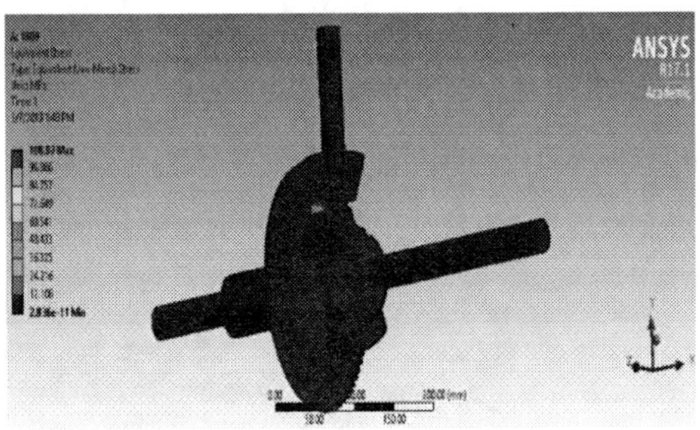

Figure19:Von-Mises Stress For Grey Cast Iron = 108.97MPa

Figure21: Von-Mises Stress For Grey Cast Iron = 435.89MPa

Figure20: Total Defomation For Grey Cast Iron= 0.030367mm

Figure22: Total Defomation For Grey Cast Iron= 0.12147mm

For The Grey cast Iron Torque = 130 N-m

For The Grey cast Iron Torque = 130 N-m

Figure23:Von-Mises Stress For Grey Cast Iron = 980.76MPa

Figure24:Total Deformation For Grey Cast Iron= 0.27331mm

Table 3: Comparison table for materials at 1800RPM

Materials	Stress MPA	Strain	Total Deformation (mm)	Density (Kg/m3)
Ni-Cr Steel	118.05	0.00074772	0.020104	7800
Epoxy E-Glass_UD	39.239	0.0032399	0.067215	7700
Aluminium Alloy	40.628	0.00065256	0.017663	2770
Grey Cast Iron	108.97	0.0011294	0.030367	7200

Table 4: Comparison table for materials considered at 3600 rpm

Materials	Stress MPA	Strain	Total Deformation (mm)	Density (Kg/m3)
Ni-Cr Steel	472.22	0.00299	0.080417	7800
Epoxy E-Glass_UD	156.96	0.01296	0.26886	7700
Aluminium Alloy	162.51	0.0026102	0.070653	2770
Grey Cast Iron	435.89	0.0045177	0.12147	7200

Table 5: Comparison table for materials considered at 5400rpm

Materials	Stress MPA	Strain	Total Deformation (mm)	Density (Kg/m3)
Ni-Cr Steel	1062.5	0.0067295	0.18094	7800
Epoxy E-Glass_UD	353.15	0.029159	0.60493	7700
Aluminium Alloy	365.66	0.0058731	0.15897	2770
Grey Cast Iron	980.76	0.010165	0.27331	7200

7. CONCLUSIONS

1. The developed Hypoid gear compared with existing gear material, which has Ni-Cr steel. The analysis was done, torque at 1800 rpm, 3600 rpm, and 5400 rpm under static loading condition.
2. The results of the static structural analysis of von-misses stress, strain, and deformations shows the aluminium alloy is a valid replaceable material in the presence of previous established gear material which has Ni-Cr steel.
3. The hypoid gear was manufactured with the help of aluminium alloy material the 50% of stress was reduced and the aluminium alloy has low density, based on low density the overall weight of the gearbox was reduced.
4. By decreasing of the weight of the gearbox the efficiency of the gear will be increased.

5. After the aluminium alloy material the next suitable material is Epoxy E-glass unidirectional composite material is suitable for manufacturing of the hypoid gear.

8. FUTURE SCOPE OF THE WORK

In the present investigation of static analysis of hypoid gear is analyzed by single composite material in as a future work, the number of composites and some advanced alloy, nano materials with different rpms was also tested for manufacturing of the hypoid gear to overcome of the gear teeth failure by applying of the excessive loads.

REFERENCES:

[1] Mayank Bansal, Nidhi Sindhu and Santosh Anand,2016, "*Structural and Model Analysis of a Composite Material Differential Gear Box Assembly*", International Journal of Engineering Sciences & Research Technology, Volume 5, No 6, June 2016, Page 575-594.

[2] Rohit Sreekumar and T. Jeyapoovan,2016, "*Design and Analysis of composite Bevel Gear in an Automobile Differential Gear Box*", International Journal of Engineering Sciences & Research Technology, Volume5, No 5 May 2016, page 248-257.

[3] C. Veeranjaneyulu, U. Hari Babu, 2012, '*Design And Structural Analysis of Differential Gear Box at Different Loads*', International Journal of Advanced Engineering Research and Studies, Vol. 1, Issue II, January-March, 2012, Page. 65-69.

[4] PSG Design Data Book,1966, PSG College of Technology Coimbatore, pp 8.1-8.64

[5] B. Venkatesh, V. Kamala, A.M.K. Prasad, 2010, *Modelling and Analysis of Aluminium A360 Alloy Helical Gear for Marine Applications*', International Journal Of Applied Engineering Research, Dindigul Volume 1, No 2, 2010, Page. 124-134.

[6] Vilmos simon, 2000,"*FEM stress analysis of hypoid gears*", Mechanism and machine theory, ELSEVIER, Volume 35,issue 9,pages 1197-1220.

[7] AnoopLega, PuneetKatyal, Vishal Gulati, 2012, '*Computer Aided Design and Analysis of Composite Gearbox Material*', International Journal of Mechanical Science and Civil Engineering (IJMSCE), Volume-1, Issue-1, December 2012, page.

[8] Chabra Pankaj, Bhatia Amit,2012, "*Design and Analysis of Composite Material Gear Box*", International Journal of Mechanical and Civil Engineering, Vol.1 (2012), Issue1, pp 15-25.

[9] Sushil B. chopade, Prof K.M Narkar, Pratik K. satav,2015, "*Design and Analysis of E-Glass / Epoxy composite monoleaf spring for light vehicle*" International Journal of Innovative Research in science engineering and technology, Vol.4, Issue-1, January 2015, page. 18801-18808.

[10] Alok Hegde , 2015, "*Tensile properties of Uni-directional glass / epoxy composites at different orientations of fibres*", Int. Journal of Engineering Research and applications, Vol. 5, Issue-3, March 2015, page. 150-153

IOP Conf. Series: Materials Science and Engineering 998 (2020) 012039 doi:10.1088/1757-899X/998/1/012039

FABRICTION OF MECHANICAL SEGWAY WITH HANDLE

M.Vishnu Vardhan[1*], Sairam Kotari[1], Mohd Sameeroddin[1*], Md. Kashif Gohar Deshmukh[1], Gunti Rohit[1]

[1] Department of Mechanical Engineering, Vardhaman College of Engineering, Shamshabad, Hyderabad, Telangana 501218, India

[*]E-mail: mukkotivishnu@gmail.com, mohdsameer9550@gmail.com

Abstract. Segway is a two-wheeled motorized personal vehicle consisting of a platform for the fleet mounted above an axle an upright post surmounted by handle with two guiding wheels, controlled by the rocker switches. Segway is mainly used for the short distance travel where there is no use of fuel and no pollution. The main objective of this project is to produce an easy way to move and a comfortable space for the user and reduce pollution as this model is eco-friendly. The present project proposes a cost effective and innovative design compared to the existing segway available in the market. The proposed project also aims to reduce the e-waste by eliminating few electronic components.

1. Introduction

This Fabrication of mechanical Segway with handle is a personal transporter which moves in short distances from place to place [1]. This Segway is a device that moves in the forward and backward directions. It has a potential to speed up to (6-7 km/h) indoor and outdoor operation, capable of turning in place, and has a weight carrying capacity of up to hundred kilograms which carries the person while riding [2]. The main principle behind the load carrying capacity is the type of material used in building the Segway that is Iron which is durable and strong which in return supports high weight carrying capacity . Generally the expenditure of these components,conjoin with the expenditure of operating experimentation and testing[3]. This Segway is a human transporter (HT) conferred an contemporary economic product for the human transportation [4]. Segway which is working on batteries balances and moves on two wheels which are guarded by microprocessors and exclusive gyroscopes[5]. The robotic mobility platform has a miniature impression , a zero turning radius ,the capability to carry over distinct terrains and the capability to bear up to the load setup. These properties made researchers experiment considering the locomotive part of their robotic systems[6]. It is furnished with a motionless T- shaped control shaft attached into a platform seated on two parallel wheels.Segway's are directed continuing up and operate according to human body dynamics [7]. The Segway has no braking system or an accelerator to boost up the speed rather it has a handgrip on the handle of Segway which is helpful for the rider while operating it also it is used while taking turns by the switches attached on to the handle.The steering mechanism of Segway is also given by the switches attached which are used to turn in opposite directions forward and backward [8]. Basically, a segway is mostly used for personal transportation in urban environment. This was invented by Dean L. Kamen. Segway is like a scooter but the wheel arrangement is placed side by side instead of parallel to each other [9]. The advantages of segway's are zero emissions, zero turn radius, no noise and higher degree of freedom. Segway is used for urban police patrolling, military surveillance, off road riding and urban sightseeing [10].

IOP Conf. Series: Materials Science and Engineering 998 (2020) 012039 doi:10.1088/1757-899X/998/1/012039

Fig. 1 Mechanical Segway

2. Literature survey

Meyer et al. [11] studied the HT segway which is regulated and powered by the consumer product safety commission rather tha the national highway traffic safety administration. The study examined the challenges for transportation by HT segway which includes types of pavements, travels on sloped surfaces, and issues on safety purposes. Brain .G.R. huges [12] investigated on the onboard computers computers to control the power directly on to the wheels to balance the forces by the rider, this study also shown that the segway balances at various conditions such as constant velocity, acceleration and deceleration. Thompson et al. [13] studied on the mechanical segway design and fabrication by using gyroscopic sensors, and electric motors driven by battery. The designed model of segway costs around Rs. 20,000/- which is very much less compared to the one which is in the commercial market which made highly cost effective. Brett Browning et al. [14] studied on the RMP segway which is a balancing robot base and it is also be able to play soccer autonomously.

The existing segway's in the market costs around Rs. 28,000/- which is most expensive and robust in design. The present project proposes a cost effective and innovative design which costs around Rs. 15,000/- which is very less compared to the existing segway available in the market [15]. The proposed project also aims to reduce the e-waste by eliminating few electronic components.

3. Components of Mechanical Segway

The following are components of mechanical Segway [16]:

3.1 Frame and Base plate
3.2 Suspension handle
3.3 Silicon tires
3.4 DC Gear motor
3.5 Pillow block bearings
3.6 Batteries
3.7 Rocker switches
3.8 Chain and Sprockets
- Gears
- Belt and Pulleys
- Crank
- Coupling

3.9 Guiding wheels

3.1 DC Gear motors

The Dc Gear Motor coverts the electrical power into mechanical power. Two Dc Gear Motors; each of 12v are used in Segway each is attached to chain sprocket system which is further connected to wheel for the movement of Segway.

IOP Conf. Series: Materials Science and Engineering 998 (2020) 012039 doi:10.1088/1757-899X/998/1/012039

Table 1 Specifications of the Dc Gear Motor

PARAMETER	VALUE
Motor	Dc 12 volts
Material	metal
Speed of the motor	30 rpm
Torque of the motor	35kgcm
Power of the motor	1.099kw

3.2 Pillow Block Bearings

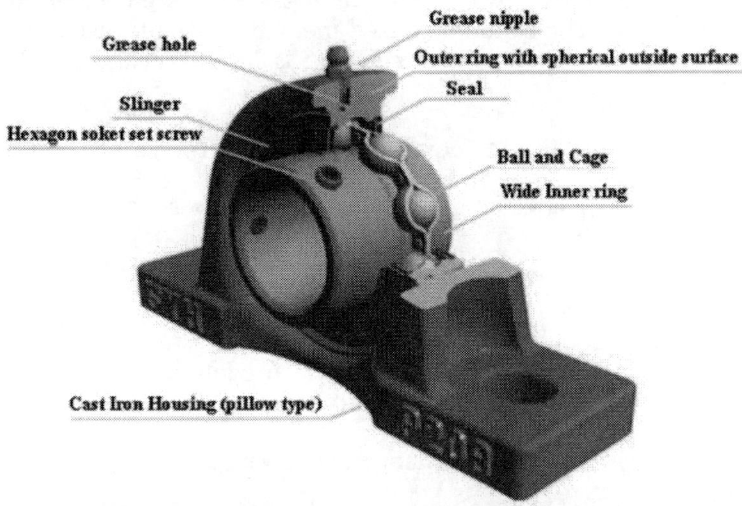

Fig. 2 Pillow Block Bearings [16]

A pillow block is a base which is used to give hold for a rotating shaft with the help of suitable bearings and different accessories. Two pillow block bearings are used in Segway each is attached to the wheel [17].

3.3 Batteries

Battery is the main power source in the mechanical Segway. There are two batteries used each is connected to Dc Gear Motor by means of cables. The power of the batteries is 12v each these are the rechargeable in both ways electric socket and solar plates.

Definite contrasting sequence of electrode materials and electrolytes are used along with lead-acid,zinc-air,nickel-cadmium(NiCad),nickel-metal hydride,(NiMH),lithium-ion(Li-ion),and lithium ion polymer. The type of batteries used in segway are rechargeable batteries which are more costlier than the disposable batteries but by the usage of this batteries the cost of ownership and environmental impact will be low. Recharging requires electricity while disposable batteries require thye change of whole battery [18].

Table 2 Specifications of Battery

PARAMETER	VALUE
Type	lead-acid
Normal Capacity	7.6 Ah
Voltage Nominal	12V

3.4 Rocker Switches

Rocker switch is a type of switch that breaks and connects the electrical circuit by" Rocking" in one direction to break the circuit, and the other direction is connected to the circuit.

These rocker switches are placed on the suspension handle the main function of these rocker switches is for the steering mechanism.

The power is given to the mechanical Segway for its movement only when the switch is pressed [18].

Fig. 3 Rocker Switches [17]

3.5 Chain and Sprockets

The power generated from the electric motors,car engines and wind generators is by the rotary Motion of the drive shaft. The commonly used method for the transmission of of motion and force from the output drive shaft in a mechanism is by the means of a component called as SPROCKET.

Fig. 4 Energy Transmission Process

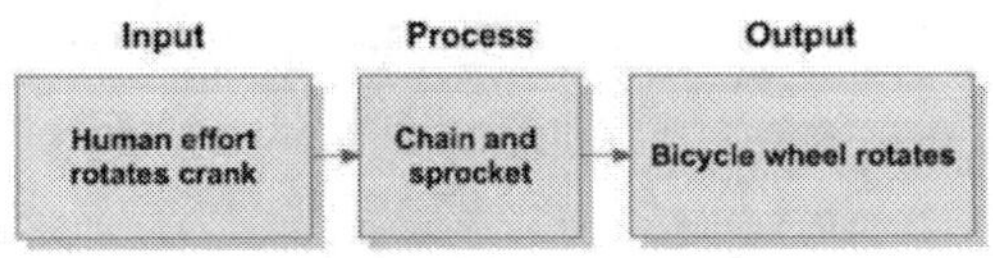

A sprocket is generally a toothed wheel used in the bicycles and motorcycles to transmit the motion from one shaft to another whereas the chains used along with it are used to transmit the motion and force from one sprocket to another one such chains are called as power transmission chains.

Fig. 5 Chain and Sprocket [18]

The mechanism that is used in the bicycle for the movement is used in mechanism with the help of Chain and Sprocket system.

Fig. 6 Chain and Sprocket drive on a bicycle

4. *Fabrication of Mechanical Segway*

4.1 Making the basic design for the Segway

First, we make a basic design for the Segway, as our requirement. This design is made as comfortable for the transporter to run the Segway. This helps the passenger to move freely at his own place. This Segway is not self-balancing so that there are two guiding wheels at front and back for the support. This device consists of switches attached to handle to be controlled by the personal transporter. This basic model is made by using CATIA software.

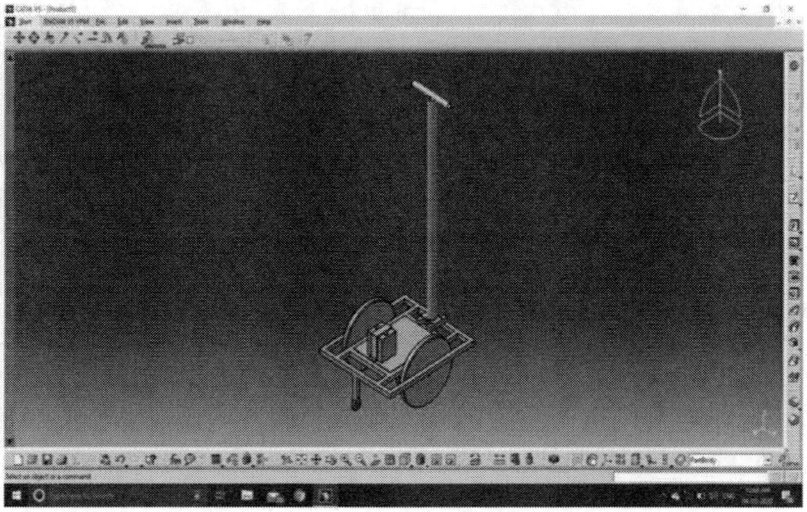

Fig. 7 Basic model of mechanical segway

In the above chapter we selected the materials foe the components required to male this project. Now we make the components ready by using some engineering techniques for the step by step assembly. First, we prepare the frame initially in which we have to attach the motors to it. This attachment is done by arc welding which we have discussed earlier and surface grinding too to get a good finish of the metal. Frame is 120 centimeters. The arc welding process and the surface grinding process is carried out accordingly. Now the motors of the 12volts DC motors are taken along with the chain and sprocket setup. Then these motors are fixed by used the arc welding technique, which is also used to attach the frame of the mechanical Segway.

Fig. 8 Fixing Motors and Base Plate

One motor is fixed in the left is established on the right direction and the other motor is fixed to the frame in the reversed direction as the turning requirements. These motors are checked thoroughly before fixing them to the frame and they must be in the center of the shaft as the chain must be fixed. The sprocket must be at the center of the shaft as the center is taken between the tire and the frame where the bearings are fixed. Now the chain and sprockets are taken according to the shaft size. These sprockets are fixed to the shaft at the time when we attach tires and bearings itself, this helps us to not to work hard, so that we have to remove again the whole frame setup to fix this. Then the chain must be fixed to the motor output and to

the sprocket according to the bits and according to the length required. We have taken 51 bits of chain to attach one side of the tire and motor which is sprocket and the motor output. These bits which are the ends ones must be fixed by the clevis pin and link plates of the chain. For this model we require a suspension so that the passenger can move easily and in case if he loose balance, he can rely on the handle which is a suspension handle and gets compressed at the bottom of the handle. Suspension handle is likely to be the whole handle of a bike. This handle is fixed by the bolts at the top and at the bottom it is welded to the frame. We can't just weld the suspension handle to the frame so that first we weld a nut and bolt slot so that the suspension handle can be fixed further and have a little movement front and back. This come onto action when the person on the mechanical Segway lose the balance, he can rely on it. These wheels are used to give support for the mechanical Segway while driving it. These wheels are fixed one at the front end and he back end of the frame so that middle silicone tires are used make front and back movements and also the turning movement. These small wheels which are attached at the front and back end are used to get the required turn additionally helping the main tires, as a result we get the turn. This works as if we want to take a left turn then we must hold the right rocker switch one for forward movement and the left rocker switch in the backward action. Similarly, if we want to take a right direction, we must hold the right rocker switch one for backward movement and the left rocker switch in the forward action.
Apart from the battery cables the connecting wires are used to connect the motor and the rocker switches in order to move the mechanical Segway. These connecting wires are used to give the electric connection from the motors, batteries and the rocker switches. The circuit connection of the mechanical Segway is as below.

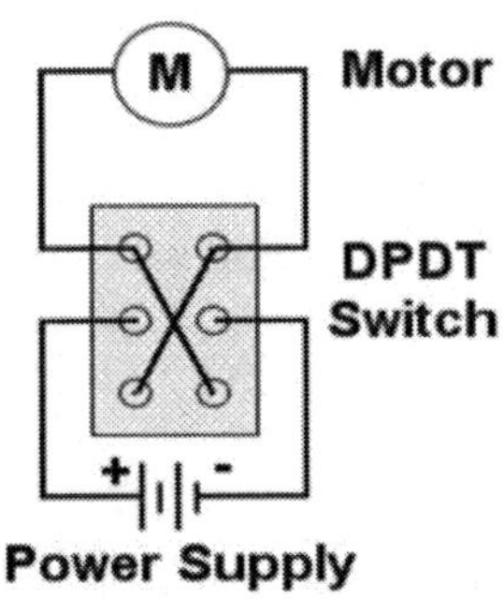

Fig. 9 Connection of Motors, Batteries with Rocker Switches [15]

when testing the mechanical segway in the initial trails the operator advantages were naturally lower as a sign of safety. This trail was made to check the system performance and its efficiency also the steering mechanism which is controlled by the rocker switches was also tested. Which worked well in taking left and right turns along with this the expected speed was attained and the safety of the operator was also as expected.

Table 3 Weights of Individual Components

Components	Quantity	Weight(Kg)
Frame	1	3.5
Tires	2	5
Motors	2	2.4
Batteries	2	2.3
Pillow block bearings	2	2
Chain and sprockets	2	1.5
Shaft	1	2
Suspension Handle	1	5
Guiding wheels	2	0.5
Other small components		1
Total		25.2

IOP Conf. Series: Materials Science and Engineering 998 (2020) 012039 doi:10.1088/1757-899X/998/1/012039

5. *Basic Modeling in CATIA V5*

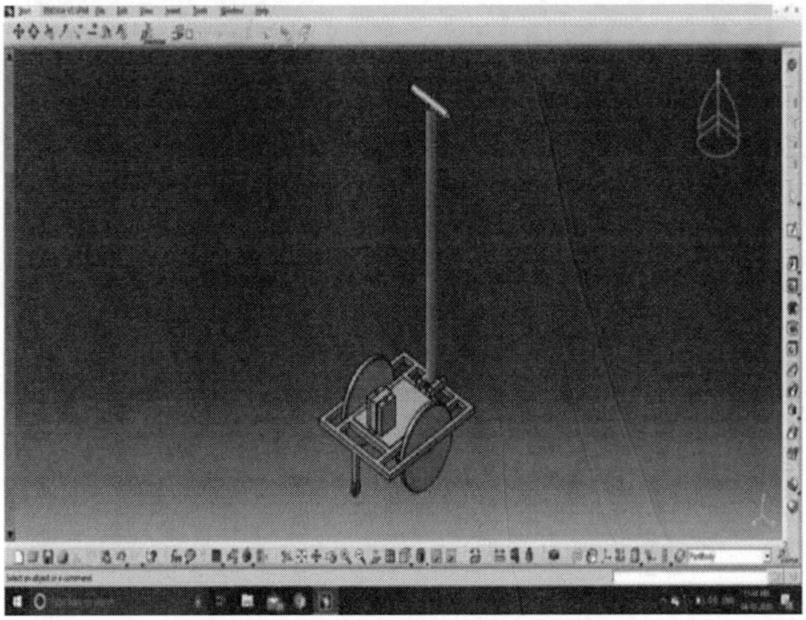

Fig. 10 Basic Design for Making the Mechanical Segway

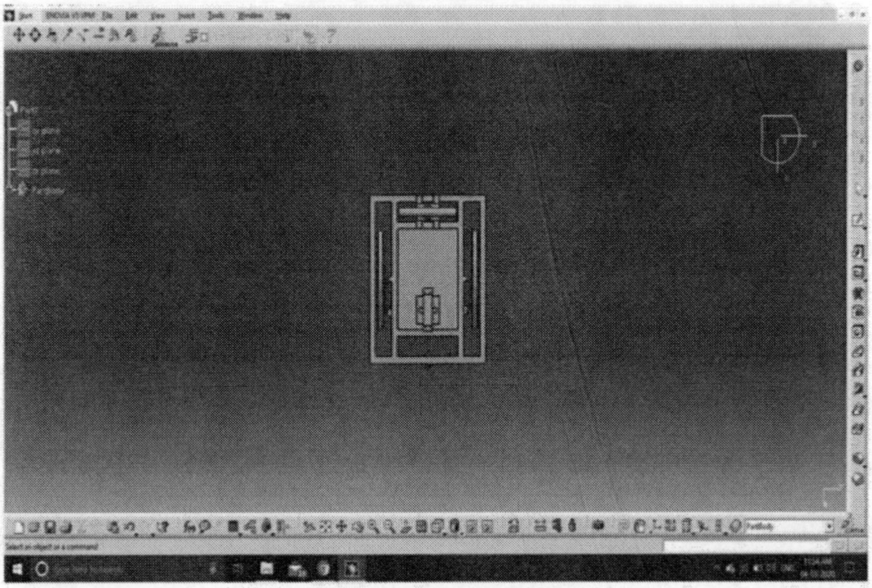

Fig. 11 Top view of Segway

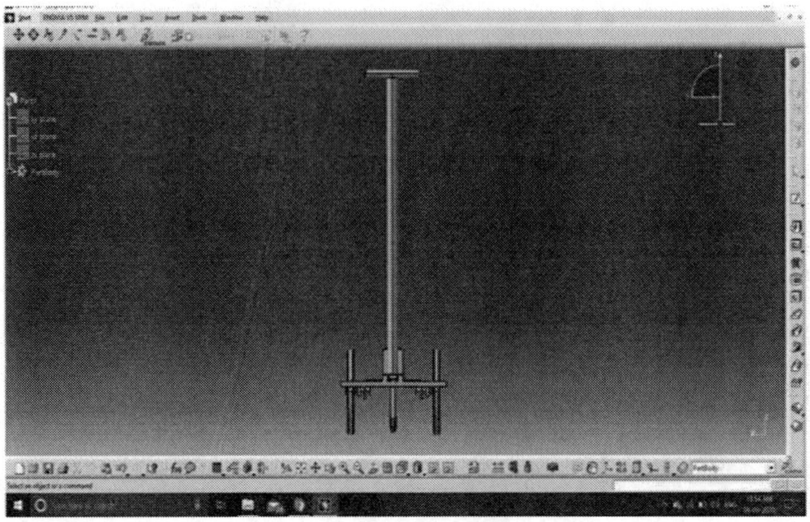

Fig. 12 Front view of the Segway

6. *Design Calculations*

Design Factors for the proposed Segway

Load Capacity = 90Kgs.
Base Width = 22"
Base Length = 26"
Mast Length = 32"
Types of motion = Forward, Backward, Left, Right
Travel Speed = 5-7 Km/h
Power Supply = 12V, 7 AH Lead-Acid Battery

Materials Used for Each part

Body of the Segway = Iron
Mast: Suspension Handle = steel
Wheels = Silicone -Eco-friendly

Power of Driving Motor

P = 2πNT/60000
Where, P = Power of Motor (KW), N = Speed (in rpm), T = Torque (Nm), T = W x R
Formula for power = 2(3.14) N*T/6000
Speed of the Motor = 30 RPM
Max Torque Produced, T = 35 kg-cm.
Therefore, the power of the motor, P = 2(3.14) (30) (35)/600 => P = 1.099KW = 1099 WATTS

Selection of the Sprockets

Number of Sprocket Teeth n = 15
Chain Pitch = 0.490m
Sprocket diameter = 40mm

Stress acting on the Base

Weight of the user = 100 kg (max)
Weight of the personal transporter = 25.2 kg (approx.)
Total weight = Weight of the user weight of the personal transporter = 125.2 kg.
Loads acting = Total weight*Gravity g=10m/sec
Thus, load acting = w*g = 125.2*10 = 1252N. P= N Area of the base, A=660.4*558.8=369032.52 mm^2.

Yield Stress

Maximum weight on the base = 125.2 kgs
Yield stress = maximum weight*gravity/area of base
Yield stress = 125.2*10/369032.52 = 0.0033928= 3.34*10 -3 n/mm^2.

Factor of Safety

FOS = ultimate stress/allowable stress

FOS = 1252/1000 = 1.252
FOS = 1.25

7. *Results and Discussions*

Firstly, it is balancing the mechanical segway without the usage of any type of sensors and programming. This is observed by some initial trails made and as per the initial driving results and why the vehicle was unable to get the expected result so we made a lot of advancements in terms of loads and motor capacity and shaft and bearing which helped us to get desirable results we expected for. The results obtained are tabulated in table 4 and table 5.

Table 4 Driving Results

Observations	Speed
Speed expected	9-10 km/h
Speed obtained	6-7 km/h

Table 5 Results observed

Observations	Result
Stopping time	1.1 seconds
Stopping distance	0.3 ft.

Initial Driving Results:

The initial driving attempts were made to run the segway by controller and resulted in positive way. This test was made to check the mechanical segway operated by the rocker switches to move forward and backward as required. After many tests and changes, the operating planning with the remaining error serving as a origin of speed, turned out to be as successful and the speed controlled by the rocker switch is very good for the forward and backward movement. Also the changes are made by replacing capacity and wires brought very good results overall.

Conclusions

In this project a mechanical Segway is modeled and fabricated which is cost effective and innovative model as compared to the existing segway available in the market.
The following conclusions can be drawn from to the present study.
1. The designed model of mechanical Segway is thoroughly tested in all the aspects and proved to be effective.
2. The designed mechanical Segway can be used for a full extent for personal transportation use and safe to use for travelling at low speeds.
3. The model which is designed is cost effective compared to the model which is available in the market.
4. The study also reduced the e-waste by eliminating few electronic components.
5. Much advancement can be made in future to this model to make it more effective by using sensors and servos to move and turn the vehicle smoothly.
6. Also the load carrying capacity of the vehicle can be changed by using heavy motors and reduce vehicle weight by using light weight durable materials for body.

References

[1] S. Betgé-Brezetz, et al, "Uncertain map making in natural environments," *Proc. IEEE International Conference on Robotics and Automation*, 1996, pp. 1048-1053.
[2] D. S. Apostolopoulos, et al, "Technology and field demonstration of robotic search for Antarctic meteorites," *International Journal of Robotics Research*, vol. 19, no. 11, pp. 1015-1032, 2000.

[3] R. Simmons, et al, "Experience with rover navigation for lunar-like terrains," *Proc. IEEE/RSJ International Conference on intelligent robots and systems*, 1995, PP.441-446.
[4] H. Seraji and A. Howard, "Behavior-based robot navigation on challenging terrain: a fuzzy logic approach," *IEEE Transactions on Robotics and Automation*, vol. 18, no. 3, 2002.
[5] I. S. Kweon and Takeo Kanade, "High-resolution terrain map from multiple sensor data," *IEEE Transactions on Pattern Analysis and Machine Intelligence*, vol. 14, no. 2, pp. 278292, 1992.
[6] E. Krotkov and R. Hoffman, "Terrain mapping for a walking planetary rover," *IEEE Transactions on Robotics and Automation*, vol. 10, no. 6, pp. 728-738, 1994.
[7] C. M. Shoemaker and J. A. Bornstein, "The demo III UGV program: a testbed for autonomous navigation research," *Proc. IEEE ISIS/CIRA/ISAS Joint Conference*, 1998, pp. 644-651.
[8] C. Ye and J. Borenstein, "A new terrain mapping method for mobile robots obstacle negotiation," *Proc. the UGV Technology Conference at the 2003 SPIE AeroSense Symposium*, Orlando, FL, April 21-25, 2003.
[9] C. Ye and J. Borenstein, "A novel filter for terrain mapping with laser rangefinders," conditionally accepted for publication in the *IEEE Transactions on Robotics and Automation*, 2003.
[10] D. B. Gennery, "Traversability analysis and path planning for a planetary rover," *Autonomous Robots*, vol. 6, no. 2, pp. 131146, 1999.
[11] Meyer, Caroline, "Rolling right along. Scooter meets few obstacles but safety issues grow", *International Journal of Engineering Science and Computing,* Vol. 6, No. 5, 2016. DOI 10.4010/2016.1233
[12] Brian G.R. Hughes, "The Unique Physics of the Segway PT Balanced at All Times", May 30,2009.
[13] M Thompson, J.Beula Julietta Mary,"Design and fabrication of failsafe segway," *International Journal of Mechanical and Industrial Technology*, vol. 2, no. 1, pp. 767-782, April 2014.
[14] Brett Browning, Jeremy Searock, Paul E. Rybski and Manuela Veloso" Turning Segways into soccer robots" *Industrial Robot: An International Journal*, Vol. 32, No. 2, 2005, pp.149–156.
[15] C. Ye and D. W. Wang, "A novel navigation method for autonomous mobile robots," *Journal of Intelligent and Robotic Systems*, vol. 32, no. 4, pp. 361-388, 2001.
[16] S. Singh, et al, "Recent progress in local and global traversability for planetary rovers," *Proc. IEEE International Conference on Robotics and Automation*, 2000, pp. 1194-1200.
[17] S. Lacroix, et al, "Autonomous Rover Navigation on Unknown Terrains: Functions and Integration," *International Journal of Robotics Research*, vol. 21, no. 10-11, pp. 917-942, 2002.
[18] H. Haddad, et al, "Reactive navigation in outdoor environments using potential fields," *Proc. IEEE International Conference on Robotics and Automation*, 1998, pp. 1232-1237.
[19] Y. Koren and J. Borenstein, "Potential Field Methods and Their Inherent Limitations for Mobile Robot Navigation," *Proc. IEEE International Conference on Robotics and Automation*, 1991, pp. 1398-1404

IOP Conf. Series: Materials Science and Engineering 998 (2020) 012040 doi:10.1088/1757-899X/998/1/012040

Experimental validation of fatigue life for thread rolling form tool material

GVR Seshagiri Rao[1], VVSH Prasad[2],M Sunil kumar [3]and VKVS Krishnam Raju [4]

[1] Associate Professor, Mechanical Engg.Dept, IARE, Hyderabad, T.S, India

[2] Professor, Mechanical Engg.Dept, IARE, Hyderabad, T.S, India

[3] Assisant Professor, Mechanical Engg.Dept, IARE, Hyderabad, T.S, India

[4] Associate Professor Mechanical Engg.Dept, IARE, Hyderabad, T.S, India

E-mail:lakshmiseshu29@gmail.com,visuvaranasi@gmail.com,snl.kmr2@gmail.com
krish.bobj@gmail.com

Abstract. A mechanical machine needs to be considering the effect of cyclic stress to avoid failures, the machine considered in this work is rotating beam fatigue tester to know the failure points. This machine present the fatigue strength values of different materials when it is subjected to repeated loads at different stress values. To establish the fatigue characteristics, various materials are tested with initial test, at a stress value that is under the material's ultimate strength. Machine components are subjected to repeated varying forces or fluctuating loads of definite magnitude and stress cycles are to be measured. In this paper it has been analyse the performance EN8 and D2 steel. These materials are used for general applications of thread rolling process which experiences weariness stacking. Further, concentrates on fatigue testing machine to improve strength quality is shown.In the same way the method is repeated with the different values of load on EN8 and D2 steel specimens and the results are plotted and to represented on S-N diagram.

1. Introduction

Rotating beam fatigue tester is utilized to find out the fatigue strength or fatigue life of a metal. 90% of the failures occur due to fatigue in a common industry. During nineteenth century, it was viewed that a fatigue fracture did not have any visible permanent deformation, this leads to a designing issue. A significant breakthrough in understanding of fatigue failure process occurred in the twentieth century with the assistance instruments like a computer, microscope by which it was not considered as a designing issue. The thought behind this work isn't to give answers to the unanswered queries yet in addition to solve the problem from alternate point of view. This incorporates describing the whimsical idea of fatigue failure by performing tests on various examples and clarifying the known methodology of fatigue test. Basically fatigue failures are expensive due to this we perform fatigue analysis because a component failing in prototype phase is more expensive than in design phase. The component failing in service can cause a launch delays, warranty claims, recall and legal liability. In the study of materials fatigue, is the phenomenon of reducing the strength of a material occurred by load application i.e. cyclic load. It is dynamic, capricious and basic harm which happens when the specimen is subjected to cyclic stacking. Compared to yield strength or ultimate strength maximum stress of the material is less. If the applied loads are greater than a certain limit value, then microscopic cracks are occurred by applying the load up to higher limit. Due to this a crack is formed and it will expand on total structure until it is failed basically Fatigue failure occurs due to formation and propagation of cracks. It is a three stage process.

IOP Conf. Series: Materials Science and Engineering 998 (2020) 012040 doi:10.1088/1757-899X/998/1/012040

American society for testing and materials (ASTM) Committee, Murakami [3] reported that in 19th century research in the field of fatigue began and is reported in number of a range of methods for fatigue computation. It suggest the number of cycles until failure N to the cyclic loading S. The components may also fail due to low level of stress in some cases.

Gracia et al. [4] has worked on structures which are performed by using hydraulic testing machines. These are capable of applying large cyclic loads up to 100KN. The fatigue life of a structure is the number of cycles taken to break the structure. It is used for creating stress strain life curve. Wilhelm August Julius Albert [2] A German mining administrator calculated and reported on breakdown of excavation elevator cables ensuing from repetitive loadings, this is known of metal fatigue. In addition that he has developed a experiment apparatus for conveyor chains used in the mine. M.N. James and G. kang [1] studied the air worthiness Standards generally require a fatigue check performed for big aircraft previous to certification to decide their secure life. VVSH Prasad et.al (5) an attempt is being made to study the forming forces in thro feed of the component at roll as shown in figure 3.0 and work intersection and to determine the optical torque requirement using ANSYS software. To generalize the thread rolling process, software code is developed which can be used for different Work –Tool combinations to estimate the torque requirements

2. Materials

Assurance of failing attributes is basically through the stress versus no. of cycles to failure curve, which is commonly known as S-N curve. The fatigue test was directed according to IS 5075: 1975. The samples for the test were machined on a CNC lathe machine from EN8 and D2 steel bars to the shape as appeared in Figure 1.

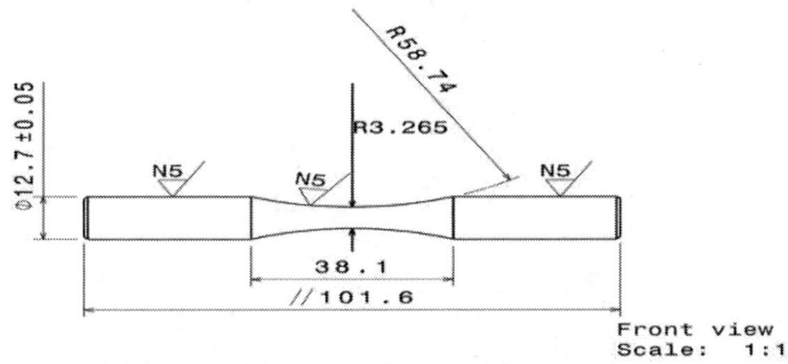

Figure 1. Test Specimen

Table 1 give an idea about the composition of the EN 8 material and Table 2 the mechanical properties of EN8.similiarly Table 3 and Table 4 show the Composition of D2 steel, Mechanical properties of D2 steel correspondingly.

Table 1. Chemical composition of EN31 Material

Element	**Min**	**Max**
Carbon	0.35	0.45%
Manganese	0.60	1.0%
Silicon	0.5	0.35%
Sulphur	-	0.6
Phosphorous	-	0.6

Table 2. Mechanical properties of EN8

Maximum Stress	700 to 850 N/mm^2
Yield Stress	465 N/mm^2
0.2% Proof Stress	450 N/mm^2
Elongation	16%
Impact KCV	28 Joules
Hardness	201 to 255 Brinell

Table.3 Chemical composition of D2 Steel

Element	**Min**	**Max**
Carbon	1.4	1.6
Manganese	-	0.6
Silicon	-	0.6
Sulphur	-	0.03
Phosphorous	-	0.03
Chromium	11	13
Molybdenum	-	0.6
Iron		Remaining

IOP Conf. Series: Materials Science and Engineering 998 (2020) 012040 doi:10.1088/1757-899X/998/1/012040

Table. 4. Mechanical properties of D2 Steel

Hardness, knoop (converted from Rockwell C hardness)	769
Hardness, Vickers	748
Izod impact unnotched	77.0J
Poisson's ratio	0.27 to 0.30
Elastic Modulus	190 a to 210 Gpa

3. Fatigue testing machine

Generally the fatigue test is to decide the life expectancy that might be normal from a material exposed to cyclic stacking. The fatigue life of material is the absolute number of cycles that a material can be exposed to a solitary stacking condition. To carry out fatigue test, a sample is loaded into a fatigue testing machine. This cycle of loading and unloading is then repeated until the end of the test is reached. The test may run to a predetermined number of cycles or until the sample has failed depending on the parameters of the test. The type of fatigue testing machine using is Rotating beam type. The specimen function is as a single beam symmetrically loaded at two points. Fatigue testing machine details and rolling process as shown in fig.2 and fig.3 respectively.

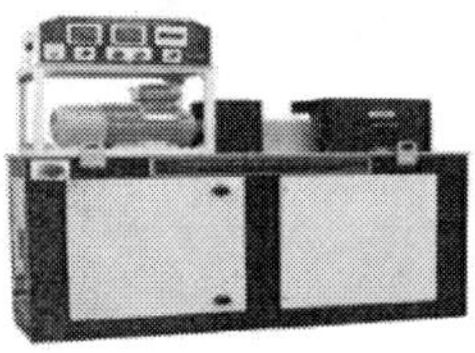

Figure 2. Fatigue testing machine **Figure 3.**Rolling

4. Mechanical and Test Specifications

Specimen material and size: Non heat treated ∅12.7 and 101.6 mm long
Collect size: ∅ 13mm
Load: 50 to 1000N (load applied with dead weights)
Loading ratio: 1:5
Specimen speed: Min 1000 rpm, Max 5000 rpm
Bending moment: 3 to 55Nm
Bending Stress: 110 to 2190Mpa
Setting block length: 40mm
Centre of profile: 19.05mm
Bending moment distance: Bending distance = loading distance + middle of profile centre
Loading distance = Loading point + Housing Covering plate + Collect length (9.9+9+17) = 35.95mm
Bending distance = 35.95+19.05 i.e. 55mm
Counters (counts): 1 to 999999

IOP Conf. Series: Materials Science and Engineering 998 (2020) 012040 doi:10.1088/1757-899X/998/1/012040

5. Procedure

The testing sample is held on machine toward one side to give rotational movement though the opposite end is connected to a heading and furthermore exposed to a heap or stress. At the point when the sample stacked between the grasps is turned about longitudinal pivot, the upper and lower portions of the sample measure length are exposed to ductile and compressive stresses. Accordingly, stress varies at the same time anytime in sample surface. The test continues until sample failure happens. The upset counter is utilized to acquire the quantity of cycles to failure relating to the stress applied. Increasing the load applied to the sample brings about a decrease in number of cycles to failure. It is essential to preparation of the samples as per the IS standard geometry. This includes cylindrical grinding up to surface roughness of 0.4 microns RA. That specimen is subjected to fatigue loading under normal ambient conditions from 5 to 1000N varying from up to 5000rpm cycles. The fatigue strength of the specimen is experimentally determining the S-N (Endurance strength-No. Of cycles) plots are drawn. Similar procedure has been adopted and subjected to thermal boundaries varying between 5000-9000c. And S-N curves are plotted and comparative studies are carried out for different materials EN8 and D2 steel. Table 5 and Table 6 shows the fatigue life calculations of EN8 and D2 steel materials.

6. Calculations

Bending Moment, M = Load applied x Bending distance Bending Stress, $\sigma b = 32M/\pi d3$
Endurance Stress, $\sigma e = Ka.Kb.Kc.\sigma e1$
Where Ka = a (σut) b, Surface Finish Factor
Kb = 1.24d-0.107, Size Factor
Kc = Reliability Factor
Log10N-3 = EF
Where EF = Fatigue strength corresponding to the N cycles

EN8

Table.5 Fatigue life Calculations of EN8

Load in N	Speed in RPM	Bending Moment in Nm	Cycles
50	1000	2.75	7.61
100	1000	5.5	12336
200	1000	11	25515
250	1000	13.75	37436

D2 STEEL

Table 6. Fatigue life of Calculations of D2 Steel

Load in N	Speed in RPM	Bending Moment in Nm	Cycles
50	1000	2.75	23555
100	1000	5.5	30394
200	1000	11	42766
250	1000	13.75	51368

IOP Conf. Series: Materials Science and Engineering 998 (2020) 012040 doi:10.1088/1757-899X/998/1/012040

7. Results and discussion:

The S-N curvature is utilized to plot the extent of a rotating stress and the number of cycles to failure for a given case material. Both stress and the quantity of cycles are shown on logarithmic scales. The exhaustion qualities of a given material are taken by its S-N graph, a case of which is appeared in fig. 7. The representation of fatigue test in which an example of material is trying to failure by applying a alternate stress of extent S comprises of spots. It is rehashed for various estimations of S and number of cycles to failure then N is plotted against S. For various materials the plot of log S against log N is shown to in straight line.

S-N Curve

S-N curvature is drawn with logarithm of number of cycles for failure by the side of x axis and yielding stress by the side of y axis. The yielding stress was calculated by means of the equation = $L \times P \times 32/\pi \times d^3$

Where, L = Distance from predetermined end of the specimen to specimen's contact point with load arrangement (mm)

d = Diameter at the narrow measurement of the sample (mm)

P = Applied load (N)

The magnitude of bending stresses (in this case alternating fully reverse alternating bending stresses are concerned therefore bending stress is indicated on y-axis different stress in axial loaded fatigue testing) and the equivalent fatigue life N (log10) i.e. on a logarithmic scale for which the graph is plotted and shown in figure 4.

EN8/C40

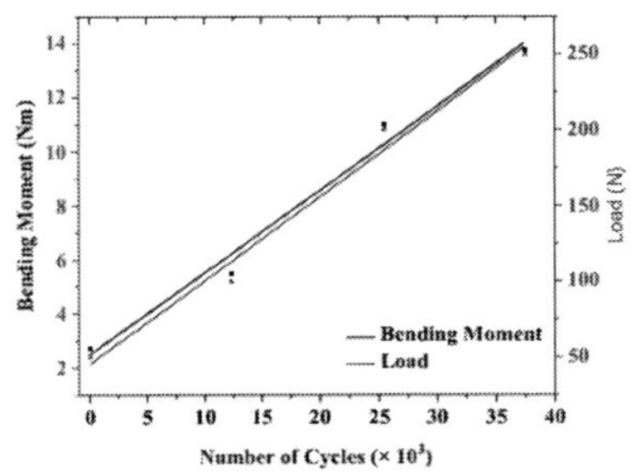

Figure 4. Bending moment and load curves of EN8 steel specimen.

D2 Steel

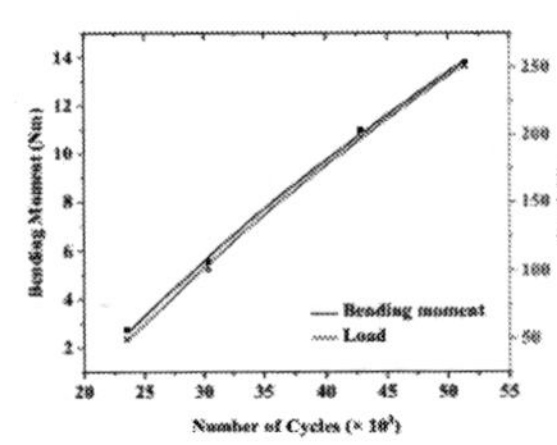

Figure 5. Bending moment and load curves of D2 steel specimen

Comparison

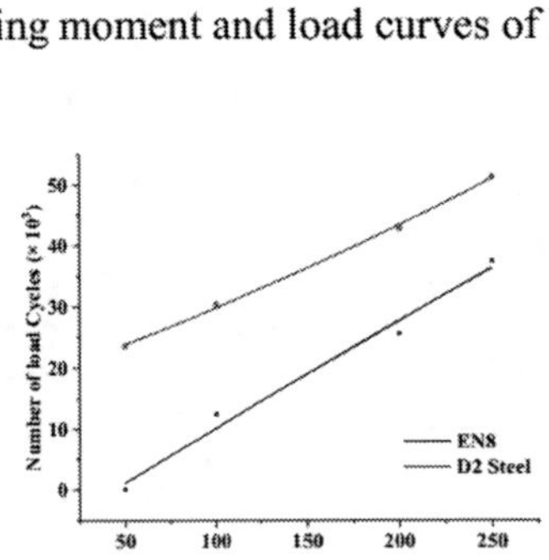

Figure 6. Curves of EN8 and D2Steel specimens

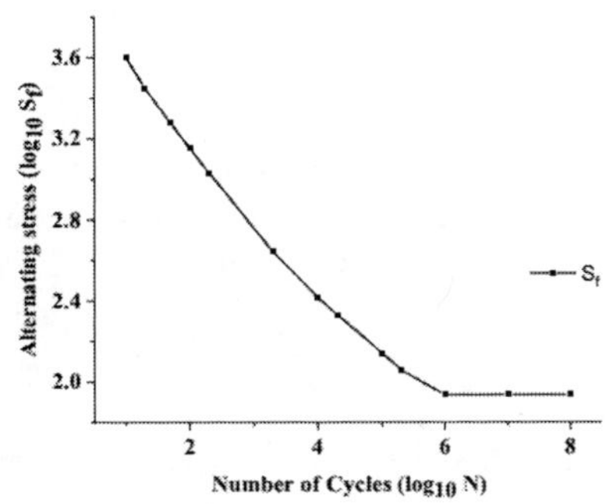

Figure 7. S-N Curve of D2Steel specimen

Figure 4,5 and figure 6 shows the bending moment and load curve of the two circumstances. One clear observable fact that can be experiential during all the curves is that, whereas the stress increase, the number of cycles for rupture reduces. Despite the fact that it very well may be notice that for the stress esteems the examples withstand higher the number of cycles before fracture. Furthermore, the D2 steel test withstands more occasions the number of cycles before fracture when contrasted with the EN8 example. Regardless of whether we think about a plan for a fixed number of cycles, state 105 cycles, while EN8 tests can withstand of 512 N/mm2, the D2 steel tests can withstand 744.3 N/mm2. Improving the fatigue life of D2 steel as it displays a higher failing life when contrasted with different cases. The results are plotted from reports with the observation of S-N curve as shown in Figure 6. During fatigue test we observed the stages of failure over the specimen materials and obtained the fatigue life by number of cycles. The components which are subjected to load changes with time can undergo failure at stresses below the material's maximum strength. This is referred to as Fatigue Failure. It accounts big majority of mechanical test.

8. Conclusions

- D2 steel specimen has considerably advanced UTS than that of the specimen of EN8 steel.
- Although D2 steel specimen endure advanced fatigue cycles prior to failure when compare to EN8 steel, the sample has extra fatigue existence than other specimen as lower bending stresses are concerned.
- At higher levels of stresses D2 steel specimen have two times the life of EN8 specimens.
- As D2 steel, steels show a advanced fatigue existence compare in the direction of others and have a enhanced extent for application concerning fatigue load
- The interest in studying and analysing of EN8 and D2 steel materials fatigue life have made an impact on us.
- For designing and dimensioning, the predictable load is compared with the capability to resist loads (fatigue strength). It means that the theoretical calculations and expressions were understood earlier than the realistic approach.
- The equipment used in this project for the fatigue characteristics of various metals is limited to standard shapes for above mentioned report.
- In future this obstacle can be overcome by introducing various shapes of metals used, which results in greater favourable results for withstanding the fatigue loads.

IOP Conf. Series: Materials Science and Engineering 998 (2020) 012040 doi:10.1088/1757-899X/998/1/012040

9. Acknowledgements

We proudly thank to Institute of Aeronautical Engineering, Hyderabad for permitting us to present this work. This Research was supported by the ministry of science and technology, Department of science & technology R & D Division, under the FIST program, project no: SR/FIST/College-2017/28(G).

10. References

1. M.N. James, G.kang Jiaotong Professor University of Plymouth, and UK Professor Southwest University school of mechanics and Engineering, China.
2. C. Barthias "There is no infinite fatigue life in metallic materials", Fatigue fract. Eng. Mater. Struct, Vol.22, no.7, pp.559-565, July 1999
3. Y Murakami, metal fatigue: effects of small defects and non metallic inclusions, 1st edition oxford, UK: Elsevier, 2002.
4. I.M. Garcia, D.G. Montiel, and C.Barthias, "Fatigue life assessment of high strength low alloy steel at high frequency", The Arabian J.Sci.Eng, Vol.33, pp.237-247, 2008.
5. VVSH Prasad, C.Labesh Kumar, Dr BDY Sunil, Dr.G.V.R.Seshagiri Rao "Determination of Optimal Torque in Thread Rolling Operation" International Journal of Engineering Trends and Technology (IJETT) – Volume 68 Issue 5 - May 2020

Modelling and Fabrication of Engine Valve To Improve Performance of An I.C Engine

P Ramakrishna Reddy[1,2], Pendam Shivanath [2] , Vittala Shandilya[2], Acha Srikanth [2], B. Balakrishna[3]

1,Research Scholar, Department of Mechanical Engineering, University college of Engineering, JNTUK, Kakinada, India

2,Department of Mechanical Engineering, Vardhaman College of Engineering, Shamshabad, Telangana, India

3 Department of Mechanical Engineering, University college of Engineering,JNTUK, Kakinada, India

E-mail:[1]er.p.ramakrishna@vardhaman.org, [3]balakrishnajntu06@gmail.com

Abstract: Modelling and fabrication of engine valves are made to increase the turbulence process by swirl. By proving swirl process to increase fast combustion phenomenon and enhance the efficiency of an engine. Engine should work at minimum speeds in order to have decrease the mechanical losses and good combustion enabling improves the combustion efficiency and mechanical efficiency (1). For this process be used to improves brake thermal efficiency and increase the usage of air-fuel mixture in combustion process which less impact pollution release to environment. The main importance of this work is to evaluate the performance of an Engine and to improve it by changing the valve design. Manufacturing of intake valves made to do an experimental work. this being done after modelling of valve and fabrication and then experimental work on test engine to compare two valves. Even if we adapt the this concept and also additional concentration required on inlet manifold.

Key Words: Turbulence, Air swirl, Mechanical efficiency, Thermal efficiency, Air-fuel ratio.

1. Introduction

The inside-cylinder flows of Internal Combustion Engine have drawn much importance to the automotive area. It is due to the fact that the flow structure developed by intake flows is related vey nearly to the design changes and enhancement engine performance parameters of the engines. The generation of turbulence of higher impact is one of the most crucial concepts for altering and stabilizing the ignition process, high movement of flame (2), mainly in case of less air-fuel mixture burn combustion. In general, two types of process are utilized in order to develop and preserve the rotational flows efficiently. These concepts are usually known as swirl process and tumble flow process, which are produced the rotations in the horizontal and vertical plane direction of the engine cylinder, res. By this process of engine performance was improved.

By adaption of these two types of fluid motion in development of turbulence i.e. swirl and tumble, we are only main focus on swirl and how to improve swirl which are mainly depends on factors like changing inlet valve design etc. after checking the previous works we implemented an idea in valve design to improve swirl and for analyzing the amount of swirl produced, we main focus on a parameter called swirl ratio. As swirl ratio is more impact on to increase the turbulence (3, 4). Main theme here is to introduce a valve with 3 mask and 3 grooves in fluid flow. This valve all ways comes, over valve head attached to valve stem this is directly introduced to develop the turbulence at the entering stage of fluid into the chamber and hence produce high swirl.

Swirl is mainly used to supply the required amount of fuel-air mixing quantity. Swirl is generated with the support of a suitable inlet port and it is amplified at the end of the compression stroke by forcing the air towards the cylinder axis into the combustion chamber. The experimental work is carried out on a single cylinder four stroke diesel engine. A diesel generator is the combination of a diesel engine with an electric generator (often an alternator) to generate electrical energy.

There are models used in the experimental analysis and compared with each other they are
1.1. Conventional Valve:
The conventional inlet valve consisting of the face which is impact to very high temperature and pressures during the process of combustion. Valve is having a sensitive area called seating which should be very clear enough in dimensions and finishing so that accurate locking and sealing improve enhance the whole Engine Performances.

Fig 1. Conventional Valve

1.2. Modified Valve (3Mask & 3Groove):
Grooving is the process of removing a small amount of piece of metal on the valve head without disturbing the valve seating area; the small cavities are called grooves (5, 6). The small built-up pieces called MASKS. In this type of Valve consisting three Grooves in numbers with small pieces build on head and a three number of small built-up pieces.

Fig 2. Modified Valve

2. Modelling of the Engine valve:

The modelling of engine valve is done in Catia-V5 software without changing dimension (as per engine valve specifications). But here we are adding a small built- up pieces called MASK. Making Grooves by removing small amount of material in order to compensate the weight and valve timing.

The Specification addition features of engine valve are

Width of the mask: - 4mm
Thickness of the mask: - 45 degrees
Width of the groove: - 3mm
Depth of the groove: - 3mm
Outer diameter of the groove: - 22mm
Inner diameter of the groove: -8mm.

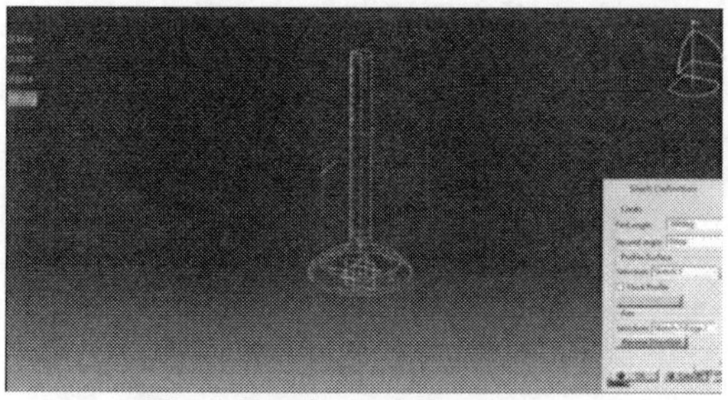

Fig 3. Preview of Sketch

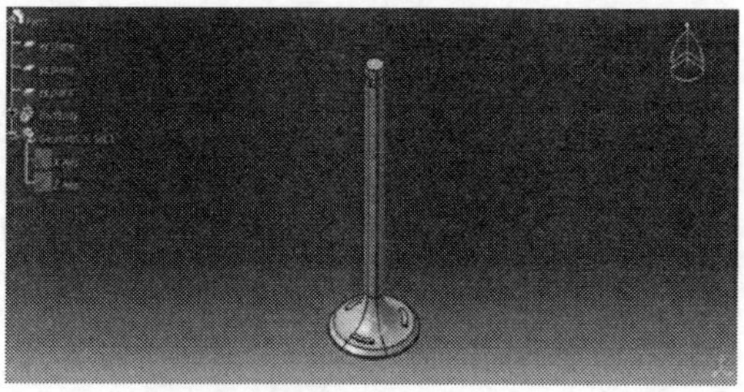

Fig 4. Engine Valve

3. Fabrication of Engine Valve:

In order to check and compare the performance of the engine, both conventional and modified valves must be used. And the modified valve needs to be separately manufactured in order to maintain the exact same dimensions and weight of the valve, so that the valve timing would not change by the effect of increase in weight. Care must be taken while producing the valve, so that the weight of the Conventional Valve and Modified Valve remain the same.

3.1. Production of Valve in CNC machine:

In this process a stainless-steel billet of 120* 40mm is taken and the dimensions of the conventional valve are fed to the CNC machine so that there is no sign of error can be made. As we feed the dimensions of the valve, the system takes over the machine and starts the operation. The tool used in this operation is called as high-speed steel (HSS). And the billet is rotated at approx. 1500rpm. As a result, the excess material is chipped off and the required shape of the valve is obtained.

3.2. Making Grooves to the Valve:

After the production of valve in CNC machine, grooves are made to the valves as per specified in the modelling. Adding grooves to the valve is done by electric discharge machining (EDM) in which copper electrode is used to cut the valve and make grooves to it. In this process spark is generated at the tip of the electrode which helps in cutting the valve accordingly. This concept is also called as Spark erosion process. The coolant which is used in this process is called spark EDM oil.

Fig 5. Manufacturing of Grooves in EDM machine

IOP Conf. Series: Materials Science and Engineering 998 (2020) 012041 doi:10.1088/1757-899X/998/1/012041

3.3. Manufacturing of Masks:

Masks are stainless steel structures which are initially made in circular shape on lathe machine and are made to cut on grinding machine(8). They are cut in such a way that each of the mask is of 120 degrees and placed in between the grooves of the valve. They are grinded through grinding process and welded to the valve. Each of the Mask is about 3mm thickness and placed at 120degree angle to the valve.

3.4. Welding of the Masks:

After manufacturing of the masks, they are welded on to the valve by a method of welding called brazing. In brazing brass is used as filler material to weld the two materials of different thickness. As the thickness of the masks is too low, and the thickness of the valve is too high, the masks cannot be welded by arc welding process, as it melts the masks(7). Therefore, the brazing process is used to weld the masks on to the valve. Brazing is done at 450degree Celsius temperature at which the filler material melts but the welding temperature still remains below the melting point of base metal. As a result, the welding is done perfectly on the valve.

Fig 6. Modified Valve

3.5. Buffing of the Valve:

Buffing is the finishing processes for smoothing a work piece's surface using an abrasive and a work wheel or a leather strop. Initially polishing refers to that use an abrasive that is glued to the work wheel, while buffing uses a loose abrasive applied to the work wheel. In order to improve the surface finish, Polishing is a more impact process while buffing is less impact on metal surface finish, which leads to a finer and good finish. For this valve buffing process is used in order to have a smooth and brighter finish of the valve.

4. Installation of Valve in the Engine:

After the buffing of the valves, the clearance of the valve is noted and checked in order to install it in the engine. Installation of the valve begins with the removal of rocker arm assembly followed by removal of engine head. After removing the engine head, old valves are replaced by the modified valves with masks on its head. After replacing the conventional valve with the modified valve, all the other components of the engine are reassembled in their original positions. The engine that is used for this experimentation is single cylinder 4-Stroke Diesel Engine, with 1500RPM and 5 BHP of horse power. The fuel used in this engine is H.S diesel.

IOP Conf. Series: Materials Science and Engineering 998 (2020) 012041 doi:10.1088/1757-899X/998/1/012041

Fig 7. Single Cylinder Diesel Engine

4.1. Experimentation:

In order to conduct the experiment, the valves should fit in the engine head perfectly. Then only the experimentation can be started. After the reassembly of the engine, the engine is started and the values of engine speed, time required for fuel consumption, voltage, current and manometer readings are taken. The values of the above are required to calculate the efficiency related engine, and Air-Fuel Ratio. For every 10ml drop of fuel, the time required for fuel consumption, voltage current, and manometer readings are taken(9,10). Initially to start the experiment, the engine should be started by rotating the Engine shaft. After starting the engine, by maintaining constant voltage of 220 volts, and by adding 5 consecutive loads, the values of manometer reading and the time required for drop of 10ml fuel is noted down for all of the 6 loads i.e. one no load conditions and 5 other consecutive loads.
Calculations are done based on the values that are obtained during the experiment and the below formulae are used to calculate the values of related efficiencies of an engine and Air-Fuel Ratio.

Fig 8. Experimental Setup

Observation Table:

Table1: Conventional Valve:

Load [N]	Engine speed [RPM]	Time for 10ml fuel consumption ion in [sec]	Generator Voltage [v]	Generator current [amps]	Manometer reading (h1) [mm]	Manometer reading (h2) [mm]	Manom eter reading h=(h1-h2) [mm]
No load	1630	71	220	0	66	73	139
1	1599	58	220	4.1	63	69	132
2	1586	47	220	8.2	62	67	129
3	1576	40	220	12.4	61	66	127
4	1556	34	220	16.5	58	64	122
5	1545	28	220	20.8	57	62	119

IOP Conf. Series: Materials Science and Engineering 998 (2020) 012041 doi:10.1088/1757-899X/998/1/012041

Table 2: Modified Valve:

Load (N)	Engine speed (RPM)	Time for 10ml fuel consumption ion in (sec)	Generator Voltage (v)	Generator current (amps)	Manometer reading (h1) (mm)	Manometer reading (h2) (mm)	Manometer reading h=(h1 -h2) (mm)
No load	1600	75	220	0	69	74	143
1	1590	62	220	4.1	66	73	139
2	1580	50	220	8.2	63	69	132
3	1569	42	220	12.4	62	67	129
4	1552	36	220	16.5	61	65	126
5	1542	29	220	20.8	60	64	124

5. Results and Discussion

Table 3: Conventional Valve:

LOAD [N]	B.P [kW]	B.S.F.C [kg/kWh]	B.M.E.P [bar]	I.P [kW]	Mech. Efficiency (%)	Brake Thermal Efficiency (%)	Volumetric Efficiency (%)	Air- Fuel Ratio
No load	0	-	0	3.75	0	0	83.69	53.77
1	1.002	0.514	1.36	4.752	21.08	15.91	83.14	42.80
2	2.004	0.317	2.742	5.754	34.826	25.80	82.87	34.29
3	3.031	0.2464	4.173	6.781	44.69	33.19	82.74	28.95
4	4.033	0.2178	5.625	7.783	51.81	37.55	82.14	24.12
5	5.084	0.2098	7.141	8.834	57.55	38.98	81.704	19.62

Table 4: Modified Valve:

LOAD (N)	B.P (kW)	B.S.F.C (kg/kWh)	B.M.E.P (Bar)	I.P (kW)	Mech. Efficiency (%)	Brake Thermal Efficiency (%)	Volumetric Efficiency (%)	Air-Fuel Ratio
No load	0	-	0	3.35	0	0	86.51	57.63
1	1.002	0.4808	1.3679	4.352	23.02	17.01	85.83	46.97
2	2.004	0.2981	2.7533	5.354	37.43	27.44	84.17	36.91
3	3.031	0.2346	4.1927	6.381	47.50	34.85	83.79	30.65
4	4.033	0.2057	5.6401	7.383	54.62	39.75	83.72	25.96
5	5.084	0.2026	7.1561	8.434	60.27	40.37	82.59	20.75

Graphs

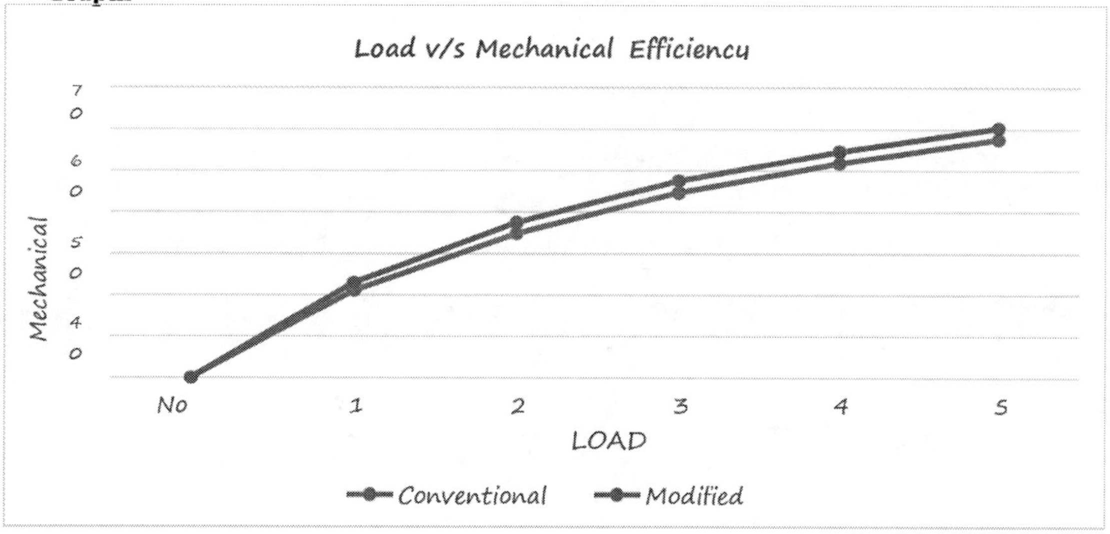

Fig 9. Mechanical Efficiency vs Load

Fig shows the variation of mechanical efficiency with load, as the load increases the mechanical efficacy increases and it decreases the nature at 5kw load.

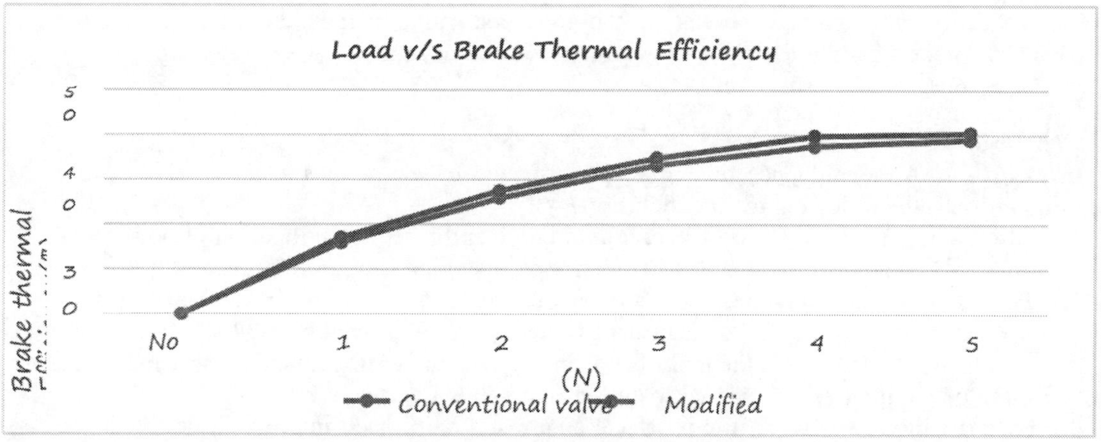

Fig 10. Brake Thermal Efficiency vs Load

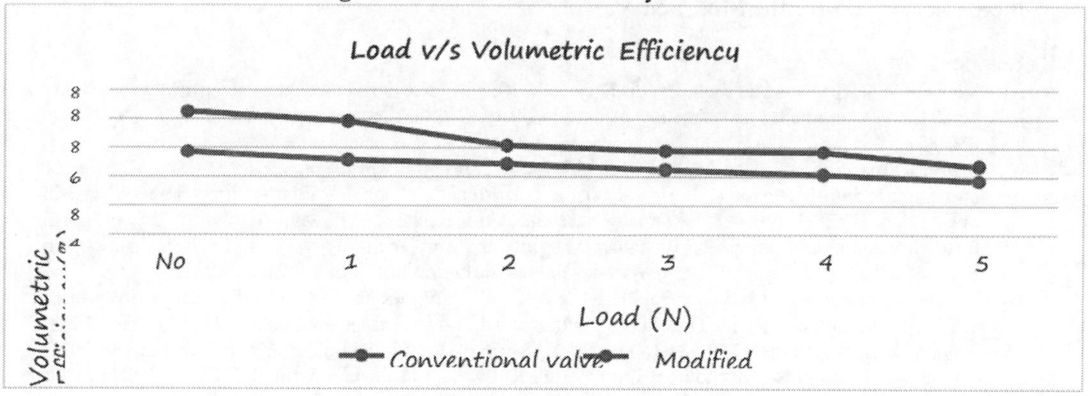

Fig 11. Volumetric Efficiency vs Load

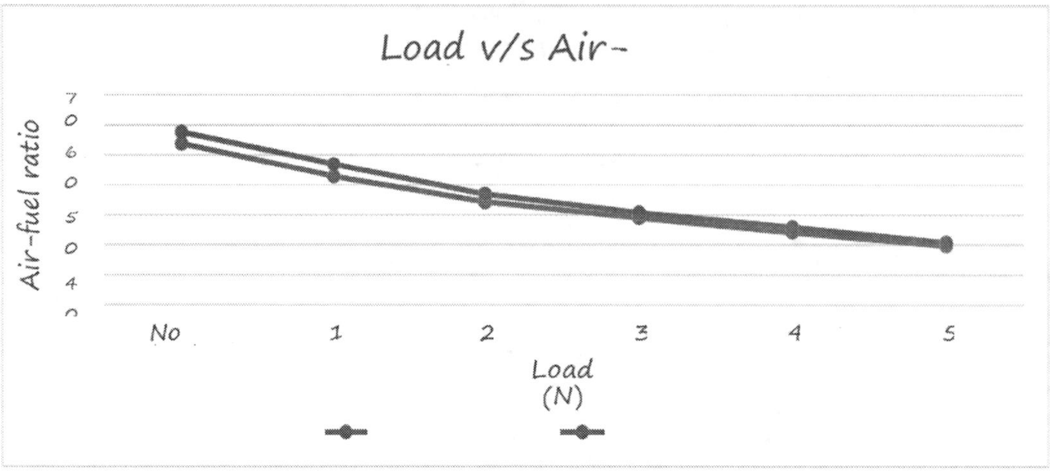

Fig 12. Air-Fuel ratio vs Load

As mentioned before, the main theme of this work is to improve the performance of an I.C engine by doing some changes in the valve design and dimensions, after drawing above graphs the expected result is achieved. The overall average values of conventional valve parameters like Mechanical Efficiency, Brake Thermal Efficiency, Volumetric Efficiency and Air-Fuel Ratio are 34.99%, 25.23%, 82.69%, 33.92 respectively.
Whereas the overall average values of the modified valve parameters like Mechanical Efficiency, Brake Thermal Efficiency, Volumetric Efficiency and Air-Fuel Ratio are 37.14%, 26.57%, 84.60%, 36.47 respectively.

Conclusions:-

Investigation of the influence of engine performance by changing the inlet valve profile configuration is the following main conclusions

1. Introducing of inlet valve profile configuration it influence the utilization of power was improved due to this Mechanical Efficiency of an engine increased by 2.15%,
2. The swirl concept improves by changing the area of profile due to this better mixing will takes place by this Brake Thermal Efficiency was improved by 1.34%,
3. Volumetric Efficiency is the main factor in engines due to this intake air was induction taken much improved compared to convention valve improved by 1.91%,
4. By changing the valve profile it effects on air fuel ratio much improved due to this the performance of an engine improves, compared to convention valve Air-Fuel Ratio is increased by 2.55 for the Modified Valve.

References

[1] Vamsidhar V, Srinivas Reddy k, Pandurangadu V 2018. International journal of scientific research and review, an experimental investigation of dimple inlet valve on diesel engine performance, Vol 7(6), p 830-837.

[2] Vinod Kumar Sharam, Man Mohan, Chandra Mouli 2017. IOP Conference Series Materials Science and Engineering, Effect of intake swirl on the performance of single cylinder direct injection diesel engine, Vol II (263), p 01-08. Dhanush Dineshan, Akhif Rasheed P E, Mihir Kulkarni, T. Venugopal 2017. International Journal of Mechanical Engineering and Technology (IJMET), Investigation On Swirl Motion Of CI Engines Through Intake Valve Configurations, Vol 8 (12), p 519-528.

[3] Pankaj N.Shrirao, Kapil B.Salve, Sachin S. Pente 2015. International Journal of Science and Research (IJSR), Swirl Induction with Dimpled Cylinder Head and its Effect on Exhaust Emission of Diesel Engine, Vol 4 (10), p 1359-1363. Dr. P. N. Shrirao, R. M. Sherekar and P. K. Bhoyar 2015. International Journal of Mechanical Engineering Research and development (IJMERD), Effect of air swirl with Dimple Cylinder head on performance characteristics of diesel engine, Vol 1 (1), p 01-07.

[4] Pavan Chandra P V, Siva Chaitanya A 2015. Int. Journal of Engineering Research and Applications

(IJERA), Design and Analysis of the Effect of a Modified Valve with Helical Guideways on Combustion Performance of an I.C Engine, Vol 5(8), p 33-42.

[5] Hiregoudar Yerrenagoudaru, Malikarjuna 2014. IOSR Journal of Mechanical and Civil Engineering, Effect of inlet Air Swirl on 4- stroke single cylinder diesel engine's performance, Vol II (4), p 59-68.

[6] C.Ramesh Kumar and G. Nagarajan 2012. APRN journal of engineering and applied science, Investigation of flow during intake stroke of a single cylinder internal combustion engine, Vol 7(2), p 180-186.

[7] Abdul Rahim Ismal & Rosli Abu Bakar Semin 2008. American journal of applied science, An investigation of vale lift effect on air flow and co-efficient of discharge of 4-strokeengine based experiment, Vol 5(8), p 963-971.

[8] M.G. Rasul and R.Glasgow 2005. Internationational conference on mechanical engineering (ICME), Performance improvement of an internal combustion engine, Vol 5(42), p 01-05.

[9] Sihun lee, Kun tong, Bryand. Quay 2000. SAE paper series, Effcet of swirl and tumble on mixture preparation during cold start of a gasoline direct injection engine. p (1-14).

[10] Tomonori Urushihara, Tsutomu Nakata, Akihiko Kakuhou, Yasuo Takagi 1996. (JSAE), Effect of swirl and tumble motion on fuel vapour behavior and mixture in lean bur engine, Vol 17, p 239-244.

Evaluation of tensile strength in riveted, adhesive and hybrid joints

K V P Chakradhar[1] , **A BalaRaju**[2] **and Anil Kumar Veerapaneni**[3]

[1]Professor, Mechanical Engineering Department, Vardhaman College of Engineering, Hyderabad, India.
[2]Associate Professor, Mechanical Engineering Department, Sreenidhi Institute of Science & Technology, Hyderabad, India.
[3]Managing Director, Aarshanano Composite Technologies Pvt. Ltd., Medikonduru, Guntur, India.
E-mail: chakradharkvp@vardhaman.org, info@aarshanano.com.

Abstract. Many automobile industries prefer hybrid joining practices in the part-assembly of advanced lightweight automobile and commercial vehicles. The main cause for this is the greater need for mixed material design. One of the combinations is composite with aluminium, like a combination of Titanium-Aluminium, CFRPs with aluminium etc. This paper deals with comparative study of mechanical strength evaluation in riveted joints, adhesive joints and hybrid single-lap joints. Hybrid joints are a combination of riveted and adhesive joints. The riveted joint produces enormous stress around the holes of the fasteners, which is detrimental to its performance. The adhesive joint allows good load distribution; however, it leads to plastic deformation that is detrimental in the long run. The hybrid joint has better load bearing capacity when compared to the riveted or adhesive joints. Thus, the study indicates that a hybrid joint has better properties suitable for structural, automobile and aerospace applications.
Keywords: Hybrid joints, Riveted joints, Adhesive joints, Single-lap joint, Carbon composite.

1. Introduction

The process of combining different joining techniques and materials is called hybrid joining. Two or more joining operations are carried out consequently, leading to enhanced properties of the joint due to synergic load bearing interface. Hybrid joining can provide merits in terms of processability and load bearing capacity [1]. Irrespective of the merits related with hybrid materials, there is unwillingness to implement them for industrial usage. This unwillingness may be due to the hampered knowledge of joining techniques with hybrid materials. In 1985, L.J.H. Smith [2] studied the performance of bonded-bolted composite joints. The study provides an insight on bonded-bolted joints and concludes that it does provide exceptional strength compared to a better designed adhesive joint. In 2005, Gordon Kelly [3] studied the load transmission effect in adhesive bonded and bolted hybrid composite joints using 3-D analysis. It gives an insight that the performance of hybrid joining technique is enhanced in comparison to adhesive bonding. In 2008, Ryosuke Matsuzaki, Motoko Shibata [4] experimentally investigated the joint strength of aluminium alloy, glass-epoxy hybrid composites fabricated using bolted/adhesive joining method. It gives an insight that the glass fibers in the hybrid joints contribute to much higher fatigue strength than that of bolted joints. In 2010, H Ngoc and Eric [5] simulated the life of adhesive bonded and bolted joints. They suggested that the mechanical fasteners reinforced the hybrid structure in case the adhesive failed. In 1971, F. Erdogan and M. Ratwani [6] analyzed the distribution of stress in smoothly tapered and stepped joint plates. Here bonded plates with a smoothly tapered joint are compared with that of stepped joint. In 2015, S. Raviraja and L. Nafeez Ahmed [7] proposed a paper on the comparative study of bolted and hybrid joints, in epoxy/glass fiber composite laminates. Here modeling, analysis of 3D models and fabrication of the composite joints were carried out. In 2006, Kweon, Jung and Kim et al. [11] analyzed the performance of aluminium and carbon composite joints. They concluded that the usage of fasteners leads to a significant increment in weight of the part. In massive parts like ship and aircraft, the cutback in the number of fasteners used, lead to weight reduction of parts. In

IOP Conf. Series: Materials Science and Engineering 998 (2020) 012042 doi:10.1088/1757-899X/998/1/012042

2009, Banea M D, Da Silva L F [12] proposed a review paper on composite materials joined by using adhesives. They concluded that bonded joints are light in weight, less expensive and damage resistant. Machining operations are not required for adhesive joints. Adhesive joints inhibit propagation of crack, leading to better performance and life of parts.

Holes drilled for mechanical fastening cause concentration of stresses which affects the strength and performance of the joint. However, for assembling parts, bolted joints are found to be more reliable and are used extensively for many engineering applications. Since the mode of failure is progressive, strength predictions in the bolted joints are easy and accurate. Mechanical fasteners have better shelf life in comparison to many adhesives.

Conventional bolted, riveted and welded joints are used for connecting various industrial parts. The reinforcement of adhesives to conventional joints enhances the strength of the joints without changing the existing design. The objective of the work is to evaluate the tensile strength of conventional and hybrid joints and decide on their suitability for making components for structural, aerospace, automobile and underwater applications.

2. Material and Methods

The materials used for the plates used for preparation of riveted, adhesive and hybrid (composite/metal) joints are Towpreg (T-700) (a composite with combination of carbon, basalt or fiberglass filament and a thermosetting resin as binder material) and Aluminum alloy. The properties of aluminium alloy and towpreg are given in Table 1 and Table 2. The adhesive used for preparing the joint is Araldite AW 106 / Hardener HV 953 U. Dimensions used for modeling the plate for analysis are given in Table 3.

Table 1. Material properties for Aluminium Alloy [8]

Material	Property	Value
Al alloy (HE-15) *(Al-90%; Cu-5.00%; Si-0.9%; Mn-1.00%; Mg-0.80%; Fe-0.50%; Zn-0.25%; Ti-0.15%; others-1.4%)*	E (MPa) υ UTS (MPa) ρ (kg/m^3)	70,000 0.33 470 2,700

Note: E-Young's modulus; υ-Poisson ratio; UTS-Ultimate tensile strength; ρ-Density.

Table 2. Material properties for Towpreg (T-700) [9]

Stiffness Property	Value
E_1 (Along fiber)	1,32,000 MPa
E_2 (Across fiber)	10300 MPa
G (Shear modulus)	3000 MPa
υ	0.3
ρ	1570 Kg/m^3
σ (Tensile)	2100 MPa
σ (Compressive)	1050 MPa
τ (Shear stress)	90 MPa

IOP Conf. Series: Materials Science and Engineering 998 (2020) 012042 doi:10.1088/1757-899X/998/1/012042

Table 3. Dimensions of the plate

S.No.	Parameter	Dimension (mm)
1	Length of plate	100
2	Width of plate	25
3	Thickness of plate	10
4	Overlapping length	40
5	Adhesive Thickness	0.1

2.1. Preparation and testing of Adhesive Joint

The carbon fiber composite and the Aluminium plates prepared and tested as per ASTM D638-02a standard. They are bonded together using with adhesive as shown in Figure 1. A layer of 0.1 mm thick adhesive applied in between the plates.

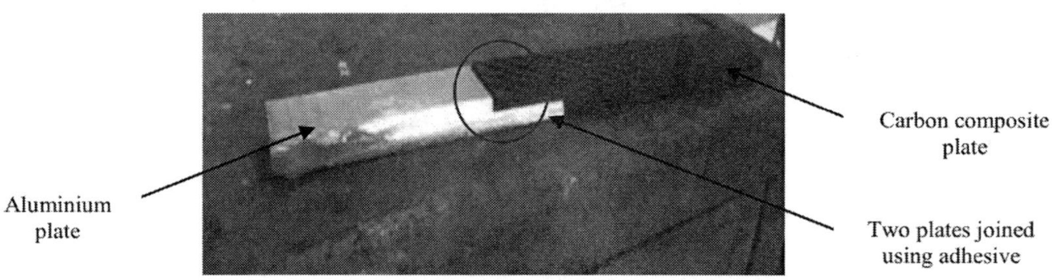

Figure 1. Carbon composite and Aluminium plates joined using adhesive

2.2. Preparation of Riveted Joint

The composite and the aluminium plates prepared as per ASTM D638-02a standard. Holes of 4.5 mm drilled on the composite and aluminium plates as per requirement. Both the plates are fastened together using a suitable fastener as shown in Figures 2a and 2b.

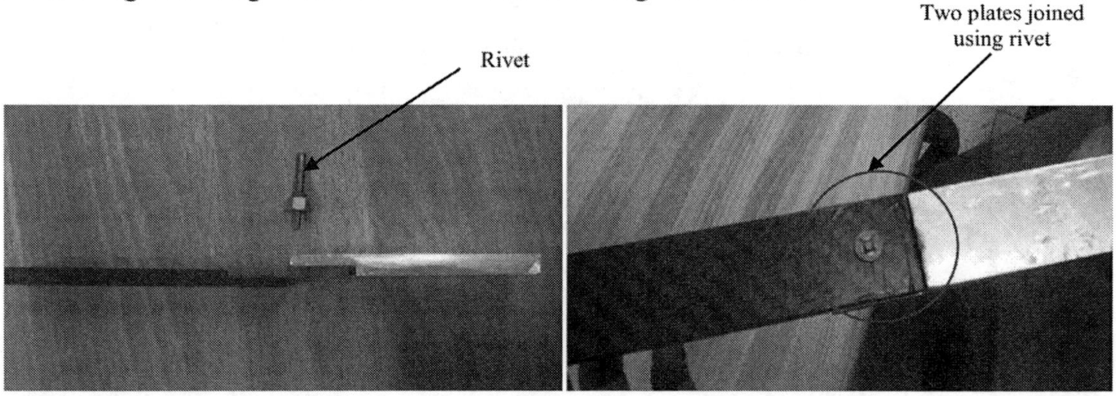

Figures 2a and 2b. Carbon composite and Aluminium plates joined using fastener.

2.3. Preparation of Hybrid Joint

The composite and aluminium plates prepared as per ASTM D638-02a standard. Holes drilled on the composite and aluminium plates as per requirement. Then the adhesive applied on the plates. Both the plates after bonded together by adhesive are fastened together using a rivet to form hybrid joint as shown in Figure 3.

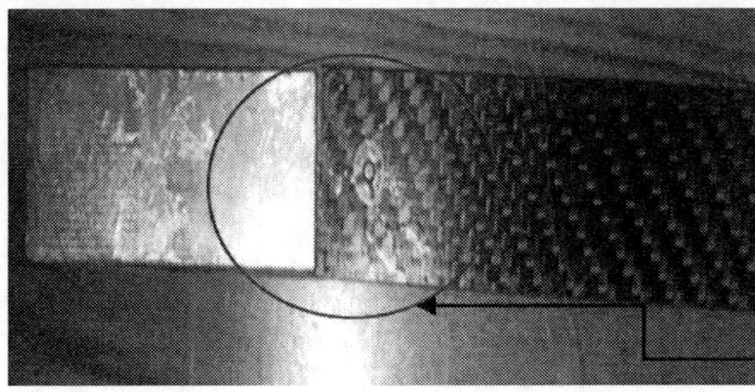

Figure 3. Carbon composite and Aluminium plates joined using adhesive and rivet

2.4. Experimental testing of Joints

A standard UTM used for performing tensile test on the specimens (plates). A load of 1000N is applied on one end of the specimen and on other end it is fixed. INSTRON 4505 machine of 10 T capacities was used for tensile testing. The test was conducted three times for each case and the mean of the three test results taken as the final value. A sample experimental setup for the test is shown in Figure 4. A failed riveted joint after testing is shown in Figure 5.

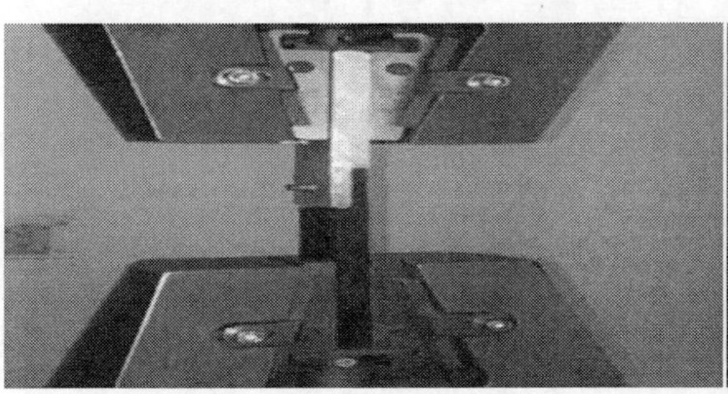

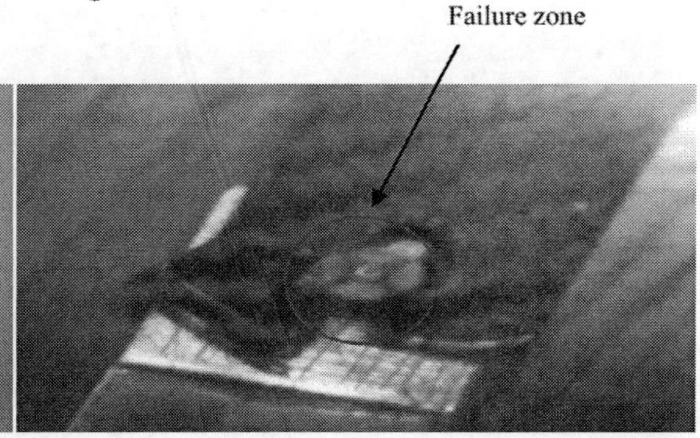

Figure 4. Experimental setup for tensile testing

Figure 5. Riveted joint plates after testing

3. Results and Discussion

The function of any joint is to produce an effective and defectless bond between the plates and transmit loads efficiently and uniformly. In the adhesive bonded joints, joint failures are generally nonexistent as they possess exceptional dissemination of loads over a broader area when compared to mechanical joints. The combination of adhesive and riveted joints as hybrid joint provides excellent load distribution and required mechanical strength to the joint [10, 13].

The tensile test conducted on the adhesive, riveted and hybrid single-lap joints gave the results as shown by the Force-Displacement curves as plotted in Figures 6-8. The Force-Displacement curves are plotted considering eight displacement values (0-0.8 mm) for different load conditions and the maximum load before failure evaluated. At 0.7 mm displacement, the plates bonded by adhesive and hybrid combination indicated that the joints could withstand maximum loads of 6.28 KN, and 9.48 KN respectively. While, the riveted joint could withstand a maximum load of 8.40 KN at 0.6 mm displacement and it failed well before it reached 0.7 mm. The study reveals that the hybrid joint (rivet and adhesive) has better load bearing capability when compared to the other two joints. The summary of ultimate tensile loads and tensile stresses taken by the joints before failure is shown in Table 4. The failure of the carbon composite plate due to tearing of the plate at the edge is shown in Figure 5. This is due to tension stress failure in the plate, with slight deformation of the rivet.

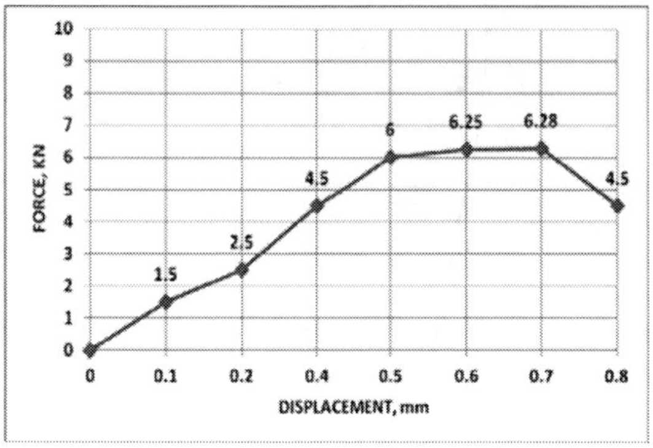

Figure 6. Force-displacement curve (adhesive joint)

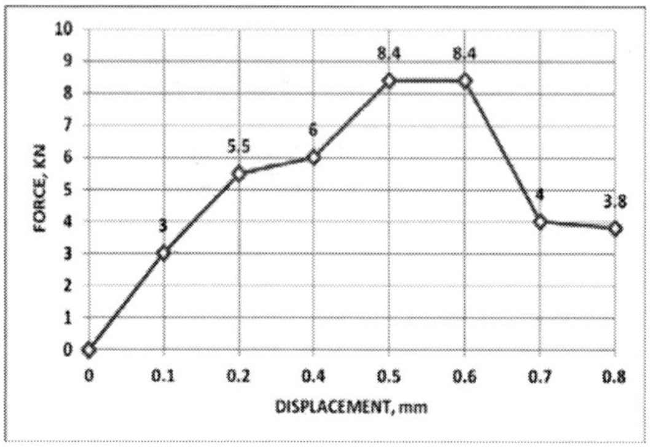

Figure 7. Force-displacement curve (rivet joint)

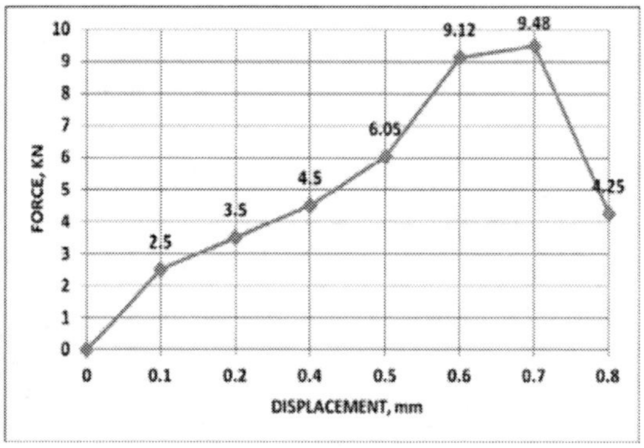

Figure 8. Force-displacement curve (hybrid joint)

Table 4. Summary of ultimate tensile loads and tensile stresses taken by the joints

Type of joint	Displacement (mm)	Ultimate tensile load (KN)	Tensile stress (N/mm^2)
Adhesive joint	0.7	6.28	25.12
Rivet joint	0.6	8.40	33.60
Hybrid joint	0.7	9.48	37.92

4. Conclusions

The tensile strength evaluation and comparison of adhesive joint, riveted joint and hybrid joint test results revealed the ensuing evidence.

1. Tensile test on aluminium and towpreg carbon composite joined plates bonded with adhesive indicated that it can withstand an ultimate tensile load of 6.28 KN and tensile stress of 25.12 N/mm^2 at 0.7 mm displacement. It failed due to interfacial bond failure.
2. Tensile test on aluminium and towpreg carbon composite joined plates bonded with single rivet indicated that it can withstand an ultimate tensile load of 8.40 KN and tensile stress of 33.60 N/mm^2 at 0.6 mm displacement and it had failed before 0.7 mm. It failed due to tearing of composite plate near the hole.
3. Tensile test on aluminium and towpreg carbon composite joined plates bonded with single rivet and adhesive (hybrid combination) indicated that it can withstand an ultimate tensile load of 9.48 KN and tensile stress of 37.92 N/mm^2 at 0.7 mm displacement. It also failed due to interfacial bond failure.
4. In comparison to adhesive joint, the hybrid joint indicated that it has 50.95% higher load bearing capacity.
5. In comparison to the riveted joint, the hybrid joint indicated that it has 12.85% higher load bearing capacity.
6. Experimental study revealed that the tensile strength of a hybrid joint is considerably superior to that of adhesive and riveted joints.
7. Experimental study also revealed that the better load transmission in hybrid joints can give improved structural performance in comparison to other joints as is also supported in literature [10].
8. Reinforcement of the adhesive joints with rivets gives the joint enhanced strength as is also supported in literature [13].

The study may pave the way for the development of light-weight materials suitable for making components for structural, aerospace, automobile and underwater applications.

Acknowledgements

The authors thank Aarshanano Composite Technologies Pvt. Ltd., Medikonduru, Guntur for providing technical assistance required for the work.

References

[1] Yadav P D and Patil M J 2014 *Journal of Engineering Research and Technology* **3** 957
[2] Hart Smith L J 1985 *Journal of Aircraft* **22** 993
[3] Gordon Kelly 2005 *Composite Structures* **69** 35
[4] Ryosuke Matsuzaki, Motoko Shibata and Akira Todoroki 2008 *Composites Part A: Applied Science and Manufacturing* **39** 154
[5] Cat-Tan Hoang-Ngoc and Eric Parroissien 2010 *International Journal of Adhesion and Adhesives* **30** 117
[6] F. Erdogan and M. Ratwani 1971 *Journal of Composite Materials* **5** 378
[7] Raviraja S and Nafeez Ahmed L 2015 *International Journal of Innovative Research in Science, Engineering and Technology* **4** 214
[8] Parthiban T, Subramanian G and Muthuraman K 2017 *Journal of Marine Science and Application* **16** 237
[9] Shukur A Hasan, Carlo Santulli, Mohd. Yazid B Yahya, Chong L Gang and Mohd. NFB Abu Bakar 2018 *FME Transactions* **46** 108
[10] Jeevi G,Sanjay Kumar Nayak and Abdul Kader M 2019 *Journal of Adhesion Science and Technology* **33** 1497
[11] Kweon JH, Jung JW, Kim TH, et al. 2006 *Composite Structures* **75** 192
[12] Banea MD and Da Silva LF 2009 *Proc IMECHE* **223** 1
[13] Ezzine M C, Amiri A, Tarfaoui M and Madani K 2018 *Advances in Aircraft and Spacecraft Science* **5** 595

Effect of Higher Transition Foundation on Dynamic Stability Boundaries of Simply Supported Beam

B.Subbaratnam[1,A], Ravinder Rao Jalagam[2,B], Shakti Prasanna Jena[3,C] and M. Satyanarayana Gupta[4,D]

[1,3] Department of Mechanical Engineering, Vardhaman College of Engineering, Shashabad-501 218, Hyderabad, Telangana, India.
[2]Director Technical, AES SERVICES LLP, Hyderabad, Telangana, India
[4]Department of Aeronautical Engineering, MLR Institute of Technology, Hyderabad, Telangana, India

[a] Corresponding Author: b_subbaratnam@yahoo.com
[b)]ravi@aesservicesllp.com, [c)]shaktipjena@gmail.com and [d)]msngupta.m@gmail.com

Abstract: Accurate closed form solution to predict the dynamic stability behavior of simply supported (SS) beam on elastic foundation, subjected to an end axial periodic load, has been derived using simple single term trigonometric function which satisfies the geometric boundary conditions. The effect of higher transition foundation on dynamic stability regions is studied. The energy method is used in the present work to evaluate the dynamic stability boundaries. Euler - Bernoulli beam theory is used for the analysis. Numerical results are presented using proper non-dimensional parameters with varying foundation parameters below and above the second transition foundation value of elastic foundation, in both the digital and analogue forms for better comprehension of the dynamic stability behavior of the beams. The effect of the higher transition foundation value of the elastic foundation on the dynamic stability behavior is brought out in the present work.

Keywords: Dynamic stability, beams, periodic loads, elastic foundation, transition foundation.

1. INTRODUCTION

Prediction of dynamic stability boundaries of structural elements like beams subjected to periodic axial loads is an important input for the structural design engineers. Earlier studies on the dynamic stability of beams/columns are studied by Timoshenko and Gere [1]. The dynamic stability of structural elements is well discussed in the classic work of V.V. Bolotin [2]. In Ref. [3] the dynamic stability of slender bars, subjected to an axial periodic load at the free ends, is investigated using the FE method. It is shown that the dynamic stability curves obtained using the non-dimensional parameters are the same, for any boundary condition, as long as the vibration and stability mode shapes are similar. In Ref.[4], the dynamic stability of structural members subjected to a periodic loads is investigated by Rao *et al.* It is shown that the boundaries/regions of the dynamic stability are obtained with suitable non-dimensional parameters are same for the structural members, where the mode shapes of the frequency and buckling are same or nearly the same, subjected to end periodic axial loads. In Ref. [5] one of author's previous work, it is demonstrated that the deviation in assumed mode shapes with reference to exact mode shapes can be accurately assessed by means of L_2 norms of associated mode shapes. In Ref. [6] of author's previous work, it is demonstrated that the deviation in assumed mode shapes with reference to exact mode shapes can be accurately assessed by means of L_2 norms of associated mode shapes.

Structural members like uniform beams and plates on elastic foundation are commonly encountered in many disciplines of engineering. The effect of elastic foundation on the regions of dynamic instability of beams is studied in Ref. [6]. An instability finding of Ref. [6] is that, as the foundation modulus (parameter) increases, the boundaries of dynamic instability are shifted away from the vertical axis and the instability region width decreased. It has been shown in Ref.[3] that the mode shapes of the buckling of columns resting on an elastic foundation depend on the value of the foundation stiffness parameter. A transition foundation value exists and beyond this transition foundation the mode shape of the buckling changes. For simply supported beam, the vibration modes shape and stability mode shape are the same (one –half sine wave), when the foundation parameter is below the first transition foundation parameter. The stability and vibration mode shapes are different (two-half sine waves for stability & one half-wave for free vibration problems respectively) when the foundation parameter is above the first transition value. For a simply supported column the value of this transition foundation parameters are found to be 4, 36, 144….. for (first, second, third…….) [1]. Furthermore, it does not exist for the vibration mode shapes of beams on an elastic foundation. Recently B. Subba Ratnam *et al* investigated the dynamic stability of beams on elastic foundation [7]. In this study, it is shown that, as the elastic foundation increases, the width of the regions of dynamic stability decreases, thus the beam less sensitive to the dynamic stability phenomenon to periodic loads. Similar studies on SS rectangular plates subjected to biaxial periodic compressive load on elastic foundation have been reported in Ref. [8].

In Ref. [9] the dynamic instability of an axially loaded beam resting on an elastic foundation with damping is investigated. It is shown that, if the foundation stiffness or damping increases, the critical dynamic load also increases and the regions of instability shifts to a higher applied frequency. The dynamic stability of a non-uniform (tapered) cantilever beam resting on an elastic foundation was examined by Lee [10]. The effect of elastic foundation on natural frequencies, static buckling loads and regions of dynamic instability of Timoshenko beams with various boundary conditions are studied by Yokoyama [11]. Subbaratnam *et al* [12], studied the two dimensional problems like rectangular plate on elastic foundation and the effect of elastic foundation on the regions of plates with uniform edge loads.

The main aim of the present proposed work is to investigate the effect of higher transition foundation on the regions of dynamic stability of SS beams resting on elastic foundation subjected to end axial periodic loads. The effect of elastic foundation, changing the mode shapes for stability and vibration on the dynamic stability boundaries, with increasing the values of foundation parameter, subjected to periodic forces is derived and discussed which highlights the simplicity and robustness of the proposed analytical work in obtaining the reasonably realistic solutions to dynamic instability problems.

2. FORMULATION

The mathematical formulation based on the variational energy principle and the evaluation of the dynamic stability boundaries is briefly discussed here for the simply supported beam resting on elastic foundation. A uniform SS beam, subjected to a concentrated axial compressive periodic load P(t), resting on an elastic foundation, is shown in Fig.1. The total potential energy Π of the beam is given by

$$\Pi = U + U_F - T - W \tag{1}$$

where U = Strain energy, U_F = Energy stored in the elastic foundation, T = Kinetic energy and W= Work done by the periodic load. The expressions for U, U_F, T and W are given, by

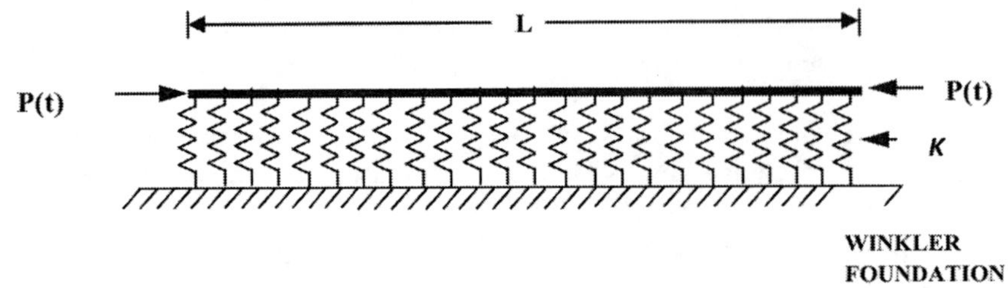

Figure.1. *Simply Supported beam resting on uniform Elastic Foundation with axial Periodic load*

$$U = \frac{EI}{2}\int_0^L w''^2 dx \tag{2}$$

$$U_F = \frac{k}{2}\int_0^L w^2 dx \tag{3}$$

$$T = \frac{\overline{m}\omega^2}{2}\int_0^L w^2 dx \tag{4}$$

and

$$W = \frac{P(t)}{2}\int_0^L w'^2 dx \tag{5}$$

where ' (prime) = first derivative, '' (double prime) = second derivative with respect to x, E = Young's modulus, I = area moment of inertia, $\overline{m}$ = mass per unit length, w = lateral displacement , P(t) = axial compressive periodic load and ω = natural frequency.

The load P(t) is taken as

$$P(t) = P_S + P_t \cos\theta t \tag{6}$$

where P_S = constant part of P(t), P_t = periodic part of P(t), θ = radian frequency of the compressive load P_t. The quantities P_s and P_t are expressed in terms of the buckling load parameter P_{cr}

Substituting, Eqs. (2) to (5) in Eq. (1), the differential equation governing the dynamic stability problem of beams resting on elastic foundation is written in terms of the energies, as

$$\Pi = \frac{EI}{2}\int_0^L w''^2 dx + \frac{k}{2}\int_0^L w^2 dx - \frac{P(t)}{2}\int_0^L w'^2 dx - \frac{\overline{m}\omega^2}{2}\int_0^L w^2 dx \tag{7}$$

As ω= θ/2, Eq. (7) becomes

IOP Conf. Series: Materials Science and Engineering 998 (2020) 012043 doi:10.1088/1757-899X/998/1/012043

$$\Pi = \frac{EI}{2}\int_0^L w''^2 dx + \frac{k}{2}\int_0^L w^2 dx - \frac{P(t)}{2}\int_0^L w'^2 dx - \frac{\overline{m}}{2}\frac{\theta^2}{4}\int_0^L w^2 dx \tag{8}$$

Substituting the expression for P*(t)* given in Eq. (6) in Eq. (8), we get

$$\Pi = \frac{EI}{2}\int_0^L w''^2 dx + \frac{k}{2}\int_0^L w^2 dx - (P_S + P_t\cos\theta t)\frac{1}{2}\int_0^L w'^2 dx - \frac{\overline{m}}{2}\frac{\theta^2}{4}\int_0^L w^2 dx \tag{9}$$

As a first approximation the condition for the existence of these boundary solutions [2] is

$$\Pi = \frac{EI}{2}\int_0^L w''^2 dx + \frac{k}{2}\int_0^L w^2 dx - \left(\alpha \pm \frac{\beta}{2}\right)\frac{P_{cr}}{2}\int_0^L w'^2 dx - \frac{\overline{m}}{2}\frac{\theta^2}{4}\int_0^L w^2 dx \tag{10}$$

where α and β are the static load parameter($=\frac{P_s}{P_{cr}}$) and dynamic load parameter($=\frac{P_t}{P_{cr}}$) respectively.

Note that in Eq. (10), two conditions are combined under the plus or minus sign, thus yielding the two boundaries of the regions of instability. In this analysis static buckling load factor of the beam is considered as the reference.

2.1. Critical (Buckling) load and frequency parameters of SS beam resting on a foundation

For SS beam the distribution of lateral deflection, in the non-dimensional form is assumes as

$$W = a\sin m\pi X \tag{11}$$

where a is the undetermined constant, m is the mode number, m_b for buckling and m_f for free vibration problem, W and X are the non-dimensional lateral deflection and axial coordinate respectively.

Substituting Eq.(11), in the non-dimensional form of the Eq.(10) by the length of the beam and taking variation w. r. t a and equating to zero. We get the buckling load parameter and the frequency parameter in terms of the mode number as

$$\lambda_b = \frac{P_{cr}L^2}{EI} = \pi^2\left(m^2 + \frac{\gamma}{m^2}\right) \tag{12}$$

and

$$\lambda_f = \frac{\overline{m}\,\omega^2 L^4}{EI} = \pi^4\left(m^2 + \frac{\gamma}{m^2}\right) \tag{13}$$

where P_{cr} is the critical load, ω is the natural frequency, γ is the elastic foundation parameter ($\gamma = \frac{kL^4}{\pi^4 EI}$), λ_b is the buckling/critical load parameter and λ_f is the natural frequency parameter ($= \frac{\overline{m}\,\omega^2 L^4}{EI}$).

2.2. Transition Values of the Foundation Parameter

The transition value γ_{Ti} (i = 1, 2, 3 . . . and corresponding to the buckling mode number m_b) of the elastic foundation parameter λ_b change of mode shape from m_b to $(m + 1)_b$ half sine waves is evaluated from the equation

$$\lambda_{bm} = \lambda_{b(m+1)}. \tag{14}$$

Then the value of γ_{Ti} as

$\gamma_{Ti} = m_b^2(m + 1)_b^2$, $(i, m_b = 1, 2, 3, 4 \ldots)$. (15)

The following two significant observations from the evaluation of the γ_{Ti} are:

a) γ_{Ti} for the buckling problem,

b) for the free vibration problems the phenomenon of transition does not exist.

The values of γ_{Ti} for the buckling problem are 4, 36, 144 . . . for (i = 1, 2, 3 . . .). In this present work the effect of the second transition foundation parameter γ_{T2} = 36 and its effect on the dynamic stability regions of SS beam is studied in detail.

2.3. Variation of λ_b and λ_f for different values of m_b and m_f

Eq. (12) and Eq. (13) are the expressions for bucking load parameter (λ_b) and frequency parameter (λ_f) for the different values of m_b, m_f and γ. Depending on these values of γ, m_b and m_f, the expressions for λ_b and λ_f are evaluated concisely and presented here;

2.3.1 For $\gamma < \gamma_{T1}$ ($m_b = m_f = 1$)

The expressions for λ_b and λ_f are

$$\lambda_b = \frac{P_{cr}L^2}{EI} = \pi^2[1+\gamma] \tag{16}$$

and

$$\lambda_f = \frac{\overline{m}\omega^2 L^4}{EI} = \pi^4(1+\gamma) \tag{17}$$

2.3.2 For $\gamma > \gamma_{T1}$ and $\gamma < \gamma_{T2}$ ($m_b = 2$, $m_f = 1$)

These expressions are

$$\lambda_b = \frac{P_{cr}L^2}{EI} = \pi^2\left[4+\frac{\gamma}{4}\right] \tag{18}$$

and

$$\lambda_f = \frac{\overline{m}\omega^2 L^4}{EI} = \pi^4(1+\gamma) \tag{19}$$

2.3. 3 For $\gamma > \gamma_{T1}$ and $\gamma < \gamma_{T2}$ ($m_b = 2$, $m_f = 2$)

The expressions for λ_b and λ_f are

$$\lambda_b = \frac{P_{cr}L^2}{EI} = \pi^2\left[4+\frac{\gamma}{4}\right] \tag{20}$$

and

$$\lambda_f = \frac{\overline{m}\omega^2 L^4}{EI} = \pi^4(16+\gamma) \tag{21}$$

2.3.4 For $\gamma > \gamma_{T2}$ and $\gamma < \gamma_{T3}$ ($m_b = 3$, $m_f = 1$)

The expressions for λ_b and λ_f are

$$\lambda_b = \frac{P_{cr}L^2}{EI} = \pi^2\left[9+\frac{\gamma}{9}\right] \tag{22}$$

and

$$\lambda_f = \frac{\overline{m}\omega^2 L^4}{EI} = \pi^4(1+\gamma) \tag{23}$$

2.3.5 For $\gamma > \gamma_{T2}$ and $\gamma < \gamma_{T3}$ ($m_b = 3, m_f = 3$)

The expressions for λ_b and λ_f are

$$\lambda_b = \frac{P_{cr} L^2}{EI} = \pi^2\left[9 + \frac{\gamma}{9}\right] \tag{24}$$

and

$$\lambda_f = \frac{\overline{m}\omega^2 L^4}{EI} = \pi^4(81+\gamma) \tag{25}$$

The expressions for λ_b and λ_f for the above cases are given for the sake of completeness, for any combination of m_b, m_f and γ the buckling and frequency parameters can be easily obtained for a given value of these three parameters, taking care of the transition value of the foundation parameter for γ.

3. DYNAMIC STABILITY FORMULAS

The dynamic stability formulas can be obtained, by applying the reference values of P_{cr} and ω for different values of γ with the consideration of the proper values of m_b and m_f. These formulas are to predict the dynamic instability boundaries of the SS beam resting on elastic foundation are:

a) Substituting the Eqs. (11), (16) and(17) in Eq. (10), half sine wave (first mode) for buckling and vibration ($m_b = m_f = 1$) as the reference values, where $P_{cr} = \frac{\pi^2 EI}{L^2}$ and $\overline{m} = \frac{\pi^4 EI}{\omega^2 L^4}$, without consideration of the effect of elastic foundation, then the dynamic stability formula is obtained, as

$$1+\gamma-\left(\alpha \pm \frac{\beta}{2}\right) - \frac{\theta^2}{4\omega^2} = 0 \tag{26}$$

where $\alpha = \frac{P_s}{P_{cr}}$ and $\beta = \frac{P_t}{P_{cr}}$.

or

$$\Omega = \frac{\theta}{\omega} = 2\sqrt{(1-\alpha)(1\pm\mu)+\gamma} \tag{27}$$

where μ is the non-dimensional parameter, $\mu = \frac{\beta}{2(1-\alpha)}$.

b) Substituting Eqs. (11), (18) and(19) in Eq. (10) for $m_b = 2$ and $m_f = 1$ then the dynamic stability formula is written, as

$$1 - \frac{\left(\alpha \pm \frac{\beta}{2}\right) P_{cr} .4 \frac{\pi^2}{L^2}}{\frac{\pi^4 EI}{L^4}\left[16 + \frac{kL^4}{\pi^4 EI}\right]} - \frac{\frac{\overline{m}\theta^2}{4}}{\frac{\pi^4 EI}{L^4}\left[1 + \frac{kL^4}{\pi^4 EI}\right]} = 0 \tag{28}$$

Substituting the values of $P_{cr} = \frac{4\pi^2 EI}{L^2}$ and $\overline{m} = \frac{\pi^4 EI}{\omega^2 L^4}$ without consideration of foundation in Eq. (28) becomes

$$\Omega = \frac{\theta}{\omega} = 2\sqrt{\left[\left((1-\alpha)(1\pm\mu)+\frac{\gamma}{16}\right)\left(\frac{(1+\gamma)}{(1+\frac{\gamma}{16})}\right)\right]} \tag{29}$$

c) In the similar way, substituting Eqs.(11), (22) and(23) in Eq.(10) for m_b =3 and m_f=1, the dynamic stability formula is obatined, as

$$1-\frac{\left(\alpha\pm\frac{\beta}{2}\right)P_{cr}.9\frac{\pi^2}{L^2}}{\frac{\pi^4 EI}{L^4}\left[81+\frac{kL^4}{\pi^4 EI}\right]}-\frac{\frac{\overline{m}\theta^2}{4}}{\frac{\pi^4 EI}{L^4}\left[1+\frac{kL^4}{\pi^4 EI}\right]}=0 \tag{30}$$

Substituting the reference values for $P_{cr} = \frac{9\pi^2 EI}{L^2}$ and $\overline{m} = \frac{\pi^4 EI}{\omega^2 L^4}$ without consideration of foundation, Eq.(30) is written, as

$$\Omega = \frac{\theta}{\omega} = 2\sqrt{\left[\left[(1-\alpha)(1\pm\mu)+\frac{\gamma}{81}\right]\left[\frac{(1+\gamma)}{(1+(\gamma/81))}\right]\right]} \tag{31}$$

The above three dynamic stability equations (formulas) are derived based on the mode numbers m_b and m_f for the values of the first and second transition foundation parameter γ lesser or greater than γ_{T1} and γ_{T2}.

4. MASTER DYNAMIC STABILITY FORMULA

Substituting the Eq.(11), the values of reference buckling load P_{cr} and frequency ω with consideration of foundation for half sine wave for stability and half sine wave vibration (m_b = 1, m_f =1), two half wave for stability and two- half wave for vibration (m_b =2, m_f=2) and three half waves for both vibration and stability problems (m_b = 3, m_f=3), in Eq. (10), we get

$$\Omega = \frac{\theta}{\omega} = 2\sqrt{(1-\alpha)(1\pm\mu)} \tag{32}$$

It may be noted that the master dynamic stability formula shown in Eq. (32) does not contain the foundation parameter explicitly, unlike the earlier formulas where in the foundation parameters γ appear explicitly.

5. NUMERICAL RESULTS AND DISCUSSION

Figure.1 shows, SS beam resting on elastic foundation, subjected to an axial periodic load. The formula developed in the present study to investigate the dynamic stability boundaries of SS beam resting on an elastic foundation, subjected to axial periodic concentrated loads is very general. In the non-dimensional form the characteristic values of the beam such as the fundamental frequency and the static buckling load parameters do not appear explicitly. However these characteristic values appear implicitly in the definition of the non-dimensional parameters μ and Ω.

IOP Conf. Series: Materials Science and Engineering 998 (2020) 012043 doi:10.1088/1757-899X/998/1/012043

Table.1 *. Values of Ω_1 and Ω_2 for $\gamma = 5.0$ ($\gamma > \gamma_{T1}$)*
(for stability two half sine waves and for vibration one half sine wave)

μ	$\alpha = 0.6$		$\alpha = 0.7$		$\alpha = 0.8$	
	Ω_1	Ω_2	Ω_1	Ω_2	Ω_1	Ω_2
0.0	3.6095	3.6095	3.3466	3.3466	3.0612	3.0612
0.1	3.5067	3.7094	3.2636	3.4276	3.0009	3.1204
0.2	3.4008	3.8067	3.1784	3.5067	2.9393	3.1784
0.3	3.2915	3.9016	3.0910	3.5840	2.8765	3.2355
0.4	3.1784	3.9942	3.0009	3.6598	2.8122	3.2915
0.5	3.0612	4.0848	2.9081	3.7340	2.7464	3.3466

Table.2. *Values of Ω_1 and Ω_2 for $\gamma = 15.0$ ($\gamma > \gamma_{T1}$)*

(for stability two half sine waves and for vibration one half sine wave)

μ	$\alpha = 0.6$		$\alpha = 0.7$		$\alpha = 0.8$	
	Ω_1	Ω_2	Ω_1	Ω_2	Ω_1	Ω_2
0.0	6.6468	6.6468	6.3935	6.3935	6.1297	6.1297
0.1	6.5467	6.7455	6.3155	6.4705	6.0756	6.1834
0.2	6.4450	6.8427	6.2366	6.5467	6.0210	6.2366
0.3	6.3416	6.9386	6.1566	6.6219	5.9659	6.2893
0.4	6.2366	7.0331	6.0756	6.6963	5.9102	6.3416
0.5	6.1297	7.1265	5.9935	6.7699	5.8541	6.3935

Table.3. *Values of Ω_1 and Ω_2 for $\gamma = 25.0$ ($\gamma > \gamma_{T1}$)*
(for stability two half sine waves and for vibration one half sine wave)

μ	$\alpha = 0.6$		$\alpha = 0.7$		$\alpha = 0.8$	
	Ω_1	Ω_2	Ω_1	Ω_2	Ω_1	Ω_2
0.0	8.9246	8.9246	8.6942	8.6942	8.4576	8.4576
0.1	8.8331	9.0151	8.6239	8.7640	8.4095	8.5054
0.2	8.7408	9.1047	8.5530	8.8331	8.3611	8.5530
0.3	8.6474	9.1934	8.4816	8.9018	8.3124	8.6003
0.4	8.5530	9.2812	8.4095	8.9699	8.2634	8.6474
0.5	8.4576	9.3683	8.3368	9.0375	8.2142	8.6942

Table.4. *Values of Ω_1 and Ω_2 for $\gamma = 35.0$ ($\gamma > \gamma_{T1}$)*
(for stability two half sine waves and for vibration one half sine wave)

μ	$\alpha = 0.6$		$\alpha = 0.7$		$\alpha = 0.8$	
	Ω_1	Ω_2	Ω_1	Ω_2	Ω_1	Ω_2
0.0	10.8117	10.8117	10.6007	10.6007	10.3855	10.3855
0.1	10.7278	10.8950	10.5366	10.6645	10.3419	10.4289
0.2	10.6433	10.9776	10.4721	10.7278	10.2981	10.4721
0.3	10.5580	11.0596	10.4072	10.7908	10.2541	10.5151
0.4	10.4721	11.1410	10.3419	10.8534	10.2100	10.5580
0.5	10.3855	11.2218	10.2761	10.9157	10.1656	10.6007

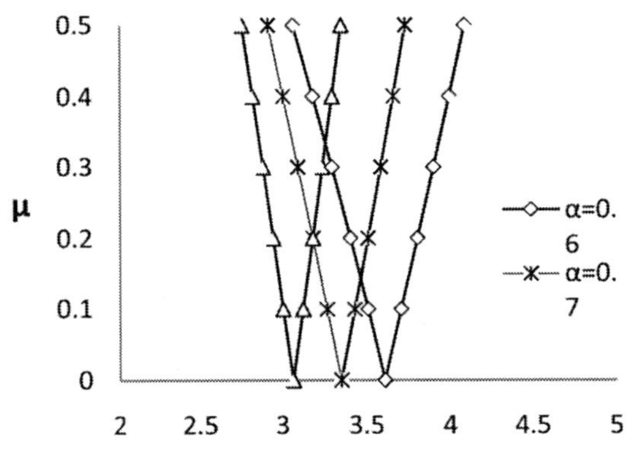

Figure.2. *Non-dimensional dynamic stability curves of SS beam resting on an foundation for $\gamma = 5.0$ ($\gamma > \gamma_{T1}$)*

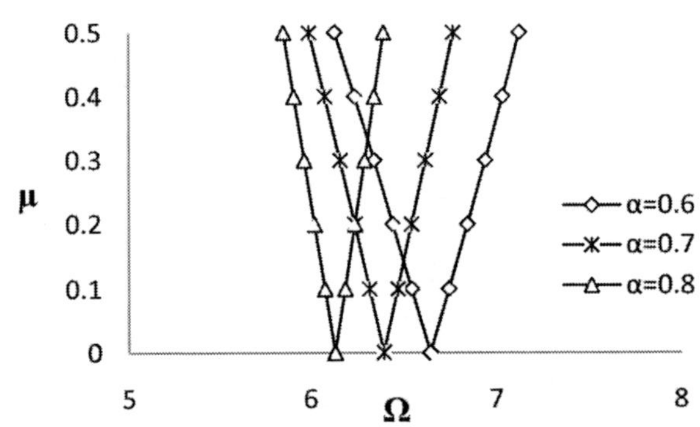

Figure.3. *Non-dimensional dynamic stability curves of SS beam resting on an foundation for $\gamma = 15.0$ ($\gamma > \gamma_{T1}$)*

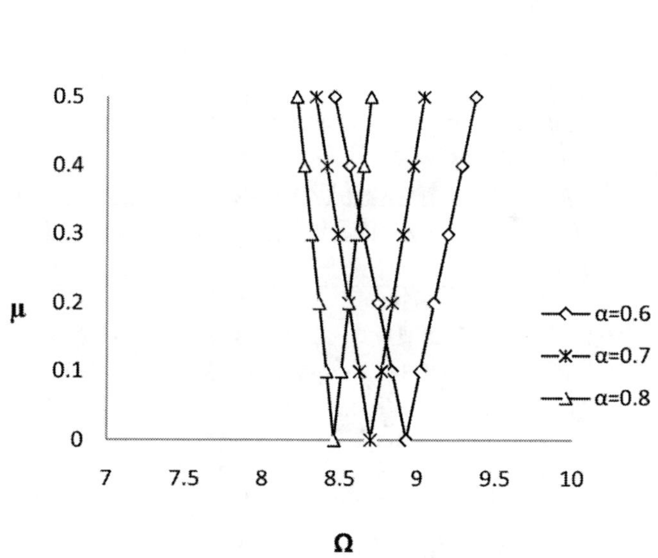

Figure.4. *Non-dimensional dynamic stability curves of SS beam resting on an foundation for $\gamma = 25.0$ ($\gamma > \gamma_{T1}$)*

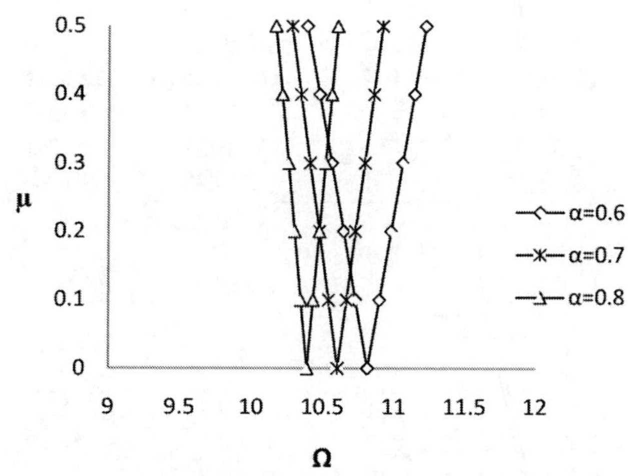

Figure.5. *Non-dimensional dynamic stability curves of SS beam resting on an foundation for $\gamma = 35.0$ ($\gamma > \gamma_{T1}$)*

Tables 1 to 4 gives the values of the stability boundaries $\Omega_1$and Ω_2, between which it is dynamically unstable with varying μ for $\alpha = 0.6$, 0.7 and 0.8 with the mode shape of two half sine wave for stability problem and one half sine wave for vibration problem for $\gamma > 4$ ($\gamma > \gamma_{T1}$).The instability boundaries $\Omega_1$and Ω_2 given in Tables 1 to 4 are for α=0.6, 0.7 and 0.8 with $\gamma = 5$ to $\gamma = 35$. Figures 2 to 5 shows the instability boundaries $\Omega_1$and Ω_2 given in Tables 1 to 4 for better visualization of the regions of dynamic stability. It can also be observed that with the increasing elastic foundation parameter, the dynamic instability regions are shifted away from the vertical axis. One significant observation in this study is the regions of dynamic instability decreases for $\gamma = 5$ to $\gamma = 35.0$ for above the first transition foundation and below the second transition foundation parameter value.

Table.5. *Values of $\Omega_1$and $\Omega_2$for $\gamma = 45.0$ ($\gamma > \gamma_{T2}$)*
(for stability three half sine waves and for vibration one half sine wave)

μ	$\alpha = 0.6$		$\alpha = 0.7$		$\alpha = 0.8$	
	Ω_1	Ω_2	Ω_1	Ω_2	Ω_1	Ω_2
0.0	10.6314	10.6314	10.0598	10.0598	9.4536	9.4536
0.1	10.4065	10.8517	9.8818	10.2346	9.3276	9.5779
0.2	10.1767	11.0675	9.7006	10.4065	9.2000	9.7006
0.3	9.9415	11.2793	9.5160	10.5757	9.0705	9.8218
0.4	9.7006	11.4871	9.3276	10.7421	8.9391	9.9415
0.5	9.4536	11.6912	9.1354	10.9060	8.8058	10.0598

Table.6. *Values of $\Omega_1$and $\Omega_2$for $\gamma = 55.0$ ($\gamma > \gamma_{T2}$)*
(for three half sine waves for stability and one half sine wave for vibration)

μ	$\alpha = 0.6$		$\alpha = 0.7$		$\alpha = 0.8$	
	Ω_1	Ω_2	Ω_1	Ω_2	Ω_1	Ω_2
0.0	11.9980	11.9980	11.4285	11.4285	10.8291	10.8291
0.1	11.7735	12.2184	11.2520	11.6023	10.7052	10.9516
0.2	11.5446	12.4348	11.0728	11.7735	10.5798	11.0728
0.3	11.3112	12.6476	10.8905	11.9423	10.4530	11.1926
0.4	11.0728	12.8568	10.7052	12.1087	10.3246	11.3112
0.5	10.8291	13.0627	10.5166	12.2728	10.1945	11.4285

Table.7. *Values of $\Omega_1$and $\Omega_2$for $\gamma = 60.0$ ($\gamma > \gamma_{T2}$*
(for stability three half sine waves and for vibration one half sine wave)

μ	$\alpha = 0.6$		$\alpha = 0.7$		$\alpha = 0.8$	
	Ω_1	Ω_2	Ω_1	Ω_2	Ω_1	Ω_2
0.0	12.6450	12.6450	12.0781	12.0781	11.4831	11.4831
0.1	12.4213	12.8648	11.9027	12.2509	11.3604	11.6046
0.2	12.1936	13.0809	11.7247	12.4213	11.2364	11.7247
0.3	11.9614	13.2935	11.5440	12.5895	11.1109	11.8437
0.4	11.7247	13.5027	11.3604	12.7554	10.9840	11.9614
0.5	11.4831	13.7088	11.1738	12.9192	10.8557	12.0781

Tables 5 to 7 gives the values of the stability boundaries $\Omega_1$and Ω_2, between which it is dynamically unstable for three-half sine waves for stability problem and one-half sine wave of vibration problem above the value of second transition foundation (γ>36) for $\gamma = 45.0$ to $\gamma = 60.0$ respectively, obtained from the Eq. (31). It is observed that, the instability regions are initially wider above the second transition foundation value and the dynamic instability regions are shifted away from the vertical axis. The other significant observation is, the dynamic instability regions, decreasing with increasing foundation value for above the second transition foundation and below the third transition foundation value. Figures 6 to 8 shows the instability boundaries $\Omega_1$and Ω_2 given in Tables 5 and 7 for better visualization of the regions of dynamic stability.

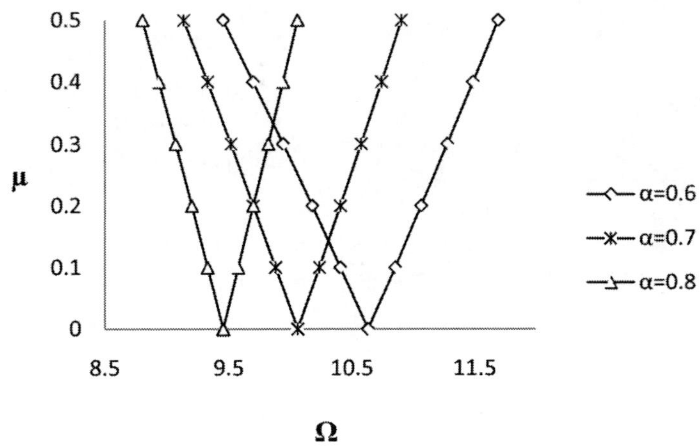

Figure. 6*. Non-dimensional dynamic stability curves of SS beam resting on an foundation for* $\gamma = 45.0$ *(*$\gamma > \gamma_{T2}$*)*

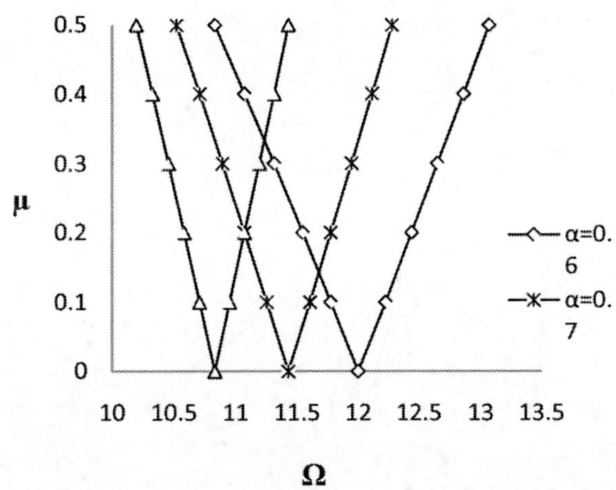

Figure. 7*. Non-dimensional dynamic stability curves of SS beam resting on an foundation for* $\gamma = 55.0$ *(*$\gamma > \gamma_{T2}$*)*

IOP Conf. Series: Materials Science and Engineering 998 (2020) 012043 doi:10.1088/1757-899X/998/1/012043

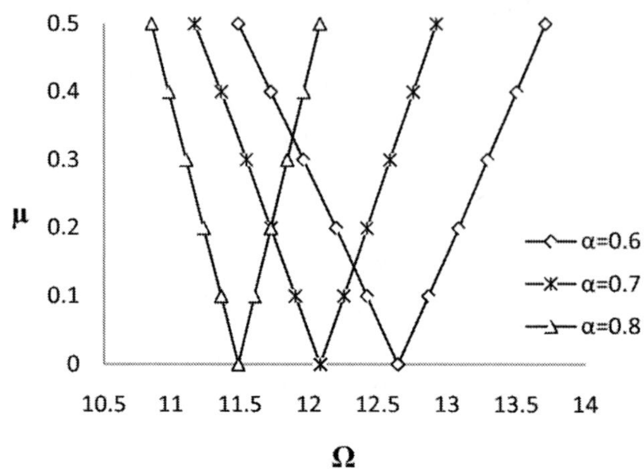

***Figure.8**. Non-dimensional dynamic stability curves of SS beam resting on an foundation for γ = 60.0 ($\gamma > \gamma_{T2}$)*

Figure 9 shows, the dynamic stability curves for two-half wave for both the stability and vibration problems, the reference critical load and reference frequency parameters are taken with consideration of elastic foundation, for the foundation parameter below the second transition value(γ<36) and three half waves for both vibration and stability problems, both reference critical load and reference frequency parameters are with consideration of elastic foundation for the below the third transition value(γ<144) are obtained from Eq. (32) are presented.

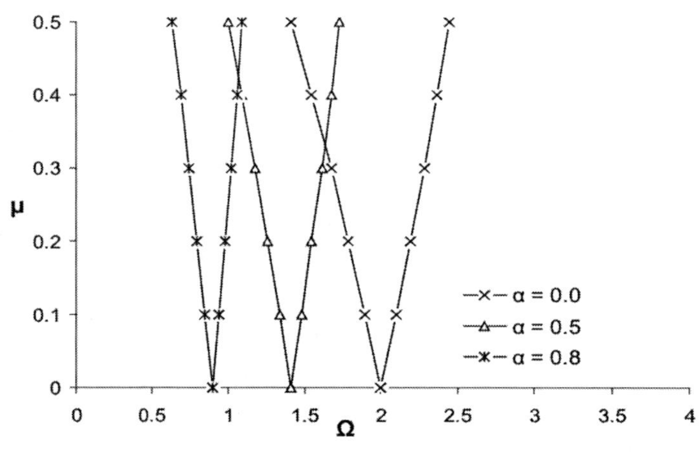

***Figure.9**. Master Dynamic stability curves*

IOP Conf. Series: Materials Science and Engineering 998 (2020) 012043 doi:10.1088/1757-899X/998/1/012043

6. CONCLUSIONS

Analytical solutions are obtained to simplify the dynamic stability behavior of simply supported beams resting on an elastic foundation subjected to an end periodic concentrated axial load. single term trigonometric admissible function, which satisfies the geometric boundary conditions, is used to obtain the feasible solution by means of classical Rayleigh – Ritz method. The effect higher transition foundation on the dynamic stability boundaries is established. The results, obtained with varying second transition foundation parameter, below and above this transition value, are presented in this work. As the elastic foundation increases, the width of the regions of dynamic instability decreases, thus making the beam less sensitive to the dynamic stability phenomenon to periodic loads. Finally, the use of the non-dimensional parameters derived in the present work is demonstrated, for the beams on elastic foundation, to obtain the master dynamic stability boundaries (curves) for the same mode for stability and vibration problems, the reference critical/buckling load and frequency parameters are obtained with consideration of foundation. No attempt is made here to validate the formulation as it has been established beyond doubt in the earlier studies of the author.

Acknowledgements

The authors are grateful to the managements of their respective organizations for their constant encouragement during the course of this work.

References

[1] Timoshenko S.P. and Gere J. M., 1962, Theory *of Elastic Stability*, Mc Graw- Hill, New York.
[2] Bolotin V.V., *Dynamic Stability of Elastic Systems*, Holden-Day, SanFrancisco, 1964
[3] Brown J. E., Hutt J. M. and Salama A. E., 1968, Finite element solution to dynamic stability of bars, *AIAA Journal*, vol.6, pp.1423-1425.
[4] Rao G.V., Ratnam B. S. and Janardhana G. R., 2008, Master Dynamic Stability Formula for Structural Members Subjected to Periodic Loads, *AIAA Journal*, vol.46, pp. 537-540.
[5] Rao G.V., Ratnam B. S., Jagadish Babu Gunda and Janardhana G. R., 2011, Master Formula for Evaluating Vibration Frequencies of Structural Members under Compressive Loads, *The IES Journal Part A: Civil & Structural Engineering,* vol.4(2), pp.79-88.
[6] Abbas B. A. H. and Thomas J., 1978, Dynamic stability of Timoshenko beams resting on an Elastic foundation, *Journal of Sound and Vibration,* vol.60, pp.33-44.
[7] Ratnam B.S., Rao G. V. and Janardhana G. R., 2011, Dynamic Stability of Beams on Elastic Foundation, *Journal of Structural Engineering*, vol.38(2), pp.150-160.
[8] Kanaka Raju K., and Venkatesara Rao G., 1993, Vibration of Initially Stressed Beams and Plates around Transition Values of Elastic Foundation Stiffness, *Journal of Sound and Vibration*, vol.161, pp.378-384.
[9] Engel R.S., 1991, Dynamic stability of Axially Loaded Beam on an Elastic foundation with Damping, *Journal of Sound and Vibration*, vol.146 (3), pp.463-477.
[10] Lee H. P.,1996, Dynamic Stability of a Tapered cantilever Beam on an Elastic Foundation subjected to a Follower Force, *International Journal of Solids and Structures*, vol.33(10), pp.1409-1424.
[11] Yokoyama, T., 1988, "Parametric instability of Timoshenko beams resting on an elastic foundation." *Computers & structures,* vol.28, pp. 207-216

[12] B. Subbaratnam, M. Harsha and S.Naresh Kumar,2019, " Dynamic Stability of Plates Resting on Elastic Foundation" *International Journal of Mechanical and Production Research and Development (IJMPERD),* vol. 9, pp.581-590.

IOP Conf. Series: Materials Science and Engineering 998 (2020) 012044 doi:10.1088/1757-899X/998/1/012044

Multi-Criteria Optimization of Aerospace Alloys in WEDM Using TOPSIS

SANJAY HERALGE[1], UMASHANKER GUPTA[2] and JAMIL DABIR[3]

[1,3]Mechanical Engineering Department, Walchand College of Engineering, Sangli, India
[2]Defence Metallurgical Research Laboratory, Hyderabad, India
Email: heralgesanjay@gmail.com

Abstract. Wire EDM is commonly used for machining of complex profiles in hard and exotic materials. In the present work the key performance of Wire EDM such as wire cutting speed, surface roughness and cut width in machining of three different aerospace class materials such as near beta titanium alloy, forged nickel base alloy and aged aluminum alloy are studied. Apart from electrical parameters, the wire with zinc coated brass and plain brass are used. TOPSIS (Technique for Order Preference by Similarity to Ideal Solution) is the Multi-Criteria Decision-Making method used for selecting the best setting of control parameters. Parameter sets are ranked from best to worst based on the performance score. Zinc coated wire outperforms the brass wire almost for every response variable in case of Ni and Ti alloy while with Al it is as good as plain wire.
Keywords. Wire EDM, TOPSIS, Cutting speed, Kerf width, Surface roughness

1. Introduction

Wire Electro-Discharge Machining (WEDM) is a non-conventional machining process, which is popularly used in industries for precision machining of difficult to machine electrically conductive materials. In Wire EDM, wire is used as tool for removing material from intended workpiece. Thousands of electrical discharges per second is generated between tool and work, which produces tiny craters by melting and evaporation on work material.

F. Nourbaksha et al. [1] carried out the Wire EDM of Ti6Al4V and found that using zinc coated wire gives better surface finish and high cutting speed. Mohapatra et al. [2] optimized gear cutting of Inconel 718 with wire EDM by using TOPSIS method. Prohaazka et al. [3] investigated the impact of Wire EDM on annealed non-alloyed steel with minimum carbon content using magnesium, zinc and tin wire. They mentioned term called as "Heat sink effect" which states that in zinc coated wire during the sparking zinc evaporates keeps core away from damage and keeps it cool. Muniappan et al. [4] analyzed the effect of machining parameters on kerf width and surface roughness using TOPSIS Multi-Criteria Decision Making (MCDM) strategy to evaluate ideal Wire EDM machining parameter for aluminum hybrid composite machining with the use of brass wire. S. Dhale et al. [5] compared the performance of plain brass wire and zinc coated wire on Inconel 718 material. MRR with zinc coated wire observed to be higher than brass wire but the increase in the machining speed results into the higher surface roughness also. Santhanakumar et al. [6] optimized micro wire electrical discharge machining (μ-WEDM) while machining of Ti alloy using TOPSIS with hybrid approach and response surface methodology. Jangra et al. [7] carried out the study with Wire EDM trim cut on materials such as HCHCr steel alloys, WC-Co composite, Monel-400 and Nimonic-90. They studied the effect of material properties on WEDM. N.Chakladar [13] used TOPSIS with AHP (Analytic Hierarchy Process) and studied their combined effect on selection of process in non-traditional machining process and also discussed how TOPSIS combined with AHP satisfies the multi-criteria decision making (MCDM). A.Singh [14] adopted TOPSIS with Taguchi method for optimization of multiple parametrs of surface roughness machined with GFRP polyester composites and concluded why TOPSIS is best method for MCDM. The approach used in TOPSIS is ranking based which further used to choose the best and worst alternatives from positive (as near as possible to the ideal solution)

IOP Conf. Series: Materials Science and Engineering 998 (2020) 012044 doi:10.1088/1757-899X/998/1/012044

and negative (as far as possible from the ideal solution) ideal solutions. TOPSIS is relatively simple to understand and implement in case Multiple Criteria Decision Making problem. Assigning weights to normalized value of outputs or responses based on application or requirement and calculating performance score, which is used to rank the alternative, this makes TOPSIS very easy to understand and implement in real life problem. Therefore, in this study also TOPSIS with FAHP (Fuzzy Analytic Hierarchy Process) is used for selection of process parameters. From the past literature, it is concluded that selection as well as optimization of proper process parameter is very important in manufacturing industry. The effect during machining would be adverse if improper machining parameters have been selected. There are very limited works in Wire EDM of aerospace material is available. Therefore, the present work focused on optimization of three aerospace alloys using TOPSIS and their performance has been compared on cutting speed, kerf width or cut width and surface roughness. The generated machining data will be useful for academic and industries involved in processing of aerospace materials.

2. Materials and Methods

The experiment was performed on 'Electronica Ultracut' wire EDM using power pulse mode on near beta Ti alloy for structure applications, forged Ni based alloy for turbine disc application and aged Al alloy for structure applications. Two types of wires, uncoated brass wire or plain wire (UC) and zinc coated wire (C) of 0.25mm diameter are used. Taylor Hobson's contact type surface roughness profilometer with 0.8mm cut off and ISO gaussian filter used for measuring surface roughness of workpiece. For reading purpose, each surface measured and total six measurements were taken and then the results are averaged. Kerf width or width of cut are measured using optical profile projector. Samples are machined for straight 15 mm length in 10 mm thickness. The parameters for study are carefully selected from literature and to cover the range of parameters specified in machine manual. Table 1 shows the parameter chosen for the study, where pulse on time (T_{ON}), current (I) and type of wire are varied and other parameters are kept constant at particular values considering minimum wire breakage. The Pulse on time is varied at three different level and current with two levels. In 'Electronica Ultracut' wire EDM machine current is available in two settings low current (11) and high current (12) and the values 11 and 12 are not current in amperes

Table 1. Machine process parameters used in the experiment.

Sr No	Parameters	Value
1	Pulse on time (μsec)	110,120 and 130
2	Pulse off time (μsec)	50
3	Servo voltage (Volts)	20
4	Wire tension (N)	15
5	Wire feed (m/min)	8
6	Water pressure (Kg/cm^2)	4
7	Servo feed (mm/min)	25
8	Current	11,12
9	Type of wire	Plain brass and zinc coated

. To understand the difference in three metal alloys selected for study their physical and mechanical properties are listed in table 2 such that one can compare the properties easily. Thermal conductivity, tensile strength, young's modulus, melting point and density these are the properties listed in table 2 for three alloys.

Table 2. Physical and mechanical properties of work materials.

Work Material	Properties of work material				
	Thermal conductivity (W/mK)	Tensile strength (MPa)	Young's modulus (GPa)	Melting point (°C)	Density (gm/cm3)
Near Beta Ti alloy	7.8	930	110	1642	4.65
Forged Ni based alloy	11.72	780	220	1320-1365	8.19
Aged Al alloy	116-170	172-476	73.8	543	2.84

3. Results and Discussion

Table 3 illustrates the experimental plan and obtained performance data. Each alloy was machined on two current settings and on each current pulse on time varied to three levels with coated and uncoated wire. Three responses cutting speed (Vc), Kerf width (KW) and Surface roughness (Ra) for all three alloys are listed in table is such way that in single table itself one can draw meaningful conclusion. Total 12 experiments are carried on each metal alloys and above three responses are measured for every experiment and thus total 36 experiments are carried out in entire experiment and all are shown in table 3.

Table 3. Experimental data obtained for Ti, Ni and Al alloy.

	Control Factors			Responses								
				Ti alloy			Ni alloy			Al alloy		
Ex. No.	Wire	I	T_{ON}	Vc (mm/min)	KW (mm)	Ra (μm)	Vc (mm/min)	KW (mm)	Ra (μm)	Vc (mm/min)	KW (mm)	Ra (μm)
1	UC	11	110	0.923	0.3424	2.3100	0.831	0.3041	2.7580	1.57	0.2915	3.0000
2	UC	11	120	1.25	0.3872	3.0248	1.1278	0.312	3.7091	1.6519	0.3343	3.9924
3	UC	11	130	1.25	0.3853	2.9410	1.2087	0.3148	3.6203	1.6538	0.339	4.1900
4	UC	12	110	1.0018	0.3479	2.4857	0.9225	0.2931	2.6206	1.6146	0.3043	3.0251
5	UC	12	120	1.49	0.4102	3.3575	1.415	0.34	3.4622	1.8248	0.3776	4.2400
6	UC	12	130	1.5	0.421	3.4075	1.4763	0.3391	3.7369	1.8292	0.3808	4.6538
7	C	11	110	1.03	0.3421	2.2337	0.95	0.2914	2.5625	1.5806	0.2947	3.0261
8	C	11	120	1.3	0.3839	3.1144	1.2	0.311	3.5165	1.6556	0.3387	3.9234
9	C	11	130	1.32	0.3879	2.9465	1.19	0.321	3.5033	1.6538	0.347	3.9575
10	C	12	110	1.12	0.3422	2.3577	1.0252	0.2929	2.6277	1.6198	0.3044	3.2460
11	C	12	120	1.6	0.4056	3.2515	1.4691	0.3399	3.6100	1.8181	0.367	4.5586
12	C	12	130	1.61	0.4081	3.2504	1.5151	0.3377	3.6527	1.8337	0.3713	4.4384

3.1 Effect of T_{ON} and type of wire on cutting Speed (V_C)

Cutting speed directly affects the productivity of machining process. Figure 1 shows the wire cutting speed of all metal alloys with coated and uncoated wire at high current setting (12). As the T_{ON} increases the Vc increases for all three metal alloys and further better with coated wire. The reason for higher cutting speed with coated wire is 'Heat sink effect' i.e. increase in conductivity within the sparking gap due to the fast melting and vaporization of the zinc particles. This leads to improved flushing condition results in increase in spark gap width, which protects the brass core from damage and also maintain the lower temperature at the core [3].

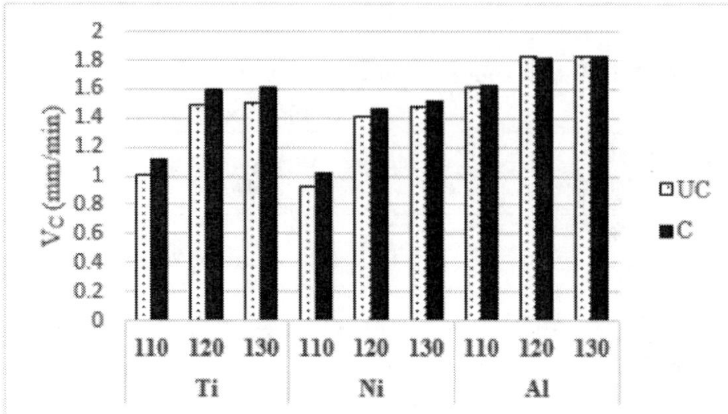

Figure 1. Effect of T_{ON} and type of wire on cutting speed.

Among three material, Al alloy exhibits higher cutting speed than other alloys. For Al alloy cutting speed with coated wire and uncoated wire is almost same. Figure 1 states that cutting speed of: Al > Ti > Ni this may be because the Al alloys has lowest melting point and highest conductivity among all three metals which leads easy and more metal removal because of intense sparking. As the behavior of cutting speed at low current is same as that of in high current only the value of cutting speed less in low current hence only data machined with high current is represented in graph which is sufficient to understand the impact of control factors on response variables. If thermal conductivity of material is high then increase in fraction of heat transfer to the most of the bulk material and therefore less melting and evaporation of work material takes place [7] which is here in case of Al alloy where little increase in cutting speed is observed with increases in T_{ON}. Whereas Ti has low thermal conductivity, which results more increase in cutting speed with increase in T_{ON}.

3.2 Effect of T_{ON} and type of wire on Kerf width (KW)

The amount of material which is wasted while machining in the gap of wire and workpiece is called as kerf width. Due to this kerf width, we cannot get desired dimensions which decreases the dimensional accuracy of machined part. Therefore, the kerf width should as minimum as possible. The kerf with is equals to wire diameter added with 2 times of spark gap. As Ton increases, the kerf width also increases as shown in figure 2. Machining with zinc coated wire results in lesser kerf width than with uncoated brass wire. In the figure 2 below, kerf width at high current setting (12) is represented.

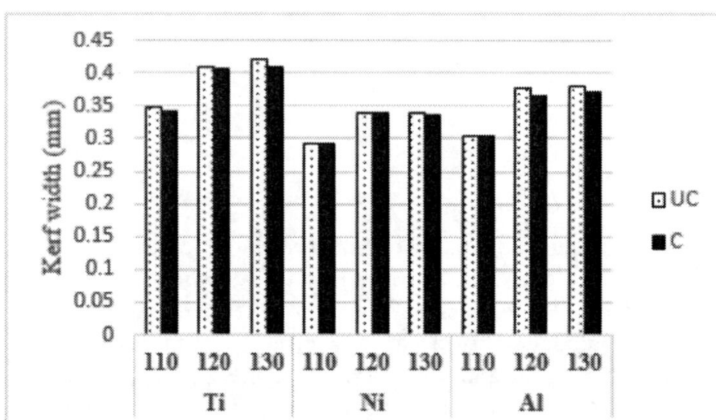

Figure 2. Effect of T_{ON} and type of wire on Kerf width.

Kerf width is also dependent on the work material thermo-physical properties such as melting point, thermal conductivity etc. As the thermal conductivity is less more material will be melted and evaporate as compare to the material with higher thermal conductivity. Hence, in Ti alloy kerf width is more than other two metal alloys.

3.3 Effect of T_{ON} and type of wire on Surface roughness (Ra)

In WEDM process surface roughness of machined surface is depends on the size and depth crater produced after melting as well as expulsion of work material. As shown in figure 3 as the T_{ON} increases, the discharge energy increases which leads to large and deep craters results in more surface damage and increased surface roughness.

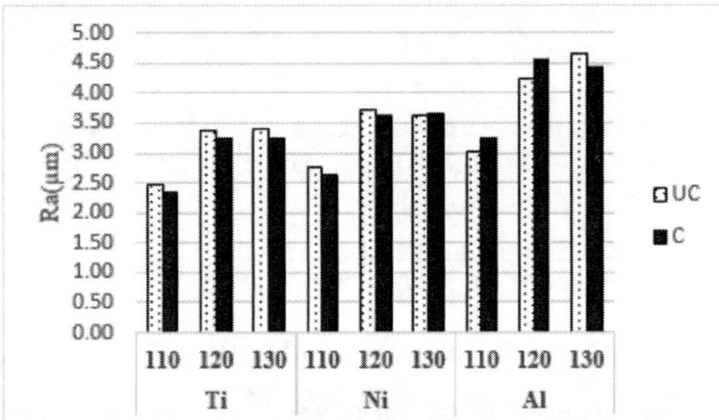

Figure 3. Effect of T_{ON} and type of wire on Surface roughness.

If the work material have low melting point, high discharge energy results into overheating and evaporation of molten metal forming gas bubble that explodes when discharge ceases. Hence Al alloy having low melting temperature has more surface roughness with both coated and uncoated wire whereas Ti and Ni alloy with high melting point has less surface roughness than Al alloy. Surface roughness when machined with coated wire is less than uncoated wire for Ti and Ni alloy but for Al alloy surface roughness with coated wire is more. As per previous literature in some cases with use of zinc coated wire surface roughness was improved [1] and, in some cases, surface roughness was not improved like in work of Kapoor et al. [10] hence there is no uniformity in results with coated wire that states that the surface roughness always improves when coated wire is used. Improved surface roughness with the zinc coated wire may be due to the existence of zinc and this provides higher strength to wire and this wire maintains itself straight under vibration [1]. In addition to this eventual evaporation of zinc coating leads to less debris in spark gap results in minimum secondary sparking.

3.4 Multi-Criteria Decision Making (MCDM)- TOPSIS Optimization

TOPSIS-Technique for Order Preference by Similarity to Ideal Solution is closest to the ideal alternative and so far from the negative ideal. Table 4 consist of summarized data from TOPSIS method where the data from every step is represented in table format for easy understanding and V_j^+ and V_j^- are listed below end of respected column for better view and understanding. Overall, all the data from TOPSIS steps are summarized in single table named table 4. The TOPSIS method consists of the following step [2, 4, 6, 13, 14, 15]:

Table 4. Multi-objective decision making using TOPSIS method.

Ex No	Wire	I	TON	Vc	KW	Ra	Si+	Si-	Pi	Rank
1	UC	11	110	0.013647	0.056969	0.162958	0.011494	0.250801	0.956178	3
2	UC	11	120	0.018482	0.064423	0.213385	0.056559	0.238572	0.80836	7
3	UC	11	130	0.018482	0.064107	0.207473	0.050689	0.239619	0.825395	5
4	UC	12	110	0.014812	0.057884	0.175353	0.019943	0.248278	0.925647	4
5	UC	12	120	0.022031	0.06825	0.236859	0.080105	0.237693	0.747937	11
6	UC	12	130	0.022178	0.070047	0.240387	0.083857	0.238113	0.73955	12
7	C	11	110	0.01523	0.05692	0.15758	0.008576	0.251537	0.967031	1
8	C	11	120	0.019221	0.063874	0.219709	0.062686	0.237695	0.791312	8
9	C	11	130	0.019517	0.06454	0.207861	0.051036	0.239659	0.824433	6
10	C	12	110	0.01656	0.056936	0.166323	0.011356	0.252187	0.956912	2
11	C	12	120	0.023657	0.067484	0.229381	0.072575	0.237585	0.766007	10
12	C	12	130	0.023805	0.0679	0.229304	0.07256	0.237638	0.766085	9
			Vj+	0.023805	0.056919	0.157579				
			Vj-	0.013647	0.070047	0.240387				

Step 1: Calculate normalized Matrix (normalized values) using below formula, X_{ij} represents the actual value of output in cell. The next is to calculate the $\sum_{i=1}^{n} X_{ij}^2$ that is sum of all the values in single column then use it in formula given below where n=1,2..12 which are number of experiment or number of alternatives in this study and repeat this for every column to prepare matrix. Considering cutting speed as beneficial attribute due to higher values while Ra and KW as non-beneficial because of smaller values.

$$\overline{X}_{ij} = \frac{X_{ij}}{\sqrt{\sum_{i=1}^{n} X_{ij}^2}} \tag{1}$$

Where i represents rows or number of experiments and j represents column or number of criteria or output responses.

Step 2: Calculate the weight for every response variable. In this study, the Fuzzy Analytic Hierarchy Process (FAHP) with Geometric mean [11] is used to calculate the weight. The inter-relations used to calculate the weight is: if Vc = x then Kerf width = 5x, if Cutting speed (Vc) = x then Surface roughness (Ra)= 7x and if kerf width (kw)= x then surface roughness= 5x. These relations can be changed based on customer requirement and area of application. In this study, these relations are taken in order to have minimum surface roughness, minimum kerf width and maximum cutting speed. The relations are managed in such way that more weightage will be assigned to surface roughness then to kerf width and less to cutting speed.

Based on this the weights (Wj) obtained for Vc= 0.0667, KW= 0.2198 and Ra= 0.7134

Step 3: Calculate the weighted Normalized matrix, by multiplying weights to normalized value

$$Vij = \overline{X}_{ij} \times Wj \tag{2}$$

Where Wj is weight calculated in step 2, $\overline{X}_{ij}$ is the normalized value in for ith row or experiment from jth column or output response from step1 calculated using equation (1) and V_{ij} is the weighted normalized value for ith row or experiment from jth column or output response. The values in the column of Vc, KW and Ra in table 4 are weighted normalized values.

Step 4: Select the ideal best (V_j^+) and ideal worst (V_j^-) value for every column in matrix or table that is from V_{ij} calculated in step 2 with help of equation (2). Ideal best is maximum value for beneficial attribute and minimum value for non-beneficial attribute whereas ideal worst is minimum value for beneficial attribute and maximum value for non-beneficial attribute from every column or output response . For example, V_j^+ for surface roughness is minimum value of V_{ij} form surface roughness column and for cutting speed it is maximum value of V_{ij} from cutting speed column.

Step 5: Calculate the Euclidian distance from the ideal best

$$S_i^+ = \left[\sum_{j=1}^{m} \left(V_{ij} - V_j^+ \right)^2 \right]^{0.5} \tag{3}$$

Where S_i^+ the Euclidian distance from the ideal best, V_{ij} is weighted normalized value in respective cell and V_j^+ is the ideal best value of respective column as shown in table 4.

Step 6: Calculate the Euclidian distance from the ideal worst

$$S_i^- = \left[\sum_{j=1}^{m} \left(V_{ij} - V_j^- \right)^2 \right]^{0.5} \tag{4}$$

Where S_i^- is the Euclidian distance from the ideal worst, V_{ij} is weighted normalized value in respective cell and V_j^- is the ideal worst value of respective column as shown in table 4. Here m=3 ie. for Vc, KW and Ra column.

Step 7: Calculate the Performance score (P_i) and rank this to take decision.

$$P_i = \frac{S_i^-}{S_i^+ + S_i^-} \tag{5}$$

Where (P_i) represents the performance score and S_i^+ and S_i^- calculated from equation (3) and (4) respectively. Larger performance score represents best choice among all alternatives

Table 4 shows the data obtained by TOPSIS method for Ti alloy where ranking is done by decreasing order of performance score obtained from equation (5) which are : 7-10-1-4-3-9-2-8-12-11-5-6 where it is understood that the experiment no. 7 which is ranked 1st is first best choice which will give minimum surface roughness, minimum kerf width and maximum cutting speed. Next one is experiment no. 10 which is ranked as 2nd is second best choice. Optimized values for Ti alloy are from the 1st best choice experiment no.7 having the highest performance score. The values are current setting 11, coated wire and T_{ON}=120. Same procedure was followed to calculate the performance score for both Ni and Al alloy. The experiment numbers, their sequence and values are same for all three alloys as given in table 3 (also in table 4) and their response output is also given in table 3. Hence individual tables for Ni alloy and Al alloy representing calculation of TOPSIS method are not shown in this paper only their experiments number are ranked based on the performance score after following same steps. For Ni alloy the experiment numbers are ranked in descending order of their Performance score as: 10-7-4-1-5-9-8-11-3-12-2-6 which indicates that experiment no. 10 is first best choice. Optimized values for Ni alloy are from the best choice experiment no.10, which are current setting 12, coated wire and T_{ON}=110. For Al alloy the experiment numbers are also arranged in descending order of their Performance score as :1-7-4-10-8-9-2-3-5-12-11-6 where experiment number 1 is first best choice. Optimized values for Al alloy are from the best choice experiment no.1, which are current setting 11, uncoated wire and T_{ON}=110

It is observed that the experiment number 7 is in first two best choices in every metal alloy whereas experiment number 6 is last best choice (worst) for all metal alloys.

4. Conclusions

In present work, an empirical study in Wire EDM has been conducted on three different metal alloys namely near beta Ti alloy, forged Ni alloy and aged Al alloy. Cutting speed, kerf width and surface roughness were the performance characteristics. With the help of TOPSIS all experiments were ranked based on their performance score where it was observed that experiment number 7 was there in top 2 best choices in every metal alloys which shows that the machining with coated wire leads to desirable results. Results shows that with increase in T_{ON} cutting speed, surface roughness and kerf width increases which is due to the increased discharge energy lead to high material removal. Coated wire due to its zinc coating that is having less melting point evaporates early which leads to increased spark gap to improve flushing efficiency results into high cutting speed and decreased kerf width. In future, work the surface integrity in wire EDM using coated wire can be taken up.

Acknowledgments
Authors would like to acknowledge Director 'Defence Metallurgical Research Laboratory, Hyderabad' for providing all the requisites to carry out this experimentation work and thanks to the staff of 'Walchand Collge of Engineering, Sangli' for their help and support in experimental work.

References

[1] Nourbaksh F, Rajkumar K P, Malshe A P and Cao J 2013 Wire Electro-discharge Machining of Titanium Alloy, *Procedia CIRP* **5** pp 12-18,
[2] Mohapatra K D and Sahoo S K 2018 A Multi Objective Optimization of Gear Cutting in WEDM of Inconel 718 using TOPSIS Method. *Decision Science Letters* **7**, pp 157-170.
[3] Prohaszka J., Mamalis A.G and Vaxevanidis N.M. 1997 The Effect of Electrode Material om Machinability in Wire Electro-Discharge Machining. *Journal of Material Processing Technology* **69**,pp 233-237
[4] Muniappan A, Jaivaakheish A P, Jaykumar V, Arunagiri A and Senthikumar A 2018 Multi Objective Optimization of Process Parameters in WEDM of Aluminum Hybrid Composite Using Taguchi and TOPSIS Technique. *IOP Conference Series: Material Science and Engineering* **402**, pp 1-10
[5] Dhale S. R and Kulkarni M.L 2016 Effect of Electrode Coating in Wire Electrical Discharge Machining of Inconel-718, *ISOR-JMCE* **13**,pp 53-59
[6] Santhanakumar M and Adalarasan R 2017 An Integrated Approach of TOPSIS and Response Surface Methodology for Optimising the Micro WEDM Parameters. *Int. J. Operational Research* **28**, pp 18-34
[7] Jangra kk, Kumar V and Kumar V 2014 An Experimental and Comparative Study on Rough and Trim Cutting Operation in WEDM of Hard to Machine Material. *Procedia Material Science* **5**, 1603-1612
[8] Baindur T, Patil M and Rajyalakshmi G 2017 WEDM Machining of Aerospace Material for Improving Material Properties. *Materials Today: Proceedings* **4**, pp 9107-16
[9] Kapoor J, Singh S and Khamba J 2010 Recent Developments in Wire Electrodes for High Performance WEDM. *Proceedings of the World Congress on Engineering* **2**

IOP Conf. Series: Materials Science and Engineering 998 (2020) 012044 doi:10.1088/1757-899X/998/1/012044

[10] Kapoor J, Khamba J and Singh S 2013 Effect of Different Electrodes on the performance of WEDM. *Material Science Forum* **751**, pp 27-34

[11] Calik A, Paksoy T and Pehlivan N.Y 2018 Comparison of Methods in FAHP with Application in Supplier Selection. *Fuzzy Analytic Hierarchy Process, Chapman and Hall/CR,* ed A Emrouznejad and W Ho (Florida),pp 45-77,

[12] Gadakh V 2012 Parametric optimization of Wire Electrical Discharge Machining Using TOPSIS method. *Advances in Production Engineering and Management* **7**, pp 157-164

[13] Chakladar N and Chakroborty S 2008 A Combination TOPSIS- AHP Method Based Approach for Non-Traditional Machining Process Selection. *Proceedings of the Institution of Mechanical Engineers, Journal of Engineering Manufacture* Part B **222,** pp 1613-23

[14] Singh A, Datta S and Mahapatra S 2011 Application of TOPSIS in the Taguchi for Optimal Machining Parameter Selection. *J. Manuf. Sci. Prod.* **11**, pp 49-60

[15] Bhat R, Mohan N, Sharma S, Kini A, Shivakumar S and Naik N 2020 Multi Response Parametric Pptimisation in Machining of Marine Application Based GFRP Composite with HSS Drill: Application of TOPSIS approach. *Materials Today: Proceedings*

IOP Conf. Series: Materials Science and Engineering 998 (2020) 012045 doi:10.1088/1757-899X/998/1/012045

Stabilization of Expansive Soil by using Jute Fiber

Dharmesh Lal[1], Pravalika Reddy[2], R. Sandeep Kumar[3], V. Govardhan Rao[4]

[1]Assistant Professor, Department of Civil, Vardhaman College of Engineering, Hyderabad, India

[2,3,4] B.Tech Student, Department of Civil, Vardhaman College of Engineering, Hyderabad, India

E-mail:dharmeshlal34@gmail.com

Abstract. The instability of structure lies mainly in the properties of soil under which it is constructed. For instance, if the soil is having weak engineering properties, there are chances of crack formation and settlement. One of the major problematic soils of this type are the Black cotton soils which are mainly classified as expansive soils. They are obviously a hindrance to the strength and stability of structures. The present study deals with the improvement of soil characteristics with the addition of Jute fiber sheet. A series of Triaxial and CBR tests were done on normal soil and Jute fiber reinforced one. The observed values indicate a significant contribution of strength improvement when fiber sheets were added.

1. Introduction

Soil can be considered as a complex material. The construction of any road or pavement consisting of number layers which includes subgrade, sub base, Base and Wearing course. Sub grade is very important in road construction. If subgrade soil is expansive soil or black cotton soil which as a tendency to swell or shrinkage characteristics. So the structures cannot be constructed on these types of soils without any stabilization. Stabilization means the process of improving the properties of soil by using suitable additives such as fly ash, cement, jute fiber, lime, plastic powder etc. In this investigation jute geotextile is used to stabilize the expansive soil or black cotton soil. The geo textile used for this investigation was Tossa jute. Jute fiber can be proved as a more superior compare to other fibers in terms of its stability or strength, functionality etc.

Black cotton soil is formed due to weathering of rocks. Black cotton soils are black in color due to the presence of titanium oxide. These soils are very hard in dry condition but losses its strength or stability in wet condition. Due to the presence of high content of montmorillonite makes the soil is more expansive. The property of high swell and shrinkage characteristics or volumetric changes the soil can be considered as dangerous in construction activities. In India these soils are found in many areas which include Andhra Pradesh, Madhya Pradesh, Maharashtra, Gujarat, Orissa etc. Enhancement in performance with the addition of Jute fibers were reported by [1-6]. In a comprehensive study by [4], it was observed that Jute fibers provide an efficient stabilization for expansive soils. It was also observed that Optimum moisture content decreased and dry density increased with the addition of Jute fiber [6].

IOP Conf. Series: Materials Science and Engineering 998 (2020) 012045 doi:10.1088/1757-899X/998/1/012045

2. Materials and Methodology

2.1 Clay

Generally clayey soils are fine grained soils, having particle size less than 2µ. Clayey soils (shown in Figure 1) are plastic in nature due to its size and geometry as well as moisture content, and these soils become hard, brittle and non-plastic upon drying. Table 1 lists the properties of clay.

Figure 1 Photograph of Clay

Table 1 Properties of soil

IOP Conf. Series: Materials Science and Engineering 998 (2020) 012045 doi:10.1088/1757-899X/998/1/012045

Property of soil	Value
Liquid limit	59%
Plastic limit	37.925 %
Plasticity index	21.075%
Specific gravity	2.125
Optimum moisture content (OMC)	24.27%
Maximum dry density (MDD)	14.915 kN/m^3
Cohesion	54.96 kPa
Angle of Internal Friction (degree)	6^0
CBR	1.05%

2.2 Jute Fiber

Jute fiber is also called as Golden Fibre. The natural jute fiber reinforcement is shown in Figure 2.

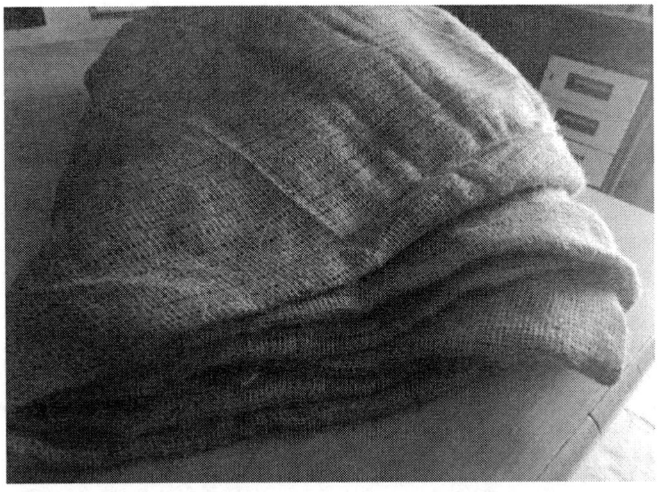

Figure 2 Photograph of Jute Fiber

2.3 Methodology

Basic tests were conducted for clay and after that a detailed elaborate study on the unreinforced and Jute fiber reinforced soil were conducted. CBR and Traxial were the main tests to identify the engineering behavior of soils. Jute fibers were laid in three layers to evaluate the performance characteristics. Triaxial and CBR tests were conducted for reinforced soil by inclusion of 1, 2 and 3 layers of jute fiber.For one layer, the fiber was placed at middle of the sample , for two layer it was at a distance of 1/3rd and 2/3rd of the sample and finally for three layer, jute fiber was located at a distance of 1/4th , 2/4th and 3/4th of the sample.

The soil samples of 38mm diameter and 76 mm height were prepared and tested under tri axial condition. Various tri axial tests were conducted on different samples with or without jute layer reinforcement.

3. Results and Discussion

A detailed and elaborate Triaxial and CBR tests wrere conducted to study the performance improvement. Stress strain graphs were plotted to find our C and Ø values of the samples. Table 2 shows the results of values obtained for reinforced and Jute fiber reinforced soil.

Figure 3 shows the results of Cohesion values with the addition of fiber. The results clearly shows an increase in shear strength properties of soil. This is mainly attributed to the multi-cellular structure of the Jute fiber, which effectively bonds and gives tensile strength to the soil. However, beyond two layers, cohesion value decreased. This clearly shows that optimum is already attained and adding of an additional layer may reduce the reinforcing effect.

However, it was also observed that there was no clear variation in angle of friction as evident from the table. Several other studies from the literature [1-3] comply with the present study results

Table 2 C and Ø values for different layers of jute fiber

No. of layers jute fiber	**C(kN/m^2)**	**Ø(degrees)**
0	54.96	6
1	58.54	6.1
2	153.52	6.3
3	72.11	6.25

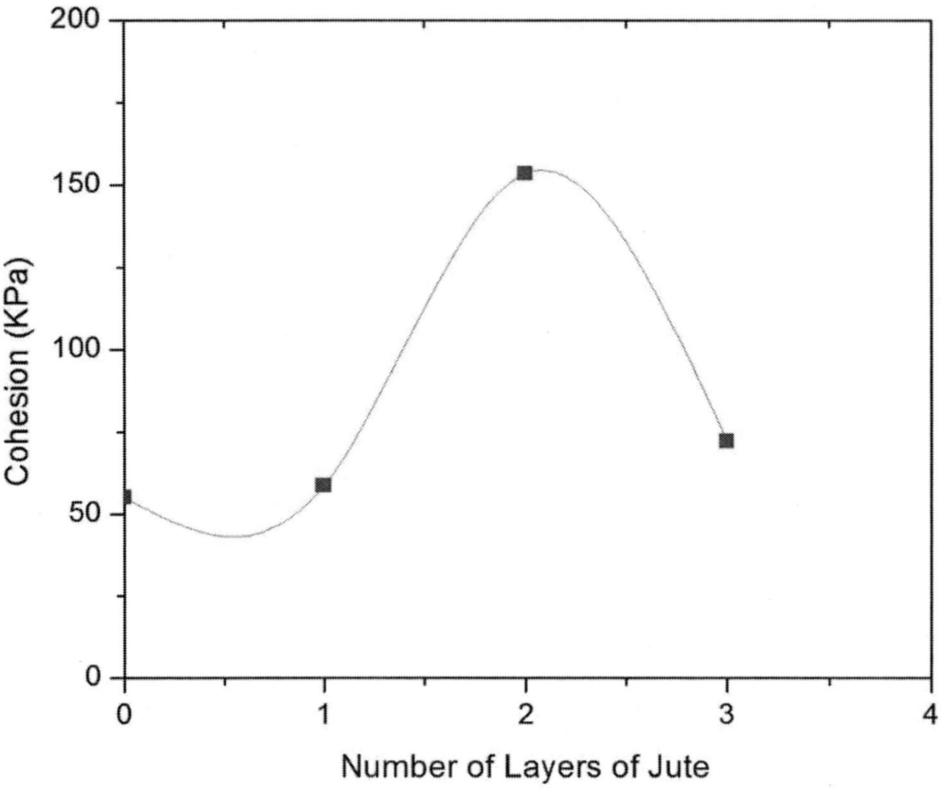

Figure 3 No.Of Jute Sheet Layers vs. Cohesion

The California bearing ratio test is conducted for both unreinforced and reinforced soil with 2 layer jute fiber sheet. It has been noted that the CBR value is increased by 21 % for 2.5mm penetration and 16.26 % for 5mm penetration.This is mainly due to the strength and stiffness enhancement happened with the soil.

4. CONCLUSION

Various tests were conducted for both reinforced and unreinforced soil to determine the strength and stiffness characteristics. In this investigation jute fiber sheet were arranged in 1, 2 &3 layers.

Based on the results obtained from this investigation, the following conclusions could be drawn:

IOP Conf. Series: Materials Science and Engineering 998 (2020) 012045 doi:10.1088/1757-899X/998/1/012045

- CBR and Triaxial tests showed that placing of jute fiber reinforcement in layers have significantly enhanced the strength and stiffness characteristics of soil.

- It was observed that placing of two layer jute fiber reinforcement is more effective compared to 1 & 3 layers.

- The C and ∅ values of the plain soil are 54.96 kPa and 6°, whereas it is 153.52 kPa and 6.3° respectively when two layers of jute fiber sheet were inserted.

REFERENCES

[1] Yagya Sharma (2019), "Improvement of Soil Properties by Using Jute Fibre as Soil Stabilizer", American Journal of Engineering Research (AJER), Vol 6, no 10, 123-129.

[2] Das T and Singh B, (2018), "Deformation and Strength characteristics of Jute Geotextile Reinforced Soils". JERAD, Vol. 8 No. 04, PP. 987-995.

[3] Singh H.P (2018) "Strength and Stiffness of soil Reinforced with Jute Textiles". IJCET, Vol.3, No.3, PP.1143-1146.

[4] BairagiHarshita, Yadav R.K and Jain R (2017), "Effect of Jute Fibres on Engineering characteristics of Black Cotton Soil". International Journal of Engineering Sciences & Research Technology, Vol 8, PP.705-707.

[5] Gray, D.H., and Ohashi, H (1983) "Mechanics of fiber reinforcing in sand". Journal of Geotechnical Engineering, 112 (8), 335-353.

[6] Aggarwal, P. and Sharma, B. (2010), "Application of Jute Fibre in the Improvement of Subgrade Characteristics" Geotextiles and Geomembranes., 28(1), 54–62.

[7] C. S., Shi, B., and Zhao, L. Z. (2010). "Interfacial shear strength of fiber reinforced soil". Geotextiles and Geomembranes., 28(1), 54–62.

[8] Islam M, Iwashita K (2010), "Earthquake resistance of adobe reinforced by low cost traditional materials". Journal of Natural Disaster and Science, Vol. 32, pp 1–21.

[9] Tera, S., and Reddy Jagannatha, H.P (2011) "International Journal of Innovation, Management and Technology, 2 (3), 186-191.

[10] Choudhary A, Gill K, Jha J; (2012), "Improvement in CBR of Expansive Soil Subgrade with a single Reinforcement layer". ECU Publications, Preceedings of Indian Geotechnical Conference, Paper No. B-214, Delhi.

[11] Singh Pardeep, Gill K.S; (2012) "CBR Improvement of Clayey soil with Geo-grid Reinforcement". IJETAE, Volume 2, Issue 6, PP. 315-318.

[12] Jagan B; 2016 "A Critical review on applications of natural Jute Fibers A Case Study". (IJCIET) Volume 7, Issue 3, 2016, pp. 200–210.

[13] Lal D, Sankar N, Chandrakaran S 2017. Surface heave behaviour of coir geotextile reinforced sand beds. J InstEng India SerA, 98(1–2), 121–125.

[14] Lal D, Sankar N, Chandrakaran S 2017. Performance of shallow foundations resting on coir geotextile reinforced sand bed. Soil Mechanics and Foundation Engineering, 54(1), 60–64.

[15] Anzar Hamid (2017), "Subgrade Soil Stabilization Using Jute Fibre as a Reinforcing Material", IJEDR.

[16] Yagya Sharma (2017), "Improvement of Soil Properties by Using Jute Fibre as Soil Stabilizer", American Journal of Engineering Research (AJER), Vol 6, no 10, 123-129.

MODEL AND MATERIAL ANALYSIS OF GOKART CHASSIS

D.V. Ramana Reddy[1,A], G. Revanth Kumar[2,B]

[1]Assistant Professor, Department of Mechanical Engineering, Vardhaman College of Engineering, Hyderabad, Telangana,

[2]PG Scholar, Department of Mechanical Engineering, Vardhaman College of Engineering, Hyderabad.

[a]Corresponding Author: ramana.mech@vardhaman.org

Abstract:
The present study of this paper is to select the best material for manufacturing the Go-Kart. To meet this designing of go-kart chassis done by using CAD software CATIA tool and simulate with ANSYS work-bench. To select the best material, strength to weight ratio plays major role on it. Steel is considering to existing material another two materials Aluminum-6061, Kevlar selected. The main reason behind choosing Aluminum-6061, Kevlar materials is two materials are having less density and high yield limit values, and this factor can increase the strength/weight ratio compare to existing material steel. In this project material selection is one of the main design decisions that greatly increase the safety, reliability and performance. The material selection is choosen after my extensive research with various parameters like strength, weight and cost. We are comparing Aluminum-6061and Kevlar materials with mild steel. In this process boundary conditions applied on each material and calculated results like deformation, stress, strain, safety factor, natural frequency values, by knowing all these results, concluded with optimum material.

Keywords: *Chassis, Go-Kart, Aluminum-6061, Kevlar, Natural Frequency.*

1. INTRODUCTION

Go-kart is a four wheeled vehicle; mostly it is used for racing purpose. Go-kart is a light weighted vehicle and it is easy to operate. Go-karts does not have suspension due to ground clearance is very small as compare to other vehicles. Mostly Go-Karts are used for racings. These Go-Karts are very low in cost and relatively safe way to introduce drivers, the main parts of Go-Kart are chassis, Engine and transmission, Steering, Breaking system, Wheels In our project the main focus is chassis and it consists of a body frame made up of pipes of various cross sections that are welded together which supports all the components, the chassis should be rigid from all the bending and torsion stress.

2. MATERIALS, MODELING AND METHODOLOGY

2.1Materials:

The materials used for our chassis are aluminum-6061 and Kevlar are taken as two materials and steel as existed material. After comparing all materials, choose these three materials because of their good properties which is suitable for Go-Kart.

IOP Conf. Series: Materials Science and Engineering 998 (2020) 012046 doi:10.1088/1757-899X/998/1/012046

Table1: Materials and their properties

Material	Steel	Al-6061	Kevlar
Yield strength	250MPa	276MPa	650MPa
Young's modulus	$2.0*10^5$MPa	$0.689*10^5$MPa	$0.83*10^5$MPa
Poissions ratio	0.29	0.33	0.29
density	7850 Kg/m^3	2700 Kg/m^3	1440 kg/m^3

2.2 MODELING:

We prepared the model by considering all the aspects and requirements. Model the GO-Kart Chassis in CAD software and then analyze it in ANSYS software, Analysis mainly depends upon the safety, comfort, cost, strength, weight, vibrations, stress, strain and deformations.

Table2: Chassis dimensions

Gokart parameters	values
Vehicle Length	50 inch (1270mm)
Vehicle wheel base	28 inch (711.2mm)
Vehicle width	20 inch (508mm)
Tube dimension	1 inch (25.4mm)
Pipe thickness	2 mm

2.3 METHODOLOGY:

The following Steps used in methodology:

- Model the GO-Kart using CATIA software tool
- Analysis with aluminum 6061, kelvar and steel materials by using ANSYS workbench
- Checking stress, strain, deformation, Natural frequencies
- Estimate the approximate weight and cost by comparing with three materials.

2.3.1 Model of chassis:

The modeling of the chassis is done in CATIA software, designed according to the requirements. Then the design is imported into the ANSYS for further analysis.

Figure1. Model of the chassis

IOP Conf. Series: Materials Science and Engineering 998 (2020) 012046 doi:10.1088/1757-899X/998/1/012046

2.3.2 Meshing of chassis model:

The meshing is the most important tool for analysis and one can expect drastic changes when the results of mesh are obtained. The chassis is divided into finite pieces so that each element is analyzed very extremely.

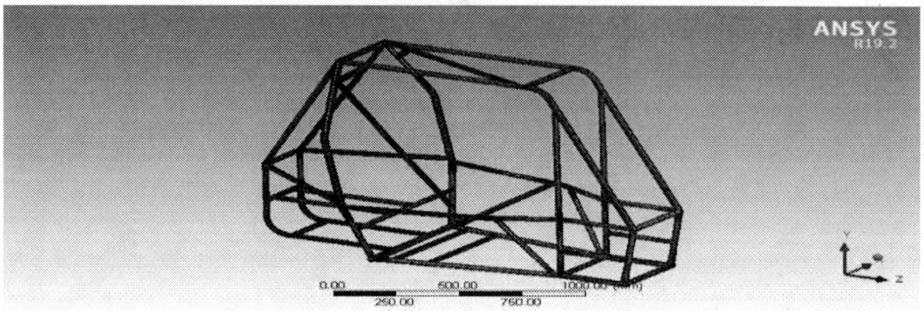

Figure 2: Mesh model of Chassis

2.3.3 Analysis parameters:

In this analysis we have to consider the vehicle which is running in real life. For that condition we have tested our chassis to verify whether the chassis is safe to sustain the condition. The value of force to be applied on the frame for this test could be obtained by one of the following methods. Impulse Method: As we know that force (F) multiplied by time of collision (t) is impulse and is equal to change in momentum of the body.

I.e. Mathematically, **F*t = m*(v-u)**

Where m is mass of the vehicle i.e. taken to be approximately 30 kg, u is the maximum velocity of the vehicle which we take as 60 kmph, and v is final velocity i.e. 0 kmph in critical condition. Here the unknown parameter is the collision time (t), which is obtained by prototype testing.

G-Force Method: This is widely used method when prototype testing is not a feasible solution, in which the force applied is calculated in F-force. Force is mass of the vehicle (m) multiplied with a (acceleration due to gravity)

Mathematically, **F = m*a = 30*9.81 = 294.3 N** (5)

Table3: forces the applied forces in different collisions

Front impact	4 G = 1177.2N
Rear impact	3 G = 882.9N
Side impact	2 G = 588.78 N
Roll-over	1.5 G = 441.45 N

2.3.4 Boundary conditions:

According to above mentioned calculations here maximum applied force could be 1.2KN approximately in any direction. So that here checking our object maximum impact force bearing capacity, here safety factor is consider to be measurement, and it should maintain minimum 1.5 to apply that load.

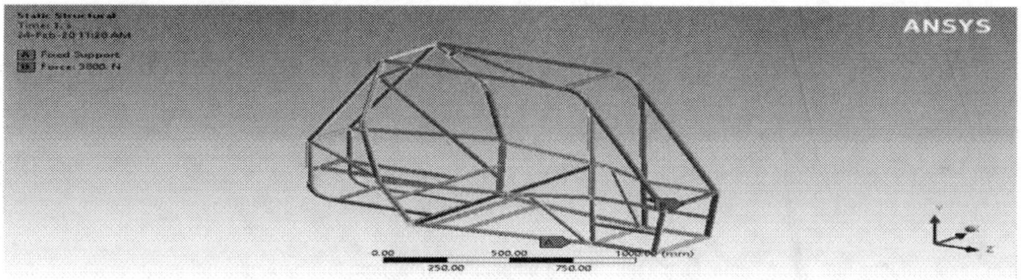

Figure3: Boundary conditions

Force →3.8KN

After completion of boundary conditions here we have to check results by solving. Just click on solve option and select results like deformation, strain, stress and safety factor values for the object.

2.3.5 Post processing:

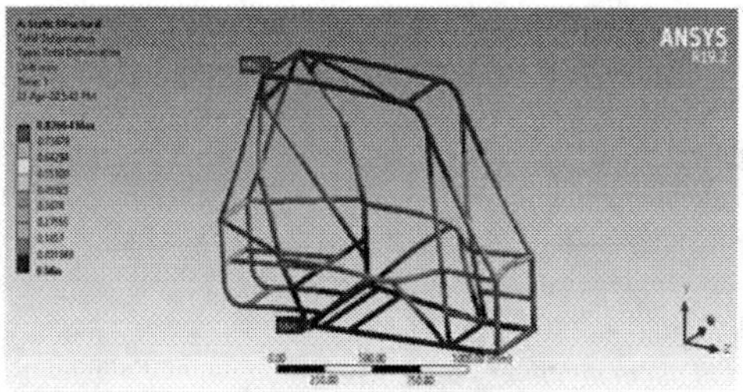

Figure 4: Total deformation of steel

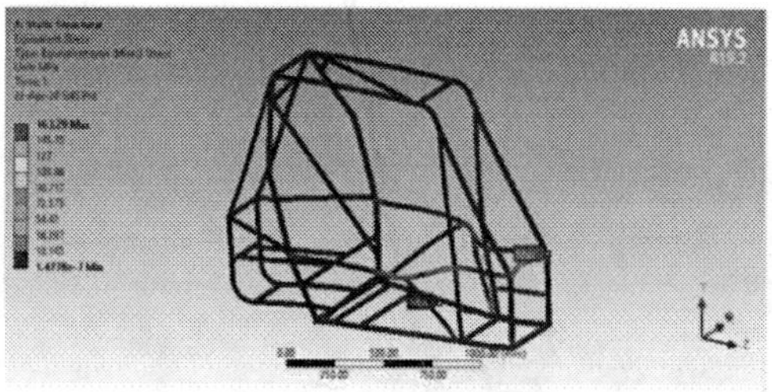

Figure5: Von-Mises stress of steel

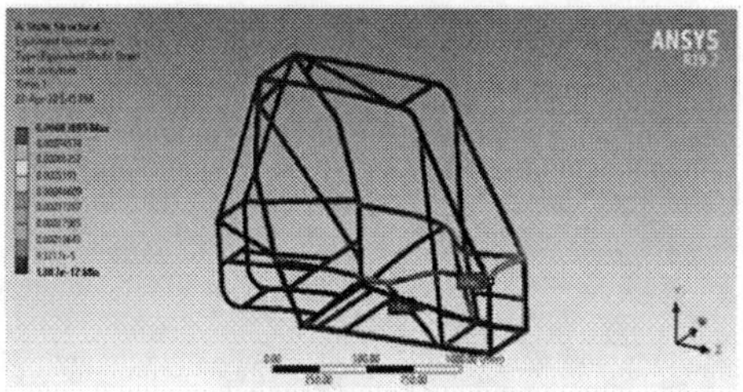

Figure 6: Total Strain of steel

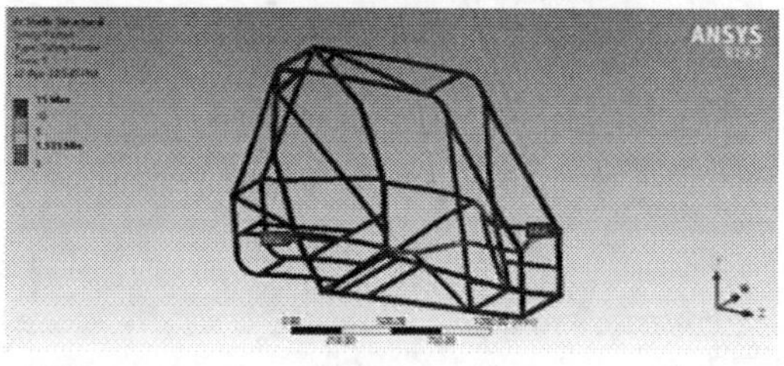

Figure7: Safety factor of steel

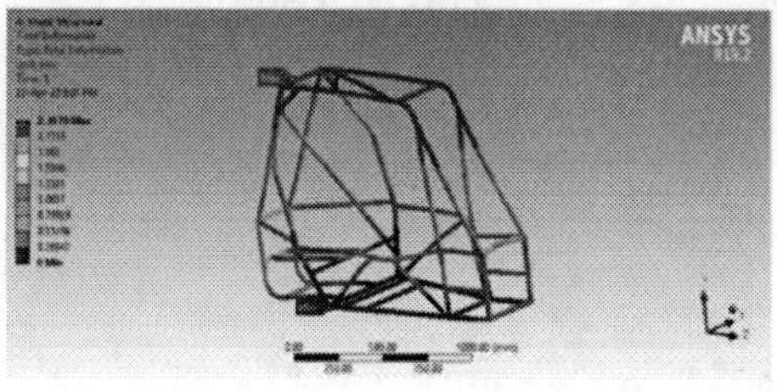

Figure8: Total deformation of Al-6061

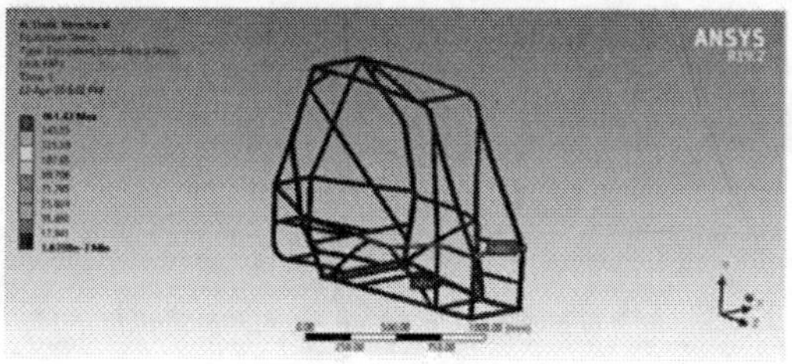

Figure9: Von-Mises stress of AL-6061

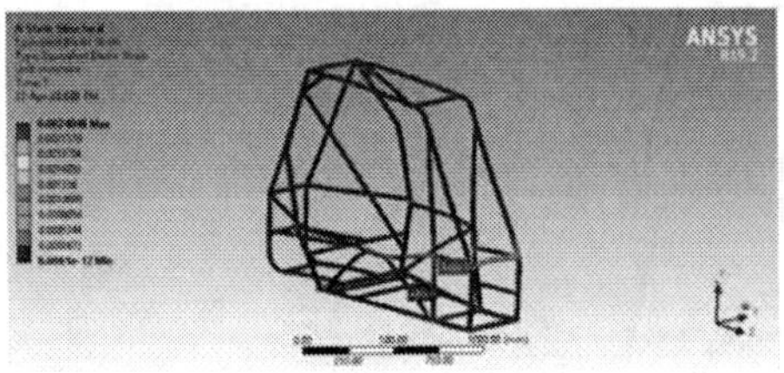

Figure10: Total strain of Al-6061

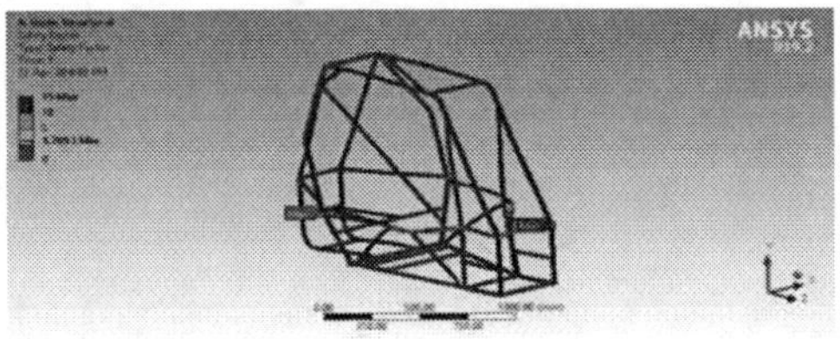

Figure11: Safety factor of Al-6061

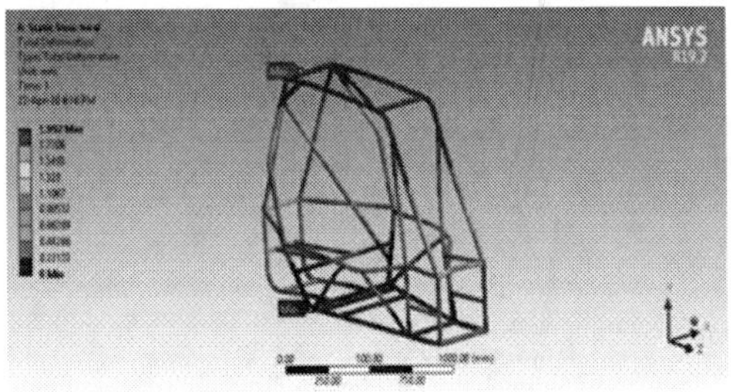

Figure12: Total deformation of Kevlar

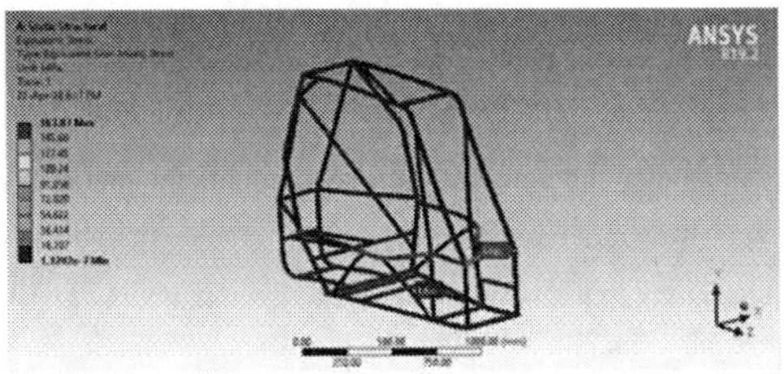

Figure13: Von-Mises stress of Kevlar

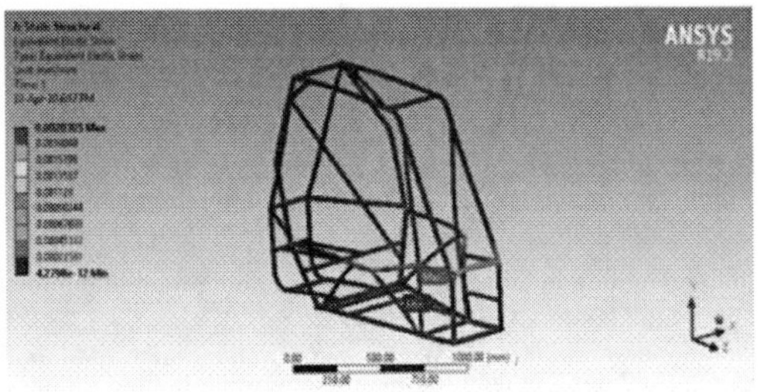

Figure14: Total strain of Kevlar

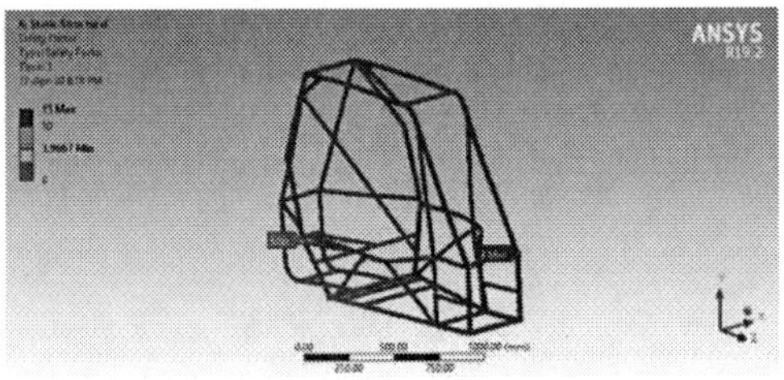

Figure15: Safety factor of Kevlar

Table4: Comparison of Deformation, stress, strain and safety factor

	Steel	Al-6061	Kevlar
Deformation (mm)	0.82664	2.3979	1.992
Stress (Mpa)	163.29	161.47	163.87
Strain	0.00083895	0.0024048	0.0020305
Safety factor	1.531	1.7093	3.9667

IOP Conf. Series: Materials Science and Engineering 998 (2020) 012046 doi:10.1088/1757-899X/998/1/012046

Modal analysis

Steel:

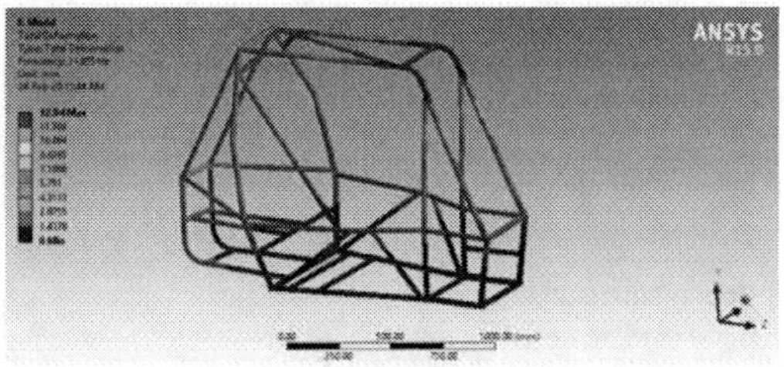

Figure16: Mode1 Deformation

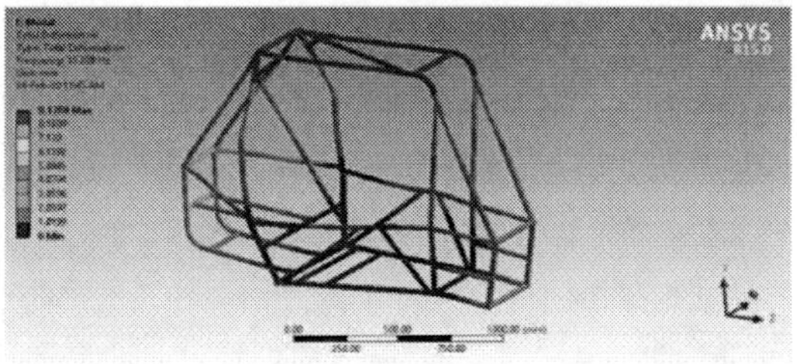

Figure17: Mode2 Deformation

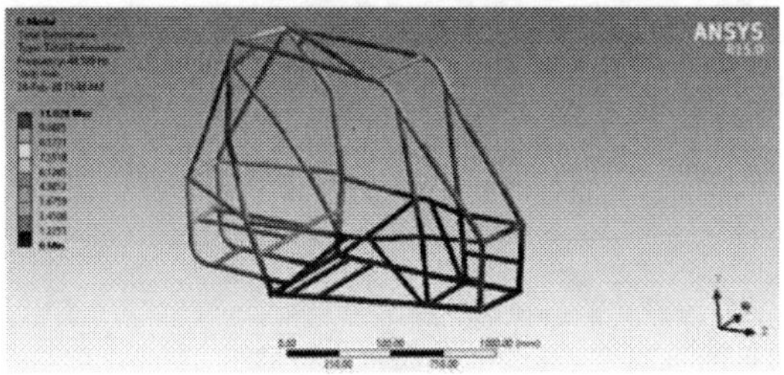

Figure18: Mode3 Deformation

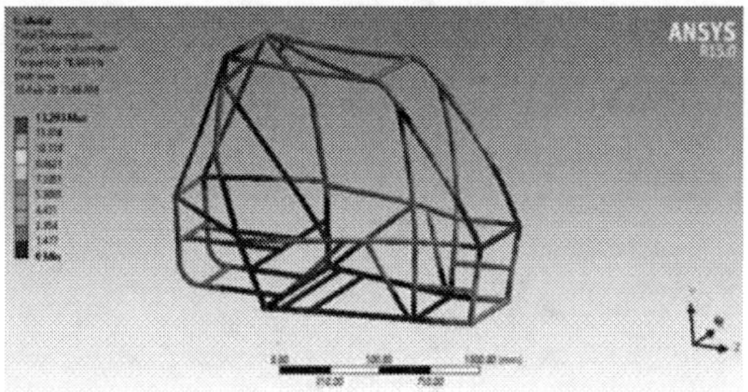

Figure 19: Mode4 Deformation

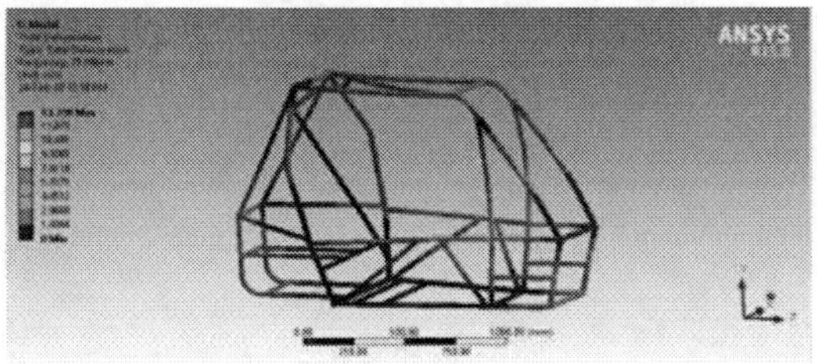

Figure20: Mode 5 Deformation

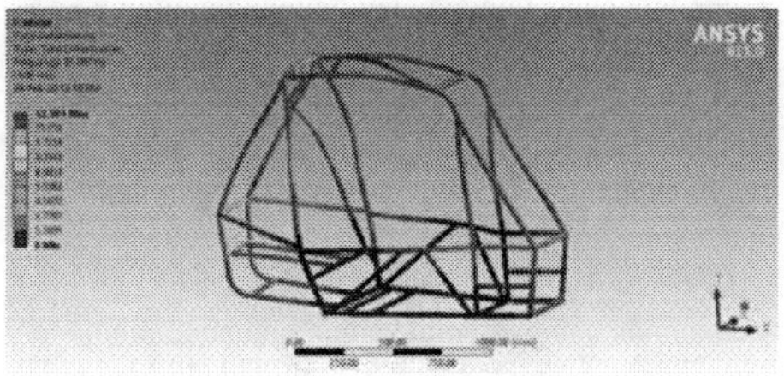

Figure21: Mode 6 Deformation

Aluminum-6061:

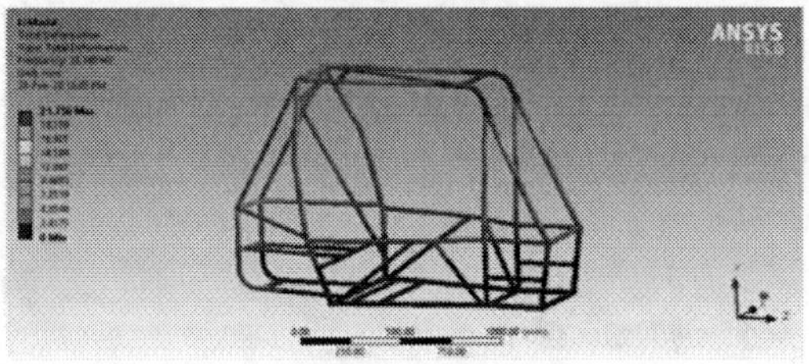

Figure22: Mode1 Deformation

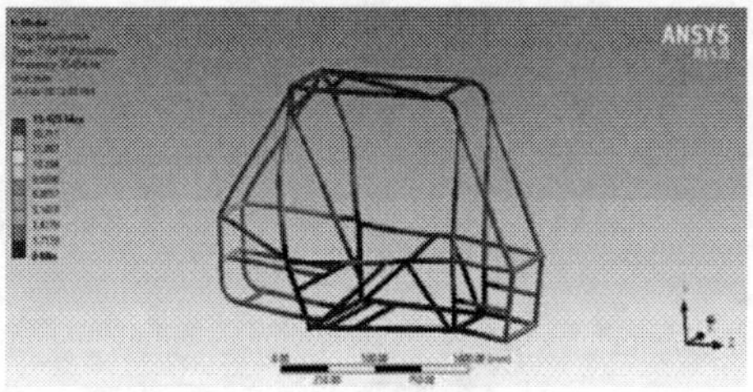

Figure23: Mode2 Deformation

IOP Conf. Series: Materials Science and Engineering 998 (2020) 012046 doi:10.1088/1757-899X/998/1/012046

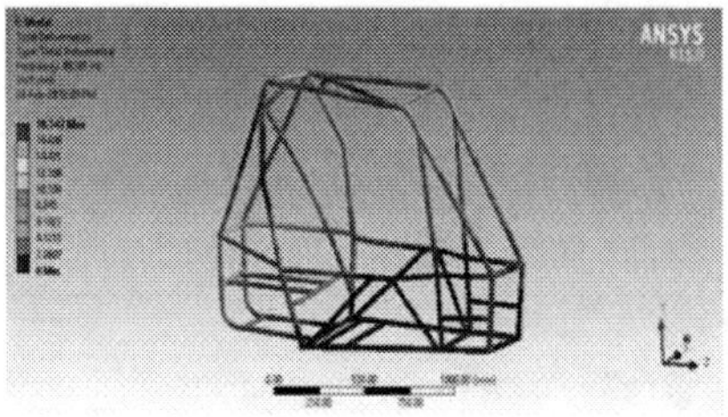

Figure24: Mode 3 Deformation

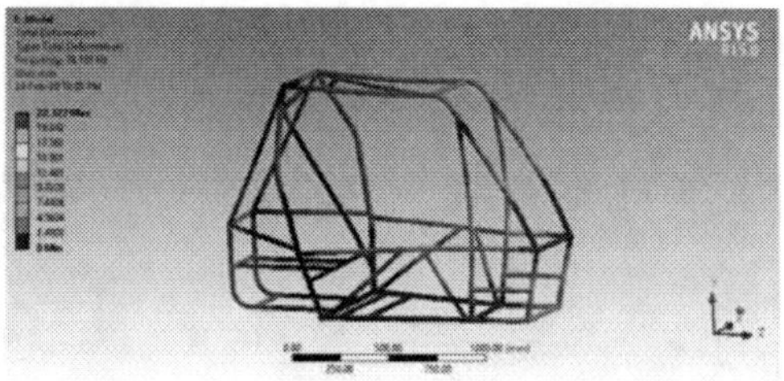

Figure25: Mode4 Deformation

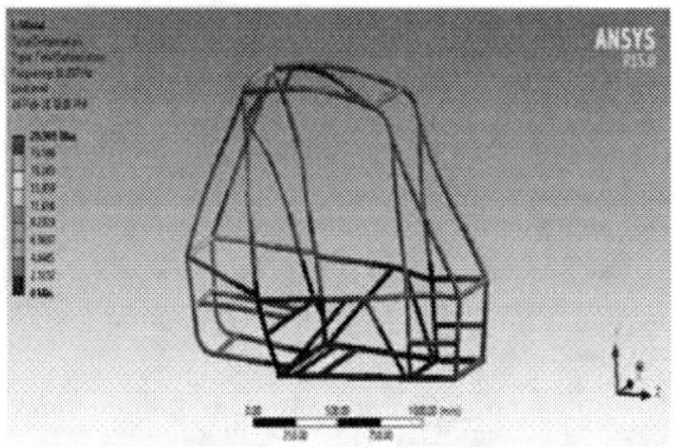

Figure26: Mode 5 Deformation

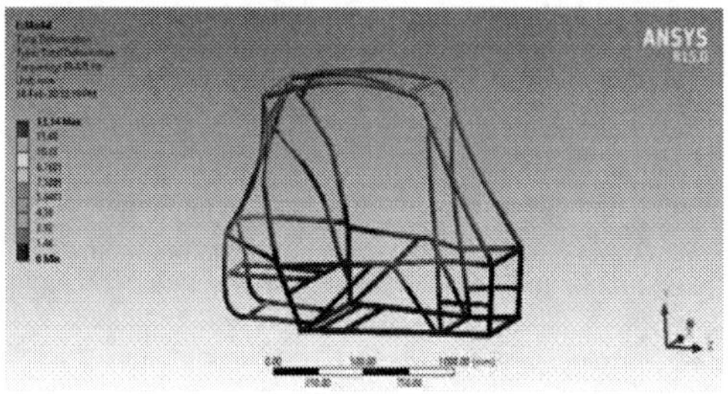

Figure 27: Mode 6 Deformation

Kevlar:

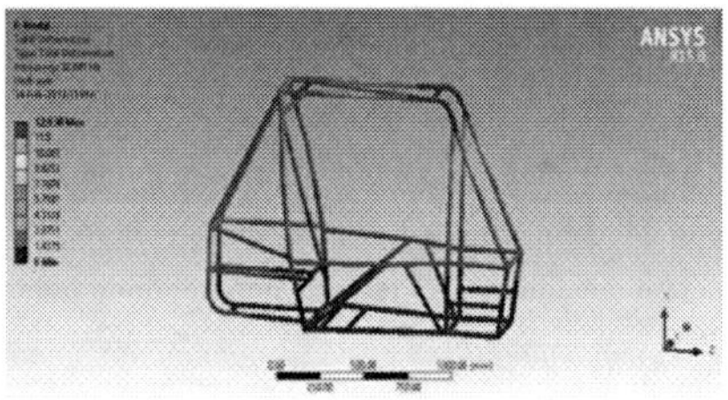

Figure 28: Mode 1 Deformation

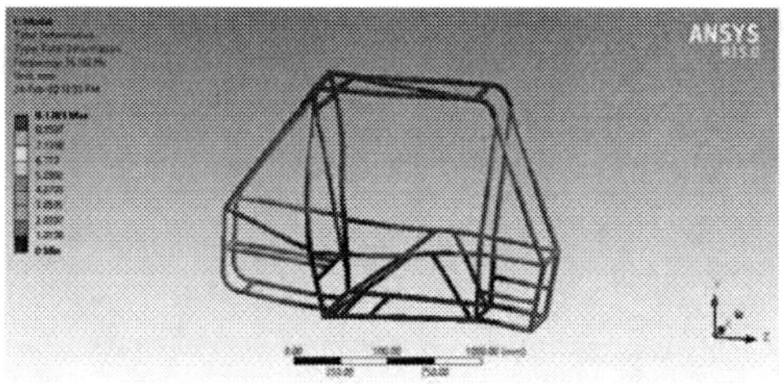

Figure29: Mode2 Deformation

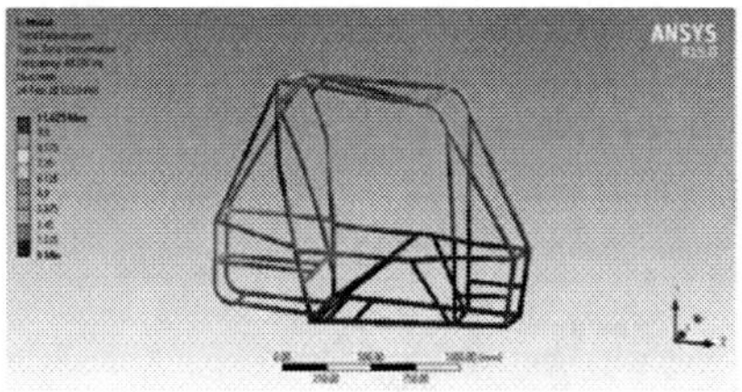

Figure 30: Mode 3 Deformation

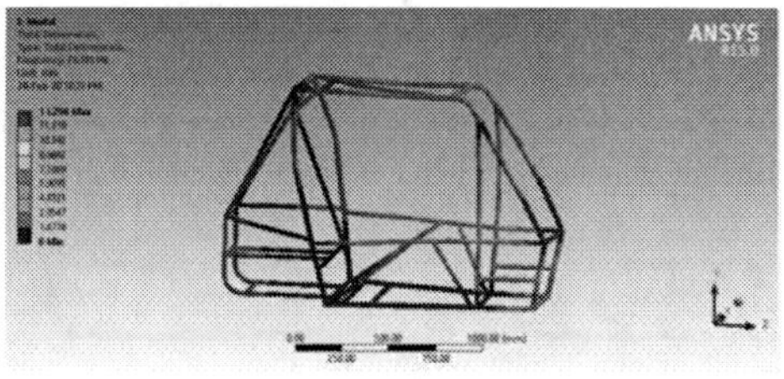

Figure31: Mode 4 Deformation

IOP Conf. Series: Materials Science and Engineering 998 (2020) 012046 doi:10.1088/1757-899X/998/1/012046

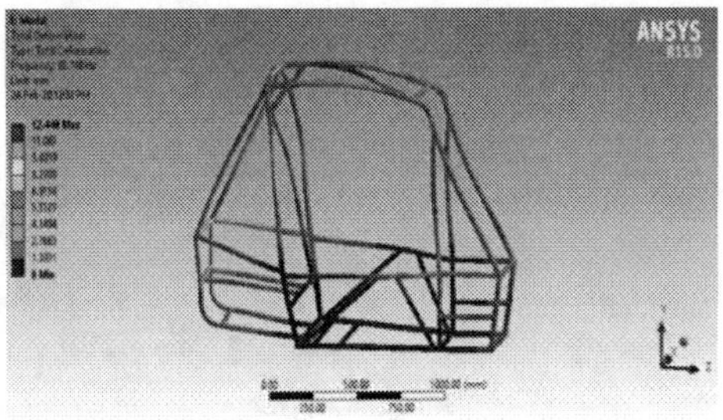

Figure32: Mode 5 Deformation

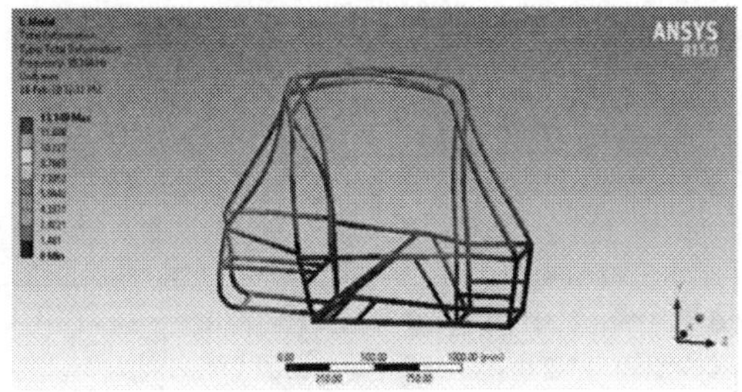

Figure33: Mode 6 Defor

Table 5: Comparison of Frequency in different modes

	Steel	Al-6061	Kevlar
Model (Hz)	31.855	32.363	32.645
Mode2 (Hz)	35.289	35.856	36.162
Mode3 (Hz)	48.589	49.383	49.787
Mode4 (Hz)	76.849	78.101	78.745
Mode5 (Hz)	77.146	82.037	82.748
Mode6 (Hz)	81.097	93.475	95.384

Weight and cost (approximate) estimation:
Steel:
Steel chassis gross weight:
Total weight = 28.21 kg
Cost of material 80/kg (approx.)
Total cost for chassis = 28.21*80= 2256.8 RS
Al-6061:
Al-6061 chassis gross weight
Total weight = 9.9904 kg
Cost of material 280/kg (approx.)
Total cost for chassis = 9.9904*280 = 2797.312 RS
Kevlar:
Kevlar chassis gross weight
Total weight = 4.23 kg
Cost of material 4501/kg
Total cost for chassis = 4.23*4501= 19039.21 RS

CONCLUSION:

In this thesis go-kart frame designed by using cad tool CATIA tool and analyzed with the help of l ANSYS workbench and in this case 3.8KN load applied as impact force, and the existing material steel can bear up to this load only, to improve the strength and also to improve the efficiency of the object by reducing total gross weight here 3 materials were chosen and analyzed with same boundary conditions.

From the results of static analysis each material has good safety factor values and all materials can withstand 3.8KN impact force, among all strength wise Kevlar c has taken 1st place, al-6061 has

IOP Conf. Series: Materials Science and Engineering 998 (2020) 012046 doi:10.1088/1757-899X/998/1/012046

2nd position, steel material have been occupied 3rd position respectively. In real time there are so many factors to be consider to select a new material, by knowing only static analysis results it is not possible to choose one good material for our object, to get more clear knowledge on each material here dynamic load boundary conditions also applied by using model analysis and calculated natural frequency results

From natural frequency values all materials have higher range frequency values than existing material and it means these all materials are good at vibrational conditions also, among all Kevlar is having good static and dynamic boundary conditions.

To get more accurate knowledge on material here cost estimation has been made approximately, and these cost estimations were give final results, from the weight and cost estimation it is clearly shown Kevlar increases 10 times material cost nearly compare to al-6061 and steel materials, previously Kevlar material is used for Ferrari F40 chassis, but in India this material wont suitable for our environment and economic conditions even though it has good strength to weight ratio values, among all steel Is cheapest material but while using al-6061 there is no much difference in cost compare to steel, and al-6061 has less density and it can decrease the total weight up to 33%, weight is reduced means indirectly fuel cost is also reduced.

Finally here project can be concluded with al-6061 material due to low weight and high strength among all materials and this material can increase the overall performance of the object and reduce the maintenance cost compare to steel.

References:

[1] Alpesh V. Mehta, Nikunj Padhiar, Jaydip Mendapara, 2011 "*Design and Analysis of Hybrid Go-kart*" IJAET/vol **II** / Issue I /277-288.

[2] Lim Wai Tuck, 2007 " *Design and Fabrication of ump Go-kart Chassis*" University of Malaysia Pahang nov, http://umpir.ump.edu.my/id/eprint/2276.

[3] Kiral Lal, Abhishek O S, 2016, "*Design, Analysis and Fabrication of go-kart*" International Journal of Scientific & Engineering Research, Volume 7, Issue 4, 429 ISSN 2229-5518.

[4] Pattanshett SV, 2016, "*Design and analysis of GO-KART chassis*", International journal of Mechanical and Industrial Technology, **4** (1) 150-164.

[5] Chow H Y, 2016,"*Go-Kart's design and construction based on theoretical and experimental findi*ngs", University of Malaysia, 2016.

[6] Michell S, Rajagopal K, Muthy V S S, Harikrishna V,2017, "*Design of a go kart vehicle*", International journal of science, engineering and technology research **6** (3) .

[7] Hajare K, Shet Y, Khot A,2016, "*Design and Analysis of a Go-Kart Chassis*", International Journal of engineering technology, management and applied science, **4** (2).

[8] Khan A S, Danish K, Kathola K, Bhonde, G Ghevande, Harshil M, 2017 "*Review on design and analysis of structure of go-kart vehicle*", International journal of research in advent technology, Special Issue National Conference 2017, April 9th 2017.

[9] Nath A, Vikram J C, Nongrum L, Marboh P, 2015, "*Design and fabrication of a Go Kart*, International journal of innovative research in science", Engineering and technology, **4**(9) .

[10] Kelkar K, Gawai S, Suryawanshi T, Ubaid S , Kharat R,2017,"*Static Analysis of Go-Kart chassis*", International journal of research in advent technology, Special Issue National Conference 2017.

CATALYTIC OXIDATION OF CARBON COMPOSITES

Dhanraj B[1, a *], Kiran Kumar N [2, b] and Rama Siva Reddy K [3, c], Dr. M. Vishnuvardhan[4,d], Anil Kumar V[5,e]

[1]Department of Mechanical Engineering, Vardhaman College of Engineering, Hyderabad-501218, India.
[2]Department of Mechanical Engineering, B.V.Raju Institute of Technology, Narsapur, Hyderabad-502313, India.
[3]Department of Mechanical Engineering, B.V.Raju Institute of Technology, Narsapur, Hyderabad-502313, India.
[4]Department of Mechanical Engineering, Vardhaman College of Engineering, Hyderabad-501218, India.
[5]Department of Mechanical Engineering, National Institute of Technology, Surat, Gujurat , India-

[*a]dhanraju2205@gmail.com, [b] kiran.nulka@bvrit.ac.in, [c] ramsivareddy.k@bvrit.ac.in
[d] mukkotivishnu@gmail.com, [e] veerapanenianilkumar@gmail.com.

Keywords: Aircraft brakes, catalytic oxidation, carbon composite, alkaline earth acetate

Abstract: A airbrake make use of multiple rotors for CC brakes, coupled with stators. A rotor is sourced by spokes, and the brake system provides protection for the stators. Pressing the brakes along with the hydraulic actuators performs braking action and the frictional torque is provided to stop aircrafts from overheating disc brakes. A friction between the rotating and stationery disc causes them to heat up to around 500 ^{0}C with a surface temperature of 2000 ^{0}C. He3nce, the materials used in this environment should show high thermal shock resistance. This paper focuses primarily on the study and evaluation of the effect of alkaline earth acetates on the catalytic oxidation of Carbon Composites aircraft brakes.

1. Introduction

1.1 Carbon Composite

Carbon/Carbons (CC) are a carbon matrix made up of carbon fibres. Composites of CC carbon matrix, with valuable mechanical and thermal properties at 2000 ^{0}C. CC Composites are made of carbon fibres, and almost all commercial carbon fibres are produced by turning a carbonaceous precursor into a fiber type[1]. Instead it connects the precursor fiber to make it fusible. Eventually, the cross-connected precursor fiber is heated in an inert atmosphere at a temperature of 1200 ^{0}C to 3000 ^{0}C.

The polymer Pepeg was kept in an oven for a period of 6 hours at a temperature of 700C to take away surplus solvents, subsequently a vacuum bag to compress the curing method for a period of 30 minutes[2]. It was made to 50mm cups by a water-cooled diamond saw, and nest treated by a post healing cycle in an air circulated oven at a temperature of 230C for a period of 8 hours. Carbon composites offer the option of a heat-resistant material capable to work at temperature above 2000C percent lighter density and a low thermal expansion coefficient, this material not only survive at high temperature, when it is heated it gets more stronger[3].

Strelko et al[5] investigated on the catalytic carbon gasification originated from electron interactions among a metal and a carbon matrix ensuing in the form of carbon oxides and metal-carbon complexes. Especially, carbon dioxide formed by the thermal break down of K2CO3 activated the carbon content efficiently in making a porous structure. Saber et al[6] examined the association among K2 13CO3 and surface oxides which is carbon-bounded with temperature controlled reaction(TPR). The investigation was done and heating mixtures of 100mg of a variety of carbon molecules and 10mg of labeled K213CO3 at temperature up to 732C in flowing helium. For raising the content of carbon the amount of CO2 released has been noted. The CO2 emitted from the K213CO3 carbon molecules was now slowly bigger than CO2 formed by carbon molecules itself. Cazorla et al [7] investigated on calcium –catalyzed carbon gasification by means of high-purity carbon that was packed with CaCO3.at 800C steady-state and step-response tests were primarily done in helium flow and then in CO2. Theoretically, it was shown that un catalyzed reaction produced

IOP Conf. Series: Materials Science and Engineering 998 (2020) 012047 doi:10.1088/1757-899X/998/1/012047

same quantity of 13CO and 12CO. the catalyzed reaction produced considerably higher concentrations of 12Co but not for increasing reaction temperature. It was also stated that a redox process relating the development of a higher oxide be supposed to the removed for the catalyzed reaction.

2. Practical setup

The experiments were conducted in the laboratory to identify the loss of carbon-carbon test samples over the specific period of duration (time).when it is subjected to conditions of oxidation simulation.

2.1 Step by step preparation

Carbon Carbon composites were impregnated with boric acid and potassium acetate and the behavior was tested in the furnace at 600 ^{0}C, 550 ^{0}C, 500 ^{0}C.

2.2 Segregation samples

DUNLOP AEROSPACE COMPANY collected small cubes of Carbon-Carbon composite samples. The samples collected from the company are used for production of aircraft brakes. Samples for observation about 30-40.the maximum and minimum weights were measured in the total sample set and were respectively 7.7436(grams) and 6.5766(grams) and average was 7.2349(grams).

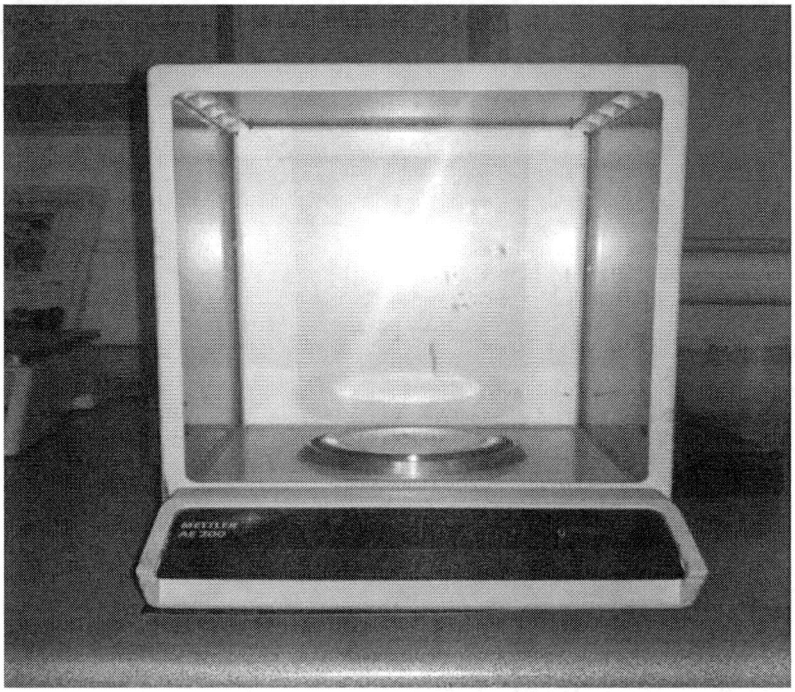

Fig.1 METTLERAE200 weighing machine

Maximum and minimum weights reported, and it is composed of ten samples in four sets each. Those sets are tested in Table.1

Table.1. Details of Classification of the Samples

SAMPLE SET	TEMPERATURE	CONDITION
SET1	600°C	WITH IMPREGNATION
SET2	600°C	WITH IMPREGNATION
SET3	550°C	WITH IMPREGNATION
SET4	500°C	WITH IMPREGNATION

3. Impregnation of Samples

Now, the test samples were able to impregnate with solution. 15 grams of boric acid, 49.1 grams potassium acetate were taken with molar solutions 0.5 and respectively. Which dissolved into 500ml distilled water. Suppose, mixture isn't fully dissolved into water and it should heated for the uniform solution creation. Dipped sample must be kept 40 minutes in solution. Subsequently, samples can be dried and keeping them circulation the preheated air oven, can be shown in figure.2 at 1100 ^{0}C at least for two hours and it were then removed from oven and place to cool down for 120seconds.the, the test sample are again weighed to identify the absorbed acetate. Likewise, boric acid of 5 grams and potassium acetate of 4 grams for next three sets of separate solutions. The aircraft brakes have endured various environmental conditions. Hence impregnated procedure was carried out to promote these conditions.

Fig.2 To cool in air passing oven

3.1 Heat treatment of test samples

Heat treatment of samples was done on regular period in order to know the oxidation of composite. The heat treatment of composite carried out in a called muffule furnaces .the complete process is explained in point wise

1. First taken given set of sample one and their weights are calculated
2. The heat treatment is carried out set one in a temperature 600 degree centigrade.
3. The sample set is cooled for two hours in a air circulated over for two hours
4. Again heat-treated for 600 degree centigrade after that weights are calculated to check oxidation.
5. This process is carried out for 48 hours continuous
6. The oxidation is checked with the help of weights
7. same steps are carried out for different set of samples

Figure.3 Heat treating in a furnace

3.2 Microscopic Investigation

Microscopic investigation is carried out how the acetate s are impregnated in the composites

3.3 *Scanning Electron Microscope(S.E.M)*

After the heat treatment process the sample tests were observed in the lens of Electron Microscope Scanning Carry JCM 5700. This test is conducted for experimental verification. The microscopic experiments were performed to learn about the structure of carbon-carbon composites, also how the composite samples had diffused the acetates and oxides. Packed with units EDS, SEM. EDS an experiment on the sample sets was carried out to micro-analyze the samples.

4. Results and Calculations:

4.1 Oxidation Test Results

Boric acid, potassium acetate were heated represents in the Table.2 for a certain amount of time at 600 ^{0}c temperature, to demonstrate the difference of the carbon – carbon samples.

Table.2 weight change of sample set 4 over period of time

Initial Weight	Weight after Impregnation	Weight after Heat Treatment Process		
		2nd Hour	3rd Hour	4th Hour
6.7257	7.0896	6.8842	6.5609	6.0396
6.7261	6.9290	6.7258	6.4530	5.9505
6.8526	7.1316	6.8587	6.6080	6.2048
6.8537	7.6087	7.1759	6.7494	6.1814
7.3400	7.6137	6.6527	6.2006	5.5097
7.3425	8.2467	7.5224	6.6389	5.5064
7.5534	7.6100	7.2396	6.8273	6.1340
7.5894	7.8399	7.1064	6.3221	5.2272
7.6914	8.0907	7.6807	7.3479	6.6611
7.6985	7.7417	7.5529	7.3498	6.9485

Table.3 weight variation of sample set 4

Initial weight	2nd hour	Weight loss	3rd hour	Weight loss	4th hour	Weight loss
6.7257	6.8842	-0.158	6.5609	0.3233	6.0396	-0.521
6.7261	6.7258	0.0003	6.453	0.2758	5.9505	0.5025
6.8526	6.8587	-0.006	6.608	0.2507	6.2048	0.4032
6.8537	7.1759	-0.322	6.7494	0.4265	6.1814	0.568
7.34	6.6527	0.6873	6.2006	0.4521	5.5097	0.6909
7.3425	7.5224	-0.179	6.6389	0.8835	5.5064	1.1325
7.5534	7.2396	0.3138	6.8273	0.4123	6.134	0.6933
7.5894	7.1064	0.483	6.3221	0.7843	5.2272	1.0949
7.6914	7.6807	0.0107	7.3479	0.3328	6.6611	0.6868
7.6985	7.5529	0.1456	7.3498	0.2031	6.9484	0.4014

Table.4 Weight variation of potassium acetate at 600°c

Initial Weight	Weight after Impregnation	Weight after Heat Treatment		
		2nd Hour	3rd Hour	4th Hour
7.2161	7.3271	4.3642	2.9054	1.6691
7.2206	7.3672	4.7269	3.2123	2.0154
7.2412	7.4239	4.2466	3.6786	2.973
7.43	7.5922	4.5864	3.5342	2.4887
7.7328	7.8780	4.1974	3.4426	2.7628

Table.5 Weight deficiency of potassium acetate at 600°c

Initial weight	7.2161	7.2206	7.2412	7.43	7.7328
2nd hour	4.36442	4.7269	4.2466	4.5864	4.1974
weight loss	2.8519	2.4937	2.9946	2.8436	3.5354
3rd hour	2.9054	3.2123	3.6786	3.5342	3.4426
Weight loss	1.4588	1.5146	0.568	1.0522	0.7538
4th hour	1.6691	2.014	2.973	2.4887	2.7628
Weight loss	1.2363	1.1969	0.7056	1.0455	0.6798

IOP Conf. Series: Materials Science and Engineering 998 (2020) 012047 doi:10.1088/1757-899X/998/1/012047

5. Analysis of results:

The below figure displays the rate of percentage carbon losses to the specific time for potassium acetates, boric oxide, potassium acetates and boric acid, respectively

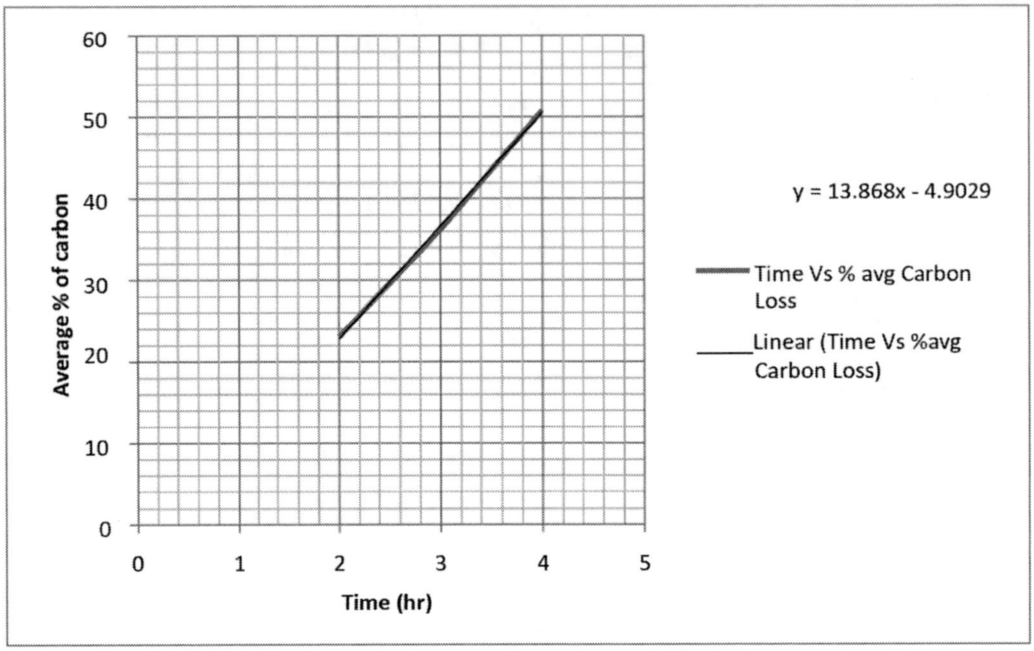

Fig.4 C vs. sample set2 at 600°

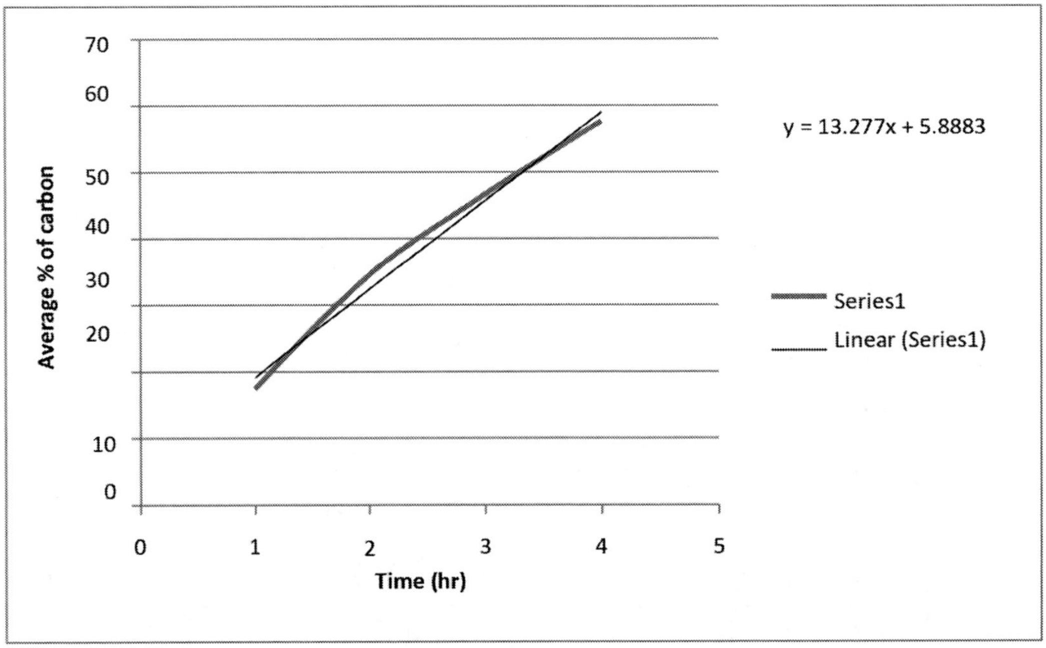

Fig.5 C vs. ko2 a at 600°c

5.1 Activation Energy Calculation

In Activation Energy shown below, the energy needed to resolve the before reactants can be converted to products. It may take very quick, if activation energy is at low temperature.

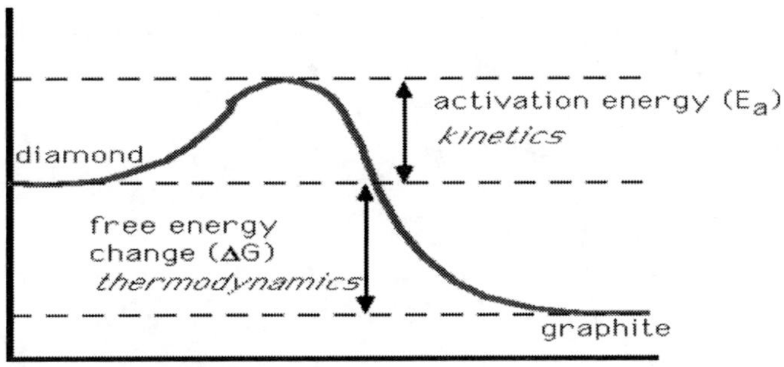

Fig.6 Activation of Energy

5.2 Reaction rate governed by Arrhenius equation

Energy of activation can be determined by intrigueln (k) with the graphs of reciprocal temperature. The rate constant derived from slope, and the % of carbon losses over the time. In figure below, the graph obtained by combining boric , potassium acetate, boric acid and potassium acetate shall be given.

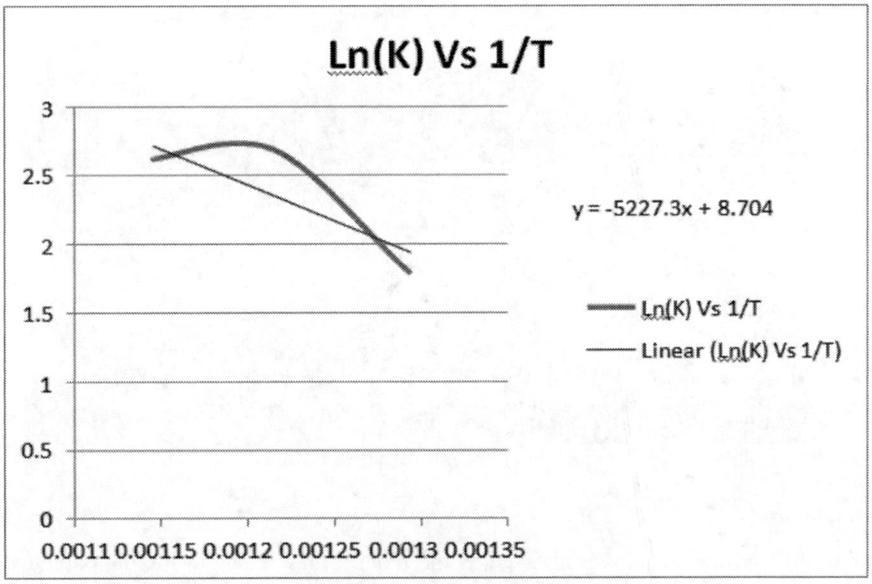

Fig.7 Arrhenius test samples- impregnation

The slope is of – Q / R, and on each Arrhenius plot was superimposed on a best-fit line for getting value for the slope.

Slope=-Q/R. Where Q=Energy Activation, R= gas constant

-5284.3=-Q/8.391

Therefore, Q=42469.72J/mole

For mixture of potassium acetate & boric oxide the activation energy is required to resolve reactions is 42469.72J/mole.

IOP Conf. Series: Materials Science and Engineering 998 (2020) 012047 doi:10.1088/1757-899X/998/1/012047

6. EDS Analysis

Under EDS analysis, the samples were checked to check whether in the c-c composites Where Diffusion of Acetates. The JCM 5700 SEM has an EDS to it. And the figure represents the different elements and how much of carbon-carbon composites samples are available, C-C samples also contained potassium, silicon, oxygen, and boron as shown in the graph below.

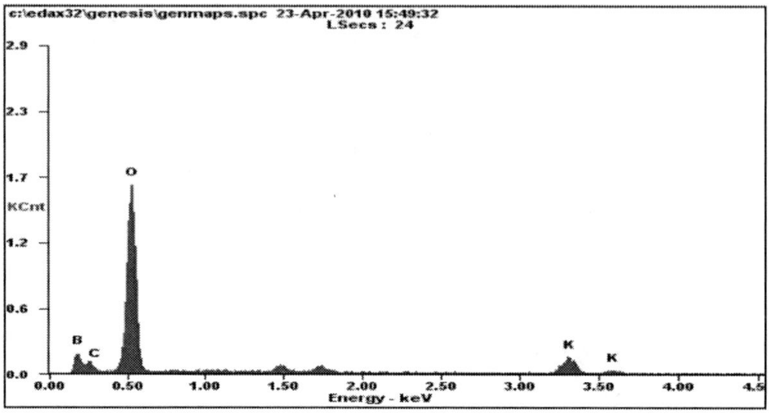

Fig.8 Visualization of CC Composites in EDS Analysis

6.1 EDS -Spectra (Observation)- CC Composite

The below figure represents the boric oxide on sample surface blocks oxygen access to fiber.

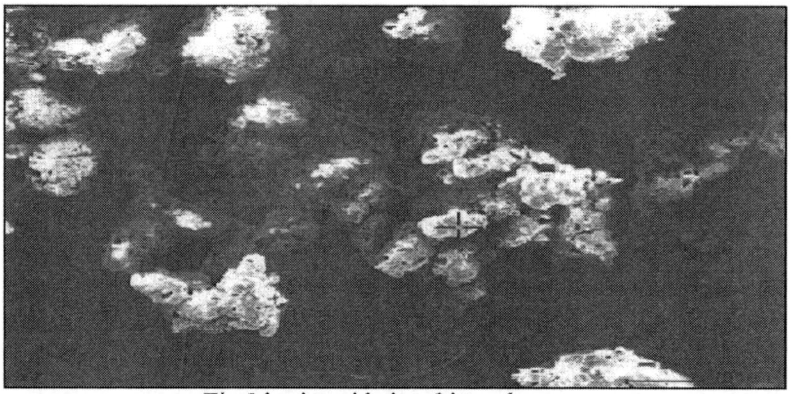

Fig.9 boric oxide in white color

7. Conclusions:

Similar to the boric acid, potassium acetate and potassium acetate alone, the catalytic effect of the potassium acetate and boric oxide mixture has stronger results from the experimental tests. Combined potassium acetate and boric oxide exhibit identical properties on C-C test samples as validate sodium acetate. And from the above graphs, it can represent that, on c-c samples the amount of carbon loss over time has more use. It has been confirmed from previous studies the boron oxide serves as an inhibits throughout oxidation reaction.

Energy activation, for potassium acetate and boric oxide a is Q=42469.72J/mole, means the reactants required to transform products to reactions are very less compared with boric oxide (Q=57390.9426 J / Mole) mixtures the magnesium and potassium acetate (Q=51587.772 J /mole) Potassium acetates needed less quantities of triggering energy which is much lesser than combined potassium acetate and boric acid to turn reactants to the products. It showed weaker catalytic effect of potassium acetate on C-C Composite test samples Oxidation. Although, it shows extraordinary non-corrosive properties.

And the potassium acetate and boric acid can be considered to serve as an inhibitor throughout oxidation reaction. Potassium acetate are good catalyst for sample oxidation and it couldn't be used in the practical defense against oxidation.
Thus, potassium acetate and boric acid efficient catalysis for the oxidation of c-c samples and therefore not a Practical defense against oxidation. But, several other alternative means of protection are used by the aircraft industry instead of using boric oxide, potassium-acetate respectively.
Future studies on C-C tests of the mixture of boric oxide and potassium acetate, potassium acetate and magnesium acetate and potassium acetate and calcium acetate will be carried out. Compared with the present experiment conducted, that could provide better results.

8. References

[1] Wu and Radovic, catalytic oxidation of carbon/carbon composite materials in the presence of potassium and calcium acetates, 2005, pp. 333-344.

[2] Wu and Radovic inhibition of catalytic oxidation of carbon/carbon composites by boron doping, 2005, pp. 1768-1777.

[3] David A.Fox and Alfred H.White Effect of Sodium Carbonate upon Gasification of Carbon and Production of Producer Gas,2002,pp.259-266.

[4]JA Moulijn, F Kapteijn, Towards a unified theory of reactions of carbon with oxygen containing Molecules,1995, pp.1155-1165.

[5] V.V. Strelko, N.V. Gerasimenko, N.T. Kartel', Low-temperature activation and oxidation of carbon Materials impregnated with potassium carbonate, Russ. J. pp. 286-289.

[6] J.M. Saber, J.L. Falconer, L.F. Brown, Interaction of potassium carbonate with surface oxides of carbon, 1968, pp.1356-1359.

[7] D. Cazorla-Amoros, A. Linares-Solano, F.H.M. Dekker, F. Kapteijn, Isotopic steady-state and step response study on carbon gasification catalyzed by calcium, 1995, pp. 1147-1154.

[8] H Zoheidi, DJ Miller, Role of oxygen surface groups in catalysis of hydrogasification of carbon black by potassium carbonate, 1987,pp. 809-819.

[9] Ushakiran Kumar sanivad Msc major project 2008/2009 (university of ertfordshire) [10]Levar Johnson final year project 2006/2007 (university of Hertfordshire)

[10] http://www.soes.soton.ac.uk/staff/pmrp/GY102/module1/m1f9.GIF

[11] http://www.chemguide.co.uk/physical/basicrates/arrheniuS1.gif

A REVIEW OF ULTRASONIC ASSISTED ELECTROLYTIC IN-PROCESS DRESSING (UA-ELID)

[1]Dhanesh B Awati and [2]Rahul M Chanmanwar

[1] PG Student, Department of Mechanical Engineering, Walchand College of Engineering, Sangli, Maharashtra, India
[2] Faculty member, Department of Mechanical Engineering, Walchand College of Engineering, Sangli, Maharashtra, India
Email: [1]dhanesh.awati@walchandsangli.ac.in, [2]rahul.chanmanwar@walchandsangli.ac.in

Abstract: This paper presents previous research work of researchers on electrolytic in-process dressing and its ultrasonic-assisted system variant. This paper also contains information about the ELID system and applications of ultrasonic-assisted ELID. This technique enables continuous dressing of wheels made using metallic bond in the grinding process itself while retaining sharp grits from the super abrasive wheels. An ultrasonic-assisted setup requires an ultrasonic power generator, piezoelectric transducer (PZT), horn assembly. An ultrasonic generator generates ultrasonic pulses, which are supplied to the piezoelectric transducer. The piezoelectric transducer then converts electrical energy to mechanical energy and vibrates either in longitudinal mode or in transverse mode. The horn system then amplifies these vibrations, and amplified vibration is supplied to the grinding tool. The relevance of research papers referred to regarding setup development of ultrasonic-assisted electrolytic in-process dressing, effect of process parameters and modeling of the ultrasonic-assisted ELID process is described in this paper.
Keywords: ELID, PZT, Ultrasonic assisted, Mathematical modeling, the Oxide layer

1. INTRODUCTION

Electrolytic in-process dressing (ELID) is a traditional dressing method employed for the uninterrupted truing of abrasive wheels. In 1990, Japanese research scholars Ohmori and Nakagawa, for the first time, developed ELID grinding for creating excellent topographic surfaces on hard and brittle materials. Then onwards, many scholars used this method for creating mirror like surfaces by employing very fine grain metallic bonded wheels [2]. Ohmori [2] stated that ELID grinding contains a basic setup which contains a power supply, an electrode as either cathode or anode, a metallic-bonded abrasive wheel, and an alkaline electrolytic-grinding fluid.

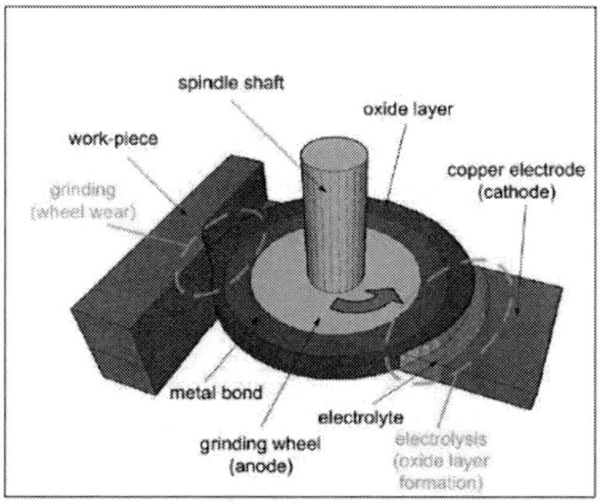

Figure 1. The schematic of ELID process. [10]

Nowadays, many researchers are interested in ELID grinding for its intelligent method of self-dressing of the grinding wheel. Many researchers have reported their findings regarding exploring the mechanism and application of the process, though comparatively less work has been reported so far to

IOP Conf. Series: Materials Science and Engineering 998 (2020) 012048 doi:10.1088/1757-899X/998/1/012048

create a smart machine tool, especially for ultrasonic-assisted ELID grinding. A newly developed technique of ELID is instrumental in resolving the limitations of traditional grinding for hard and brittle materials. This technique offers continuous sharpening of wheels made using metallic bond in the grinding process itself while retaining sharp grits from the super abrasive wheels. The schematic of the ELID process is illustrated in figure 1 above.

The continuous sharpening of the wheel involves the basics of the electrolysis process. Here electrical energy gets converted into chemical energy. Due to electrical energy, the dissolution of ions arises in the electrolytic solution, and the cations and anions in the solution travel towards cathode and anode, respectively. This will trigger a chemical reaction at the anode, which results in the formation of metal oxide. This oxide layer will either dissolve in an electrolyte or quickly warn out during grinding allowing new grit protrusion.

2. UA-ELID PROCESS

ELID works on the principle of electrochemistry. After the pre-dressing, the bonding material, which acts as an anode, gets dissolved, and the oxide layer of the bonding material is generated on the wheel surface. This layer acts as an insulating covering. This is the reason because of which the increase in oxide layer thickness decreases the electrical conductivity of the abrasive wheel surface and thus the electrolysis current reduces. The oxide layer will be effortlessly machined during the grinding process. As a result of this, the increment in current takes place which in turn accelerates the anodic dissolution. This makes ELID process as self-regulating and a dynamic equilibrium process [31]. The repetition of this cycle enables the process to achieve stable grinding.

Ultrasonic vibrations are generated by providing electrical power to piezoelectric plates. These plates are used in the vibrator. With the help of the excitation of different piezoelectric plates by high-frequency electrical signals, the vibrator was resonated. The transformation of electrical energy into mechanical vibration is done through the piezoelectric transducer (PZT). As the amplitude of oscillation generated at the surface of PZT was incapable to achieve a sufficient cutting rate, the tool-horn assembly is proposed to intensify the amplitude of vibration. Figure 2 illustrates bi-modal vibration produced by PZT systems, i.e., longitudinal vibration mode and transverse vibration mode [17].

In longitudinal mode the vibrations generated are in line with the axial (Z) direction, and in transverse type, vibrations produced are in the plane orthogonal to the axis. For generating longitudinal vibrations, single bunch of circular piezoelectric plates is required as illustrated in figure 2 (a). After the application of a voltage which is having a sine wave like structure to the transducer, the piezoelectric plates start broadening and shrinking, so that the vibrator is reverberated and the tooltip which is adhered to the tail end of the horn pulsates longitudinally along its axis. "The amplitude of vibration depends on the applied voltage, the property of piezoelectric material, and the arbor structures. The pulsation amplitude is boosted by the horns such that its maxima developed at the tooltip. As shown in figure 2 (b), if two semicircular shaped piezoelectric plates are situated on the PZT and two sinusoidal voltages with 180° phase difference are applied to the piezoelectric plates, the two piezoelectric plates will broaden and shrink alternately, finally causing the tool adhered to the end of the horn to vibrate in the transverse mode in the XY plane" [17].

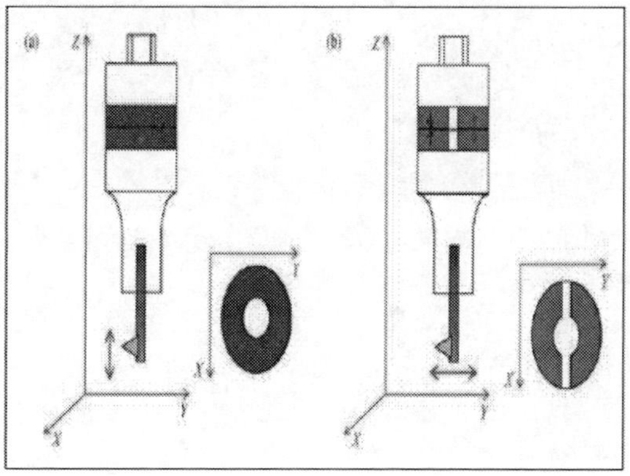

(a) Longitudinal mode

(b) Transverse mode

Figure 2. Types of PZT system for ultrasonic vibration. [17]

Concerning ultrasonic horn design, Y. Choi et al. [23] have designed a tool horn length of 130.8 mm using FEM analysis. They have also found that the maximum amplitude of vibration can be obtained at an ultrasonic frequency of 20 kHz [23,42]. For fabricating the ultrasonic unit, A. Zahedi et al. [42] have selected lightweight, reasonable strength, large processability, and large sonic efficacy, aluminum alloy 7075 for the fabrication of the ultrasonic module. The intended modelu consists of two parts: i) booster-sonotrode, which establishes the assosiation among the transducer and the workpiece, and ii) the housing, which retains the vibrating system on the grinding machine spindle. Figure 3 illustrates a diagramatic representation of the ultrasonic system. It is advantageous to have a pulsation mode in the center plane of the booster so that the casing will be seperated from the oscillating sequence (transducer, booster, workpiece). At the same time the maxima of oscillation should be positioned at the top of the workpiece [42].

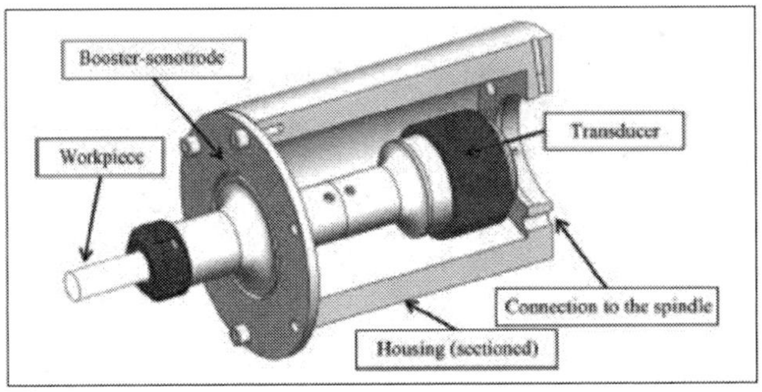

Figure 3. Diagrammatic representation of the ultrasonic system with elements. [42]

3. EXPERIMENTATION

This chapter covers various materials, experimental conditions used during experimentations, and the effect of various process parameters in ELID and UA-ELID processes.

3.1 Materials in experimentations

Different materials like workpiece material, grinding wheel material, grinding wheel grit sizes, electrolyte, concentration, and operational situations are tabulated in Table 1 below.

Table 1. Gist of materials and experimental conditions used in ELID.

Workpiece Material	Grinding Wheel Material	Wheel Diameter (mm)	Grit Sizes (#)	Electrolyte	Concentration	Experimental Conditions
Reaction Bonded Silicon Nitride (RBSN) [1]	Synthetic Diamond 400 – R-100-BX 1614 [1]	75 [4,24,31, 41]	400 [1]	CIMTERIA L HD 90 [1]	15:1 [1,24]	Voltage (V) 20,40 [7] 60 [2,3,4,5,7,11,31, 37,38,39,40], 70 [36] 80 [7], 90 [3,16,28,32,34, 38,39], 100 [7,24]

IOP Conf. Series: Materials Science and Engineering 998 (2020) 012048 doi:10.1088/1757-899X/998/1/012048

Silicon infiltrated Reaction Bonded Silicon Nitride (SRBSN) [1,2]	Cast Iron Fiber Bonded Diamond (CIFB) [2,3,4,5,11, 16,24,33,36, 37,39,40,41]	150 [3,26]	325 [2,4, 29, 37, 39]	Noritake AFG-M [2,3,39,40]	1:50 [2,3,4,5,31, 37,39,41]	Peak Current (A) 0.3 [24], 0.5 [38], 1[32,38,40], 2[7,38,40], 3,4 [38], 5[11,36,39], 7[11], 9 [11,37], 10 [2,4,5,28], 13 [37],16 [3], 20 [39], 24[3]
Sintered Silicon Nitride (Si_3N_4) [2,5]	WA60G8V [25]	200 [2,5,38]	600 [2]	CG-7: NUXES [4,5,24,31, 41]	1:20 [11,32,33,38]	Pulse on time (μs) 2 [36,39], 4 [3,40], 5 [2,4,5,16,26, 28,31,34], 8,10,14 [7] 30 [7, 24]
BK 7 glass [4,7,24,28,2 9,41]	Copper bonded Diamond [26]	305 [11,37]	1200 [24, 29, 40,41]	TRIM 229 [11,37]		Ultrasonic frequency (kHz) 20 [23], 20.6 [42], 28 [14], 35 [16], 100 [31]
Silicon Wafers [5,29]	Cobalt Cast Iron Compound Diamond [26,40]	23 [14]	2000 [2]	TRIM C270 [11,19,33]		Axial Ultrasonic Amplitude (mm) 0.008 [14] 0.0082 [16] 0.02-0.025[42]
Zirconia [8,42]	Bronze bonded Diamond [31]	25 [16]	4000 [2,4,7, 11, 24,29, 39]	TJMX-V [31,34,38]		Tool-w/p Gap (mm) 0.1-0.3[34,37], 0.25[41] 0.4 [31]
Sapphire [11]	CBN [32,38]	205 [25]	170 [3]	Kyodo Oil [40]	50:1 [40]	Pre-dressing (min.) 5 [40] 10-15 [3,34] 20 [41] 61.87 [37]
Ti6Al4V [23]	Tungsten bonded CBN [34]	80,100 [26], 400 [26,42] 325,500 [26,40]	8000 [5,36]			Feed rate (mm/min) 30, 50, 70 [40] 0.3, 0.6, 0.9, 1.2 [42]
Carbon steel [25]	D151C118 Diamond wheel [42]	211 [33]	280 [16, 31]			
GCr15 [32]		60 [34]	1000 [33]			
SiC [33]		2.5 [40]	20000			
GaN [36]			[36]			

3.2 Effect of process parameters

Process parameters are key variables on which the functioning of the process relies as explained in section 2 of this paper. The impact of different operational variables like wheel velocity, depth of cut, feed velocity, oxide layer thickness, inter-electrode gap, selection of cathode and anode, vibration frequency and amplitude and mode of oscillation is studied by different researchers [16,22,32,38,42].

1) Wheel velocity

As the abrasive wheel speed increases, the surface roughness decreases as shown in figure 4 (a) this is because higher the wheel speed lesser the time of engagement of the abrasives with the workpiece and as the engagement time is less, there is a less chance of cutting of oxide layer by chips. The linear behaviour of oxide layer enhances the uniform surface topography [16].

2) Depth of cut

The increase in depth of cut raises the surface roughness because of the indentation fracture mechanics. Depth of grinding increases the extrusion resistance offered by the workpiece. This resistance between the wheel and the workpiece can expand the cracks thus the surface becomes highly nonuniform and thus the quality of the surface decreases as shown in figure 4 (b) [16].

3) Feed velocity

Feed velocity is the rate at which abrasive wheel is moving in the axial direction. Again, rise in the feed velocity rises the surface roughness as the rate of extrusion resistance increases and the fundamental of indentation fracture mechanics applies. The influence of feed velocity on surface roughness is shown in figure 4 (c) [16].

4) Oxide layer thickness

Oxide layer plays an essential role in generating the surface quality. Oxide layer acts as an insulating covering so, increase in oxide layer depth decreases the electrolysis current and vice a versa. Formation of oxide layer lowers down the actual depth of cut of abrasives. Therefore, rise in the electrolytic control current reduces the oxide layer thickness and in turn raises the actual depth of cut of abrasive particles. Furthermore, this leads to the rise of undeformed chip thickness of abrasive particles and thus the surface roughness and waviness rise as illustrated in figure 4 (d) [38].

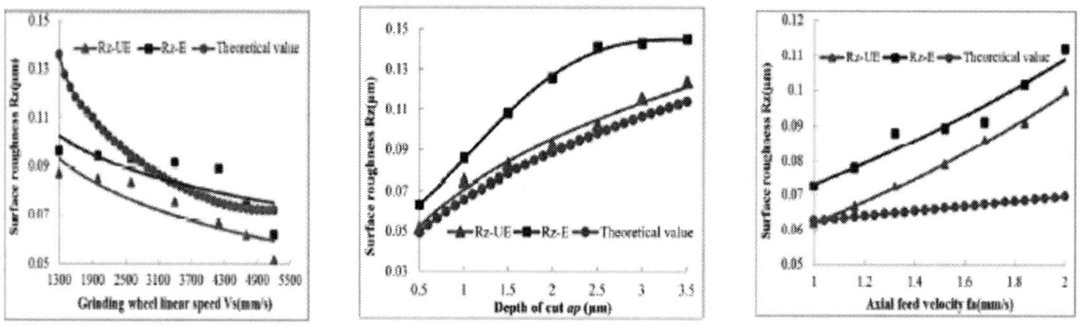

(a) [16] (b) [16] (c) [16]

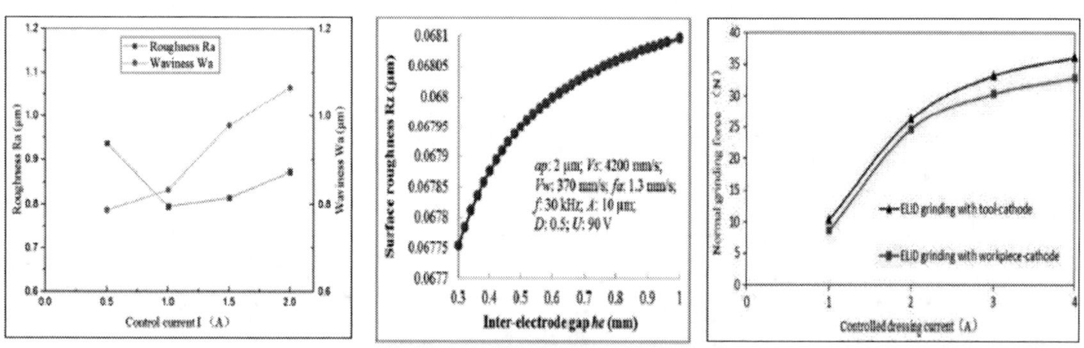

(d) [38] (e) [16] (f) [32]

Figure 4. Graphs of effect of various process parameters.

5) Inter-electrode gap

Inter-electrode gap is a gap maintained between abrasive wheel and a workpiece surface. The rise in this gap increases the electrode potential due to which the tendency of electric spark discharge rises. This phenomenon develops the corroded pits on the workpiece surface as a result of which the surface quality decreases as shown in figure 4 (e) [16,38]

6) Selection of cathode and anode

The importance of electrode is explained in section 2 of this paper. Figure 4 (f) shows the effect of selection of tool as a cathode and anode on normal grinding force. As discussed earlier, if dressing current increases then there will be reduction in the oxide layer thickness due to which actual grinding depth rises. Also, with the reduction in layer thickness, no. of protruding grains increases and this results in the rise of grinding force. Normal grinding force in ELID with workpiece as cathode is less than that of ELID with tool as cathode [32].

7) Vibration frequency and amplitude

Energy ratio decreases with the combined effect of increase in ultrasonic pulsation frequency and the amplitude which is also illustrated in figure 5. The decrease in the grinding energy ratio inherently increases the surface quality [42].

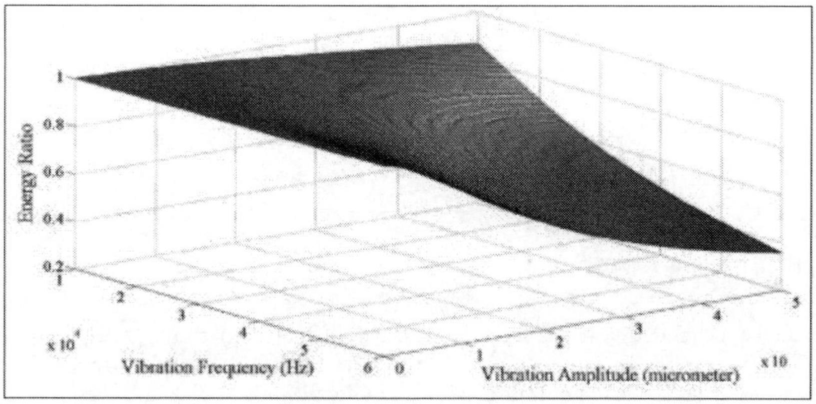

Figure 5. Impact of oscillation frequency and amplitude on the energy ratio. [42]

8) Mode of oscillation

Vibrations can be applied in two modes i.e., longitudinal vibrations (orthogonal to the grinding direction) and transverse vibrations (along the grinding direction) figure 6 illustrates these vibrations schematically. Amplitude of vibration is less in longitudinal mode than the transverse mode also, longitudinal deepens the valley height whereas transverse reduces the peak height. This is the reason why longitudinal vibrations increases the surface roughness and transverse vibrations smoothens the surface roughness [22,42].

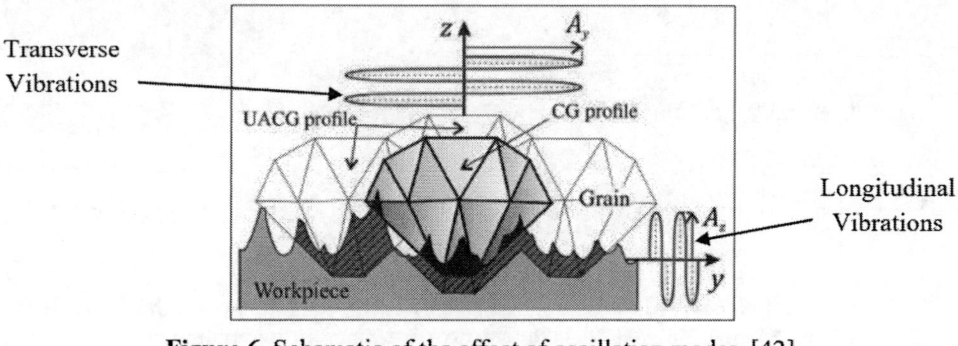

Figure 6. Schematic of the effect of oscillation modes. [42]

4. MATHEMATICAL MODELING

In this chapter, various mathematical models regarding oxide layer formation and surface quality prediction in ELID as well as UA-ELID are explained.

4.1 Oxide layer formation

An oxide film is formed by the electrolysis process as explained in chapter III of this paper. The formation of an oxide film significantly contributes in improving the topography of the surface in ELID grinding. When the oxide film is too dense, the work surface has remote possibility of being machined by an abrasive wheel surface which consists of pointed abrasives. When the oxide layer is too lean, the metal bond is eroded quickly and the grinding wheel fade out at higher rate. Therefore, the depth of the oxide layer affects the operational state instantly. Many researchers worked on developing the mathematical model of oxide layer thickness formation [12,14,19].

X. Jia et al. [14] have studied and formulated the mathematical model of oxide layer thickness in UA-ELID multiplicate grinding. In addition to electrolysis and Faraday's law, they have incorporated the effect of ultrasonic vibrations and removal of the oxide layer due to crack propagation. Worn out grains are being squeezed out by workpiece particles. This squeezing cause regional softening of the oxide layer and surface fractures got enhanced by residual stress of permanent distortion. The unit grinding force on single grain (F_p) can be expressed as per Eq. (1) and the average normal force per grain (F_{ng}) can be expressed in Eq. (2).

$$F_p = \frac{6\lambda a_p v_w}{\pi \cos\beta N_m (v_s+v_w)} \Bigg/ \oint \sqrt{1+\left(\frac{2A_0\pi f}{v_s+v_w}\cos\left(\frac{2\pi f}{v_s+v_w}.x+\varphi_0\right)\right)^2}\,dx \tag{1}$$

$$F_{ng} = F_p\left(\frac{a v_w}{v_s}\right)^2 \frac{a_p}{D}\sin\beta\,\tan\beta \tag{2}$$

Where "λ is a constant related to material, a_p is plastic deformation per unit grain, v_w is workpiece feed speed, v_s is the velocity of grinding wheel, β is the half conical angle of a single grain, N_m is grain number of unit area, A_0 is the amplitude of ultrasonic vibration, f is the frequency of ultrasonic vibrations, φ_0 is initial phase angle and a is the distance between continuous cutting edges" [14].

"When F_{ng} > holding force for a single grain, the macroscopic cracks are created on the contact interface between oxide layer and grains. By the principle of indentation fracture, the volume of the oxide layer fall of per grain (α) can be written" [14] as Eq. (3) and thus the thickness of the oxide layer (h) at any time (t) can be expressed in Eq. (4).

$$\alpha = \alpha_0\left(F_{ng}\right)^{\frac{9}{8}}(K_c)^{\frac{1}{2}}(H)^{\frac{5}{8}}\left(\frac{E}{H}\right)^{\frac{4}{5}} l_g \tag{3}$$

$$h = \sqrt{\frac{\eta AUS_e\alpha_1 R_c t}{2\left(1-\sqrt[3]{\pi/6}\,\kappa^{(2/3)}\right)\rho_y\rho nF\pi^2 DB} + \frac{\rho_d^2\sigma^2}{4\rho_y^2}} - \frac{\rho_d\sigma}{2\rho_y} \quad (t \le t_0) \tag{4}$$

$$h = \sqrt{\frac{\eta AUS_e\alpha_1 R_c t}{2\left(1-\sqrt[3]{\pi/6}\,\kappa^{(2/3)}\right)\rho_y\rho nF\pi^2 DB} + \frac{\rho_d^2\sigma^2}{4\rho_y^2}} - \frac{\rho_d\sigma}{2\rho_y} - \frac{\left(6\kappa/\pi d_g^3\right)^{(2/3)}(v_s+v_w)\,\alpha.\alpha_2 R_c(t-t_0)}{2\left(1-\sqrt[3]{\pi/6}.\kappa^{(2/3)}\right)\pi^2 D} \quad (t > t_0)$$

Where, "η is current efficiency, A is atomic mass, U is effective voltage, S_e is cathode active area, α_1 is the wrap angle of layer formation area, R_c is duty ratio, t is time, κ is the volume ratio of grain, ρ_y is the resistivity of an oxide layer, ρ is the density of the oxide layer, n is no. of valence electrons, F is

Faraday's constant, D is the diameter of the grinding wheel, B is outside diameter width of the grinding wheel, σ is electrode gap, d_g is the diameter of diamond grain, α_2 is the wrap angle of the grinding area and t_0 is the time when the oxide layer just begins to shed" [14].

4.2 Surface quality prediction

UA-ELID process is generally used for the mirror-like surface finish of brittle materials. Bo Zhao et al. [16] have analyzed the interrelationship between abrasives and workpiece surface by using kinematic mechanics. Impact of ultrasonic vibration on single abrasive grain of grinding wheel, the effect of oscillations on two adjoining abrasive grains and effect of ELID process parameters has been cumulatively studied. They have finally developed the predictive mathematical model of surface roughness in ultrasonic-aided ELID internal grinding as shown in Eq. (6). The true grinding depth (a_p') can be modeled as shown in Eq. (5).

$$a_p' = a_p - \frac{\eta_2 . M . D . U . A_c}{z . F . \rho . A_a(\rho_e . h_e + \rho_f . h)} \tag{5}$$

Where "a_p is the nominal grinding depth, η_2 is the current efficiency, M is the molecular weight of the metal bond, D is the duty ratio, U is the electrode voltage, A_c is the effective cathode area, z is the valence of metallic element, F is the Faraday's constant, ρ is the density of metal bond, A_a is the effective conducting area at the anode, ρ_e is the electrolyte resistivity, h_e is the inter-electrode gap, ρ_f is the resistivity of the oxide layer and h is the thickness of the oxide film" [16].

After adding the effect of ultrasonic vibrations and the effect of ELID parameters, finally, Bo Zhao et al. [16] have developed the mathematical model of surface roughness (R_z) for internal UA-ELID as shown in Eq. (6).

$$R_z = \sqrt{2} . \sqrt[3]{\frac{\pi}{6V_g}} . \frac{\eta_1 . f_a^{\zeta_3} . d_{gavg}}{f^{\zeta_1} . A^{\zeta_2} . V_s} . \sqrt{a_p' . \left(\frac{R_w - R_s}{R_s . R_w}\right) . \left\{(R_w + a_p')^2 + \left[\frac{R_w . f_a}{V_s - V_w} + 2 . A . \pi . f . \frac{R_w}{V_s - V_w}\right]^2\right\}} \tag{6}$$

Where, "V_g is the volume fraction of grains, η_1, ζ_1, ζ_2, ζ_3 are the correlation coefficients and indexes for considering the effect of adjacent grits on the workpiece surface, f_a is the axial feed velocity, d_{gavg} is the average diameter of the abrasive particle, f is the frequency of ultrasonic vibrations, A is the amplitude of ultrasonic vibrations, V_s is the velocity of the grinding wheel, R_w is the radius of the workpiece, R_s is the radius of the grinding wheel and V_w is the velocity of the workpiece" [16].
Bo Zhao et al. [16] have also found that values of surface roughness predicted using the above model of surface quality they had developed were in good agreement with the actual surface roughness values by the experiments.

5. APPLICATIONS

5.1 Fine grinding of bearing components

In the bearing industry, it is foremost essential to enhance the surface quality of roller surfaces, which assists in decreasing disturbances and fluctuations during the functioning. SiC is usually used as the material of bearing rollers, however, by the virtue of its large hardness, it is hard to obtain a mirror-like surface by machining. For obtaining remarkable bearing quality, the surface of a roller is required to be ground, lapped, and then polished with a conventional finishing method, which increases time as well as cost. With the ELID grinding of Sic rollers, we can achieve surface roughness of 30-70 nm depending on load [33].

"The surface roughness of the bearing raceway is considered as a critical parameter of transmission accuracy. A relatively rough surface indicates the possibility of deeper damage since rough surface on the raceway can be correlated to subsurface damage, while a relatively fine surface will exert an adverse effect on the fatigue life of the finished ball bearing. Therefore, it is desired that the surface roughness

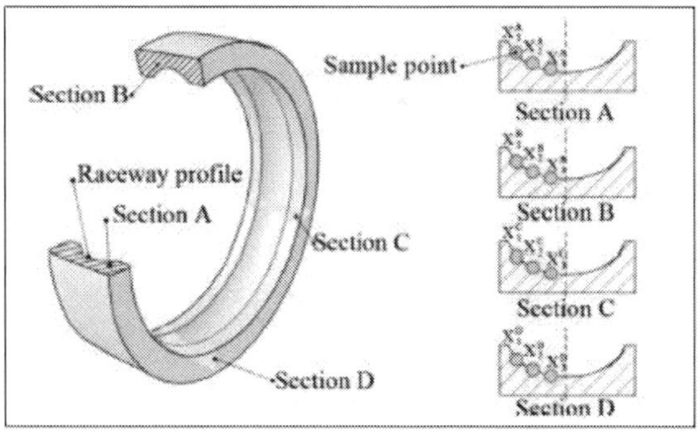

Figure 7. Bearing raceway profile with sections Studied. [34]

Figure 8. Mirror surface generation on Si. [10]

on the raceway is constant in an acceptable range. With the grinding time extended from 2 to 30 min, the surface roughness on the raceway was kept stable between 0.43 and 0.53 μm constantly at the feed rate from 0.0012 to 0.006 mm/min" [34]. The section-wise raceway profile is shown in figure 7 [34]. They also concluded that the ELID groove grinding combined with intermediate truing has a considerable scope of application in accurate fabrication of bearing raceway.

5.2 Grinding and thinning of Silicon Wafers

Silicon wafers are largely employed in integrated circuits. In order to achieve high quality and low cost, J. Liu et al. [35] have used the ELID finishing of silicon wafers. They found that the surface roughness of about 1.85 nm and subsurface crack of less than 1 μm was achieved [35]. During the thinning of silicon wafers by ELID, unstable forces were observed [6]. So, M. Islam et al. [6] have newly developed Injection Electrode assisted ELID and have effectively thinned down silicon wafers from "750 μm to 70 μm" [35]. Also, they have observed a high surface quality of about 6 nm.

5.3 Light Emitting Diodes (LED)

The use of hard and brittle materials in the light-emitting diodes has been increased in several years. Silicon carbide (SiC), gallium nitride (GaN), and sapphire ($\alpha - Al_2O_3$) are used as support material for LEDs. These materials are complicated to machine by virtue of mechanical properties they possess also these materials are chemically stable so, difficult to erode as well. H. Lee et al. [36] have used the sequential process of ELID and chemical mechanical polishing (CMP) to fine machining of these materials. They have observed surface roughness (R_a) for Sic 0.4 nm, for sapphire wafer 0.8 nm and GaN 1 nm [36].

5.4 Optical Components

Micro-optical components play an important role in applications like fiber optics, optical storage systems, and portable information devices. Surface finishing of micro components needs smaller grinding wheels, small grinding velocity, and adequate wheel-workpiece stiffness. Mirror surface finish of optical components as shown in figure 8 is desirable. With ELID grinding, microlens can be finished with great quality as ELID grinding wheels have high wheel-workpiece stiffness. Most of the optical components require Si wafers with high finish [3,10]

5.5 Micro hole machining

In many industrial fields there is an ample demand of the micro hole machining of advanced ceramics. Manufacturing of micro hole in hard and brittle materials has problems such as abrasive wheel is prone to wear more and loaded due to low grinding velocities. The process is slightly unstable also removing of grinding chips is difficult and precision as well as surface quality of machined holes are not excellent.

C. Zhang et al. [40] have used ELID – II system for micro-hole machining and have found that grinding forces reduced and micro-hole entrance and exit didn't have crack marks [40]. Micro holes of 250 μm were finished on ceramic materials [26].

6. Conclusions

This paper gives the gist of literature referred for the ELID and UA-ELID processes. After the study of literature, the following are concluded

i. UA-ELID enhances the quality of the surface than the normal ELID process. Also, UA-ELID works in ductile regime hence is more suitable for nano composite ceramics than ELID.
ii. Longitudinal vibrations reduce the surface topography whereas transverse vibrations increase the quality of the surface.
iii. Recommended values of electric current, voltage and inter-electrode gap are 0.5-1A, 60-90V and 0.1-0.3 mm respectively. If current is less than 0.5A, then corrosion pits are developed on the surface due to electric spark and if current is greater than 1A, then it reduces the depth of oxide layer thus actual grinding depth increases which in turn reduces the surface quality.
iv. Many researchers had worked in the mathematical modeling of oxide layer thickness formation while some researchers had worked in developing surface quality prediction models for both ELID and UA-ELID. These models were used again by some researchers for developing precision controlling and online monitoring systems.
v. While much of the work has been done for ELID – I & ELID – II systems, less work has been reported in ELID – III & ELID – IV systems. So, the authors take this opportunity to highlight this area for further developments in this area.

Acknowledgements
The authors acknowledge the support provided by Walchand College of Engineering Sangli. The authors would also like to acknowledge the work of all the researchers in the field of ELID grinding.

References

[1] Bandyopadhyay B 1995 The effects of grinding parameters on the strength and surface finish of two silicon nitride ceramics, *Journal of Materials Processing Technology* **53**, pp 533-543.
[2] Ohmori H, Takahashi I and Bandyopadhyay B 1996 Ultra-precision grinding of structural ceramics by electrolytic in-process dressing (ELID) grinding, *Journal of Materials Processing Technology* **57**, pp 272-277.
[3] Bandyopadhyay B, Ohmori H and Takahashi I 1997 Efficient and stable grinding of ceramics by electrolytic in-process dressing (ELID), *Journal of Materials Processing Technology* **66**, pp 18-24.
[4] Fatima K, Senthil Kumar A, Rahman M and Lim H 2003 A study on wear mechanism and wear reduction strategies in grinding wheels used for ELID grinding, *Journal of Wear* **254**, pp 1247-1255.
[5] Islam M, Senthil Kumar A, Balakumar S, Lim H and Rahman M 2006 Performance evaluation of a newly developed electrolytic system for stable thinning of silicon wafers, *Journal of Thin Solid Films* **504**, pp 15-19.
[6] Jiao Y, Ping Hu and Pei Z 2005 Rotary ultrasonic machining of ceramics: Design of experiments, *International Journal of Manufacturing, Technology, and Management* 7, pp 192-207.
[7] Saleh T, Sazedur Rahman M, Lim H and Rahman M 2007 Development and performance evaluation of an ultraprecision ELID grinding machine, *Journal of Materials Processing Technology* **192**, pp 287-291.
[8] Manicone P, Iommetti P and Raffaelli L 2007 An Overview of Zirconia ceramics: Basic properties and clinical applications, *Journal of Dentistry* **35**, pp 819-826.
[9] Gao G, Zhao B, Xiang D and Kong Q 2009 Research on the surface characteristics in ultrasonic grinding nano-zirconia ceramics, *Journal of Materials Processing Technology* **209**, pp 32-37.
[10] Rahman M, Senthil Kumar A and Biswas I 2009 A Review of Electrolytic In-Process Dressing (ELID) Grinding, *Journal of Key Engineering Materials* **404**, pp 45-59.

IOP Conf. Series: Materials Science and Engineering 998 (2020) 012048 doi:10.1088/1757-899X/998/1/012048

[11] Peidong Hang 2009 A study on the electrolytic in-process dressing (ELID) grinding of Saffire with acoustic emission monitoring, *The University of Toledo Digital Repository*.
[12] Zhang H, Kuai J and Zhang F 2010 Modeling of Thickness of the Oxide Film in ELID Grinding, *Journal of Advanced Materials Research* **135**, pp 376-381.
[13] Jiang J, Ge P, Bi W, Zhang L, Wang D and Zhang Y 2013 2D/3D ground surface topography modeling considering dressing and wear effects in grinding process, *International Journal of Machine Tools and Manufacture* **74**, pp 29-40.
[14] Jia X, Zhao B and Bian P 2013 Predictive modeling of oxide layer thickness in ultrasonic ELID multiplicate grinding, *Journal of Advanced Materials Research* **815**, pp 86-92.
[15] Zhu Z, Dhokia V, Nassehi A and Newman S 2013 A review of hybrid manufacturing processes – State of the art and future prospectives, *International Journal of Computer Integrated Manufacturing* 2013.
[16] Bo Zhao, Chen F, Jia X, Zhao C and Wang X 2017 Surface quality prediction model of nano-composite ceramics in ultrasonic vibration-assisted ELID mirror grinding, *Journal of Mechanical Science and Technology* **31**, pp 1877-1884.
[17] Shaolin X, Kuriyagawa T, Shimada K and Mizutani M 2017 Recent advances in ultrasonic-assisted machining for the fabrication of micro/nano-textured surfaces, *Front. Mech. Eng*. **12**, pp 33-45.
[18] Kushan M, Orak S and Uzunonat Y 2017 Ultrasonic Assisted Machining Methods: A Review, *International Journal of Advanced Engineering Research and applications* **3**, pp 294-302.
[19] Alqahtani B, Zhang M, Marinescu I, Bafakeeh O and Sofyani S 2019 Microscopic characterization and modeling of the oxide layer for electrolytic in-process dressing (ELID) grinding with a focus on voltage, electrode-wheel gap, and coolant flow, *The International Journal of Advanced Manufacturing Technology*.
[20] Azarhousang B and Tawakoli T 2011 Development of a novel ultrasonic unit for grinding of ceramic matrix composites, *International Journal of Advanced Manufacturing Technology* **57**, pp 945-955.
[21] Wos P and Michalski J 2011 Effect of Initial Cylinder Liner Honing Surface Roughness on Aircraft Piston Engine Performances, *The Journal of Tribol Lett* **41**, pp 555-567.
[22] Kuo Kei-lin and Tsao Chung-Chen 2012 Rotary ultrasonic-assisted milling of brittle materials, *Transactions of Nonferrous Metals Society of China* **22**, pp 793-800.
[23] Choi Y, Park K, Hong Y, Kim K, Lee S and Choi H 2013 Effect of Ultrasonic Vibration in Grinding; Horn Design and Experiment, *International Journal of Precision Engineering and Manufacturing* **14**, pp 1873-1879.
[24] Sazedur Rahman M, Saleh T, Lim H, Son S and Rahman M 2008 Development of an on-machine profile measurement system in ELID grinding for machining aspheric surface with software compensation, *International Journal of Machine Tools & Manufacture* **48**, pp 887-895.
[25] Furutani K, Hieu N, Ohguro N and Nakamura T 2003 Automatic compensation for grinding wheel wear by pressure based in-process measurement in wet grinding, *The Journal of Precision Engineering* 27, pp 9-13.
[26] Rahman M, Senthil Kumar A, Lim H and Fatima K 2003 Nano finish grinding of brittle materials using electrolytic in-process dressing (ELID) technique, *Sadhana* **28**, pp 957-974.
[27] Wang Y, Lin B, Wang S and Cao X 2014 Study on the system matching of ultrasonic vibration-assisted grinding for hard and brittle materials processing, *International Journal of Machine Tools and Manufacture* **77**, pp 66-73.
[28] Fatima K, Rahman M, Senthil Kumar A and Lim H 2007 Modeling of Ultra-Precision ELID Grinding, *Transactions of the ASME* **129**, pp 296-302.
[29] Fatima K, Schinhaerl M, Geiss A, Rascher R and Sperber P 2010 A Knowledge-based feed-back control system for precision ELID grinding, *The Journal of Precision Engineering* **34**, pp 124-132.
[30] Kumar M, Kanmani S S, Vamsi K P and Venugopal A 2014 Vibration Assisted Conventional and Advanced Machining: A Review, *The Journal of Procedia Engineering* **97**, pp 1577-1586.
[31] Klocke F, Klink A and Henerichs M 2009 ELID dressing behavior of fine-grained bronze bonded diamond grinding wheels, *International Journal of Abrasive Technology* **2**, pp 358-367.

[32] Wang Z, Ren C, Chen G, Zhang L and Deng X 2018 A Comparative study on the state of the oxide layer in ELID Grinding with tool-cathode and workpiece-cathode, *International Journal of Advanced Manufacturing Technology* **94**, pp 1299-1307.

[33] Zhao L, Bafakeeh O, Liu H and Marinescu I 2016 ELID Fine Grinding of SiC Bearing Rollers, *The Journal of Key Engineering Materials* **686**, pp 204-211.

[34] Wu M, Ren C and Zhang K 2015 ELID groove grinding of ball-bearing raceway and the accuracy durability of the grinding wheel, *International Journal of Advanced Manufacturing Technology*.

[35] Liu J, Pei Z and Fisher G 2007 ELID grinding of silicon wafers: A literature review, *International Journal of Machine Tools & Manufacture* **47**, pp 529-536.

[36] Lee H, Kasuga H, Ohmori H, Lee H and Jeong H 2011 Application of electrolytic in-process dressing (ELID) grinding and chemical mechanical polishing (CMP) process for emerging hard-brittle materials used in light-emitting diodes, *Journal of Crystal Growth* **326**, pp 140-146.

[37] Pavel R, Pavel M and Marinescu I 2004 Investigation of pre-dressing time for ELID grinding technique, *Journal of Materials Processing Technology* **149**, pp 591-596.

[38] Wang Z, Ren C, Chen G, Deng X and Ji C 2019 Influence of oxide layer on grinding quality in ELID grinding bearing outer ring raceway with workpiece-cathode, *The International Journal of Advanced Manufacturing Technology*.

[39] Yin S, Ohmori H, Uehara Y, Chen F, Zhu Y and Wang Y 2009 Ultra-precision fabrication of large stamping die of Wolter mirror for an X-ray telescope by ELID grinding, *Proceeding of SPIE* **7282**.

[40] Zhang C, Ohmori H and Li W 2000 Small-hole machining of ceramic material with electrolytic interval-dressing (ELID - II) grinding, *Journal of Materials Processing Technology* **105**, pp 284-293.

[41] Biswas I and Senthil Kumar A 2010 Experimental study of wheel wear in electrolytic in-process dressing and grinding, *International Journal of Advanced Manufacturing Technology* **50**, pp 931-940.

[42] Zahedi A, Tawakoli T and Akbari J 2015 Energy aspects and workpiece surface characteristics in ultrasonic-assisted cylindrical grinding of alumina-zirconia ceramics, *International Journal of Machine Tools and Manufacture* **90**, pp 16-28.

★★★

IOP Conf. Series: Materials Science and Engineering 998 (2020) 012049 doi:10.1088/1757-899X/998/1/012049

Superhard Alloys: A Review of Empirical Modeling and Computational Strategies

Ch.Chandramouli[1,2], B.Venkatesh[1], M.Manzoor Hussain[3]

[1]Mechanical Engineering Department, Vardhaman College of Engineering, Hyderabad, India
[2]Research scholar, Mechanical Engineering Department, University college of Engineering, JNTUH, Hyderabad, India
[3]Mechanical Engineering Department, University college of Engineering, JNTUH, Hyderabad, India
E-mail: chandramouli@vardhaman.org,b.venkatesh@vardhaman.org.

Abstract. Superhard alloys has gained major research interest in recent past as it represents an effort to create man made alloys with hardness on par with the hardest substance available to us i.e. a naturally occurring diamond .They are virtually incompressible solids with high electron density and high bond covalency. Present paper aims to review the empirical methods and computational techniques developed by employing them using published literature, in order to attain a better understanding of complexities involved in synthesis of such compounds as a first step towards developing a new superhard metallurgical composition with similar mechanical attributes.
Keywords: Superhard alloys ,Ultrahard alloys, Computational Metallurgy, Numerical Modelling, Metallurgy Guided ML, High Entropy Alloys, Quaternary Carbides.

1. Introduction

The ability to resist deformation under loading is defined as "hardness". This is often measured using indentation depth under specific load which is then used to evaluate material performance against standardized scales like Mohr's, Brinell, Rockwell and Vickers[1][2][3].If the measured value on Vickers scale in above 40 GPA or 80 GPA, then elements or alloys with such property are said to exhibit Super hardness or Ultra hardness respectively. Besides high hardness they are generally chemically inert with good compressional strength, high bulk modulus and melting point[4][5].Such Materials have a wide range of applications in fields like automotive, oil and gas, aviation, defence, space exploration etc[6][7]. In 20th century a number of experiments were carried out aiming to design and synthesis super hard materials for industrial applications, which resulted in two famous compounds namely diamond [8],cubic boron nitride (c-BN) and zinc blended boron nitride (zb-BN)[9]. It is estimated that about 3000 million carats are produced every year as they are predominantly used as super abrasives[10].The superhard diamond is not a very good abrasive material, as it is not suitable for high speed machining and forms iron carbides when used to machine ferrous alloys making c-BN as only alternative which has only half the hardness of that of a diamond due to which recent research has seen a rise in search of materials other than carbon allotropes which are chemically and thermally stable with many focussed on achieving it using nano crystalline tweaking[11].

2. Current Research Trends

High hardness values are a contribution of three dimensional network formed by base elements of the alloy which is generally made up of short and long covalent bonds. The rigidity of these bonds are crucial for hardness values. For example graphite has layered hexagonal structure called graphene along crystallographic direction 'c' with weak van der Waals forces resulting in a very soft material whereas graphene bond energies are equal to that of diamond rendering it as one of the toughest materials[12], from which on can infer that composition, crystal structure along with pressure - temperature phases are important parameters in design of super hard materials. As carbon and boron are two hardest periodic elements know to us most of the super and ultra-hard materials developed using compounds wherein they are present in different phases[13]. A number of carbon allotropes have been experimented using such ab initio approach [14], predicting phases such as monoclinic M-carbon[15]. Similarly the rich phases of boron promise a large family of superhard compounds[16]. Systems such as boron carbide B_4C and boron superoxide B_6O with α- rhombohedral boron (α-B_{12}) present high hardness along with chemical inertness, wear resistance and melting point[17]. Recently boron superoxide system with β- orthorhombic boron has been predicted which exhibits hardness values higher than α-B_6O[18]. A number of such hypothetical systems like tetragonal B_4CO_4, Metastable $B_6C_2O_5$& B_2CO_2have been predicted with high hardness values[19]. Transition metal borides like WB_4 [20]and ReB_2 [21] have been considered as potential super hard material but their high reported values have been criticised[22].Nitrides formed by reacting metals with nitrogen gas like Re_2 $(N_2)(N_2)$ and similar compounds[23] in extreme pressure and temperature conditions is a promising approach but the high price is an inconvenience. Clathrates like polymerised partially decomposed fullerenes are superhard[24], similar polymerization is observed in carbon nano tubes[25]. Besides extreme hardness tetrahedral clathrate structures suggest existence of carbon clathrate framework which are not only harder than diamond but also present other properties like super conductivity. Recently predicted C_3B_3 clathrate structure that can host Sr atom is one such example[26]. Figure 1 presents an illustration in research trends of Hard, Superhard & Ultrahard Materials [27]

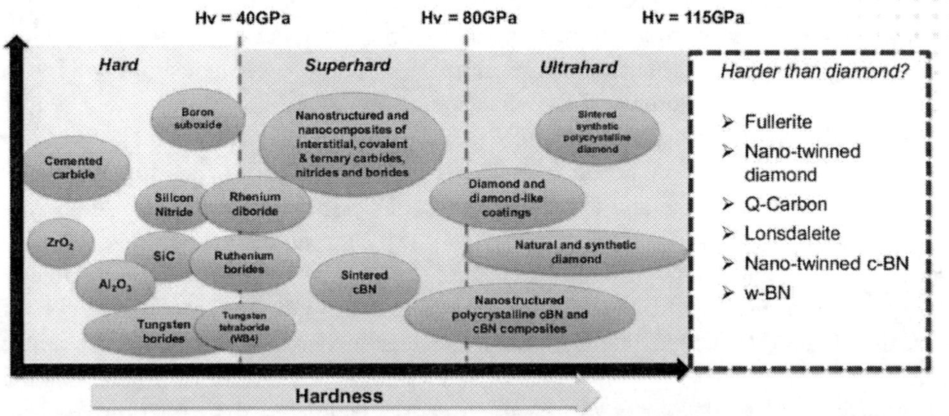

Figure 1: Hard, Superhard & Ultrahard Materials

However a major part of research is focussed on borides, carbides and nitrides of transitional metals like Rhenium [28]. Molybdenum[29], Chromium[30], Tungsten[31][32]Etc. Compounds like TiN[33] and WC[34] with hardness of 22 and 30 GPA respectively are already widely used in many industries. However these transitional metal complexes present a unique challenge as it is difficult to evaluate their compositions and crystal structure using regular experimental techniques like X-ray diffraction for which new methods to predict crystal structure have been developed and used to predict novel super hard materials.[35][36]. One such scenario where the super hardness of rhenium boride was predicted to be ~48 GPAbut was later was refuted[37]. Sum of the transitional metals Carbides and Nitrides being perused as the potential Superhard material are presented in figure 2[38].

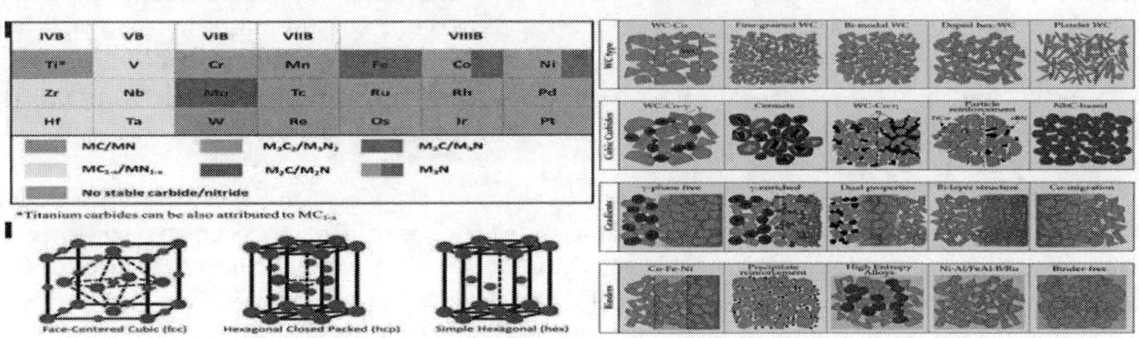

Figure 2: Transition Metal Carbides and Nitrides.

3. Empirical Modeling

Predicting mechanical properties using crystal structure is one of common approaches used in quest of novel super hard materials and a number of different techniques have been developed for the same[39][40]. A number of empirical models have been proposed which use crystal structures, compositions and material physical properties as an input to predict the hardness, like in the Chen model assumes that indentation size and dimensions are related to shear and bulk modulus[41]whereas Lyakhov-Oganov model uses crystal structure and chemical compositions[42] but it is often preferred to compare the model predictions with experimental data for no one model can accurately predict for all types of material combinations and phases.

3.1. Thermodynamic Phase Change

Besides the empirical methods the general strategies of predicting new chemical systems with superhard properties can be distinguished into two methodologies which are dependent on phases formed under high pressure conditions. (I) phase change under high pressure like in case of graphite to diamond conversion, (II) phase change induced densification of heterogeneous complexes under high pressure. A third approach has also gained great interest in recent times, it focuses on structural rearrangement at nano scale in systems made of unusual compositions of multiple elements under high pressure. An example of case (I) is orthorhombic γ-B_{28} discovered in 2008 has changed our understanding of allotropies of boron and its ability to form new phases under high pressure[43]. Till 2008 only three phases of known 16 phases of boron namely α-B_{12} rhombohedral at low temperature, β-B_{106} rhombohedral at high temperature and low pressure and t-B_{192} tetragonal at high temperature and high pressure were considered as promising systems [44][45].With predicted hardness above 50GPA this discovery has renewed interest in different metastable phases of boron recently a nonicosahedral β-B allotrope of boron have been identified to format 2100 K and 115 GPA[46].

As per literature Case (II), where in phase change leads to densification can be achieved either by pressurising preformed low density complexes to induce phase change in turn densification (d-BC_5) or initiating chemical reaction at high pressure which causes estimated densification ($B_{13}N_2$). In practice these two approaches are mutually inclusive and can occur in same compound (c-BC_2N). With near diamond like structure d-BC_5 is synthesised by dissolving boron in carbon under controlled conditions of temperature and pressure in order to avoid initiation of any phase change and then using high pressure to convert the unusual stoichiometric solution in to boron 'diamond' with an estimated hardness of 70GPA[47]. The in-situ study of B-BN system made the discovery of rhombohedral sub nitride $B_{13}N_2$. With a unit cell structure similar to BC_4it exhibits super hardness with hardness value of 41 GPA. Unlike boron carbide it is thermodynamically stable and at 5GPA it melts to form a eutectic solution and at further high pressures it form boron rich phases namely rhombohedral $B_{13}N_2$ and tetragonal $B_{50}N_2$. At further increase of pressure the $B_{50}N_2$ decomposes tetragonal t-B_{52} and $B_{13}N_2$ [48][49]. With hardness of 76GPAcubic BC_2Nfills the gap between cubic boron nitride and diamond and is synthesised by phase transition of $(BN)_{0.48}C_{0.52}$ at high pressures. A macroscopic sample have been produced using a large-volume multi-anvil apparatus with a 5000-ton press at Bayerisches Geoinstitut. The super cell has a combination of high density phase C-B-N layers like in graphite and non-dense layers distributed across the structure leading to formation of zb-BC_2N phase contributing to high resistance to deformation under uniaxial loading[50].

3.2. Nano Structures and Hall Petch Effect

In later half of nineteenth century a phenomenon was proposed claiming significant increase in hardness of polycrystalline materials when there microstructure was made of nano sized grains and was called 'Hall Petch Effect' which has again gained popularity as the search of superhard material shifted its focus towards high pressure induced nanostructuration[51].A simplified way to explain this effect is through dislocation pile-up because in polycrystalline material the grain boundaries act as barriers to crystal dislocation. Under applied stress, the more mismatch in the grain layout would mean more energy required to change the dislocation direction and move in to adjacent grain. Hence as the grain size is decreased the number of grain boundaries per unit volume increases and also rises the number of dislocations occurring in an instance which are piling up against each other which causes an increase in energy required to overcome them. This means more stress is needed to be applied to overcome these dislocations in finely grained material in comparison to coarsely grained one thus making it harder[52][53]. This relation between yield strength and grain size is defined by Hall-Petch as

$$\sigma = \sigma_0 + {}^{k}/_{d^2}$$

However this equation is observed to be inaccurate as grain size decrease beyond a critical diameter (around 10nm) where in a decrease in hardness is observed causing 'Inverse Hall Petch Effect'[54][55]. This inverse effect is not completely understood yet it is understood that at critical grain size the dislocation controlled plasticity shifts towards grain boundary controlled plasticity where in sliding and grain boundary diffusion become prominent causing a reduction in hardness. A new generalised equation has been proposed considering these factors

$$\sigma = \sigma_0 + \frac{k}{d^2} + k_1 + \frac{B_0}{d} + Bd^3$$

Where in k_1 and B_0 are constants and B is obtained using strain vs temperature data terms represent diffusion controlled dislocation mechanism. For large grain sizes first terms are dominant and for extremely small grain sizes the last three terms are important to explain inverse effects. Another issue observed in nano structures is as one increases the temperature due to grain growth the size is no longer in nano scale. This phenomenon is well observed in 'nano scaled diamond' and 'nano c-BN'. It has been observed that this can be easily overcome by maintaining the quality of starting material and controlling the temperature and pressure during synthesis process.

4. Computational Metallurgy

Two new approaches namely in situ and ab initio have gained immense popularity in recent years Advances in X-Ray diffraction along with improved optical and tomographic techniques and Synchrotron radiation sources for studying extreme conditions have propelled the in situ studies whereas next generation artificial intelligence and machine learning engines have pushed the ab initio beyond the frontiers of regular methodologies used in turn paving way for new generation of superhard materials in form of high entropy transition metal carbides[56][57].

4.1. In Situ and Ab Initio approaches

As observed by many it is not always possible to study superhard material using experimental techniques like X-ray diffraction studies alone, a hybrid approach of first principle equation for predicting mechanical properties from crystallography validated using experimental data is needed for better prediction is needed. A number of superhard materials are formed due to metastable phases at ambient pressures and temperatures and there synthesis requires controlled smooth transition monitored using constant feedback. Numerous microscopic models use parameters such as crystal geometry, atomic proximity and connectivity, valence electron density, electronegativity are available to predict hardness value[58]. Such crystal structure prediction based algorithms have gained prominence in predicting new superhard composition of elements in recent years. Modified versions of first principle equations have been proposed by many which can used to optimise the hardness of such predicted crystal structures[59][60]. A major drawback in such a system is that most of these models predict very high hardness values for structures with non-interacting forces like van der Waals forces which are needed for holding three dimensional shape of compound, this is overcome by models such as CALYPSO and UPEX which chemical graph theory to identify atomic bonds[61][62]. We can estimate hardness using elastic, bulk and shear moduli but such a macroscopic approach are not employed to predict potential material compositions as it is very expensive to calculate these properties using first principle methods hence instead they are used to evaluate only promising set of permutations which are predicted from initial search settings[63][64][65][66].

4.2. Numerical Modelling
A number of models have been proposed to explain the phenomena of plastic deformation due to applied load using microscopic crystallographic concepts such as slip and twinning of grain structures.

4.2.1. Taylors Model
It is based on the assumption that the deformation field of all the grains in the specimen under load is equal disregarding the shape and size of grains in local neighbourhood also does not consider kinematic coupling where in the rotation of a grain due to external loading effects rotation of grains around it.It calculates the total deformation using slip, twinning and lattice distortion in given crystal system using power law based Cauchy stress equations and Taylor's assumptions [67].

4.2.2. Viscoplastic Self Consistent Model (VPSC)
Similar to Taylor model it also doesn't consider the shape and rotational effects of neighbouring grains under external loading but allows individual strain response for each grain in the system and their difference is dependent on stiffness between grain and surrounding homogenous media calculated relatively. It averages the microscopic behaviour of grains uniformly to match the overall macroscopic loading. In VPSC, each grain does not interact with its neighbouring grain instead its interaction is with entire aggregate treated as a unit equivalent medium [68].

4.2.3. Crystal Plasticity Finite Element Model (CPFEM)
It also considers geometric distortions caused due applied load. The change in shape and rotation are calculated and because of this feature it can predict material failure more accurately than the Taylor's and VPSC models. Where they are able to predict failure due to textural effects, CPFEM has potential to evaluate localised deformations as direct interaction with neighbouring grains is considered and accordingly calibrated while calculating the final state of given specimen under load as illustrated in figure 3[69].

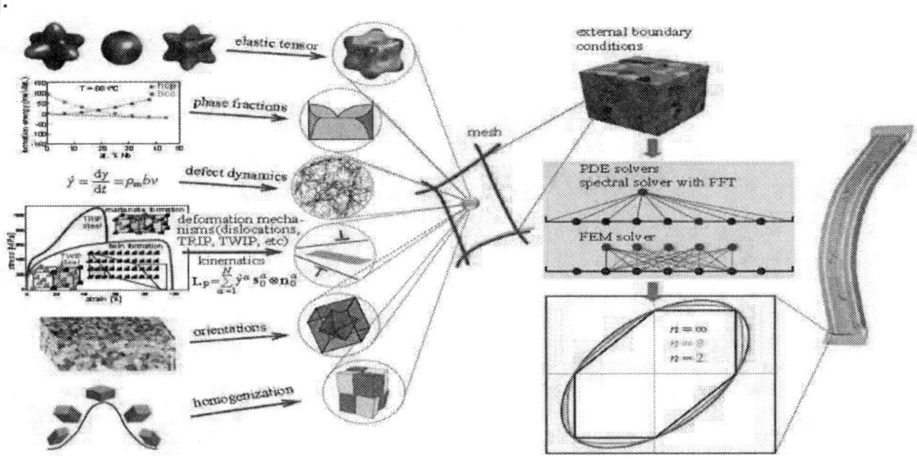

Figure 3: Crystal Plasticity Finite Element Methods

Using such approaches a number of material databases coupled with Machine Learning models have been developed which can predict a whole range of elemental combinations along with their properties like supercritical temperature, melting point, electronic bands, elastic properties for urinary, binary, ternary, quaternary and quinary materials. Databases such as Open Quantum Materials Database (OQMD[70], Automatic Phonon Library (APL), Automatic Gibbs Library (AGL), Automatic Elastic Library (AEL) containing vibrational properties, thermal and thermo-mechanical properties such as entropy, enthalpy, bulk and shear modulus along with Search API's(Application Program Interface) like RESTful and AFLUX[71] allow AFLOW[72], CALYPSO, CALPHAD, UPEX ,XtalOpt[73], RandSpg[74] and other similar algorithms to access more than 2 million entries for estimation in a single run which have predicted compounds such as R3m-BC_2N, Pnnm-CN, cubic BC_3,Cr_3C_2, Cr_7C_3,$Cr_{23}C_6$,

CrB, CrB_2, CrB_4, Cr_2B, Cr_3B_4, Cr_5B_3, Pmmn WB_5, $Re_{0.5}W_{0.5}C$ and $ReWC_{0.8}$ which exhibit superhard property[75][76][77].

Further using optimisation techniques such as Mendelevian Chemical Spacing, Energy Filtering and Pareto Technique combined systems like B_xC_y, B_xO_Y, C_xN_Y, Cr_xC_Y, Re_xB_Y, W_xB_Y, Si_xC_Y, W_xC_Y , Al_xO_Y, Ti_xC_Y , Si_xN_Y , Ti_xN_Y , Be_xO_Y , Ru_xO_Y, Os_xO_Y , Rh_xB_Y , Ir_xB_Y, Os_xB_Y and Ru_xB_Y which have been long argued to possess high hardness values were found to be true, further promising systems such as Mo_xB_Y ,B_xP_Y,V_xB_Y,Fe_xB_Y,Mn_xB_Y,Mn_xH_Y,$(Hf_{0.25}Ta_{0.25}Zr_{0.25}Nb_{0.25})C$,$(Cr_{0.2}Nb_{0.2}Ti_{0.2}V_{0.2}W_{0.2})C$,$(Zr_{0.2}Hf_{0.2}Mo_{0.2}Ta_{0.2}W_{0.2})C$ were predicted[78][79][80]. This kind of behaviour of material can be easily represented using Ashby Map of Fracture Toughness vs. Yield Strength as illustrated figure 4[81].

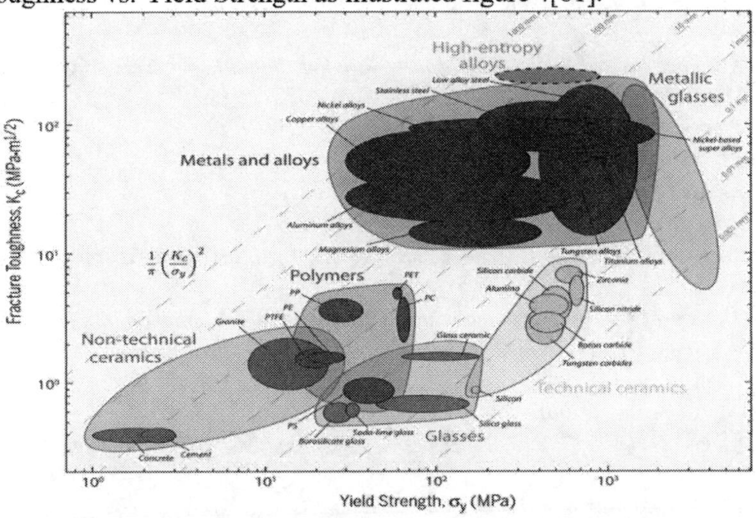

Figure 4: Ashby Map of Fracture Toughness vs. Yield Strength.

5.Summary

Super hard alloys like any other super material were developed as a solution for drawbacks in existing compounds and are constrained by the expensive and complex synthesis methodologies. During early stages of this quest major focus was on structures similar to that of diamond, which is the hardest known material available to us. As part of which a range of new allotropes of boron and nitrogen with potential of bonding capabilities similar to carbon were identified. With recent advances in modern experimental techniques, both at micro and macro level along with miniaturization of computational resources have opened a range of new compositions of elements utilizing in situ and ab initio techniques. A majority of these are built on the foundation of the idea that the hardness value is correlated to indentation dimensions which is in turn effected by the physical and chemical properties such as grain size, shape, valency, affinity, thermodynamic phases etc. Computational models such as VPSC, CPFEM coupled with normalisation algorithm like particle optimisation along with AI and ML models driven by AGL, AEL and other similar open databases has an ability to scan two million entries in a single run generating a plethora of potential configurations. The only short coming of such an approach is that it is based on published literature and the data isn't as accurate as it is believed to be, proved as in case of inverse Hall Petch recently established. Also all the models published either generalise or let user control the parameters while evaluating the values thus effecting the final solution. From current review we understand that a combination of CPFEM with CALYPSO is a popular method of proposing new type of super hard materials especially while considering transitional metal based complexes. This new advancement in metallurgical studies is not yet fully understood and needs more resourceful systems for further research.

References

[1] D.Tabor 1951 *The Hardness of Metals.* Clarendon : Oxford.

[2] Argon, F. A. McClintock and A. S. 1966, *Mechanical Behavior of Materials.* Reading : Addison-Wesley.

[3] Wilkinson, F. A. Cotton and G. 1972, *Advanced Inorganic Chemistry, 3rd ed.* New York : Wiley.

[4] W.D.Sproul,1996 *,New routes in the preparation of mechanically hard films,* Science 273(5277):889.

[5] D.M.Teter,*1998,Computational alchemy: the search for new superhard materials,* MRS Bull 23(01), pp. 22-27.

[6] Vepˇrek, 1999,*The search for novel, superhard materials*, J Vac Sci Technol A 17(5), pp. 2401–2420.

[7] Haines J, Leger J, Bocquillon G,*2001, Synthesis and design of superhard materials*, Annu Rev Mater Res 31(1), p. 1.

[8] *F. P. Bundy, H. T. Hall, H. M. Strong, and R. H. Wentorf,1957,Nature176, pp. 51–55,1955.*

[9] *Wentorf, H. R. J., Chem. Phys.26, pp. 956–960.*

[10] *QYResearch. 2019, Global (United States, European Union And China) Super Hard Material Market Research Report 2019-2025. s.l. : QY RESEARCH.*

[11] *Solozhenko, V. V. Brazhkin and V. L 2019, J. Appl. Phys.125(13), 130901.*

[12] *K. Li, P. Yang, and D. Xue, 2012,Acta Mater.60(1), pp. 35-42.*

[13] *Kurakevych, O. O. 2009, J.Superhard Mater.31(3), pp. 139-157.*

[14] *Q. Zhu, Q. Zeng, and A. R. Oganov, 2012, Phys. Rev. B85(20), 201407.*

[15] *Q. Li, Y. Ma, A. R. Oganov, H. Wang, H. Wang, Y. Xu, T. Cui, H.-K. Mao and G. Zou., 2009,Phys. Rev. Lett.102(17), 175506.*

[16] *V. A. Mukhanov, O. O. Kurakevych, and V. L. Solozhenko, 2010,J. Superhard Mater 32(3), pp. 167-176.*

[17] *H. Dong, A. R. Oganov, Q. Wang, S.-N. Wang, Z. Wang, J. Zhang M. M. D. Esfahani, X.-F. Zhou, F. Wu, and Q. Zhu, 2016, Sci. Rep.6, 31288.*

[18] *O. O. Kurakevych, V. L. Solozhenko, 2011, J. Superhard Mater. 33(6), pp. 421-428.*

[19] *S. Wang, A. R. Oganov, G. Qian, Q. Zhu, H. Dong, X. Dong, and M. M. Davari Esfahani,2016 Phys. Chem. Chem. Phys.18(3), pp. 1859–1863.*

[20] *R. Mohammadi, A. T. Lech, M. Xie, B. E. Weaver, M. T. Yeung, S. H. Tolbert and R. B. Kaner, 2011 Proc. Natl. Acad. Sci. U.S.A.108(27), pp. 10958–10962.*

[21] *H.-Y. Chung, M. B. Weinberger, J. B. Levine, A. Kavner, J.-M. Yang,S. H. Tolbert, and R. B. Kaner, 2007 s.l. : John Wiley & Sons, Science316(5823), pp. 436-439.*

[22] *N. Dubrovinskaia, L. Dubrovinsky, and V. L. Solozhenko, 2007 Science318(5856) 1550.*

[23] *M. Bykov, S. Chariton, H. Fei, T. Fedotenko, G. Aprilis, A. V. Ponomareva,F. Tasnádi, I. A. Abrikosov, B. Merle, P. Feldner, S. Vogel, W. Schnick,V. B. Prakapenka, E. Greenberg, M. Hanfland, A. Pakhomova, H.-P. Liermann,T. Katsura, N. Dubrovinskaia. 2019 s.l. : Wiley, Nat. Commun.10(1), 2994.*

[24] *V. V. Brazhkin, A. G. Lyapin, V. L. Solozhenko, V. I. Bugakov S. N. Dub, O. O. Kurakevych, M. V. Kondrin, and E. L. Gromnitskaya, 2008 Fullerenes Nanotubes Carbon Nanostruct.16(5–6), pp. 475–485.*

[25] *M. Hu, Z. Zhao, F. Tian, A. R. Oganov, Q. Wang, M. Xiong, C. Fan, B. Wen J. He, D. Yu, H.-T. Wang, B. Xu, and Y. Tian, 2013 Sci. Rep.3, 1331.*

[26] *L. Zhu, H. Liu, R. E. Cohen, R. Hoffmann, and T. A. Strobel. 2017 arXiv:1708.03483(2017). [Online].*

[27] *Kanyanta, Valentine. 2016 "Hard, Superhardand UltrahardMateriaMicrostructure-Property Correlations for Hard, Superhard, and UltrahardMaterials. s.l. : Springer International Publishing, ISBN 978-3-319-29289-2.*

[28] *M.-M. Zhong, X.-Y. Kuang, Z.-H. Wang, P. Shao, L.-P. Ding, and X.-F. Huang), 2013 J. Phys. Chem. C117, 10643.*

[29] *Y. Liang, X. Yuan, Z. Fu, Y. Li, and Z. Zhong, 2012 Appl. Phys. Lett.101, 181908.*

[30] *L. Han, S. Wang, J. Zhu, S. Han, W. Li, B. Chen, X. Wang, X. Yu, B. Liu,R. Zhang, Y. Long, J. Cheng, J. Zhang, Y. Zhao, and C. Jin, 2015 Appl. Phys. Lett.106,221902.*

[31] *Q. Gu, G. Krauss, and W. Steurer, 2008 Adv. Mater.20, 3620.*

[32] *X.-Y. Cheng, X.-Q. Chen, D.-Z. Li, and Y.-Y. Li, 2014 Acta Crystallogr. C70,85.*

[33] *D. S. Stone, K. B. Yoder, and W. D. Sproul, 1991 J. Vac. Sci. Technol. A9, 2543.*

[34] *M. P. Groover. 2010 Fundamentals of Modern Manufacturing: Materials, Processes and Systems. s.l. : John Wiley & Sons.*

[35] *Oganov, A. R. 2011 Modern Methods of Crystal Structure Prediction. s.l. : Wiley.*

[36] *A. R. Oganov, G. Saleh, A. G. Kvashnin. 2018 Computational Materials Discovery. s.l. : Royal Society of Chemistry.*

[37] *A. R. Oganov, C. J. Pickard, Q. Zhu, and R. J. Needs, 2019Nature Rev. Mater.4,331.*

[38] *Yu Zhong, Xinhui Xia ,Fan Shi , Jiye Zhan , Jiang ping Tu ,Hong Jin Fan.2016 Transition Metal Carbides and Nitrides in Energy Storage. 5, s.l. : Wiley, Vol. 3. 1500286.*

[39] *A.R.Oganov. 2011 Modern Methods of Crystal Structure Prediction. s.l. : Willey.*
[40] *C.J.Pickard, Q.Zhu, R.J.Needs, A.G.Oganov, 2019 Nature Rev.Mater, Vols. 4,331.*
[41] *X.Q.Chen, H.Niu, D.Li, Y.Li, 2011 Intermetallics, Vols. 19,1275.*
[42] *A.O. Lyakhov, A.R.Oganov. 2011 Phys,Rev.B, Vols. 84,092103.*
[43] *V.L.Solzhenko, O.O.Kurakevych, A.R,Oganov, 2008 Superhard Mater, Vol. 30(6), pp. 428-429.*
[44] *P.Macchi, 2011 Superhard Mater, Vol. 33(6), pp. 380-387.*
[45] *Y.L.Godec, A.R.Oganov, V.L.Solozhenko, 2012 Superhard Mater, Vols. 34(1), 74.*
[46] *I.Chuvashova, E.Bykova, K.Glazyrin, M.Mezouar, V.Prakapenka, 2017 Phys.Rev.B, Vols. 95(18), 180102.*
[47] *A.Ritcher, R.Ries, C.P. Daghlian, 2006 Diam.Relal.Mater, Vol. 15(11), pp. 2019-2023.*
[48] *O.O.Kurakevych, V.L.Solozhenko, 2014 Superhard.Mater, Vol. 36(4), pp. 270-278.*
[49] *V.Bushlya, V.L.Solozhenko, 2017 Superhard.Mater, Vol. 39(6), pp. 422-426.*
[50] *S.K.Sharma, H.W.Hubble, I.Kudryashov, 2004 Raman Spectrosc, Vol. 35(10), pp. 822-825.*
[51] *N.J.Petch, 1953 Iron.Steel.Inst, Vol. 174, pp. 25-28.*
[52] *T.Irifune, A.Kurio, S.Sakamoto, H.Sumiya, 2003 Nature, Vol. 421, pp. 599-600.*
[53] *A.K.Mukerjee, D.Wolf, S.R.Philpot,V.Yamakov, H.Gleiter,2004 Nat.Mater, Vols. 3,43.*
[54] *C.S.Pande, K.P.Cooper, 2009 Prog.Mater.Sci, Vol. 54(6), pp. 689-706.*
[55] *Z.Zhao, B.Xu, Y.Tian, 2016 Annu.Rev.Mater.Res, Vol. 46(1), pp. 383-406.*
[56] *J.Philippe, M.Berg, G.Bromiley, F.Bergame, J.P.Perrilat, M.Alvarez-Murga, M.Morand, R.Atwood, A.King, S.Reigner, Y.Le Godec,M.Mezourar, 2016 High Pressure Res, Vol. 36(4), pp. 512-532.*
[57] *A.Courac, Y.Le Godec, V.L.Solozhenko, 2019 J.Appl.Phys, Vols. 126, 151102.*
[58] *F.Gao , 2003 Hardness of Covalent Crystal, Phys.Rev.Lett, Vols. 91, 015502.*
[59] *A.Simuneck, J.Vackar, 2006 Hardness of Cvalent and Ionic Crystal:First Principles Calculations, Phys.Rev.Lett, Vols. 96, 085501.*
[60] *K.Li, X.Wang, F.Zang, D.Xue,2008 Electronegativity Identification of Novel Superhard Materias, PhysRev.Lett, Vols. 100, 235504.*
[61] *O.A.Lyakhov, A.R.Oganov, 2011 volutionary Search for Superhard Materials: methodology and applications to form carbon and TiO2, Phys.Rev.B, Vols. 84, 092103.*
[62] *X.Zhang, 2013 First Principle Structural Design of Superhard Material, Chem.Phys, Vols. 138, 114101.*
[63] *D.M.Teter, 1998 Computational alchemy: The search for new superhard materials, MRS Bull, Vol. 23, pp. 22-27.*
[64] *. F.M.Gao, L.H. Gao, 2010 Microscopic models of hardness, J. Superhard Mater, Vol. 32, pp. 148-166.*
[65] *Chen, Q.Xi, H.Niu, D.Li, Y. Li, 2011 Modeling hardness of polycrystalline materials and bulk metallic glasses, Intermetallics, Vol. 19, pp. 1275-1281.*
[66] *Y.Tian, B.Xu, Z. Zhao, 2012 Microscopic theory of hardness and design of novel superhard crystals, Int. J. Refract. Met. H, Vol. 33, pp. 93-106.*
[67] *Izadbakhsh, 2012 Crystal plasticity based finite element modelling of large strain deformation in AM30 magnesium alloy, IOPScience, Vol. 20.*
[68] *Hutchinson, John W, 1976 Bounds and Self-Consistent Estimates for Creep of Polycrystalline Materials. London : Proc. Royal Soc. Lond, Vol. 101. A 348, 101-127.*
[69] *Franz Roters, Philip Eisenlohr, Thomas R. Bieler, Dierk Raabe. 2011 Crystal Plasticity Finite Element Methods. s.l. : Wiley, 9783527642090, 3527642099.*
[70] *J.E.Saal, S. Kirklin, M. Aykol, B. Meredig, C.Wolverton, 2013 Materials design and discovery with high-throughput density functional theory: The open quantum materials database (OQMD), JOM, Vol. 65, pp. 1501–1509.*
[71] *F.Rose, 2017 AFLUX: the LUX materials search API for the AFLOW data repositories, Comput. Mater. Sci, Vol. 137, pp. 362-370.*
[72] *C.Toher, 2017 Combining the AFLOW GIBBS and elastic libraries to efficiently and robustly screen thermomechanical properties of solids, Phys. Rev. Mater, Vols. 1, 015401.*
[73] *Z.Falls, P.Avery, E.Zurek, 2018 XtalOpt version r11: an open-source evolutionary algorithm for crystal structure prediction, Comput. Phys. Commun, Vol. 222, pp. 418-419.*
[74] *P.Avery, E.Zurek, 2017 RandSpg: an open-source program for generating atomistic crystal structures with specific spacegroups, Comput. Phys. Commun, Vol. 213, pp. 208–216.*
[75] *A.Y. Liu, M.L Cohen , 1989 Prediction of new low compressibility solids, Science, Vol. 245, pp. 841-842.*
[76] *C.M. Sung, M.Sung,1996 Carbon nitride and other speculative superhard materials, Mater. Chem. Phys, Vol. 43, pp. 1-18.*
[77] *Landis, D.D, 2012 The computational materials repository, Comput. Sci. Eng, Vol. 14, pp. 51-57.*

[78] *E.Gossett, 2018 AFLOW-ML: a RESTful API for machine-learning predictions of materials properties, Comput. Mater. Sci, Vol. 152, pp. 134-145.*

[79] *A.M.Tehrani, 2018 Machine learning directed search for ultraincompressible superhard materials, J. Am. Chem. Soc, Vol. 140, pp. 9844–9853.*

[80] *Pranab Sarker, Tyler Harrington, Cormac Toher, Corey Oses, Mojtaba Samiee, Jon-Paul Maria, Donald W. Brenner, Kenneth S. Vecchio, Stefano Curtarolo, 2018 High-entropy high-hardness metal carbides discovered by entropy descriptors, Nat. Commun, Vols. 9, 4980.*

[81] *Robert O Ritchie, Mark A Meyers,Shiteng Zhao, Zezhou Li, 2019 Mechanical properties of high-entropy alloys with emphasis on face-centered cubic alloys, s.l. : Progress in Materials Science, Elsevier, Vol. 102.*

Effect of Surface Modification with Aluminium and Copper on the Performance of Stainless Steel 304 Fin

[a, b] **Sreedhar Vulloju and** [c] **P.S. Kishore**

[a] *Research scholar, Department of Mechanical Engineering, College of Engineering Andhra University, Visakhapatnam, India.*

[b] *Assistant Professor, Department of Mechanical Engineering, Vardhaman College of Engineering, Shamshabad, Hyderabad, India.*

[c] *Professor, Department of Mechanical Engineering, College of Engineering, Andhra University, Visakhapatnam, India.*

www.sreedharvulloju@gmail.com

www.psrinivaskishore@gmail.com

Abstract: In this present work, experiments were performed on coated Stainless Steel 304 fin to understand the effect of coating on heat transfer parameters of fin. Aluminium and Copper were chosen as coating materials and these were coated on the Stainless steel 304 substrate by using Thermal Spray Method. Conducted experiments on different coated and none coated fins at various heat inputs using Fin-Pin apparatus in our heat transfer laboratory and necessary readings were noted to calculate Reynolds Number, Nusselt Number, Heat transfer coefficient, Thermal conductivity and fin efficiency. The value of thermal conductivity of fin without was obtained as 52.3 W/m K, coated with Aluminium as 166.75 W/m K and coated with Copper as 280.6 W/m K. The improvement of percentage of fin efficiency in case of fin coated with Aluminium was obtained as 73.6% compared with fin without coating, but in case of fin coated with Copper was obtained as 81.7% compared with fin without coating and 4.6% compared with fin coated with Aluminium.

Keywords: Heat flow rate, stainless steel fin, coating, thermal conductivity, Fin efficiency.

1. Introduction

Fins can be used to enhance the performance of heat exchangers with different types of fin profiles. Internal fin can give better option to increase the cooling of electronic systems rather than other methods (Chiang Ko-Ta et al [1]). Position of extended surface, fin shape and cost may play important role to get uniform flow of heat from the system (B. Moshfegh et al. [2]. Rectangular geometry of fin was best suitable profile to get more heat transfer among various geometrical shapes due to its simple construction, easy manufacturing and low cost. Generally two fin configurations like horizontally based vertical fins and vertically based vertical fins are used in various applications. Though, horizontal orientation was not recommended due to its low quantity of heat transfer from the system (Welling J. R. et al.[3]). Enhance the heat flow with fins were most suitable method because of its easy construction, most economical and convenient. Fins can be used to increase the heat flow by increase the surface area to heat flow to obtain required rate of heat transfer from the system. It is required to maintain minimum the gap between fins and base material surface as to restrict to fall of the heat transfer using fins even if surface area increases with number of fins. Fins may also create narrow the flow passage and afford boundary layer interferences which may result on the performance of heat transfer from the system (F. Harahap et al. [4]). Improving surface area with fin is most suitable due to its less weight, less space occupancy and most economical. Extended surfaces were used in automobiles and various engineering applications to achieve required heat flow (D.Q. Kern et al. [5]). It was noticed that more heat transfer was obtained using nano fluids compared to other fluids. Study the nano fluid properties like as viscosity, thermal conductivity and specific volume during the 2000. (Sadik Kacak et al. [6]). It was noticed in his experimental work, volume fraction (0.003 % by volume)

IOP Conf. Series: Materials Science and Engineering 998 (2020) 012050 doi:10.1088/1757-899X/998/1/012050

of copper oxide (CuO) nano particles were mixed with de-ionized water dispersed to form nano fluid. He studied the performance of convective heat transfer at the steady state by passing nano fluid through a copper tube and concluded that achieve better results compared to conventional fluids (Asirvatham et al. [7]). Studied the performance of shell and tube heat exchangers with Segmental type and Disc-Doughnut Type Baffles and it was concluded that heat exchangers with Disc-Doughnut Type Baffles transfers more heat compared to Segmental type baffles (V.Sreedhar et al. [8]). Brass is coated on Stainless steel using Laser Engineered Net Shaping (LENS) technique to analyze thermal conductivity of Stainless Steel. They observed that at temperature of 100°c, coated Stainless Steel shows better improvement in thermal conductivity which was increased by 65% and low thermal resistance was observed between coating material and substrate by reduce the dilution with defect free sound interface. They observed that best future based thermal coating can develop with metal matrix composites which should had low thermal expansion to improve the heat flow characteristics by using laser processing (Felix A. Espana et.al [9]). They conducted experiments on substrate coated with nano materials coated and suggested the importance of nano material coating on the parent material become futuristic parameter in the field of thermal applications (Prabhu et al. [10]). If fluid velocity increases, then heat transfer co-efficient also increases and vice versa. However pump takes more power as to supply fluid with high velocity. To obtain required heat transfer flow with enhancing the heat transfer coefficient is not better option by increasing the fluid velocity. In other way, rate of heat flow can be increased by reducing the sink temperature and it was practically impossible to decrease sink temperature. Finally, rate of heat flow between the fluids can be improved by increasing the surface area in case of convective heat transfer which was achieved by adding fin (extended surface) to the base material. The performance of fin can be improved by using surface modification with high thermal conductivity materials on the substrate (Prabhu et al. [11]). Caron nano tubes were coated on Aluminium fins with the help of PVD method to analyze the performance of fin and examined the otheat flow characteristics between with and without coated Aluminium fin. It was observed that efficiency of fin was improved by 5% with coated Aluminium fin than without coated Aluminium fins (R. Senthilkumar et al. [12]). Stainless steel fin was made in elliptical shape and coated with multi walled carbon nano tubes .They were conducted experiments on with and without coated fin at different surface temperatures to study the effect of coating on the subastrate. Estimated the various heat flow parameters like temperature distribution, thermal conductivity, heat transfer coefficient, efficiency of fin, shaped tube efficiency at various heat inputs and studied the results with coated and non coated Stainless Steel fin. It was obtained from the results that thermal conductivity of coated Stainless Steel fin was increased by 21.1% compared with non coated Stainless Steel because of decrease in surface temperature. Also the convective heat transfer coefficient of coated Stainless steel fin was improved by 7% compared to without coated Stainless Steel fin. The shaped tube efficiency of coated Stainless Steel was enhanced by 6.2% compared to without coated Stainless Steel fin. The fin effectiveness of coated Stainless Steel fin was increased by 21.8% compared to without coated Stainless Steel fin (J.Nagarani et al. [13]). Brass and Aluminium were chosen as coating materials and stainless steel fin was selected as substrate. Fin was coated with 250 micron of thickness by using Twin wire coating technique and conducted experiments on with and without coated fin at different heat inputs to analyze the effect of coating on the performance of fin. It was noted that Nusselt number of S.S coated with Aluminium was improved by 1.36% compared to coated with Brass and 2.1% compared to without coated fin and there was rise in efficiency of fin coated with Brass and Aluminium materials by 14-73% compared to without coated fin (Sreedhar Vulloju et.al [14]).

Form the above information, it was concluded that coating can effect the rate of heat transfer through the substrate. We chosen stainless steel 304 fin as substrate and it was coated with Aluminium and Copper using thermal spray coating method. We conducted experiments on coated and non coated fin using pin fin apparatus at different surface temperatures to study the influence of coating on the performance of fin.

2. Substrate and Coating Materials

Generally coating is applied on the surface of substrate to protect from corrosion and increasing lifetime but here the coating on substrate can also to boost heat transfer characteristics. Coating method is used for either in permitting or diathermic of heat flow based on its application. Thermal Barrier Coatings are coating material which not allows the heat transfer from or into a system. High thermal conductivity materials are selected as coating materials causes to improve the heat flow performance of the systems. In our work Stainless steel 304 is considered as substrate. Aluminium and Copper are chosen as coating materials of 250 micron thickness. The coating material wires and different coated fins were shown in fig. 1 and fig. 2.

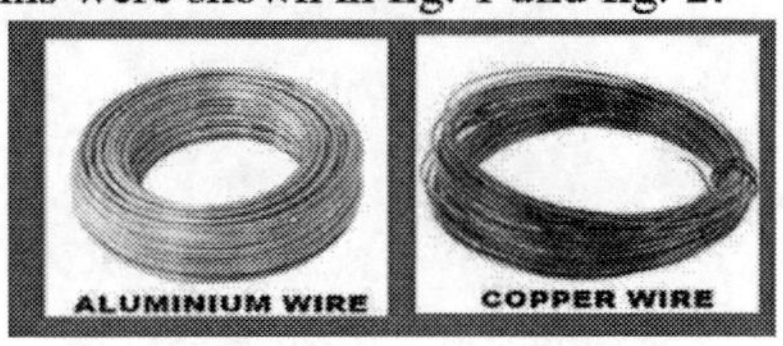

Fig 1. Coating Material Wire

Fig 2. Coated Stainless Steel

3. Microstructure of Coated Fin

The coating was provided on the stainless steel fin with Aluminium and Copper with thickness of 250 microns using twin wire coating method. After the coating process, the specimens were dried individually for 3-6 days depends on thickness of coating on the substrate. It is required to know the bonding of coating material on stainless steel fin as to examine the effect of coating on the performance of fin. If coating is done on the substrate without any defects so that heat loss can be decreases through the coating surfaces. The microstructure of coated fin was observed through the microscope integrated with computer and its image had been taken as shown in the fig 3 & 4.

3.1 Stainless Steel coated with Aluminium

Etchant Used: Polished

Observation: Microstructure shows (fig 3.) finely distributed intermetallic particles on the substrate meanwhile some porosities on coating were observed. No bonding defects were observed. Surface of coating material on substrate was appeared as smoothly and no cracks were noted. This information implied that Aluminium was coated on substrate properly using thermal spray coating method. Air is occupied in the porosities which may leads to decline the heat flow from the fin to surroundings.
(Porosities- It is a void space on a material by formation of holes)

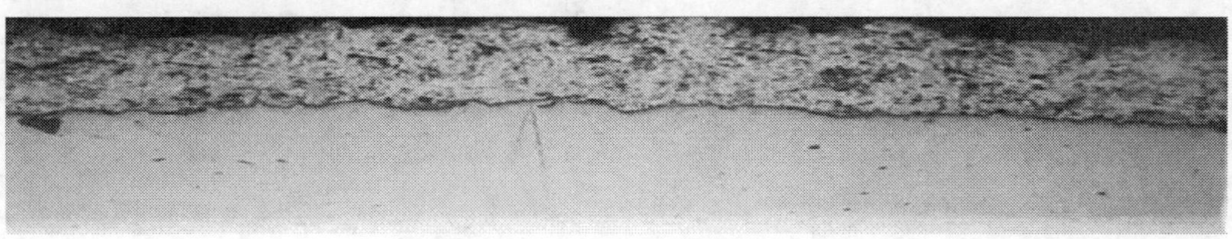

Fig 3.Microstructure of Stainless Steel coated with Aluminium

3.2 Stainless Steel coated with Copper

Etchant Used: Polished

Observation: Microstructure shows (fig 4) coating particles were completely bonding with the base metal. Little porosities observed on the coating were observed. No bonding defects were observed.

Surface of coating material on substrate was appeared as smoothly and no cracks were noted. Thickness of coating material of 250 microns over the entire surface of stainless steel fin was maintained as uniformly and no uneven coating surfaces on the substrate were observed. This information implied that Copper was coated on base material properly using thermal spray coating method. It was noted that formation of porosities in Copper coating is somewhat less compared to Aluminium coating.

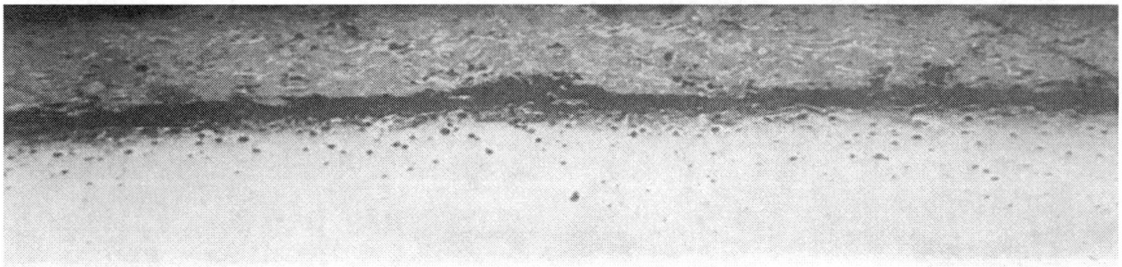

Fig 4.Microstructure of Stainless Steel coated with Copper

4. Thermal Analysis

Pin Fin Apparatus in our heat transfer laboratory was used to conduct experiments on coated fin at different heat inputs as shown in the Fig.5. It consists of duct, test section, blower, measuring flow meter, thermocouples and control panel. Blower is used to supply the air through the duct. Discharge of air through the duct was measured by Orifice meter. Dimmer stat was arranged on control panel which was used to supply the required voltage to the fin surface. Voltmeter, Ammeter and Digital temperature indicator were used to measure voltage ,current and temperature respectively.

Table 1.Specifications of fin and test setup:

Diameter of Fin :14.3mm	Blower capacity: 0.24 HP, 2800 rpm, 180 watts
Length of fin : 180mm	Range of Temperature Indicator 0-100°c
Dimensions ofDuct = 150 mm x100 mm x 700 mm	Range of Voltmeter: 0-230 V
Diameter of Orifice= 14 mm	Range of Ammeter: 0-2 Amp
Coefficient of discharge C_d= 0.64	Dimmer stat: Open type, wire wound

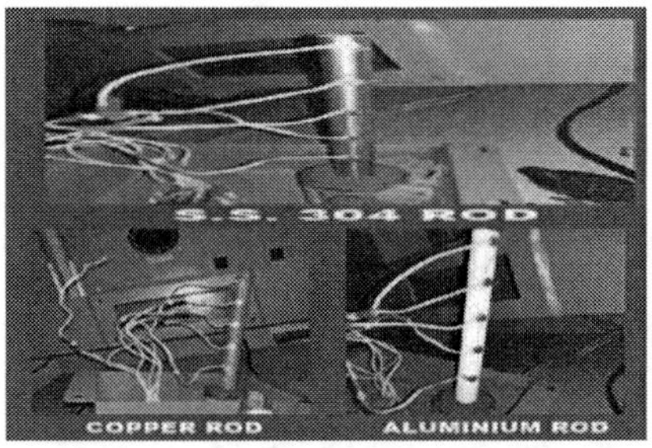

Fig 5. Fin arranged with Thermocouples

Fig 6. Pin Fin Experimental Set Up.

Five holes were drilled at equal distance on Stainless Steel Fin as to attach thermocouples to measure average surface temperature of fin (ta) as shown in fig.

$$\text{Average Surface Temperature ta} = \frac{(t_1+t_2+t_3+t_4+t_5)}{5} \qquad (1)$$

Here, t_1 °c is the surface temperature of fin at hole1 and similarly t_2, t_3, t_4 and t_5 are the surface temperature of fin at holes 2, 3, 4 and 5 respectively as shown in fig 6. Ta is required to calculate mean film temperature of air (t_{mf}).

Mean film temperature of air is obtained is expressed in the equation (2)

$$t_{mf} = t_a + t_d \qquad (2)$$

Where, Duct temperature (td) which is measured with thermocouple arranged inside the duct. The required properties of air are obtained based on the value of T_{mf}.

Velocity of air at orifice (V_0) is obtained is expressed in the equation (3)

$$V_0 = C_d \sqrt{\frac{2gh\rho_w}{\rho_a}} \quad \text{m/s} \qquad (3)$$

Where, g= acceleration due to gravity = 9.81 m/s^2, h= Manometer reading in m

ρ_w= density of water = 1000 kg/ m^3, ρ_a= density of air kg/m^3

C_d = 0.64 orifice meter coefficient

Velocity of air through the duct (V_d) is calculated from the continuity equation and it is expressed by an equation (4)

$$V_d = \frac{\left[V_0 \times \pi/4 \times d_0^2\right]}{[w \times b]} \quad \text{m/sec} \qquad (4)$$

Where, d_0 = diameter of the orifice in m, w = width of duct in m, b = breadth of duct in m

Reynold's Number (Re) is given by equation (5)

$$Re = \frac{VD}{\nu} \qquad (5)$$

Where, D =Diameter of the Fin in m, V = Velocity of air through duct in m/s, ν = kinematic viscosity of the air at t_{mf}.

Nusselt Number [N_u] is expressed by an equation (6) (Sreedhar Vulloju et.al [19])

$$\text{Nusselt Number: } N_u = 0.683(Re)^{0.466}(Pr)^{0.333} \quad 40<Re<4000 \qquad (6)$$

Heat Transfer Coefficient of air (h) is calculated by an expression (7)

$$\text{Heat Transfer Coefficient (h)} = \frac{(Nu \times K_{air})}{D} \qquad (7)$$

Where, K_{air} =Thermal conductivity of air at t_{mf}, Fin parameter [m] as given in equation (8)

$$m = \sqrt{\frac{(hp)}{(KA)}} \qquad (8)$$

Here, P = perimeter of the fin m, A = Cross sectional area of Fin m^2, K = Thermal conductivity of Fin material

Efficiency of Fin (η) is expressed by an equation (9)

$$\eta = \frac{Tan(mL)}{mL} \qquad (9)$$

Where, L= Length of the Fin in m

5. Results and Discussions

Experiments were conducted at steady state conditions on coated and none coated fin using Pin Fin equipment at various voltage ranges between 50 to 75 watts shown in fig.6 and noted necessary

IOP Conf. Series: Materials Science and Engineering 998 (2020) 012050 doi:10.1088/1757-899X/998/1/012050

readings to estimate heat flow parameters like Reynolds number, Nusselt number, Thermal conductivity, Heat transfer co-efficient and fin efficiency.

Table2. Result parameters of Fin Material

Fin Material	Reynolds Number	Nusselt Number	Heat Transfer Coefficient(W/m ^{2}K)	Thermal Conductivity(W/m K)	Fin Parameter(m)	Efficiency (%)
Stainless Steel	164.87	6.55	12.015	52.3	10.96	50.4
S.S Coated With Al	176.19	6.69	12.1	166.75	4.42	87.5
S.S Coated With Cu	182.82	6.86	12.81	280.6	3.51	91.6

Reynolds Number shows an important parameter in the flow of heat transfer rate as its value increases which causes to improve Nusselt Number as well as Heat transfer through the fin The value of Reynolds Number of Stainless Steel 304 without coating was obtained as 164.87, coated fin with Aluminium as 176.19 and coated fin with Copper as 182.82. It was pointed out that coating can effect on the value of Reynolds Number as shown in fig 7 and it increases the Reynolds number in different way with different coating materials. It was noted that Reynolds Number of Copper coated fin was more than other two cases. The value of Reynolds Number of Fin coated with Aluminium was found as 6.8% more than the fin without coating but in case of fin coated with Cu, it was obtained as 10.8% more than the fin without coating.

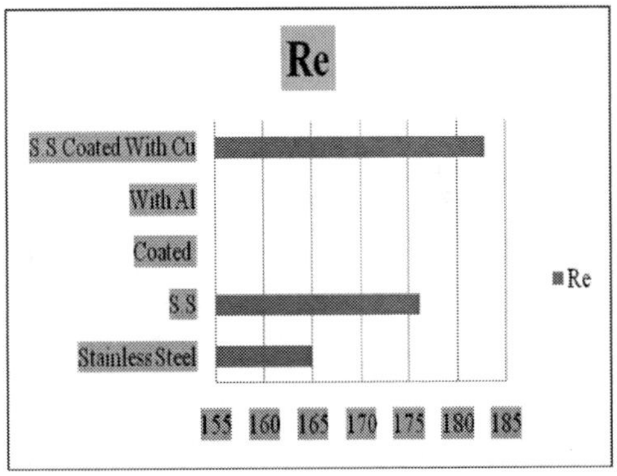

Fig 7. Fin Material V/S Reynolds Number

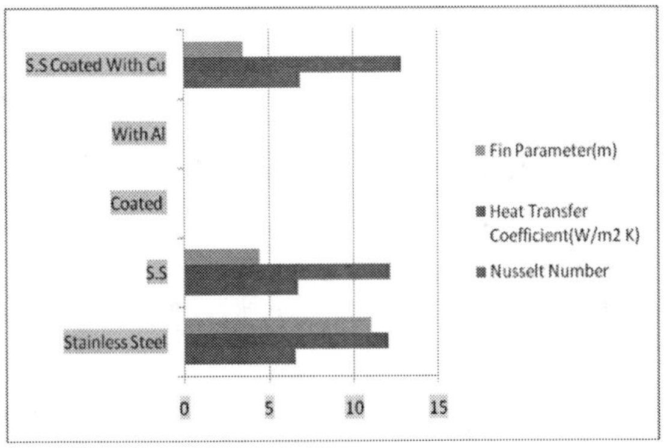

Fig 8. Fin Material V/S Fin Parameter,Heat Transfer Co-Efficient & Nu

Nusselt Number in fin without coating was found as 6.55, coated with Aluminium as 6.69 and coated with Copper as 6.86. From these results, it was observed that coating can effect the value of Nusselt Number and Copper coated fin has the Nusselt Number more compared with Aluminium coated fin and fin without coating as shown in fig 8. It was noted that coating on the substrate can increase Nusselt Number. Heat transfer co-efficient (h) plays a important role in the convection mode of heat transfer through the fin. It was calculated for different types of fins by an expression given in equation (7). The value of heat transfer co-efficient in case of fin without coating was found as 12.015 W/m^{2}K, coated with Aluminium as 12.1 W/m^2 K and coated with Copper as 12.81 W/m^2 K. It was noted that Coating

IOP Conf. Series: Materials Science and Engineering 998 (2020) 012050 doi:10.1088/1757-899X/998/1/012050

can effect the value of heat transfer co-efficient and its value was obtained more in fin coated with Copper compared with other two cases as shown in fig 8.

Thermal conductivity (K) is the property of material which influences the flow of heat transfer through the conduction mode and it was determined for different types of fin with the principle of Fourier equation. The value of thermal conductivity of fin without was obtained as 52.3 W/m K, coated with Aluminium as 166.75 W/mK and coated with Copper as 280.6 W/mK as shown in table 2. We noted that coating can influence the value of thermal conductivity of fin and its value may vary with respective to thermal conductivity of coating material as if coating material has more thermal conductivity causing to increasing the value of thermal conductivity of substrate and vice versa as shown in fig 9. So that we found that fin coated with Copper was obtained as high value of thermal conductivity compared to fin coated with Aluminium. It was examined that the value of thermal conductivity of fin coated with Aluminium was found as 3.18 times more than the fin without coating , but in case of fin coated with Copper was given as 5.36 times more than the fin without coating and 1.68 times more than the fin coated with Aluminum. Fin parameter (m) was calculated using an equation (8) and its value was utilized in the determination of fin efficiency as equation of fin efficiency is the function of fin parameter given by expression(9). We noted that the value of fin parameter of fin without coating as 10.96, coated with Aluminium as 4.42 and coated with Copper as 3.51as shown in the table 2. The less value was noted in fin coated with Copper compared to other cases. So that coating with high thermal conducting can decrease the value of fin parameter and vice versa. From the result table 2, we observed that fin efficiency of fin without coating as 50.4%, fin coated with Aluminium as 87.5% and fin coated with Copper as 91.6% and the coating can effect the fin efficiency as shown in fig 10. It shows that the value of fin parameter of fin coated with Copper

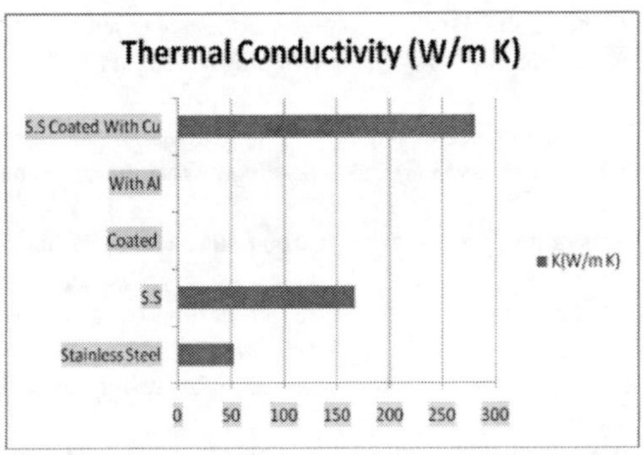

Fig 9.Fin Material V/S Theraml Conductivity

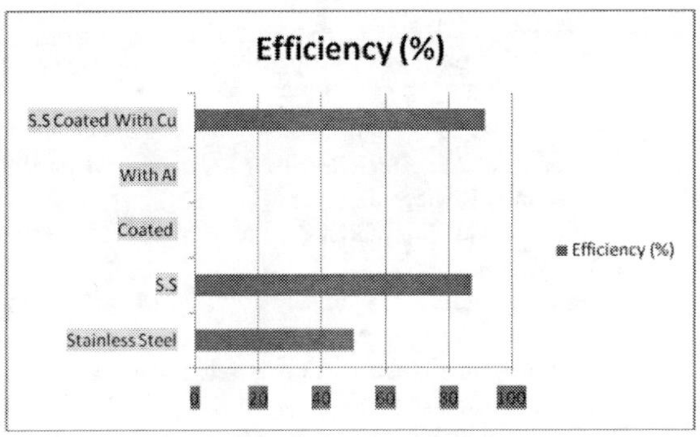

Fig 10.Fin Material V/S Fin Efficiency

has less and it has more efficiency corresponding to Al coated fin and none coated fin. The improvement of percentage of fin efficiency in case of fin coated with Aluminium was obtained as 73.6% compared to fin without coating, but in case of fin coated with fin coated with Copper was obtained as 81.7% compared with fin without coating and 4.6% compared with fin coated with Aluminium.

6. Conclusions

Experiments were conducted on coated and none coated Stainless Steel 304 using Pin Fin Apparatus at various heat inputs as to understand the effect of coating on heat flow characteristics of fin. From experimental results, it was noted that fin coated with Aluminium and Copper has effective performance than the fin without coating. It was concluded that the coating on fin can improve the thermal conductivity of fin material rather than heat transfer coefficient even increase little surface area so that heat dissipation from the fin through the mode of conduction is more compared with convection heat transfer. Surface area of coating material on substrate was not influence more on heat transfer as thickness of coating was very small(250 microns).The coating layer on the surface of fin absorbs heat passing through the fin material and transfer to surrounding . Porosities were formed on surface of fin during the coating process but air fills in the porosities which leads to decrease the heat transfer through the fin. Copper was coated on stainless steel 304 fin more effectively so that less porosities were observed in copper coated fin compared to Aluminium coated fin. So fin coated with Copper shows better performance results compared to fin coated with Aluminium.

References

[1] Chiang Ko-Ta 2005 Optimization of the design parameters of parallel-plain fin heat sink module cooling phenomenon based on the Taguchi method, *Int. Commun. Heat Mass Transfer* **32** 1193e2120.
[2] Moshfegh B and Nyiredy R 2004 Comparing RANS Models for Flow and Thermal Analysis of Pin Fin Heat Sinks *15th Australasian Fluid Mechanics Conference*,The University of Sydney, Sydney, Australia 13e17.
[3] Welling J R and C N Wooldridge 1965 Free convection heat transfer coefficients from vertical fins, *J. Heat Transfer* **87** 439-444.
[4] Harahap F and McManus H N 1967 Natural convection heat transfer from horizontal rectangular fin arrays, *J. Heat Transfer* **89** 32-38.
[5] Kern D Q and Kraus A D 1972 Ext. Surface Heat Trans, McGraw-Hill, New York.
[6] Kakac S and Pramuanjaroenkij A 2009 Review of convective heat transfer enhancement with nanofluids, *Int. J. Heat Mass Transfer* **52** 3187e3196.
[7] Asirvatham LG, et al 2009 Experimental study on forced convective heat transfer with low volume fraction of CuO/Water nanofluid. Energie 2:97–119
[8] Sreedhar V et al2017 Experimental Investigation on Shell and Tube Heat Exchanger Using Segmental and Disc-Doughnut Type Baffles *International Journal of Mechanical Engineering and Technology* Scopus Indexed Journal,Volume **8** Issue PP-975-984.
[9] Felix A et al 2009 High Thermal Conductivity Coatings via LENS for Thermal Management Applications *W. M. Keck Biomedical Materials Research Laboratory School of Mechanical and Materials Engineering* Washington State University Reviewed.
[10] Prabhu S and Vinayagam BK 2010 Nano surface generation of grinding process using carbon nano tubes. Sadhana **35**:747–760
[11] Prabhu S and Vinayagam BK 2011 AFM surface investigation of Inconel 825 with multi wall carbon nano tube in electrical discharge machining process using taguchi analysis. *Archiv Civil Mech Eng* 9:149–170
[12] Senthil Kumar R et al 2013 Analysis of natural convective heat transfer of nano coated aluminium fins using Taguchi method *Heat and Mass Transfer* **49**:55–64 DOI 10.1007/s00231-012-1063-1
[13] Nagarani J et al 016 Heat transfer enhancement of the AISI stainless steel 304 elliptical annular fin coated with the carbon nanotube *International Journal of Ambient Energy* Vol. **37**, No. 1, 55–63.
[14] Sreedhar Vulloju et al 2019 Enhancement of Heat Transfer Characteristics of Plain Fin Coated with Brass and Aluminium *International Journal of Recent Technology and Engineering (IJRTE)* ISSN: 2277-3878, Volume-**8** Issue-4.

PERFORMANCE OF FLY ASH AND GGBS BASED GEOPOLYMER CONCRETE USING SINGLE ALKALINE ACTIVATOR SOLUTION AND ITS COST ANALYSIS

Dr. G. Mallikarjuna Rao[a], Y. Anil Kumar[b], Dr.Jagadish Vengala[c] Dr. M. Venu[d]

[a] Associate Professor, Deapartment of Civil Engineering, Vardhaman College of Engineering, Shamshabad, Hyderabad.
[b] PG Student, Deapartment of Studies in Civil Engineering, Vardhaman College of Engineering, Shamshabad, Hyderabad.
[c] Associate Professor, Deapartment of Civil Engineering, PVP Institute of Technology, Kanuru, Vijayawada.
Assistant professor, Department of Civil Engineering, Vardhaman College of Engineering, Shamshabad, Hyderabad.

Abstract

Concrete is the most durable, versatile and reliable building material in the world. Next to water, concrete is the most used material. Ordinary Portland cement production accounts for 5-7% of total greenhouse gas emissions. It also consumes a lot of energy. Today, Geo- polymer concrete (GPC) has become a more sustainable concrete compared to conventional concrete. This research work focuses on the performance of Fly Ash and GGBS based geo- polymer concrete cured in outdoor conditions using a single alkaline activator solution (Geo activator). The variations of Fly Ash and GGBS were (FA100-GGBS0; FA90-GGBS10; FA70-GGBS30; FA50-GGBS50). In this study, only the Geo activator is used to activate Fly Ash and GGBS in the mixtures. Geo polymer concrete of size 100 mm x 100 mm x 100 mm was used in this research work to measure compressive strength at different ages of 7 days and 28 days. The results reveal that the increase in the proportion of GGBS content in the mix increases compression strength of concrete. The cost analysis was also carried for geopolymer concrete, compared with conventional concrete and the results reveals that for the lower grades, the cost will be the same as for conventional concrete and for the higher grades, the use of geo polymer concrete can reduce the cost of construction up to some extent.

Keywords: Geopolymer concrete, Fly Ash, GGBS, Compressive Strength, Geo-activator and Cost Analysis:

1. INTRODUCTION

Concrete is the most widely used building material in the world after water, and Portland cement is the main ingredient used in concrete. Cement production releases large amounts of carbon dioxide (CO_2) into the atmosphere that contribute significantly to greenhouse gas emissions. It is estimated that a ton of carbon dioxide is released into the atmosphere when a ton of OPC is produced. World cement production per year is 2.6 billion tons. Cement production requires a large amount of limestone that is now rapidly depleting. On the other hand, fly ash is a by-product produced by the combustion of coal. And a large amount of fly ash is deposited in landfills that affect groundwater. Therefore, we can use Fly Ash as a pozzolanic material to replace cement and we can reduce waste and we can also improve concrete performance. Therefore, products such as fly ash and GGBS need to develop alternatives to existing cements that use the cement properties of the industry. GGBS is a byproduct of steel making process plants. Availability of Fly Ash and GGBS provide you with more opportunities to use more products worldwide. Davidovits developed a binder called Geo-polymer in 1978 to describe alternative cementitious materials. Geopolymer concrete is alternative to conventional concrete where cement is completely replaced by Fly Ash, GGBS, Metakolin.etc. Which are rich in silica and alumina and an ecological concrete that does not emit greenhouse gases during the polymerization process. And the reaction between materials a binding property can be achieved through the use of alkaline solutions. The geo-polymers are made of raw materials with silicon (Si) and aluminum (Al) content, so the cement can be completely replaced with fly ash and GGBS(GGBS).

2. LITERATURE REVIEW

Davidovits [1999] proposed that binders could be produced by a polymer reaction of alkaline liquids with silicon and aluminum in materials of geological origin or by-products such as fly ash and rice husk ashes. He called these binders as geo polymers. **Wang SD et al. [1995]** reported that NaOH and Na_2SiO_3 are commonly used in combination to form an alkaline solution. This solution activates silica and alumina in fly ash to form alumina silicate hydrate and forms calcium silicate hydrate (C-S-H) by reacting with calcium in GGBS. **Pinto, A [2004]** reported that the molar ratio of Na_2SiO_3 / NaOH is a factor that affects the properties of geopolymers. After several investigations, he suggested that the 2.5 ratio provides maximum compressive strength with a constant binder content.
Mallikarjuna Rao et al. [2015] studied source material of geopolymers (fly ash and GGBS) and alkaline activator consisting of sodium metasilicate and sodium hydroxide of different molarities (8,12,16M). the ratio of Meta silicate sodium to sodium hydroxide considered in this study is 2.5. From the results it has been indicated that the combination of fly ash and GGBS results in a reduced final setting time and greater compressive strength. **B.Rajini [2014]** To study the effect of fly ash and GGBS on the mechanical properties of GPC with different proportions from 0 to 100%. Sodium silicate and sodium hydroxide are used as alkaline activators. Study of compressive strength, tensile strength divided for 7, 28, 56 and 90 days and at room temperature. **J. Thaarrini [2016]** studied the cost of production of $1m^3$ of OPC and GPC. Calculated based on the market prices. Cost of OPC and GPC is marginally same for lower grades. And the cost of OPC is higher than the GPC for higher grades. From this paper the cost of GPC is 1.7% higher than OPC for small grades M30. And the cost of GPC is 11% lower than the OPC for higher grade like M50.

3. EXPERIMENTAL STUDY

The present research paper aim is to determine the compressive strength of Fly Ash and GGBS based geo- polymer concrete using single alkaline activator solution and cured at ambient temperature and its cost analysis.

3.1 Materials

Fly ash: fly ash is a byproduct of coal obtained from the La Mangada Thermal Power Plant (NTPC). In this experiment we use class F flyash as the source material.
GGBS: The crushed blast furnace slag is a byproduct of the steel processing plant that replaces the cement obtained from Toshali Cement Pvt. Ltd Bayyavaram India.

Table 1: Chemical composition of fly ash and GGBS (% by mass)

Chemical Composition	Fly ash	GGBS
SiO2	60.11	34.06
Al2O3	26.53	20
Fe2O3	4.25	0.8
SO3	0.35	0.9
CaO	4.00	32.6
MgO	1.25	7.89
Na2O	0.22	NIL
LOI	0.88	NIL

The fine aggregates: used according to 383: 1970 according to Zone II are obtained from the sand of

IOP Conf. Series: Materials Science and Engineering 998 (2020) 012051 doi:10.1088/1757-899X/998/1/012051

nearby rivers.
Coarse aggregate: The substance that remains above the aggregate size of 4.75 mm in the BIS sample is called coarse aggregate. We use locally available coarse aggregate with a maximum size of 16 mm. In this experiment, aggregates of various sizes were used.

16 mm -12.5 mm - 30%

12.5 mm - 10 mm -40%

10 mm - 4.75 mm -30%

Geo activator: In this experiment, only a single alkaline liquid was used, namely geo activator, and no sodium hydroxide and sodium silicates were used. Geo activator Solutions of silica modulus SiO_2:Na_2O (M_s) =2.92:1 with 28.98% SiO_2, 9.92% Na_2O by weight, which was produced from KIRAN GLOBAL Ltd, Chennai India.
Super plasticizer: to improve the workability of fresh concrete water, the super plasticizer based on naphthalene was reduced and added to the mixture.

3.2 Mix Design
In this experiment, we designed four different mix proportions for conventional concrete: Mix A (20MPa), Mix B (30MPa), Mix C (40MPa) and Mix D (50MPa) and the same mix proportions were used for geo- polymers concrete. For Mix A the ratio was set to 1: 1.5: 3: 0.50. And of Mix B 1: 1.51: 2.52:0.43, Mix C 1: 2.56: 3.26: 0.40, Mix D 1: 1.09: 3.52: 0.36. In geopolymer concrete, the mix design is the same as conventional concrete, where only a single alkaline solution was used. And the cement is completely replaced by fly ash and GGBS in four different mixtures at different proportions. And water is completely replaced by geo-activator. The alkali / binder ratio is equal to the water / cement ratio. The amount of water we use in conventional concrete is the same amount of geoactivator solution used in geopolymer concrete.

Table 2: Mix proportions of conventional concrete

Mix no	Binder (kg/m^3)	Alkaline/binder	Fine aggregate (kg/m^3)	Coarse aggregate(kg/m^3)	Water (kg/m^3)	SP(kg/m^3)
MIX A	396	0.50	608	1216	192	Nil
MIX B	438	0.43	660	1104	190	Nil
MIX C	350	0.40	896	1140	140	2.64
MIX D	427	0.36	466	1504	130	3.42

Table 3: Mix proportions of Geo-polymer concrete

Mix no	Binder (kg/m^3)	FA:GGBS	Alkaline/binder	Fine aggregate (kg/m^3)	Coarse aggregate (kg/m^3)	Alkaline liquid (kg/m^3)	SP(kg/m^3)
Mix A	396	100:0	0.50	608	1216	198	Nil
Mix B	438	90:10	0.43	660	1104	190	Nil
Mix C	350	70:30	0.40	896	1140	140	2.64
Mix d	427	50:50	0.36	466	1504	130	3.42

IOP Conf. Series: Materials Science and Engineering 998 (2020) 012051 doi:10.1088/1757-899X/998/1/012051

4. RESULTS AND DISCUSSIONS

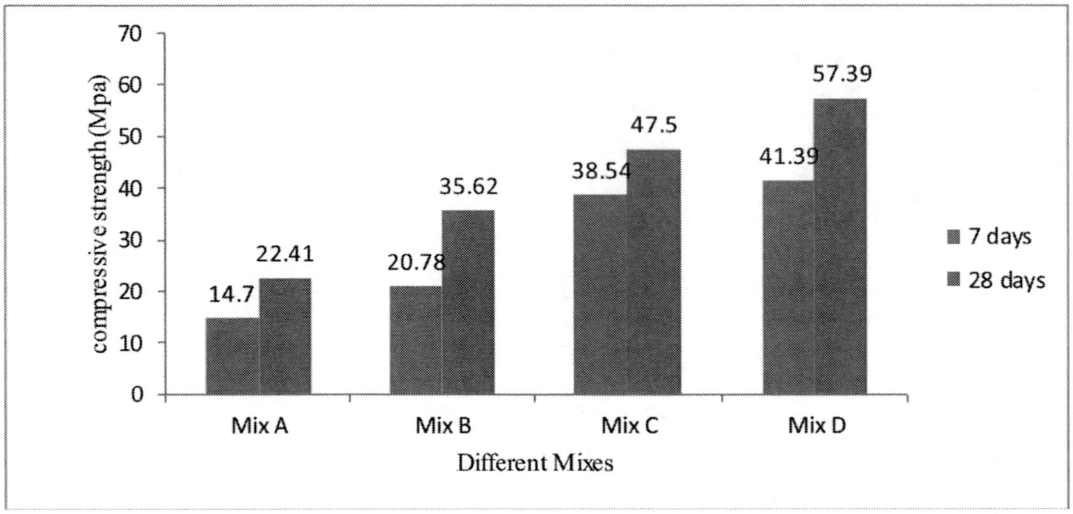

Figure 1: compressive strength of conventional concrete between 7 and 28days.

Figure 1 shows the compressive strength of conventional concrete at 7 and 28 days. Mix A is of 20MPa, Mix B is 30MPa, Mix C is 40MPa and Mix C is 50Mpa. Strength after 28 days obtained as 22.41, 35.62, 47.5, and 57.39MPa. With increase in the cement content for different grades of concrete the strength was also improved. The increase in the strength is due to formation of the calcium silicate hydrate gel. The gain of strength of concrete improved with increase in the age of the concrete for different grades. The gain of the strength can be observed from figure 1

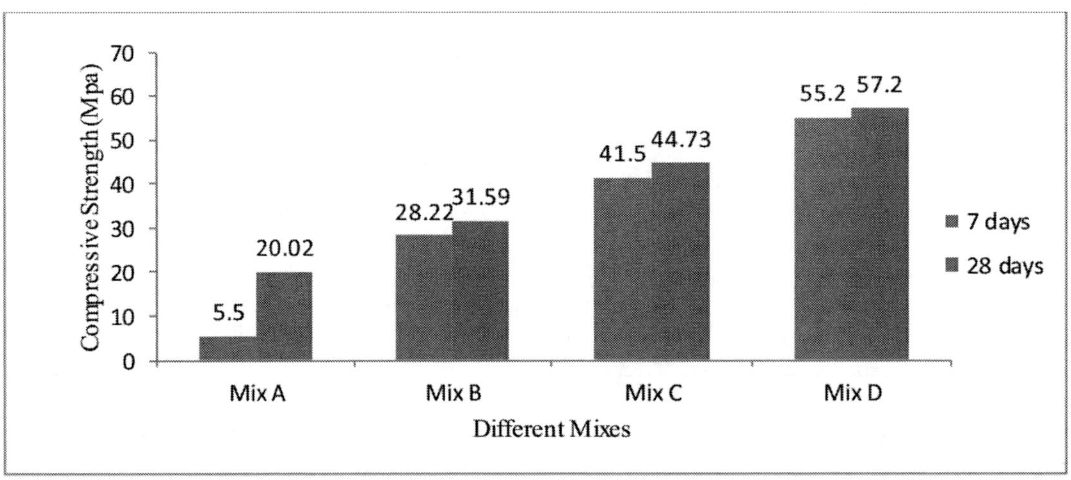

Figure 2: compressive strength of GPC between 7 and 28 days

From figure 2, we can observe the compressive strength of geopolymer concrete cubes tested for 7 and 28 days. The compressive strength of GPC mix A shows the lesser strength compared with other mixes. The strength of geopolymer mainly depends up on the fly and GGBS content in the mixes. For mix A only fly ash content was taken as the source material it shows very less strength due to slow polymerization process. Whereas for other mixes the strength gain for 7 days is high and up to 28 days it was continued. The reason for more strength gain is due to faster polymerization process and dissolution of the GGBS particles were high and the form of the C-A-S-H gel is more for Mix B, Mix C and Mix D.

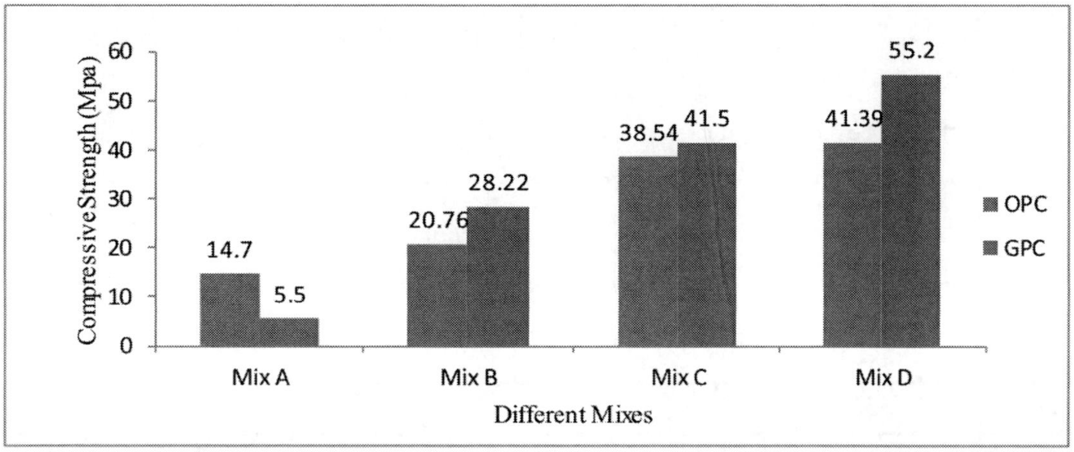

figure 3: comparison of 7 days compressive strength between Conventional Concrete and GPC

From figure 3, we compared the strength between 7 days for conventional concrete and Geopolymer concrete. The strength of GPC was less compared with conventional concrete. In conventional concrete the strength gain is due to the formation of C-S-H gel whereas for the geopolymer concrete the strength of geopolymer specimens was due to polymerization process and it was the slow process and the strength gain was very less compared with conventional concrete specimens. The polymerization process was also depends up on the alkaline activator and particle size distribution of the source materials.

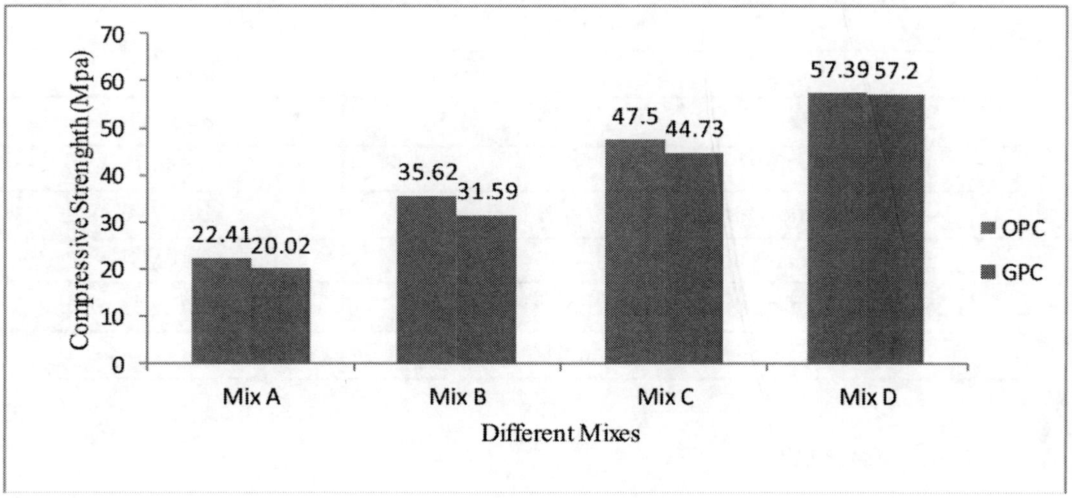

Figure 4: Comparison of 28 days compressive strength between conventional Concrete and GPC

From Figure 4 we can see the compressive strength for 28 days between the conventional concrete and the geopolymer concrete. The strength gain for 28 days is less geopolymer concrete specimens compared with conventional concrete specimens. For the same mix proportions, the strength of the conventional concrete specimens and geopolymer concrete specimens were compared at 28 days. The same conventional strength can be achieved without cement content. In our study we tried to replace the total cement content with fly ash and GGBS and water with the geo-activator solution. The strength was almost similar at 28 days curing period and hence for in-situ conditions also this concrete will be most suitable. In previous studies the geopolymer concrete specimens were prepared with sodium silicate and sodium hydroxide solution and cured under oven curing at 60^0 for 24 hours, these mixes were not suitable at insitu conditions and they were quite costlier because of the source materials and activator solutions. Now in the present study the mixes were prepared with single activator solution and the cost comparison was also done, these mixes were helpful to cast geoploymer concrete under insitu condition with required strength.

Table 4: shows the binder content, alkaline/binder ratio and compressive strength of OPC and GPC.

Mix NO	Binder kg/m^3	FA: GGBS	Alkaline/binder	Conventional concrete (MPa)	Geo-polymer concrete (MPa)
Mix A	396	100:0	0.50	22.41	20.02
Mix B	438	90:10	0.43	35.62	31.59
Mix C	350	70:30	0.40	47.50	44.73
Mix D	427	50:50	0.36	57.39	57.20

4.1 Cost Analysis

The components of GPC and OPC and their cost are shown in Table 5. Buy bulk materials from suppliers, you can get them at low cost.

Table 5: Ingredient Materials and their cost

S.NO	MATERIALS	Rate in Rs. /per
1	Cement	330/50kg
2	Fly ash	800/MT
3	GGBS	1.5/kg
4	Fine aggregate	1000/MT
5	Coarse aggregate	700/MT
6	Sodium silicate (Na_2Sio_3)	16/kg
7	Super plasticizers (conplast sp430)	250/kg

4.2 Cost analysis for OPC and GPC

IOP Conf. Series: Materials Science and Engineering 998 (2020) 012051 doi:10.1088/1757-899X/998/1/012051

Table 6: shows the cost of producing 1 m³ of OPC in grades 20MPa and 30MPa is calculated based on the amount of material.

S. No	Materials	Rates in Rs	unit	Mix A (20MPa)		Mix B (30MPa)	
				Quantity	Amount	Quantity	Amount
1	Cement	330	50 kg	396	2614	438	2891
2	Fine Aggregate	1000	MT	608	608	660	660
3	Coarse Aggregate	700	MT	1216	851	1104	773
	Total				4073/-		4324/-

Table 7: shows the cost of producing 1 m³ of OPC in grades 40MPa and 50MPa is calculated based on the amount of material.

S.NO	Materials	Rates in Rs	Units	Mix C(40MPa)		Mix D(50MPa)	
				Quantity	Amount	Quantity	Amount
1	Cement	330	50kg	350	2310	427	3245
2	Fine aggregate	1000	MT	896	896	466	466
3	Coarse aggregate	700	MT	1140	1628	1504	1052
4	Super plasticizer	250	kg	2.64	660	3.42	855
	Total				5494/-		5618/-

Table 8: The cost of production of 1m³ of GPC of grade 20MPa and30MPa.

S.NO	Materials	Rates in Rs	Units	MIX A(20MPa)		MIX B(30MPa)	
				Quantity	Amount	Quantity	Amount
1	Fly ash	800	MT	396	316.8	394	315
2	GGBS	1.5	1 kg	0	0	44	66
3	Fine aggregate	1000	MT	608	608	660	660
4	Coarse aggregate	700	MT	1216	851	1104	773
5	Geo activator	16	1 kg	198	3168	190	3040
	Total				4943.8/-		4854/-

Table 9: The cost of production of 1m³ of GPC of grade 40MPa and50MPa.

S.NO	Materials	Rates in Rs	Units	MIX C(40MPa)		MIX D(50MPa)	
				Quantity	Amount	Quantity	Amount
1	Fly ash	800	MT	245	196	213.5	170.8
2	GGBS	1.5	1 kg	105	158	213.5	320.25
3	Fine aggregate	1000	MT	896	896	466	466
4	Coarse aggregate	700	MT	1140	798	1504	1052.8
5	Geo Activator	16	1 kg	140	2240	130	2048
6	Super plasticizer	250	1 kg	2.8	700	3.4	850
	Total				4988/-		4758.4/-

Table 10: Savings in cost of production for GPC and OPC concrete.

Grade of concrete	Cost of production of 1m³ of OPC	Cost of production of 1m³ of GPC	Savings in cost (Rs)	Savings in %
MIX A (20MPa)	4073/-	4943.8/-	-870.8/-	-17.61%
MIX B (30MPa)	4324/-	4854/-	-530/-	-10.92%
MIX C (40MPa)	5494/-	4988/-	506/-	9.21%
MIX D (50MPa)	5618/-	4758.4/-	859.6/-	15.30%

The above calculations show that the cost of producing OPC is higher than the cost of producing GPC concrete for higher grade. For 20MPa grade concrete of GPC, the production cost is slightly higher (17.61%) than the small grade OPC. For the 30MPa grade of GPC, the production cost is higher than OPC of the same grade (10.92%). For 40MPa grade OPC, production costs are higher than GPC of the same grade (9.21%). For the 50MPa grade of concrete, the OPC cost is 15.30% higher than the GPC of the same grade. Therefore, it was concluded that cost savings could be maximized in the production of high- grade than the low-grade Geo-Polymer concrete.

5. CONCLUSION

- The combination of fly ash and GGBS can be used in the development of GPC without the need for heat curing to satisfy the compressive strength.
- The compressive strength of geopolymer concrete with 100% fly ash provides 25% strength, or 25% strength, for 7 days and provides more strength as 100% increases in 28 days.
- The compressive strength of geopolymer concrete increases with the increase in the GGBS replacement

IOP Conf. Series: Materials Science and Engineering 998 (2020) 012051 doi:10.1088/1757-899X/998/1/012051

rate of fly ash. The increase in intensity is 7 days more than 28 days.

- The combination of fly ash and GGBS is the solution to develop geopolymer concrete under outdoor hardening conditions.
- The dry density for the combination of fly ash and GGBS varies from 2400-2500 Kg / m^3, similar to ordinary concrete.
- The maximum density observed in the geopolymer sample represents a maximum compressive strength of 7 days of age for different alkali to binder ratios and different combinations of fly ash and GGBS.
- Depending on the change in binder content, the compressive strength was increased in different combinations.
- In the cost analysis, we observe that the production of OPC is higher than the cost of producing high-grade GPC. And the cost of GPC is slightly higher than the cost of OPC for a small grade like M20.

6. REFERENCES

1. Davidovits, "Synthetic mineral polymer compound of silico aluminate family and preparation process," US patent 4472199, 1978
2. Wang, SD. Pu, XC. Scrivener KL & Pratt, "Alkali-activated slag cement and concrete: a review of properties and problems," Advanced Cement Research, 27:93–102, 1995
3. Roy, DM, "Alkali-activated cements: opportunities and challenges," Cement and Concrete Research, 29:249–54,1999.
4. Komnitsas, K. Zaharaki, "Geopolymerisation: A review and prospect for the mineral industry," Miner. Eng. 20, 1261–1277, 2007.
5. Villa, C. Pecina, E.T. Torres, R. Gomez, "Geopolymer synthesis using alkaline activation of natural zeolite," Constr. Build. Mater.24, 2084–2090, 2010.
6. Hardjito, D. Wallah, SE. Sumajouw, DMJ and Rangan, "On the Development of Fly ash - Based Geopolymer Concrete," ACI Materials Journal, 101(6):467-472, 2004.
7. Pinto, "Alkali-activated metakaolin based binder," PhD Thesis. University of Minho.2004.
8. DjwantoroHardjito, Chua Chung Cheak& Carrie Ho Lee Ing, "Strength and Setting Times of Low Calcium Fly Ash-based Geopolymer Mortar." Modern Applied Science, Vol. 2, pp. 3-11, 2008.
9. Deb PS, Nath P, Sarker PK, "The effects of ground granulated blast-furnace slag blending with fly ash and activator content on the workability and strength properties of geopolymer concrete cured at ambient temperature," Materials & Design; 62(0):32-9,2014.
10. Nath, P and Sarker, "Effect of GGBS on setting, workability and early strength properties of fly ash geopolymer concrete cured in ambient condition," Construction Building Materials, 66:163-171, 2014.
11. Raijiwala DB and Patil HS, "Geopolymer Concrete: A Concrete of Next Decade," JERS, 2(1):19-25, 2011.
12. Alonso S and A Palomo, "Alkaline activation of metakaolin and calcium hydroxide mixtures : influence of temperature, activator concentration and solids ratio," Materials Letter, 47(1-2):55-62, 2001.
13. Xu, H. van Deventer, J.S.J, "The geopolymerisation of alumino-silicate minerals," Int. J. Miner. Process, 59, 247–266, 2000.
14. Barbosa, V.F.F. Mackenzie, K.J.D and Thaumaturgo, "Synthesis and Characterization of Materials Based on Inorganic polymers of Alumina and Silica: Sodium Polysiliate," Polymers. International Journal of Inorganic Materials V.2, pp.309-317,2000.
15. A. Palomo, M.W. Grutzek, M.T. Blanco, "Alkali-activated fly ashes. A cement for the future", Cement and Concrete Research, 29 (8), pp. 1323- 1329, 1999.
16. UbollukRattanasak, PrinyaChindaprasirt, "Influence of NaOH solution on the synthesis of fly ash geopolymer", minerals engineering, 22(12):1073-1078 · October 2009.
17. G. Mallikarjuna Rao, Gunneswara Rao, "Final Setting Time and Compressive Strength of Fly Ash and GGBS-Based Geopolymer Paste and Mortar", Arabian Journal for Science and Engineering, 11; 3067-3074,2015.

Investigating the Effects of Welding Parameters on Mild Steel by SMAW Technique

ROHIT PAGARE[1], DHANESH AWATI[2], SHRINATH MANE[3], VIJAY TELI[4] and AMAR BHANDARE[5]

[1,2,3,4,5]Mechanical Engineering Department, Walchand College of Engineering, Sangli, India
Email: pagarerohit27@gmail.com

Abstract. The Shielded Metal Arc Welding (SMAW) is a joining technique in which melting and joining of metals with the help of an arc between the workpiece and the filler rod. The core purpose of the paper is to examine the mechanical characteristics of welded metal under various standalone parameters like type of filler material, electrode diameter, and current for SMAW. This paper employed the half factorial Design of Experiment (DOE) using Taguchi L4 array. Varieties of welding filler materials used were E6013 and E7016, with an electrode diameter of 3.15mm and 4mm and the welding current of 90A and 110A. The desired output variables were tensile strength and Rockwell hardness of the welded joint. This paper found that when electrode diameter increases, the deposition rate increases, and hence HAZ increases, which decreases the weld strength. Using the welding filler E7016, we can obtain the maximum tensile strength for welded metals.
Keywords. SMAW, Design of Experiments, Tensile Strength, Hardness, HAZ

1. Introduction

Welding is a continual process for joining two or more pieces of metal using regional amalgamation derived from a suitable blending of pressure, temperature, as well as metallurgical condition. An electric arc ensures proper fusion and joining of the materials by providing essential heat. The welding process is extensively applied in construction because of its easy applicability, minimum arrangement of equipment and also is inexpensive compared to other processes. However, this paper concentrated primarily on the Shielded Metal Arc Welding Process and the core purpose of the paper is to examine the mechanical characteristics of welded metal under various standalone parameters like type of filler material, current and electrode diameter for SMAW. This paper implemented the half factorial design for the Design of Experiment (DOE). The schematic of the SMAW is shown in figure 1.

The process of melting and joining metals with the help of heat generated by an arc between workpiece and coated electrode, is called as Shielded Metal Arc Welding (SMAW) also known as Stick Welding. The arc generation is assisted by the flux i.e. the outer electrode coating and it also prevents adulteration by enabling slag layer and shielding gas. The weld joint is result of depositing filler material from the core of electrode along the desired area at constant speed.

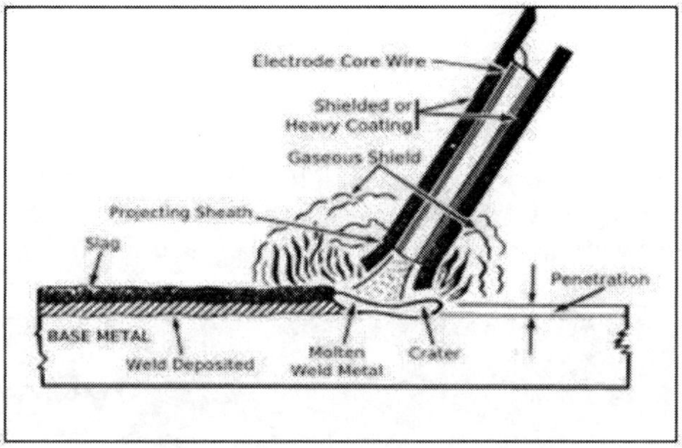

Figure 1. Schematic of the SMAW process.

Kchaou et al. [1] found that the welded joint of material UNS N08028 is ductile in the impact test and tensile test with changing the parameters such as the filler material diameter, number of passes, current, and voltage. Pamnani et al. [2] discussed that the inclusions content of the welded joint of material DMR 249A varies with the welding process, voltage, current speed (mm/s), heat input (kJ/min), number of passes, heat input for final passes. Singh et al. [3] analyzed that vibration applied on the weld pool successfully enhances welded joint mechanical properties of 6mm thick MS plate with varying parameters such as welding speed, frequency of vibration, current, hardness, S/N ratio for hardness. Ganesh et al. [4] analyzed the residual stresses as well as thermal cycles in DMR-249A steel welded joints. Sadeghi et al. [5] analyzed that the probability of crack formation reduces after studying the microstructure in the study of investigating corrosion behavior and abrasive wear of surface repair of grey cast iron of SMAW technique with preheating of electrodes. Tahir et al. [6] found that for optimum tensile strength and optimum hardness, E7016 and E7018 welding filler are responsible respectively. Welding current was also kept as a parameter in the inspection of the welded material mechanical properties.

2. Materials and Methods

The samples were welded initially using the arc welding technique, and parameters varied during welding were combinations of filler material, welding current & electrode diameter. The experiments were designed by using the Taguchi Method. The L4 array of design of the experiment is shown in Table 1 below. After the welding of 4 samples as L4 array is used, Specimens for the tensile test were made according to the ASTM E8M-04 standard. Tensile testing on specimens was carried out, and parameters responsible for the highest tensile strength and lowest tensile strength were noted down. For these parameters, we again welded two specimens and, on these specimens, Rockwell hardness testing was carried out.

Table 1 L4 array of design of experiment.

Experiment No.	Parameters		
	Current (A)	Filler Material	Electrode diameter (mm)
1	90	E6013	3.15
2	90	E7016	4
3	110	E6013	4
4	110	E7016	3.15

IOP Conf. Series: Materials Science and Engineering 998 (2020) 012052 doi:10.1088/1757-899X/998/1/012052

3. Experimental Setup and Procedure

3.1 Sample Preparation

MS Plates were marked and cut with dimensions (100*50*5 mm). Edge preparation of plates was done by using files. Bevel angle of 45^0 and root face of 2 mm is maintained. The portion near to the weld zone was polished by filing operation to avoid the inclusion of rust in the weld zone. The prepared sample is shown in figure 2. Ten such samples were prepared for welding.

Figure 2. Sample Preparation.

3.2 Experimental Setup

Experiments were designed as per the Taguchi L4 Array shown in Table 1. Root opening of about 1 mm is maintained and welding of specimens was carried out as per the conditions required. Two spot welds at ends were made so that plates will not move during welding. After spot welding was done, actual welding was carried out. Size of the weld is 5 mm and the length of the weld is 50 mm. Four specimens were prepared for tensile testing. After each welding, Specimens were labeled for easy identification for later analysis. The welded specimen is shown in figure 3.

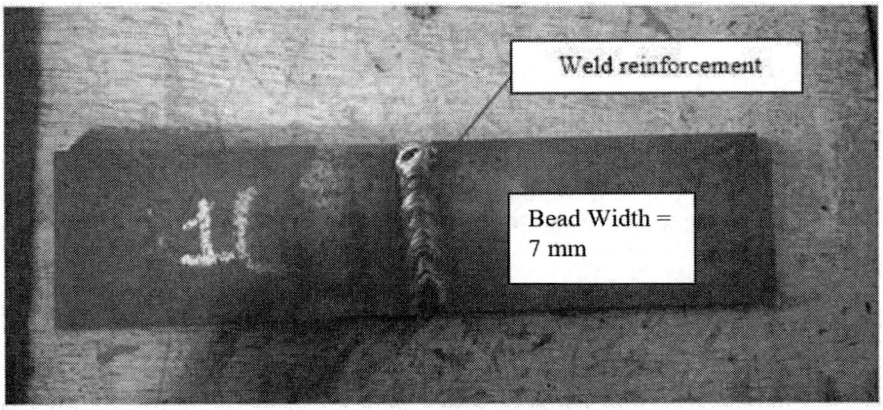

Figure 3. Welded Specimen.

3.3 Tensile specimen preparation

After welding was carried out, samples were allowed to air cool. Making of tensile test specimen according to the ASTM E8M-04 standard for the tensile test was made. As per the dimension, the material was cut by hand cutter and milling machine. Excellent finishing and grinding of weld reinforcement were done by Grinding Machine. Tensile specimens are shown in figure 4 below.

3.4 Tensile Testing

To determine the Ultimate Tensile Strength (UTS), we used the universal testing machine (UTM) machine. Specimen first fitted in the bottom holder, and then it was fitted in the crosshead. In the software, tensile specimen type – PLATE was selected. Width – 20mm, Gauge length – 100mm, and thickness – 5mm was selected. Testing was started after pressing the hydraulic valve. The pressure Relief valve was pressed after breaking of the specimen. The plot of stress vs. strain was plotted by software. Broken specimens after tensile testing were shown in figure 4 (b) below.

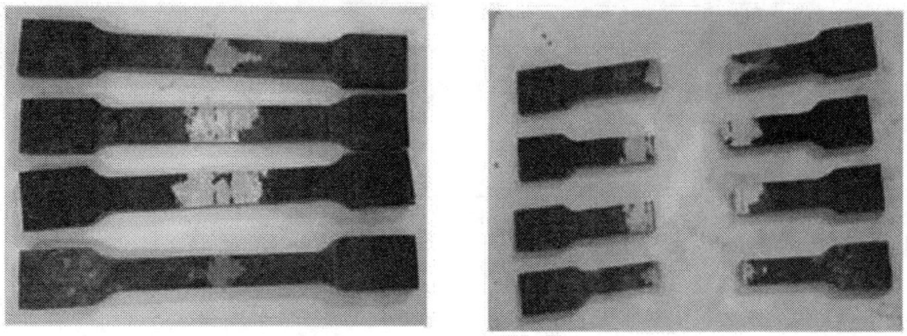

(a) (b)

Figure 4. (a) Tensile samples before testing, and (b) Tensile samples after testing.

3.5 Hardness Testing

Parameters responsible for the highest strength and lowest strength were selected for hardness testing, i.e., parameters for experiment no. 3 and 4 were selected. According to selected parameters, the welding of samples was done. As per ASTM E-18 guidelines, hardness testing on the sample is carried out. For Rockwell's hardness test, specimen dimensions of (50*25*5 mm) were marked on the sample, and hardness specimen were prepared by fine finishing and grinding of reinforcement by grinding machine and the top part of the weld as seen in figure 3 was fine finished. Superfinishing of the specimen is done on a rotary wheel. Scale "B" with load 100 kgf or 980.7 N and 1.588 mm diameter ball indenter was used. With major and minor load hardness testing is carried out on hardness tester. Six readings of hardness by indentation were taken both in fusion zone and in base metal. Average of these readings is calculated for final hardness. Hardness testing specimen after testing is shown in figure 5.

Figure 5. Hardness specimen testing.

4. Results and Discussion

4.1 Tensile Testing

Tensile testing of all four specimens according to the L4 array was done. Table 2 illustrates the results of all the experiments.

Table 2 Tensile testing result of specimens.

Experiment No.	Parameters			UTS (MPa)
	Current (A)	Filler Material	Electrode diameter (mm)	
1	90	E6013	3.15	293.4
2	90	E7016	4	373.6
3	110	E6013	4	229.4
4	110	E7016	3.15	400.28

Stress vs. strain graphs for experiment no. 3 and 4, i.e., minimum and maximum strength, is illustrated in figure 6 and figure 7 below.

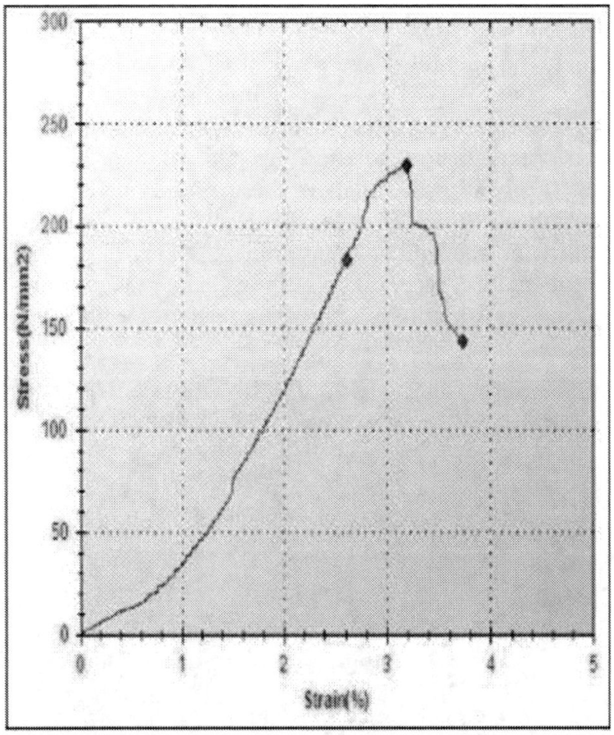

Figure 6. Stress vs. strain graph of specimen 3.

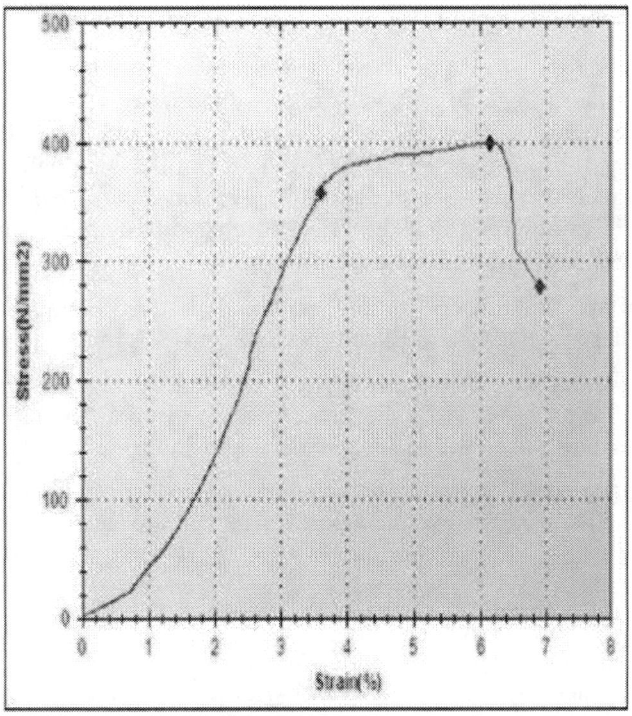

Figure 7. Stress vs. strain graph of specimen 4.

4.2 Hardness Testing

Rockwell hardness testing for specimens having the highest tensile strength and lowest tensile strength in tensile testing was carried out. Data on hardness testing is illustrated in Table 3 below. Hardness testing was carried out in the fusion zone and near base metal. An average of six readings of hardness testing was taken.

Table 3 Hardness testing result of specimen.

Experiment No.	Parameters			Rockwell Hardness (HRB)	
	Current (A)	Filler Material	Electrode diameter (mm)	Fusion zone	Base metal
L	90	E6013	3.15	88	71
H	90	E7016	4	92	69

The difference in the fusion zone hardness and the base metal hardness is because of the temperature gradient. In the fusion zone, sudden heating and cooling takes place due to which time of heat dissipation is less and as temperature gradient is more, hardness value in the fusion zone is greater than the hardness value at the base metal. Also, it is much evident that due to low amount of hydrogen, tendency of hydrogen embrittlement decreases which inherently increase the strength of the weld joint and hardness in the fusion zone.

IOP Conf. Series: Materials Science and Engineering 998 (2020) 012052 doi:10.1088/1757-899X/998/1/012052

5. Conclusions

The research findings demonstrated that the welding filler, welding current, and electrode diameter substantially affect the tensile strength as well as hardness of the welded metal. This experiment found that as electrode diameter increases, the deposition rate increases, and hence HAZ increases, which decreases the weld strength. In this experiment, with E7016 as filler material, it was observed that a low amount of current might cause insufficient deposition of filler material, which causes a decrease in the weld strength. It was also observed that the effect of electrode diameter on tensile strength was more prominent than the current.

1) In tensile testing, optimum parameters for highest tensile strength using SMAW technique are: Current – 110A, Filler material – E7016 and Electrode diameter – 3.15mm

2) Hardness testing reveals that for an increase in the joint's tensile strength, the weld hardness also rises. The hardness at the fusion zone was higher than the base metal as the low-temperature gradient at base metal, and rapid heating and cooling take place at the fusion zone.

3) The basic Low Hydrogen coated formulation ensures a defect-free quality weld. The superb and well-established flux formulation ensure excellent performance of the electrode in AC.

References

[1] Kchaou Y, Haddar N, Henaff G, Pelosin V and Elleuch K 2014 Microstructural, compositional and mechanical investigation of Shielded Metal Arc Welding (SMAW) welded super austenitic UNS N08028 (Alloy 28) stainless steel, *Material and Design* **63**, pp 278-285.

[2] Pamnami R, Jayakumar T, Vasudevanb M and Sakthivel T 2016 Investigations on the Impact Toughness of HSLA Steel Arc Welded Joints, Journal of Manufacturing Process,21, 75-86, 2016.

[3] Singh P., Patel D., Prasad S., Optimization of Process Parameters During Vibratory Welding Technique Using Taguchi's Analysis, *Perspectives in Science* **8**, pp 399-402.

[4] Pamnami R, Vasudevan M, Jayakumar T, Vasantharaia P and Ganesh K 2016 Numerical simulation and experimental validation of arc welding of DMR-249A steel, *Defence Technology* **12**, pp 305-315.

[5] Sadeghi A, Moloodi A, Golestanipour M and Shahri M 2017 An investigation of abrasive wear and corrosion behavior of surface repair of gray cast iron by SMAW, *Journal of Materials Research & Technology* **6**, pp 90-95.

[6] Tahir A, Lair N and We F 2018 Investigation on mechanical properties of welded material under different types of welding filler (shielded metal arc welding), *AIP conference proceedings* **1958**, pp 1-7.

DESIGN FAILURE MODE AND EFFECT ANALYSIS – CASE STUDY

[1]Adila R. Nadaf, [2]Rahul M. Chanmanwar

[1] PG Student, Department of Mechanical Engineering, Walchand College of Engineering, Sangli, Maharashtra, India

[2] Faculty member, Department of Mechanical Engineering, Walchand College of Engineering, Sangli, Maharashtra, India

E-mail: [1]Adila.nadaf@gmail.com, [2]Rahul.chanmanwar@walchandsangli.ac.in

Abstract. Design-Failure mode and effect analysis (D-FMEA) is very important tool used in industries to detect failure during design stage. D-FMEA is a technique to find out causes of failure and their effects on the performance of a system(parts) during design phase. In this paper, the D-FMEA of critical components of designed automatic tilting mechanism trolley is carried out. The critical components of automatic tilting mechanism trolley are identified and analyzed by the D-FMEA tool. This paper provides risk priority number of components of trolley. The component with higher risk priority number has higher chances of failure. The risk priority number of critical components is reduced by incorporating corrective actions. This paper provides probable causes of failure, effects of failure and corrective actions to be undertaken during design phase of Automatic tilting mechanism trolley. The D-FMEA tools helps in reducing accidents.

Keywords. Design failure mode and effect analysis, Risk priority number, Severity, Occurrence, Detection.

1. INTRODUCTION

There are various operational and safety issues concerning with material handling trolleys due to which the operator faces various health disorders like musculoskeletal disorder (MSD) that is why study of the material handling trolleys is essential. In this paper burr material handling trolley is considered. In industries normally, the burr handling trolleys are transferred manually from one place to other. Manual pushing and pulling of trolleys causes fatigue to operator. Normally 800-1000 kg of burr is transferred over long distance by manual pushing. In manufacturing industries burr material handling is transferred manually over long distance. In this paper a better design of material handling trolley is prepared to reduce load carried per unit off human efforts. The transferring of trolley is not done by pushing or pulling but it is transferred through forklift and hence due to which entire manual movement of trolley is eliminated.

Hence a trolley with automatic tilting mechanism is designed named as automatic tilting mechanism trolley as shown in Fig. 1. Trolley with tilting mechanism is designed to reduce human efforts and increase safety in material handling. In this paper, D-FMEA of critical component of automatic tilting mechanism trolley is carried out.

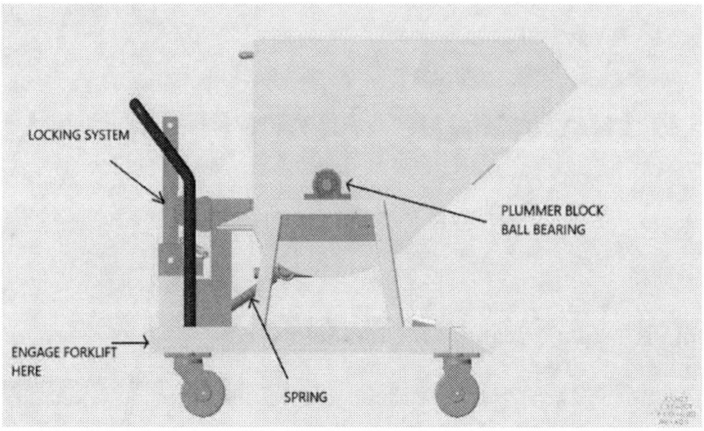

Fig. 1 Trolley image

2. AUTOMATIC TILTING MECHANISM TROLLEY DESIGN

In this new design of trolley automatic tilting mechanism is incorporated. The centroid of trolley is calculated by using Creo software for full load and no-load conditions. Figure 2 shows center of gravity of empty and full bin. The approx. value are given below.
Center of gravity of empty bin (CS0)
Xcs0=407mm Ycs0=438mm
Center of gravity of full bin (CS1)
Xcs1=437mm Ycs1=439mm

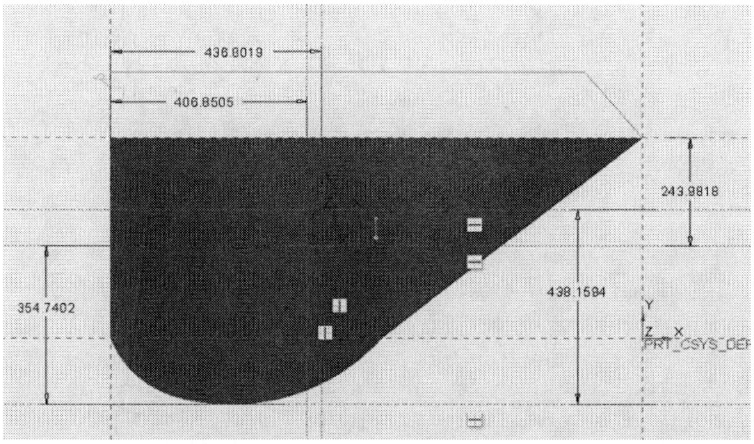

Fig. 2 Center of gravity of trolley

Centroid of trolley is shifted by 180mm (as shown in Fig. 3) so that the trolley will get automatically tilted. At this shifted centroid point Plummer block ball bearing are provided. The trolley rotates about this point. Helical extension springs are used to avoid the sudden jerk which the trolley will undergo while tilting. It acts as a shock absorber. By sitting in the forklift, the lever of locking system can be pulled by using rope. When the latch is disengaged the trolley will unload automatically. Fig. 4 shows the tilted trolley image. The manual movement of trolley is eliminated, and trolleys are taken to unloading area by means of forklift.

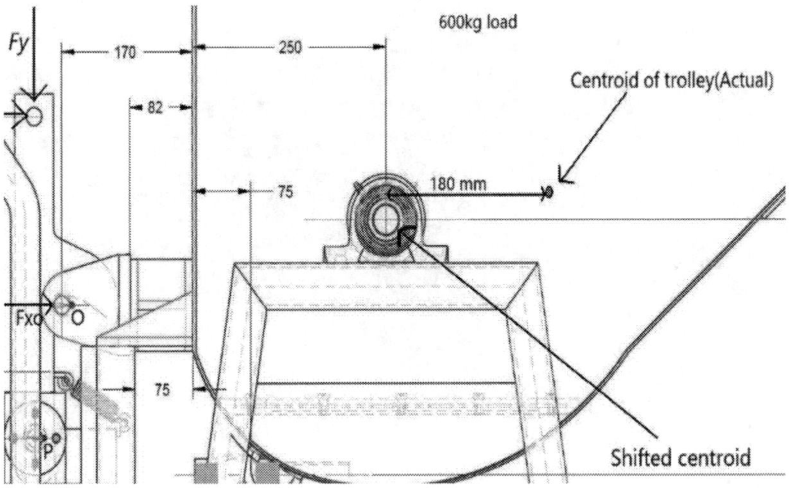

Fig. 3 Shifted centroid

2.1. Design of shaft: Material Selection - Mild steel is used as shaft material due to its properties like high tensile strength, Shear strength, Good modulus of elasticity and rigidity, High toughness and less cost compared to other steel materials.
Total load inside the trolley - 800kg
One bearing load carrying capacity - 400kg
Weight - 400 X 9.81 N
The shaft diameter is calculated from above data by bending stress equation and is 35 mm.
2.2. Design of bearing Since the radial load acts on the shaft, ball bearings are used for this trolley with automatic tilting mechanism.
For 35 mm shaft diameter the bearing selected from bearing catalogue is, 6007.
2.3. Design of Spring The helical extension spring is used as a shock absorber. The material of spring used is standard.
Spring material: En747 (Quench and temper 32-36 Rc)
For the given application, estimated maximum spring force (P) and the corresponding required deflection (δ) of the spring.
P - 40kg
δ - 180mm - Deflection of spring
k - Spring stiffness
Shear stress - 525 N/mm^2
Wire diameter - d - 3.5 mm
Coil diameter - D - k X d = 3.5 x 7 = 24.5 mm
Number of coils - N - 62
Spring length - L - 217mm

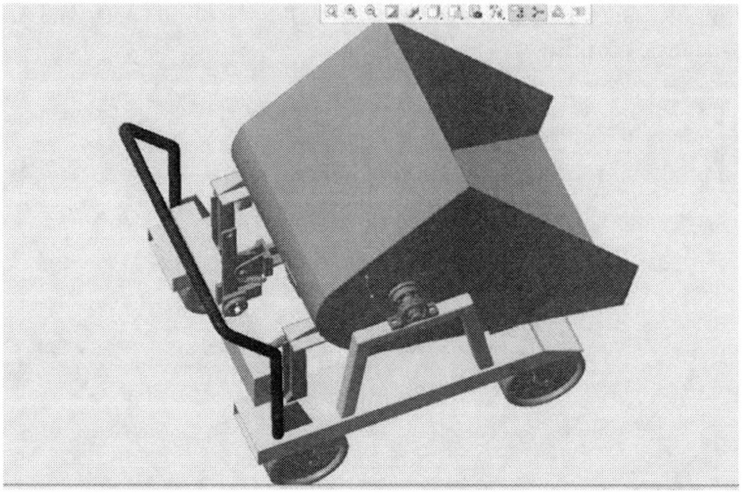

Fig. 4 Tilted trolley image

The Fig. 4 shows the tilted trolley image or tilted position of trolley.

3. D-FMEA

D-FMEA is a technique to find out causes of failure and their effects on the performance of a system(parts). It helps in analysing for possible failure modes and to prevent them from occurring by incorporating corrective actions. D-FMEA is a analysing technique of new design prior to implementation. This will help to reduce the chances of failure after implementation. D-FMEA is one of the most efficient analysis techniques.

"Severity (S) is the fact or condition being severe to the individual component or system as a whole. Rating from 1 to 10 is given, with 1 means no effect to the system and 10 for very hazardous effect on the system. The severity rating criteria along with its description is given in table no. 1" (Pattnaik,2015).

"Occurrence (O) means how frequently a failure or incident will happen. The rating of 1 to 10 is given where, 1 means "very unlikely to happen" and 10 means "failure is inevitable". The occurrence rating criteria along with its description is given in table no. 2" (Pattnaik,2015).

"Detection (D) means chances of failure detection. Similar to severity and occurrence, a numeral rating of 1 to 10 is given where 1 means "the failure will be detected very easily" and 10 means "the failure is not possible to detect with current controls". The detection rating criteria along with its description is given in table no. 3" (Pattnaik,2015).

Risk Priority Number (RPN) is multiplication of the Severity, Occurrence and Detection. RPN is given below [4,5],

$$RPN = (\text{Severity Rating}) \times (\text{Occurrence rating}) \times (\text{Detection rating})$$
$$RPN = S \times O \times D$$

Table 1. Severity Rating Criteria

SR. NO.	EFFECTS	RANKING	DESCRIPTION
1	Hazardous without warning	10	Failure occurs without warning. Machines cannot be repaired, and they are inoperable. Affects safe operation without warning.
2	Hazardous with warning	9	Failure with warning. Affects safe operation with warning.
3	Very High	8	Loss of primary function and system is inoperable. It involves major repair and rework. It makes product inoperable.
4	High	7	Makes product operable at reduced performance (customer dissatisfaction).

5	Moderate	6	No loss of primary function. Only a part of a system is failed and that part can be made operable by rework.
6	Low	5	Components is working but at reduced comfort. The working of system is affected. Efficiency of operation is reduced.
7	Very Low	4	By minor changes failures can be avoided. System or components are not affected majorly.
8	Minor	3	There is no performance loss but cause little disturbance. By minor rework it can be reworked.
9	Very Minor	2	Components is operable with minor loss or no loss. Rework or repair is not required.
10	None	1	No effect.

Table 2. Occurrence Rating criteria

SR. NO.	EFFECTS	RANKING	DESCRIPTION
1	Very High: Inevitable failures	10	Failure in every third component (1:3)
2	High: Failures occur almost as often as not.	9	Failure in every sixth component (1:6)
3	High: Repeated failures	8	Failure in every ninth component (1:9)
4	High: Failures occurs often	7	Failure in every 50 components (1:50)
5	Moderately High: Frequent failures	6	Failure in every 150 components (1:150)
6	Moderate: Occasional failures	5	Failure in every 800 components (1:800)
7	Moderately Low: non-frequent failures	4	Failure in every 4500 components (1:4500)
8	Low: Relatively low failures	3	Failure in every 30000 components (1:30000)
9	Very Low: Failures are few and far between	2	Failure in every 150000 components (1:150000)
10	Remote: Failure is unlikely	1	Failure in every 1.5 million components (1:1.5million)

Table 3. Detection Rating criteria

SR. NO.	EFFECTS	RANKING	DESCRIPTION
1	Uncertainty	10	Unable to detect a cause of failure by controls.
2	Very Remote	9	Very low chances.
3	Remote	8	Poor chance of detection of failures by controls.
4	Very Low	7	Very low chances of detection.
5	Low	6	Possibility of detection of failure mode by controls.
6	Moderate	5	Moderate chances of detection.
7	Moderately High	4	Failures have good chances of detection.
8	High	3	High chances of detection of failure i.e. automatic detection.
9	Very High	2	Very high chances of detection.
10	Almost certain	1	Controls certain to detect. Error proofed by product design.

4. D-FMEA COMPONENTS:

The D-FMEA was applied on different components of the Automatic tilting mechanism trolley.
Components of D-FMEA:

- **Failure of bearing shaft:** The total load in the trolley bin acts on the shaft hence it is the most important component of D-FMEA. The load acting on trolley is carried by bearing shaft. Hence failure of bearing shaft may result in loss of primary function i.e. tilting.
- **Failure of wheels:** The wheels are used for manual movement of trolley which is important to locate trolley in any area of shop floor.

- **Failure of bearing resting frame:** The load is transferred from shaft to bearing and from bearing to resting frame. Hence failure of resting frame may result in loss of primary function(tilting).
- **Failure of pin:** The overall trolley rotating torque is taken by locking pin. Hence failure of pin may result in loss of primary function(tilting). Also sudden failure of pin can lead to accident.
- **Failure of helical extension spring:** It is used as a shock absorber to avoid sudden jerk on trolley bin.
- **Failure of forklift Square section:** The trolley is taken to unloading area by means of forklift and forklift is engaged in forklift square section. The total lifting load is acting on the section. Hence failure of forklift square section can damage entire trolley.

5. D-FMEA IMPLEMENTATION:

5.1. Failure of bearing shaft If shaft failure is caused by first two cases, then the severity rating is 10 since the trolley is inoperable. Also, sudden failure causes damage to whole trolley and hence the repair is almost impossible. It may result in safety incidents. By carrying out analysis of shaft for given load, the severity rating is reduced to 8 from 10[10]. The detection rating is 7 in first two cases, since controls have poor chance of detection. But it is reduced to 6 by time to time inspection.

In third case, welding failure the severity rating is 10. By providing flange at the end of shaft (as shown in Fig. 5) the severity rating is reduced to 4. The detection rating is 6 since controls have good chance of detection. By providing flange the welding failure is reduced. Hence, here detection rating is reduced from 6 to 4. The occurrence rating given here is 3 in all cases. By assuming failure in every 30000 components. Occurrence rating is reduced to 3 from 2.

Table.4: D-FMEA of bearing shaft

Sr. No.	Components	Potential failure mode	Effects of failure	S	Cause of failure	O	Current process control	D	RPN	Recommended action(s)	S	O	D	RPN
1	Bearing shaft	• Bending of shaft, Fatigue failure	• May damage bearing. • Entire mechanism failure may occur.	10	• Stress exceeds yield stress of material due to excess radial load and impact load. • Fatigue failure of shaft due to cyclic loading.	3	NA	7	210	• Design shaft with proper specification • Analysis of shaft is carried out to find maximum permissible stress it can carry. • Time to time preventive maintenance and inspection.	8	2	6	96
		• Undersize shaft diameter	• Failure of shaft. • Wrong bearing selection. • Bearing failure. • Entire mechanism failure may occur.	10	• Shaft not designed with proper specification. • Wrong inputs regarding load carried by shaft.	3	NA	7	210	• Design shaft with proper specification • Analysis of shaft is carried out to find maximum permissible stress. • Time to time preventive maintenance and inspection.	8	2	6	96
		• Welding failure between shaft and bin	• Mechanism failure. • Failure of bearing and trolley parts.	10	• Less welding surface area of shaft. • Also, difficulty in welding of shaft with thin sheet of bin.	3	NA	6	180	• Flange is provided at the end of shaft were weld is to be made to increase surface area of welding.	4	2	4	32

The analysis of shaft is carried out, allowable and von-mises stresses are compared [10].

Fig. 5 Von mises stresses in shaft

Material used: Mild steel
Yield strength of material: 250MPa Assume FOS: 2
Allowable stress: 125MPa
Load on shaft: 400 kg
Von-mises stress: 93.52MPa (from analysis as shown in Fig.4)
Maximum principal stress: 100.079MPa (from analysis)

5.2 Failure of wheels If wheel failure is caused by wheel undersize, then the severity rating is 6 since the trolley is operable and manual movement of trolley cannot be performed. By replacing wheel, the trolley can be used same as before.

The detection rating is 7, since controls have poor chance of detection. Due to use of preventive maintenance of wheels the detection rating is reduced to 4.

The occurrence rating given here is 3 in all cases. By assuming failure in every 30000 components. Occurrence rating is reduced to 3 from 2 by performing time to time preventive maintenance.

Table.5: D-FMEA of wheels

Sr. No.	Components	Potential failure mode	Effects of failure	S	Cause of failure	O	Current process control	D	RPN	Recommended action(s)	S	O	D	RPN
2	Wheels	• Wheel undersize	• Failure of wheels • Failure of castor • More force required for pushing/pulling.	6	• Wheels not selected properly from the catalogue. • Less load carrying capacity of wheels.	3	NA	7	126	• Analysis of wheel for given radial load is carried out. • Also, wheels having high material yield strength is selected. • Preventive maintenance of wheels to be carried out.	6	2	4	48

5.3. Failure of bearing resting frame If failure of bearing resting frame is caused by buckling, then the severity rating is 10 since the trolley is inoperable. Also, sudden failure causes damage to bearing and other trolley parts and hence the repair is almost impossible. By carrying out analysis of bearing resting frame for given load, the severity rating is reduced to 8 from 10[10].The detection rating is 7 in first case, since controls have poor chance of detection.

In second case too, severity rating is 10 since the trolley is inoperable. Also, sudden failure causes damage to bearing and other trolley parts and hence the repair is almost impossible. And detection rating in this case is 6 since by visual inspection controls have moderate chance of detection. Since bending has chances of detection by visual inspection.

The occurrence rating given here is 3 in all cases. By assuming failure in every 30000 components. Occurrence rating is reduced to 3 from 2. This can be reduced by providing rib to the resting frame.

If locking pin failure is caused by shear or bending, then severity rating given is 8. Here primary function failure, trolley will remain in tilted position due to which automatic tilting mechanism will not be performed. By designing pin properly by considering proper design standard the severity rating can be reduced and alternate lock is provided to bin for support. Due to which the severity rating is reduced to 5, alternate lock will perform the function of pin in case of failure of pin. The detection rating is 7, since controls have poor chance of detection.

Table.6: D-FMEA of Bearing resting frame and Pin

Sr. No.	Components	Potential failure mode	Effects of failure	S	Cause of failure	O	Current process control	D	RPN	Recommended action(s)	S	O	D	RPN
3	Bearing resting frame	• Buckling	• Automatic tilting mechanism failure. • May damage bearing/shaft.	10	• Aspect ratio not defined properly. • Less thickness of column.	3	NA	7	210	• Rib provided to avoid buckling. • Analysis is carried out. • Thickness is high enough to sustain the load.	6	2	7	84
		• Bending and structural failure.	• Automatic tilting mechanism failure. • Bearing and shaft failure.	10	• Repeated high cyclic loading. • Stress exceeds yield stress of material due to excess radial load and impact load.	3	NA	6	180	• Rib provided to avoid buckling. • Analysis is carried out.	6	2	6	72
4	Pin	• Pin failure Bending or shear failure	• Trolley will remain in tilted position i.e. mechanism will not work.	8	• Pin not designed according to specification. • More load acting on pin. • High torque since all rotating load acts on pin. • Stress exceeds yield stress of material due to excess radial load and impact load.	3	NA	7	168	• Design pin properly. • Alternate lock is provided. • Analysis of pin is carried out.	5	2	4	40

5.4 Failure of helical extension spring In all the cases of failure of helical extension spring, the severity rating is 6 since the trolley is operable and primary function can be performed. Spring corrosion can be eliminated by providing anti corrosive coating. Hence life of spring can be increased. The severity rating is reduced to 1 from 6 due to coating.

The detection rating is 7 in first two cases, since controls have poor chance of detection. In first case detection rating is reduced to 4, this is be done by carrying out analysis and time to time inspection of spring. In second case too detection rating is reduced to 4, this is done by providing mark on trolley above which no material should be filled in trolley. The detection rating is 4 in third case, since corrosion can be easily detected by eyes.

The occurrence rating given here is 3 in all cases. By assuming failure in every 30000 components. Occurrence rating is reduced to 3 from 2.

Table.7: D-FMEA of Helical extension spring

Sr. No.	Components	Potential failure mode	Effects of failure	S	Cause of failure	O	Current process control	D	RPN	Recommended action(s)	S	O	D	RPN
5	Helical extension spring	• Fatigue failure of spring	• Sudden jerk on trolley. • Unbalance of loads.	6	• Due to faulty choice of springs • Spring fails due to loads exceeding the yield stress of the spring material. • During tilting spring must extend and compress. Hence it is a cyclic phenomenon which may result in fatigue failure.	3	NA	7	126	• Choose springs according to trolley loads and other specification. • Visual inspection.	6	2	4	48
		• Spring relaxation	• Spring mechanism failure.	6	• When load carried more than specified limit of spring.	3	NA	7	126	• Mark given on trolley above which no material should be filled.	6	2	4	48
		• Spring corrosion	• Less life. • Sudden failure of spring.	6	• Material selection not properly done.	3	NA	4	72	• Material selection should be done properly.	1	2	4	8

5.5. Failure of forklift Square section In all the cases of failure of forklift square section by bending, the severity rating is 10 since the trolley is inoperable and primary function cannot be performed i.e. tilting of trolley. It is reduced to 8 by carrying out analysis and designing the part properly. If failure is by improper welding the severity rating is 6 it is reduced to 4 by efficient welding and time to time inspection. If failure is by corrosion the severity rating is 6 it is reduced to 3 by time to time painting of trolley.

The detection rating is 6 in first two cases, since controls have good chance of detection. In third case detection rating is reduced to 3, since this can be easily detected by eyes. The detection rating is 4 in third case, since corrosion can be easily detected by eyes.

The occurrence rating given here is 3 in all cases. By assuming failure in every 30000 components. Occurrence rating is reduced to 3 from 2.

Table.8: D-FMEA of Forklift square section

Sr. No.	Components	Potential failure mode	Effects of failure	S	Cause of failure	O	Current process control	D	RPN	Recommended action(s)	S	O	D	RPN
6	Forklift square section	• Bending	• Primary function degradation and movement of trolley by forklift will not be possible. • Sheet bending due to more load.	10	• More lifting load than specified. • Less strength of material.	3	NA	6	180	• Analysis is carried out to find maximum permissible stress.	8	2	6	96
		• Broken welds	• Bending and weakening of frame. • All mounting parts gets weakened.	6	• Fabrication Defects	3	NA	6	108	• Efficient welding. • Visual inspection of welds.	4	2	4	32
		• Corrosion	• Less life of part. • Sudden failure of part due to weakening.	6	• Paint removal. • Used in high moisture environment.	3	NA	4	72	• Proper time to time painting of trolley.	3	2	4	24

6. STRESSES COMPARISON:

The Table. 9 shows the allowable and von-mises stress comparison.

Table.9: Stress comparison

Components	Material	Load acting(kg)	Yield strength(MPa)	Allowable stress(MPa)	Vonmises stress(MPa)	Max Principal stress(MPa)
Bearing shaft	Mild steel	400	250	125	93.52	100.079
Wheels	Ultra-high molecular weight polyethylene (UHMW-PE)	200	39.98	19.98	3.79	3.86
Bearing resting	Mild steel	400	250	125	36.95	40.12

frame						
Pin	Mild steel	600	250	125	21.2	24.85
Helical extension spring	En747 steel	400	280	140	35.11	38.22
Forklift square section	Mild steel	800	250	125	25.64	41.67

7. PRIORITY GRAPH:

Priority Graph is a representation of risk priority numbers of the system/components on the graph. From the above evaluation, the graph is shown below with RPN on Y axis and parts of trolley on X axis. The Fig.6 shows components of trolley with higher and lower RPNs. The recommended corrective/preventive measures are undertaken for critical components of trolley to reduce chances of failure. Fig.6 shows the priority graph.

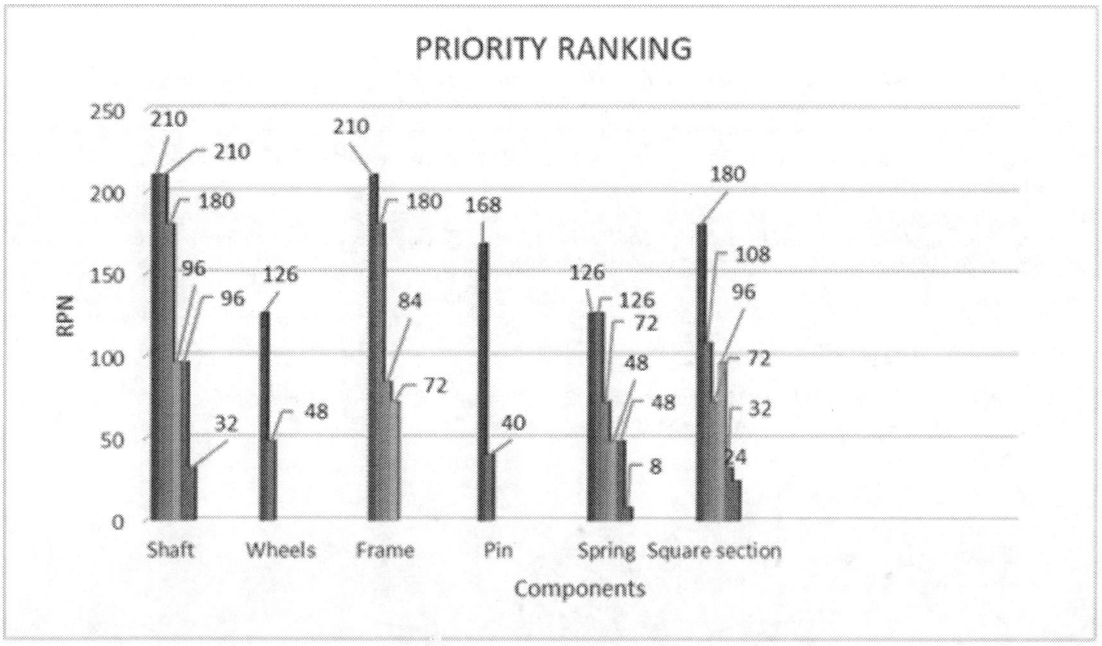

Fig.6 Priority graph of components

CONCLUSION

The D-FMEA technique was successfully applied to the automatic tilting mechanism trolley components. The analysis helped us to find the critical components of the mechanism. The critical components of trolley are shaft, frame, pin and square section which has highest risk priority number. The RPN rating is reduced for critical components by incorporating the corrective actions. Required preventive measures were undertaken as mentioned in above respective tables and incorporated during design and fabrication stage of the automatic tilting mechanism trolley.

IOP Conf. Series: Materials Science and Engineering 998 (2020) 012053 doi:10.1088/1757-899X/998/1/012053

References

[1] American Society of Quality (ASQ), Failure Modes and Effects Analysis (FMEA); *http://asq.org/learn-aboutquality/process-analysis-tools/overview/fmea.html.*

[2] S. P. Raut, L. P. Raut, *A Review of Various Techniques Used for Shaft Failure Analysis*, International Journal of Engineering Research and General Science, Vol. 2, Issue 2, Feb-Mar 2014.

[3] M. Kolich, *Using Failure Mode and Effects Analysis to design a comfortable automotive driver seat*, Applied Ergonomic, Vol.45, pp. 1087-1096, 2014.

[4] S. Pattnaik, *Design failure mode effect analysis of all-terrain vehicle*, International Journal of Research in Engineering and Technology, Vol.04, Issue 06, pp. 2319-1163, June-2015.

[5] G. J. Pawar, *Application of Design Failure Modes and Effect Analysis (DFMEA) to Vertical Roller Mill Gearbox,* International Journal of Engineering Research, Vol. 4, Issue 12, pp. 663-667,2015.

[6] N. Sellappan, D. Nagarajan, K. Palani, *Evaluation of Risk Priority Number (RPN) in Design Failure Modes and Effects Analysis (DFMEA) using Factor Analysis*, International Journal of Applied Engineering Research ISSN, Vol. 10, Issue 14, pp. 34194-34198, 2015.

[7] W. H. Skewis, *Support system technology corporation*, Failure Modes of Mechanical Springs.

[8] S. M. Muzakkir, K. P. Lijesh, Harish Hirani, *Failure Mode and Effect Analysis of Journal Bearing*, International Journal of Applied Engineering Research, Vol.10, pp. 36843-36850,2015.

[9] J. Singh, S. Saha, *Static analysis of alloy wheel using ansys15.0,* International Journal of Research in Engineering and Technology, Vol. 04, Issue 07, July-2015.

[10] P. P. Graceraj, *A novel DFMEA Model for High Power Diesel Engine Design*, International Journal of Engineering Research & Technology (IJERT), Vol. 4, Issue 05, May-2015.

[11] M. Behera, B. Behera, *Design Failure Modes and Effects Analysis (DFMEA) of a Human Powered Recumbent Vehicle*, International Journal of Engineering Research & Technology (IJERT), Vol. 5, Issue 04, April-2016.

[12] A. A. Nannikar, D. N. Raut, R. M. Chanmanwar, S. B. Kamble, *FMEA of Manufacturing and Assembly process*, International Conference on Technology and Business Management, 26-28, 2012.

Experimental Investigation of YSZ Coated Piston Crown on Performance and Emission Features of LHR Diesel Engine with Mahua Biodiesel

R Rajesh[1] Dr.C.Vijaya Bhaskar Reddy[2] and Dr. B.Durga Prasad[3]

[1]Research Scholar, Department of Mechanical Engineering, JNTUA(A), Anantapur, Andhra Pradesh, India.

[2]Professor, Department of Mechanical Engineering, SVCET(A), Chittoor, Andhra Pradesh, India.

[3]Professor, Department of Mechanical Engineering, JNTUA(A), Anantapur, Andhra Pradesh, India.

E-mail:mr.rajesh301@gmail.com, bhaskarreddychalla@gmail.com, mukdhajntu@gmail.com.

Abstract. Researchers have carried out numerous studies on low heat rejection (LHR) engines. Among the studies reported, some are experimental and many are theoretical in nature. The thermal barrier coatings (TBC) play a prominent role in LHR engines. Ceramic coatings are TBCs, they shield the base material and high temperature protect the source. The present work is focused on Yttria-stabilized zirconia (YSZ) material with a thickness of 500 microns. Is a bond coat which is used a plasma spray Tanique for costed on the required surface. Experimental analysis is carried out on a diesel engine with diesel and Mahua biodiesel with different loading conditions. This study results to expose that brake thermal efficiency is bigger along with lessening of BSFC. The LHR engine reduces the Hydro carbon (HC) and carbon monoxide (CO) emissions drastically but the Nitrogen oxide (NOx) emission is increased.

Keywords: LHR Engine, TBC, YSZ, BSFC, Emissions, HC, CO, NOX

1. Introduction

Various impressions of low heat rejection (LHR) engines are established to using A new techniques like ceramic coating in the preferred components, avoid the air gap in the piston and the other apparatuses etc. It is also found that the coatings on condition that on the engine components are simple with progressive coating techniques.

The following journals are referred which forms the basis to carry out the project. Parker et al. [1] developed a monolithic reaction bonded silicon nitride piston for accomplishing low heat rejection to the coolant throughout the load range and reported higher efficiency. Stang et al. [2] bolted the crown, which consisted of a top supported with a conventional aluminum base made of ceramic to the aluminum body of a piston, separated by a very high insulating layer. However, this piston was operated only for 264 hours after which mechanical problems like loosening of the bolt were encountered and further testing had to be abandoned. Krishnan et al. [3] worked on monolithic ceramic piston and evaluated the performance of single cylinder engine and found a drop in brake specific fuel consumption (BSFC) of about 3% with 9% reduction in frictional horse power (FHP). Timoney et al. [4] reported an increased frictional horsepower with insulated engine, with plasma spray coated zirconia piston, with different degrees of air swirl achieved by masked valves, though it improved the fuel economy. Wade et al. [5] reported an improvement in the fuel consumption for un-cooled single cylinder diesel engine, in the range of 4 to 7% with ceramic coated cylinder head, valves, piston and liner in the area above the piston rings, to the baseline water cooled engine. Havstad et al. [6] found small improvements in fuel consumption in the range of 5 to 9% with ceramic insert in the piston and in the liner above the piston rings, when compared with water cooled engine. A constant air/fuel ratio was

IOP Conf. Series: Materials Science and Engineering **998** (2020) 012054 doi:10.1088/1757-899X/998/1/012054

maintained for both versions of the engines to ensure similar combustion in both cases. Cheng et al. [7] conducted tests for single cylinder operation on 8-cylinder engine revised, employing a ceramic-coated piston and head. BMEP and airflow rate remained constant, however air–fuel ratio was not kept constant. This resulted in deterioration of the engine performance with insulation, increasing BSFC upto around 17%.Ciniviz et al. [8] clarified that Associated with a normal diesel engine, the power of engine was enlarged by 2%, the engine twisting was enlarged by min 1.5% and max. 2.5 %, and brake sfc (B.S.F.C.) was diminished by min. 4.5 and max. 9 %.

2. Methodology

It was divided into different sections namely, Plasma spray Coating, fabrication of an insulated piston, Experimental programme for pure diesel, experimental investigation with LHR engine with mahua biodiesel. The plasma spray coating on piston top shown in Fig.1, and the Normal piston, TBC Insulated piston and Piston de-accumulate and accumulate in Engine are shown in fig .2.1(a & b) and 2.2 (c & d).

Fig. 1 Plasma spray coating on piston top

(a). Normal Piston (b). TBC Insulated Piston

Fig. 2.1 Difference of before TBC and after TBC of piston

(c). Normal piston dis-assemble in the engine (d). Coated piston was assembled in the engine

Fig. 2.1 LHR engine replace the TBC piston with place of Normal Piston

3. Diesel Procedure on Predictable Engine and LHR Engine

Photographic View of the Investigational setup shown in Fig 3 (a & b). used for the surveys of LHR diesel engine with MAHUA diesel is shown in Fig 3 (c), and the Fig 3. (d, e &f) shows reading taken by use of computer. The predictable engine had an aluminium alloy piston with a bore diameter of 87.5 mm and a length of stroke is 110 mm. The esteemed yield of the engine is 5.20 kW at a frequency speediness of engine 1500 rpm. The compression ratio was 17.5:1. The fuel injector required 3 holes of 0.25 mm size.

A direct injection type is involved as a combustion chamber with no different research for swirling motion of air. Electric dynamometer is associated with the engine for finding its BP. Finding fuel consumption of the engine here used Burette method to cut off. Air- consumption of the engine was restrained by the air-box method.

Fig.3 (a). Investigational Setup of LHR Engine

Fig.3 (b). Investigational Setup for LHR Engine Operation

Fig.3 (c). preparation of MAHUA bio-diesel

Fig 3 (d). Reding taken by the computer

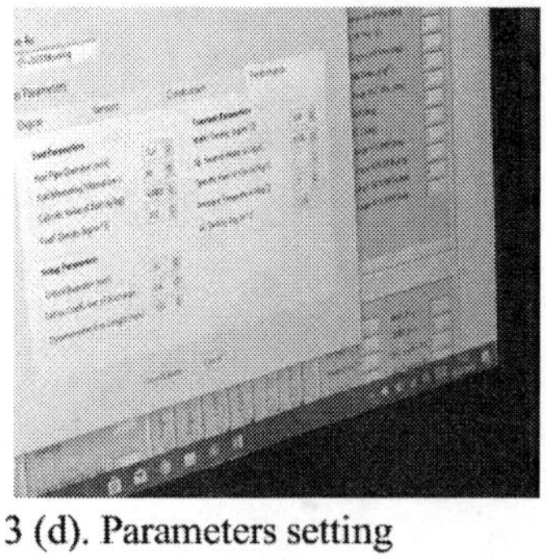

Fig 3 (d). Parameters setting

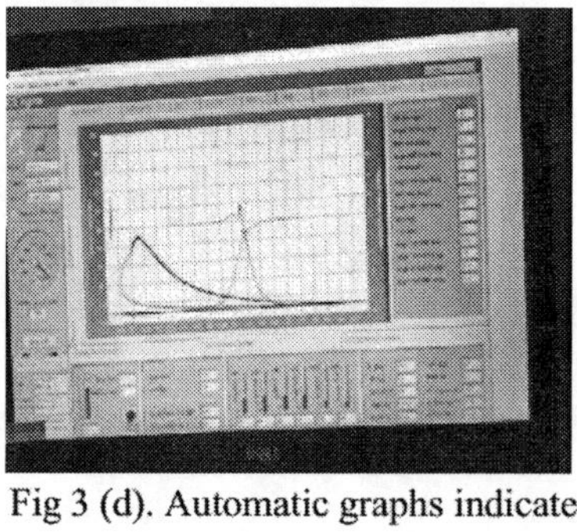

Fig 3 (d). Automatic graphs indicated

IOP Conf. Series: Materials Science and Engineering **998** (2020) 012054 doi:10.1088/1757-899X/998/1/012054

4. Performance Parameters

LHR engine consisted of a ceramic coating (Yttria stabilized zirconium) of 500 microns on the piston head. Plasma spray coating technique is used in this process. In this a conventional single cylinder CI engine with diesel is used in

Test 1 and with mahua is Test 2 then convert the diesel engine to a LHR Engine by coating its piston crown with a 500-micron layer of YSZ by a plasma spray method for Test 3, in Test 4 mahua oil is used with LHR Engine. From the table 1 to 7, Engine parameters, namely Bth η, BSFC, Mech η, vol η, power & emission characteristics were measured. Then include specifications of the coating, specifications of the engine and engine conditions.

Table 1. Specifications of Plasma spray coating.

Coating parameters	Specifications
Plasma gun	3MB plasma spray gun
Nozzle	GH Type nozzle
Pressure of organ gas	100–120 PSI
Flow rate of organ gas	80–90 LPM
Pressure of hydrogen gas	50 PSI
Flow rate of hydrogen gas	15–18 LPM
Powder feed rate	40–45 g per minute
Spraying distance	3–4

Table 2. Test Engine specifications.

Type	Specifications
Engine	Four Stroke, Single Cylinder water-cooled Diesel Engine
Rated Speed	1500 rpm
Bore Diameter	87.5mm
Stroke	110 mm
Compression ratio	17.5:1
Loading	Electric loading
Orifice Diameter	29.6 mm
Coefficient of Discharge	0.6

Table 3. Testing Engines and Conditions.

Testing Engine	Condition
Base Line Engine (Diesel)	Test-1
Base Line Engine (Mahua Biodiesel)	Test-2
LHR Engine With YSZ Coat (diesel)	Test-3
LHR Engine YSZ Coat (MahuaBiodiesel)	Test-4

Table 4. Conventional Engine with Diesel.

Load (kg)	Speed (rpm)	BP (kW)	IP (kW)	FP (kW)	TFC (kg/hr)	SFC (kg/ kWh)	BThEff (%)	IThEff (%)	Mech Eff(%)	VolEff (%)	HBP (%)	HJW (%)	HGas (%)	HRad (%)
1559	0.05	0.01	2.68	2.67	0.35	23.53	0.36	65.98	0.55	86.87	0.36	9.64	15.78	74.22
1537	4.51	1.32	3.76	2.45	0.50	0.38	22.66	64.76	35.00	86.45	22.66	18.69	17.59	41.35
1520	9.11	2.63	4.89	2.26	0.75	0.28	30.19	56.07	53.84	85.58	30.19	19.36	16.54	33.90
1507	13.52	3.87	5.97	2.10	1.00	0.26	33.32	51.39	64.84	85.84	33.32	20.67	17.12	28.89
1490	17.78	5.03	6.88	1.85	1.25	0.25	34.66	47.38	73.16	85.07	34.66	21.82	20.14	23.38

Table 5.LHR engine with YSZ coating (diesel)

Load (kg)	Speed (rpm)	BP (kW)	IP (kW)	FP (kW)	TFC (kg/hr)	SFC (kg/ kWh)	BThEff (%)	IThEff (%)	Mech Eff(%)	VolEff (%)	HBP (%)	HJW (%)	HGas (%)	HRad (%)
1565	0.09	0.03	2.38	2.35	0.31	11.57	0.74	65.84	1.13	88.68	0.74	12.69	19.80	66.77
1553	4.69	1.38	3.76	2.38	0.52	0.37	22.98	62.52	36.76	88.62	22.98	13.22	18.43	45.36
1526	9.03	2.62	5.16	2.54	0.72	0.28	31.08	61.19	50.80	88.11	31.08	15.29	18.50	35.13
1518	13.45	3.88	6.04	2.16	0.98	0.25	33.93	52.79	64.28	86.57	33.93	15.18	18.57	32.32
1500	17.91	5.11	7.19	2.08	1.19	0.23	36.87	51.91	71.03	85.87	36.87	16.66	20.60	25.86

Table 6. Conventional Engine with Mahua Biodiesel

Load (kg)	Speed (rpm)	BP (kW)	IP (kW)	FP (kW)	TFC (kg/hr)	SFC (kg/ kWh)	BThEff (%)	IThEff (%)	Mech Eff(%)	VolEff (%)	HBP (%)	HJW (%)	HGas (%)	HRad (%)
1560	0.05	0.01	2.35	2.33	0.35	23.83	0.36	57.61	0.63	88.85	0.36	1.11	19.29	79.23
1540	4.79	1.40	3.75	2.35	0.56	0.40	21.89	58.60	37.35	88.29	21.89	12.03	17.54	48.55
1528	8.99	2.61	4.87	2.26	0.81	0.31	28.05	52.27	53.66	87.43	28.05	14.38	17.11	40.46
1516	13.60	3.92	6.11	2.19	1.06	0.27	32.06	49.95	64.18	86.14	32.06	16.06	17.64	34.25
1496	17.92	5.09	6.99	1.90	1.36	0.27	32.42	44.50	72.86	85.50	32.42	15.84	18.01	33.72

Table 7. LHR engine with YSZ coating (mahua)

Load (kg)	Speed (rpm)	BP (kW)	IP (kW)	FP (kW)	TFC (kg/hr)	SFC (kg/ kWh)	BThEff (%)	IThEff (%)	Mech Eff(%)	VolEff (%)	HBP (%)	HJW (%)	HGas (%)	HRad (%)
1552	0.06	0.02	2.18	2.16	0.35	19.11	0.44	52.01	0.85	88.81	0.44	7.46	18.60	73.49
1540	4.45	1.30	3.62	2.32	0.66	0.51	16.75	46.56	35.97	88.40	16.75	10.89	14.66	57.70
1520	9.22	2.66	4.82	2.16	0.91	0.34	24.76	44.82	55.24	87.26	24.76	12.00	14.57	48.68
1509	13.67	3.92	5.99	2.07	1.16	0.30	28.51	43.56	65.46	86.31	28.51	12.76	15.65	43.07
1492	17.95	5.09	7.14	2.05	1.52	0.30	28.38	39.80	71.30	84.89	28.38	13.00	15.80	42.83

5. Results and Discussions

Investigational result shows that a lack of heat refusal to the chilling medium needed to result in an increase in exhaust energy all load periods. Hence, the TBC engine, increasing the vol η of the engine at dissimilar loadings and speed situations. The parameters from the Graph 1 to 7 have different parameters with corresponding to the tables 1 to 7. The discrepancy in brake thermal efficiency vs brake power. TBC coated engine with Test 3 has higher efficiency than Test 1, Test 2, and Test 4 engines. This is due to the low thermal conductivity of YSZ, in the piston crown the thermal resistance rises, hence it cannot consent the heat energy to the coolant and further standard. Then below graphs show differences of related characteristics.

Graph.1, 2 shows the variation in brake thermal efficiency vs brake power. TBC coated engine with Test 3 has higher efficiency than Test 1, Test 2, Test 4 engines. The maximum brake thermal efficiency obtained for TBC coated engines Test 4 and Test 3 respectively. The Test 4 and Test 3 due to the reduction in fuel burning and the power output is same to helps increasing the brake thermal efficiency. Graph 3. shows the variation in volumetric efficiency of baseline engines and TBC coated engines in Test 3 under different load conditions. Graph 3. shows the comparison of thermal balance sheet is made for the baseline engine and TBC coated engines under different loads. It is also observed that, the heat energy available at the exhaust is increased at all loading conditions.

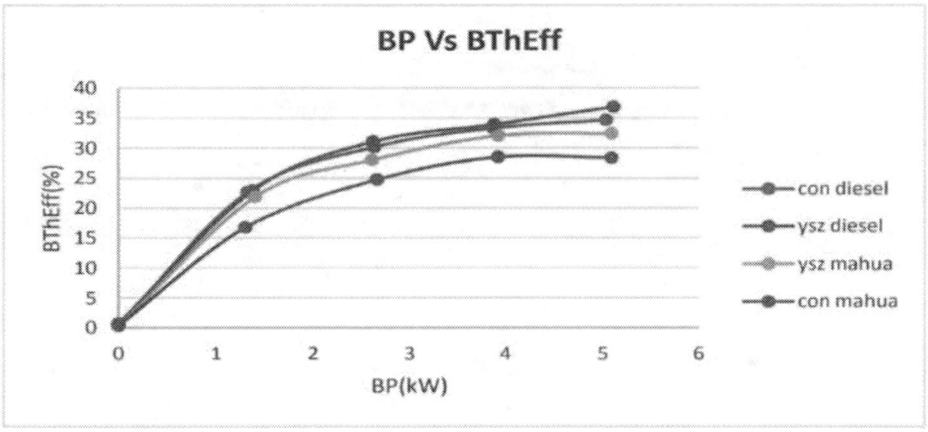

Graph 1. Difference of Bth η of reference engine & TBC engine under BP

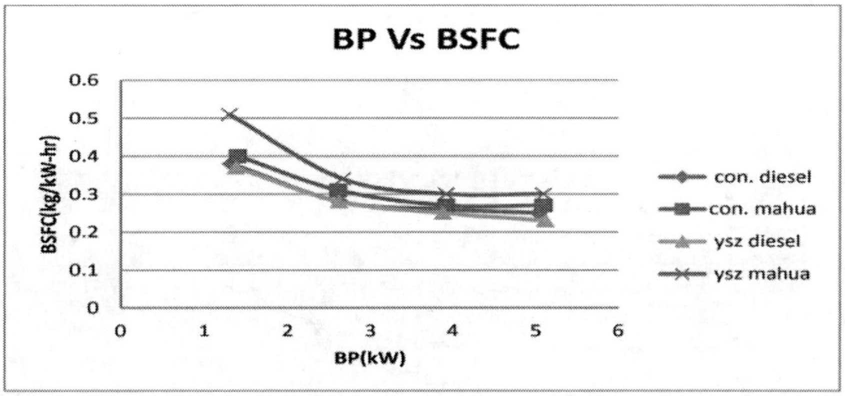

Graph 2. Difference of Brake Sp. Fuel Consumption of reference engine & TBC engine under B P

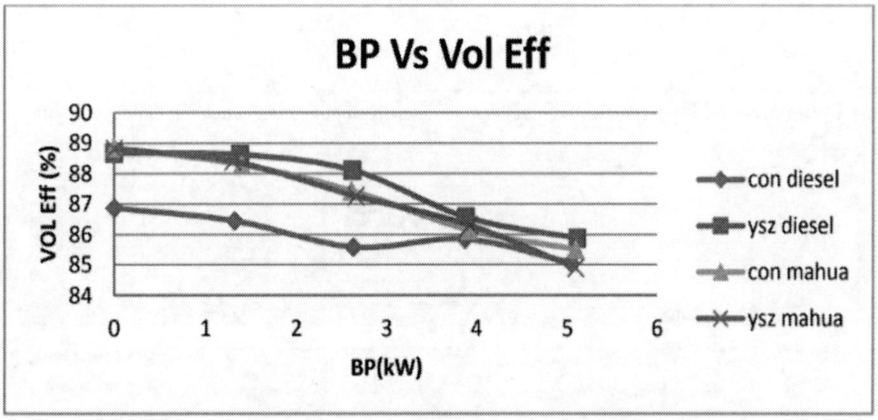

Graph 3. Difference of Volumetric Efficiency of reference engine & TBC engine under B P

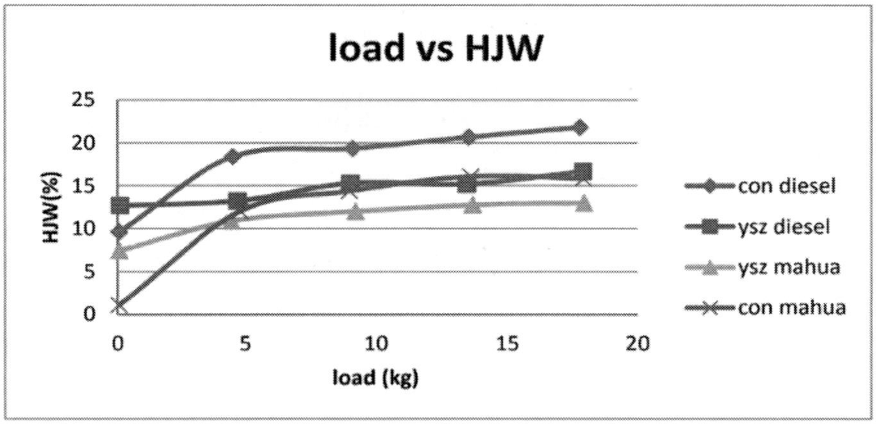

Graph 4. Difference of HJW under different Loads in Thermal Balance

6. Emission Features

The appliance is verified the record of ISO 8178-4 ''C1'' 8 Methods of trial sequence for off-road automobiles, and its emission features. After the investigational results, the emission of HC, CO and NOX emissions is considered. It is initiate of CO and HC emissions are condensed in Test 3 and Test 4 in TBC engines when related to the reference engine in Test 1 and Test 2. The NOX emissions are increased in Test 3 and Test 4 in the TBC coated engines.Graph. 5 shows variations in HC emissions with load and Graph 6 shows CO with load.

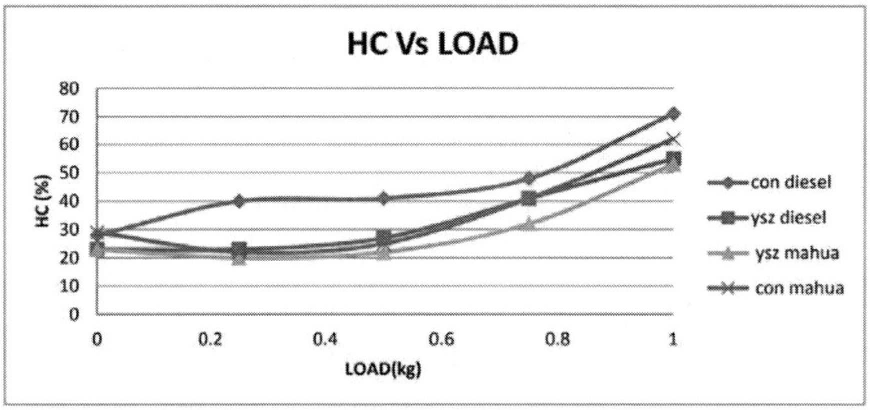

Graph 5. Difference of Hydrocarbon Emission of reference engine and TBC engine under changed Loads

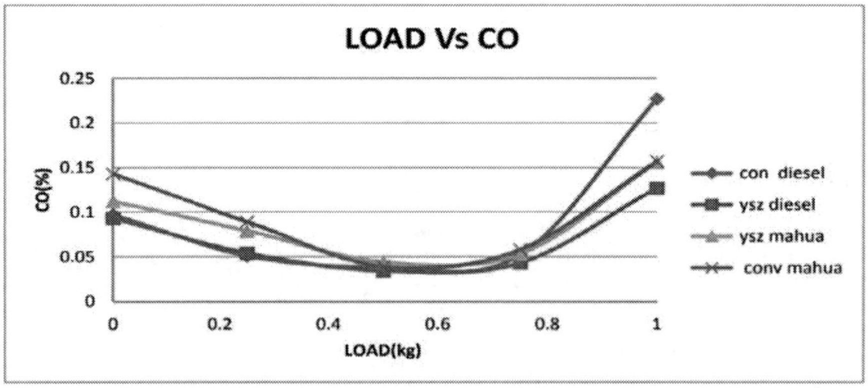

Graph 6. Difference of Carbon monoxide emission of reference engine and TBC engine under changed Loads.

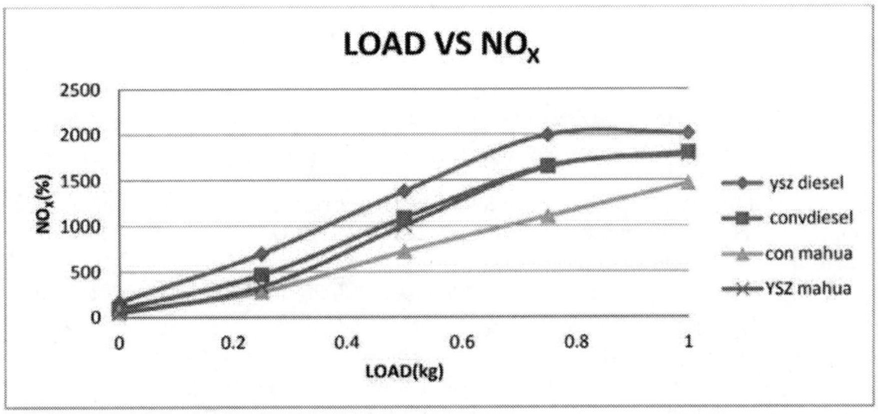

Graph 7. Difference of Nitrogen oxide emission of reference engineand TBC engine under changed Loads.

In general NOX level is different straight values with in the temperature of inside the cylinder in engine. NOX emissions higher un base line engine compared to TBC engine. Smoke levels increases by 12% and NOX levels increases by 19% in TBC coated engine. In past investigation explained that LHR engine are generally higher NOX emissions. It is to be higher combustion in temperature and curation of combustion is longer. It was experimentally determined that TBC coated engines Test 3 engine higher in NOX emission at various loads. Graph 7 Variation of Nitrogen oxide emission of baseline engines and TBC engines under different Loads.

7. Scope for Future Work

There is sufficient scope for carrying out further work on the following aspects of other materials with low thermal conductivities and high strength can be tried out as crown materials.

- Endurance Tests on the LHR engine may have to be carried out.
- The effect of varying compression ratios on LHR engine can be studied in detail.
- The naturally aspirated engine can be changed to supercharged engine and its effect on the performance of LHR engine can be studied.
- For controlling NOX in LHR engine, catalytic converters can be designed. For reducing smoke levels in LHR engine with pure diesel operation suitable cyclone separators can be designed.
- The multi insulation LHR engine can be used to study its effects in alcohols. 9. Heat Released Rates can be studied.

8. Conclusion

A conventional single cylinder CI engine with diesel is used in Test 1 and with mahua is Test 2 then convert the diesel engine by using coated piston to an LHR Engine with 500-micron film of YSZ ceramic coating a plasma spray method for Test 3, in Test 4 mahua oil is used with LHR Engine. Engine limits, namely Bth η, BSFC, Mech η, vol η, power and emission remained measured to inspect the properties of YSZ, and its performance and emission features of the engine. The succeeding conclusions canister be tired from the investigational outcomes. Brake thermal efficiency is enhanced at all loads in the TBC coated engines in Test3 condition. In Test3 brake thermal efficiency is an extreme increase due to low thermal conductivity of YSZ. TBC coated engines in the Test 3 condition reduce the specific fuel consumption, when compared to the reference engines in Test 1. HC emissions were condensed drastically in the TBC coated engines in Test 3, whereas Carbon monoxide emission is condensed in Test 2, Test 3 and Test 4 conditions separately when compared to the baseline engines in Test 1. Smoke levels increases by 12% and NOX levels increases by 19% in TBC coated engine. Finally, Exhaust gas temperature was enhanced charging conditions within chance to improve the NOX emissions of the ceramic treated engines.

Acknowledgements this experimental work was supported by the SVCET, Chittoor. We are very thankful Dr. C. Vijaya Bhaskar Reddy and Dr. B Durga Prasad for helping to carry out the experimentation process.

References

[1] Parker, D.A. and Smart, R.E., "An evaluation of silicon nitride diesel pistons", The mechanical Engineering properties and applications of ceramics, Editor D.J. Godfrey, Proceedings of the British ceramic Society, 167-181,(1978).

[2] Stang John, H., "Designing adiabatic engine components", Proceedings of SAE International Congress & Exposition, Detroit, Michigan, Paper No.780069,(1978).

[3] Krishnan, D.B., Raman, N., Narayanaswamy, K.K and Rohtagi, P.K., "Performance of an Al-Si graphite particle composite piston in a diesel engine", Transactions of Wear, 60,205-215, (1980).

[4] Kadoli, F.G., "Ceramic thermal barrier coatings for the improvement of diesel engine performance", PhD thesis, I.I.T., Bombay, (1984).

[5] Wade, W.R., Havstad, H.,Ounsted, E.J.,Trinker, F.H. and Garvin, I.J., "Fuel economy opportunities with an un-cooled diesel engine", SAE Paper No. 841286,(1984).

[6] Havstad, P.H., Gervin, I.J. and Wade ,W.R.,"A ceramic insert un-cooled diesel engine",SAE Paper No. 860447,(1986).

[7] Cheng, W.K., Wong, V.W. and Gao,F., "Heat transfer measurement comparison in insulated and non-insulated diesel engines", SAE Paper No. 890570,(1989).

[8] Ciniviz, M., Hasimoglu, C., Sahin, F. and Salman, M. S. (2008). Impact of thermal barrier coating application on the performance and emissions of a turbocharged diesel engine. Proceedings of The Institution of Mechanical Engineers Part D-Journal Of Automobile Engineering, 222 (D12), 2447–2455, Published: DEC 2008.

Influence of Alkaline ratios on strength properties of Fly ash- Ground Granulated Blast Furnace Slag Based Geopolymer Mortars

M. Venu[a], G. Mallikarjuna Rao[b], Y. Anil Kumar[c] Sri Rama Chand Madduru[d], and Ramamohana Reddy Bellum[e]

[a]*Assistant Professor, Department of Civil Engineering, Vardhaman College of Engineering, Shamshabad, Telangana, India.*

[b]*Associate Professor, Department of Civil Engineering, Vardhaman College of Engineering, Shamshabad, Telangana, India.*

[c]*PG Stdent, Department of Civil Engineering, Vardhaman College of Engineering, Shamshabad, Telangana, India*

[d]*Assistant Professor, Department of Civil Engineering, Sree Chaitanya College of Engineering, Karimnagar, India.*

[e]*Research Scholar, Department of Civil Engineering, Vignan's Foundation for Science, Technology and Research, Vadlamudi, Guntur, India.*

Abstract: Geopolymer concrete is a new and alternative to cement concrete, thus decreasing greenhouse emissions and leading to better construction practices. This can be produced by complete replacement of cement with pozzolanic materials such as Fly Ash (FA), Ground Granulated Blast Furnace Slag (GGBS) and activation through alkaline activator solution. The present study aims to investigate the effect of 8M sodium hydroxide solution for the chemical ratios of Na_2SiO_3 to NaOH are 1.5, 2.0, 2.5, and 3 with different combinations of fly ash and GGBS. The cast specimens were outdoor cured and tested at the age of 1, 3, 7, and 28 days for compressive strength, followed by XRD analysis to identify the mineralogical variations with a change in parameters. The obtained results concluded that GGBS content with 8M concentration of sodium hydroxide in the mix increases the compressive strength of geopolymer mortar. The optimum dosage of Na2SiO$_3$/NaOH was 2.5 based on the compressive strength of mortar. The compounds responsible for strength were identified through XRD analysis as N-A-S-H gel and C-A-S-H gel in mixes with fly ash and GGBS as binders.

Keywords: Compressive strength, Outdoor curing, Polymerization, geopolymer mortar, X-Ray Diffraction analysis.

1. Introduction

Enormous quantities of by-products are generated due to industrial activities, which affect the atmosphere harmfully. A massive amount of carbon dioxide (CO2) is emitted during the production processes of cement, which are energy-intensive **[1]**. Geopolymer (GP) binder is an alternative to the Ordinary Portland Cement (OPC) binder, which is produced by activation the industrialized by-products rich in alumino-silicates like Fly Ash (FA), Ground Granulated Blast Furnace Slag (GGBS), zeolite, etc. with activator liquid **[2]**. Worldwide several studies have been done on FA-GGBS blended GP binders. FA (Class-F) based binders typically need high curing temperature and high alkali-activator levels to enhance the properties. These alkali liquids are accountable for the higher emissions of carbon dioxide, and manufacturing expenses of GP, as using greater alkali content leads to an improvement in both of these **[3]**.

On the other hand, FA-GGBS based binders require relatively lower activator dosages, which can reduce both issues like CO_2 emissions and cost. Combination FA and GGBS allow at ambient curing to enable a compact and denser microstructure; as a result, enhanced mechanical properties **[4]**. Despite around 35-40 years of research studies **[5]**, comparatively few reports concentrated on the optimistic mix proportions in terms of source materials and their influence on the amount of alkaline liquid, fresh and hardened properties and recognition of the microstructural products. The majority of the reports available on pastes [6,7] and only some of the investigations are focused on the applications of FA-GGBS based mortar and concretes [8–10]. The significant outcomes from these literature reports were that the replacement of GGBS enhanced the mechanical properties but reduced the consistency and workability. Considering the structure of GP can assist in designing of more reliable and durable constructions [11]. The activator liquid plays an essential role in attaining the strength of GP materials. The solution activates SiO2 and Al2O3 in FA to form N-A-S-H gel and C-A-S-H gel by reacting with calcium in GGBS [12,13].

On the other hand, GPs are formed by treating aluminosilicate materials (such as FA, GGBS, etc.) with alkali liquids. Normally, FA consists of a homogeneous mix of alumina-silicate and quantities of crystalline products like quartz, mullite, magnetite, and hematite [14]. In the formation of GPs, with the combination of FA contributes reactive species through dissolution in the alkaline environment with different $Na_2SiO_3/NaOH$, the residue is characterized using X-ray analysis [15].

In this study, an attempt has been made to prepare GP mortars with FA and GGBS, and hardened properties such as compressive strength were determined. Mineralogical variations were studied through XRD analysis. To estimate the effect of alkaline ratios, the mass ratio of $Na_2SiO_3/NaOH$ solution was varied as 1.5, 2, 2.5, and 3.0. The molarity of sodium hydroxide used in this study was 8M. All the specimens were outdoor cured.

2. Materials

FA and GGBS were used according to ASTM C 618 [16] and ASTM C989 [17], respectively, as binders in the present experimental work. Locally available FA and GGBS were used and mineral composition was shown in Fig. 1. Oxide components of FA and GGBS were presented in Table 1. The local river sand was used as fine aggregate conforming to Zone-II as per BIS: 383-1970 [18] and the specific gravity was 2.64.

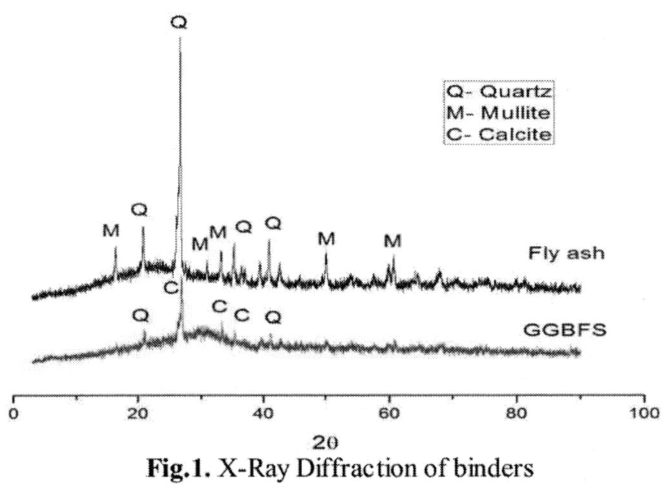

Fig.1. X-Ray Diffraction of binders

Table 1. Chemical composition of binders

Oxide	SiO_2	Al_2O_3	Fe_2O_3	SO_3	CaO	MgO	Na_2O	LOI
Fly ash	60.11	26.53	4.25	0.35	4.00	1.25	0.22	0.88
GGBS	34.06	20	0.8	0.9	32.6	7.89	-	-

2.1. Preparation of Alkaline Solution

The alkaline liquid used in the present experimental work was a blend of NaOH and Na_2SiO_3 liquids. However, the activation can be done using a NaOH or Na_2SiO_3 liquid alone, but the reaction process is slower in such a case. The molarity of the NaOH plays a vital role in attaining the strength of the GP samples. The molarity of the NaOH is taken as 8M in the experimental program. For the preparation of the NaOH liquid of 8M, 320 grams (8x40) of NaOH pellets were dissolved in one litre of potable water. The prepared NaOH liquid was allowed to cool, and then the NaOH liquid was blended with Na_2SiO_3 liquid, and the ratios of Na_2SiO_3/NaOH were varied as 1.5, 2, 2.5, and 3.

2.2. Mixed Proportion

Five mixes were used in this paper with different proportions of FA-GGBS. The ratio of Na_2SiO_3/NaOH was varied as 1.5, 2.0, 2.5, and 3.0. The density of GP samples was around 2000-2200 kg/m^3. The mix proportioning of all the materials were presented in Table 2.

Table 2: Mix Proportions of geopolymer Mortar for different ratios of Na_2SiO_3/NaOH

Mix ID	Label	FA	GGBS	Fine Aggregate	NaOH	Na2SiO3	Alkaline liquid	Na2SiO3/NaOH
M1	F100G0	880	0	880	176	264	440	1.5
M2	F75G25	660	220	880	176	264	440	1.5
M3	F50G50	440	440	880	176	264	440	1.5
M4	F25G75	220	660	880	176	264	440	1.5
M5	F0G100	0	880	880	176	264	440	1.5
M6	F100G0	880	0	880	146.66	293.32	440	2
M7	F75G25	660	220	880	146.66	293.32	440	2
M8	F50G50	440	440	880	146.66	293.32	440	2
M9	F25G75	220	660	880	146.66	293.32	440	2
M10	F0G100	0	880	880	146.66	293.32	440	2
M11	F100G0	880	0	880	125.7	314.28	440	2.5
M12	F75G25	660	220	880	125.7	314.28	440	2.5
M13	F50G50	440	440	880	125.7	314.28	440	2.5
M14	F25G75	220	660	880	125.7	314.28	440	2.5
M15	F0G100	0	880	880	125.7	314.28	440	2.5
M16	F100G0	880	0	880	110	330	440	3
M17	F75G25	660	220	880	110	330	440	3
M18	F50G50	440	440	880	110	330	440	3
M19	F25G75	220	660	880	110	330	440	3
M20	F0G100	0	880	880	110	330	440	3

2.3. Casting and Curing

The binders and fine aggregates were mixed uniformly for 2-3 min in a pan mixer. Next, the activator solution was added and continued blending for another 3-4 min. Consequently, fresh GP mortar placed in the moulds and placed in a vibration table for 1-2 min to remove the air voids. After 24 hours, all the samples were demoulded and placed in outdoor curing for different ages. Fig.2 illustrates the outdoor curing process of GP mortar samples.

Fig.2. Geopolymer samples left for Outdoor Curing

3. Results and Discussions

3.1 Compressive Strength

The compressive strength of GP mortars with dissimilar proportions of FA and GGBS were considered in the experimental study. The strength of mortar cubes prepared with dissimilar ratios of Na_2SiO_3 to NaOH ranged from 1.1 MPa to 71.78 MPa for all the combinations of FA and GGBS. This clearly shows the effect of activator combinations on the compressive strength of FA and GGBS based GP mortars. The specimen prepared with F100-G0 has deficient strength compared with all other mix combinations of FA and GGBS. With an increase in the GGBS levels, the compressive strength of GP was also enhanced. The early strength was attained with greater percentages of GGBS. The mixture of NaOH to Na_2SiO_3 liquid plays an essential role in gaining the strength. With an increase in the Na_2SiO_3/NaOH ratio from 1.5 to 2.5, the compressive strength increases, and beyond 2.5, the compressive strength decreases. The mineralogical variation and the minerals which are responsible for the strength contribution for all mixes of FA and GGBS were observed through XRD analysis.

The effect of calcium content present in the different combinations of FA and GGBS was studied in the experimental program. In the FA based mixes, the silica and alumina are the main minerals which are responsible for the strength contribution. The dissolution of these minerals helps in attaining the strength through the polymerization process. With the addition of GGBS level in the mix helps in achieving more strength compared with FA based samples. A combination with F100-G0 attained very less strength compared to all the mixes due to less calcium content. The highest Ca/Si ratio was observed in F0-G100 compared with the mixes shows the higher compressive strength, which was observed from Figs. 3-6. The presence of more amounts of Ca results in additional C-A-S-H gel formations, it leads to enhancement of strength. The attainment of the strength in GP mortar mainly depends on the GGBS content in the mix.

The chemical composition present in the source materials and their combinations is responsible for the strength contribution of GP mortars. The replacement of silica and alumina with some amount of calcium content is helpful for the formation of the new reaction products like albite, microcline, and anorthite observed through XRD analysis. These minerals are other forms of the alkali aluminosilicate GP gel (N-A-S-H) and calcium alumina silicate hydrate gel. These minerals are responsible for the strength attainments of the GP mortars prepared with dissimilar proportions of FA and GGBS.

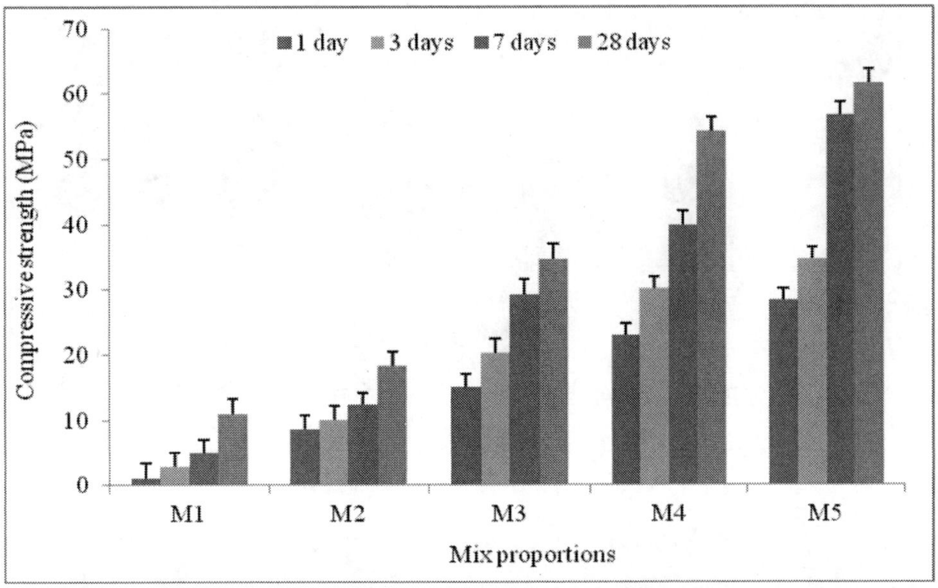

Fig. 3. Compressive Strength of Geopolymer Mortar with $Na_2SiO_3/NaOH$ -1.5

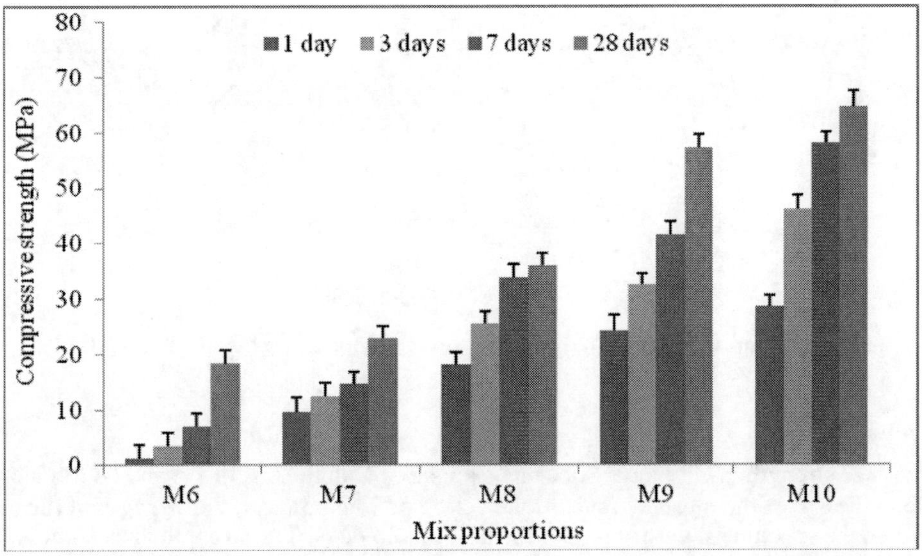

Fig. 4. Compressive Strength of Geopolymer Mortar with $Na_2SiO_3/NaOH$ -2

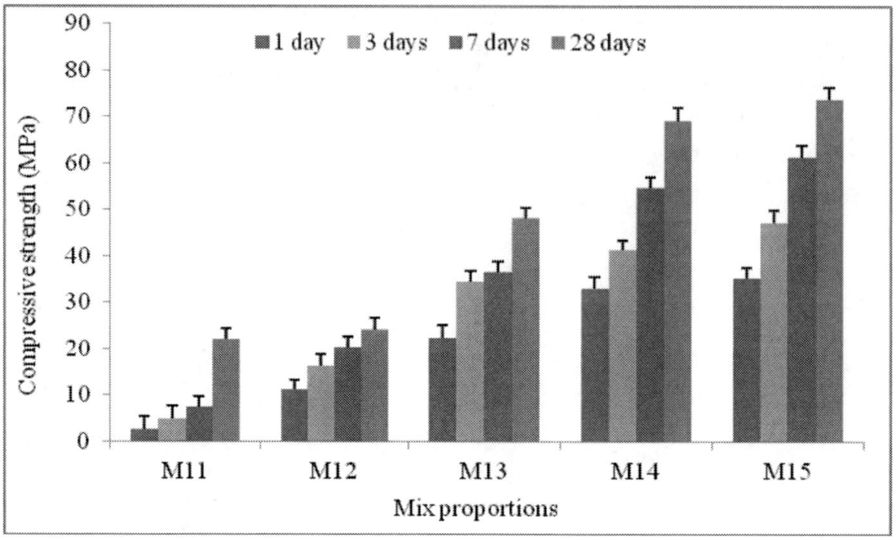

Fig. 5. Compressive Strength of Geopolymer Mortar with Na_2SiO_3/NaOH -2.5

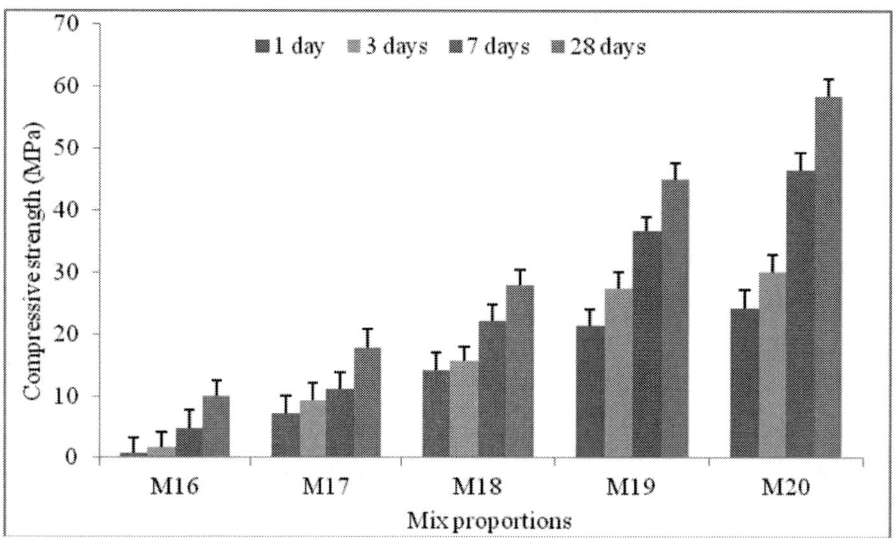

Fig. 6. Compressive Strength of Geopolymer Mortar with Na2SiO3/NaOH -3

3.1.1. Effect of Age:

The compressive strength of GP mortar specimens prepared with the mix of FA and GGBS increases with an increase in the age of the mortar cubes. At the period of 1 and 3 days, the strength of the mortars was very less when the specimens were compared with 7 and 28 days. The strength increment was observed for the samples prepared with more amount of GGBS content. The maximum strength among all the samples was observed for the mix of F0-G100 at the age of 28 days. The ratio of Na_2SiO_3 to NaOH plays a crucial role in attaining the strength of GP mortars. The strength was lower for the mixes prepared with sodium silicate to sodium hydroxide ratio 1.5. As the ratio of Na_2SiO_3 to NaOH increases the strength of the blends with dissimilar proportions of FA and GGBS also increases up to 2.5. After that, the strength decreases due to the large quantity of sodium silicate solution, which will affect the polymerization process. Age plays an essential role in influencing the polymerization process, and as the age increases, the strength increases with the formation of C-A-S-H gel.

3.2. XRD analysis

The fragments of mortar cubes after the experiment were ground to a fine powder and were used for XRD analysis. The major reaction products formed during the activation are alumina-silicate products in an amorphous state or poorly crystalline state. The XRD patterns of mortar samples for different combinations of FA and GGBS for different alkali ratios are shown in Fig.7. The minerals identified in XRD analysis are Quartz (SiO_2), Mullite (Al_2O_3) and Calcite (CaO), Hematite (Fe_2O_3), Alumina (Al_2O_3), Alite (Ca_3SiO_5), Belite (Ca_2SiO_4) and Portlandite ($Ca(OH)_2$) were observed in all the samples irrespective of mix proportions. Fig.7 was evident that the main peak detected as quartz at $2\theta = 24.5°$ having N-A-S-H/C-A-S-H, and calcite as associate peaks.

As seen from Fig.7, the reaction products in mix 100FA-0GGBS were N-A-S-H gel. With an increase in GGBS content, the calcium content increases and results in an increased amount of C-A-S-H gel. It is reported in the literature that the Ca^{2+} ions replace all the Na^{+} ions in the mix in favor of C-A-S-H formation until available Ca^{2+} is exhausted **[19]**. The increase in compressive strength at higher dosages of GGBS is due to the structure of C-A-S-H and C-S- H gel along with N-A-S-H gel, which improved the strength of mortar.

The use of higher alkaline ratios increases the silica content resulting in a greater quantity of aluminosilicate gel, which on reaction with available Na^{+} or Ca^{2+} results in N-A-S-H or C-A-S-H gels. The presence of higher amounts of silicates reacts among themselves, forming oligomeric silicates, which on reaction with alumina results in poly (sialate-siloxo) or poly (sialate-disiloxo) polymer structures. These polymer structures are comparatively more durable than poly (sialate) structures; thereby the strength is improved at higher alkaline ratios. The main reaction product for F100-G0 is N-A-S-H gel, and whereas in F0- G100, the main reaction product was C-A-S-H gel.

For different combinations of FA and GGBS, the peak of C-A-S-H was observed at $2\theta = 24.5°$ approximately. The C-A-S-H was formed due to the polymerization of FA and GGBS with alkali activators, which also has led to the transformation of the amorphous phase to the crystalline phase. The quartz and calcite were the main minerals that are responsible for the strength contribution for the samples prepared with F0-G100 content along with N-A-S-H gel. The strength was much improved for F0-G100 samples and this strength was contributed by the mineral anorthite which was observed through XRD analysis and amorphous compounds dissolve easier than crystalline compounds during the first step of polymerization, yielding greater amounts of reactive SiO2 and Al2O3 to combine during the geopolymeric reaction product, which is responsible for the higher strengths **[20]**.

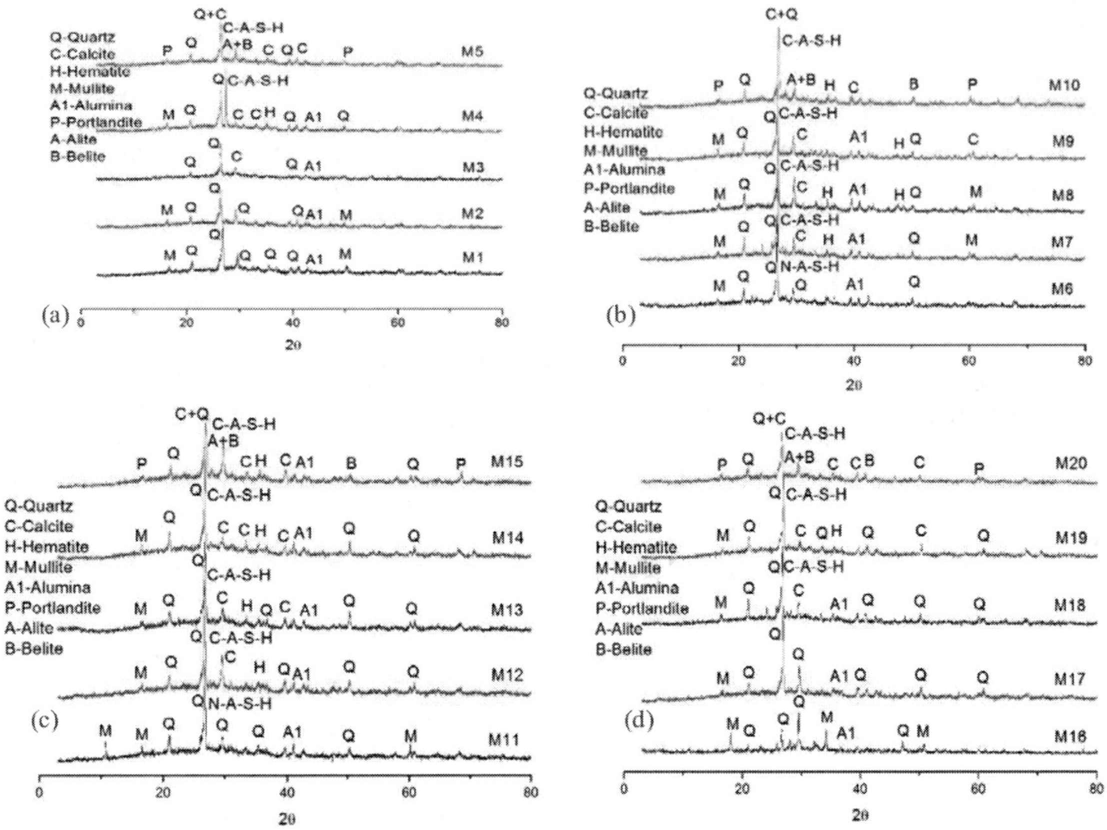

Fig. 7. XRD analysis of GPC (a) M1-M5, (b) M6-M10, (c) M11-M15, (d) M16-M20

Calculations

This paper presents the effect of Na_2SiO_3 to NaOH ratios (1.5, 2, 2.5, and 3) on strength properties of geopolymer concrete with different proportions of fly ash and GGBS. From the experimental data the following conclusions were drawn,

- Compressive strength of geopolymer mortar increases with increase in percentage replacement of fly ash with GGBS.
- To develop geopolymer concrete under outdoor curing condition, combination of fly ash with GGBS is a relevant solution.
- The Ca/Si ratio plays an important role in affecting the strength of the geopolymer mortar as there increase in Ca/Si ratio the strength of the geopolymer mortar also increases.
- For the combinations of fly ash and GGBS, the minerals quartz and calcite were responsible for improving the compressive strength of fly ash and GGBS based geopolymer mortars.
- Presence of CaO has a positive effect on strength of mortar due to formation of additional C-A-S-H gel along with N-A-S-H gel.

References

[1] M.C.G. Juenger F. Winnefeld, J.L. Provis, J.H. Ideker, Advances in alternative cementitious binders, Cem. Corcr. Res. 41 (12) (2010) 1232-1243, http://dx.doi.org/10.1016/j.cemconres.2010.11.012.

[2] J.L. Provis, Green concrete or red herring? – Future of alkali-activated materials, Adv. Appl. Ceram. 113 (8) (2014) 472–477.

[3] J.L. Provis, A. Palomo, C. Shi, Advances in understanding alkali-activated materials, Cem. Concr. Res. 78 (2015) 110–125.

[4] S. Kumar, R. Kumar, S.P. Mehrotra, Influence of granulated blast furnace slag on the reaction, structure and properties of fly ash based geopolymer, J. Mater. Sci. 45 (3) (2009) 607–615.

[5] M.A. Smith, G.J. Osborne, Slag/fly ash cements, World Cement Technology 8 (6) (1977).

[6] W.C. Wang, H.Y. Wang, M.H. Lo, The fresh and engineering properties of alkali activated slag as a function of fly ash replacement and alkali concentration, Constr. Build. Mater. 84 (2015) 224–229.

[7] X. Gao, Q.L. Yu, H.J.H. Brouwers, Reaction kinetics, gel character and strength of ambient temperature cured alkali activated slag – fly ash blends, Constr. Build. Mater. 80 (2015) 105–115.

[8] I. Ismail, Durability as a Function of Microstructure of Alkali-activated Slag/Fly Ash Binders, Doctoral Dissertation, University of Melbourne, Australia, 2013.

[9] P.S. Deb, P. Nath, P.K. Sarker, The effects of ground granulated blast-furnace slag blending with fly ash and activator content on the workability and strength properties of geopolymer concrete cured at ambient temperature, Mater. Des. 62 (2014) 32–39.

[10] P. Nath, P.K. Sarker, Effect of ggbfs on setting, workability and early strength properties of fly ash geopolymer concrete cured in ambient condition, Constr. Build. Mater. 66 (2014) 163–171.

[11] Wang SD, Pu XC, Scrivener KL and Pratt PL, Alkali-activated slag cement and concrete: a review of properties and problems. Advanced Cement Research, 27 (1995) :93–102.

[12] Bellum RR, Muniraj K, Madduru SRC (2000). Influence of activator solution on microstructural and mechanical properties of geopolymer concrete, Materialia, March 19, 2020. DOI: https://doi.org/10.1016/j.mtla.2020.100659.

[13] C.K. Yip, G.C. Lukey, J.L. Provis, J.S.J. van Deventer, Effect of calcium silicate sources on geopolymerization, Cem. Concr. Res. 38 (2008) 554–564, http://dx.doi.org/10.1016/j.cemconres.2007.11.001

[14] Song S, Sohn D, Jennings H.M and Mason, T.O, Hydration of Alkali-Activated Ground Granulated Blast Furnace Slag, Journal of Materials Science 35 (2000): 249-257.

[15] Mallikarjuna Rao G and Gunneswara Rao TD (2015) Final Setting Time and Compressive Strength of Fly Ash and GGBS-Based Geopolymer Paste and Mortar. Arabian Journal for Science and Engineering 40:3067–3074.

[16] ASTM C 618, Standard Specification for Coal Fly Ash and Raw or Calcined Natural Pozzolan for Use in Concrete, ASTM International, West Conshohocken, PA, 2008.

[17] ASTM C989 / C989M-18a, Standard Specification for Slag Cement for Use in Concrete and Mortars, ASTM International, West Conshohocken, PA, 2018, www.astm.org.

[18] BIS (Bureau of Indian Standards) (1970) IS 383-1970: Specification for coarse and fine aggregates from natural sources for concrete. Bureau of Indian Standard, New Delhi, India.

[19] Garcia-Lodeiro, A. Palomo, A. Fernandez-Jimenez, D.E. Macphee, Compatibility studies between N-A-S-H and C-A-S-H gels. Study in the ternary diagram Na2O–CaO–Al2O3–SiO2–

IOP Conf. Series: Materials Science and Engineering 998 (2020) 012055 doi:10.1088/1757-899X/998/1/012055

H2O Cement and Concrete Research 41(2011): 923–931.
[20] Fernandez-Jimenez and Palomo, Alkali-activated fly ashes: properties and characteristics. 11th International Congress on the Chemistry of Cement (Durban, South Africa), Vol. 3 (2003), pp. 1332-1340.

IOP Conf. Series: Materials Science and Engineering 998 (2020) 012056 doi:10.1088/1757-899X/998/1/012056

Optimization of Helical Coil Tube Heat Exchanger: A Systematic Review

Shaik Rafi[1], Dr.C.Sivarajan[2], Dr.Mallikarjun.Ch[3]
[1]Research Scholar, Department of Mechanical Engineering, Annamalai University
[2]Guide, Associate Professor, Department of Mechanical Engineering, Annamalai University
[3]Co-Guide, Principal, Department of Mechanical Engineering, A. M. Reddy Memorial College of Engineering and Technology

E-mail: shaikrafi340@gmail.com

Abstract. Helical coil heat exchanger is used in the food industries, air cooling industries etc.., due to its advantage of structural compact and high heat transfer coefficient. Many existing researches have been carried to improve the heat transfer efficiency and pressure drop. Mathematical model of the heat exchanger has been used to validate the performance of the heat exchanger. This paper involves in review the researches related to the optimization of helical coil tube heat exchanger. Some of the optimization methods such as Taguchi method, Genetic Algorithm (GA) and Multi-Objective Genetic Algorithm (MOGA) is used. From the analysis, Artificial Neural Network (ANN) and GA method has efficient performance in modelling and optimization of heat exchanger. Although, some methods involved in economic optimization of heat exchanger and these method has lower performance.

1. Introduction

Helically coils are vastly used as an efficient heat exchanger due to their compactness and high heat transfer coefficients. The helically coiled heat exchangers are used in numerous applications such as air conditioning systems, refrigeration systems, thermal power plants systems, chemical processing, food processing and medical equipment [1].The study of the thermal performance of these types of heat exchangers was extensively performed in the past years. Many researchers have studied thermal performance and frictional characteristics of these types of heat exchangers and proposed various correlations for Nusselt numbers and friction factors of both sides [2].Optimization of energy consumption is one of the most important subjects in the engineering and process design. Therefore, investigations on heat transfer enhancement (HTE) methods are quite important [3].The experimentally analyzed the effects of the geometrical parameters on heat transfer performance. And they found that the structural parameters have significant effects on flow and heat transfer [4].With the development of CFD technology and computers, the numerical investigations on heat exchangers are more and more popular [5].

Recently, various researchers have studied the thermal characteristics of these kinds of heat exchangers for use with different applications [6].They obtained the impact of various operational and geometrical parameters and proposed correlations for estimation of the Nusselt numbers, thermal and exergy efficiency, optimal and critical values of operational and geometrical parameters and entropy generation rate [7]. With advantages of larger heat transfer area, higher heat transfer coefficient, and more uniform heating, helically coil-tube heat exchangers can enhance heat transfer performance [8]. This phenomenon is not consistent with the entropy generation minimization principle. Thus, the optimization with only entropy generation rate as the optimization objective function has its limitations [9, 10]. In this paper, the researches related to the Helical Coil Tube Heat exchanger were reviewed. The analysis shows that the ANN and GA method has the higher efficiency in modelling and optimization of heat exchanger.

The paper is organized as review on Helical Coil Tube Heat exchanger is provided in section 2, comparative analysis of optimization of Helical Coil Tube Heat exchanger is provided in section 3 and conclusion is provided in section 4.

2. Review on Helical Coil Tube Heat exchanger

Helical Coil Tube Heat exchanger has been used in many industrial applications such as air conditioning system, food industries etc. Many researches have been carried out to improve the helical coil tube heat exchanger. These methods involves in applying mathematical model to validate the performance of heat exchanger. Artificial Neural Network is used to model the heat exchanger to analyse the effect of input parameters. The optimization methods such as GA, PSO is applied to find the optimal performance of the heat exchanger. Computational Fluid Dynamics (CFD)is applied to simulate the heat exchanger to evaluate the performance. The major objectives of the method are to find the optimal design parameter to provide higher heat transfer efficiency and lower pressure. The general block diagram of the helical coil heat exchanger is shown in Figure 1.

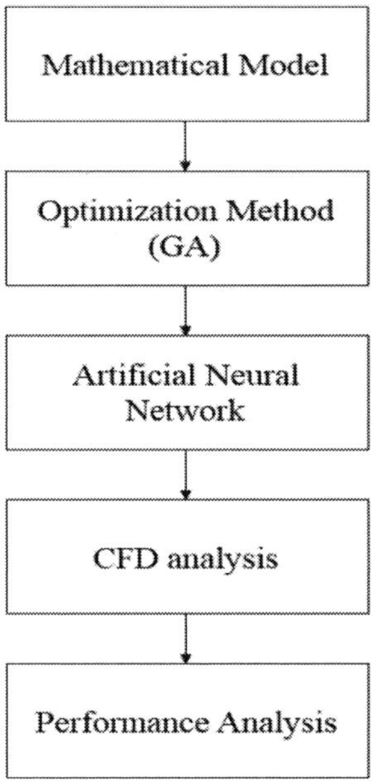

Figure 1. The general block diagram of optimization of Helical coil tube heat exchanger

Baqir, et al. [11] applied the optimization method to enhance the thermal performance and reduce the cost of the helical coil heat exchanger with air injection technique. The economic optimization is based on the optimal air flow and shell-side flow rate of the exchanger. The experimental result shows that parameters such as injected air flow, shell-side flow and coil-side flows has the impact on effectiveness. The effectiveness of the exchanger is increased in this method and has considerable economic performance. The efficiency of the method can be improved by using optimization methods such as Genetic algorithm and Particle Swarm Optimization.

Campet, et al. [12] optimize the internal shape of single-started helical coil tube heat exchanger. In order to improve heat exchanger efficiency, the inner surface of channel feature rounded rib. The Gaussian process regression and efficient global optimization are used in this method for optimization. The optimization method involves in improve the heat transfer efficiency for similar pumping power. The experimental result shows that the method increases the heat transfer efficiency. The mathematical model has been used in this research to analyse the performance. The optimization performance can be improved by efficient optimization method.

Sharifi, et al. [13] applied Artificial Neural Network (ANN) and Genetic Algorithm (GA) for the optimization of the heat exchanger. The ANN method was used to analyse the performance of the heat exchanger for various input variables. The GA is used to find the optimal performance of the heat exchanger in the method. The experimental result shows that the developed method has the higher performance in the optimization of heat exchanger. The experimental analysis shows that selection of wrong wire insertion affects the overall performance of heat exchanger. The optimization performance can be increased by using PSO method.

Mohapatra, et al. [14] uses the three fluid heat exchanger to analyze the performance of the design

IOP Conf. Series: Materials Science and Engineering 998 (2020) 012056 doi:10.1088/1757-899X/998/1/012056
parameters. Taguchi based optimization method was used to find the optimal parameter for the higher efficiency and lower pressure. The experimental result shows that flow rate of hot water is the important parameter for heat transfer and tube size is the important parameter for pressure drop. Effective optimization methods like GA can be used to improve the performance.

3. Comparative Analysis

Helical Coil Tube Heat exchanger has been used in many industries such as food, refrigeration, aerospace etc., due to its compact structure, ease of manufacturing and high heat transfer efficiency. Some of the researches involves in optimizing the Helical coil tube heat exchanger to improve the efficiency and reduces the cost. The recent research in the optimization of helical coil tube heat exchangers are reviewed in this section. The comparative analysis of the recent researches in helical tube heat exchanger is shown in Table 1.

Table 1. Comparative analysis of optimization method in Helical Coil Tube Heat Exchanger

Author(s)	Methods	Advantages	Limitations
Han, et al. [15] (2020)	Multi-Objective Genetic Algorithm (MOGA)	Design parameters are analysed under equal Reynolds number. Provides the better performance than existing method.	The energy saving is required to improve in the design.
Wang, et al. [16] (2019)	Two-layered Multi-Objective optimization	The experimental and numerical analysis was carried out. The analysis shows that the performance is improved in the method. The optimal structural parameter is provided in this method.	The performance of the optimization is required to be improved.
Gord, et al. [17] (2016)	The optimal geometry and operational conditions are analysed using second law of thermodynamics	Four important parameter are consider in this method for the optimization. To optimize the geometry, the fixed optimal condition is considered in this method.	The economic analysis is also can be considered to reduce the cost.
Alimoradi, et al. [18] (2017)	Numerical investigation of heat transfer intensification via installing annual fins on outer surface of the helical coil.	Two methods were used to analysis the design. The numerical analysis shows that the method has the considerable performance. For prediction of heat transfer performance, optimal cases and condition has been achieved.	Efficient optimization method can be applied to improve the performance.
Wang, et al. [19] (2018)	Numerical analysis has been made to study the effect of fin geometry and inlet mass flow rate.	The analysis shows that the developed method has the considerable performance.	The performance of the method is low and various parameters are required to be considered.
Miansari, et al. [20] (2019)	Numerical analysis has been carried out to evaluate the effect of circular grooves on exchanger. The grooves are created in different height for the analysis.	The geometric parameter of qualitative and quantitative analysis shows that the performance is increased. The optimal design parameter is analysed in this method.	Pressure drop is slightly affected by this method.

	Taguchi method was used to analyse the different parameters.		
Etghani, et al. [21] (2017)	Numerical analysis has been carried to analyse the exchanger. Four important parameter are considering to improve the performance. Taguchi method was used to find the optimal design performance.	The numerical analysis shows that the important parameter for optimal performance.	The performance of optimization is need to be increased.
HAN, et al. [22] (2019)	Multi-objective optimization is applied using Markowitz effective boundary theory. The Mathematical analysis has been carried out to evaluate the method.	The developed method has the higher performance than existing method. The error value of the developed method is low.	The performance of the optimization is low and need to consider various parameters.
Alimoradi, et al. [23] (2017)	Numerical model has been carried out to analysis the effect of parameters.	The method provide the optimal design for the model.	The performance of optimization is low and various parameters is required to consider.
Kareem, et al. [24] (2016)	The optimization and CFD is analysed in this method. The various cross-section is analysed in this method.	The simulation result shows that the developed method has the higher performance.	Pressure drop is required to be considered in this method.

4. Conclusion

Helical Coil Tube Heat Exchanger has been used in food industries, air conditioning industries etc., due to its advantages of structural compact and high heat transfer coefficient. Many researches have been carried out to improve the heat transfer efficiency and reduces the pressure in the heat exchanger. This paper review the researches involves in the optimization of the helical coil heat exchanger with its advantages and limitations. From the analysis, the methods such as ANN and GA has higher efficiency in modelling and optimization of heat exchanger. Some researches involved in economic optimization of heat exchanger to reduce the cost and improve the efficiency. These methods have the lower efficiency in the optimization and requires effective method for higher performance.

5. References

[1] Nada, S.A., Elattar, H.F., Fouda, A. and Refaey, H.A., 2018. Numerical investigation of heat transfer in annulus laminar flow of multi tubes-in-tube helical coil. *Heat and Mass Transfer* **54**715-726

[2] Alimoradi, A. 2017. Investigation of exergy efficiency in shell and helically coiled tube heat exchangers. *Case Studies in Thermal Engineering*, **10**1-8.

[3] Beigzadeh, R., Rahimi, M. and Parvizi, M. 2013. Experimental study and genetic algorithm-based multi-objective optimization of thermal and flow characteristics in helically coiled tubes. *Heat and Mass Transfer*, **49**1307-1318

[4] Wang, S., Jian, G., Xiao, J., Wen, J. and Zhang, Z., 2017. Optimization investigation on configuration parameters of spiral-wound heat exchanger using Genetic Aggregation response surface and Multi-Objective Genetic Algorithm. *Applied Thermal Engineering*, **119**, 603-609

[5] Kumar, P.M. and Chandrasekar, M. 2019. CFD analysis on heat and flow characteristics of double helically coiled tube heat exchanger handling MWCNT/water nanofluids*Heliyon***5**e02030.

[6] Gholamalizadeh, E., Hosseini, E., Jamnani, M.B., Amiri, A. and Alimoradi, A. 2019. Study of intensification of the heat transfer in helically coiled tube heat exchangers via coiled wire inserts. *International Journal of Thermal Sciences***141** 72-83.

[7] Kurnia, J.C., Sasmito, A.P., Shamim, T. and Mujumdar, A.S. 2016. Numerical investigation of heat transfer and entropy generation of laminar flow in helical tubes with various cross sections *Applied Thermal Engineering***102**849-860.

[8] Ferng, Y.M., Lin, W.C. and Chieng, C.C. 2012. Numerically investigated effects of different Dean number and pitch size on flow and heat transfer characteristics in a helically coil-tube heat exchanger. *Applied Thermal Engineering***36** 378-385.

[9] Shokouhmand, H. and Salimpour, M.R. 2007. Optimal Reynolds number of laminar forced convection in a helical tube subjected to uniform wall temperature. *International Communications in Heat and Mass Transfer***34**753-761.

[10] Wang, G., Dbouk, T., Wang, D., Pei, Y., Peng, X., Yuan, H. and Xiang, S., 2020. Experimental and numerical investigation on hydraulic and thermal performance in the tube-side of helically coiled-twisted trilobal tube heat exchanger. *International Journal of Thermal Sciences***153**106328.

[11] Baqir, A.S., Mahood, H.B. and Kareem, A.R., 2019. Optimisation and evaluation of NTU and effectiveness of a helical coil tube heat exchanger with air injection. *Thermal Science and Engineering Progress***14**100420

[12] Campet, R., Roy, P.T., Cuenot, B., Riber, É. and Jouhaud, J.C., 2020. Design optimization of an heat exchanger using Gaussian process. International Journal of Heat and Mass Transfer, 150, p.119264.

[13] Sharifi, K., Sabeti, M., Rafiei, M., Mohammadi, A.H., Ghaffari, A., Asl, M.H. and Yousefi, H. 2020. A good contribution of computational fluid dynamics (CFD) and GA-ANN methods to find the best type of helical wire inserted tube in heat exchangers. *International Journal of Thermal Sciences***154** 106398.

[14] Mohapatra, T., Padhi, B.N. and Sahoo, S.S., 2019. Analytical investigation and performance optimization of a three fluid heat exchanger with helical coil insertion for simultaneous space heating and water heating. *Heat and Mass Transfer*, **55**1723-1740.

[15] Han, Y., Wang, X.S., Zhang, H.N., Chen, Q.Z. and Zhang, Z. 2020. Multi-objective optimization of helically coiled tube heat exchanger based on entropy generation theory. *International Journal of Thermal Sciences***147** 106150.

[16] Wang, G., Wang, D., Deng, J., Lyu, Y., Pei, Y. and Xiang, S. 2019. Experimental and numerical study on the heat transfer and flow characteristics in shell side of helically coiled tube heat exchanger based on multi-objective optimization. *International Journal of Heat and Mass Transfer***137**349-364.

[17] Farzaneh-Gord, M., Ameri, H. and Arabkoohsar, A. 2016. Tube-in-tube helical heat exchangers performance optimization by entropy generation minimization approach. *Applied Thermal Engineering*, **108**1279-1287.

[18] Alimoradi, A., Olfati, M. and Maghareh, M., 2017. Numerical investigation of heat transfer intensification in shell and helically coiled finned tube heat exchangers and design optimization. *Chemical Engineering and Processing: Process Intensification*, 121, pp.125-143.

[19] Wang, J., Hashemi, S.S., Alahgholi, S., Mehri, M., Safarzadeh, M. and Alimoradi, A., 2018. Analysis of Exergy and energy in shell and helically coiled finned tube heat exchangers and design optimization. *International Journal of Refrigeration***94** 11-23.

[20] Miansari, M., Valipour, M.A., Arasteh, H. and Toghraie, D., 2020. Energy and exergy analysis and optimization of helically grooved shell and tube heat exchangers by using Taguchi experimental design. *Journal of Thermal Analysis and Calorimetry***139**3151-3164.

[21] Etghani, M.M. and Baboli, S.A.H., 2017. Numerical investigation and optimization of heat transfer and exergy loss in shell and helical tube heat exchanger. *Applied Thermal Engineering***121**294-301.

[22] Han, Y., Wang, X.S., Zhang, Z. and Zhang, H.N., 2020. Multi-objective optimization of geometric parameters for the helically coiled tube using Markowitz optimization theory. *Energy***192**116567.

[23] Alimoradi, A. and Veysi, F., 2017. Optimal and critical values of geometrical parameters of shell and helically coiled tube heat exchangers. *Case Studies in Thermal Engineering***10**73-78.

[24] Kareem, R.2017. Optimisation of double pipe helical tube heat exchanger and its comparison with straight double tube heat exchanger. *Journal of The Institution of Engineers (India): Series C*, **98**587-593.

IOP Conf. Series: Materials Science and Engineering 998 (2020) 012057 doi:10.1088/1757-899X/998/1/012057

Optimization of Time and Cost for a Research Project by Project Crashing Method

S Sharma[1], **N Bedi**[2] **and VK Sukhwani**[3]

[1] Department of Mechanical Engineering, Ujjain Engineering College, Ujjain, India
E-mail: shivamsharma1107@yahoo.com

Abstract. The aim of current study is to access trade off between time and cost for the duration of a research project. A linear programming model (LPP) was developed and used for the same purpose. Normal and Crashing Costs for the project activities were obtained from the principal investigator of the project and were used for conducting the analysis in the excel solver. Maximum and minimum duration for the project was obtained through critical path method and crashing method respectively. The result of the study indicated significant reduction in project duration with normal increase in project cost. A small increase in the project cost with reduction of nearly 45 days in the project interval was observed. Nearly 5% increase in cost of the project was noticed with reduction of 15% in total project duration.

1. Introduction

Project completion within minimum time incurring minimum cost is desirable in every project. Heuristic and mathematical programmings are used to achieve the same purpose. Genetic Algorithm, Fuzzy multi objective genetic algorithm (FMOGA) were observed to be Heuristic Approach whereas Linear Programming, Integer Programming and Dynamic programming were considered as Mathematical techniques [1]. The most reliable approach to analyze time-cost tradeoff is LP Solver technique was suggested by Ragsdale [2]. The use of excel solver to facilitate identification of activities to be crashed was also demonstrated by Mantel et al. [3].

Based on the past study and previous information about crashing project activities using Linear Programming model with excel solver, we had applied this technique to reduce project duration for a research project- *Development of heat transfer Correlation in micro channels* (a TEQIP Phase III sponsored project). In the current study, time cost tradeoff for the research project was conducted using Linear Programming method. The Critical Path and activities of the research Project was first identified. LP equations were prepared and solved using excel solver to obtain the optimal cost for reducing the project duration. The increase in cost due to crashing was also presented [4].

The technique used here is easy to understand and simple for implementation as well as simple to solve if large numbers of decision variables are present which is quite difficult to solve, if we solve them manually.

2. Activity Time-Cost Tradeoff

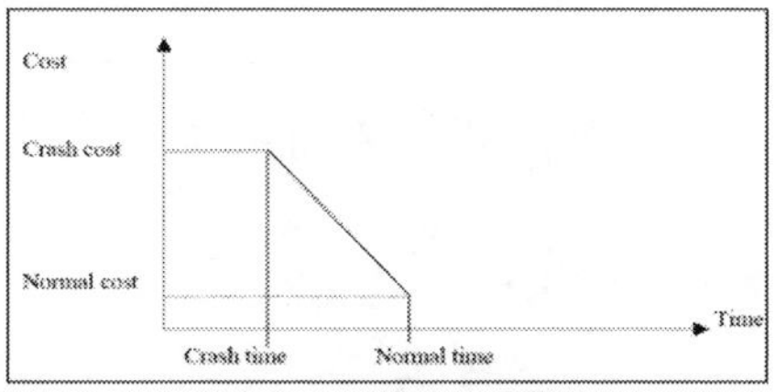

Figure 1: Time - cost trade-off for an activity.

From Fig.1 Slope of the activity can be defined as the slope of the line which is joining from Normal Time and Crash Time [5]. Mathematically, Slope of the activity or Cost Slope can be written as

$$\text{Cost Slope} = \frac{CrashingCost - NormalCost}{NormalTime - CrashingTime} \quad (1)$$

3. Problem Formulation for considered Prototype Example

Consider the following project data given in Table 1, obtained from our considered Project by discussion with the Project Manager. Here, the main objective is to Crash the Project duration from its earlier expected completion time. The following table gives data on Normal Time, Normal Cost, Crash Time and Crash Cost of our considered Project.

Firstly The Gantt Chart of Project activities is drawn as shown in Figure 2 for formulation of LPP model as stated by Perera S. [6]. In Gantt Chart along x-axis Dates (in dd/mm format) are represented and along y-axis activities are represented.

Table 1 : Data for Normal and Crash Times and their associated costs.

Activity Name	Activity	Predecessor	Normal Time (Day)	Crash Time (Day)	Normal Cost (Rs.)	Crash Cost (Rs).
Literature Reading	A	-	180	150	1500	4000
Formulation of Problem	B	-	15	10	300	400
Initialization of Procurement of Instrument	C	B	60	55	150000	180000
Further Procurement of Instruments during Setup	D	C	90	50	250000	275000
Setup of Experimental Apparatus	E	C	105	60	60000	72000
Validation of Experimental Setup	F	D,E	45	38	6000	7200
Data Generation	G	A,F	60	45	2500	3200
Conclusion & Results	H	G	15	15	0	0

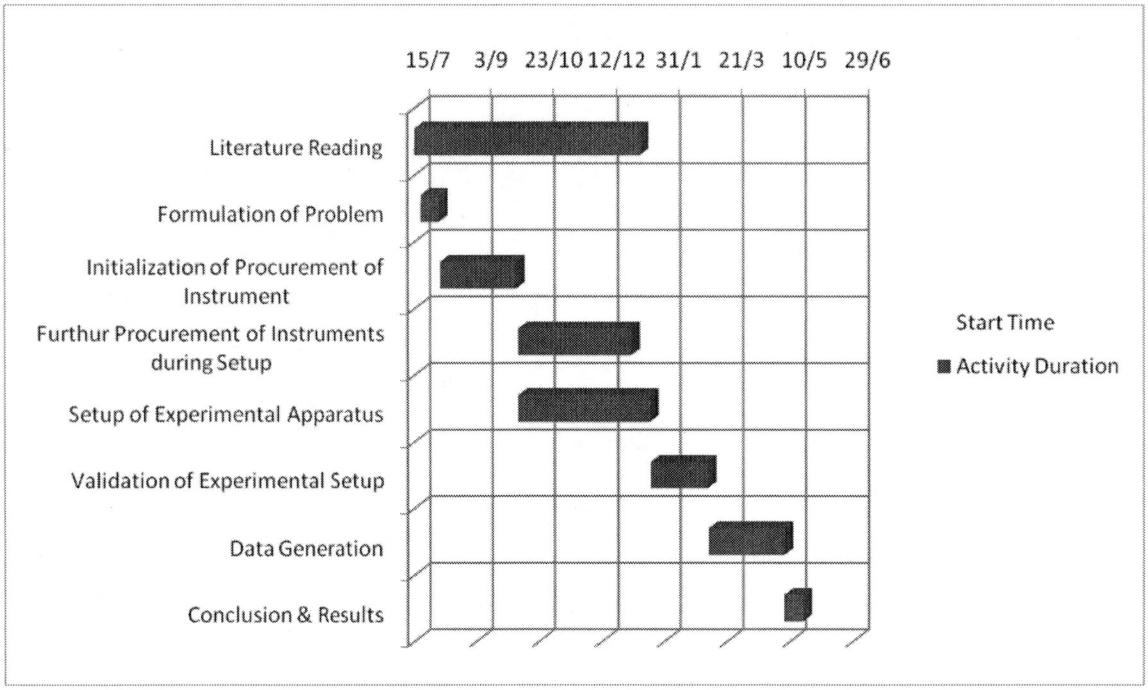

Figure 2: Gantt Chart of Project Activities.

By drawing the Project Network Diagram and Obtaining Earliest Start Time, Earliest Finish Time, Latest Start Time, Latest Finish Time, and Critical Path Length or Project Duration we had formulated LPP model. This data is represented in Table 2. The Project Network diagram is shown in Figure 3.

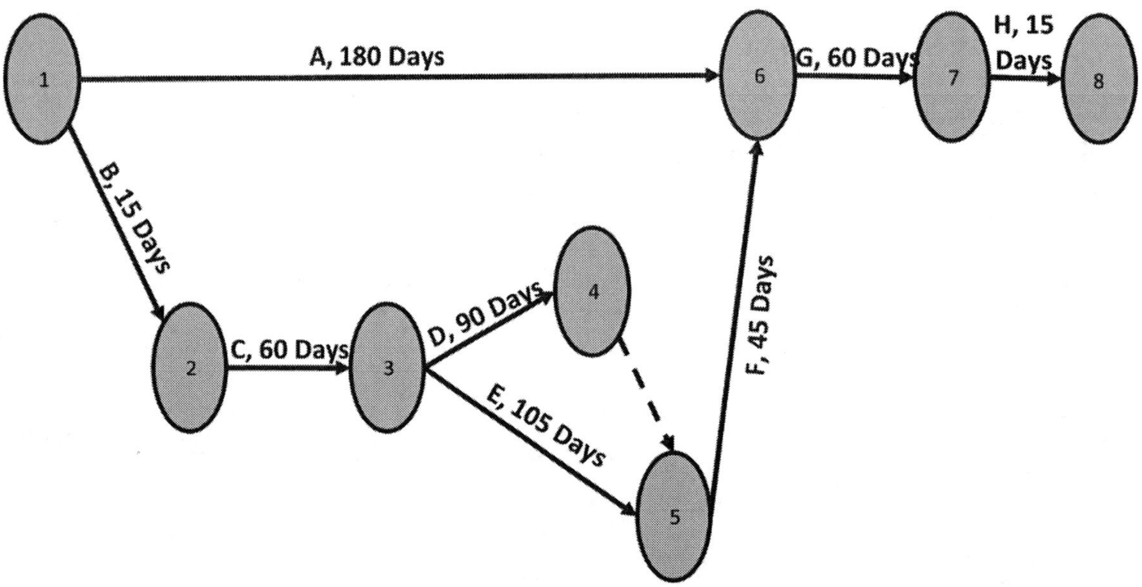

Figure 3: Network Diagram of Project Activities

Table 2: Calculation of EST, EFT, LST, LFT, and Slack.

Activity	Duration	Earliest Start Time	Earliest Finish Time	Latest Start Time	Latest Finish Time	Slack	Critical Path
A	180	0	180	45	225	45	
B	15	0	15	0	15	0	CP
C	60	15	75	15	75	0	CP
D	90	75	165	90	180	15	
E	105	75	180	75	180	0	CP
F	45	180	225	180	225	0	CP
G	60	225	285	225	285	0	CP
H	15	285	300	285	300	0	CP

From the data obtained in Table 2 and Fig. 3, we get Critical Path. The *Critical Path* of the Project is *B-C-E-F-G-H*. The observed duration of Project or *Length of Critical Path* is *300 days*.

4. Project Crashing by using Linear Programming Technique

In this technique we first construct a mathematical model of our project crashing problem by using Linear Programming for minimizing the total cost of crashing activities. If we estimate the project duration by drawing network diagrams like CPM or in Management software with exact activity duration we get it 300 days (i.e. before crashing).

Now, we are defining the decision variables, Let X is the time when an event occurs, it is measured from starting (or beginning) of the project. Therefore,

X_1 = time at which event 1 occurs
X_2 = time at which event 2 occurs
X_3 = time at which event 3 occurs
X_4 = time at which event 4 occurs
X_5 = time at which event 5 occurs
X_6 = time at which event 6 occurs
X_7 = time at which event 7 occurs
X_8 = time at which event 8 occurs
Y_A = No. of days activity A can be crashed.

Similarly, Y_B, Y_C, Y_D, Y_E, Y_F, Y_G, Y_H No. of days activity B,C,D,E,F,G,H can be crashed respectively.

IOP Conf. Series: Materials Science and Engineering 998 (2020) 012057 doi:10.1088/1757-899X/998/1/012057

Table 3: Maximum Crash Time and Cost-Time Slope of the activities.

Activity	**Normal Time**	**Crash Time**	**Normal Cost**	**Crash Cost**	**Max. Crash Time**	**Crash Cost per day**
A	180	150	1500	4000	30	83.33
B	15	10	300	400	5	20
C	60	55	150000	180000	5	6000
D	90	50	250000	275000	40	625
E	105	60	60000	72000	45	266.67
F	45	38	6000	7200	7	171.43
G	60	45	2500	3200	15	46.67
H	15	15	0	0	0	0

Objective Function:
The starting time of each activity is dependent on the starting time and duration of its immediate predecessors as we seen in Fig. 3. We also get crash cost per day (or cost slope) as shown in Table 3 by using equation (1). Using this cost slope and decision variable Y (i.e. number of days an activity should be crashed) we have developed the following objective function :

Minimize Crash Cost = $83.33Y_A + 20Y_B + 6000Y_C + 625Y_D + 266.67Y_E + 171.43Y_F + 46.67Y_G + 0Y_H$
Constraint: Based on data in Table 3 we have several Constraints.

Crash Time Constraint:

$Y_A \leq 30$
$Y_B \leq 5$
$Y_C \leq 5$
$Y_D \leq 40$
$Y_E \leq 45$
$Y_F \leq 7$
$Y_G \leq 15$
$Y_H \leq 0$

Start Time Constraint:

For 1st event, $X_1 = 0$
For 2nd event, $X_2 \geq 15 - Y_B$
For 3rd event, $X_3 \geq 60 - Y_C + X_2$
For 4th event, $X_4 \geq 90 - Y_D + X_3$
For 5th event, $X_5 \geq 105 - Y_E + X_3$
$X_5 \geq X_4$
For 6th event, $X_6 \geq 180 - Y_A$
$X_6 \geq 45 - Y_F + X_5$
For 7th event, $X_7 \geq 60 - Y_G + X_6$
For 8th event, $X_8 \geq 15 - Y_H + X_7$

Project Completion Constraint:

$X_8 \leq 255$ (Specified Deadline)

Non-Negativity Constraint: $X_1 \geq 0, X_2 \geq 0, X_3 \geq 0, X_4 \geq 0, X_5 \geq 0, X_6 \geq 0, X_7 \geq 0, X_8 \geq 0$
Similarly, $Y_A, Y_B, Y_C, Y_D, Y_E, Y_F, Y_G, Y_H \geq 0$.

5. Solution of the LPP using Excel Solver

We have defined the objective function and all the constraints for the problem stated in Kunpeng li et al. (2012) [7]. To determine the minimum crash cost we have to minimize the objective function. We can get the solution for this LP model by the help of Excel Solver. For the solution of LP model using Excel Solver, enter the objective function and constraints as shown in the Figure 4. This LP model can also be solved by MATLAB, STORM, LINDO or by other Linear Programming computer programs which are available in the market.

Decision Variable	y_A	y_B	y_C	y_D	y_E	y_F	y_G	y_H	x_1	x_2	x_3	x_4	x_5	x_6	x_7	x_8				
Value	0	4.758	0	19.13	39.337	6.3035	0	0	0	10.433	70.433	141.3	141.3	180	240	255		Objective Function		
Objective	83	20	6000	625	266.67	171.43	46.67	10000000	0	0	0	0	0	0	0	0		23622.07		
Constraint																				
Crash Time Constraint																				
Const. 1	1																	0	<=	30
Const.2		1																4.758019	<=	5
Const. 3			1															0	<=	5
Const. 4				1														19.13	<=	40
Const. 5					1													39.33719	<=	45
Const. 6						1												6.303478	<=	7
Const. 7							1											0	<=	15
Const. 8								1										0	<=	0
Start Time Constraint																				
Const. 9		1								1								15.1915	>=	15
Const. 10			1							-1	1							60	>=	60
Const. 11				1							-1	1						90	>=	90
Const. 12					1						-1		1					110.2072	>=	105
Const. 13												-1	1					0	>=	0
Const. 14	1													1				180	>=	180
Const. 15						1							-1	1				45	>=	45
Const. 16							1							-1	1			60	>=	60
Const. 17								1							-1	1		15	>=	15
Project Completion Constraint																				
Const. 18																1		255	<=	255
Non-Nagativity Constraint																				
Const. 19	1																	0	>=	0
Const. 20		1																4.758019	>=	0
Const. 21			1															0	>=	0
Const. 22				1														19.13	>=	0
Const. 23					1													39.33719	>=	0
Const. 24						1												6.303478	>=	0
Const. 25							1											0	>=	0
Const. 26								1										0	>=	0
Const. 27									1									0	>=	0
Const. 28										1								10.43348	>=	0
Const. 29											1							70.43348	>=	0
Const. 30												1						141.3035	>=	0
Const. 31													1					141.3035	>=	0
Const. 32														1				180	>=	0
Const. 33															1			240	>=	0
Const. 34																1		255	>=	0

Figure 4: LPP Model using Excel Solver (Screenshot of LP solution Procedure in Excel).

IOP Conf. Series: Materials Science and Engineering 998 (2020) 012057 doi:10.1088/1757-899X/998/1/012057

6. Results Obtained by Solving LPP

The Linear Programming Problem is solved by Excel Solver to determine the Minimum Crashing cost for reducing project duration. Figure 5 is the graph between activities on abscissa and duration of these activities on ordinates. This graph is representing the duration of activity before crashing and after crashing. The main aim of our calculations is to reduce the project duration with minimum crashing cost. The Solution for model is shown in table 4:

Objective	Final Value
Z	**23622.065**
Variable	Final Value
Y_A	0
Y_B	4.76
Y_C	0
Y_D	19.13
Y_E	39.33
Y_F	6.30
Y_G	0
Y_H	0
X_1	0
X_2	10.43
X_3	70.43
X_4	141.30
X_5	141.30
X_6	180
X_7	240
X_8	255

Table 4
Solution of the LP Problem using Excel Solver.

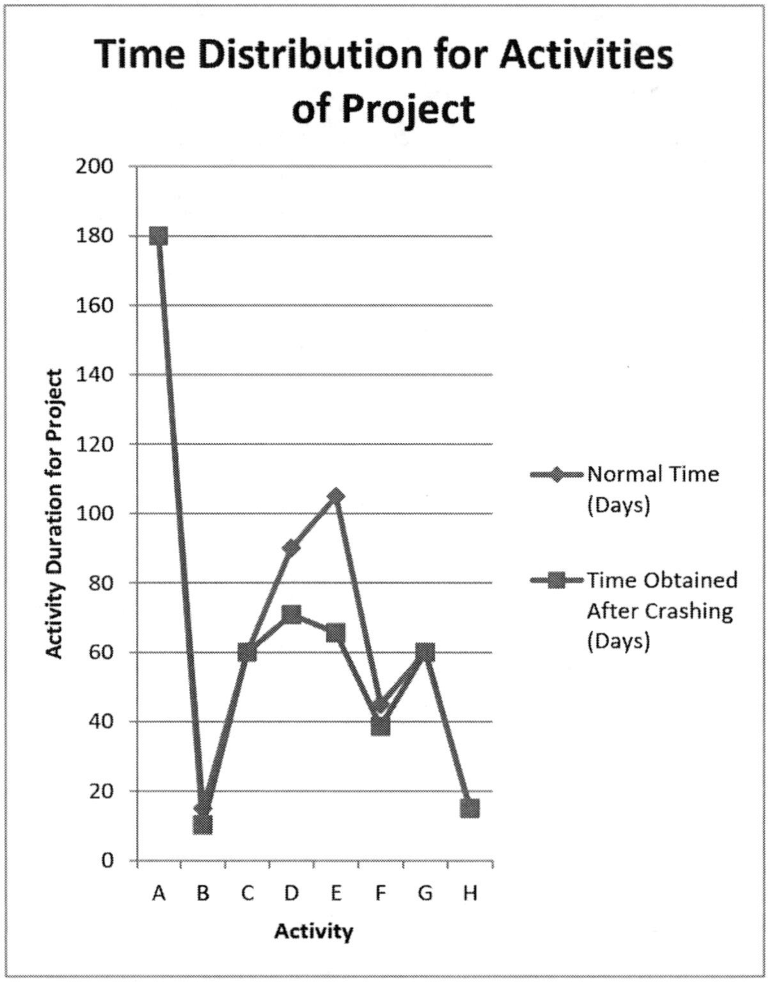

Figure 5: Time Distribution for each activity of Project

Our findings suggest that if project is completed without crashing i.e. in 300 days then the total cost of the project is Rs 470300 while if it is completed in 255 days by using crashing method then total cost of the project increases to Rs. 493922. The following Table 5 is showing these findings.

IOP Conf. Series: Materials Science and Engineering 998 (2020) 012057 doi:10.1088/1757-899X/998/1/012057

Table 5: Total Cost Before and After Crashing.

Project Duration	Direct Cost	Cost incurred due to Crashing	Total Cost
300 Days	470300	-	**470300**
255 Days	470300	23622	**493922**

7. Conclusion

From the above data, we can conclude that *Project Duration is reduced by 45 days* with the expenditure of Rs. 23622 in order to crash the activities of Project. The Project Duration can be further reduced but we have to spent more money to achieve that task.

Here, The Percentage Reduction in Project Duration is 15 % which is a significant value obtained on the expenditure of just Rs.23622. This increased crash cost is not affecting too much because here crashing cost is increasing but subsequently indirect cost (or running cost) associated with the project activities reduces for the 45 days. Therefore, the net cost required for reduction in project duration by crashing is optimum.

Further, The Project activities duration can be varied due to several external factors. So in future studies, we can convert this CPM Problem into PERT by taking activities duration (or Normal Time) as most likely occurring time, Optimistic times as 70% of normal duration and Pessimistic times as 125% of normal duration [8]. The PERT Technique follows Normal Distribution Curve. And then Analyze the above Problem in PERT. In PERT first we determine the new Critical Path and standard deviation in duration of each activity, and then measures Standard deviation (SD) of the project. By using this SD and Critical Path Length we can measure the Probability of the Completion of the project on the scheduled date.

8. References

[1] Karmaker CL and Halder P 2017 Scheduling Project Crashing Time Using Linear Programming Approach: Case Study *International Journal of Research in Industrial Engineering (In press)* **Vol 6 No. 4** 283-292

[2] Ragsdale C 2003 A new approach to implementing project networks in spreadsheets *INFORMS Transaction on Education* **3(3)** 76-85

[3] Mantel S, Meredith J, Shafer S, and Sutton M 2011 Project management in practice *Hoboken NJ: John Wiley & Sons Publishing*

[4] Diaby M, Cruz JM, Nsakanda AL 2011 Project crashing in the presence of general nonlinear activity time reduction costs *International Journal of Operational Research* **12** 318

[5] Vrat P, Kriengkrairut C 1986 A goal programming model for project crashing with piecewise linear time-cost trade-off *Engineering Costs and Production Economics* **10** 161–172

[6] Perera S 1980 Linear programming solution to network compression *J. Constr. Div.* **106** 315–326

[7] Li K, Shao B, Zelbst P 2012 Project Crashing Using Excel Solver: A Simple AON Network Approach *International Journal of Management & Information Systems (IJMIS)* **16(2)** 177-182

[8] Liu S-T 2003 Fuzzy activity times in critical path and project crashing problems *Cybern. Syst.* **34** 161–172

Applications of Lean tools for Compressor Assembly Line

Sanket Borgave[1] and Sagar Sapkal[2]
Mechanical Engineering Department, Walchand College of Engineering, Sangli, India
sanketborgave2484@gmail.com [1], sagarus1201@gmail.com [2]

Abstract. In last decades, competition between several industries to deliver high quality products is marginally increasing in terms of costs and performance. Lean manufacturing is one of those activities, which focus on the reduction of cost by eliminating non value adding activities. The general idea behind Lean manufacturing is to eliminate non value adding activities which again leads to eliminate waste. Higher flexibility, better quality and effective production system the manufacturer requires new ideas on production line. All the Lean tools are used to implement newer ideas into production line which improves the overall production process. This Paper addresses the application of lean manufacturing philosophy to the compressor assembly line of a well-known industry. In this paper, Failure Mode Effective Analysis, Value Stream Analysis and Takt time are the lean tools effectively integrated resulting in the improvement in terms of cycle time, space available, distance travelled by worker and productivity which leads to delivering the customer demand within time.
Keywords. Lean, Failure Mode and Effects Analysis (FMEA), Takt time, Cycle time, Kaizen

1. Introduction

The basic goal of any business is to satisfy the customer demand. Customer demand in the industry manufacturing the natural gas compressors is increasing day by day as the CNG market right now is on the peak. The present Gas Packaging and Process (GPP) line is unable to cope up with the increased demand due to its so many losses like more non-value-added time, old method related to material handling, improper material flow, less space and flexibility, more distance travelled by a worker in assembling station and bad ergonomics etc.

Lean manufacturing is a production practice which adds the value to a process and removes waste [1]. There are many non value adding processes on production line which can be overlooked, which increases the overall lead time of a production activities [2]. Due to competition between businesses, companies are facing more challenges and complexities. In order to respond to the changes by customer, to strengthen the product value in the organization which may be service oriented or manufacturing oriented, systematic production line with minimum non value adding processes is desired [3]. So, there is need of more value adding processes than that of non value adding processes to achieve this target. Therefore, Lean manufacturing implementation is becoming a need for any type of organizations to sustain [4].

Failure Modes and Effects Analysis (FMEA) is a technique used in product development and operations management for analysis of failure modes within an organization to classify and rate the processes in three factors as severity, detection and occurrence of the failures [5]. Risk Priority Number (RPN) is calculated where the failure occurs every time and based on these numbers, necessary action is taken.

IOP Conf. Series: Materials Science and Engineering **998** (2020) 012059 doi:10.1088/1757-899X/998/1/012059

Implementation of lean tools and its principles leads to change in organizational cultures, higher efficiency, better productivity, cost reduction and customer satisfaction. Also we use FMEA to avoid any failures in production processes, Also to detect the problem and its intensity and deciding the priorities of the failures which helps the organization to eliminate these failures in the most economical way[6]. Today the organizations are not depending upon their single way of solution the are interested in the integrations of theses tools so the efforts are being made by various researchers to integrate the Lean tools and FMEA [7]. The same efforts are made in this paper in order to make the assembly line leaner and deliver the customer demand.

DeSouza and Carpinetti [5] explained a proposal of adaptation of the FMEA method which analyses wastes and define priorities for the actions aiming at minimization or elimination of these wastes based on the criteria of occurrence, severity, and detection. Shekari and Fallahian [6] implemented Lean and FMEA together, in such a way that not only specific ability of each method, but also combination of these methods raised the overall performance of the organization. Amin and Karim [7] developed a method which evaluates the applications of lean tools selected to reduce production wastes within the assemblers time constraints. Nguyen and Do [8] transformed a traditional production line model to lean one of electronics assembly line with the help of mapping current state value stream and drawing future state value stream mapping in order to establish and implement plan, resulting in reduction of nearly 40% manpower with the same output daily, also saved 30% floor area on the line and reduced delivery time from 7.5 days to 3 days. Sapkal and Jadhav [9] have implemented Kanban and VSM in Residual Current Circuit Breaker manufacturing company and concluded that this approach helps to analyse loopholes in the current system. By implementation of this approach, they have achieved lead time reduction from 16.824 days to 8.59 days and about 18% reduction in processing time

By studying various literature regarding the lean implementation techniques, it has been found that for manufacturing and processing industries, possible and easy way is to determining the non value-added activities and implementation of effective lean tools like FMEA and VSM, which eliminate bottlenecks from the system. Researchers focused on removing non-value-added activities from every department of the organization like Engineering, Production, warehouse and the outsourcing. The current GPP line is facing similar kinds of problems that have addressed by researchers in their work. Therefore, integration of lean implementation tool with effective use of FMEA can be used for solving such problems

2. Data Collection

In the factory of compressor assembly Atlas Copco, there are total 5 stations for assembly line out of which station 1 is only for block assembly where there are 2 fixtures. And other 4 stations are for full package assembly where at each station no. 2,3,4,5 separate compressor assembly carried out manually. Data has been collected at each station in terms of Cycle time, Flexible space available at each station and distance travelled by worker. The time studies of each assembly process were taken in terms of videography analysis with the help of hand and with support of company staff and operators. This video analysis is used to analyse the task times. So, the time analysis has been done for the block area using Avix software. Table 1 shows the detailed data collected from the industry in terms of cycle time, value added time, non value added time, space available and average distance travelled by worker, for block assembly and package assembly.

2.1 Value adding activities

Value adding activities are activities, which transform the product towards completion, particular step is done right for the first time, the customer generally pays for this step to be done [10].

In CNG compressor assemblage of special parts like Block, coolers, vessels, instruments, priority panel, driving unit etc. adds the value to the machine for which customer is paying for the So, the value adding time is more precise time in the assembly system. So, there must be no compromise for this time. The worker has to take this time to complete the assembly.

IOP Conf. Series: Materials Science and Engineering **998** (2020) 012059 doi:10.1088/1757-899X/998/1/012059

2.2 Non Value-Adding Activities

Non value adding activities are activities that actually do not add value to the machine. These are further divided in 3 parts as Required, Wait, and Loss. Addition of all these times is considered as non-value adding time.

2.2.1 Required - These activities are the one which must be done, but they do not actually add value to the customer. These activities are cleaning of the parts, fitting the nut, bolt, washer, Studs. Also, the time required for handling the material.

2.2.2 Wait - Waiting caused by resetting, bad process flow or internal balancing losses. Waiting occurs when the operator waits for the product to be assembled to be arrived or the material handling equipment to be arrived. This is one of the most conspicuous wastes and it does not add value to the customer.

2.2.3 Loss-These are due to excess motion of the material handling equipment like forklift, overhead crane. Also, worker searching the part to be assembled and the tool finding to complete the assembly.

In order to implement the lean tools in an organization the main target should be to eliminate or to reduce the Non value adding time which helps the factory to deliver the goods to customer within time [10]. Table 1 shows the data collection at initial stages of the assembly line. Cycle time is the time required actually to complete the process or step. Which again is divided into value adding and non value adding time which helps us to understand how much actually non value adding time which we are going to reduce or eliminate. This step is done in AVIX software where the videography of all the steps is imported and divided accordingly.

Table 1. Data Collection

	Cycle time (min.)	VA time (min.)	NVA time (min.)	Space available (m^2)	Average dist. Travelled by worker (m)
Block Assembly	2086	565.63	1520.75	37.66	982
Package Assembly	4320	1606	2374	12.51	660

3. Takt Time Calculation

Takt time is a function of available production time and product volume[10]. In this, it is used to calculate the rate at which we are supposed to produce the product in order to satisfy the customer demand. It is determined by using formula [11]:

$$\text{Takt time} = \frac{\text{Available time}}{\text{Customer demand per day}} \quad (1)$$

The takt time and cycle time are different concepts and should not be mixed with each other. Takt time remains same for all stations on an assembly line, which means time for which work piece should stay at a station. Cycle time is process or operation time of product that is completed while the workpiece is at the station.

Customer is demanding 12 machines per month so demand for Package station will be 12 for all 4 stations. From this the demand per station per month is calculated as 3. Similarly, for block station for preparing 12 machines 12 blocks are required so demand for the block i.e. station 1 is 12 machines per month. On this station there are two fixtures on which two workers works on two different blocks so, demand per fixture per is 6. Table 2 shows the detailed takt time calculation with regarding the customer demand per month which gives the rate at which we are supposed to produce the product.

Table 2. Takt time calculation

	Station 1: Block		Station 2: Package	
Customer demand per month	12		12	
Customer demand per station/fixture	6		3	
Demand per day Considering average 24 working days per month	$\frac{6}{24} = 0.25$		$\frac{3}{24} = 0.125$	
Available Time			Hours	minutes
	Total working hours Per day.		8.5	510
	Contractual time per day		1.67	70
	Available working time per day		7.3	440
Takt time	$\frac{440}{0.25} = 1760$ min		$\frac{440}{0.125} = 3520$ min.	

4. Comparison of Takt and cycle time

As shown in figure 1, the takt time compared with cycle time and is found that the takt time is more for both block and package station which is not desired. In order to fulfil customer demand, the cycle time must be equal to or less than that of takt time.

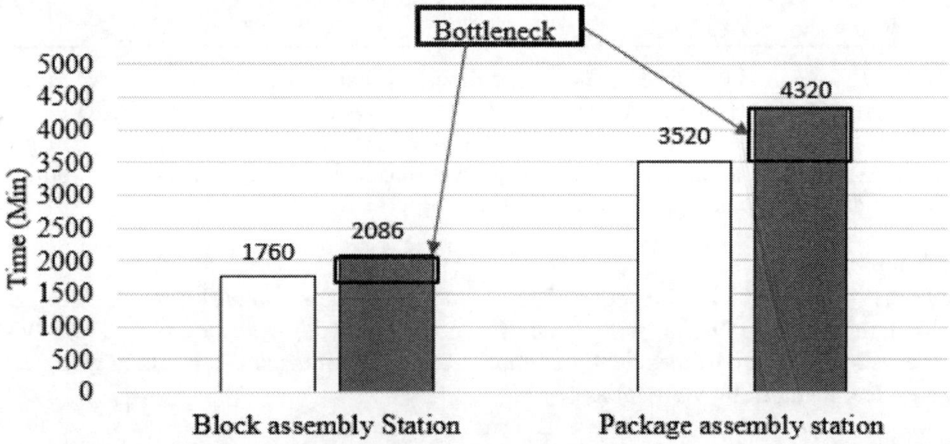

Figure 1. Comparison of Takt time and Cycle time

5. Implementation of Lean Tools

After Collection of data collection and finding the bottlenecks in the current system, the lean tool to is be implemented in order to remove the bottlenecks from the system to satisfy customer demand. The FMEA is tool is used to do so. Risk Priority Number (RPN) is calculated to prioritize the cause of failure and wates. It develops the FMEA solution and the action which is necessary, based on the RPN data.

$$\text{Severity x Occurrence x Detection} = \text{RPN} \tag{2}$$

IOP Conf. Series: Materials Science and Engineering **998** (2020) 012059 doi:10.1088/1757-899X/998/1/012059

5.1 Rating Scales

Severity: Rating 1 to rating 10 = Not Severe to Very Severe
Occurrence: Rating 1 to rating 10 = Not Likely to occur to Very Likely occur
Detection: Rating 1 to rating 10 = Likely to Detect to Not Likely to Detect

All three above factors are based on the importance of the product to the customer and how much a failure affects the quality of the product and is decided in co-ordination with managing team of the industry [2].

Table 3. FMEA Analysis for Package assembly

Sr .No	Assembly Process	Severity	Occurrence	Detection	RPN Number
1	Keeping the Frame, Driver unit and block	7	4	4	112
2	**Alignment of the Engine and block**	**10**	**9**	**6**	**540**
3	Driver accessories	6	6	3	108
4	**Vessel Mountings with fittings**	**8**	**9**	**7**	**504**
5	Cooler mountings with fittings (4 coolers)	4	6	5	120
6	**Plenum mounting with fan and motor**	**9**	**9**	**5**	**405**
7	Actuator with their Fittings	3	6	7	147
8	Instruments with their Fitting	4	9	5	180
9	Gauge Panel and gauges	2	5	6	60
10	Canopy with their Fitting	6	5	7	210
11	Priority Panel with their fittings	5	5	4	100
12	Factory Assured Test	8	8	6	384

Table 3 shows the detailed assembly process for block station and the factors allowed to each of the activities. The next step is prioritizing cause of failure and is done by calculating RPN number. The Process having higher Risk Priority Number are targeted first to deliver the product of good quality within the time. Activities at sr. no. 2, 4, and 6 in the table 3 show the higher priority number and are targeted first and Kaizen to be implemented for these processes.

5.2 Kaizen Implementation

Once the priorities decided with the help of FMEA different feasible solutions to the different targets in order to implement kaizen, which will reduce the non-value adding time are decided. Best solution amongst them considering cost, efficiency is implemented for all targeted activities with permission of the managing team which shown in table 4.

Table 4. Kaizen Implementations

Process	**Kaizen**
Alignment of Engine and Block	A LASER alignment machine is implemented instead of previous dial gauge alignment method which reduced time for alignment from 470 min. to 180 min.
Vessel Mounting	A Fixture is developed which carries all 6 vessels in it instead of 6 different pallets for different vessel resulting in time reduction to travel the vessel and space reduced of pallets
Plenum Mounting with fan and motor	An assembly of Plenum, Fan, Motor is outsourced, instead different parts assembling within factory. This reduced assembly time and space occupied within station.

IOP Conf. Series: Materials Science and Engineering **998** (2020) 012059 doi:10.1088/1757-899X/998/1/012059

Also, movable Cluster trolley implemented in place of big lineside racks which resulting in reduction in distance travelled by worker and that of space lineside for block as well as packaging stations. Also, wooden pallets are been replaced with movable steel pallets with steps helps reduction in space.

6. Results and Discussion of Lean Implementation

Once the Kaizen is implemented for all the targeted activities there is again need to collect the data in order to determine the kaizen results, so again data is collected as previous. Table 5 shows the data collection after kaizen and lean implementation.

Table 5. Data collection after Lean implementation

	Cycle time (Min)	VA time (Min)	NVA time (Min)	Space available (m^2)	Average dist. Travelled by worker (m)
Block Assembly	1542.02	563.62	978.40	51.84	487
Package Assembly	3291	1606	1705	30.08	320

6.1 Improvement in Cycle time

As shown in figure 2, the takt time compared with cycle time after the process improvement by lean and is found that the cycle time is less than that of takt time for both block and package station which fulfill the factory to customer demand. Also, there is some buffer time which the factory can sustain for further increase in customer demand.

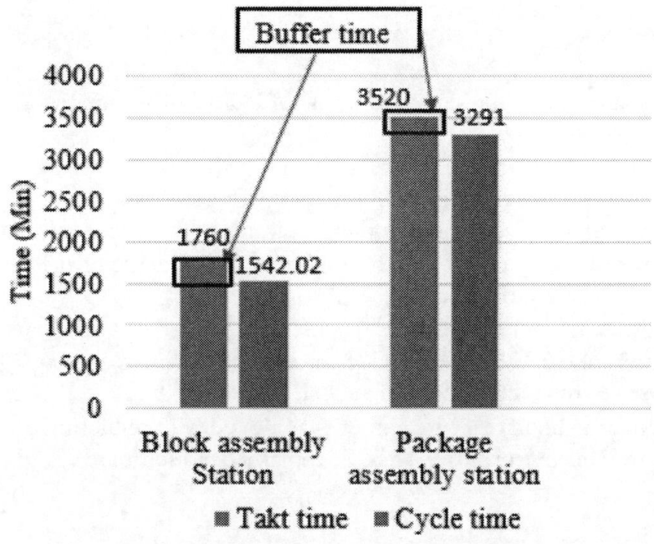

Figure 2. Improvement in Cycle Time.

6.2 Improvement in other Factors

Other factors such as non-value adding time, flexible space available, Avg. distance travelled by worker before and after lean implementation is analyzed and resulting in marginal improvement in all these factors.

Table 6. Improvement in Factors

Factor	Block Section		Package Section	
	Before implementation	After implementation	Before implementation	After implementation
Takt time (min)	1760	1760	3520	3520
Customer demand (per Month)	12	12	12	12
Cycle time(min)	2086.5	1542.02	4340	3291
Non-Value adding time(min)	1520.75	978.40	2374	1705
Flexible space available(m2)	37.66	51.84	12.51	30.08
Dist. Travelled by worker(m)	982	487	660	320

Table 6 shows the improvement factors under consideration in both block assembly and package assembly. This shows 26% reduction in cycle time this is due to reduction in non-value adding activities which is achieved with effective kaizen. This non-value adding time is marginally reduced by 35%. Also, by putting effective fixtures for material handling and putting cluster trolley for bins, flexible space is improved by 37% and distance travelled by worker is reduced by 50%. Improvement in all these factors helped the product to be delivered within time.

7. Conclusions

Lean manufacturing philosophy implemented in a GPP line of compressors to eliminate non-value adding activities.

- By using Lean tools like Value Stream Analysis, FMEA, Takt time and effective use of kaizen, the performance of the organization is found to be improved in terms of cycle time, flexible space available, and distance travelled by worker resulting in the improvement in productivity.
- FMEA tool is successfully used to analyse waste and define the priorities for the processes aimed at minimizing the wastes effectively.
- For Block section and Package section, Cycle time reduced by 25.8% and 24.1%, NVA activities reduced by 35.6% and 28.1% , Flexible space available is increased by 38% and 140%, Distance travelled by worker reduced by 50.4% and 51.5%.
- Monthly customer demand of 12 machines is achieved with reduction in cycle time by comparing it with takt time and extra buffer time is remained which can sustain for further increase in demand.

Acknowledgement

Authors are thankful to industry Atlas Copco for providing the facilities and necessary support to carry out this research work.

IOP Conf. Series: Materials Science and Engineering **998** (2020) 012059 doi:10.1088/1757-899X/998/1/012059

References

[1] Bhamu J and Sangwan K 2014 Lean manufacturing: literature review and research Issues, *International Journal of Operations & Production Management* **34 (7)** pp 876-940

[2] Kjersem K, Halse L, Kiekebos P and Emblemsvag J 2015 Implementing Lean in Engineer-to-Order Industry: A Case Study, *International Federation for Information Processing- Part II* **460** pp 248–255

[3] Vinodh S and Chethan Kumar A G 2015 A Case Study on Lean Product and Process Development', J.P. Davim (ed.), *Management and Industrial Engineering*, pp 17-30

[4] Rahani A R and Al-Ashraf M 2012 Production Flow Analysis through Value Stream Mapping: A Lean Manufacturing Process Case Study, *Procedia Engineering* **41** pp 1727 – 1734

[5] B. de Souza R and Carpinetti L R 2014 A FMEA-based approach to prioritize waste reduction in lean implementation, *International Journal of Quality & Reliability Management* **31** pp 346-366

[6] Shekari A and Fallahian S 2007 Improvement of Lean methodology with FMEA, *POMS 18th Annual conference Dallas. Texas, U.S.A.* **007-0520**

[7] Al Amin M & Karim M A 2013 A time-based quantitative approach for selecting lean strategies for manufacturing organisations, *International Journal of Production Research* **51(4)** pp 1146-1167

[8] Nguyena M, Do N 2016 Re-engineering Assembly line with Lean Techniques *Procedia CIRP* **40** pp 590-595

[9] Sapkal S and Jadhav R 2015 Productivity Improvement by using Value Stream Mapping in Manufacturing Company, *3rd International Conference on Industrial Engineering* **1** pp 496-50

[10] Sundar R, Balaji A N and Satheesh Kumar R M 2014 A Review on Lean Manufacturing Implementation Techniques, *Procedia Engineering* **97** pp 1875 – 1885

[11] Rajenthirakumar D and Sridhar R 2014 Development of Lean Assembly Line – A Case Study, *International Journal of Lean Thinking* **5 (1)**

An Emphasis of Geopolymer Concrete with Single Activator and Conventional Concrete with Recycled Aggregate and Data Analyzing using Artificial Neural Network

G. Mallikarjuna Rao[1], K. Satheesh Kumar[2], Kiran Kumar Poloju[3], Kota Srinivasu[4]

[1] Vardhaman College of Engineering, Department of Civil Engineering, Shamshabad, Hyderabad, 501218, India, Email: gmyadav25@gmail.com

[2] Vardhaman College of Engineering, Department of Civil Engineering, Shamshabad, Hyderabad, 501218, India, Email: sateeshkumar851@gmail.com

[3] Middle East College, Department of Civil Engineering, Al Rusayl, Muscat, 124, Oman, Email: kpoloju@mec.edu.om

[4] NRI Institute of Technology, Guntur, 522006, India, Email: kota.srinivasu@gmail.com

ABSTRACT

The emphasis of concrete in the advancement of construction industry is essential. Due to the rapid increase in the development of infrastructure, there is a considerable demand for the construction materials. The construction materials drained gradually due to continue use in construction and there is a degree for the advancement of these development materials. Therefore, in this context, a rapid increase in the price of conventional construction materials and various governments have put restrictions on fine aggregates. In this circumstance, the investigation attempted to determine the appropriateness of the substitute materials in concrete development because these materials are easily available at very low cost as compared to conventional fine and coarse aggregates. The effect of recycled aggregates on the strength of conventional concrete and Geo -polymer concrete was studied. The natural aggregates are replaced by the crushed aggregates of intervals of 10%,20%, 30%, 40 %,and 50% in both geo-polymer and conventional concrete. The different strengths of 20MPa, 40MPa, and 60MPa werestudied in this research work. The conventional concrete cubes were cured in normal water and geopolymer concrete cubes cured in ambient room temperature. The coarse aggregates used were of size 20mm and 10mm sieved and the recycled aggregates contains a cement mortar attached to it. The outcomes show that a slight decline in the strength of conventional concrete and high strength found with 30% substitution of coarse aggregate and eventually the prediction of data was analyzed using Artificial Neural Network

Keywords: *Geopolymer concrete, Fly Ash, GGBS, Compressive Strength, Geo-activator and Artificial Neural Network*

1. INTRODUCTION

The reused coarse aggregate acquired by dismantled concrete were utilized for concrete production. Four distinctive reused total aggregates were delivered; made with 0%, 25%, 50%, and 100%, respectively **[1]**. The blend extents of the four concretes were structured to accomplish a similar compressive strength. The shape and surface of the aggregates can likewise influence the functionality of the concrete **[2]**. Crushed concrete as partial replacement accomplishes a high level of reused coarse aggregate without followed by mortar. Concrete made with 100% of coarse reused aggregate requires highmeasures of cement to accomplish a high compressive strength and subsequently is not a monetary recommendation as it is not practical. Concrete made with 100% of reused coarse total has

IOP Conf. Series: Materials Science and Engineering 998 (2020) 012060 doi:10.1088/1757-899X/998/1/012060

20 to 25% less pressure quality than traditional cement at 28 days, with a similar viable w/c proportion and concrete amount. **[3],** learned about certain boundaries like compressive strength, ultrasonic velocity, shrinkage, water absorption tests. The outcomes demonstrated a diminishing compressive strength towards the significant level of the reused aggregate substance that is because of the low quality of the followed mortar. It was seen from the analysis that reused total cement was "good" as per UPV measurable values.**[4]** considered the mechanical properties, durability studies, and the basic presentation of reused aggregates for around 10 years (1996-2011). He discussed the acquired outcomes and the consequences of ordinary cement. The perceptions uncovered that the aggregate of – concrete grid interfacial zone of reused aggregate composed of free and permeable hydrate. Regarding the toughness properties, the carbonation and chloride resistance were lower when contrasted and the ordinary cement. For the components, for example, shrinkage and creep, indicated increased for the conventional concrete. **[5]** Consider the designing and strength properties of fly ash based geopolymer reused aggregate concrete and the outcomes were introduced in this paper. The increase of reused coarse aggregate (RCA) recovered from construction and demolition(C&D) indicated promising capacity in development industry as an option in contrast to regular aggregates. It monitors gigantic amounts of common assets and decreases the space required for the landfill removal of C&D squanders. In this examination and augmentation of 25% halfway substitutions by weight of characteristic aggregates with reused aggregates in geopolymer concrete, up to 100% substitutions were considered. The new and mechanical properties of all the over four concrete blends have been researched. The results showed that the functionality of geopolymer solid reduction than normal concrete and it took over 24 hours to set. Geo polymer-based reused aggregate concrete shows preferable quality and solidness execution over standard reused total cement. **[6].** The product not just gives general capacity to strength examination and improvement alongside an intelligent easy-to-use interface, yet additionally incorporates one of a kind of highlights, for example, artificial neural network modelling, network &multithreaded computing, and user-accessible programming modules. Visual Gene Developer is helpful for the two analysts who need to rapidly examine and improve strength, and the individuals who are keen on creating and testing new calculations in bioinformatics. **[7]** Presents an investigation of the impact of reused aggregate on the strength and durability properties of fly ash-based geo-polymer concrete. Geo-polymer concrete examples were set up from high calcium fly ash, sodium silicate and sodium hydroxide arrangement, fine aggregate from river sand, and two types of coarse aggregate, and the results demonstrated that reused aggregate can be utilized as a coarse aggregate in high calcium fly ash geo-polymer cement with the 7-day compressive strength of 30.6–38.4 MPa which were marginally lower than those of high calcium fly ash geopolymer concrete with demolished limestone. **[8]** tested the consequences of reused concrete aggregate and the results are contrasted and the characteristic RCA. The fine aggregate utilized in the concrete is both recycled and conventional aggregate. The outcomes shows the compressive, tensile, and flexural strength of reused aggregate are normal from 85% to 95% of the characteristic aggregate. **[9]** Artificial neural systems (ANNs) are generally considered as software that can assist with examining cause-impact connections in complex frameworks inside a major information structure. In this circumstance, a specific neural system apparatus, which can deal with little datasets of exploratory or observational information is shown that can help in distinguishing the principal causal elements prompting changes in some factors. Clear details of ANN are provided, its applications to specific study are discussed. There is an extensive use of concrete. Thus, the consumption of natural sand is also very high globally. Generally, in developing countries, the demand for natural sand is comparatively very high to satisfy the rapid growth in infrastructure, whereas, in the same situation, India being a developing country is lacking better quality natural sand. Mainly in India, depletion of the deposits of natural sand causes a serious threat to the society and the environment. As the extraction of natural sand from the riverbeds goes on increasing, it creates many problems. Some of

the problems include loss of vegetation on the riverbanks, loosing water, retaining sand strata, causing landslides due to the deepening of the river courses. This also affects the aquatic life and the agriculture as the level of the underground water is lowered largely. In the earlier decades, the cost of construction used to be relatively high due to the variable cost of the natural sand that is used as a fine aggregate in concrete. Considering this situation, a research has been initiated to find out an alternative substance to the natural sand, which is comparatively less expensive and available easily. In extension to this, the data prediction using Artificial Neural Network(ANN) is also studied. Data analysis and data prediction is a game changer in this modern world where ample work is needed to be done in a short time. Based on the previous statistics and data available, we can generate the new data of our needs. Which saves time, energy, and cost. Based on the algorithm of ANN, we can predict the data and compare the experimental values and analytical values. Moreover, single activator in geopolymer mortar is a highly effective, innovative and environmentally friendly material that could affect strength of geopolymer mortar compared to alkaline activator [10] The Molarity increases the amount of sodium hydroxide increases and resulted increase in strength of mortar. Few specimens were casted with 1:3 proportion using single activator and combined solution (Mix of sodium hydroxide and sodium silicate solution of ratio 2.5 with 8 Molarity) and tested compressive strength after 1, 3, 7 and 28 days under ambient curing. Besides, UPV test is done for the specimens and compared. The test results displayed that a single solution cube specimens produced better early strength properties compared to other specimens. Since there are lack of literature on the topic, less study are discussed in this paper

2. EXPERIMENTAL PROGRAMME

Materials Used in the Investigation

Fly ash: fly ash is a byproduct of coal obtained from the La Mangada Thermal Power Plant (NTPC). In this experiment, we use class F fly ash as the source material.

GGBS: The crushed blast furnace slag is a byproduct of the steel processing plant that replaces the cement obtained from Toshali Cement Pvt. Ltd Bayyavaram, India.

Fine aggregate (stone dust): The fine aggregate used in this investigation is stone dust which is grey in colour, with a specific gravity of 2.6 conforming to IS 2386 (part-3)-1963. Water absorption for stone dust is 0.6% conforming to IS 2386 (part-3)-1963

Coarse aggregate: Natural granite aggregate having specific gravity of 2.8 and water absorption of 0.1%conforming IS 2386 (part 3). Recycled aggregate demolished from buildings having specific gravity of 2.19 and water absorption of 0.2% conforming IS 2386 (part 3). Aggregate passing through 12mm IS sieve and retained on 10mm IS sieve is taken in the entire investigation as shown in Table 1.

Table 1: Specific gravity and Water Absorption Tests for RCA and Natural aggregate

TESTS	RECYCLED AGGREGATE	NATURAL AGGREGATE
Specific Gravity	2.35	2.73
Water Absorption (%)	2.89	2.19

Geo activator: In this experiment, only a single alkaline liquid was used, namely, geo activator replacing alkaline activator (Mix of sodium hydroxide and sodium silicate). Geo activator Solutions of silica modulus SiO_2:Na_2O (M_s) =2.92:1 with 28.98% SiO_2, 9.92% Na_2O by weight, which was produced from KIRAN GLOBAL Ltd, Chennai India.

Super plasticizer: To improve the workability of fresh concrete naphthalene based superplasticizer was added to the mixture.

Mix Proportions

Different grades of concrete of MIX (A)20MPa, MIX(B)40MPa, and MIX(C)60MPa were performed with replacement of coarse aggregate with crushed aggregate of 0%, 10%, 20%, 30%40% and 50% as shown in Table 2 and Table 3

Table 2: Mix proportions of Conventional Concrete

Mix	Cement(kg/m^3)	Coarse aggregate(kg/m^3)	Fine aggregate(kg/m^3)	Water (kg/m^3)
Mix A (20MPa)	396	1216	608	198
Mix B (40 MPa)	450	1380	428	180
Mix C (60 MPa)	687	866	474	170

Table 3: Mix proportions of Geo-polymer Concrete

Grade	Fly ash(kg/m^3)	GGBS (kg/m^3)	Coarse aggregate(kg/m^3)	Fine aggregate (kg/m^3)	Geo activator (kg/m^3)
Mix A (20MPa)	396	0	1216	608	198
Mix B (40 MPa)	315	135	1280	528	180
Mix C (60 MPa)	412	275	474	966	170

3.RESULTS AND DISCUSSIONS

Figure 1 shows the compressive strength of recycled aggregate concrete for 7 and 28 days. The aggregate was replaced up to 50% at an interval of 10%. From figure 1, it can be observed that the compressive strength was decreased with increase in the replacement levels of coarse aggregate. The maximum compressive strength was observed at 0% replacement level at was 28.51 MPa after 28 days, whereas for 50% replacement level of coarse aggregate it was observed as 22.31 MPa. The compressive strength decrement was maximum observed at 30% replacement of coarse aggregate.

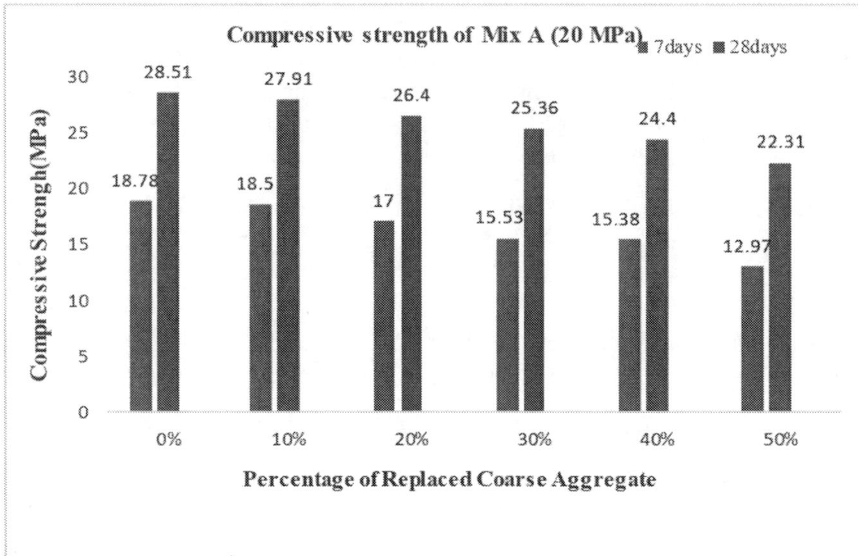

Figure 1: Compressive strength of conventional concrete cubes of Mix (A) 20 MPa

Figure 2 shows the compressive strength of conventional concrete of 40MPa for replacement of RCA from 10%-50% at the intervals of 10. For the age of 7 days and 28 days. We can see from the result the RCA shows good compressive strength for 40MPa in 10% and 20% replacement of CA and after then the strength decreases slightly when we increase RCA dosage. The replacement of RCA in concrete will decrease the compressive strength. On this context, the compressive strength for 0% replacement of RCA was 47.39 MPa and for 50% replacement was 41.31MPa

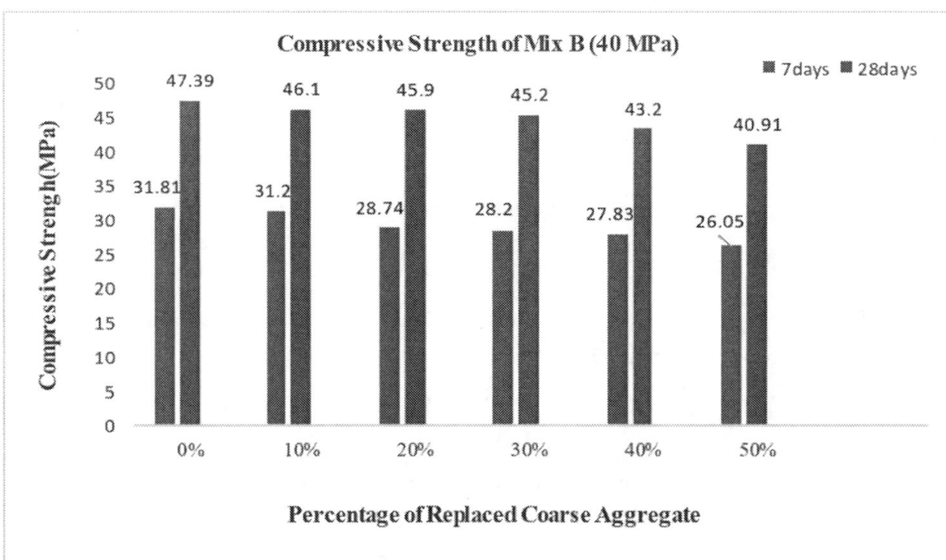

Figure 2: Compressive strength of conventional concrete Mix (B) 40Mpa

Figure 3 shows the compressive strength of conventional concrete of 60MPa for replacement of RCA from 10% to 50% at the intervals of 10 at the age of 7 days and 28 days. We can see from the result, the RCA shows good compressive strength for grade 40 MPa in 10% and 20% replacement of CA and after then the strength decreases slightly when we increase RCA dosage. The replacement of RCA in concrete will decrease the compressive strength. On this context, the compressive strength for 0% replacement of RCA was 65.13MPa and for 50% replacement was 59.78MPa

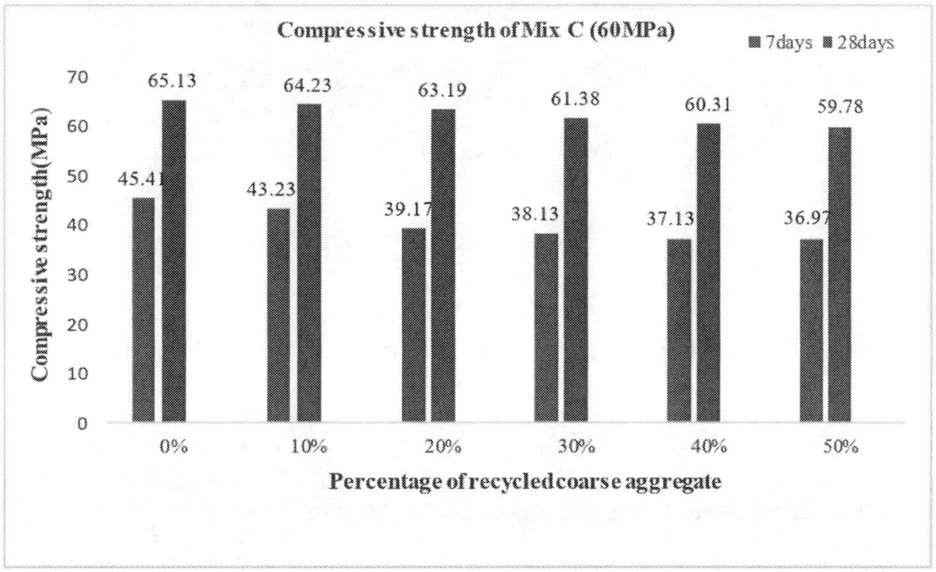

Figure 3: Compressive strength of conventional concrete cubes of Mix(C) 60Mpa

Figure 4shows the compressive strength of geopolymer concrete of 20MPa for replacement of RCA from 10% 50% at the intervals of 10 at the age of 7 days and 28 days. The replacement of RCA in concrete will decrease the compressive strength. On this context, the compressive strength for 0% replacement of RCA was 27.65MPa and for 50% replacement was 23.46MPa.

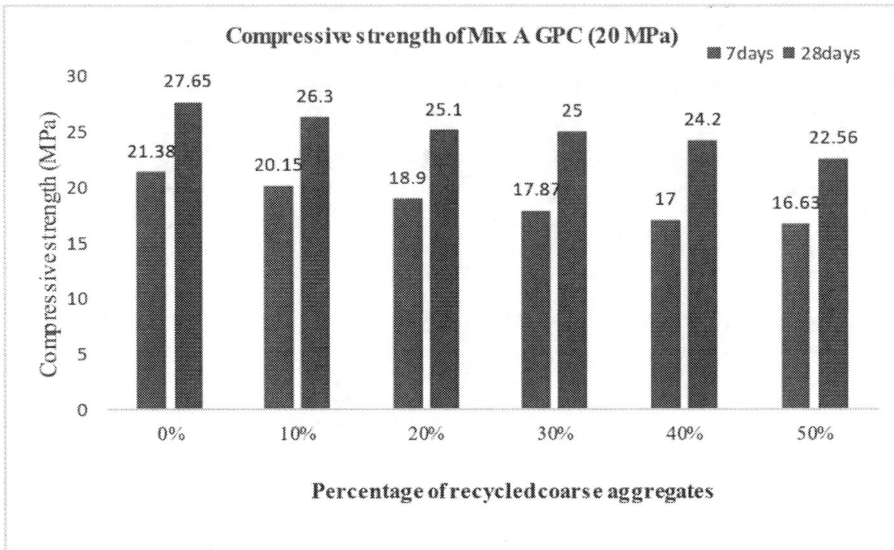

Figure 4: Compressive strength of geopolymer concrete cubes of Mix A (20 MPa)

Figure 5 shows the compressive strength of geopolymer concrete of 40MPa for replacement of RCA from 10%-50% at the intervals of 10 at the age of 7 and 28 days. The replacement of RCA in concrete will decrease the compressive strength. On this context, the compressive strength for 0% replacement of RCA was 48.4MPa and for 50% replacement was 44.13MPa

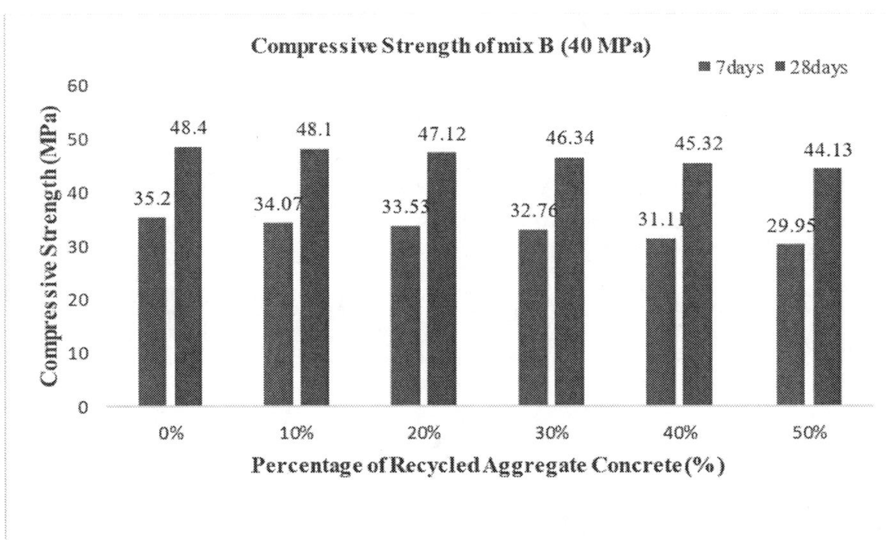

Figure 5: Compressive strength of geopolymer concrete cubes of 40 MPa

Figure 6 Shows the compressive strength of geopolymer concrete of 60MPa for replacement of RCA from 10%-50% at the intervals of 10 at the age of 7 and 28 days. The replacement of RCA in concrete will decrease the compressive strength. On this context, the compressive strength for 0% replacement of RCA was 66.83MPa and for 50% replacement was 61.50 MPa.

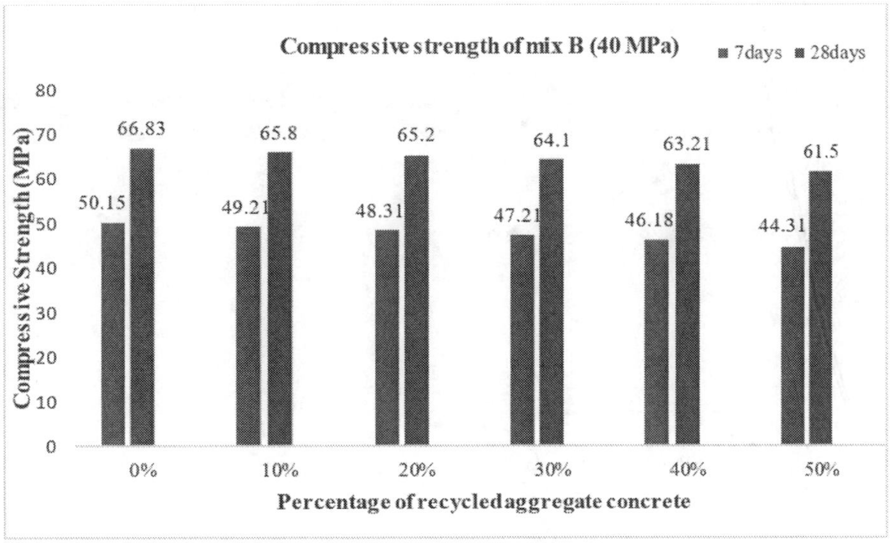

Figure 6: Compressive strength of geopolymer concrete cubes of 60 MPa

Results for data prediction using ANN

The above results obtained in experimental work for conventional concrete and geopolymer concrete were taken as input data for training in Artificial Neural Network for the grades of 20MPa, 40MPa,and 60 MPa and the data is predicted for the same grades for replacement of RCA from 0%-50% at the interval of 10%. Table 4shown the composition of materials used and the data is normalized to be less than 1.

Table 4: Maximum and Minimum range of total values

Mixes	Fly ash	GGBFS	Fine aggregate	Coarse aggregate	Alkaline solution	Recycled Coarse aggregate	Compressive strength (MPa)
MIX 1	366	0	608	1094	198	121.6	28.32
MIX 2	366	0	608	972	198	243	27.21
MIX 3	366	0	608	852	198	364	26
---	---	---	---	---	---	---	---
---	---	---	---	---	---	---	---
---	---	---	---	---	---	---	---
MIX 61	377	250	574	817	170	349	65.23
MIX 62	377	250	574	700	170	466	63.21
MIX 63	377	250	574	583	170	583	61.21

Minimum	315	0	528	583	170	0	21
Maximum	377	250	608	1242	198	690	69.83

The Figures 7, 8, and 9 shows the comparative difference between data predicted for strengths of 20MPa,40MPa and 60MPa by training data of grades 20MPa,40MPa and 60MPa.Fig 1 shows the regression curve for training data set, Fig 4.8.2 shows the regression curve for prediction data set. With the values replacement of 0% to 50% of RCA. The obtained R^2 value for training and prediction data set should be 0.99 and 0.98.

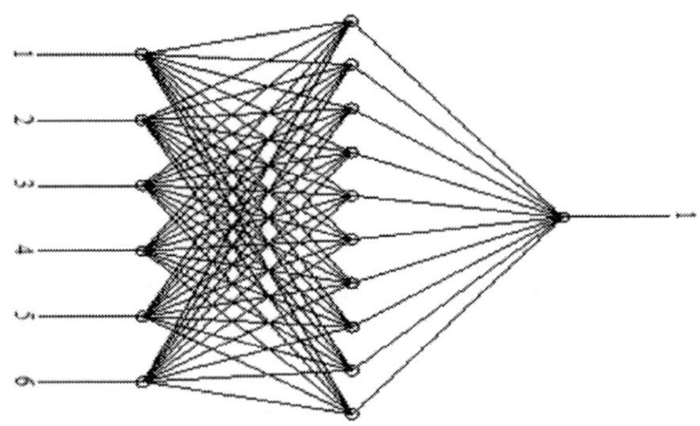

Fig 7: Network model

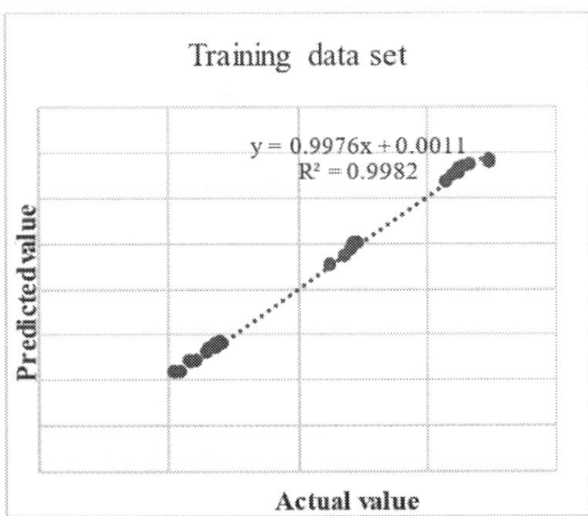

Fig: 8 Regression curve for training data set

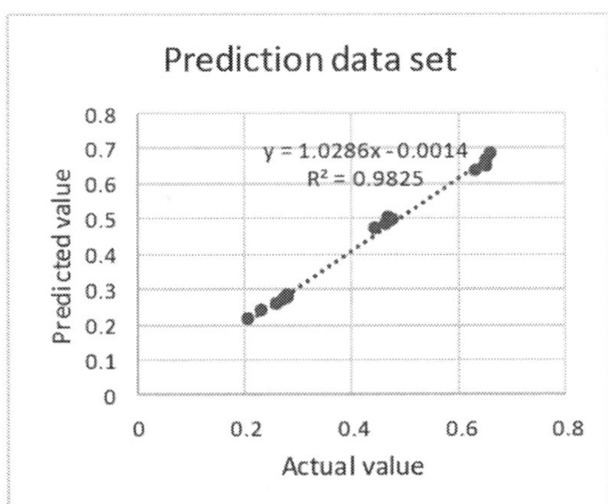

Fig 9: Regression curve for prediction data set

CONCLUSIONS

- In conventional concrete, the acceptable strength of RCA is found up to 30% replacement in all the mixes.
- In geopolymer Concrete, the early strength within 7 days is observed to be of desired strength due to high calcium content in GGBS used.
- Based on the research result for Conventional concrete, the decrease in compressive strength with 10% replacement of RCA is 1-2% and with 50% of RCA, the decrease is 18-20% and for Geo-Polymer concrete. The decrease in compressive strength with 10% replacement of RCA is 1-2% and with 50% of RCA, the decrease is 12-16%.
- Bythis, we can conclude that the use of RCA in concrete can be replaced up to 30 %, which showed minimum fluctuations, and further increase will decrease the strength
- The data prediction using Artificial Neural Network by Visual Gene Developer shown good results and the regression analysis shows very less error. We can save the time by using ANN rather than working on trial mixes which consumes cost and time.

REFERENCES

1) Etxeberria, Miren & Vázquez, E. & Mari, A. (2006). Microstructure analysis of hardened recycled aggregate concrete. Magazine of Concrete Research - MAG CONCR RES. 58. 683-690. 10.1680/macr.2006.58.10.683.

2) Kiran Kumar Poloju, Vineetha Anil, Ram Kishore Manchiryal. (2017). Properties of Concrete as Influenced by Shape and Texture of Fine Aggregate. American Journal of Applied Scientific Research. Vol. 3, No. 3, 2017, pp. 28-36. ISSN: 2471-9722 (Print); ISSN: 2471-9730 (Online). http://www.sciencepublishinggroup.com/journal/paperinfo?journalid=395

3) Wai Hoe Kwan, Mohiuddin Ramli, Mohd Zailan Suleiman (2011), Influence of the amount of recycled coarse aggregate in concrete design and durability properties, Construction and Building.

4) Xiao, J., Li, W. & Poon, C. Recent studies on mechanical properties of recycled aggregate concrete in China—A review. *Sci. China Technol. Sci.* **55**, 1463–1480 (2012).

5) Kumar, P. & Dhinakaran, Govindasamy. (2012). Effect of Admixed Recycled Aggregate Concrete on Properties of Fresh and Hardened Concrete. Journal of Materials in Civil Engineering. 24. 494-498. 10.1061/(ASCE) MT.1943-5533.0000393.

6) Jung, S. K., & McDonald, K. (2011). Visual gene developer: a fully programmable bioinformatics software for synthetic gene optimization. *BMC bioinformatics*, *12*, 340.

7) Nuaklong, P., Wongsa, A., Sata, V., Boonserm, K., Sanjayan, J., & Chindaprasirt, P. (2019). Properties of high-calcium and low-calcium fly ash combination geopolymer mortar containing recycled aggregate. *Heliyon*, *5*(9), e02513.

8) Bansal, Shishir & Singh, S. (2015). SUSTAINABLE HANDLING OF CONSTRUCTION AND DEMOLITION (C & D) WASTE International Journal of Sustainable Energy and Environmental Research. International journal of Sustainable Energy and Environmental Research. 2015. 22-48.

9) Pasini, Antonello (2015), Artificial neural networks for small dataset analysis, Journal of Thoracic Disease; Vol 7, No 5 (May 2015).

10) Kiran Kumar Poloju, Kota Srinivasu and Mallikarjuna Rao (2020), Study on Mechanical Characterization of Geopolymer Cement Mortar with Single Solution and Combined Solution, Journal of Xi'an University of Architecture & Technology, ISSN No : 1006-7930, Volume XII, Issue VIII, 2020, Page No: 481-487.

EXPERIMENTAL STUDIES OF STELLITE-6 HARDFACED LAYER ON FERROUS MATERIALS BY TIG SURFACING PROCESS

C Dinesh Chandra[1*], B Rushikesh[2], Mohammed Numan[3], B Venkatesh[4]
[4]Department of Mechanical Engineering, Vardhaman College of Engineering, Hyderabad, Telangana, India.

[1,2,3] UG Students, Department of Mechanical Engineering, Vardhaman College of Engineering, Hyderabad, Telangana, India.

E-mail:*dineshchandra962@gmail.com

Abstract: Hardfacing is the cheapest technique and it is the deposition of coatings of hard, wear resistant materials on a worn or new component that is subjected to wear, impact, high temperature, and oxidation. Weld hard facing is used to deposit thick layers of hard material by Tungsten Inert Gas (TIG) welding technique. This layer acts as a sacrificial layer at the point of impact and safeguards the base metal from wear out and corrosion. It is also used to fill the worn out part of a component there by increasing its durability and lifetime. A thick layer of stellite-6 of width 1mm is deposited on the base metals (i.e mild steel and EN-8) varying the weld currents as 120A, 140A, 160A. Vickers Hardness testing machine is used to measure the hardness of the hardfaced components and the variation of the hardness with varying weld current for both substrates. The specimens are subjected to oxidation at elevated temperature of 973K by muffle furnace for 5, 10, 15 hours duration and to study the oxidation rate with respect to different processing currents. The microstructural analysis is performed using optical microscopy on hardfaced samples. The aim of the paper is to study the effect in rate of oxidation, hardness, and microstructure with respect to varying parameters in stellite-6 hardfacing on ferrous materials.

Keywords: Hardfacing layer, Morphology, Microstructural analysis, Oxidation process, TIG welding.

1. INTRODUCTION:

In recent day's protection of the structural materials of different motor drive, cranes, crushers, bearing components, brakes, bushes, and pipeline assemblies against wear is very essential. Hence the metallurgical and mechanical engineers have carried out lot of research on wear. It became evident during the survey that wear of metals was a prominent topic in large number of the responses regarding some future priorities for research in tribological studies. Wear and oxidation are the most leading factors, which influence the life factor of the machine parts and assemblies. These hardfacing procedures can be applied in number of components like cranes, crushers, pump and fan impellers, crane wheels, turbine blades, pipeline assemblies, bucket teeth, screw conveyors, bulldozer teeth, blades, worn railway crossings, etc. MX Yao et al., have conducted study on wear, corrosion and cracking resistance of some W-or Mo-containing Stellite hardfacing alloys, and stated that the W-containing stellite alloys have better corrosion resistance in oxidizing acid. The Mo-containing Stellite alloys exhibit unusual combination of excellent wear and corrosion resistance in reducing environments[4]. Yucel Birol has investigated that the high temperature wear performance of Inconel 617 and Stellite-6 alloys and compared with that of the X32CrMoV33 hot work tool steel [12]. JL Otterloo and De Hosson have demonstrated that electron microscopy, pin-on-disc wear experiments and hardness measurements were carried out on a laser-coated cobalt based stellite alloy [9]. Buchely et al., have studied the microstructures of as-cast HS111 and as-SHSed Stellite-6 by SEM, XRD and EPMA [11]. Bingley and Schnee have studied that the abrasive

IOP Conf. Series: Materials Science and Engineering 998 (2020) 012061 doi:10.1088/1757-899X/998/1/012061

resistance is related to carbon content and secondary effect is based on hardness and chromium percentage [1]. Lemaire and Calvar have proposed the influence of addition of Nb, Cr, Mo and C on wear behavior of nickel based alloy [8]. Fontalvo and Mitterer have projected that the Co-based alloy composition having a relatively small lanthanum addition and relatively large carbon content provided remarkable oxidation resistance and wear resistance at high temperatures [7]. Bortoleto et al., have developed high Cr and V based hardfacing alloy and found that the wear resistance of high Cr and V based hard facing alloy on pin on disc is on the level of standard of NiCrBSi + 60% fused tungsten carbides and 13 times better than that of Stellite-6[2]. The objective of the hardfacing technique is to make the critical components for longer life by depositing on them with metal alloys in order to resist scratching and indentations, corrosion, erosion, abrasion, wear, impact etc.

2. METHODOLOGY:

Hardfacing is a metalworking technique and it is resistance to wear, it is an application of build-up of deposits of special alloys on the surface of base metals [3]. TIG technique deposits can functionalize surfaces and reclaim components extending their durability. The service life of a part increases by hardfacing and thereby extend the effective, efficient utilization of machinery equipments viz. core components such as crushers, dozer blades, bearings are exposed to heavy wear and require efficient surface protection measures to avoid costly down times and to reduce cost for expensive spare parts.

3. EXPERIMENTATION:

Experimental Details and Procedure for Hardfacings

Hardfacing is one of ancient technique used to improve mechanical properties, majorly surface properties of contacting surfaces and reduces the internal stress [5]. Hardfacing is performed by a TIG welding setup [6]. It was operated at 15V and a processing current of 120A, 140A, and 160A (according to AWS welding handbook). The setup is schematically shown in Figure 1. Initially scales and dust are removed from the samples. Then the sample was fixed on a suitable fixture and the argon pressure of 5kg/cm^2 was adjusted. Hardfacing was performed at different welding speeds shown in Table.1, which were measured by means of stopwatch. Straight away after hardfacing, the plate was quenched in an oil. The details of hardfacing operation is listed in Table 1. EN 8 and mild steel is used as substrate material, having the dimension of 60x30x6mm^3. Electrode for hardfacing was stellite-6. Samples sizes of 30X30mm^2 were cut from hardfaced plate whose hardfaced surface was prepared with fine finish. The composition of substrate materials is mentioned in table 2 and table 3.

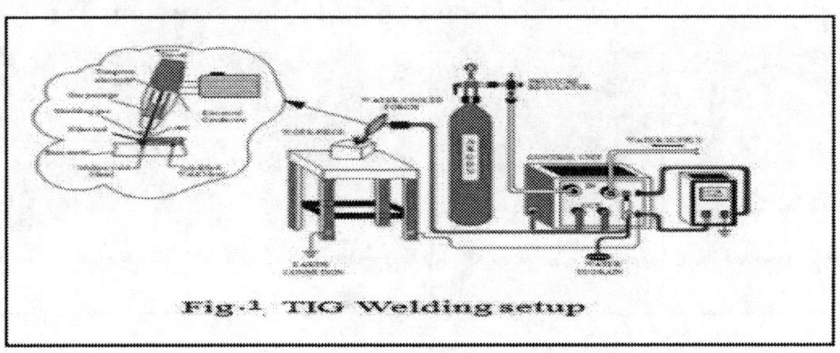

Fig.1 TIG welding set-up

IOP Conf. Series: Materials Science and Engineering 998 (2020) 012061 doi:10.1088/1757-899X/998/1/012061

Table.1 Hard-facing conditions

Parameter	**Value**
Applied voltage (V)	15
Applied current (A)	120, 140, and 160
Weld speed (mm/s)	2.25, 2.43, 2.63
Gas (argon) pressure (kg/cm^2)	5
Heat input (kJ/mm)	0.8, 0.864, 0.912

Table.2 Material composition of EN-8

Element	**Composition (%)**
C	0.34-0.40
Mn	0.60-1.0
P	0.05
S	0.005
Si	0.10-0.40

Table.3 Material composition of mild-steel

Element	**Composition (%)**
C	0.23
S	0.55
Mn	1.00
P	0.40
Si	0.35

Deposition of stellite layers on base metals:

The hardfacing alloy (i.e Stellite-6 rod) is shown in Fig.4 and which is deposited on two selected base materials i.e EN-8 and mild steel) by TIG welding process and the hardfaced samples of substrate are given in Fig.2 and Fig.3. The material composition of stellite-6 as shown in Table.4.

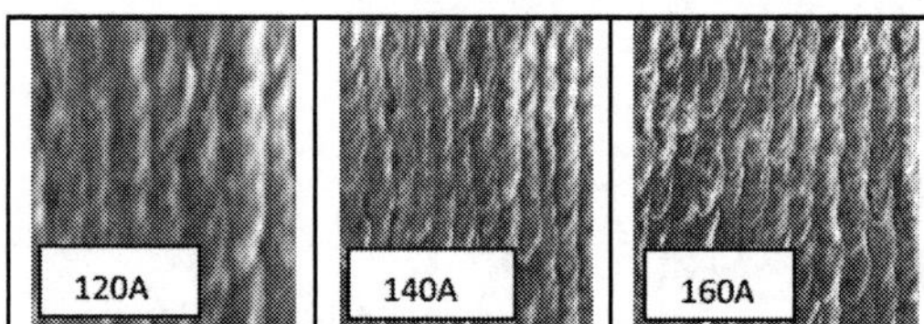

Fig.2 Hardfaced components of EN-8

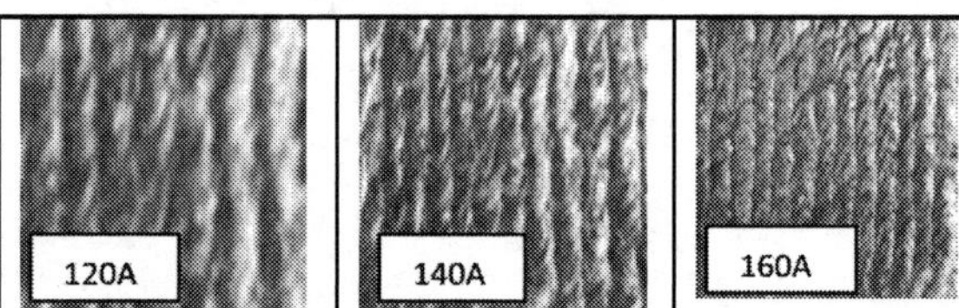

Fig.3 Hardfaced components of mild-steel

Table.4 Material Composition of Stellite-6

Element	Composition (%)
Chromium	30 %
Tungsten	4-5 %
Carbon	1.2 %
Nickel	< 3 %
Molybdenum	<1 %
Iron	<3 %
Silicon	<2 %
Cobalt	Balance

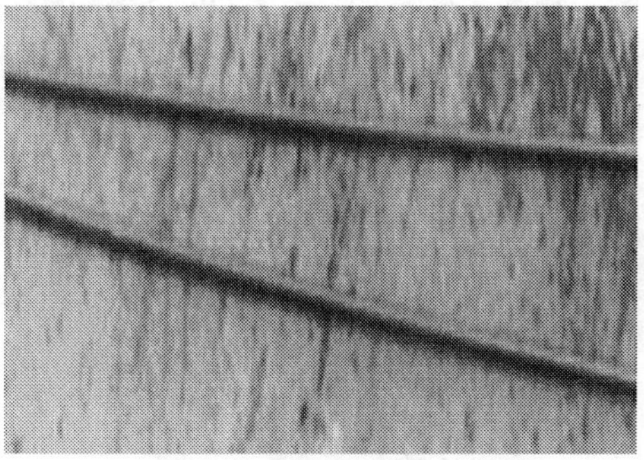

Fig.4 Stellite rods of (dia 3.2mm)

4. RESULTS AND DISCUSSION:

4.1. Hardness Test

The Vickers hardness test was performed at Mechanical Engineering department, Vardhaman Engineering College, Hyderabad. The Micro hardness test procedure as per ASTM E-384 and specified a range of light loads using a diamond intender to make an indentation, which is measured and converted to a hardness value [10].

Table.5 Hardness of Mild-Steel

Current(A)	Hardness number
120	310
140	353
160	361

Table.6 Hardness of EN-8

Current(A)	Hardness number
120	331
140	365
160	371

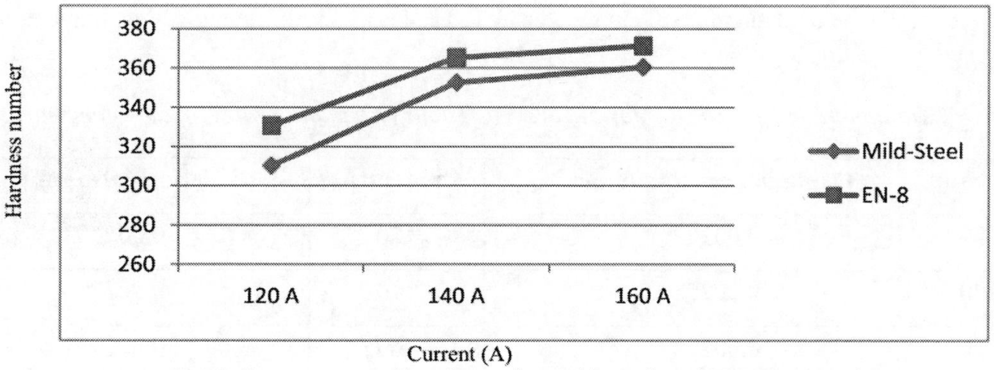

Fig.5 Comparison of hardness for mild-steel and EN-8 material

4.2. Oxidation Test at Elevated Temperature

The prepared samples are coated with the alumina on all the sides, except one side so that which enable to measure the oxidation of its hardfaced layer. These coated samples are weighed before oxidation test. The

Fig.6 Alumina Coated Samples

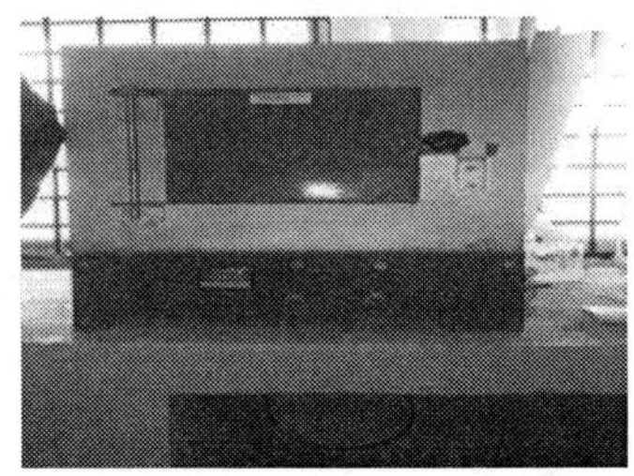

Fig.7 Muffle furnace

Table.7 Test details

Sample size	15x15x6mm	Oxidation rate= Weight gain (g)/surface area (mm^2)
Coating	Alumina	Weight gain = Wt. before oxidation(w_1) - Wt after oxidation (w_2)
Test time	5,10,15hours	
Test temperature	700^0C (973^0K)	Surface area = $15x15mm^2$

oxidation test is performed by Muffle furnace at Vardhaman College of Engineering, Hyderabad. The three samples are kept in furnace, and then heated to 700^0C and the soaking time is maintained in furnace for 5hours duration at 700°C. Repeat the same process for 10 and 15hours of soaking time at constant temperature of 700^0C. The respective samples were taken-out from furnace after cooling and weighed them. Determined the oxidation rate of alumina coated samples as per formula given in above table.7. It is found that the oxidation

Table.8 Oxidation rates of mild steel for 5hours

Current(A)	Oxidation rate (gm/mm^2)
120	0.002692
140	0.002580
160	0.002237

Table.9 Oxidation rates of EN-8 for 5hours

Current(A)	Oxidation rate (gm/mm^2)
120	0.002603
140	0.002296
160	0.002163

IOP Conf. Series: Materials Science and Engineering 998 (2020) 012061 doi:10.1088/1757-899X/998/1/012061

rate of mild steel is decreased from 0.002692gm/mm^2 to 0.002580gm/mm^2 if welding current is varies from 120A to 140A and the oxidation rate is decreased from 0.002580gm/mm^2 to 0.002237 gm/mm^2 when welding current is varies from 140A to 160A shown in table.8. It is also observed that oxidation rate for EN-8 decreases from 0.002603gm/mm^2 to 0.002296gm/mm^2 and decreased from 0.002296 gm/mm^2 to 0.002163gm/mm^2 for corresponding welding currents vary from 120A to 140A as shown in table.9.

Table.10 Oxidation rates of mild steel for 10hours

Current(A)	**Oxidation rate (gm/mm^2)**
120	0.005384
140	0.005156
160	0.004474

Table.11 Oxidation rates of EN-8 for 10hours

Current(A)	**Oxidation rate (gm/mm^2)**
120	0.005206
140	0.004592
160	0.004326

Oxidation rate is decreased from 0.005384 gm/mm^2 to 0.005156 gm/mm^2 to 0.004474 gm/ mm^2 if welding current vary from 120A to 140A to 160A as given in table.10. The rate of oxidation is recorded as 0.005206, 0.004592 and 0.004326 gm/mm^2 for the corresponding welding currents (i.e120A, 140A and 160A) and the oxidation rate is decreasing as shown in table.11. It is observed that the same trend is followed as the oxidation rates of mild-steel and EN-8 are decreasing for varying welding currents as tabulated in table.12 and table.13.

Table.12 Oxidation rates of mild steel for 15hours

Current(A)	**Oxidation rate (gm/mm^2)**
120	0.008076
140	0.007740
160	0.006711

Table.13 Oxidation rates of EN-8 for 15hours

Current(A)	**Oxidation rate (gm/mm^2)**
120	0.007809
140	0.006888
160	0.006489

IOP Conf. Series: Materials Science and Engineering 998 (2020) 012061 doi:10.1088/1757-899X/998/1/012061

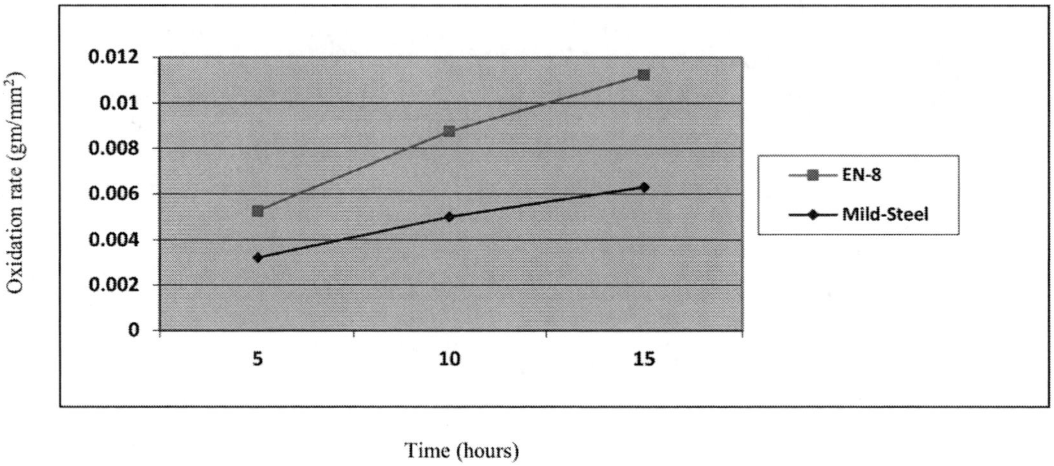

Fig.8 Variation of oxidation rates of Mild-steel and EN-8 for different times

Fig.8 shows the comparison of oxidation rates of mild steel and EN-8 for different time in hours. Rate of oxidation of mild steel is moderate at 5hours, slightly more at 10hours and very high at 15hours of soaking compared with EN-8.Thus oxidation rate is directly proportional to soaking time duration of samples in furnace at constant temperature (i.e 700^0C).

4.3 Micro-structural Analysis

The microstructure of each sample is captured using optical microscope with a magnification of 100X at the weld interface. Each sample was checked for complete fusion, freedom of cracks, porosities and other linear defects. There complete fusion of metals in each sample and defects were not found on the surface examined.

Etchant: Oxalic acid, Specimen orientation: Cross section, Magnification: 100X, Material: EN-8

Microstructure of EN-8	Observations
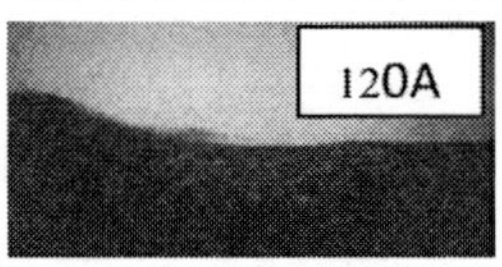	The fine grain particles are formed. The fine grain possess low strength at high temperatures because of more grain boundaries and at high temperatures grain boundary atoms break bonds easily and slide over the other grain atom which is a cause for low strength.
	Both coarse and fine grain particles are formed. The fine grains possess low strength and coarse grain particles possess high strength at high temperatures. As both coarse and fine grains are formed, its strength lies in between 120A and 160A specimens.

IOP Conf. Series: Materials Science and Engineering 998 (2020) 012061 doi:10.1088/1757-899X/998/1/012061

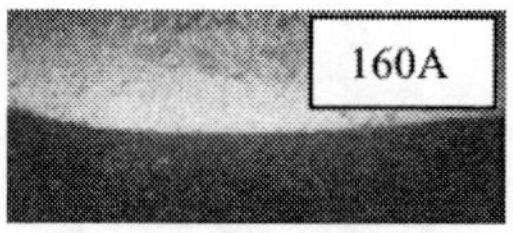	The coarse grain particles are formed. The coarse grain possess high strength at high temperatures because of less grain boundaries and no sliding deformation takes place.

Fig.9 Microstructures of EN-8 Material with different processing currents

Etchant: Oxalic acid, Specimen orientation: Cross section, Magnification: 100X, Material: mild-steel

Microstructure of mild steel	Observations
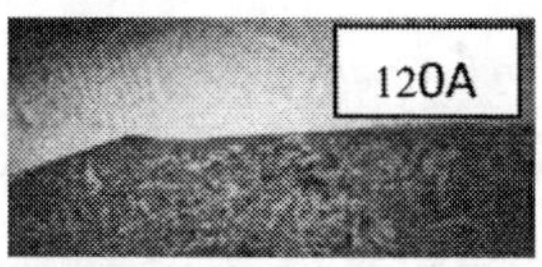	The fine grain particles are formed. These are formed due to fast cooling rate. It possess low strength at high temperatures because of more grain boundaries and at high temperatures boundary atoms breaks bonds easily and slide over the other grain atoms which is a cause for low strength.
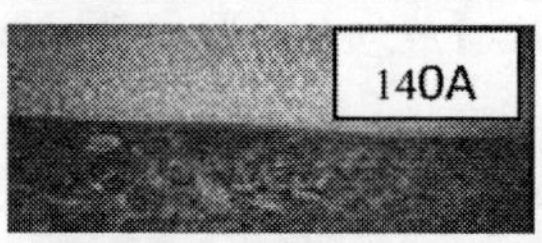	Both coarse and fine grain particles are formed. The fine grains possess low strength and coarse grain particles possess high strength at high temperatures. As both both coarse and fine grains are formed, its strength lies in between 120A and 160A specimens.
	The coarse grain particles are formed. These are formed due to slow cooling rate. The coarse grain possess high strength at high temperatures because of less grain boundaries and no sliding deformation takes place.

Fig.10 Microstructures of mild steel Material with different processing currents

5. CONCLUSION:

The hardness of the hardfaced component is greater than the hardness of the base metal. Hardness of hardfaced EN-8 material is higher in comparison with hardfaced mild-steel material. The oxidation rate of hardfaced component increases with increase in soaking time. The oxidation rate is decreasing as processing weld current increases. Hardfaced EN-8 material has good oxidation resistance in comparison with hardfaced mild-steel material. Micro-examination of the surface of the components shown complete fusion and freedom from cracks, porosities and other linear defects.

IOP Conf. Series: Materials Science and Engineering 998 (2020) 012061 doi:10.1088/1757-899X/998/1/012061

References

[1] Bingley M S, and Schnee, S 2005 "*A study of the mechanisms of abrasive wear for ceramic material under wet and dry three-body conditions*" Wear, **258** 50-61.

[2] Bortoleto E M, Rovani A C, Seriacopi V, Profito F J, Zachariadis D C, Machado I F, Sinatora A and Souza R M (2013) "*Experimental and numerical analysis of dry contact in the pin on disc test*" Wear, Vol. **301**, pp. 19-260.

[3] Venkatesh B, Sriker K, Prabhakar Vattikuti S V, Deepak K (2015) "*Wear characteristics of hard facing alloys: State-of-the art* "Elsevier Journal of Procedia Material Science 10, 527 – 532.

[4] Yao M X, Wu J B C and Xie Y (2005) "*Wear, corrosion and cracking resistance of some W- or Mo-containing Stellite hardfacing alloys,*" Materials Science and Engineering, vol. **407**, no. 1-2, pp. 234–244.

[5] Venkatesh B, Sriker K, Vijaybabu T & Parthasarathi G (2019) "Wear *Behavior of Fe-Cr-V Based Hard Facing Alloy on ASTM A105 Steel*" International Journal of Mechanical and Production Engineering Research and Development (IJMPERD), Vol. **9**, Issue 4, 995-1004.

[6] Venkatesh B, Bharti Malvi, Manish Roy and Pallab sarkar (2018) "*Effect of Welding Conditions on Erosive Wear of Hard-faced Co-Based Alloy Layer*" Journal of Engineering Tribology, Vol **232**,Issue 11, 1377 – 1389.

[7] Fontalvo G A and Mitterer C(2005) "*The effect of oxide-forming alloying elements on the high temperature wear of a hot work steel,*" *Wear*, vol. 258, no. **10**, pp. 1491–1499.

[8] Lemaire and Le Calvar M (2001) "*Evidence of tribocorrosion wear in pressurized water reactors,*" *Wear*, vol. **249**, no, 5-6, pp.338-344, 2001.

[9] De Mol Van Otterloo J L, De Hosson J M (1997) "*Microstructural features and mechanical properties of a cobalt-basedlaser coating,*" Acta Materialia 45 (3), 1225-1236.

[10] Venkatesh B, Vijaybabu T, Seshappa A, Sriker K (2017) "*Wear behavior of multilayer over mono layer of Fe-C-Cr hardfacing alloy on ASTM A 105 Steel*", International Journal of Mechanical Engineering and Technology (IJMET),Volume **8**, Issue 12, 999–1006.

[11] Buchely, Gutierrez G C, LLeon M and Toro A (2005) "*The effect of microstructure on abrasive wear of hardfacing alloys*", Wear, Vol. **259**, pp. 52-61.

[12] Yucel Birol (2010) "*High temperature sliding wear behaviour of Inconel 617 and Stellite 6 alloys,*" Wear 269 (9-10), 664-671.

A detailed second law (excergic) analysis approach of H_2O-LiBr vapour absorption cooling system

Arshi Banu P. S*, Micha Premkumar T, Seralathan Sivamani , Vijayabalan P

Department of Mechanical Engineering, Hindustan Institute of Technology & Science, Chennai-603103, India.

*Corresponding author: psarshib@hindustanuniv.ac.in

Abstract. Vapour absorption technology is generally known as a potential environmental friendly strategy to use solar heat energy reliably for space cooling applications. The thermodynamic analysis is used as an approach for determining efficiency in these systems. Two methods are the first law (FLA) and the second law (SLA) analyses in it. FLA uses the energy conservation of each component and gives the COP of the cycle. Comparing vapour absorption cooling systems on the basis of COPs is not justified. Hence second law is used for a better understanding of the performance. Second law analysis uses exergic method, thereby providing information on the losses in each component of the system which will indirectly influence the COP. During the present work on a single-effect H_2O-LiBr vapor absorption cooling system of a ton capacity, a comprehensive SLA has been carried out. Enthalpy, entropy, exergic and dead state values at each state have been found. Exergic values have been compared with the literature values. The reason for the differences in these values has been discussed. The importance of choosing dead state properties of a solution and the proper approach to find it has been provided. Among four key components of a vapour absorption system, an absorber is found to be the component with the largest energy losses.

Keywords: second law analysis, exergy, entropy, irreversibility, dead state.

1. Introduction

Vapour absorption cooling systems utilize low-grade energy, for example, solar heat, as input and operating fluids are eco-friendly materials. Therefore, it has the least global warming and ozone depletion consequences compared to vapour compression systems. H_2O-LiBr is found to be the most viable option at moderate temperatures for space cooling applications (Mc Neely 1979, Pattorson & Perez Blanco 1988, Lee *et al.* 1990, Jeter *et al.* 1992, Lenard *et al.* 1992), hence more suitable for solar energy-based space cooling applications (Jian Sun *et al.*2012, Jaruwongwittaya & Chen 2010), Hassan & Mohamad 2012). Enhancing COP is the major challenge faced by vapour absorption systems. However, complete performance evaluation is not only to enhance COP but also to overcome the drawbacks of crystallization and corrosion issues associated with H_2O-LiBr and thereby influencing the size and cost of the system. Thermodynamic analysis needs to be performed to analyze the above factors (Badarinarayana *et al.* 1982).

Methods to evaluate performance are are First Law Analysis (FLA) and Second Law Analysis (SLA) for vapour absorption cooling systems. Detailed FLA has been discussed in our previous works (Arshi Banu PS *et al.* 2018(a-b)). A second law analysis calculates the exergy-based system performance that represents the thermodynamic irreversibility. Exergy testing is defined as the maximum amount of work potential in relation to the environment of a material or a form of energy. The losses thereby represent the

IOP Conf. Series: Materials Science and Engineering 998 (2020) 012062 doi:10.1088/1757-899X/998/1/012062

real loss of work in an exergy study. Dissipation (friction), heat transfer under temperature difference is essential to irreversible processes contributing to these losses.

Some of the literary works on SLA have been discussed here. Koehler *et al.* (1988), Olivia *et al.* (1994) performed the exergic analysis on a single-effect (single heat source based) vapour absorption refrigeration system. Found enthalpy, entropy and exergy at each state. The authors suggested the different approaches in finding the dead states of the solution state points. Talbi *et al.* (2000) performed the exergic analysis on single-effect vapour absorption refrigeration system. Found enthalpy, entropy and exergy at each state. Sedighi *et al.* (2007) considered of single effect vapour absorption refrigeration system with SHX (solution heat exchange), with RHX (refrigerator heat exchanger) and with both RHX and SHX. Variations of COP with generator temperature have been discussed in each cycle. Entropy has not been calculated at all state points of the cycle. Raynaldo *et al.* (2010) considered the vapour absorption refrigeration, heat pump and heat transfer for his studies. He found physical exergy as well as chemical exergy at each state point of the above cycles. He also clearly represented the properties of the H_2O-LiBr solution, such as enthalpy, entropy and crystallization temperature in both graphical and empirical form. Detailed SLA has been performed in the present study by considering the internal circuit of the single-effect vapour absorption system. Enthalpy, entropy and exergic values of refrigerant and solution have been calculated at each state point of the cycle. The Irreversibilities of each component has been evaluated. Importance of the reference dead state and procedure to find it, for refrigerant and solution has been discussed. Exergic values have been compared with various literature values. The reasons for the differences in these values also discussed.

2. SLA of single-effect H_2O-LiBr vapour absorption cooling system

The system used in the second law analysis is an H_2O-LiBr single-effect absorption cooling system with a capacity of one ton shown in Figure 1. The four key elements are; generator (G), condenser (C), evaporator (E) and absorber (A). The generator (G) comprising a strong salt solution is supplied with heat externally. The steams evaporated from solution flows into the (C) and get condensed. Due to the condensed liquid expansion, (E) has the cooling effect. Eventually, vapour dissolved into (A) and maintains the poor salt solution. The salt solution circulates between (A) and (G) functions as a thermal compressor. LiBr is known as an absorbent solution in this sense. For heat exchanger (SHX), the solution of the absorber is preheated by its recovery before reaching the reactor by using the power from the hot fluid exiting the generator. Details of pressure, temperatures, phases at various state points of the cycle and the assumption made during analysis have been discussed in our previous work (Arshi Banu PS *et al.* 2018(b)).

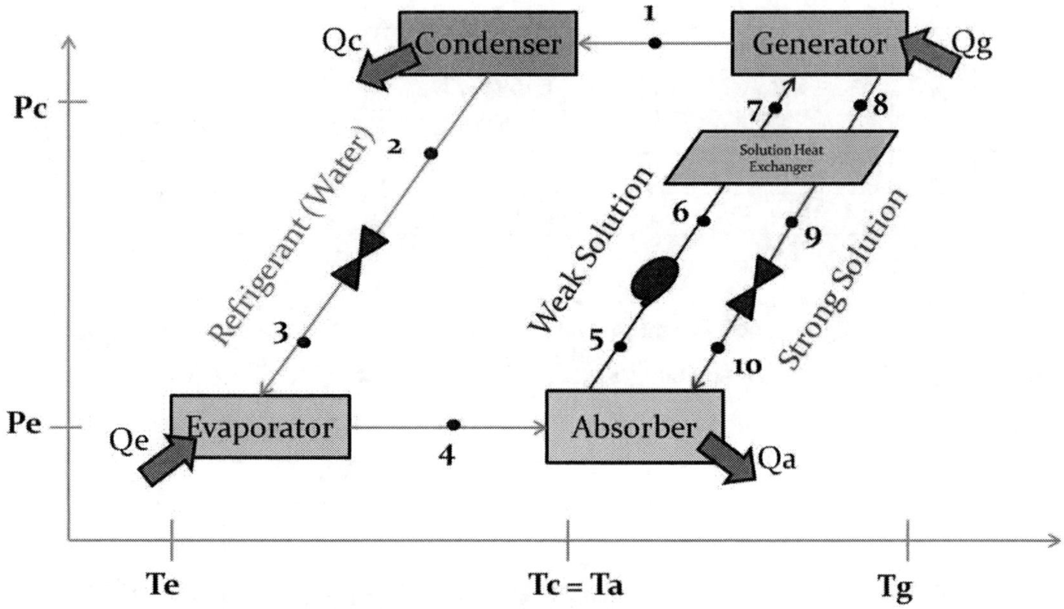

Figure 1 H_2O-LiBr vapour absorption cooling system considered for SLA

To evaluate this system according to the second law, Entropy and enthalpy values are needed at each state of the system. Enthalpy, entropy, exergy and availability values of water (refrigerant) and H_2O-LiBr solution (absorbent) have been calculated as follows.

2.1 Enthalpy and Entropies of Refrigerant (water)

Enthalpies and entropies of refrigerant (water) circuit (state point '1' to '4') can be obtained from property empirical equations or in-built functions of EES (Engineering Equation Solver) software of pure water. At state 1, water is at a super-heated steam state from the generator at temperature (T_g), enthalpy and entropy of super-heated steam are considered at temperature (T_g). Super-heated steam is condensed to liquid water in the condenser. The enthalpy and entropy of liquid water are considered at state 2. Liquid water extracts heat in evaporator and vapour is formed. Hence water vapour enthalpy and entropy considered at evaporator temperature and pressure at condition state 4.

2.2 Enthalpies of H_2O-LiBr solutions:

An h-T-X diagram analytical equation gives specific enthalpy values for H_2O-LiBr solutions for various concentrates (X) and solutions (T_s).

$$h_n = \sum_0^4 A_n X^n + T_s \sum_0^4 B_n X^n + T_s^2 \sum_0^4 C_n X^n \quad (1)$$

n	A_n	B_n	C_n
0	-2024	18.2829	-3.7008214E-2
1	163.309	-1.1691757	2.8877666E-3
2	-4.88161	3.248041E-2	-8.1313015E-5
3	6.302948E-2	-4.034184E-4	9.9116628E-7
4	-2.913705E-4	1.8520569E-6	-4.4441207E-9

2.3 Entropies of H_2O-LiBr solution:

Empirical equation (2) gives specific entropy values in terms of solution concentrations (X) and solution temperatures (T_s) for H_2O - LiBr solution (Kaita *et al.* 2001).

$$s_n = \sum_{i=0}^{3}\sum_{j=0}^{3} A_{ij}\, X^j T_s^{\,i} \qquad (2)$$

j	A_{i0}	A_{i1}	A_{i2}	A_{i3}
0	5.127558E-1	-1.393954E-2	2.924145E-5	9.035697E-7
1	1.226780E-2	-9.156820E-5	1.820453E-08	-7.991806E-10
2	1.364895E-5	1.068904E-7	-1.381109E-9	1.529784E-11
3	1.021501E-8	0	0	0

2.4 Exergy calculation

The exergy of water (refrigerant) circuit (1-2-3-4) is given by:

$$\psi_n = [h_n - h_o] - T_o[s_n - s_o] \qquad (3)$$

h_o and s_o are enthalpy and entropies of water at dead state considered at atmospheric pressure (p_o) = 1bar and temperature (T_o) = 25°C.

The exergy of a binary mixture solution is generally given by:

$$\psi_n = [h(T_n, X_n) - h_o(T_o, X_{sat})] - T_o[s(T_n, X_n) - s_o(T_o, X_{sat})] \qquad (4)$$

h_o and s_o are enthalpy and entropies of dead state for the solution is considered at temperature T_o and concentration as X_{sat}. Saturated concentration (X_{sat}) is the concentration of LiBr in water obtained at pressure (p_o) = 1 bar and temperature (T_o) = 25°C.

2.5 Availability Loss

The availability loss of a system is calculated by:

$$\Delta\psi = \sum m_i\psi_i - \sum m_e\psi_e - Q\left(1 - \frac{T_o}{T}\right) - W \qquad (5)$$

The first word on the right of the equation is the amount of input exergy. The second is the amount output exergy production, whereas the third term is heat exergy Q, transmitted at constant temperature 'T'. The final term is the transfer of mechanical work to or from the system (Arzu *et al.* 2005).

2.6 The availability loss in each component is calculated by:

Availability loss in condenser: $\Delta\psi_c = m_1\psi_1 - m_2\psi_2 - Q_c\left(1-\frac{T_o}{T_c}\right)$ (6)

Availability loss in Refrigerant Expansion Valve: $\Delta\psi_{ev} = m_2\psi_2 - m_3\psi_3$ (7)

Availability loss in Evaporator: $\Delta\psi_e = m_3\psi_3 - m_4\psi_4 - Q_e\left(1-\frac{T_o}{T_e}\right)$ (8)

Availability loss in Generator: $\Delta\psi_g = m_7\psi_7 - m_8\psi_8 - m_1\psi_1 - Q_g\left(1-\frac{T_o}{T_g}\right)$ (9)

Availability loss in Absorber: $\Delta\psi_a = m_4\psi_4 - m_5\psi_5 - m_{10}\psi_{10} - Q_a\left(1-\frac{T_o}{T_a}\right)$ (10)

Availability loss in Solution Heat Exchanger: $\Delta\psi_{hx} = m_6\psi_6 - m_7\psi_7 + m_8\psi_8 - m_9\psi_9$ (11)

Availability loss in Solution Expansion Valve: $\Delta\psi_{sev} = m_9\psi_9 - m_{10}\psi_{10}$ (12)

Availability loss in Solution Pump: $\Delta\psi_{sp} = m_5\psi_5 - m_6\psi_6 - W$ (13)

Based on these sets of equations, exergic analysis is performed and heat losses that occurred in each component of VACS have been found. The result obtained from the second law analysis/exergic analysis has been tabulated in Table 2.

3. Results obtained from SLA

Results obtained from SLA using exergic method on a single-effect H_2O-LiBr vapour absorption system of one ton capacity has been presented here. For the analysis the values of parameters considered are: generator temperature (T_g) = 87.8°C, condenser temperature (T_c) = 37.8°C, absorber temperature (T_a) = 37.8°C, evaporator temperature (T_e) = 7.2°C, ambient temperature (T_o) = 25°C and effectiveness of solution heat exchanger is assumed as 0.8. The reason for selecting the parameter values are, these are obtained from real experimental values and also similar readings have been considered by various authors, hence it will be convenient in validating and comparing current work with literature. The details are discussing in the next section. Using these parametric values SLA has been performed. At each state point of the cycle, enthalpy, entropy values were calculated and tabulated in Table 1. State 1 to 4 represents pure refrigerant values obtained from water property charts. State 5 to 7 represent week solution values and state 8 to 10 represent strong solution values. These are obtained from equation (1-2). Entropy values for refrigerant and solution are obtained from equation (3) and equation (4) respectively.

Table 2 results show the irreversibilities contributed by each component of the absorption system calculated from equation (6-13). It also represents the difference between inlet and outlet exergy and the percentage of exergic losses by each component of the vapour absorption cooling system. They have been given in the higher to lower order in the percentage of contribution. It is found that with a change in flow rates of refrigerant, weak and strong solution, the values of irreversibilities will keep changing. Whereas the percentage of irreversibilities will remains constant for the given set of component temperatures. For external circuit, irreversibilities depend on the real state point data such as inlet & outlet temperatures and mass flow rates occurring in each component. Deciding and comparing the irreversibilities considering the external circuit is not consistent and will vary case to case. Hence in the present work, internal circuit has been considered for finding the irreversibilities occurring in each component.

IOP Conf. Series: Materials Science and Engineering 998 (2020) 012062 doi:10.1088/1757-899X/998/1/012062

Table 1 Enthalpy, Entropy and Exergic data obtained from SLA

State Point	Temperature (°C)	Pressure(kPa)	Concentration (%)	Enthalpy (kJ/kg)	Entropy (kJ/kg.K)	Exergy (kW/kg)
1	87.8	7.5	0	2664	8.58	112
2	37.8	7.5	0	158.3	0.54	1.01
3	7.2	0.84	0	158.3	0.57	1.01
4	7.2	0.84	0	2514	8.97	-154.2
5	37.8	0.84	55.86	92.22	0.229	26
6	37.8	7.5	55.86	92.22	0.229	26
7	76.53	7.5	55.86	175.3	0.405	32.8
8	87.8	7.5	62.75	220	0.479	81.89
9	47.8	7.5	62.75	146	0.29	71.92
10	47.8	0.84	62.75	146	0.29	71.92

It is found that the absorber contains the highest availability losses. The reasons are the differences in temperatures accruing between absorber and surroundings. These losses can be avoided by increasing the heat transfer area of the absorber. However, this promotes the cost of absorber. In order to increase the overall performance of the system, care must be taken in designing the components to reduce the irreversibilities that occur in it. The details of absorber design to overcome these losses have been discussed in our work (Arshi Banu PS *et al.* 2018(c)). The exergic losses of generators are next to the absorber. These losses are due to refrigerant evaporation in the generator from a strong solution. By selecting appropriate working fluids and heat source combination these losses can be avoided.

Table 2 Exergic losses of H_2O-LiBr vapour absorption system

Component	Exergy Loss($\Delta\psi$) (kJ)	% of Exergy Loss
Absorber	60.53	40.9
Generator	50.57	34.2
Evaporator	7.843	5.3
Heat Exchanger	21.53	14.5
Condenser	7.403	5
System	26.24	100

4. Comparison with literature values and identification of dead state of solution

Exergic vales obtained in the present study has been compared with the literature works of three authors Koehler *et al.*(1988), Oliveir *et al.*(1994), Raynaldo *et al.* (2010) as shown in Table 3. Single effect vapour absorption refrigeration system shown in Figure 1, has been considered with 10 state points for comparison. Common temperature, pressure and flow rates have been maintained at all state points. Input parameters considered for the comparison are generator temperature as 87.8°C, condenser and absorber temperature as 37.8°C and evaporator temperature at 7.2°C. Calculated enthalpy, entropy and exergy values with these inputs have been compared with literature. It is found that there is an enormous difference in the values of exergy by various authors even though there is a similarity in enthalpy and entropy values at all state points. The major reasons identified for the differences in the exergic values are: Lack of clarity in choosing the dead state reference parameters such as pressure, temperature and concentration of solution; Differences in choosing the methodology for finding the dead state enthalpy and entropy of the solution; Variations in choosing the property values of water as well as H_2O-LiBr solution from literature.

According to Koehler *et al.* (1988), temperature and concentration are the major parameters in deciding the dead state properties whereas pressure is also a vital parameter in defining dead state, which has not been clearly represented in the work. He only considered the reference temperature as 25°C and corresponding concentration to find the dead state enthalpy and entropy of solution. Oliveira *et al.* (1994) considered pressure as 100 kPa and temperature as 25°C and concentration as 20% to find the dead state enthalpy and entropy for solution circuit. The reason behind considering the 20% concentration and not been discussed. Arzu Sencana *et al.* found dead state enthalpy and entropy (h_o and s_o) at temperature 298.5 K. Talbi *et al.* (2000), Sedighi *et al.* (2007) & Azhar *et al.* (2019 & 20202) mentioned environmental condition as the dead state reference properties. Specific pressure, temperature and concentration have not been mentioned in the paper. pressure and concentration details of the solution to find the dead state properties have not been discussed. Reynaldo *et al.* (2010) considered temperature as 25°C and pressure as one atmosphere. The concentration of the solution has been considered corresponding to the above pressure and temperature. The author mentions the corresponding saturation concentration as 55.4 % for calculations.

In the present study, dead state properties of the solution have been found at the temperature of 25°C and pressure as one atmosphere. Reference concentration has been considered at these pressure and temperature to find the dead state properties h_o, s_o for the solution circuit (For state points '5' to '10'). Exergic values obtained from the present study have similar behaviour with the Reynaldo *et al.* (2010)'s values in solution circuit state points. But the author mentions the saturation concentration as 55.47% whereas for the present study saturation concentration is 13.86 %. Refrigeration circuit (state point '1' to '4') values are matching with the values of Oliveira *et al.* (1994).

Table 3 Comparison of exergic values of present study with literature at each state point of H_2O-LiBr cycle

St. Point	Temperature (°C)	Pressure (kPa)	Mass flow rate	Concentration (%)	Enthalpy (kJ/kg)	Entropy (kJ/kgK)	Exergy (kW/kg) (Present Study)	Koehler *et al.*	Oliveir *et al.*	Raynaldo *et al.*
1	87.8	6.558	1	0	2664	8.58	112	731.9	110.8	160.7
2	37.8	6.558	1	0	158.3	0.54	1.01	622.2	1.02	50.98
3	7.2	1.016	1	0	158.3	0.57	1.01	615.2	-5.9	44.08
4	7.2	1.016	1	0	2513.8	8.97	-154.2	465.7	-155.4	-105.5
5	37.8	1.016	9.035	55.42	91.54	0.229	26	32.56	114.9	501.4
6	37.8	6.558	9.035	55.42	91.54	0.229	26	32.57	114.9	501.4
7	66.2	6.558	9.035	55.42	148.95	0.405	32.8	37.35	119.7	506.2
8	87.8	6.558	8.035	62.32	221.2	0.479	81.89	10.42	180.4	608.7
9	53.08	1.016	8.035	62.32	156.64	0.29	71.92	1.88	171.8	600.2
10	53.08	1.016	8.035	62.32	156.64	0.29	71.92	1.88	171.8	600.2

5. Conclusions

Second law analysis using exergic method of single-effect H_2O-LiBr vapour absorption system of one ton capacity has been performed and conclusions obtained from the analysis have been discussed here.

- Enthalpy, entropy and exergy values of each state point of the vapour absorption system have been calculated.
- Using the exergic values, exergy losses, irreversibilities and the percentage of losses contributed by each component of the VACS have been identified.
- It is found that the absorber contains the highest availability losses followed by the generator. In order to increase the overall performance of the system, care must be taken in designing these components, selecting appropriate working fluids and heat source to reduce the irreversibilities that occur in it.
- Exergic values have been compared with various authors. The reasons for the differences in these values have been discussed.
- The importance of choosing dead state properties of the solution and the clear procedure to find them has been discussed.

References

[1]. McNeely LA. Thermodynamic properties of aqueous solutions of LiBr. ASHRAE Transactions 1979; 85:413–34.
[2]. Patterson MR & Perez-Blanco H. Numerical fits of properties of lithium-bromideH_2O solutions. ASHRAE Transactions 1988; 94:2059–77.
[3]. Lee RJ, DiGuilio RM, Jeteter SM & Teja AS. Properties of LiBr–H_2O solutions at high temperatures and concentrations. Part II. Density and viscosity ASHRAE.Transactions 1990; 96:709–14.
[4]. Jeter SM, Moran JP & Teja AS. Properties of LiBr–H_2O solutions at high temperatures and concentrations. Part III. Specific heat. ASHRAE Transactions1992; 98:137–49.
[5]. Lenard JLY, Jeter SM & Teja AS. Properties of LiBr–H_2O solutions at high temperatures and concentrations. Part IV. Vapor pressure. ASHRAE Transactions 1992; 98:167–72.
[6]. Jian Sun, Lin Fu & Shigang Zhang. A review of working fluids of absorption cycles. Renewable and Sustainable Energy Reviews 2012; 16:1899–1906.
[7]. Jaruwongwittaya T, Chen G. A review: renewable energy with absorption chillers in Thailand. Renewable and Sustainable Energy Reviews 2010; 14:1437–44.
[8]. Hassan HZ & Mohamad AA. A review on solar cold production through absorption technology.Renewable and Sustainable Energy Reviews 2012; 16:5331–5348.
[9]. Badarinarayana K, Srinivasa Murthy S & Krishna Murthy MV. Thermodynamic analysis of R21-DMF vapour absorption refrigeration systems for solar energy applications. International Journal of Refrigeration 1982; 5(2):115-119.
[10]. Arshi Banu, P.S. & Sudharsan, N.M., Review of water-based vapour absorption cooling systems using thermodynamic analysis. Renew. Sustain. Energy Rev 2018; 82(3), pp: 3750–3761.
[11]. Arshi Banu P.S., Balaji Dhanapal, 2018, Parametric behavior of vapor absorption cooling system using H_2O-LiBr as working fluid for solar energy applications. International Journal of Mechanical Engineering and Technology (IJMET), Volume 9, Issue 13, pp. 1537–1548. ISSN Print: 0976-6340.
[12]. Koehler WJ, Ibele WE & Soltes J, 1988, 'Availability simulation of a lithium bromide absorption heat pump'. Heat Recovery Systems and CHP; vol.8, pp.157–171.
[13]. Oliveira S & Le Goff P. Analyse exerge´tique desprocessus de se´paration et de me´langeage—applicationaux thermo transformateurs. Revue Ge´ne´rale deThermique 1994; 385:21–29.
[14]. Talbi, MM & Agnew, B, 2000, 'Exergy analysis: an absorption refrigerator using lithium bromide and water as the working fluids', Applied Thermal Engineering, vol. 20, pp. 619-630.
[15]. K Sedighi, M Farhadi, & M Liaghi, Exergy analysis: parametric study on lithium bromide–water absorption refrigeration systems',2007, J. Mechanical Engineering Science, DOI: 10.1243/09544062JMES604.
[16]. Reynaldo Palacios-Bereche, R. Gonzales & S. A. Nebra, Exergy calculation of lithium bromide–water solution and its application in the exergetic evaluation of absorption refrigeration systems LiBr-H_2O,.International Journal of Energy Research (2010), (wileyonlinelibrary.com). DOI: 10.1002/er.1790.
[17]. Kaita, Y 2002, 'Simulation results of triple-effect absorption cycles', International Journal of Refrigeration, vol. 25, pp. 999-1007.
[18]. Arzu Sencana, Kemal A Yakuta & Soteris A Kalogirou 2005, 'Exergy analysis of lithium bromide/water absorption systems. Renewable Energy, vol. 30, pp. 645-657.
[19]. Arshi Banu, 0P.S., Sudharsan, N.M., 2018, 'Experimental heat and mass transfer studies on horizontal falling film absorber using water-lithium bromide' Thermal Science, (https://doi.org/10.2298/TSCI180307208A).
[20]. Md.Azhar, Altamush Siddiqui M. First and Second Law Analyses of Double Effect Parallel and Series Flow Direct Fired Absorption Cycles for Optimum Operating Parameters. J. Energy Resour. Technol 2019, 141(12), pp: 124501 (7 pages).
[21]. Md.Azhar, Altamush Siddiqui M. Comprehensive exergy analysis and optimization of operating parameters for double effect parallel flow absorption refrigeration Cycle. Thermal Science and Engineering Progress 2020, 16, pp: 100464.

Nomenclature:

Q	-	Heat load of components (W)
T	-	Temperature (°C)
h	-	Enthalpy (kJ/kg)
s	-	Entropy (kJ/kgK)
ψ	-	Availability (kJ/kg)
$\Delta\psi$	-	Exergic loss /Availability (kJ/kg)
X	-	Solution concentration (%)
FLA	-	First law Analysis
SLA	-	Second Law Analysis
RHX	-	Refrigerant Heat Exchanger
SHX	-	Solution Heat Exchanger
COP	-	Coefficient of Performance
Suffix		
a	-	Absorber
c	-	Condenser / Cooling water
e	-	Evaporator
g	-	Generator
r	-	Refrigerant (Water)
w	-	Weak solution
o	-	Dead state
s	-	Strong solution
1 to 10	-	State points

IOP Conf. Series: Materials Science and Engineering 998 (2020) 012063 doi:10.1088/1757-899X/998/1/012063

Effect of cathode end plate on passive direct methanol fuel cell performance

K Deepak
Professor, Mechanical Engineering Department, Vardhaman College of Engineering, Hyderabad-INDIA
E-mail:deepak045@gmail.com

Abstract. Direct methanol fuel cell (DMFC) belongs to the class of fuel cells that use polymer electrolyte membrane and operate on methanol-water mixture. The low overall performance of passive DMFC compared to active DMFC is due to the slow anode reaction combined with inefficient cathode reaction. In this work, an attempt was made to augment the passive DMFC performance by modifying the structural design of the cathode end plate (CEP). The passive DMFC was fabricated and the performance was investigated experimentally. Three different CEP duct configurations were studied using 1 to 5 M methanol mass fraction. The results indicated higher performance for square duct CEP configuration, with 5 M methanol mixture.

1. Introduction

DMFC is an energy conversion device that directly converts the stored chemical energy of methanol into electrical energy. Owing to ease in storage and delivery, high energy density, and longer life, DMFC is considered as a viable alternative to batteries in portable device applications [1]. DMFCs are of two types, active and passive [2]. The active type DMFC functions with the provision of an auxiliary pump and a blower to supply methanol on the anode side and air on the cathode side. In passive systems, methanol is supplied without an external device and depends on utilizing atmospheric air. The active DMFC has high reliability, yields high performance and easy to regulate the operating parameters such as temperature, methanol mass fraction and flow rate. In passive type DMFC, fuel supply depends on the fuel diffusion in the built-in tank while the atmospheric air is utilized for oxidant. The passive system is more compact and offers a unique feature for fuel cell size reduction.

Structural parameters influence the performance of DMFCs significantly. The use of high open ratio current collector (CC) exhibits high performance due to enhanced methanol flow rate from the fuel chamber to the gas diffusion film [3]. Methanol concentration is one of the vital parameter that has substantial effect on the fuel cell performance. Structural adaptions influence the degree of methanol concentration and the open circuit voltage (OCV) characteristics [4]. Methanol mass fraction, cell positioning and temperature influence the OCV values. Horizontal cell positioning is ideal for higher methanol mass fraction and vertical positioning is chosen for lower mass fraction to limit methanol crossover and carbon dioxide gas blockage [5]. Structural parameters such as current collector open ratio, gasket, and assembly shape influence the contact resistance effect between the current collector and gas diffusion film [6]. Light weight CCs were developed and feasibility for using in a 3-cell DMFC module was demonstrated, that has full possibilities for a variety of applications. Multiple cells can be incorporated to produce high voltage [7]. The Faraday efficiency of DMFC can be increased substantially by modifying the system layout and operating strategy [8]. The passive DMFCs are associated with the challenges such as methanol crossover, species management, heat and water management, slow kinetics, stability and durability. Various methanol transport barriers are used to control the methanol crosswise flow from anode side to cathode.

IOP Conf. Series: Materials Science and Engineering 998 (2020) 012063 doi:10.1088/1757-899X/998/1/012063

Working temperature boosts the electrochemical action and hence the fuel cell performance. It is important to enhance the air and water flow on the cathode side for reducing water flooding [9]. The cathode CC design is very crucial for efficient functioning of the fuel cell. A good cathode CC facilitates efficient oxygen flow for reaction; evacuate water bubbles besides electron collection [10].

2. Methodology

2.1. Cell Fixture and Membrane Electrode Assembly (MEA)

The experiment was conducted on passive DMFC. A single cell arrangement was designed that has 45 cm^2 effective cross-sectional area. Fig. 1 shows detailed view and assembled image of the passive DMFC module. The cell was molded in a polymethyl methacrylate (also known as acrylic glass) frame, and was positioned between a set of CC and diffusion layer. In this system, grooves were provided on the end plates for methanol mixture and oxygen flow. The provisions to tap current and voltage measurement were provided. A Nafion 117 electrolyte membrane that has good proton conductivity, no electron conductivity and good mechanical strength was used in MEA preparation. The polymer electrolyte film separates the anode and the cathode. Fig. 2 shows the proposed CEPs with three different configurations.

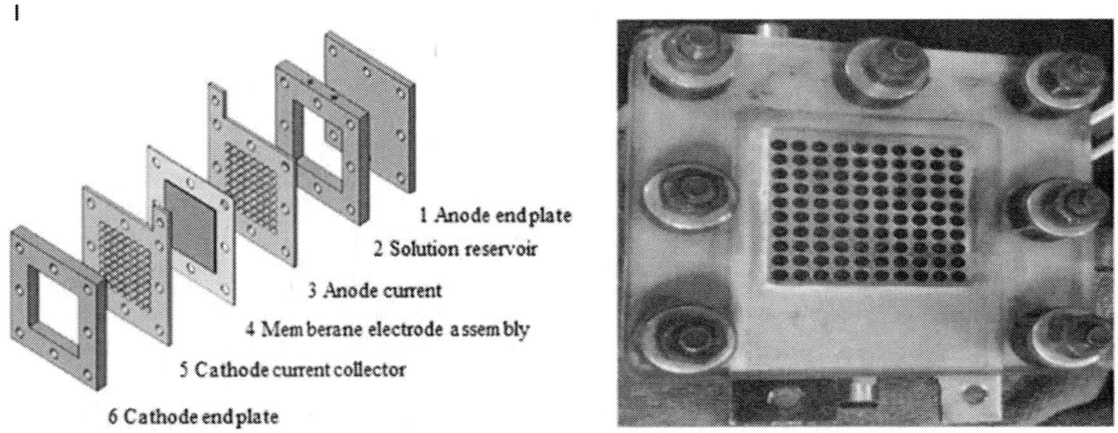

Fig. 1 Passive DMFC Module

(a) (b) (c)

Fig. 2 CEP with (a) Square duct; (b) Hexagonal duct; (c) Circular duct

IOP Conf. Series: Materials Science and Engineering 998 (2020) 012063 doi:10.1088/1757-899X/998/1/012063

2.2. Operation and Parameters

Methanol mixture was filled into the fuel chamber of the cell adjacent to anode; the area adjacent to cathode was kept exposed to the atmospheric air. The complete reactions were taken place leading to the formation of water and carbon dioxide as reaction by-products. Methanol mixture with 1 to 5 molarities was used in the experiment and the performance parameters were calculated for three different CEP configurations. Following reactions occur in the DMFC.

Anode: $CH_3OH + H_2O \rightarrow CO_2 + 6H+ + 6e -$ E_0=0.02V

Cathode: $3/2O_2 + 6H+ + 6e + \rightarrow 3H_2O$ E_0=1.23V

Complete: $CH_3OH + 3/2O_2 \rightarrow CO_2 + 2H_2O$ E_0=1.21V

2.3. Measurement Procedure

A DC power load bank was used for loading and the voltage-current data was noted. For stable voltage values, 1 minute time interval was considered for two consecutive readings. Methanol mixturewas filled using a pipette on the anode side. The atmospheric oxygen was made to flow naturally to reaction sites on the cathode side. Before beginning the experiment, a fresh MEA was activated using 1 M methanol mixture for 16 hours. The experiments were conducted at atmospheric conditions. Three observations were noted for each CEP configuration using 1 to 5 M methanol solution. The average values were plotted.

3. Results and Discussion

To determine the power density of DMFC for different CEP configurations, voltage is measured corresponding to different current density values with 10 mA/cm2 increment. The experiments were conducted using 1 to 5 M methanol mixture separately and the results of 5 M methanol mixture were plotted. The voltage decreases with current density. As depicted in Fig. 3, voltage is highest for the square configuration followed by the hexagonal configuration. The power density variation with current density, illustrated in Fig. 4 shows highest power density for the square configuration and least for the circular configuration.

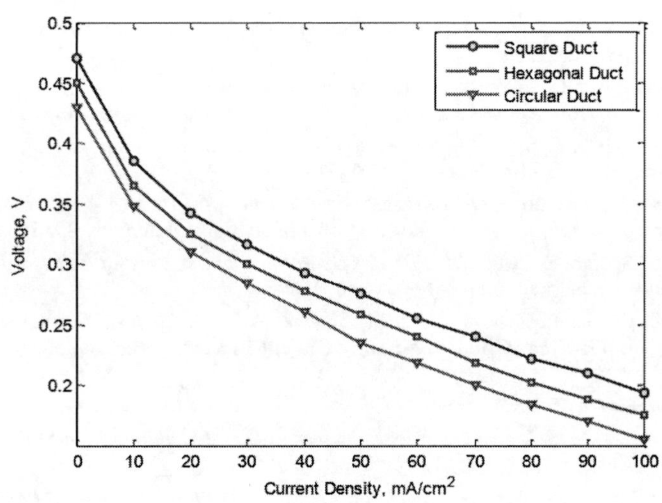

Fig. 3 Variation of voltage with current density using 5 M methanol solution

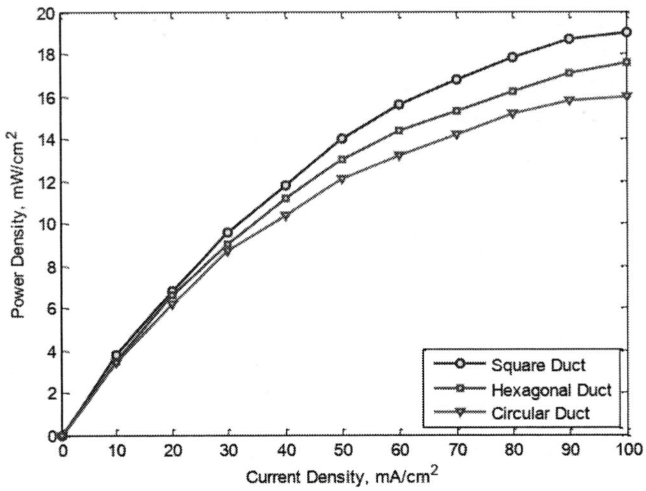

Fig. 4 Variation of power density with current density with 5 M methanol solution

4. Conclusion

This experimental work deals with the performance of a passive DMFC, with different CEP configurations. From the experimental results, it is depicted that, the CEP with square duct gives high performance. For this duct configuration, the methanol concentration for high performance is determined experimentally. The following conclusions were drawn based on the experimental results.

- The use of CEP improves the cell performance.
- The CEP with square duct gives high cell performance in comparison with hexagonal and circular ducts.
- Methanol mass fraction influences the cell performance significantly. For DMFC with square duct CEP, 5 M methanol mixture yields high performance.

References

[1] Zhao T S, Chen R, Yang W W and Xu C, 2009, *Small direct methanol fuel cells with passive supply of reactants* (Journal of Power Source, Vol.191, issue 2) p.185

[2] Kamarudin, S K, Achmad F, and Daud W R W, 2009, *Overview on the application of direct methanol fuel cell (DMFC) for portable electronic devices* (International Journal Hydrogen Energy, Vol. 34, issue 16) p.6902

[3] Lai Q, Yin G P, Wang Z, 2009, *Effect of Anode Current Collector on the Performance of Passive Direct Methanol Fuel Cells* (International Journal of Energy Research, vol. 33) p.719

[4] Yuan W, Tang Y, Wana Z, Pan M, 2011, *Operational Characteristics of a Passive Air-Breathing Direct Methanol Fuel Cell under various Structural Conditions* (International Journal of Hydrogen Energy, vol. 36) p.2237

[5] Yousefi S, Shakeri M and Sedighi K, 2014, T*he Effect of Operating Parameters on the Performance of a Passive DMFC Single Cell* (World Applied Sciences Journal, vol. 22, issue 4) p.516

[6] Borello D, Calabriso A, Cedola L, Zotto L D, 2014, *Simone Giovanni Santori, Development of Improved Passive Configurations of DMFC with Reduced Contact Resistance* (Energy Procedia, vol. 61) p.2654

[7] Kuan Y D, Lee S M and Sung M F, 2014, *Development of a Direct Methanol Fuel Cell with Lightweight Disc Type Current Collectors* (Energies vol. 7) p.3136

[8] Na Y, Zenith F and Krewer U, 2015, *Increasing Fuel Efficiency of Direct Methanol Fuel Cell Systems with Feedforward Control of the Operating Concentration* (Energies vol. 8) p.10409

[9] Munjewar S S, Thombre S B, Mallick R K, *Approaches to Overcome the Barrier Issues of Passive Direct Methanol Fuel Cell – Review* (Renewable and Sustainable Energy Reviews vol. 67) p.1087

[10] Boni M, Rao S S, Srinivasulu G N, 2020, *Performance Evaluation of an Air Breathing–Direct Methanol Fuel Cell with Different Cathode Current Collectors with Liquid Electrolyte Layer* (Asia-Pacific Journal of Chemical Engineering, vol. 15, issue 4) e.2465.

IOP Conf. Series: Materials Science and Engineering 998 (2020) 012064 doi:10.1088/1757-899X/998/1/012064

Effect of Irregularities on Seismic Performance of High Rise Structures

Prasanna Allena[1], Bhavani Chowdary T[2]

[1]Post graduate student, Department of civil, Vardhaman College of engineering, Hyderabad.
[2]Assistant professor, Department of civil, Vardhaman college of engineering,Hyderabad.

bhavanitalasila@vardhaman.org

Abstract: This Paper describes about effect of mass irregularity in regular and irregular high rise structure under seismic loads. These Irregularities refer to abrupt change of geometry, mass, strength and stiffness as well. This leads to uneven distribution of forces and their corresponding deformation over the height of structure. Performing dynamic analysis and comparing the responses of Mass irregularity taken at various stories of regular and irregular buildings. Various aspects such as Modal periods, Modal frequencies, Story drifts, and Response of Time history graphs are drawn while these structures are subjected to Mass irregularity and seismic loads.

Keywords—Irregularities, High rise structure, Non- linear Dynamic Analysis, Modal periods& frequencies etc.,

I Introduction

The word earthquake is used to express any seismic occurrence whether natural or caused by human that can produce seismic influence around any area. Earthquakes are caused generally by rapture of geological falls inside the earth, but also by other events such as volcanic movement, landslides, mine blasts, atomic test. Irregularities referring to sudden change of strength, stiffness, geometry and mass. This results in irregular distribution of forces or deformation over the height of the building. The behaviour of these types of buildings is something different. There is need of more work to be done in this regard. So, this research work is an attempted to reach on more accurate conclusion to be reduce their effect on the structure

1.1 Earth Quake: Shaking of the ground due to sudden release of energy and 95% of earthquakes are caused by plate tectonics. 5% of earthquakes are caused due to other disturbances like volcanic eruptions, rock falls, landslides and explosions

1.2 Structure Irregularities: The irregularity in the building structures may be due to irregular distributions in their mass, strength & stiffness along the height of building when such buildings are constructed in high seismic zones, the analysis and design becomes more complicated. The types of irregularities are.

a) Stiffness Irregularity
b) Mass Irregularity
c) Geometry Irregularity

As this project is dealing with Dynamic analysis for mass irregularity and geometry irregular structures of high rise building. (Where as dynamic analysis is one of the part in seismic analysis).

IOP Conf. Series: Materials Science and Engineering 998 (2020) 012064 doi:10.1088/1757-899X/998/1/012064

1.3 Mass Irregularity: Mass irregularity shall be considered to exit where the seismic weight of any story is more than 150 percent of that of its adjacent stores. In case of roof irregularity need not to be considered.

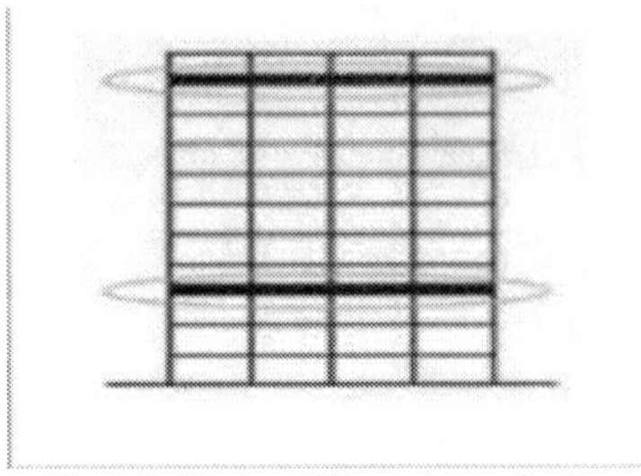

Fig 1 indicates stiffness irregularity

1.4 Geometry Irregularity: A structure is considered to be geometry irregular when the horizontal dimension of the lateral force resisting system in any story is more than 150 percent of that in its adjacent story.

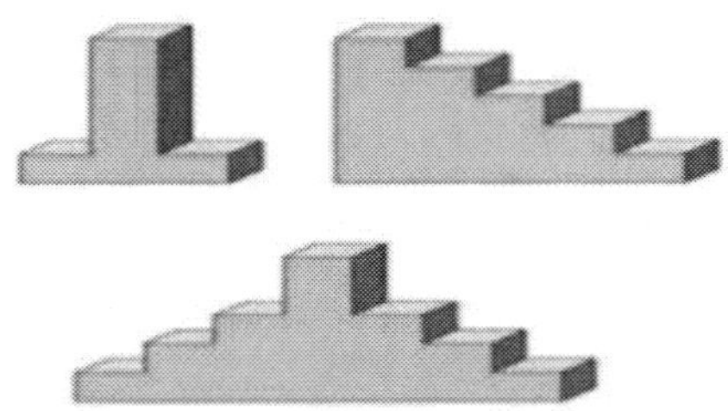

Fig 2 represents types of Geometric Irregularrity

1.5 Seismic Analysis: Analysis of earthquake vibrations and its crust. Subsets of Seismic Analysis-

a) Static Analysis
   - a) Linear Static
   - b) Non Linear Static

b) Dynamic Analysis
   - a) Linear Dynamic
   - b) Non Linear Dynamic

1.6 Non- Linear Dynamic Analysis: Time history analysis is also called as nonlinear dynamic analysis. This method is adopted when the time history is recorded of a particular location is available. In time history analysis, the maximum seismic response of a building is determined with respect to the ground motion records.

1.7 Linear Dynamic analysis: Linear dynamic analysis can be performed by response spectrum method. Response spectrum analysis is a method which is used to fin the structural response to non-deterministic and transit dynamic events. This method provides maximum seismic response of a building by measuring the natural mode of vibration.

1.8 Response spectrum analysis: The response spectrum is the best method to perform the dynamic analysis of seismic resistant building without having the time history data this method is best suited to determine peak seismic response of the building and also to find the accurate results for structural design. The maximum response is plotted against undamped natural frequency. The response spectrum is determined by time or frequency domain analysis and then for a given time period maximum response is picked in this method. The maximum response or maximum design base shear is calculated using the formula

IOP Conf. Series: Materials Science and Engineering 998 (2020) 012064 doi:10.1088/1757-899X/998/1/012064

V = k * x max
Where,
V = maximum base shear.
k = stiffness.
x max = maximum displacement

2.OBJECTIVES

- To study the nonlinear dynamic response in Geometric structures and Non Geometric structures.
- To study the variations in modal periods and frequencies when structures subjected to dynamic loads.
- To study the effect of mass irregularity at different stories when exposed to the nonlinear dynamic response of structures.
- Compare the nonlinear dynamic response of a regular structure to irregular structures when both the structures subjected to mass irregularity.

3.METHODOLOGY

Methodology used in this project to analyze the high rise structural behaviour of regular and geometry irregular with mass irregularity is:

- Selection and design plan of building before heading towards software design
- Plan the high rise structure according to NBC
- Manual Calculations are necessary to give value inputs to build structure
- Modeling and result analysis in ETABS
- Comparison of analysis and its behavior patterns in various aspects, as we are going to discuss it further going deep in topic

Specifications	
No. of stories	G+18
Storey height	3m
Material properties	
Grade of Concrete	M40
Grade of steel	Fe 500
Beam dimensions	300*600 mm
Column Dimensions	1000*1000mm,800*800mm,600*600 mm
Thickness of slab	120mm
Thickness of wall	230mm
Zone factor, Z	0.36
Importance factor, I	1.00
Response reduction factor, R	5.00
Thickness of slab	120mm
Thickness of wall	230mm

IOP Conf. Series: Materials Science and Engineering 998 (2020) 012064 doi:10.1088/1757-899X/998/1/012064

3.1 Models for Analysis

Regular Structure Models:

- Regular Structure with Mass irregularity at 6th Floor(R6MR)
- Regular Structure with Mass irregularity at 12th Floor(R12MR)
- Regular Structure with Mass irregularity at 18th Floor(R18MR)
- Regular Structure with Mass Irregularity at 6th 12th 18th Floor(R)

Geometry Irregular Structure Models:

- Geometry irregularity with mass irregularity at 6th floor(GIR6MR)
- Geometry irregularity with mass irregularity at 12th floor(GIR12MR)
- Geometry irregularity with mass irregularity at 18th floor(GIR18MR)
- Geometry irregularity with mass irregularity at 6th 12th 18th floor(GIR)

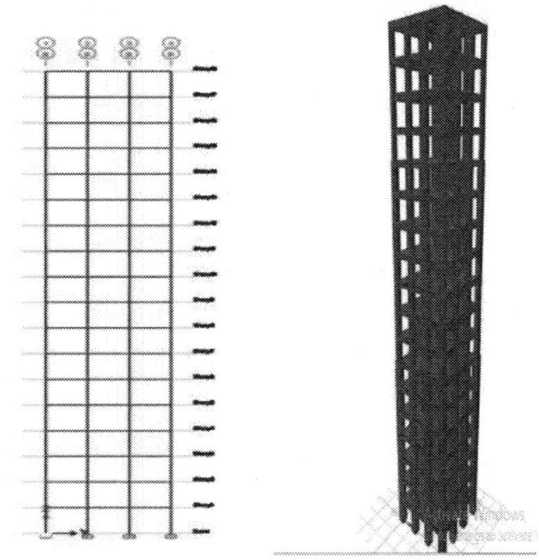

Fig 3 shows regular building with mass irregularity

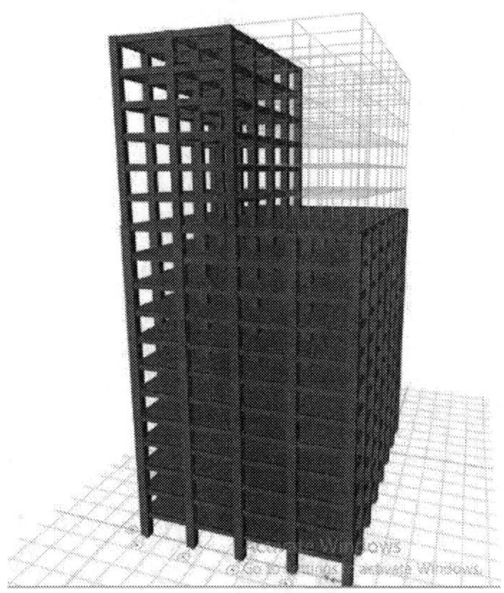

Fig 4 shows geometric irregular building with mass irregularity

IOP Conf. Series: Materials Science and Engineering 998 (2020) 012064 doi:10.1088/1757-899X/998/1/012064

Non linear dynamic analysis have performed to predict the responses of the structure, modal analysis has been simulated in order to verify the natural modes and frequencies and observations has been taken into consideration for storey drifts.

4. Results and discussion

4.1 Modal Time Periods & Frequencies

Comparison of time period and frequency of regular model with variations of mass irregularity at various levels of the high rise structure.

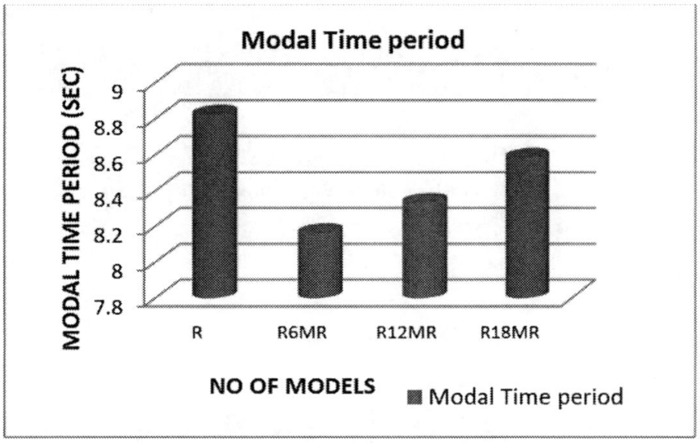

Graph 1 indicates modal time periods

As time period increases frequency decreases.

Comparison of time period and frequency of Geometric irregular model with variations of mass irregularity at various levels of the high rise structure.

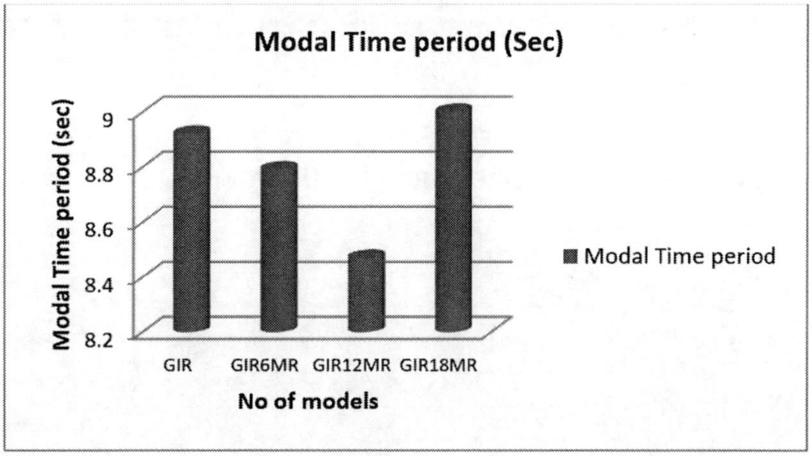

Graph 2 specifies modal time period for geo metric irregular structure

4.2 Storey Drifts

Maximum drift is observed at regular building at 18th floor mass irregularity and minimum drift at regular building at 6th ,that may be caused due to excess mass mounted at the top storey.

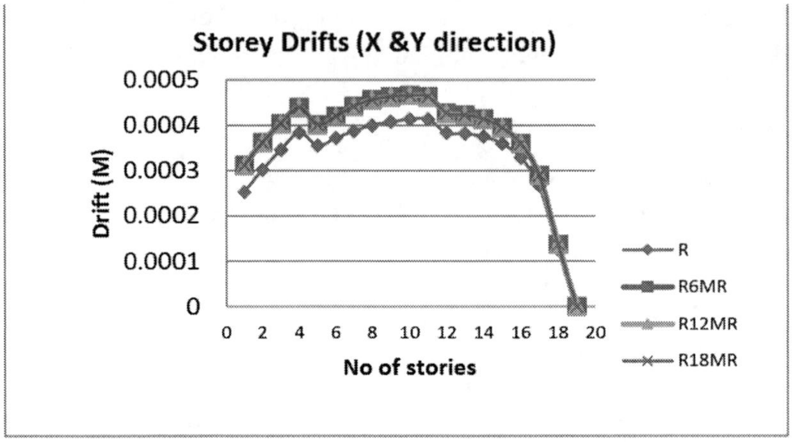

Graph 3 depicts storey drifts in X & Y direction for regular models

Comparison of storey drifts in X and Y direction of geometric irregular structure with variations of mass at 6, 12,18th floors.

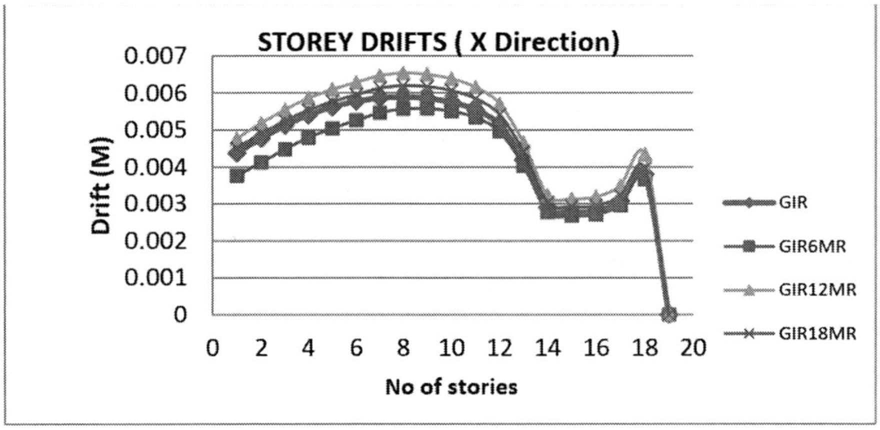

Graph 4 describes storey drifts in X direction for geometric irregular structure

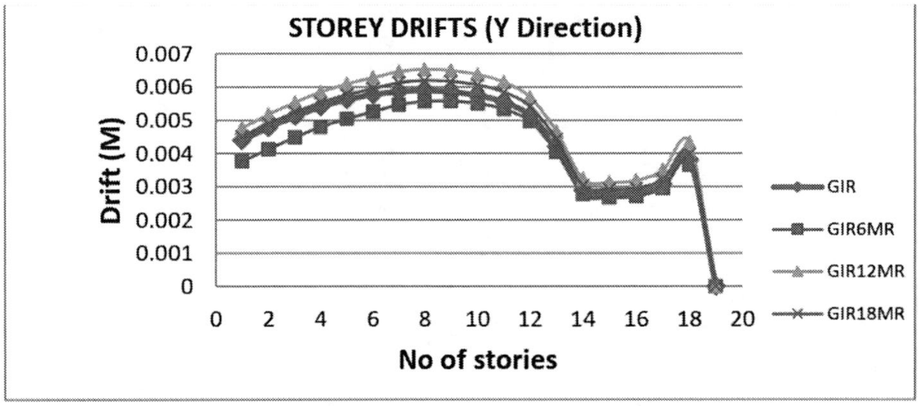

Graph 5 shows storey drifts in Y direction for geometric irregular structure.

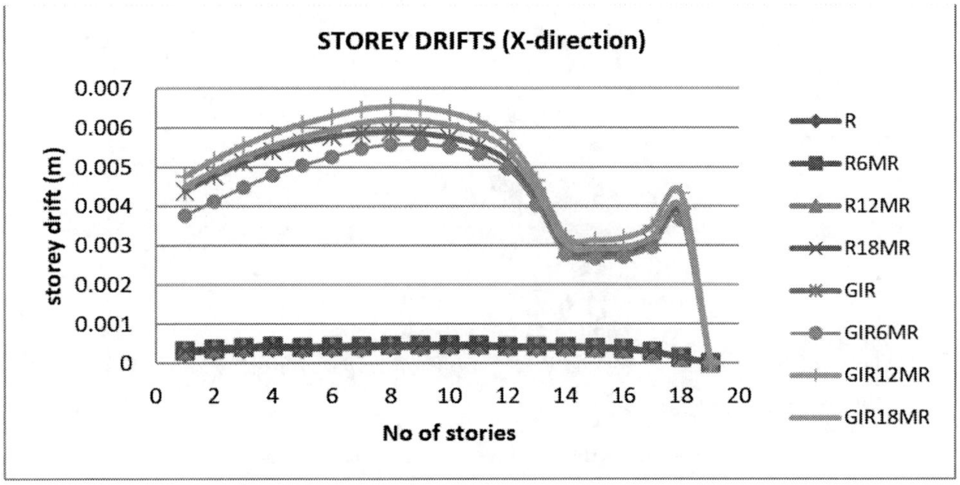

Graph 6 describes storey drifts in X direction for all models.

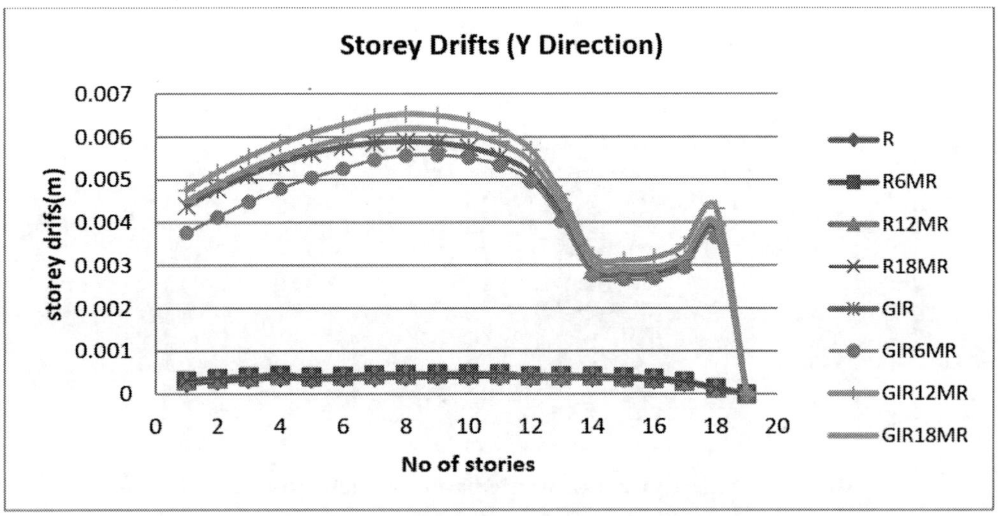

Graph 7 represents storey drifts in X direction for all models.

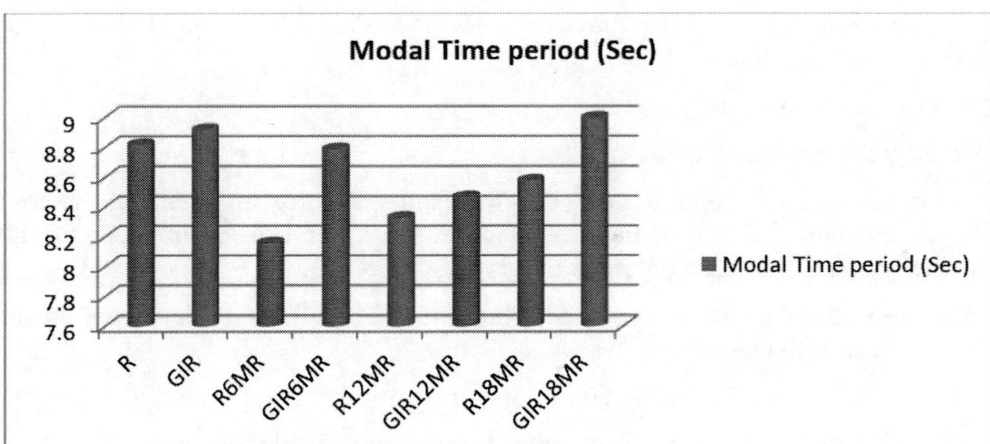

Graph 8 indicates the time period for all models

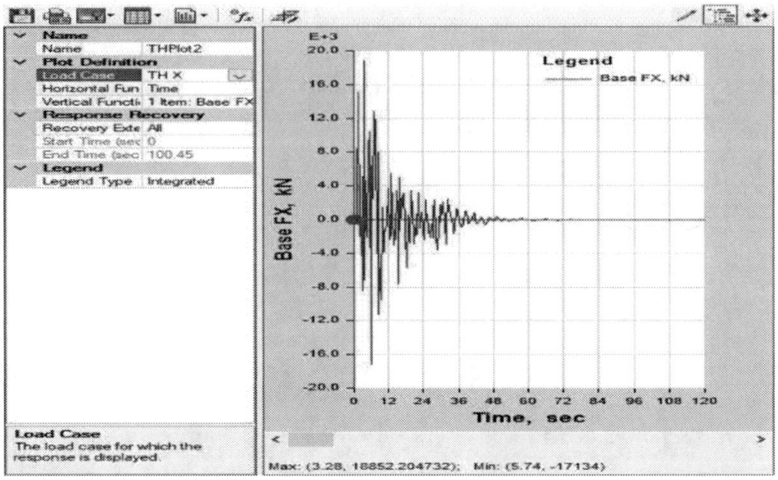

Graph 9 shows the response of regular structure@ 18th floor.

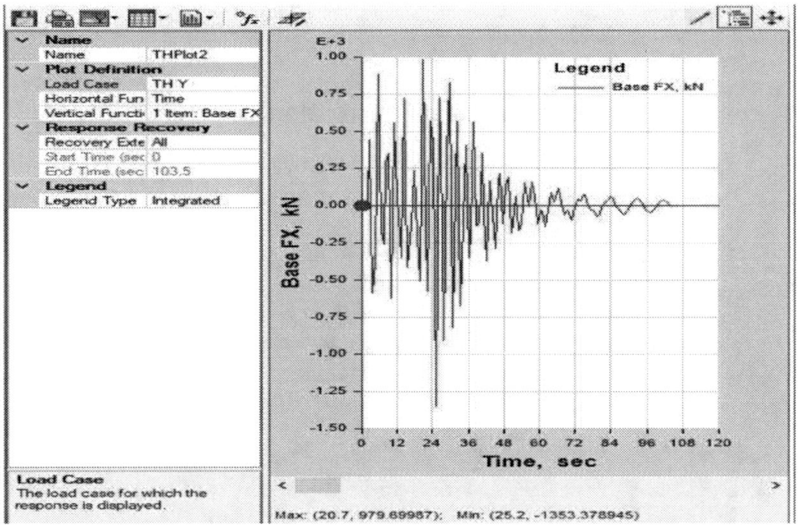

Graph 10 illustrates the response of geometric irregular structure@ 18th floor.

From graph 9 & 10, the responses have been observed that, a non geometric structure shows less resistance under seismic loads.

5. Observations of the study

5.1 Modal frequencies and time period:

- In comparison with both structure regular and geometry irregular high rise structure modal periods and frequencies followed the same trend, from 6th floor to 18th floor enhancement of the time period is observed.
- Maximum time period is observed when mass irregularity is developed on the 18th floor of the both models.
- First mode values are considered as it indicates primitive response of the structure,
- As mass of the structure of structure increases frequency decreases.
- Taking into consideration development of mass irregular in the geometric irregular structure time period varies in between 8.47 to 9.00 sec.

IOP Conf. Series: Materials Science and Engineering 998 (2020) 012064 doi:10.1088/1757-899X/998/1/012064

- In regular structure, time period varies in between 8.16 to 8.82sec.

5.2 Storey drifts

- When two models Regular, Geometry Irregular are compared, maximum storey drifts are observed in GIR i.e., 9mm drift is observed in the Y direction.
- In R models, the maximum drift observed is about 0.46mm in both X &Y directions.
- In GIR models, maximum drift is 6mm and 9mm in X &Y directions respectively.

5.3 Responses in Time History Analysis

- In R model optimum non linear responses are observed in the graph between pseudo spectral acceleration (mm/sec^2) with period (sec), the response of time history in X direction is max 8.828e10 mm/sec^2 at a period of 1.59 sec and is max 6.13e10 mm/sec^2 at a period of 1.59 sec.
- In GIR model peak non linear responses are observed in the graph between pseudo spectral acceleration (mm/sec^2) with period (sec), the response of time history at 18th floor is max 9.00e10 mm/sec^2 at a period of 2.9 sec in X direction and 9.00e10 mm/sec^2 at a period of 2.33 sec in Y direction
- By studying the non linear responses of R models, mass irregularity at 6th floor has less response; hence we can conclude that increased mass makes the model rigid at base level and exhibit greater resistance to fail.
- By analyzing the time history responses of GIR models ,mass irregularity at 12th floor has less response, since the model is stepped ,mass changes from 12th onwards ,it acts as a tunned mass damper

6. Conclusions

From the all the parameters considered in R & GIR models, when the center of gravity and center mass coincides; mass irregularity does not cause major change in frequency reduction. Increments in masses at lower stories make the model much stiffer and exhibit resistance. Here mass it acts as a damper.

References

[1] N.Anvesh , Dr. Shaik Yajdani, K. Pavan kumar "Effect of Mass Irregularity on Reinforced Concrete Structure Using Etabs" Vol. 4, Issue 10, October 2015

[2] Devesh P. Soni and Bharat B. Mistry "Qualitative review of seismic response of vertically irregular building frames" Vol. 43, No. 4, December 2006, pp. 121-132.

[3] V.K. Sadashiva, G.A. MaacRae and B.L. Deam "Determination of Irregularity limits" The 14th World Conference on Earthquake Engineering.

[4] S.Varadharajan, V.K. Sehgaland B.Saini "Review of different Structural irregularities in buildings" Vol. 39, No. 5, December 2012 - January 2013 pp. 393-418.

[5] Mr. Gururaj B. Katti and Dr. Basavraj S. Balapgol "Seismic Analysis of Multistoried RCC Buildings Due to Mass Irregularity By Time History Analysis" Vol. 3 Issue 7, July – 2014.

[6] Hema Mukundan, S.Manivel "Effect of Vertical Stiffness Irregularity on Multi-Storey Shear Wall-framed Structures using Response Spectrum Analysis" Vol. 4, Issue 3, March 2015.

[7] Pralobh S. Gaikwad, Prof. Kanhaiya K. Tolani "Study Of Dynamic Effect On Unsymmetrical Building (Rcc & Steel)" Volume 3, Issue 3, JuneJuly, 2015.

IOP Conf. Series: Materials Science and Engineering 998 (2020) 012064 doi:10.1088/1757-899X/998/1/012064

[8] Dileshwar Rana, Prof. Juned Raheem "Seismic Analysis of Regular & Vertical Geometric Irregular RCC Framed Building" Volume: 02 Issue: 04 | July-2015.

[9] Thi Thi Hein Lwin, Kyaw Lin Htat "Study on Effect of High Rise Steel Building with Different Masses" Volume 4, Issue 7, July 2015.

[10] Poncet, L. and Tremblay, r. "Influence of mass irregularity on the seismic design and performance of multi-storey braced steel frames" 13th world conference on Earthquake Engineering Vancouver, B.C., Canada August 1- 6, 2004 paper no. 2896.

A Post-processing Fusion Framework for Deep Learning Models for Crop Disease Detection

Subodh Bansal[1], Anuj Kumar[2]

[1]Department of Computer Science and Engineering, University Institute of Engineering and Technology, Panjab University, Sector 25, Chandigarh
[2]Department of Computer Science and Applications, Panjab University, Sector 14, Chandigarh
Subodhbansal991@gmail.com

Abstract. The paper proposes a Post-processing Fusion Framework (PFF) for crop disease detection. The framework combines the prediction matrices produced by the multiple deep learning models and increases the overall classification accuracy of the system. The model is tested on the Plant Village dataset covering 14 plants and 26 diseases in 54,305 images, divided into 38 classes. Five deep learning models are used in the classification stage. Further, the PFF is applied to the results of the classification stage. Post-processing enhanced the efficiency of the system and achieved an accuracy of 99.33%. It secures 99.99%, 99.61% and 99.42% scores for top-5 accuracy, Mean Reciprocal Rank and Mean Average Precision values, respectively. Additionally, the classification rate of the approach is less than one second per test image. The high speed and accuracy make the technique highly suitable for real cultivation early warning systems.

Keywords. Post-Processing, Deep Convolutional Neural Networks, Deep Learning, Image Classification, Plant disease detection.

1. Introduction

Crop loss due to diseases is a significant cause for the reduction in crop yield. Every year, over 20% of the agriculture output is lessened due to pathogenic infections caused to crops [1]. It results in a reduction in the availability of food, thus making it a significant challenge in achieving the Goal-2: 'Zero Hunger' set by the United Nations Organisation [2]. Therefore, a mechanism for mitigating the same is the need of the hour.

Crop loss due to diseases can be prevented or reduced by its prompt detection and correct identification and early treatment. So, a crop disease detection and classification system (CDDS) is required. The field of computer vision and artificial intelligence can be used for the development of the system.

Many researchers have made efforts to develop effective CDDSs. Several approaches ranging from the traditional pattern matching to the modern-day state-of-the-art deep learning (DL) models have been applied for its development. The steps involved in conventional and modern approaches are listed in figure 1(a) and 1(b), respectively.

IOP Conf. Series: Materials Science and Engineering 998 (2020) 012065 doi:10.1088/1757-899X/998/1/012065

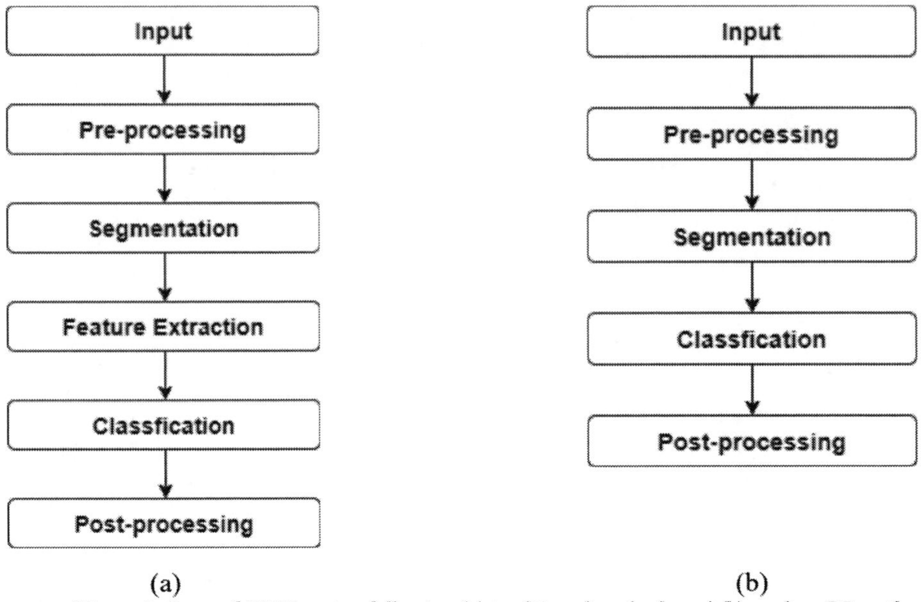

Figure 1: Steps of CDSS system following (a) traditional methods and (b) modern DL methods

In the traditional approaches, the main focus of the researchers was to handpick the best features (symptoms) for detection of the disease. Whereas in the modern approaches, testing every possible orientation of the black box DL models to maximise the classification accuracy is the focus. Still, in majority of the researches, regardless of the approach, work is done on the stages starting from pre-processing to classification. The post-processing stage is often neglected.

In the present investigation, a Post-processing Fusion Framework (PFF) is proposed to increase the accuracy of the DL models after the classification stage. It is implemented on the Plant Village [3] dataset. For validation of the framework, five DL models are used. The paper contains an in-depth literature review of the existing CDDS approaches, in section 2. Section 3 presents the proposed PFF. The experimentation of the investigation is detailed in section 4. The obtained results are discussed in section 5. Finally, the conclusions drawn, and the possible future directions of the research field are discussed in section 6.

2. Related work

The efforts made in the field of computer vision and machine learning for the development of CDDS are summarised in this section. First, the traditional methods are discussed followed by the modern methods.

2.1. Traditional methods

S. Kai et al. [4] proposed an approach for crop disease detection using Grey Level Co-occurrence Matrix (GLCM) [5] for feature extraction and Back Propagation Neural Networks (BPNN) [6]. The researcher applied the technique on a self-collected dataset of maize leaf images and achieved an accuracy of 98.00%. Z. Zhang et al. [7] experimented with variety of combinations of Genetic Algorithms (GA) [8] and Support Vector Machines (SVM) [9]. Experimentations on a self collected dataset concluded in an accuracy of 90.25%.

B. Jiang et al. [10] experimented on water stress for Maize using Super-Pixel (SP) and GLCM for feature extraction and SVM for classification. Application of the methodology on a Self constructed dataset of 1297 leaf mages (in black background), was able to achieve an accuracy of 98.97%. X. E. Pantazi et al. [11] implemented Local Binary Patterns (LBPs) for feature extraction and One-Class Classification (OCC) technique for classification. The researcher identified downy mildew, Powdery mildew and black rot diseases from crop leaf images.

IOP Conf. Series: Materials Science and Engineering 998 (2020) 012065 doi:10.1088/1757-899X/998/1/012065

2.2. *Modern methods*

The parallel computation power of computers have increased a lot in the recent years, especially with the introduction of computationally powerful Graphical Processing Units (GPU) having hundreds of computing cores. This has helped the field of Artificial Intelligence to achieve seen exponential advancement. Execution of the computationally expensive deep learning models (DL) (once thought to be just theoretical concepts) is now possible [12]. Following are some of the advancements of the same in the area of crop disease detection.

G. Wang et al. [13] used (Video Geometry Group) VGG-16, VGG-19 [14], Inception-v3 [15], and Resnet [16] DL models for detection of Black Rot disease in Apple leaves, in the Plant Village dataset achieving an accuracy of 93.1%. M. Brahimi et al. [17] used AlexNet [18] and GoogLeNet [19] models for multiple diseases of tomato leaves achieving accuracy, precision, recall and F1 score of 99.18%, 99.53%, 98.53% and 98.52%, respectively. K. Yamamoto et al. [20] applied structurally constrained Recurrent Neural Networks [21] on the same dataset achieving the accuracy of 95%.

J. Gwan et al. [22] used VGG-A DL model on radish leaf images captured using Unmanned Aerial Vehicle (UAV) for the detection of *Fusarium wilt* and scored an accuracy of 97.40%. J. Lu et al. [23] used VGG-FCN for disease detection for Wheat leaves on Wheat Disease Database 2017 (WDD2017) achieving accuracy of 97.95%. C. Dechant [24] developed a CNN model for the detection of Northern Corn Leaf Blight in Maize leaves and correctly identified the same, 96.70% of times.

A. Fuentes et al. [25] applied various (AlexNet, ZFNet, VGG- 16, GoogLeNet, ResNet-50, ResNet-101, ResNetXt-101 [26], Faster RCNN [27], R-FCN [28] and SSD) DL models on a self extracted dataset of 5000 images covering 9 diseases of tomato plant leaves from Korea, achieving an average precision of 83.06%. A. C. Cruz et al. [29] developed a modified DL X-FIDO (*Xylella FastIdiosa* Detector for *O. europaea L.*) model for olive quick decline syndrome disease of Olive plant, achieving an accuracy value of 98.60%.

A. Ramcharan et al. [30] worked on cassava plant leaf disease detection using transfer learning with Inception v3 and ImageNet DL models on a self extracted dataset of 11,670 images achieving an accuracy of 93%. B. Liu et al. [31] worked with self captured dataset of 1053 images of Apple crop using AlexNet, GoogLeNet, VGG-16 and ResNet-20 DL models achieving classification accuracy of 97.62%. J.Amara et al. [32] worked with Banana crop's self captured dataset containing 147 images of healthy and infected leaves. Further experimentation with LeNet model resulted in achievement of accuracy, precision, recall and F1-score of 98.61%, 98.67%, 98.60% and 98.64% respectively.

S. Ghosal et al. [33] worked on Soybean using self-captured dataset of 25,000 images with AlexNet DL model, achieving classification accuracy of 94.31%. M. Sibiya et al. [34] conducted experiments with CNN using an aggregation approach for Maize leaves for detection of fungal diseases from Plant Village Dataset [3] and achieved the success rate of 92.85%. J. C. A. Barbedo [35] used DL approaches on individual lesions and spots of diseased leaf instead of whole leaf which increased the accuracy by 12%.

R. V. D. Vijver et al. [36] worked on high-resolution canopy reflectance images of 32 potato plants. The researcher used spectral classifiers like partial test squares discriminant analysis (PLS-DA) and SVM to detect the presence of the early blight disease caused by *Alternaria solani*. The experiment achieved an accuracy of 92%. W. Kim et al. created a disease monitoring system for onion farms for the detection of downy mildew disease. The images were used to train a DL model developed by B. Zhou et al. [37] and achieved an accuracy of 87.20%.

3. Post-processing fusion framework

From the modern techniques discussed in the previous section, it is seen that majority of the researches end the experimentation on the classification stage, and the post-processing stage is often neglected. Therefore a post-processing technique is proposed. The working and architecture of the technique are discussed further.

The classification stage consists of one or more DL models. A DL model is a neural network having multiple hidden layers. Its last layer is called the output layer. The output layer consists of nodes equal to the number of classes present in the test dataset. For each test image, the output layer produces a percentage score at every node in the output layer. This percentage score depicts the confidence of the DL model in the test image belonging to the corresponding class. The node having the maximum prediction score is considered as the final prediction of the model.

The percentage scores list obtained at the output layer of a DL model can be written as a list of key-value pair (class name/node, percentage score). The list of key-value pair is known as the prediction matrix.

After running many simulations of the existing DL models on the standard annotated crop disease datasets such as the Plant Village dataset and analyzing the prediction matrices produced, some useful findings are noticed:

i. Many a times, a test image misclassified by one DL model is accurately classified by another.
ii. A DL model achieving higher classification accuracy also shows misclassifications that are correctly classified by the inferior models.
iii. In majority of the cases, when the prediction value is compared, the model showing the correct classification of the test image often score better than the best prediction score of model, misclassifying the image.

All the above findings are analyzed, and a general equation (1) is formulated. The formula fuses the outputs of the existing DL models to achieve better accuracy that is not possible using DL models in isolation. The technique is named Post-processing Fusion Framework or PFF in short.

$$P_{fm} = \max \Pr ob(DP_1, DP_2, DP_3, ..., DP_n) \quad (1)$$

In equation 1, P_{fm} signifies the prediction score of the experimentation after using PFF on the test image and DP_n depicts the highest prediction score of the participating DL models.

4. Experiment

The steps of the experimentation are illustrated as a flowchart in figure 2 and listed ahead.

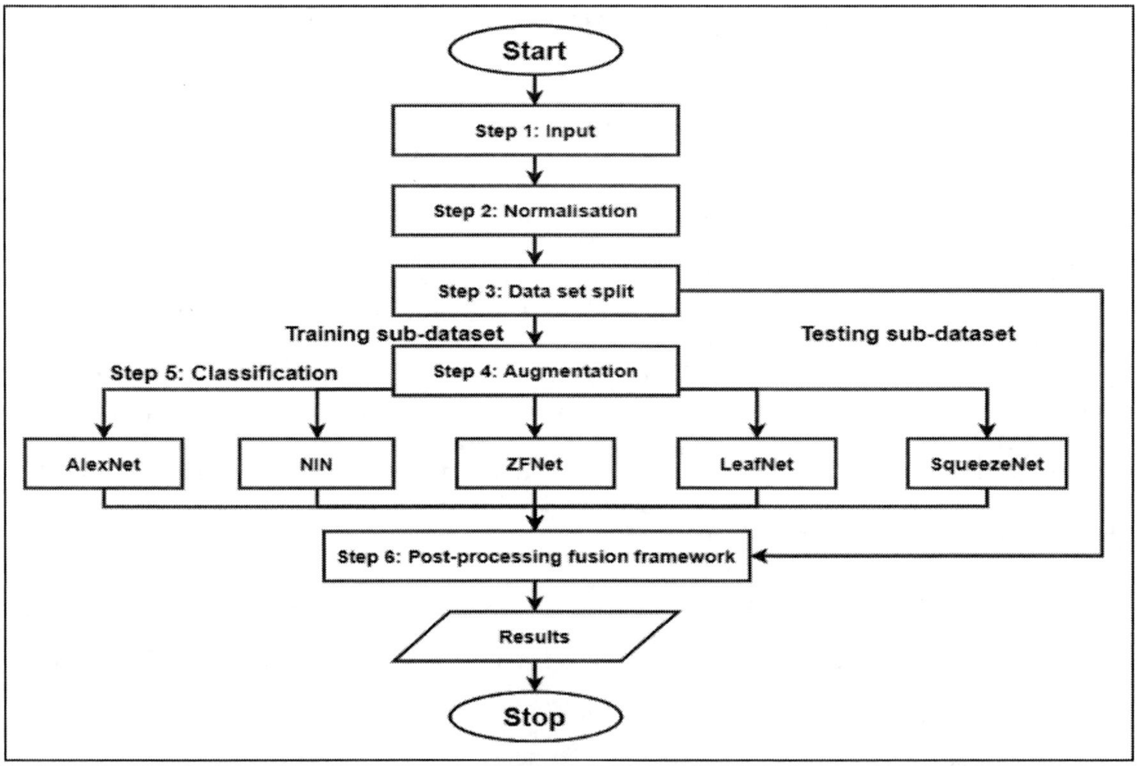

Figure 2: Methodology

IOP Conf. Series: Materials Science and Engineering 998 (2020) 012065 doi:10.1088/1757-899X/998/1/012065

Step 1: Input: the dataset is obtained and prepared for the experimentation.
Step 2: Normalization: the images of the dataset are resized and cropped for the experimentation.
Step 3: Split: the dataset is split into training and testing sub-sets.
Step 4: Augmentation: the training subset is augmented to reduce the risk of over-fitting.
Step 5: Classification: DL models are trained and tested on the training and testing subsets, respectively and the prediction matrices are prepared.
Step 6: Post-processing: the prediction matrices obtained from the classification step are post-processed using PFF.

4.1. Dataset

An open access dataset containing 54,305 annotated images provided by D. P. Hughes et al. [3] is used. It covers 14 plant species and 26 diseases. Overall, the dataset is classified in 38 classes.

4.2. Dataset normalization

The dataset is normalized, i.e. resized and cropped to 256x256 pixels. Table 1, shows that the DL models considered for the experiment use images of size 256x256 or a little less, hence it is the most optimal size. Further, the images can be reduced to this size without much loss of necessary information. The steps used for normalization are:

Step 1: Detection of the smaller dimension, length or width
Step 2: Reduction of the smaller dimension to 256 pixels while maintaining on aspect ratio
Step 3: Centring of the image to 256x256 pixels canvas, chopping the remaining area.

4.3. Training and testing datasets

The image dataset is then split into two subsets using the 80:20 approach. 80% images per class are used for neural network training and the remaining 20% for the testing purpose. The pictures are selected at random for the subsets. The numbers of images in the training and testing subsets are 43,456 and 10,849, respectively.

4.4. Dataset augmentation

After normalization, the training subset images are subjected to augmentation. Augmentation is performed to obtain better results, on a relatively small dataset. Additionally, it reduces the risk of over-fitting of the model on the training set. In the present experiment, each image is augmented into 29 images, increasing the total number of training images from 43,456 to 12,60,224. The augmentation strategy has four main elements:

- Image rotation – it is helpful to mimic alternate angles from which an image can be captured.
- Scaling – it is helpful in mimicking different distances from which an image can be captured.
- Brightness – it is helpful in mimicking varying lighting conditions in which an image can be captured, like over and underexposure especially during noon and dusk, respectively.
- Clipping – it is helpful in mimicking the event when the area of interest is only a small part of the actual image.

Additional pre-processing does not improve the classification performance in case of DL models [38]. Hence, it is not considered. Further, as the DL models using Convolutional Neural Networks automatically identifies the significant and insignificant features in images, so the complicated step of segmentation is also avoided.

4.5. Model training

After augmentation, the next step is neural network training. Five prominent deep neural networks are selected for the purpose: (i) AlexNet [18], (ii) NIN [39], (iii) ZFNet [40], (iv) LeafNet [41], and (v) SqueezeNet [42]. These models are trained, tested and implemented using Caffe DL framework using python programming language on a machine with NVIDIA GTX1080Ti GPU card, using CUDA parallel programming platform on Linux machine (Ubuntu 16.04 operating system).

The input size of the images required for each neural network is shown in table 1. So the images of the training subset are cropped accordingly.

Table 1: Neural network input image size

Neural Network	Input Image Size
AlexNet [18]	227x227x3
NIN [39]	224x224x3
ZFNet [40]	225x225x3
LeafNet [41]	256x256x3
SqueezeNet [42]	227x227x3

The hyper-parameters of the DL models are optimized using the random search approach. The hyper-parameters used in the final experimentation are listed in table 2. The learning rate starts from 1.5625 x 10^{-2} and is annealed by a factor of one half after every one-tenth of the total number of mini-batches, i.e. 7,876 images, and rested at 3.0518 x 10^{-5}.

Table 2: CNN parameters and solver training hyper-parameters

Parameter	Value	Parameter	Value
Epoch Size	12,60,224	Gamma (ϒ)	0.5
Epochs	2	Weight Decay	0.0002
Batch Size	32	Base Learning Rate (α)	1.5625 x 10^{-2}
Batches / Epoch	78,764	Learning Rate Decrease Policy	Step
Momentum (μ)	0.5	Step Size	7,876

Training of all the five DL models is performed on the augmented training dataset and the models weight biases are recorded and cached.

4.6. Post-processing fusion framework

The test images are passed one by one from the trained DL models and the probability matrices are recorded. Further, the fusion framework is applied and the final result i.e. the final prediction is obtained and recorded. The results are discussed in the following section.

5. Experimentation Results

The results of the experiment are recorded at two levels: (a) after classification, and (b) after PFF. The performance of the methodology is evaluated using four performance metrics: top-1 accuracy (O_1), top-5 accuracy (O_5), Mean Reciprocal Rank (MRR) [43] and Mean Average Precision (MAP) [44]. The results obtained are discussed further.

5.1. Obtained results

At level *a*, i.e. after classification, the individual trained models are validated on the test sub-dataset. At level *b*, the trained model's outputs are fused using PFF. The results obtained from the both levels are recorded and tabulated in table 3 and plotted in figure 3.

Table 3: Performance of CNN model architectures for classification on test dataset

Level	Model	O_1 (Value %, no. of Iterations)	O_5 (Value %, no. of Iterations)	MRR (Value %, no. of Iterations)	MAP (Value %, no. of Iterations)	Misclassifications
a	AlexNet	96.00 (63008)	99.95 (78760)	97.85 (63008)	96.87 (55132)	434
a	LeafNet	96.24 (78762)	99.86 (70884)	97.92 (78762)	97.03 (78762)	408
a	NIN	94.14 (78760)	99.90 (78760)	96.80 (70884)	95.40 (78760)	636
a	SqueezeNet	89.66 (78760)	99.44 (78760)	93.98 (78760)	91.61 (78760)	1121
a	ZFNet	96.01 (78760)	99.94 (55132)	97.88 (55132)	96.92 (55132)	433
b	PFF	99.33 (78760)	99.99 (55132)	99.61 (70884)	99.46 (78760)	70

It can be clear from the table 3 that the LeafNet DL model scores the highest average top-1 accuracy (O_1) of 96.24% at 78762 iterations. The best score in average top-5 accuracy (O_5) is achieved by AlexNet, with a score of 99.95% at 78760 iterations. The best Mean Reciprocal Rank (MRR) and

Mean Average Precision (MAP) are again achieved by LeafNet at 97.92% and 97.03%, respectively at 78762 iterations.

After implementation of PFF, the scores at all the performance metrics increases. The value of O_1 increases from 96.24% (LeafNet) to 99.33%; O_5 increases from 99.95% (AlexNet) to 99.99%; MRR increases from 97.92% (LeafNet) to 99.61%; and finally, the value of MAP increased from 97.03% (LeafNet) to 99.46%. Additionally, the fusion framework results in decreasing the number of misclassifications from 3.76% (408 images) to 0.67% (70 images).

Further, after investigating the misclassified images, the following reasons may be stated for these misclassifications:

i. Similar background noise in multiple images, like the hand of the image capturer.
ii. The picture has small Region of Interest (ROI).
iii. Various diseases have similar symptoms, even confusing the human eye.

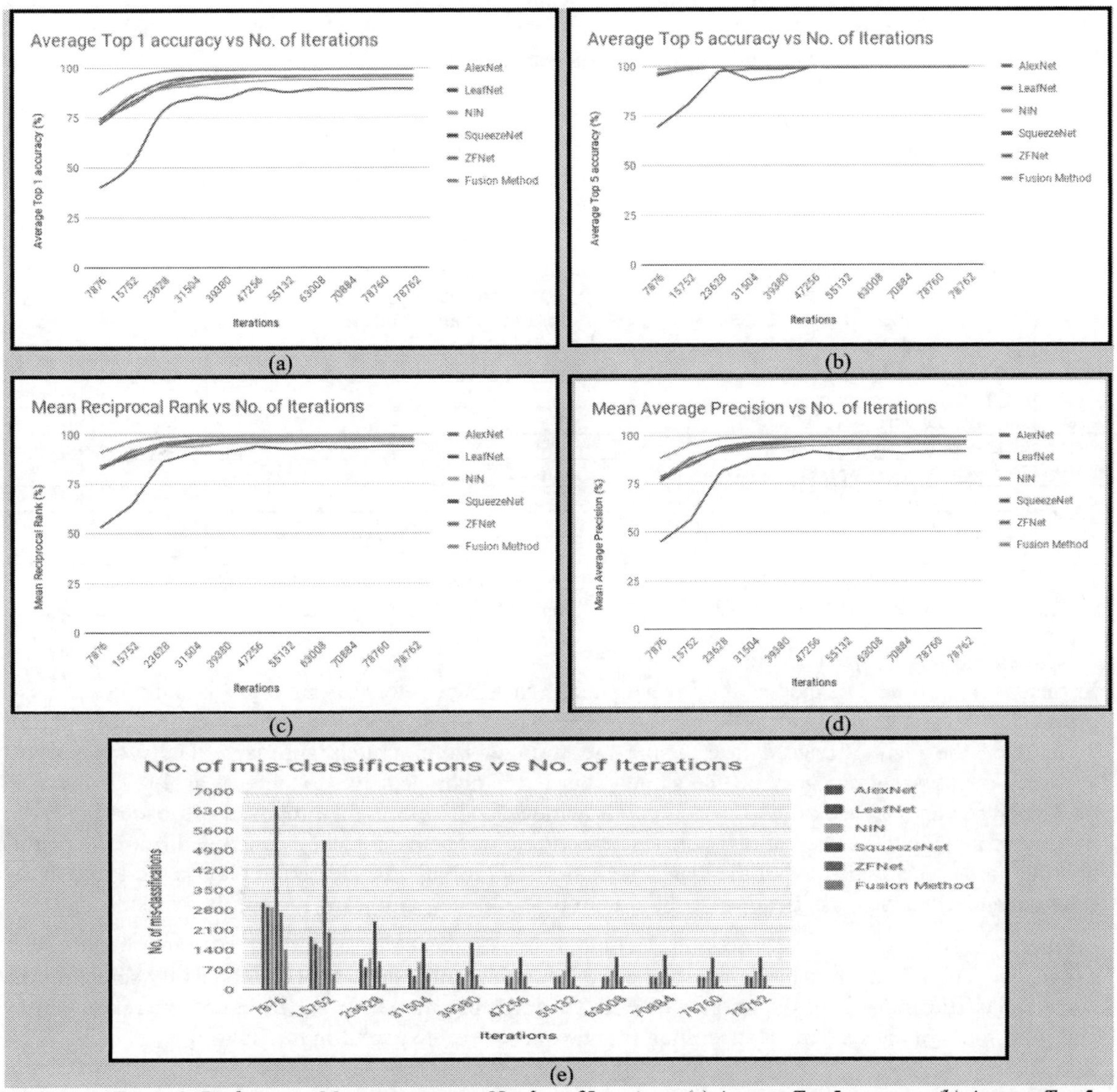

Figure 3: Performance Measurement over Number of Iterations: (a) Average Top 5 accuracy, (b) Average Top 5 accuracy, (c) Mean Reciprocal Rank, (d) Mean Average Precision and (e) Number of misclassifications

5.2. *Importance of number of iterations*

The number of iterations is an important metric to keep the space and time complexity in check. Using an excessive number of iteration may waste time in model training, this time can be utilised for optimisation of the model or its use in the field. Therefore, keeping the number of iterations in check is an important parameter.

In this experiment, the number of training images are 43,456, and are augmented to 12,60,224. The batch size is kept at 32 (optimised using the random search algorithm). The maximum number of iterations possible are 78,762 (epoch size * number of epochs/batch size, 12,60,224 * 2/32) per epoch. For comparison of iterations, the trained models are screenshot at every one-tenth of the total number of iterations, i.e. at every 7876 iterations. The results are plotted in various graphs, shown in figure 3. The results show a consistent trend in all the five graphs (3(a) to 3(e)), i.e. for all four performance metrics (O1, O5, MRR and MAP). The performance initially enhances with an increase in the number of iterations. But, after reaching a threshold (47,256 for AlexNet, 55,132 for NIN, 47,356 ZFNet, 39,380 for LeafNet, 63,008 for SqueezeNet, and 55,132 for PFF), the increase in the number of iterations became insignificant and the performance stagnates.

5.3. *Comparison with existing approaches*

The results of the existing methods and the present experimentation are shown in table 4. From the results, it is evident that the proposed approach outperformed the existing methods by a fair margin.

Table 4: Comparison of the proposed PFF technique with the existing approaches

Year	Diseases covered	Accuracy	Reference
2020	Apple scab, Bacterial spot, Black rot, Cedar apple rust, Cercospora leaf spot, Common rust, Early blight, Esca (Black Measles), Gray leaf spot, Haunglongbing (Citrus greening), Late blight, Leaf blight (Isariopsis Leaf Spot), Leaf Mold, Leaf scorch, Northern Leaf Blight, Powdery mildew, Septoria leaf spot, Spider mites, Target Spot, Tomato mosaic virus, Tomato Yellow Leaf Curl Virus	99.33%	This study
2020	Early Blight	92%	[36]
2020	Downy mildew	87.20%	[37]
2019	Common rust, Northern leaf blight	88%	[45]
2019	Nitrogen deficiency, phosphorous deficiency, potassium deficiency	90%	[46]
2019	Northern Corn Leaf Blight, Common Rust, Grey Leaf Spot	92.85%	[34]
2018	Drought stress	98.97%	[10]
2017	Northern Leaf Blight	96.7%	[24]
2015	Exserohilum Turcicum, Brown spot, Grey spot, Curvularia Lunata, Round spot and Corn rust	92.82%	[7]
2015	Leaf blight, brown spot, grey leaf spot, rust spot, Curvularia leaf spot, small spot	94.46%	[47]

6. Conclusion and future scope

In this work, various DL models are investigated, hyper-parameters are optimized, and a novel Post-processing Fusion Framework (PFF) is proposed. The PFF achieved the classification accuracy of 99.33% on the Plant-Village dataset, using the 'training from scratch' approach. The previous best accuracy in the same category on the same dataset was achieved by Mohanty et al. [38] in 2016 at 98.37%. Thus an increase of 0.96% accuracy is achieved. Therefore, the proposed approach appears to be very promising to be used in real-cultivation conditions. Hence, it can be concluded from the results that the post-processing stage of the experimentation for crop disease detection plays a vital role in the efficient and effective classification. It should not be excluded while carrying similar works.

Further, the proposed post-processing technique is simple to use and is computationally less expensive. The approach helps to overcome the limitations of individual DL models by selecting the correct classifications from the outputs of each participating model. It enhances the overall accuracy and robustness of the system. Hence, such frameworks are quite useful in the field.

6.1. *Future suggestions*

Future research may focus on the experimentation of the proposed framework using techniques like transfer learning. Moreover, the initial learning of transfer learning approach may be executed on the

IOP Conf. Series: Materials Science and Engineering 998 (2020) 012065 doi:10.1088/1757-899X/998/1/012065

augmented Plant-Village [3] dataset produced as a by-product of the present investigation. Plant disease dataset should be used instead of popular image-net [48] for pre-training for crop disease detection. It may prove to be more successful for this specified task as the pre-trained model will be quite proficient in the domain and will require very less fine-tuning. Thus, it may result in the realization of enhanced results on a relatively small dataset, like digipathos [49]. Further, the final model may be combined with a pesticide suggestive system and implemented as a smartphone app. The resulting system may assist the farmers with real-time remedial measures for crop security, against diseases. Thus, it will help farmers in maximizing crop yield and minimizing crop loss due to diseases. Consequently, improving the overall crop-yield and resulting as a positive step towards a very healthy agrarian future.

References

[1] R. N. Strange and P. R. Scott, "Plant Disease: A Threat to Global Food Security," *Annu. Rev. Phytopathol.*, vol. 43, no. 1, pp. 83–116, 2005, doi: 10.1146/annurev.phyto.43.113004.133839.

[2] "Goal 2: Zero Hunger – United Nations Sustainable Development." https://www.un.org/sustainabledevelopment/hunger/ (accessed Aug. 18, 2020).

[3] D. P. Hughes, M. Salathé, and M. Salathe, "An open access repository of images on plant health to enable the development of mobile disease diagnostics," *CoRR*, vol. abs/1511.0, 2015, doi: 10.1111/1755-0998.12237.

[4] S. Kai, Z. Liu, H. Su, and C. Guo, "A research of maize disease image recognition of corn based on BP networks," in *Proceedings - 3rd International Conference on Measuring Technology and Mechatronics Automation, ICMTMA 2011*, 2011, vol. 1, pp. 246–249, doi: 10.1109/ICMTMA.2011.66.

[5] B. Sebastian, A. Unnikrishnan, and K. Balakrishnan, "GREY LEVEL CO-OCCURRENCE MATRICES: GENERALISATION AND SOME NEW FEATURES," *Int. J. Comput. Sci. Eng. Inf. Technol.*, vol. 2, no. 2, 2012, doi: 10.5121/ijcseit.2012.2213.

[6] M. Cilimkovic, "Neural Networks and Back Propagation Algorithm."

[7] Z. Zhang, X. He, X. Sun, L. Guo, J. Wang, and F. Wang, "Image Recognition of Maize Leaf Disease Based on GA-SVM," 2015, doi: 10.3303/CET1546034.

[8] S. Mirjalili, "Evolutionary Algorithms and Neural Networks," *Soft Comput. Intell. Syst.*, vol. 780, pp. 111–136, 2019, doi: 10.1007/978-3-319-93025-1.

[9] B. Schlkopf, A. J. Smola, and F. Bach, *Learning with Kernels: Support Vector Machines, Regularization, Optimization, and Beyond*. The MIT Press, 2018.

[10] B. Jiang, P. Wang, S. Zhuang, M. Li, Z. Li, and Z. Gong, "Detection of maize drought based on texture and morphological features," *Comput. Electron. Agric.*, vol. 151, pp. 50–60, Aug. 2018, doi: 10.1016/j.compag.2018.03.017.

[11] X. E. Pantazi, D. Moshou, and A. A. Tamouridou, "Automated leaf disease detection in different crop species through image features analysis and One Class Classifiers," *Comput. Electron. Agric.*, vol. 156, no. July 2018, pp. 96–104, Jan. 2019, doi: 10.1016/j.compag.2018.11.005.

[12] Y. Lecun, Y. Bengio, and G. Hinton, "Deep learning," *Nature*, vol. 521, no. 7553, pp. 436–444, 2015, doi: 10.1038/nature14539.

[13] G. Wang, Y. Sun, and J. Wang, "Automatic Image-Based Plant Disease Severity Estimation Using Deep Learning," *Comput. Intell. Neurosci.*, 2017, doi: 10.1155/2017/2917536.

[14] A. Z. Karen Simonyan, "Very Deep Convolutional Networks for Large-Scale Image Recognition," *CoRR*, vol. abs/1409.1, pp. 1–14, 2015.

[15] C. Szegedy, V. Vanhoucke, S. Ioffe, and J. Shlens, "Rethinking the Inception Architecture for Computer Vision."

[16] J. S. Kaiming He, Xiangyu Zhang, Shaoqing Ren, K. He, X. Zhang, S. Ren, and J. Sun, "Deep Residual Learning for Image Recognition," *CoRR*, vol. abs/1512.0, pp. 1–12, 2015, doi: 10.1016/0141-0229(95)00188-3.

[17] M. M. M. Brahimi, B. Kamel, T. H. Boumediene, K. Boukhalfa, and A. Moussaoui, "Deep Learning for Tomato Diseases: Classification and Symptoms Visualization," *Appl. Artif. Intell.*, vol. 31, no. 4, pp. 299–315, 2017, doi: 10.1080/08839514.2017.1315516.

[18] B. A. Krizhevsky, I. Sutskever, and G. E. Hinton, "ImageNet Classification with Deep Convolutional Neural Networks," *Commun. ACM*, vol. 60, no. 6, p. 2017, 2012.

[19] A. R. Christian Szegedy, Wei Liu, Yangqing Jia, Pierre Sermanet, Scott Reed, Dragomir Anguelov, Dumitru Erhan, Vincent Vanhoucke, "Going Deeper with Convolutions," *CoRR*, vol. abs/1409.4, pp. 1–9, 2014, doi: 10.1002/jctb.4820.

[20] K. Yamamoto, T. Togami, and N. Yamaguchi, "Super-resolution of plant disease images for the acceleration of image-based phenotyping and vigor diagnosis in agriculture," *Sensors (Switzerland)*, vol. 17, no. 11, pp. 1–13, 2017, doi: 10.3390/s17112557.

[21] T. Mikolov, A. Joulin, S. Chopra, M. Mathieu, and A. Ranzato, "LEARNING LONGER MEMORY IN RECURRENT NEURAL NETWORKS."

[22] J. G. Ha *et al.*, "Deep convolutional neural network for classifying Fusarium wilt of radish from unmanned aerial

vehicles," *J. Appl. Remote Sens.*, vol. 11, no. 04, pp. 04262 (1–14), 2017, doi: 10.1117/1.JRS.11.042621.

[23] J. Lu, J. Hu, G. Zhao, F. Mei, and C. Zhang, "An in-field automatic wheat disease diagnosis system," *Comput. Electron. Agric.*, vol. 142, no. 1, pp. 369–379, 2017, doi: 10.1016/j.compag.2017.09.012.

[24] C. DeChant *et al.*, "Automated Identification of Northern Leaf Blight-Infected Maize Plants from Field Imagery Using Deep Learning," *Phytopathology*, vol. 107, no. 11, pp. 1426–1432, Nov. 2017, doi: 10.1094/PHYTO-11-16-0417-R.

[25] A. Fuentes, S. Yoon, S. C. Kim, and D. S. Park, "A robust deep-learning-based detector for real-time tomato plant diseases and pests recognition," *Sensors (Switzerland)*, vol. 17, no. 9, 2017, doi: 10.3390/s17092022.

[26] S. Xie, R. Girshick, P. Dollár, Z. Tu, and K. He, "Aggregated Residual Transformations for Deep Neural Networks," Nov. 2016, [Online]. Available: http://arxiv.org/abs/1611.05431.

[27] S. Ren, K. He, R. Girshick, and J. Sun, "Faster R-CNN: Towards Real-Time Object Detection with Region Proposal Networks," Jun. 2015, [Online]. Available: http://arxiv.org/abs/1506.01497.

[28] J. Dai, Y. Li, K. He, and J. Sun, "R-FCN: Object Detection via Region-based Fully Convolutional Networks." Accessed: Aug. 20, 2020. [Online]. Available: https://github.com/daijifeng001/r-fcn.

[29] A. C. Cruz, A. Luvisi, L. De Bellis, and Y. Ampatzidis, "X-FIDO: An Effective Application for Detecting Olive Quick Decline Syndrome with Deep Learning and Data Fusion," *Front. Plant Sci.*, vol. 8, no. October, pp. 1–12, Oct. 2017, doi: 10.3389/fpls.2017.01741.

[30] R. Ramachandran, D. C. Rajeev, S. G. Krishnan, and P. Subathra, "Deep learning in neural networks: An overview," *Neural Networks*, vol. 61, pp. 85–117, 2015, doi: 10.1016/j.neunet.2014.09.003.

[31] B. Liu, Y. Zhang, D. J. He, and Y. Li, "Identification of apple leaf diseases based on deep convolutional neural networks," *Symmetry (Basel).*, vol. 10, no. 1, 2018, doi: 10.3390/sym10010011.

[32] J. Amara, B. Bouaziz, and A. Algergawy, "A Deep Learning-based Approach for Banana Leaf Diseases Classification," in *BTW*, 2017, pp. 79–88.

[33] S. Ghosal, D. Blystone, A. K. A. A. K. Singh, B. Ganapathysubramanian, A. K. A. A. K. Singh, and S. Sarkar, "An explainable deep machine vision framework for plant stress phenotyping," *Proc. Natl. Acad. Sci.*, vol. 115, no. 18, pp. 4613–4618, 2018, doi: 10.1073/pnas.1716999115.

[34] M. Sibiya and M. Sumbwanyambe, "A Computational Procedure for the Recognition and Classification of Maize Leaf Diseases Out of Healthy Leaves Using Convolutional Neural Networks," *AgriEngineering*, vol. 1, no. 1, pp. 119–131, Mar. 2019, doi: 10.3390/agriengineering1010009.

[35] J. G. Arnal Barbedo, "Plant disease identification from individual lesions and spots using deep learning," *Biosyst. Eng.*, vol. 180, no. 2016, pp. 96–107, Apr. 2019, doi: 10.1016/j.biosystemseng.2019.02.002.

[36] R. Van De Vijver *et al.*, "In-field detection of Alternaria solani in potato crops using hyperspectral imaging," *Comput. Electron. Agric.*, vol. 168, no. March 2019, p. 105106, Jan. 2020, doi: 10.1016/j.compag.2019.105106.

[37] B. Zhou, A. Khosla, A. Lapedriza, A. Oliva, and A. Torralba, "Learning Deep Features for Discriminative Localization." Accessed: Aug. 20, 2020. [Online]. Available: http://cnnlocalization.csail.mit.edu.

[38] S. P. Mohanty, D. P. Hughes, and M. Salathé, "Using Deep Learning for Image-Based Plant Disease Detection," *Front. Plant Sci.*, vol. 7, no. September, pp. 1–10, Sep. 2016, doi: 10.3389/fpls.2016.01419.

[39] M. Lin, Q. Chen, and S. Yan, "Network In Network," *CoRR*, vol. abs/1312.4, 2013, [Online]. Available: http://arxiv.org/abs/1312.4400.

[40] M. D. Zeiler and R. Fergus, "Visualizing and Understanding Convolutional Networks," *CoRR*, vol. abs/1311.2, p. 2013, 2012.

[41] P. Barré, B. C. Stöver, K. F. Müller, and V. Steinhage, "LeafNet: A computer vision system for automatic plant species identification," *Ecol. Inform.*, vol. 40, no. December 2016, pp. 50–56, 2017, doi: 10.1016/j.ecoinf.2017.05.005.

[42] J. Hu, L. Shen, S. Albanie, G. Sun, and E. Wu, "Squeeze-and-Excitation Networks," *CoRR*, vol. abs/1709.0, pp. 1–14, 2017.

[43] H. Goëau *et al.*, "The imageCLEF Plant Identification Task 2013," in *Proceedings of the 2Nd ACM International Workshop on Multimedia Analysis for Ecological Data*, 2013, no. i, pp. 23–28, doi: 10.1145/2509896.2509902.

[44] H. Goëau *et al.*, "Plant Identification in an Open-world (LifeCLEF 2016)," in *CLEF 2016 - Conference and Labs of the Evaluation forum*, Sep. 2016, no. LifeCLEF, pp. 428–439, [Online]. Available: https://hal.archives-ouvertes.fr/hal-01373780.

[45] A. S. Deshapande, S. G. Giraddi, K. G. Karibasappa, and S. D. Desai, "Fungal disease detection in maize leaves using haar wavelet features," in *Smart Innovation, Systems and Technologies*, 2019, vol. 106, pp. 275–286, doi: 10.1007/978-981-13-1742-2_27.

[46] N. Leena and K. K. Saju, "Classification of macronutrient deficiencies in maize plants using optimized multi class support vector machines," *Eng. Agric. Environ. Food*, vol. 12, no. 1, pp. 126–139, 2019, doi: 10.1016/j.eaef.2018.11.002.

[47] J. Luo, S. Geng, C. Xiu, D. Song, and T. Dong, "A curvelet-SC recognition method for maize disease," *J. Electr. Comput. Eng.*, vol. 2015, Jan. 2015, doi: 10.1155/2015/164547.

[48] O. Russakovsky *et al.*, "ImageNet Large Scale Visual Recognition Challenge," *Int. J. Comput. Vis.*, vol. 115, no. 3, pp. 211–252, 2015, doi: 10.1007/s11263-015-0816-y.

[49] A. Johannes *et al.*, "Automatic plant disease diagnosis using mobile capture devices, applied on a wheat use case," *Comput. Electron. Agric.*, vol. 138, pp. 200–209, 2017, doi: 10.1016/j.compag.2017.04.013.

DAMAGE DETECTION IN COMPOSITE MATERIALS USING LAMB WAVEMETHOD

SNaresh Kumar[1], K Srikanth[2*], B Subbaratnam[3], Shakti P Jena[4]

[1, 3, 4]Department of Mechanical Engineering, Vardhaman College of Engineering, Hyderabad, Telangana, India.
[2]PG Scholar, Department ofMechanical Engineering, Vardhaman College ofEngineering, Hyderabad, Telangana,India.

Email:[*]srikanthkadhireni96@gmail.com

ABSTRACT:A composite material is a material which is made up of two or more constituent materials with variation in theirphysical or chemical properties that, when combined it produce a material withdifferent characteristics from the individual components. The individual components remain separate and distinct within the finished structure, differentiating composites from mixtures and solid solutions.Glass fiber is a material which consisting of numerous extremely fine fibers of glass. In this project both E-glass and S-glass materials were analyzed by using CAE tool abacus to detect damages of each material, and calculating results like stress, displacements, for both objects with and without damage, by knowing all these results with suitable tables and graphs project can be concluded e-glass and s-glass material behavior.

Keywords: *S-glass, E-glass, composite, lamb wave propagation.*

1. INTRODUCTION:

The propagation of Lamb waves are in solid plates or spheres. The waves are elastic whose particle motion lies in the plane that contains the direction of wave propagation and the plane normal (the direction perpendicular to the plate). In the year 1917, the Horace Lamb is an English mathematician who published the classic analysis and description of acoustic waves of this type. Their turned out properties are to be quite complex. An infinite medium supports just two wave modes traveling at unique velocities; but plates support two infinite sets of Lamb wave modes, whose velocities which depend on the relationship between length of wave and thickness of plate.

Theusage and understanding of Lamb waves has advanced greatly Since 1990s, the availability of computing power is rapidly increased. Theoretical formulations of Lamb's hadgave practical application substantially, particularly in the field of nondestructive testing.

The word Rayleigh–Lamb waves enclose the Rayleigh wave, a kind of wave which propagates on a single surface. Both Rayleigh and Lamb waves are constrained by the elastic properties of the surface(s) that guide them.

A brief review about the Lamb wave-based damage detection literature is given here:

Cesnik et al. "gave an overview of damage prognosis, description of the basic methodology of guided wave SHM, and reviewed developments from the open literature of this multidisciplinary field"[1]. Zhongqing et al. "provided a comprehensive review on the Lamb wave-based damage identification approaches for composite structures" [2]. Beadle et al. "studied the interaction of the first antisymmetric A0 mode of Lamb wave in an aluminium plate with small surface notches" [3]. Mofakhami et al. "studied the Lamb wave propagation in an aluminium plate containing a circular hole with edge notches, both theoretically and experimentally" [4]. Willberg et al. "studied the use of Lamb wave for damage detection and nondestructive evaluation" [5]. Silva et al. "studied a piezoelectric transducer network based system to be applied to aluminium and composite plates" [6]. Kessler et al. "presented the experimental and analytical survey of candidate methods for in situ damage detection of composite materials" [7]. Yamada et al. "studied the development of a new source location method using Lamb wave on anisotropic carbon fibre reinforced plastic plates" [8]. Liu et al. "studied a signal analysis method using a WT for nondestructive damage detection in the life cycle management of wind energy converters" [9]. Zabel et al. "studied wavelet analysis for applying to the numerous problems within the general field of SMH" [10]. Giurgiutui et al. "presented the results of a systematic theoretical and experimental investigation of the fundamental aspects using PW active sensors to achieve embedded ultrasonic in thin gage beam and plate structures" [11]. Finally, Gangadharan et al. "proposed a geodesic-based approach using Lamb wave to locate a damage in an isotropic metallic plate"[12].

1.1 Lamb's characteristic equations: Generally elastic waves in strong materials are guided by the limits of the media in which they propagate. A way to deal with guided wave spread, generally utilized in physical acoustics, is to look for sinusoidal answers for the wave condition for straight versatile waves subject to limit conditions speaking to the auxiliary math. This is an exemplary eigenvalue issue.

Waves in plates were among the main guided waves to be dissected along these lines. The examination was created and distributed in 1917 by Horace Lamb, a pioneer in the numerical material science of his day.

Lamb's conditions were inferred by setting up formalism for a strong plate having interminable degree in the x and y headings, and thickness d in the z course. Sinusoidal answers for the wave condition were hypothesized, having x-and z-directions of the structure

$$\xi = A_\alpha f_\alpha(z)e^{i(wt-kx)} \quad (1)$$

$$\zeta = A_\alpha f_\alpha(z)e^{i(wt-kx)} \quad (2)$$

This structure speaks to sinusoidal waves engendering in the x heading with frequency 2π/k and recurrence ω/2π. Uprooting is an element of x, z, t just; there is no dislodging in the y heading and no variety of any physical amounts in the y bearing.

The physical limit condition for the free surfaces of the plate is that the segment of worry in the z heading at z = +/ - d/2 is zero. Applying these two conditions to the above-formalized answers for the wave condition, a couple of trademark conditions can be found.These are:

$$\frac{\tanh(\beta d/2)}{\tanh(\alpha d/2)} = \frac{4 \propto \beta k^2}{(k^2+\beta^2)^2} \quad (3)$$

For symmetric modes and

$$\frac{\tanh(\beta d/2)}{\tanh(\alpha d/2)} = \frac{(k^2+\beta^2)^2}{4 \propto \beta k^2} \quad (4)$$

For asymmetric modes, where

$$\alpha^2 = k^2 - \frac{w^2}{c_l^2} \quad and \quad \beta^2 = k^2 - \frac{w^2}{c_t^2}$$

Inalienable in these conditions is a connection between the precise recurrence ω and the wave number k. Mathematical techniques are utilized to discover the stage speed cp = fλ = ω/k, and the gathering speed cg = dω/dk, as elements of d/λ or fd. cl and ct are the longitudinal wave and shear wave speeds separately.

The arrangement of these conditions additionally uncovers the exact type of the molecule movement, which conditions (1) and (2) speak to in conventional structure as it were. It is discovered that condition (3) offers ascend to a group of waves whose movement is balanced about the midplane of the plate (the plane z = 0), while condition (4) offers ascend to a group of waves whose movement is antisymmetric about the midplane. Figure 1 delineates an individual from every family.

2. MATERIALS:

The material used for our project are Glass fiber. Itis a material comprising of various amazingly fine filaments of glass.Glassmakers from the beginning of time have explored different avenues regarding glass strands, however mass assembling of glass fiber was just made conceivable with the innovation of better machine tooling.

Table1: Materials and their properties.

MATERIAL	S-glass	E-glass
Young's modulus	86e9 Pa	72e9 Pa
Poisson's ratio	0.21	0.21
Density	2485 kg/m³	2550 kg/m³

3. METHODOLOGY: The following Steps used in methodology:

- Pre-processing or modeling: In modeling it creates an input file which contains a proper design for a finite-element analyzer (it is also known as "solver").

IOP Conf. Series: Materials Science and Engineering 998 (2020) 012066 doi:10.1088/1757-899X/998/1/012066

- Processing or finite element analysis: In finite element analysis it gives a visual file as output.
- Post-processing or producing report, picture, animation, and etc., from the system as output file: In post processing it can visualize the analysis results approximately.

4. CAE Modeling:

The modelling of the material done in ABAQUS Software according to the requirements.

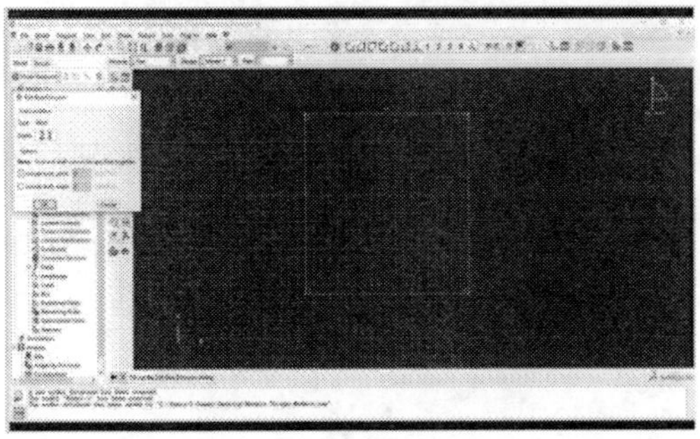

Figure 1: Before Extrusion

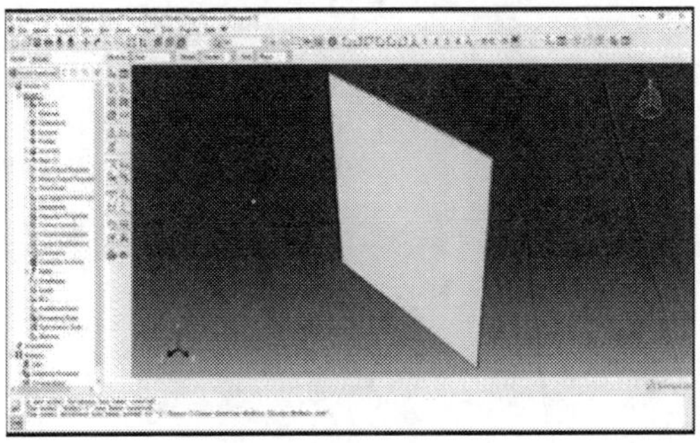

Figure 2: Extruded model.

4.1Meshing:

Now to create sensor and exciter here we need to divide complete object into small parts, to do this here partition option were used, to do partition either there should be plane or axis should have, and here planes were chosen and created in "XY" plane direction with a distance of 10mm and repeated this process in all 4 directions at a distance of 10mm from edge

Figure 3: Meshing results for without cracking plate.

Figure 4: Meshing results for with cracking plate

Now select the job module to solve the analysis results,

And now click on monitor to see the time increment all frequency values, and then click on results option to see the results, and those results are given below and discussed it,

5. RESULTS AND DISCUSSION:

5.1 Without crack result of S-glass.

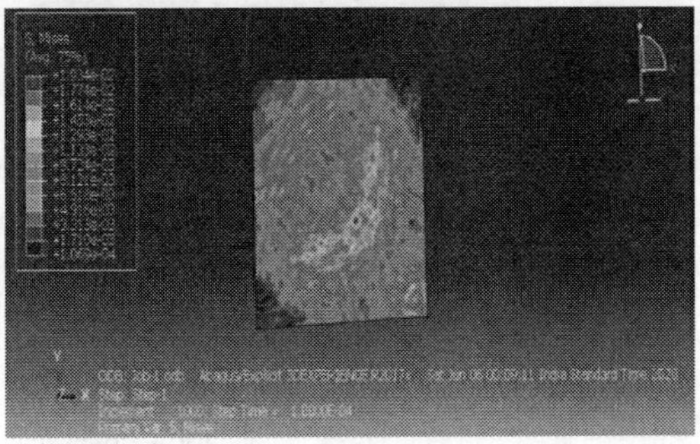

Figure 5: Von misses stress

Figure 6: Equivalent pressure stress.

Displacement vs time graphs in U1, U2 and U3 directions

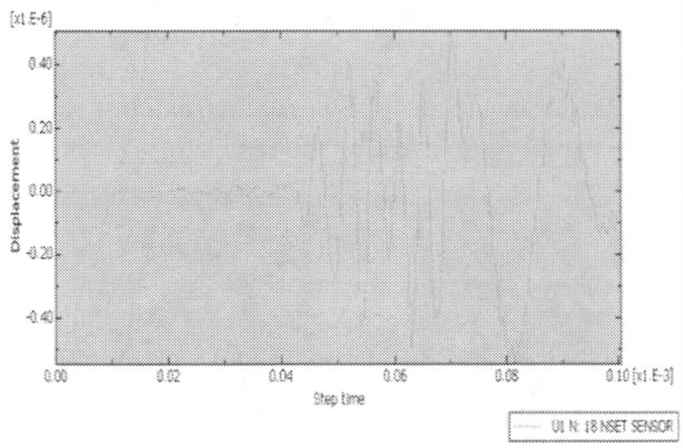

Figure 7: Displacement vs time in U1 direction.

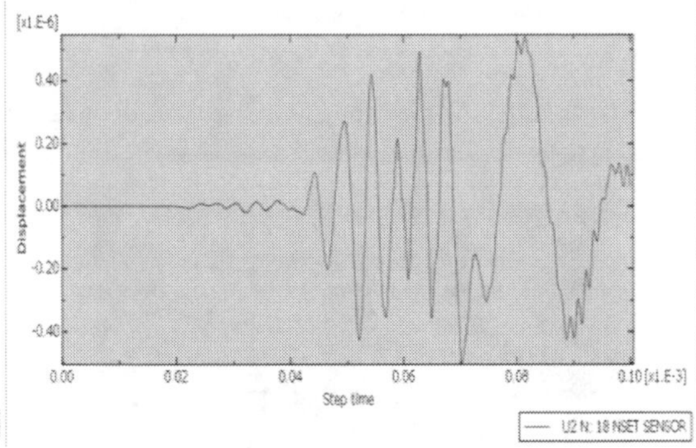

Figure 8: Displacement vs time in U2 direction.

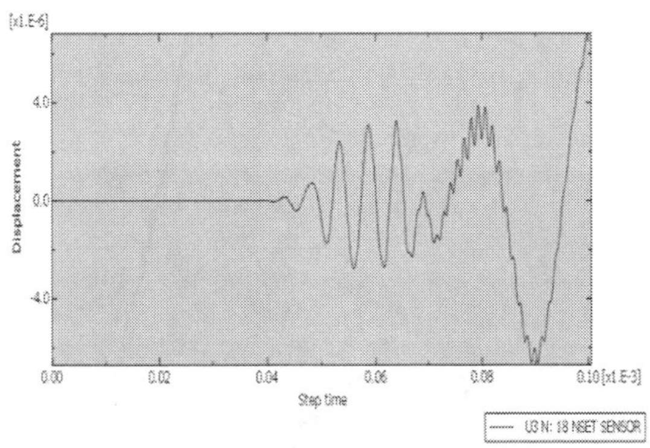

Figure 9:Displacement vs time in U3 direction

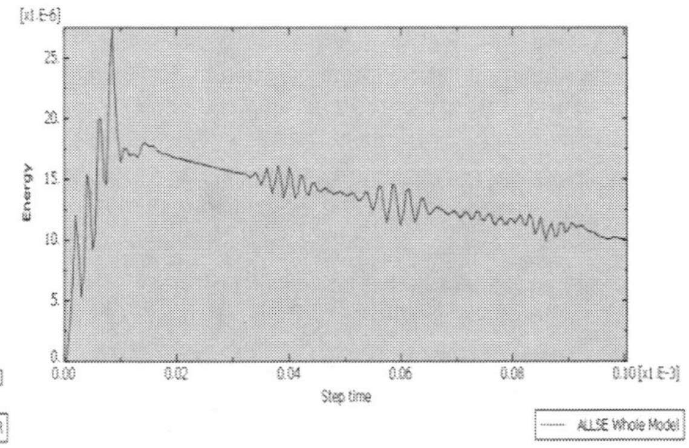

Figure 10: Strain energy vs time.

5.2 Without crack result of E-glass.

Figure 11: Von misses stress.

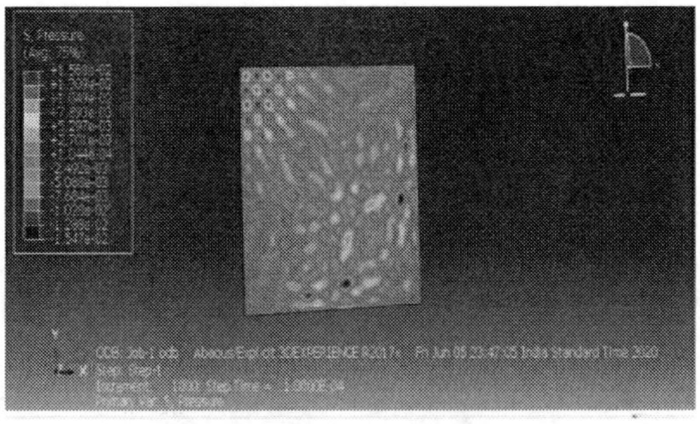

Figure 12: Equivalent pressure stress.

IOP Conf. Series: Materials Science and Engineering 998 (2020) 012066 doi:10.1088/1757-899X/998/1/012066

Displacement vs time graphs in U1, U2 and U3 directions

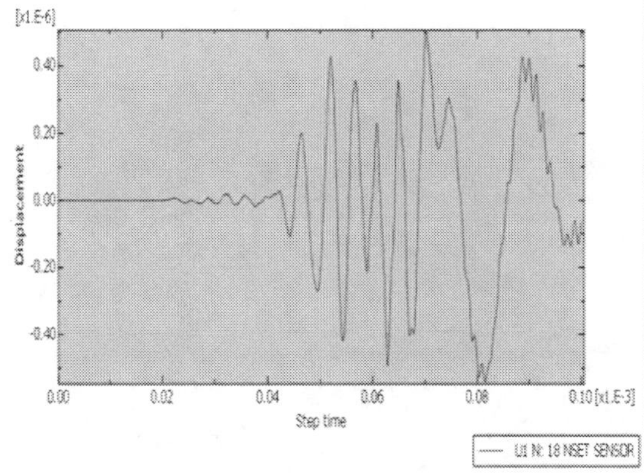

Figure 13: Displacement vs time in U1 direction.

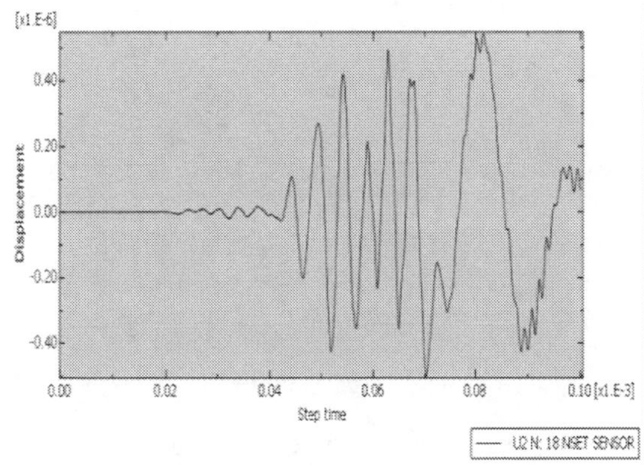

Figure 14: Displacement vs time in U2 direction.

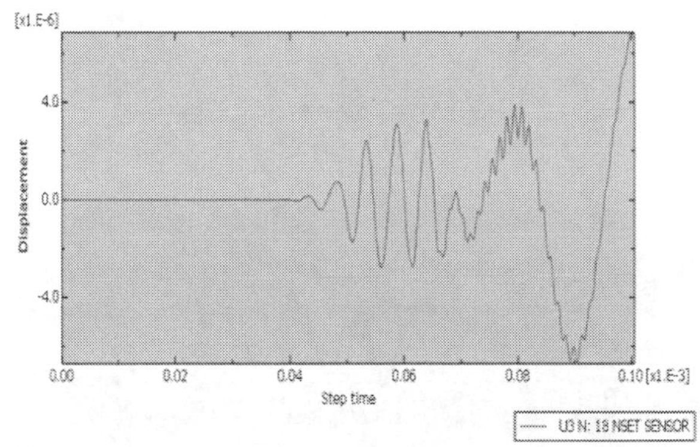

Figure 15: Displacement vs time in U3 direction.

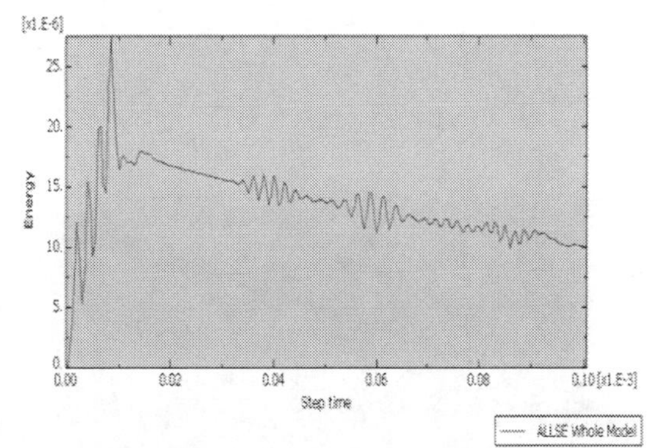

Figure 16: Strain energy vs time.

Discussion: The above results are of S and E glass without crack materials. Both the materials have different properties, based on their properties the results are obtained. Through FE simulation it is found to be good and there is no damage in the material. The below graphs are plotted by comparing the results like von misses stress, pressure stress, displacement in U1, U2 and U3 directions. Comparing both the results S-glass has less stress values and E-glass has more stress values which shown in below graphs.

Graphs: Below graphs are for both S-glass and E-glass of without crack plate

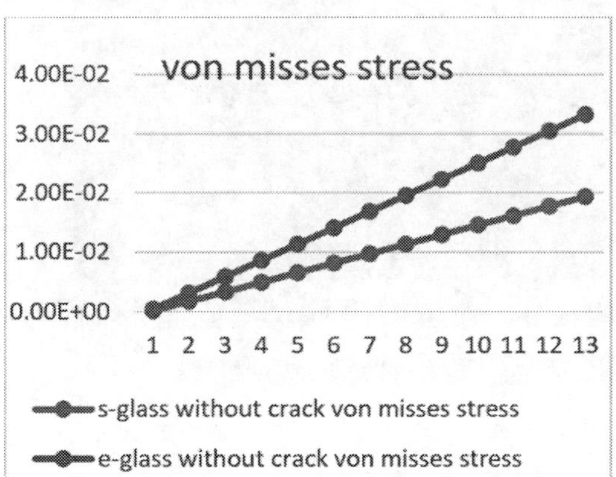

Figure 17: Von misses stress.

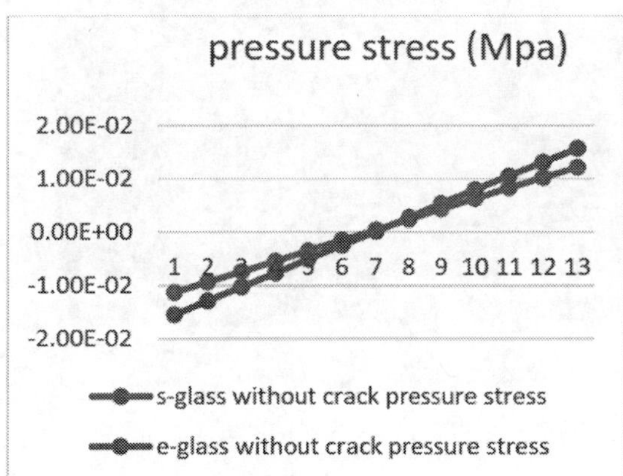

Figure 18: Pressure stress.

IOP Conf. Series: Materials Science and Engineering 998 (2020) 012066 doi:10.1088/1757-899X/998/1/012066

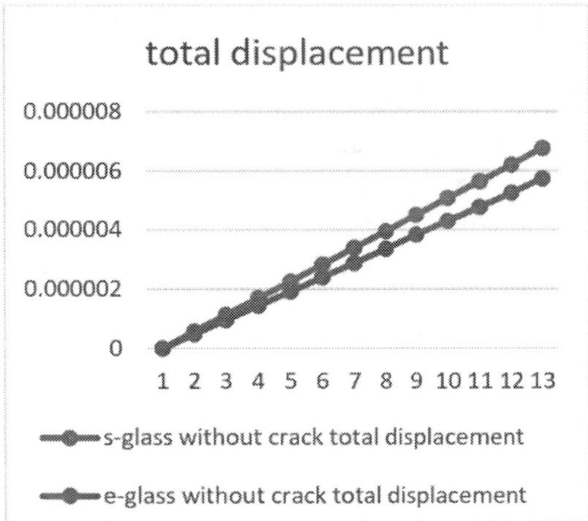

Figure 19: Total Displacement.

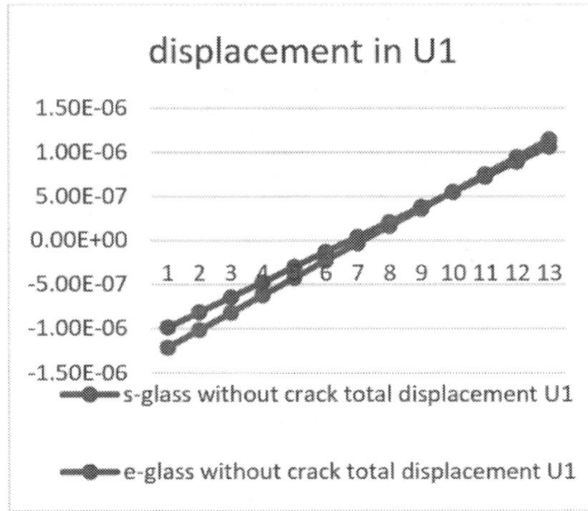

Figure 20: Displacement in U1

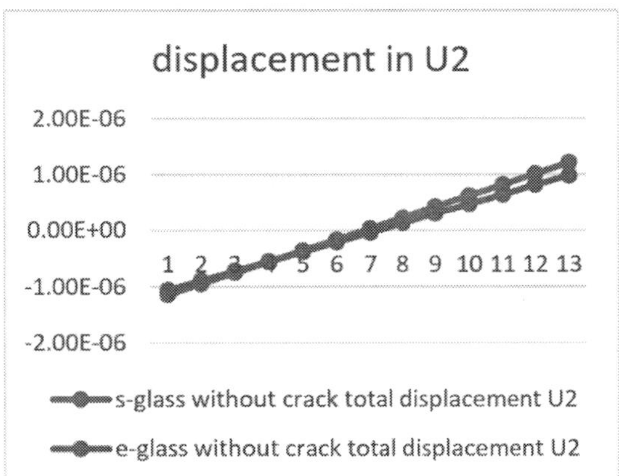

Figure 21:Displacement in U2

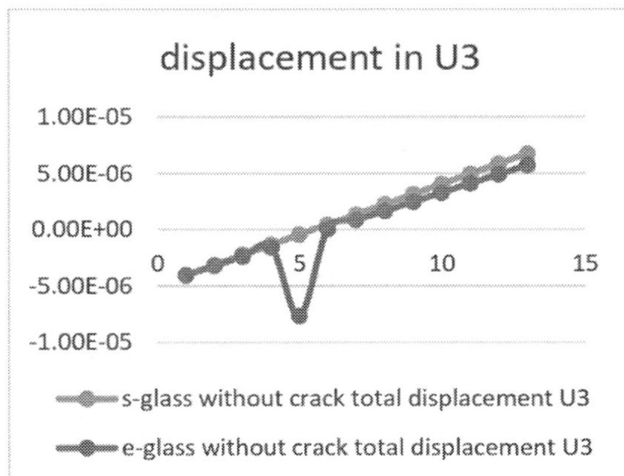

Figure 22:Displacement in U3

5.3 Lamb wave propagation with crack result of S-glass.

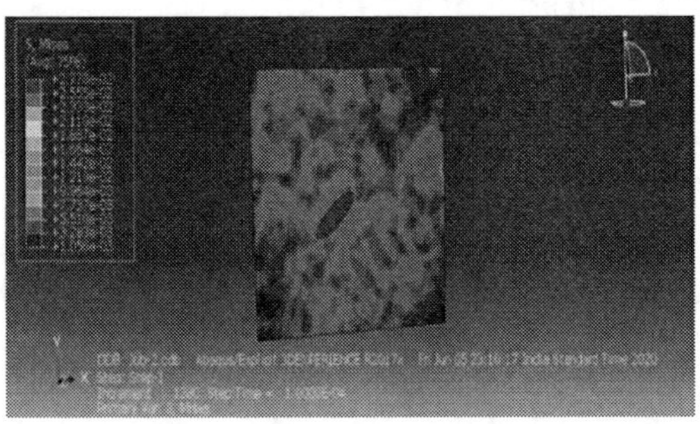

Figure 23: Von misses stress.

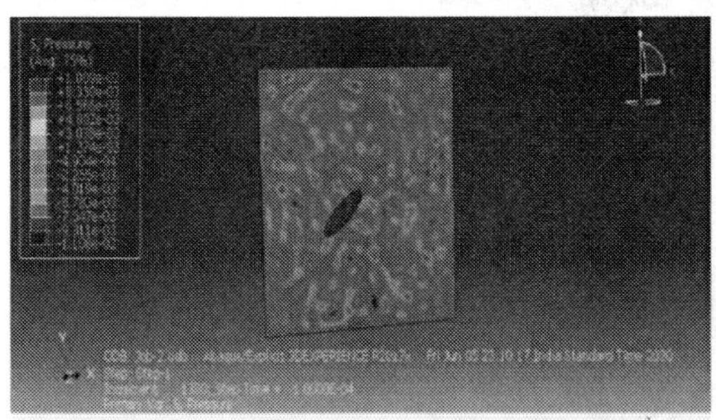

Figure 24: Equivalent pressure stress.

IOP Conf. Series: Materials Science and Engineering 998 (2020) 012066 doi:10.1088/1757-899X/998/1/012066

Displacement vs time graphs in U1, U2 and U3 directions

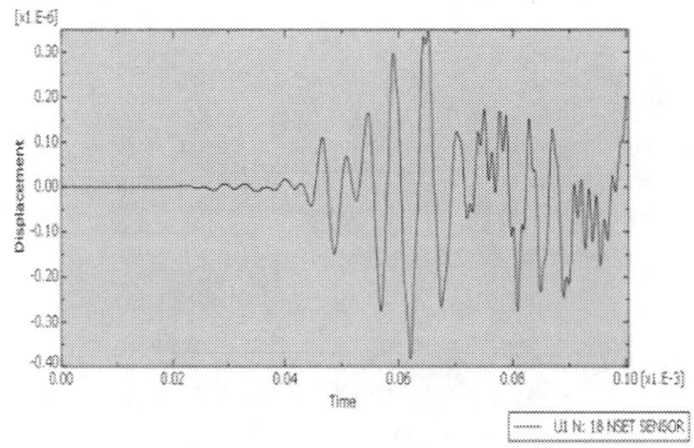

Figure 25: Displacement vs time in U1 direction.

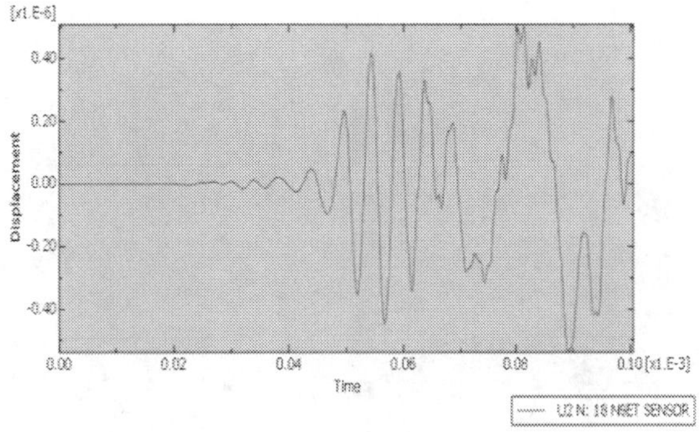

Figure 26: Displacement vs time in U2 direction.

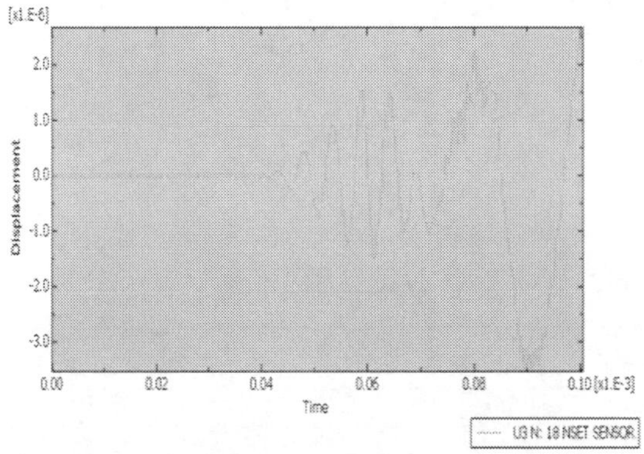

Figure 27: Displacement vs time in U3 direction.

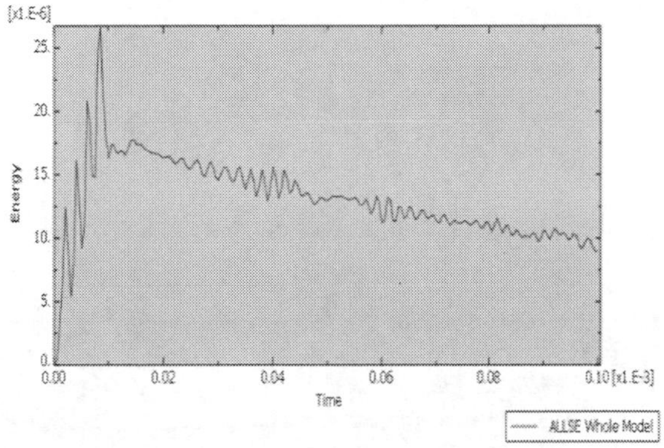

Figure 28: Strain Energy v/s time

5.4 Lamb wave propagation with crack results of E-glass.

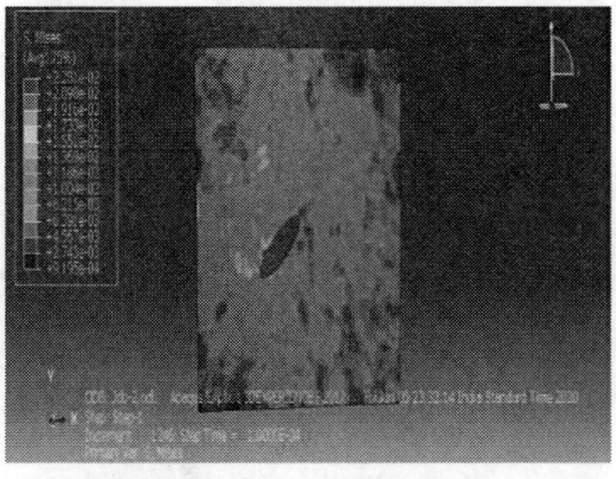

Figure 29: Von misses stress.

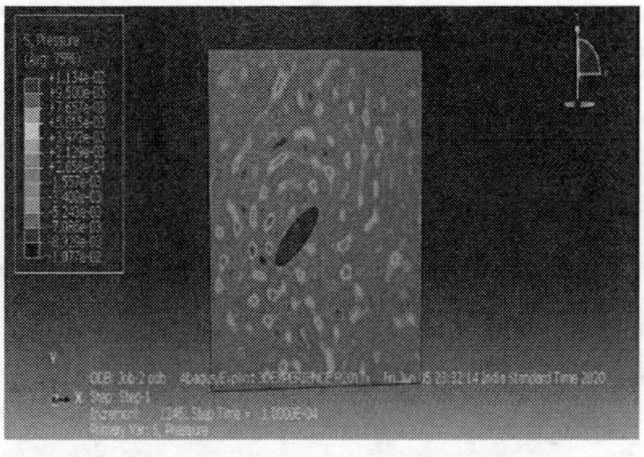

Figure 30: Equivalent pressure stress.

Displacement vs time graphs in U1, U2 and U3 directions

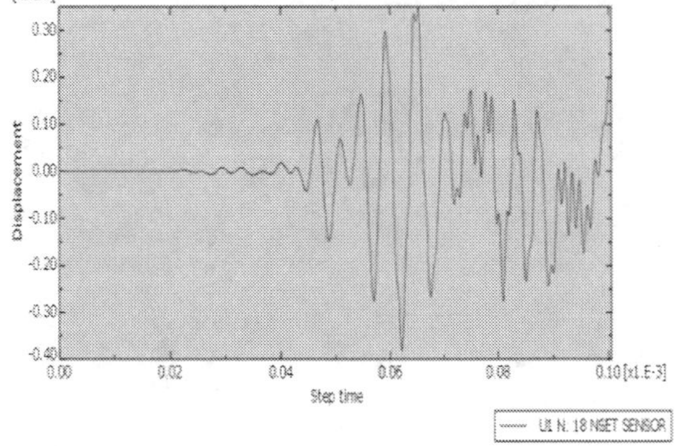

Figure 31: Displacement vs time in U1 direction.

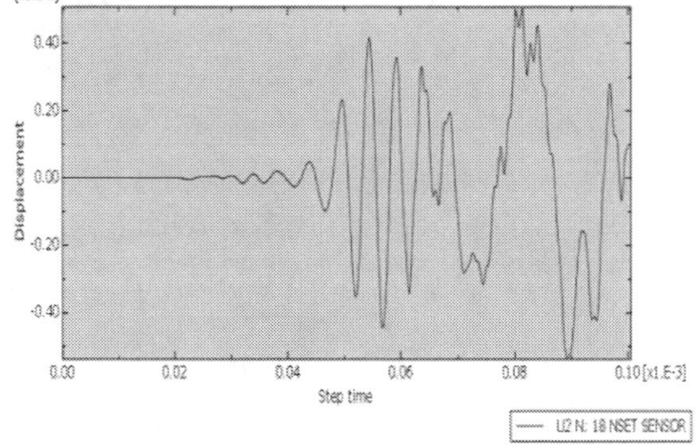

Figure 32: Displacement vs time in U2 direction.

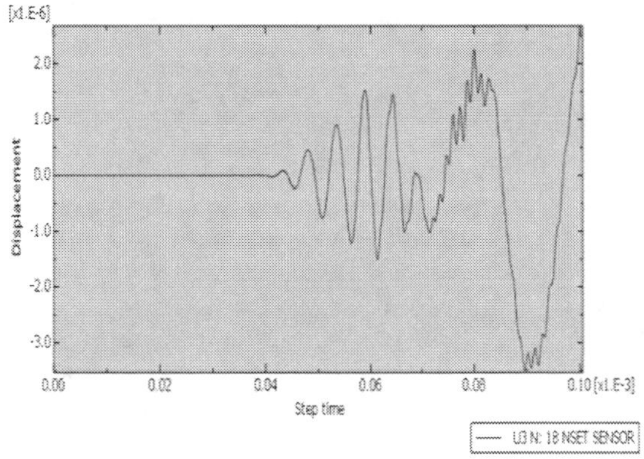

Figure 33: Displacement vs time in U3 direction.

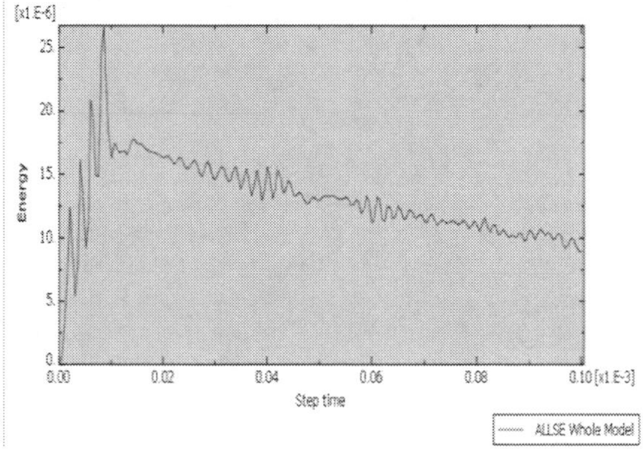

Figure 34: Strain Energy v/s time.

Discussion: The above results are of S and E glass with crack material. Both the glasses have different properties, based on their properties the results are obtained. Through FE simulation it is found to be known that the damage occurred in the material. Below graphs are plotted by comparing the results like von misses stress, pressure stress, displacement in U1,U2 and U3 directions. The stress value in S-glass has high and E-glass has less which compare to without crack, which leads to be known that there is a damage in the material.

Graphs: Below graphs are for both S-glass and E-glass of with crack plate.

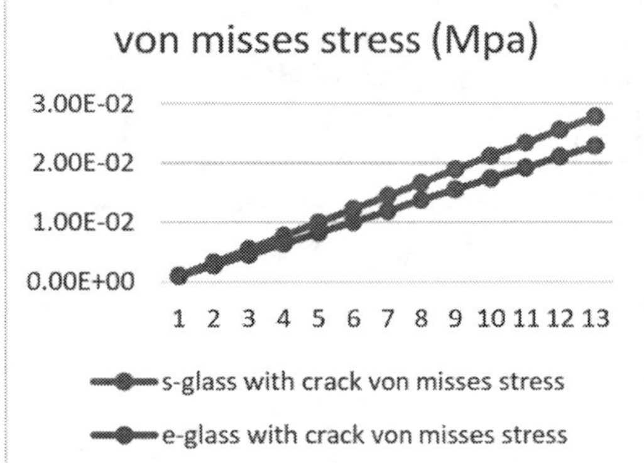

Figure 35: Von misses stress

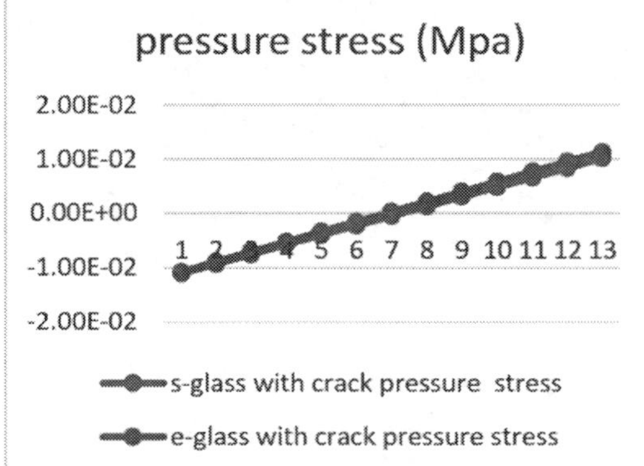

Figure 36: Pressure stress

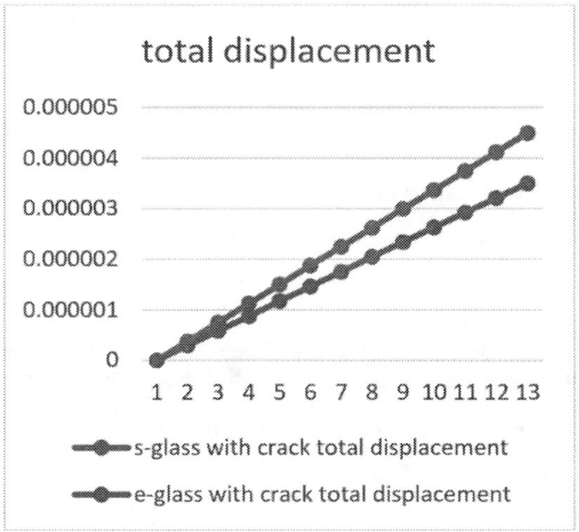

Figure 37:Total Displacement

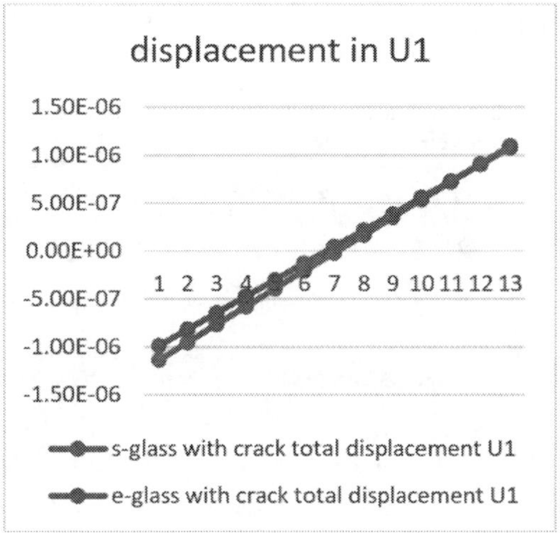

Figure 38: Displacement in U1

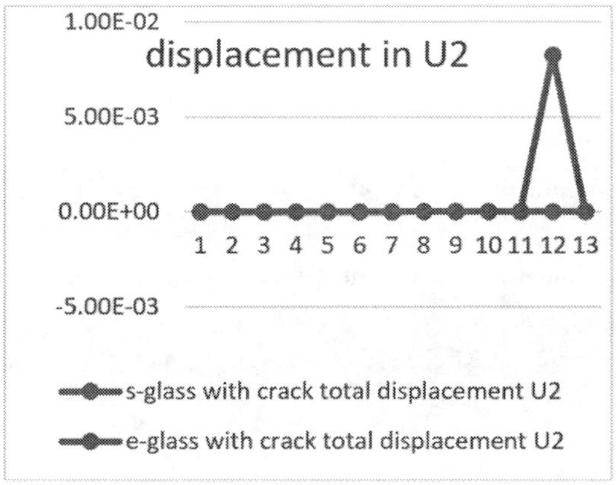

Figure 38: Displacement in U2

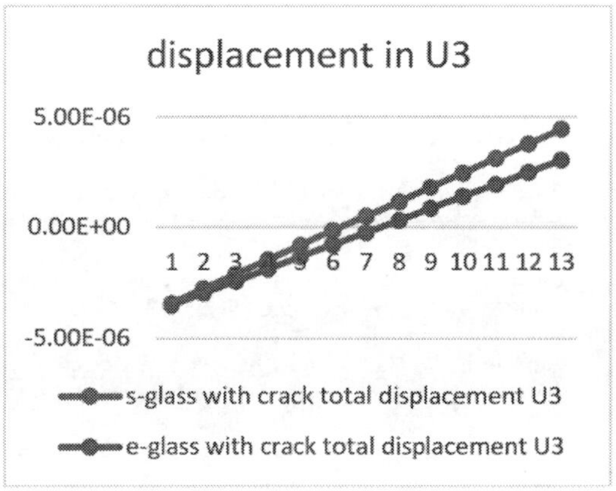

Figure 39: Displacement in U3

6. CONCLUSIONS:

In this article calculating displacement and stress values for e-glass and s-glass materials by using lamb wave method with the help of CAE tool Abaqus, in this process it is observe that S glass has less stress values on the plate, and high displacement results compare to E glass materials,

From the results of both models here it is observe that the strain energy value is increasing while increasing time period and it has maximum value at 0.08e-3 time period, similarly it has maximum displacement value at 0.08e-3 time period.

If there is any damage on material, the stress values will increases for both materials, compare to without crack object, and the displacement of each material will be decreases, among 2 materials, e-glass material is having less stress values with and without crack when compare to s-glass material, it shows the results as, less stress can increases the object performance in durability, and their fatigue conditions.

IOP Conf. Series: Materials Science and Engineering 998 (2020) 012066 doi:10.1088/1757-899X/998/1/012066

REFERENCES:

[1] Gangadharan R, Murthy C R L, Gopalakrishnan S, Bhat M R and MahapatraD R 2010 International Journal on Smart Sensing and Intelligent Systems,*Characterization of cracks and delaminations using PWAS AD Lamb wave based time-frequency methods,* 3(4).

[2] Raghavan A and Cesnik C E S 2007 The Shock and Vibration Digest,*Review of guided-wave structural health monitoring,* 39(91).

[3] Zhongqing S, Lin Y, and Ye L 2006 A review, Journal of Sound and Vibration,*Guided Lamb waves for identification of damage in composite structures:* 10 1016.

[4] Beadle B M, Hurlebaus S, Jacobs L J and Gaul L 2005 In Proceedings of the 23rd International Modal Analysis Conference (IMAX XXIII), *Detection and localization of small notches in plates using Lamb waves,*96.

[5] Mofakhami M R and Boller C 2009 Materials forum,*Scattering of Lamb waves from edge cracks in circular holes for understanding damage detection in more complex structures,* 33.

[6] Willberg C, Perez J M, Ahmad Z and Gabbert U 2009Proceedings in Applied Mathematics and Mechanics,*Simulation of piezoelectric induced Lamb wave in plates,* 9.

[7] Silva C, Rocha B and Suleman A 2010 2nd International Symposium on NDT in Aerospace, *Guided Lamb waves based structural health monitoring through a PZT network system,*

[8] Kessler S S and Spearing SM 2002 Smart Materials and Structures, *Damage detection in composite materials using lamb wave methods,* 11, 269–278.

[9] Yamada H, Mizutani Y, Nishino H, Takemoto M and Ono K.2000Proceedings of EW-GAE 2000 European Working Group on Acoustic Emission, *Lamb wave source location of impact on anisotropic plates,*77–83.

[10] Li J and Liu S 200817th World Conference on Non-destructive Testing, *The application of time-frequency transform in mode identification of Lamb waves,*25(28).

[11] Zabel V and Brehm M 2005 Proceedings of the SAMCO Summer Academy,*Wavelet analysis in structural health monitoring and damage detection,*5(9).

[12] Giurgiutiu V, Bao J and Zhao W 2003 Society for Experimental Mechanics, *Piezoelectric wafer active sensor embedded ultrasonics in beams and plates,*43(4), 428–449.

IOP Conf. Series: Materials Science and Engineering **998** (2020) 012067 doi:10.1088/1757-899X/998/1/012067

Mechanical Characterization and Machining of Mg-AZ91D/SiC/Al_2O_3MMC by Squeeze Cast Method

Prudhvi Raj K [*1,a], **Seshappa A** [*2,b], **Anjaneya Prasad B** [3,c]

*1Department of Mechanical Engineering, Mahatma Gandhi Institute of Technology, Hyderabad-500075, India
*2Department of Mechanical Engineering, Vardhaman College of Engineering, Hyderabad-501218, India
1,2,3 Department of Mechanical Engineering, Jawaharlal Nehru Technological University Hyderabad, College of Engineering, Hyderabad-500 085, India.

*a kprajlecturer@gmail.com, b seshurambollimi@gmail.com, c baprasadjntu@gmail.com
Keywords: Squeeze Casting, Hybrid Metal Matrix Composites, Machining, Mechanical Characterization.

Abstract: In the automotive industry, the high surface finish and superior mechanical property materials are prepared with the help of the squeeze casting process. The paper investigates on the chosen materials of MMC cast using magnesium alloy (Mg-AZ91D) reinforced with various %wt of silicon carbide (SiC) also aluminum oxide (Al_2O_3) of nano particulates (20 nm). The preferred squeeze casting technique produces magnesium matrix composites along with silicon carbide (SiC) and aluminum oxide (Al_2O_3) particulates direct to an absolute drenched of inside the liquefy magnesium of hybrid metal matrix composite. Using CNC lathe and orthogonal array (L9), Magnesium metal matrix composites (Mg-AZ91D) are machined. To determine Metal Removal Rate (MRR) and Surface roughness (Ra) with various input parameters having the CNC lathe. Magnesium metal matrix composites (Mg-MMC) also, comparing the results with the unreinforced magnesium alloy with reinforcement of silicon carbide (SiC) and aluminum oxide (Al_2O_3) of nano particulates, Mechanical properties of the Magnesium metal matrix composite (Mg-MMC) and optimal machining parameters are evaluated.

1. Introduction

Magnesium based Hybrid metal matrix composites (Mg-HMMC) especially having sound fatigue resistance which is elementary or base materials for Automotives, because of its advanced assets-like superior wear, corrosive resistance, high strength to weight ratio, lightweight, low density, high hardness, high temperature, thermal shock resistance, high specific modulus, and high fatigue strength. Turn to liquid also stirring techniques is used to make Hybrid metal matrix composites (HMMCs). This new edition has been greatly enlarged and updated to provide both scientists and engineers with a clear and comprehensive understanding of composite materials. In describing both theoretical and practical aspects of their production, properties, and usage, the edition crosses the borders of many disciplines. Incessant progressions contain amalgamated substance for the period of additional become wider your horizons function. In the present scenario more than 200 composite materials are available instead of as an average of 1600 engineering materials in the market, so the importance of composite materials for engineering applications day to day life increases [1].

In various industries like aircraft, aerospace, and automotive rapid demands for composite materials, because composites are wonder material with lightweight, high strength to weight ratio and stiffness property has come a long way in replacing the conventional materials like metals, wood, etc. Composite materials are promising predominantly into response toward unique consider as of expertise suitable in the direction in quickness progressive proceedings into engineering. When the ceramic particles added in the materials significant mechanical properties of composite have been improved and comparing many traditional engineering materials such as metals, have a low specific gravity that makes their properties particularly superior in strength and modulus. While a consequence of concentrated learning interested in the essential landscape of resources also enhanced appreciative of their composition possessions correlation, there is a possibility in the direction of making wider novel amalgamated possessions among enhanced properties are

IOP Conf. Series: Materials Science and Engineering **998** (2020) 012067 doi:10.1088/1757-899X/998/1/012067

mechanical also physical. Prepared novel concentrations are ceramic matrix composites, metal matrix composites having and Polymer matrix composites are having high performance amalgamated composites [2]. The selected works of Mg-HMMCs are manufacturers employing compress transmit along with its possessions also machining aptitude concert is premeditated.
M.R. Amin et. Al. the author's research work of unstable characteristic liquid run throughput into practice Prandtl combination distance end to end hypothesis. Individual foremost limitation consists of its effort be too into accumulation just before the constraints having of 2-D instead of 3-D model, it is applicable for a lightweight material like pure aluminum for the results cannot exist functional on behalf of aluminum alloys somewhere spongy region grown for the duration of solidification [3].
M. Hasan, L.Begum completed a follow a line of investigation taking place magnesium alloy AZ91 with A 3-D arithmetical simulation of an industrial-sized block caster. A magnesium alloy AZ91 has been carried out for the steady-state operational phase of the caster. While using of open-top liquefies discharge method built-in among a permeable sieve next to the hot-top with a simulated model, with a, superheat of 640 °C, a fixed inlet all parametric studies were performed, with help of temperature and velocity the results are obtainable illustratively. All final results are coming based on the crucible depth, thickness, mold shell thickness and axial temperature also considered and its speed of the casting processing through its homogeneous mixture of composite material. [4].
M. Mounib et. al completed his follow a line of investigation scheduled presentations on MMCs (metal matrix composites) truly powerfully scheduled the allocations of a constituent part in the metal matrix while moreover lying on the substance responses are occurs next to the phase transformations stage. The present learning the substance responses within the selected base materials constituent part of (Al_2O_3) and (Al_2O_3) through Mg-AZ91D along with EL21, correspondingly, moreover present work was the distributions of the selected reinforcement in the composites. Dissimilar processe s a r e differential scanning calorimetric (DSC), as well as XRD be bringing into play en route for the best part of the selected substance responses with near recognize yield. Transmission electron microscopy (TEM), [5].
R. Sunil al. completed his follow a line of investigation going on SMMCs (Surface metal matrix composites) be a cluster of advanced materials for engineering anywhere the external surface of metals are customized with scatter inferior stage within the outline of constituent part otherwise filaments with the central part are not affected of chemical composites of materials along with its structure. These materials are exactly suitable for the automobile field, airbuses, power industries also recent trends for biomedical. Newly, (FSP) friction stir processing practices have been in advance broad fame into manufacturing exterior amalgamated into unyielding circumstances. Selected AZ19D life forms complex in the direction of procedure metals and contains effective progression with friction stir processing practices (FSP) in the direction of manufactured surface metal matrix composites (SMMCs) [6].
S. Aravindan et. al. completed his follow a line of investigation going on (Mg-AZ91D) amalgamated resistant employing (SiC) silicon carbide constituent part among wt% based or volume percentage at different sizes be manufacture through stir casting process consist of two steps based on the various sizes reinforced particles are added in the Mg-AZ19D, to evaluate the characteristics of the MMCS in while transmit also (T6) temperature environment. Obtained results while compared with available standards properties. Allocations of the selected constituent part also cracked exterior be deliberate during the SEM images [7].
A. Dey and K. M. Pandey. completed his follow a line of investigation based on the composite materials with Magnesium be probable resources designed for different appliances like aerospace with protection association suitable in the direction of along with low down compactness, good quality physical along with mechanical properties. Advancements are wear behavior, specific strength, damping behavior, stiffness, creep employing low energy possessions be considerably predisposed in adding particles together interested in the metal matrix composites also conventional engineering materials. Present investigation summary lying on magnesium with different

IOP Conf. Series: Materials Science and Engineering **998** (2020) 012067 doi:10.1088/1757-899X/998/1/012067

combinations of reinforcement and its alloy along with limitations and advantages [8].
Finished lines of investigation on amalgamated are fabricated employing the squeeze casting method, Mg-AZ91D matrix material. The selected reinforcement constituent part must be alive physically dominant as well as rigid than the matrix material, as a result, headed for turn out the estimated intensification consequence. But, the mixture of substance, type, and size along with wt% of the reinforcements as well as its interactions with the matrix is essential to obtain desirable properties. Among high-quality thermal stability, ceramic particles as reinforcements are the most desirable due to high hardness, strength, and melting point. Among suitable concern on cost as well as functions are on mind SiC, Al_2O_3 reinforcements are fashionable into magnesium composites. Huge research information is available on SiC particle reinforced magnesium MMC. If SiC considers in magnesium composites, having highly brittle by nature. While Al_2O_3 particles have the advantages of low cost, exhibit high specific stiffness and excellent oxidation resistance. Al_2O_3 particles can exist in many crystalline phases and produce the most stable hexagonal alpha. The machining procedure is completed through a computer-controlled lathe machine (CNC). With help of Surface roughness testing equipment evaluate the Surface roughness of the machined surface. Parametric study be accomplish intended for the material removal rate (MMR), surface roughness (Ra). Brinell hardness machine is used to know the Hardness number, with help of a universal tensile test rig to know the tensile strength.

2. Materials and Methods

A base material magnesium, the most widely used die-cast alloy (AZ91D), having a tremendous amalgamation of its oxidization confrontation, mechanical properties, and casting ability. The Mg-AZ91D of magnesium alloy having a lesser concentration, higher specific modulus and unit weight to load-carrying capacity situate good heights, also used for huge potency are not essential. Whereas higher stiffness or a thick, light desired form are needed. As the exact quantity of three metallic impurities like iron, copper, and nickel are inbuilt in the Mg-AZ91D of magnesium alloy, So oxidization confrontation is attained with putting into effect extremely severe restrictions. For the production of this alloy, it is essential to bring into play principal magnesium next to a very squat echelon adding together. Instances of complex parts manufacturing are used in the designing of aircraft or housing in addition to reciprocating or rotary machine components.
Silicon carbide (SiC), it occurs in nature as the extremely rare mineral moissanite. A small particle of silicon carbide can tie simultaneously employing hard pressing with sintering machine set up headed for extremely structured rigid ceramic objects. Those are extensively employing into relevance's needs elevated continued existence, such seeing that automobile coaster brake, vehicle clutches also ceramic plates into bullet evidence upper part body particles, besides that acknowledged while carborundum(hard block solid), be a complex as a chemical formula SiC of silicon and carbon.
Aluminum oxide (Al_2O_3) (or) aluminum oxide is a chemical compound of aluminum and oxygen with the chemical formula Al_2O_3. It is the most commonly occurring of several aluminum oxides. It occurs naturally in its crystalline polymorphic phase α-Al_2O_3 as the mineral corundum, as an abrasive owing to its hardness, and as a refractory material owing to its high melting point. It is also used in microdermabrasion, both in the machining process available through dermatologists and estheticians, and as a manual dermal abrasive used according to manufacturer directions.

2.1. *Squeeze casting of Mg-AZ91D/SiC/ Al_2O_3 Composites -*

As liquid metal forging is a Squeeze casting, be a part of the grouping of transmitting as well as falsify course of action. At the bottom of the test set up having pre-heated die through this molten metal is poured into the chosen die. When the molten metal gets start solidification, the pressure is applied on the upper half die and it closes with help of hydraulic pressure in terms of tons, it happens at the solidification process. The applied pressure or force is slightly less than the forging force, along with components having enormous specify be capable of subsisting are formed. Mechanical properties are enhanced when the metal matrix composite is having less porosity. Through this transmit and pressure applied method like squeeze techniques en route for manufacture the ferrous and non-ferrous materials.

Table 1.Chemical composition of the Mg-AZ91D matrix alloy

%	Al	Cu	Fe	Ni	Zn	Mn	Si	Mg
Mg-AZ91D	8.3-9.7	0.03	0.005	0.002	0.35-1.0	0.15-0.5	0.1	Bal

2.2. *Mg-AZ91D/SiC/ Al_2O_3 Composites Fabrication Procedure-*

The aluminum oxide (Al_2O_3), Silicon carbide (SiC) with a Size of 20 (nm), and magnesium alloy AZ91D be stimulated inside a crucible shown in Figure 1. At 800^0c magnesium, alloy AZ91D is heated in the furnace beginning. In Pre-heater at 400^0c alumina in addition to boron carbide are as well preheated.

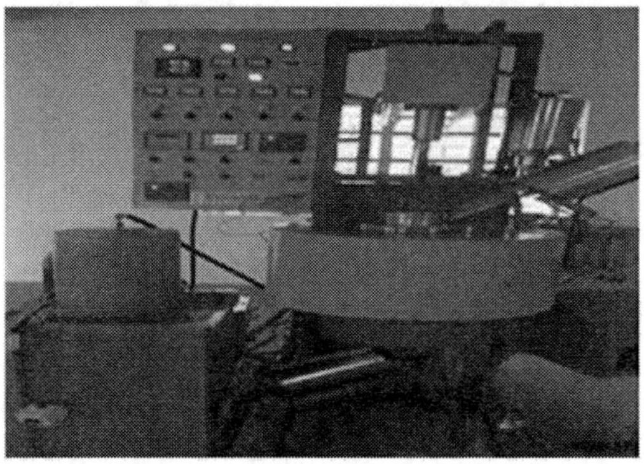

Fig.1. Stir casting setup with Squeeze

With a suitable load of 400KN applied on the molten with help of a hydraulic die having a mixture of molten metal is coming from the crucible with the bottom gate. Table 2 shown the wt% of materials is solidification the essential profile, as well as the dimension of the hybrid metal matrix composite materials, are obtained.

Table 2. Samples prepared

Samples	Materials in terms of wt.%		
	Mg-AZ91D %	SiC %	Al_2O_3 %
1	100	0.0	0.0
2	94	3.0	3.0
3	91	6.0	3.0
4	88	9.0	3.0

2.3. *Orthogonal array (L9)*

The orthogonal array (L9) is intended on behalf of realizing the consequence of 4 autonomous dynamics every one contains 3 dynamic rank standards. At the same time as here be various regular orthogonal ranges presented, each of the ranges is preordained on behalf of a particular numeral indent of autonomous intend unpredictable's also ranges. At the same time as in various cases, there is nothing communication representation hypothesis is valid, present exist various cases someplace here be a plain indication of communication. The selections of array imagine to here be zero communication in between selected factors of any two. The communication connecting the temperature also material properties would be alive a classic interaction bond. The below table 3 shown obtained values with the selected orthogonal array of 3 variables. In support of illustration, condition individual requirements en route for carrying out experimentation in the direction of identity with the authority of four dissimilar autonomous changeable employing every changeable contains three lays down standards, after that orthogonal array (L9) may be the accurate selection.

Table 3. Orthogonal Array (L9)

CNC Machine	Mg-AZ91D/SiC/$Al_2O_3$of materials								
Inputs Variables	S-1	S-2	S-3	S-4	S-5	S-6	S-7	S-8	S-9
Spindle speed(rpm)	350	550	750	550	750	350	750	550	350
Depth of cut	0.2	0.3	0.4	0.2	0.3	0.4	0.2	0.3	0.4
Tool Feed rate	0.1	0.1	0.1	0.2	0.2	0.2	0.3	0.3	0.3

2.4. CNC Machining

The below figure is shown as a machined and unmachined workpiece with help of a CNC Lathe machine. The main aim to study the amount of material removal rate and amount of surface roughness of the Magnesium metal matrix composite of Mg-AZ91D/SiC/ Al_2O_3, if Al_2O_3 is constant wt.% in all samples and SiC % varying at wt.% in twice in sample 3 and sample 4.

Fig.2. Mg-AZ91D/SiC/Al_2O_3

3. Results with Discussion

3.1. Metal Removal Rate (MRR)

The Desirable MRR valves are shown in table 4. There are two ways to evaluate the amount of materials removal rate of Mg-AZ91D/SiC/ Al_2O_3, one is a weight basis and the other one is volume-based. The Material Removal Rate (MRR) also known as the immediate material removal rate as the rate at which the cross-section area of material being removed moves through the workpiece. Another way is the measurement of how much material is removed from a part in a given period.

Table 4. Mg-AZ91D/SiC/Al_2O_3 Metal Removal Rate and Surface Roughness values

CNC Machine	Mg-AZ91D/SiC/$Al_2O_3$of materials								
Inputs Variables	S-1	S-2	S-3	S-4	S-5	S-6	S-7	S-8	S-9
Spindle speed(rpm)	350	550	750	550	750	350	750	550	350
Depth of cut	0.2	0.3	0.4	0.2	0.3	0.4	0.2	0.3	0.4
Tool Feed rate	0.1	0.1	0.1	0.2	0.2	0.2	0.3	0.3	0.3
MRR$\times 10^{-8}$*(m/Sec)*	2.19	3.42	5.22	4.41	6.42	8.21	7.34	4.11	1.32
Ra*(μm)*	0.45	0.56	0.58	1.89	1.78	1.92	4.23	5.39	6.42

3.2. Ra(Surface Roughness) test

The Ra is deliberate by using the SURFTEST machine set up shown in figure 3. Surface Roughness (Ra) is a constituent of exterior consistency. Surface roughness is enumerating through the variations into the track of the customary vector of a real exterior from the surface's ultimate outward appearance. However variations are big, the exterior surface as of rough; meanwhile, they are little, the exterior face is smooth.

IOP Conf. Series: Materials Science and Engineering **998** (2020) 012067 doi:10.1088/1757-899X/998/1/012067

Fig.3. The test set up for Surface Roughness

Table.5 represents of measured values of the samples' surface roughness through the surface testing machine. Surface roughness (Ra) is characteristically believe headed for being present high occurrence, short-wavelength elements of surface measurements' (observe exterior metrology). Other than, during the observation surface roughness is frequently essential headed for recognizing together the frequency and amplitude in the direction of making certain with the purpose of an exterior be in shape intended for an intention.

Table.5. Surface Roughness Test Experimental values

CNC Machine Inputs Variables	Mg-AZ91D/SiC/$Al_2O_3$of materials								
	S-1	S-2	S-3	S-4	S-5	S-6	S-7	S-8	S-9
Spindle speed(rpm)	350	550	750	550	750	350	750	550	350
Depth of cut	0.2	0.3	0.4	0.2	0.3	0.4	0.2	0.3	0.4
Tool Feed rate	0.1	0.1	0.1	0.2	0.2	0.2	0.3	0.3	0.3
Ra(μm)	0.45	0.56	0.58	1.89	1.78	1.92	4.23	5.39	6.42

3.3. *Hardness*

The Brinell scale characterizes the indentation hardness of materials through the scale of penetration of an indenter, loaded on a material test-piece. It is one of several definitions of hardness in materials science.

The volume fractional percentage of SiC/ Al_2O_3 nano particulates fashioned as increasing the hardness value of metallic matrix composite results are shown in Fig.4a. The percentage of SiC/ Al_2O_3 of in Mg-AZ91D had to escalate the proportion of SiC/ Al_2O_3 through the stiffness assessment enhances in the same way. Whereas pure Mg-AZ91D (magnesium alloy) had not had such hardness while evaluates among Magnesium/SiC/ Al_2O_3 (MMC) amalgamated.

3.4. *Tensile*

With the consequence of SiC/ Al_2O_3 wt% in Mg-AZ91D MMC, the tensile strength is scrutinized, along with the results are plotted in Fig.4b. Tensile strength increases with an increase in the weight fraction of reinforced nanoparticles of SiC/Al_2O_3. The Tensile strength of fabricated hybrid Magnesium/ SiC/Al_2O_3 (MMC) also increases with an increase in weight fraction of reinforced particles SiC/Al_2O_3. The tensile strength of fabricated 9 wt% SiC+ 3wt% Al_2O_3 – Mg-AZ91D MMC is 224MPa which is 1.3 times of Mg- Matrix.

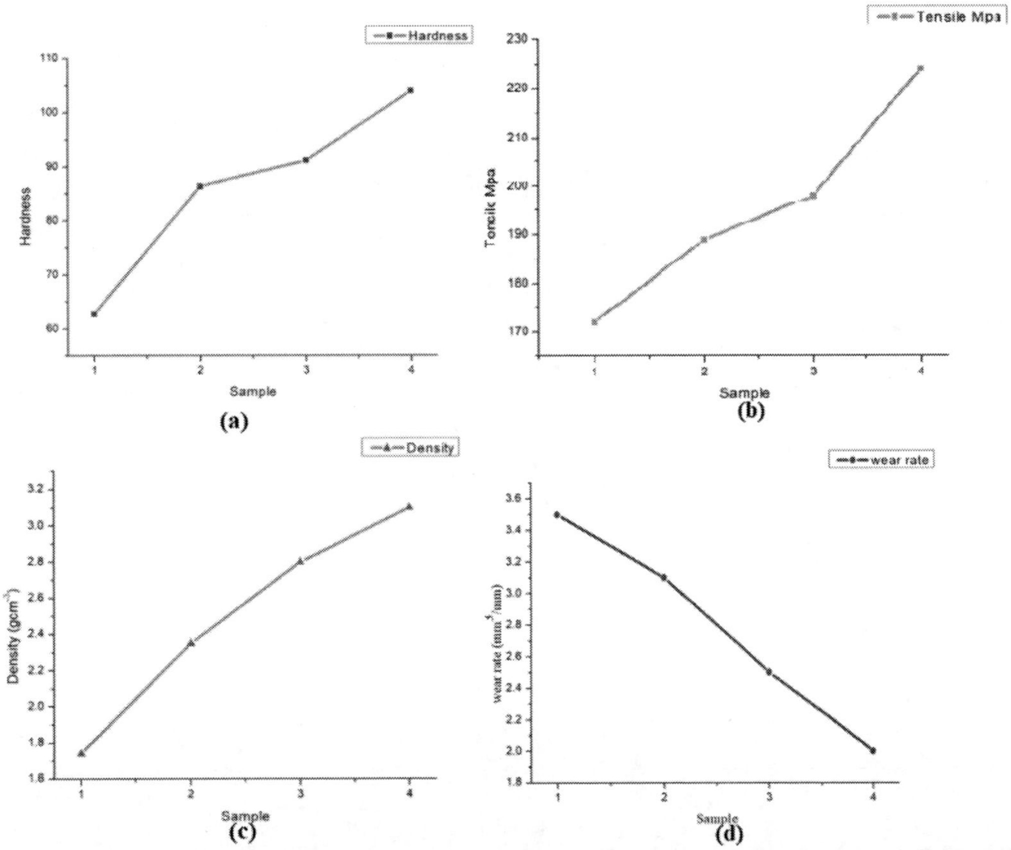

Fig.4. (a) wt% SiC/Al_2O_3 in Mg-AZ91D V/S Hardness of composites, (b) wt% SiC/Al_2O_3 in Mg-AZ91D V/S Tensile Strength, (c) wt% SiC/Al_2O_3 in Mg-AZ91D V/S Density, (d) wt% SiC/Al_2O_3 in Mg-AZ91D V/S Wear

3.5. Density

Fig.4c, Represents the density of the prepared samples determined utilized the relation as follows: $= w1\,(\rho w)/w1-w2$, where ρw is the density of water, w1=sample weight in the air, w2= sample weight in water, ρ = density of prepared cast Mg-MMC sample. Based on Archimedes's principle, the density of the prepared SiC/Al_2O_3 wt% in Mg-AZ91D MMC samples are determined.

3.6. Wear

By using the pin on disc test rig the selected or sample specimens are performed wear tests. In Fig.4.d stated lower wt% of SiC/Al_2O_3 added in Mg-AZ91D as a reinforcement the extent of reduction is higher if higher wt% of SiC/Al_2O_3 added in Mg-AZ91D wear rate decreases because the wear rate decreases with increases the SiC/Al_2O_3 wt% in Mg-AZ91D as reinforcement, for the reasons is thermal stability and hardness of the reinforcement are more than the Mg-AZ91D alloy.

3.7. Performances Measures with the effect of Machining Parameters

With considering the process parameters of spindle speed, Depth of cut(DOC), Tool feed rate, Metal Removal Rate (MRR), Surface Roughness (Ra) also reinforcement proportions (%Ref) be considered to drown the graph based on the experimental assessment.

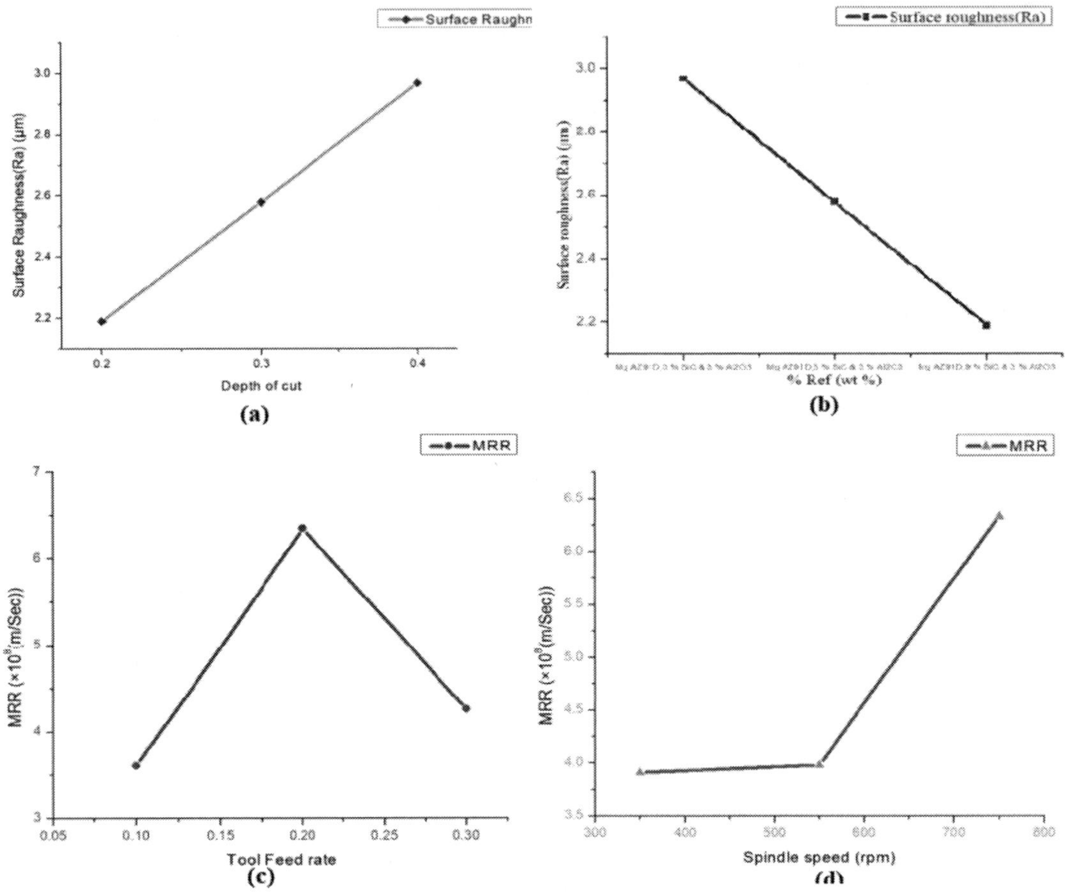

Fig.5. (a) DOC v/s Surface Roughness (Ra), (b) %Ref v/s Surface Roughness (Ra), (c) Feed v/s Material Removal Rate (MRR), (d) Speed v/s Material Removal Rate (MRR)

From Fig.5.a, the depth of cut increases, the surface roughness values are to some extent increased also there are no identical variations due to tool feed rate and spindle speed. Fig.5.a stand for the chart contains an X-axis depth of cut whereas Y-axis consists of surface roughness. From Fig.5.b X-axis taken as the reinforcement (%Ref) added in the Mg-AZ91D of 3%, 6%, and 9% of Sic moreover 3% of Al_2O_3 in addition to the Y-axis consists of surface roughness. Its experimental results observed that % of reinforcements are increased in Mg-AZ91D; it increases the hardness of the composite material, so the surface roughness value decreases with the good surface finish of the components. However, the friction-free components are fabricated and it becomes suitable for the automobile industry. Fig.5.c contains the X-axis tool feed rate and Y-axis Material removal rate (MRR). Concerning tool feed rate increases up to a certain level the material removal rate also increased. Further tool feed rate increased the material removal rate decreased. From Fig.5.d consists of speed to materials removal rate, X-axis consists of speed, and Y-axis consists of material removal rate. When spindle speed increases to obtained more material removal rate. It's not considered DOC and tool feed rate.

4. Conclusion

In this investigation, squeeze castings techniques are used to build magnesium metal matrix (Mg-MMC), made-up with make use of for the fabrication of hybrid Mg-AZ91D/SiC+ Al2O3-MMC.

The dissimilar weight fractions of hard reinforced particulates SiC and Al_2O_3 are utilized to fabricate hybrid Mg-MMC samples. The different tests are conducted on cast hybrid Mg-MMC samples to identify the various physical and mechanical properties.
The following conclusions are based on the fabrication and machining of Mg-AZ91D/ SiC/Al_2O_3 composite material.

- With help of the squeeze casting technique, Mg-AZ91D with reinforcement's nanoparticles of SiC/Al_2O_3 distributions are easily done.
- The density of hybrid Mg-MMC increases with an increase in weight fraction of SiC/Al_2O_3 nano particulates as SiC/Al_2O_3 both have a higher density over Mg-matrix. Investigation results reveal that the density of Mg-MMC with 9wt% SiC + 3% Al_2O_3 is about 1.77 times of Mg-AZ91D. The hardness increases with an increase in weight fraction of hard reinforced nano particulates with SiC/Al_2O_3 (8 & 3 wt %) respectively - MMC is 104HV, whereas the hardness of Mg-AZ91D is only 62.8HV.
- Tensile strength increases with an increase in weight fraction of reinforced nano particulates SiC/Al_2O_3. The incorporation of SiC/Al_2O_3 nano reinforcement particulates to Mg-AZ91D alloy increases the wear resistance of the composite.
- The purposes of L9 array selection to reduce the number of tests are conducted through the CNC Lathe machine. It generates great values of measurable performance conducted on Mg-AZ91D/ SiC/Al_2O_3 composite materials.
- Since the reinforcement percentages increase in magnesium composites of Mg-AZ91D/SiC/Al_2O_3, the surface roughness are decreases. The observed variations concerning tool feed and spindle speed.
- They both selected output responses purely depend on the depth of cut whereas compared to other input process parameters in CNC lathe.

5. References

[1] L. M. Manocha, A. R. Bunsell, Advances in composite materials, Pergamon Press, Oxford, 2 (1980), 1233-1240.
[2] G. Fisher, Composite: Engineering the ultimate material, American Ceramics Society, 63 (2005), 360-364.
[3] M.R.Amin, A. Maharajn, Modeling of turbulent heat transfer during the solidification process of continuous casting, I. Mater. Process. Tech. 174(2006) 155-166.
[4] M. Hasan, L. Begum, Modelling of magnesium DC casting process using a 3D turbulent CFD model, Advanced Material Processing, 28(2012), 145.
[5] M. Mounib, M. Pavese, C. Badini, W. Lefebvre, H. Dieringa, Reactivity, and microstructure of Al2O3-reinforced magnesium-matrix composites, Advances in Materials Science and Engineering, 39 (2013), 1-7.
[6] B. R. Sunil, Developing surface composites: A comparative survey, International Journal of Advances in Materials Science and Engineering, 4(3) (2015) 1-8.
[7] S. Aravindan, P. V. Rao, K. Popanna, Evaluation of physical and mechanical properties of AZ91D/SiC composites by two-step stir casting process, Advanced Engineering Material, 6(2015), 1-11.
[8] Dey, K. M. Pandey, Magnesium metal matrix composites-a review, Indian Foundry Journal, 57(2015), 35-40.

DYNAMIC BEHAVIOR OF ROPEWAY STRING

SHAKTI P JENA[1,a], MOHD IBRAHIM[2,b*], B. SUBBARATNAM[3,c], S NARESH KUMAR[4,d]

[1, 3, 4]Department of Mechanical Engineering, Vardhaman College of Engineering, Hyderabad, Telangana, India.

[2]PG Scholar, Mechanical Engineering Department, Vardhaman College of Engineering, Hyderabad, Telangana, India.

E-mail: [*b)]shaazibrahim641@gmail.com [a)]shaktipjena@vardhaman.org
[c)]b_subbaratnam@gmail.com [d)]snk.vce@gmail.com

Abstract: In this paper a new approach for dynamic behavior of a string is aimed to study lateral vibrations induced in an elastic ropeway string. We are going to deal with the dynamic behavior of a string. In hilly areas we have problem for traveling from one place to another place. There we are using a rope drive system using a string as a rope. A ropeway is a type of maritime lifting gadget used to move light stores and hardware across waterways. Here, multiple ropeways are designed by using solid works and then analyzing with static, transient and dynamic boundary conditions. To suggest optimum materials for optimum ropeway string, 4 materials were chosen (SAE 1020, SAE1137, Al-17075, Al-7475) to analyze and study the dynamic behavior of ropeway string by knowing each material results, comparing with each other material results and then showing suitable tables and graphs, finally concluding thesis with optimum material with optimum ropeway string to improve the performance and their durability.

1. Keywords: String, Solid Works, Ansys Workbench, Natural Frequency.

2. Introduction: Since few years, the scientific interest for the dynamics of structures travelled by moving loads has growing considerably because of the design and the efficient construction of systems are able to reach high velocities. There are many literature mainly focused on the analysis of the dynamic response of beam and on elastic foundations when loaded under different points. A.J. Wallis Tayler [1] has studied the aerial or wire ropeways, their construction and management. Walter G. Booth [2] has designed and studied applications of the aerial ropeways. James M. W. and Brown john [3] determined the dynamics of an aerial cableway system. Jacques Dubuisson, Michel Cantin [4] examined slipping resistance of monocable aerial ropeway carrier grips. Andreas Pichler [5] has examined the dynamic behavior of the grip and the terminal equipment of a detachable monocable ropeway at the terminal entry. Mark Löhr, Adams [6] has performed the simulation for ropeway technology. Stephan Liedl [7] have computed the motions and forces in the rope system of aerial ropeways during operation. Alexander BorisoffKazakof [8] have proposed advances in engineering software for lift transportation systems. Marta Knawa and Danuta Bryja [9] have developed modeling problems of steeply inclined cableway subjected to moving load. Danuta Bryja and Marta Knawa [10] have explored Computational model of an inclined aerial ropeway and numerical method for analyzing nonlinear cable-car interaction. But dynamic behavior of string when it is subjected to moving mass is also important to study. In few areas mainly hill stations even for the short distances, there is a problem of electricity and travelling from one place to another. To overcome this difficulty, we can use rope drive (using string carrying mass attached to it) When a string is used as a rope then its dynamic behavior when carrying mass with it is to be studied carefully otherwise it may not give exact results. In this we are going to deal mainly with dynamic behavior of string under moving mass, required stiffness and vibrations of string etc.,

3. Problem formulation:

3.1 Arbitrary loads of ropeways: The below image shows the Single span circular ropeway span and the Loading parameters among brackets is indicated with Ls and the altitude difference is indicated with hs.

IOP Conf. Series: Materials Science and Engineering 998 (2020) 012068 doi:10.1088/1757-899X/998/1/012068

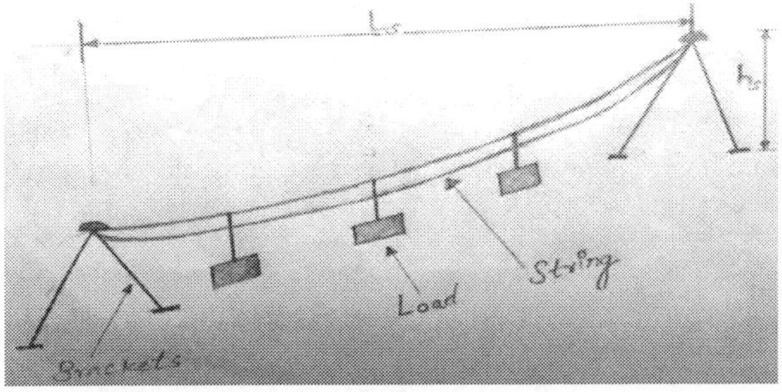

Figure 1 Ropeway of single span circular

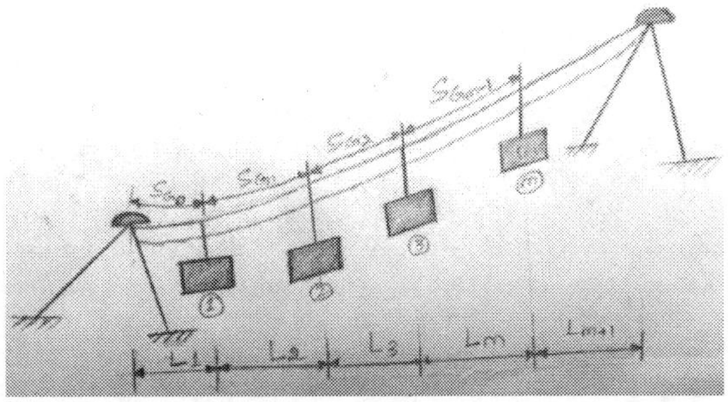

Figure 2 Loads parameters

The total length of the bearing cable is indicate with the term of sl_0 and the total length of traction cable is denoted with st_0 and it can be calculated completing the erection of the ropeway section formula, $V_{Bi}=V_{Ai}+qs_i$ (i=1...m+1)
Total length weight of cable is denoted with q here. Here the altitude difference & spans of every section values are shown. In this process the cross section area & modulus of elasticity is denoted with EA_0, here superscripts are denoted with 1 or t represents the bearing cable or traction variables.

$$L_i = \frac{H_i}{q}\left[\frac{qs_i}{EA_0} + asinh\frac{V_{Bi}}{H_i} - asinh\frac{V_{Ai}}{H_i}\right] \quad (1)$$

$$h_i = \frac{V_{Bi}^2 - V_{Ai}^2}{2qEA_0} + \frac{H_i}{q}\left(\sqrt{1+\frac{V_{Bi}^2}{H_i^2}} - \sqrt{1+\frac{V_{Ai}^2}{H_i^2}}\right) \text{ (i=1...m+1)} \quad (2)$$

where L is length of the cable, H is height of bracket on which string is mounted and q is the total length weight of string.

3.2 Traction cable & bearing cable analysis: The combine interactivity between bearing cable & traction cable of one & only pass over ropeway is mostly cerebrate from the below steps

$$L_i^l - L_t^l = 0 \ , h_i^l - h_i^t = \text{(i=1...m+1)} \quad (3)$$

The below equation shows return traction cable

$$L_{m+2}^t - L_s = 0 \ , h_{m+2}^t - h_s = 0 \quad (4)$$

It is not necessary to know the loads movement from where to where in a one only pass over ropeway it is constant the value of total altitude variation & span, the addition of spans portion is same to complete horizontal span & the addition of altitude difference portion is same to complete altitude difference value

$$\sum_{j-1}^{m+1} L_j^l - L_s = 0 \ \ \sum_{j-1}^{m+1} h_j^l - h_s = 0 \quad (5)$$

3.3 Constant loads of Space: Here we have m+1 equation due to unchanged values of traction cable when it is pulled by load

$$\sum_{j=s_i+1}^{2+m} S_j^t - S_i^G = 0 \ , \ \sum_{j=g_m+1}^{s_i+1} S_j^t - S_m^G = 0 \text{ (i=0...m-1)} \quad (6)$$

4. Materials and methodology:

4.1 Materials:

Table 1 Material properties

Material :	SAE1020	SAE1137	Al-7075	Al-7475
Young's modulus	2*10^11 Pa	200*10^9 pa	71.7*10^9 Pa	70.3*10^9 Pa
Poisson's ratio	0.29	0.29	0.33	0.33
Yield strength	394.70 Mpa	500 Mpa	503 Mpa	450 Mpa
Density	7870 Kg/m³	7800 Kg/m³	2810 Kg/m³	2810 Kg/m³

5. Methodology:

5.1 Solid works process: In this thesis solid works software is used to develop the both ropeway strings, in this process the mean diameter of total number of strings is consider as 12mm. To create ropeway string here extrude option were used to generate, and here 1000mm is consider to be total length of the string, and after creating the sketcher need to check whether it is completely closed or not if there is any open ends need to fix them to closed loop to convert the sketcher into 3D object, and here need to remember one more thing that is the sketcher is don't have any overlapping it means the sketcher should not pass on other sketcher. After creating all these sketches the image shows the final image of the circular straight string and helical twisted string.

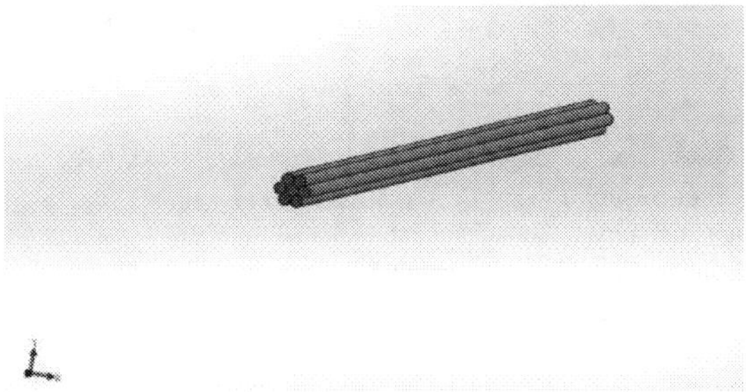

Figure 3 Straight string model

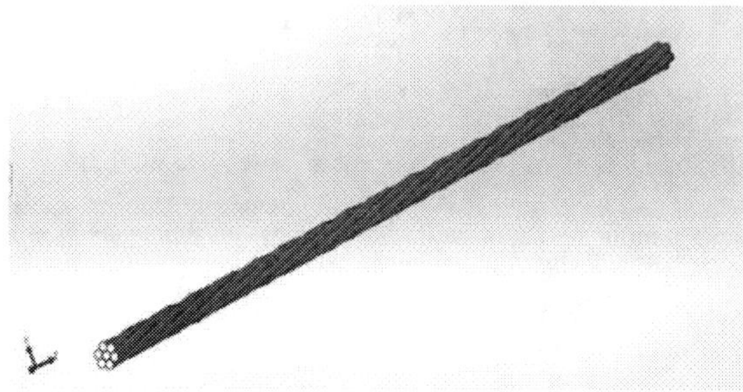

Figure 4 Helical twisted string model

5.2 Ansys process: Select ANSYS workbench and then click static structural module to do analysis process which is shown in below image. After selecting static analysis module now enter the material properties of SAE 1020, SAE 1137, Al 7075, Al7475 and those values are given below. After completing design in solid works now the final model were imported into ANSYS workbench and the model is showing in below image, to import into ANSYS workbench here the object should be in either .iges or .step file format only.

5.2.1 Straight ropeway string Meshing: To calculate the results here need to apply boundary conditions on it, before applying boundary conditions first of there should be a contact between each string so that the load will transfer from one end to other end, this meshing will create intersections and cross connection in between them so that by that contact the load can transfer throughout the object.

5.2.2 Boundary conditions: Here force is consider to as boundary condition on it and here Fixing both ends by using fixed support option then after apply Force 430 N apply then solve the results of deformation, stress, strain, safety factor, this 430N is maximum limit to apply on circular string, at this loading the safety factor is nearby 1.5 only, so that it is consider to be maximum allowable load on this circular string without any breakage on it.

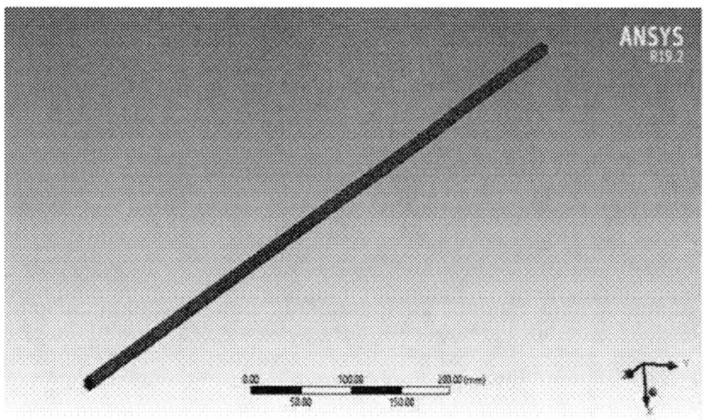

Figure 5 Straight rope way string meshing

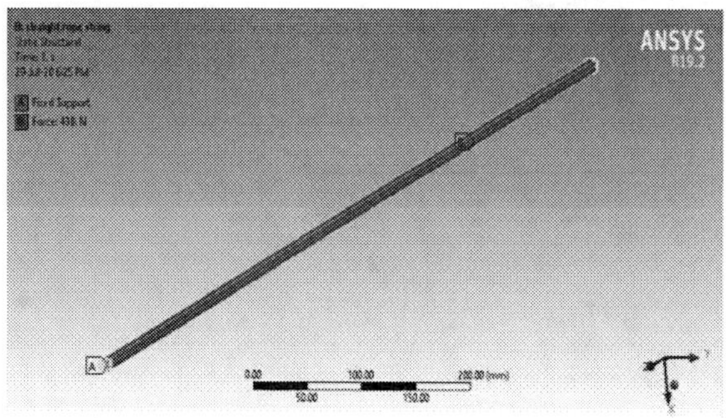

Figure 6 Straight rope way string boundary conditions

5.3 Straight ropeway string results:

5.3.1 Material SAE1020

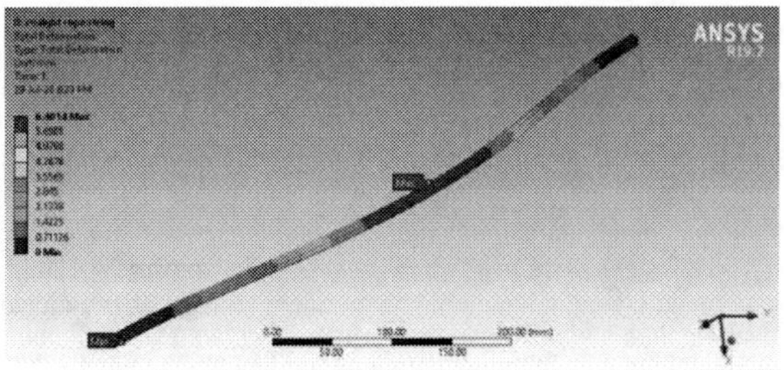

Figure 7 Material SAE1020 Deformation

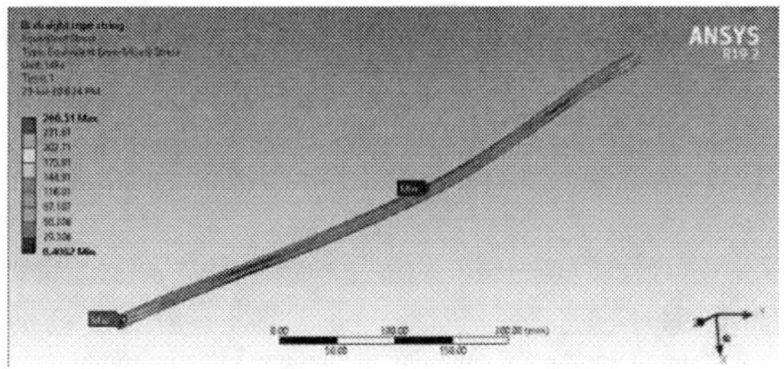

Figure 8 Material SAE1020 Stress

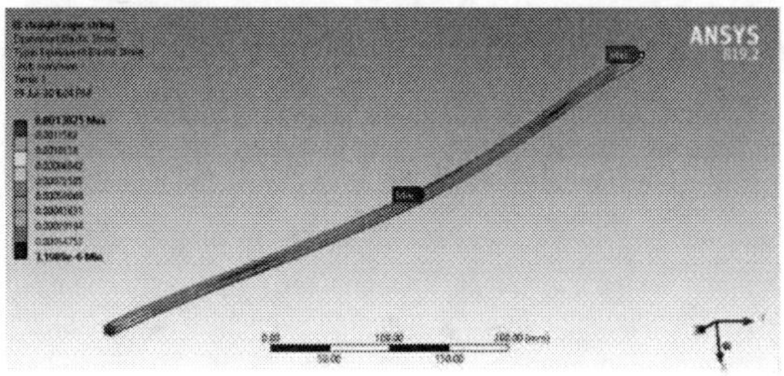

Figure 9 Material SAE1020 Strain

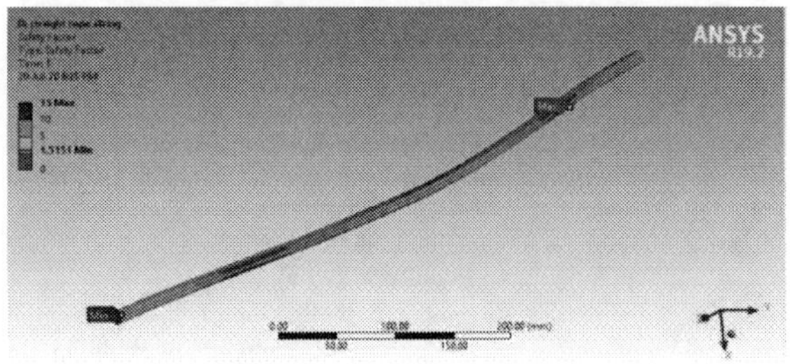

Figure 10 Material SAE1020 Safety factor

5.3.2 Material Al-7075

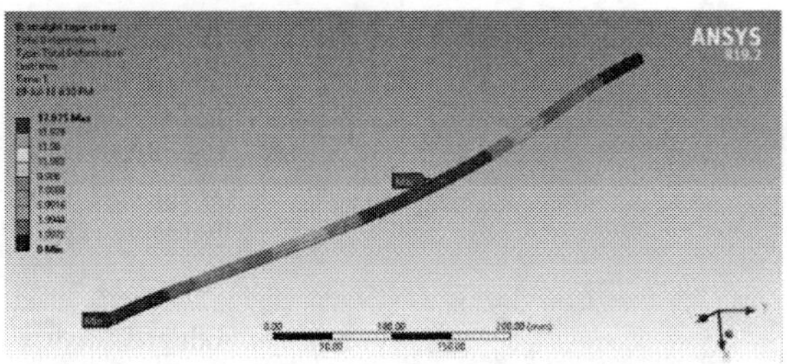

Figure 11 Material Al-7075 Deformation

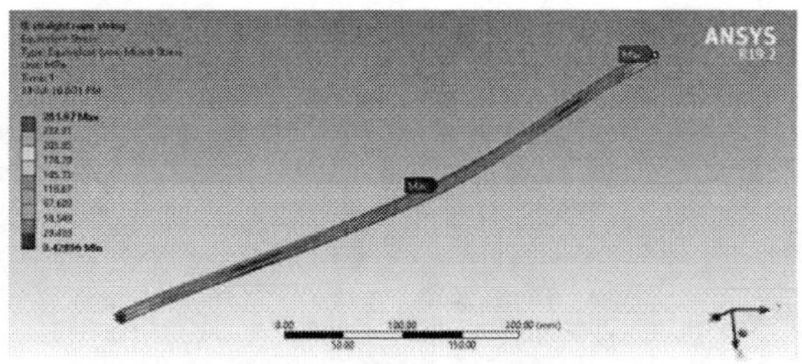

Figure 12 Material Al-7075 Stress

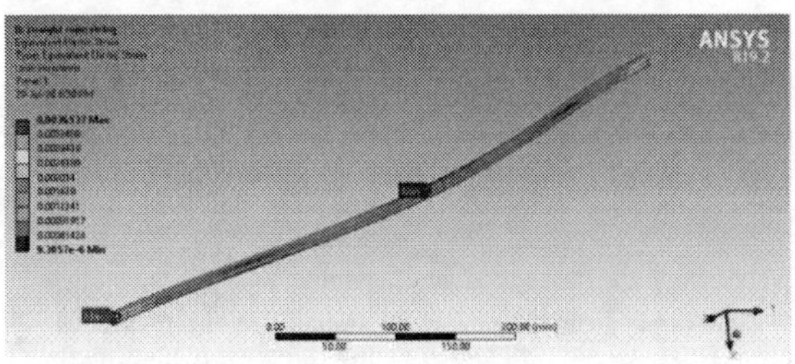

Figure 13 Material Al-7075 Strain

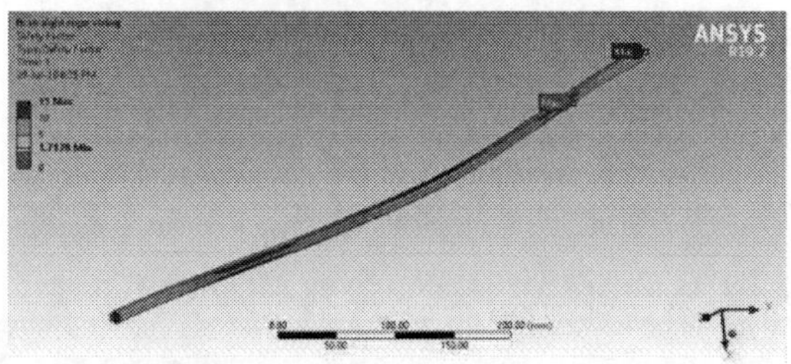

Figure 14 Material Al-7075 Safety factor

5.4 Tables: The following table shows the results of straight ropeway string for different materials

Table 2 Straight ropeway string results

Material	SAE 1020	SAE 1137	AL7475	AL7075
Deformation (mm)	6.401	6.4014	18.333	17.975
Stress (Mpa)	260.51	260.5	261.97	261.9

Strain	0.0013025	0.0013025	0.0037265	0.0036537
Safety factor	1.5151	1.9193	1.7178	1.9201

5.5 Graphs: The following graphs illustrates the results of Straight ropeway string for different materials

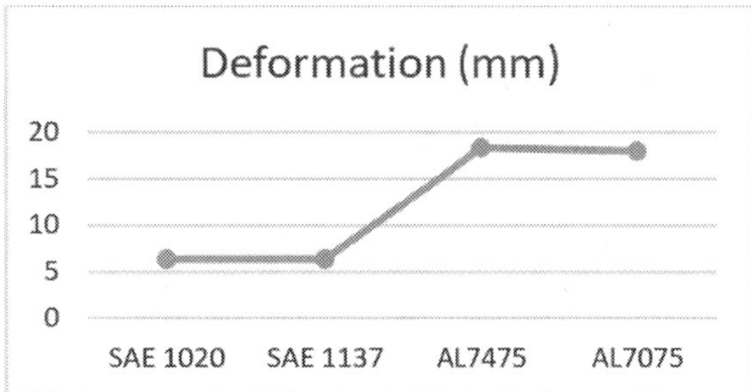

Graph-1 Straight ropeway string deformation

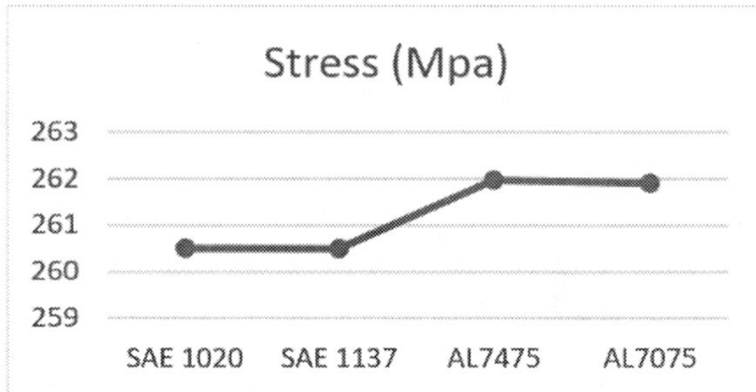

Graph-2 Straight ropeway string stress

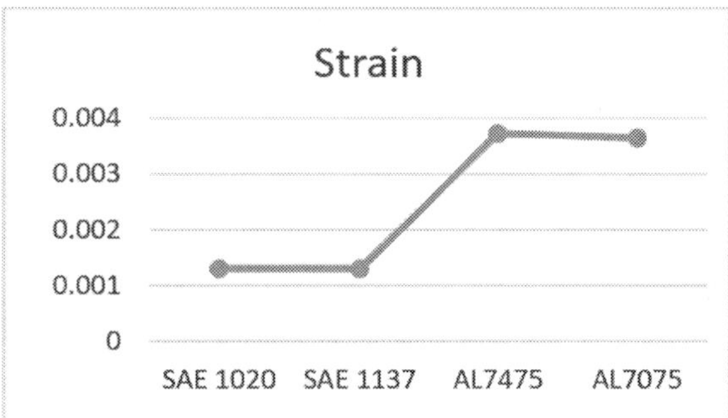

Graph-3 Straight ropeway string strain

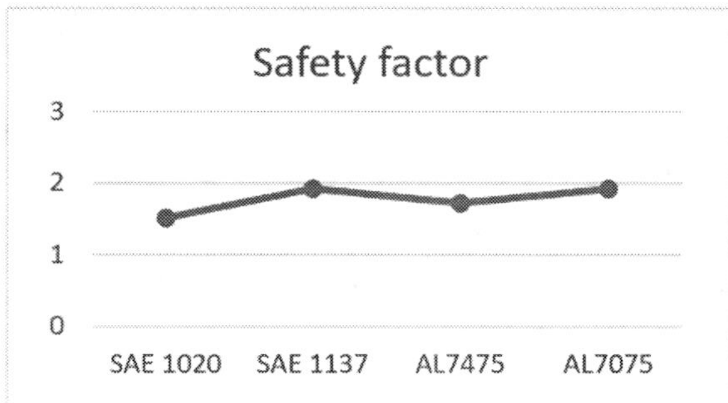

Graph-4 Straight ropeway string Safety factor

6. Helical twisted ropeway string:

6.1 Helical twisted rope string meshing: To calculate the results here need to apply boundary conditions on it, before applying boundary conditions first of there should be a contact between each string so that the load will transfer from one end to other end, this meshing will create intersections and cross connection in between them so that by that contact the load can transfer throughout the object.

6.1.1 Helical twisted ropeway string boundary conditions: Here force is considered as boundary condition on it and here Fixing both ends by using fixed support option then after apply Force 450N apply then solve the results of deformation, stress, strain, safety factor, this 450 N is maximum limit to apply on helical twisted string, at this loading the safety factor is nearby 1.5 only, so that it is consider to be maximum allowable load on this helical twisted string without any breakage on it.

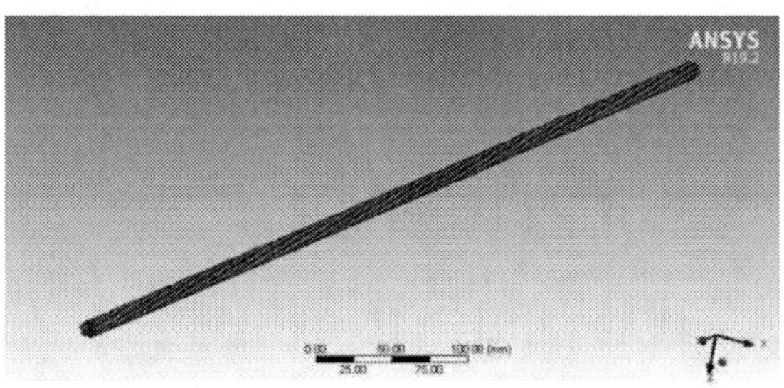

Figure 15 Helical twist rope string Meshing

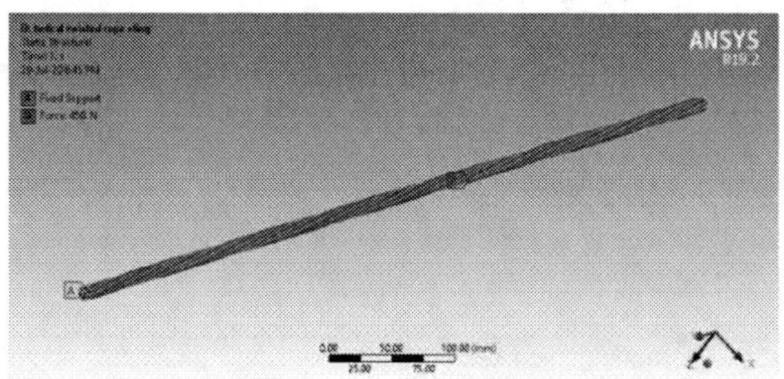

Figure 16 Helical twist rope way string Boundary conditions

6.2 Helical twisted string ropeway results:

6.2.1 Material SAE1020

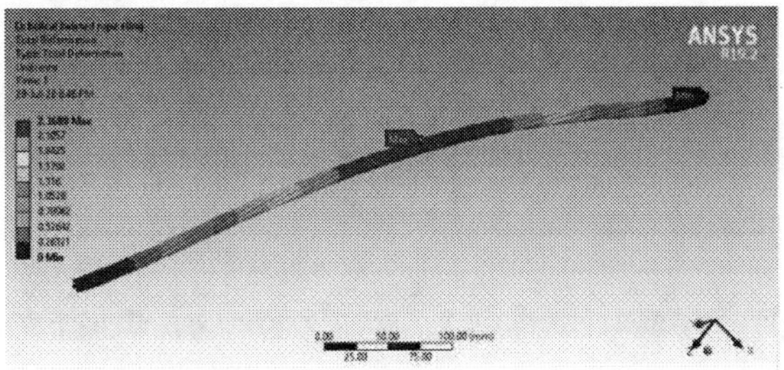

Figure 17 Material SAE1020 Deformation

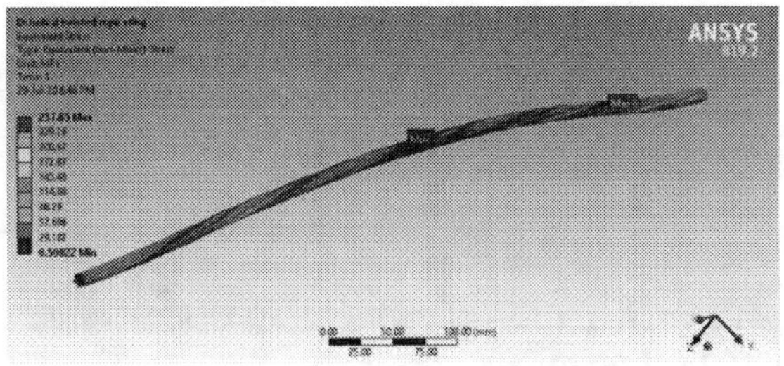

Figure 18 Material SAE1020 Stress

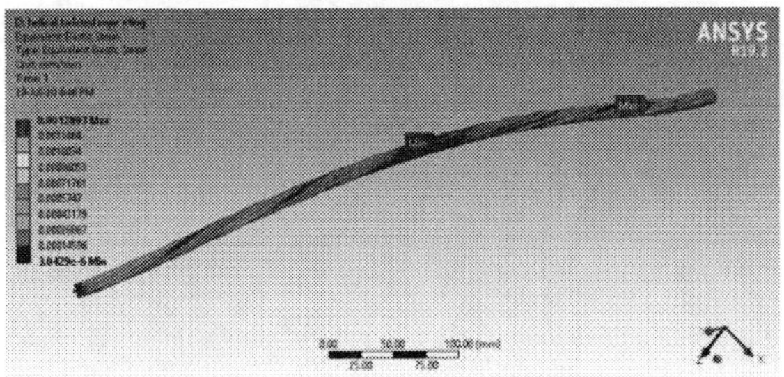

Figure 19 Material SAE1020 Strain

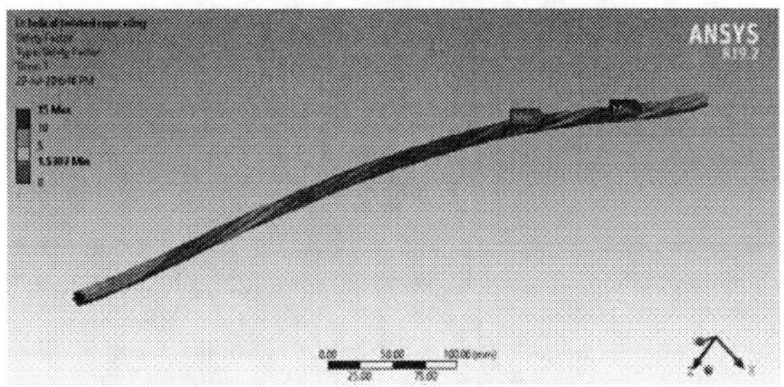

Figure 20 Material SAE1020 Safety factor

6.2.2 Material Al-7075

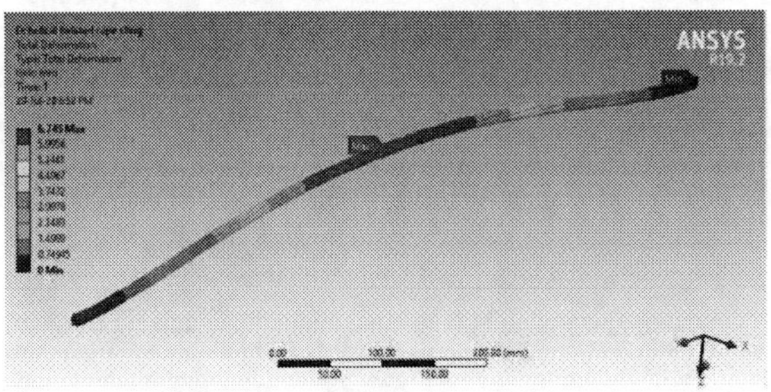

Figure 21 Material Al-7075 Deformation

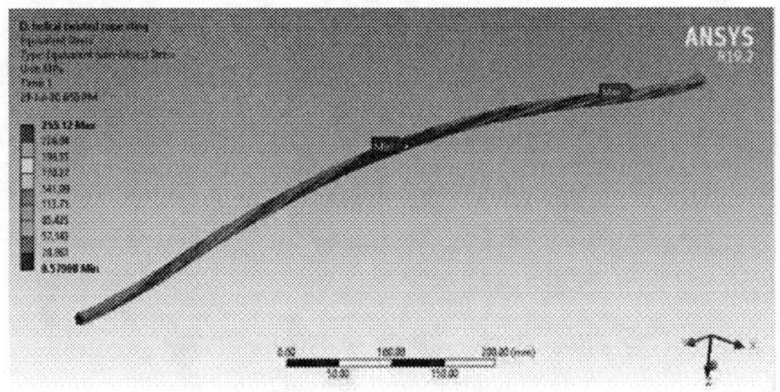

Figure 22 Material Al-7075 Stress

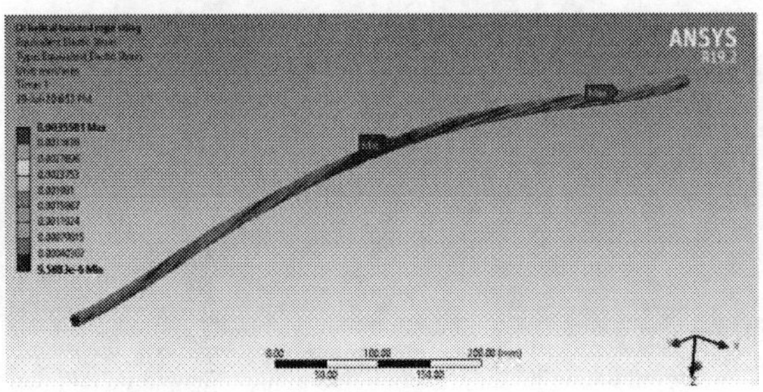

Figure 23 Material Al-7075 Strain

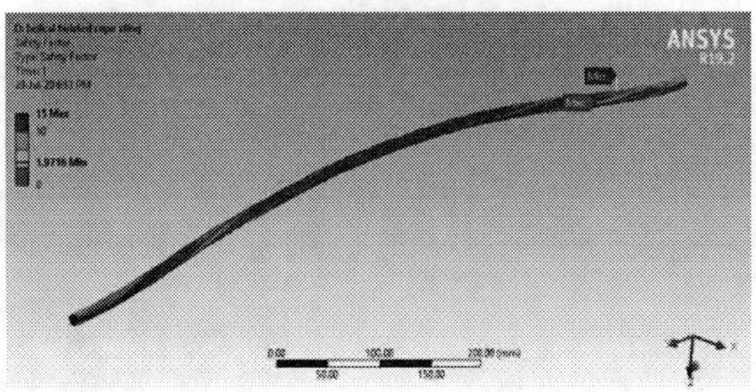

Figure 24 Material Al-7075 Safety factor

6.3 Tables: The following table shows the results of helical twisted ropeway string for different materials

IOP Conf. Series: Materials Science and Engineering 998 (2020) 012068 doi:10.1088/1757-899X/998/1/012068

Table 3 Helical twisted ropeway string results

Material	SAE 1020	SAE 1137	AL7475	AL7075
Deformation (mm)	2.3689	2.3689	6.8794	6.745
Stress (Mpa)	257.85	257.85	255.12	255.21
Strain	0.0012893	0.0012893	0.003629	0.0035581
Safety factor	1.5307	1.9391	1.7639	1.9716

6.4 Graphs: The Following graphs illustrates the results of helical twisted ropeway string for different materials

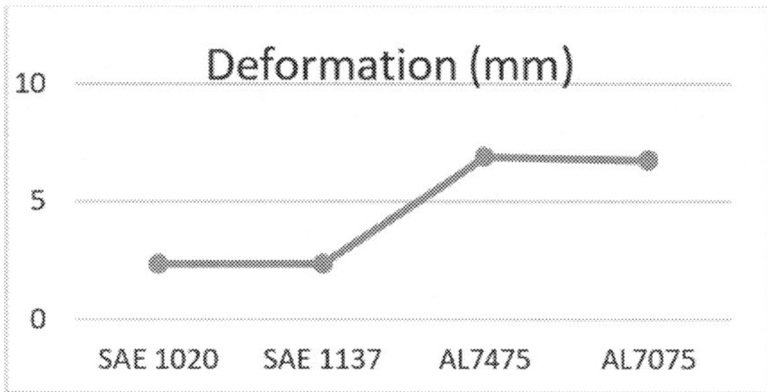

Graph-5 Helical twisted string deformation

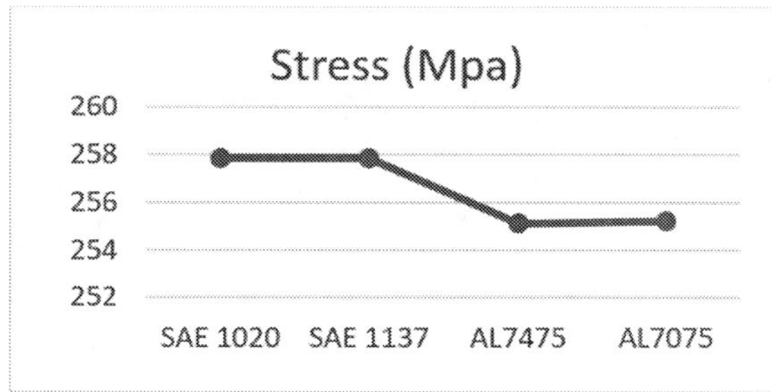

Graph-6 Helical twisted string stress

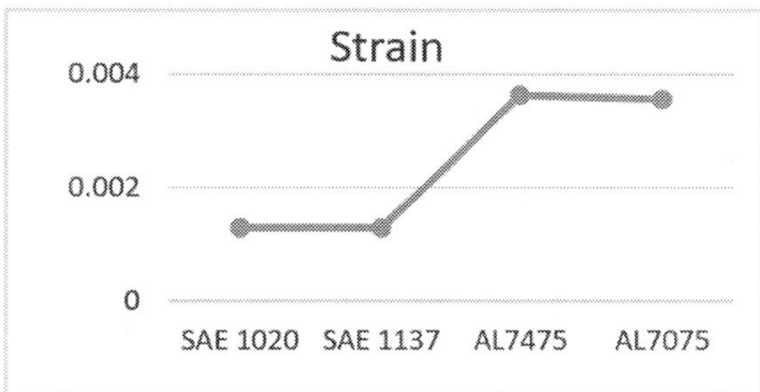

Graph-7 Helical twisted string strain

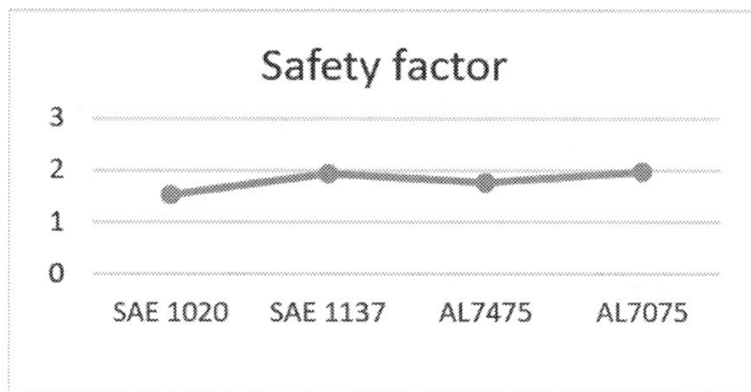

Graph-8 Helical twisted string safety factor

7. Transient analysis:

7.1 Transient analysis boundary conditions: In this transient analysis module different type of load values applied at different time period, here 43 seconds of time is taken to apply 430N of load on it, and each step is increment of 5secs in time period, and 50N in load steps.

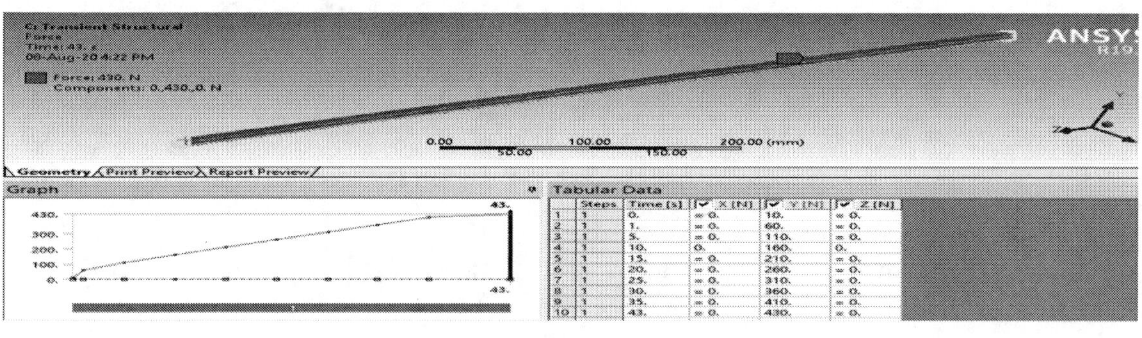

Figure 25 Transient analysis

7.2 Straight ropeway transient analysis results:

7.2.1 Material SAE1020

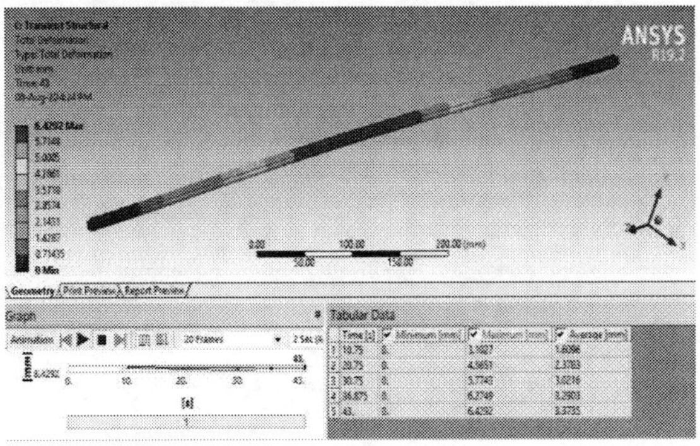

Figure 26 Material SAE1020 Deformation

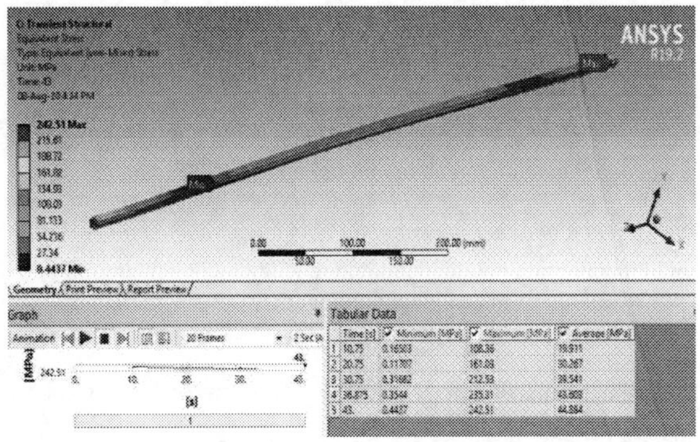

Figure27 Material SAE1020 Stress

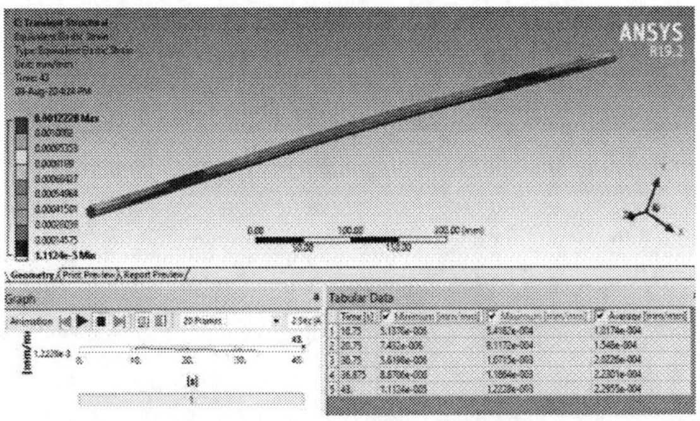

Figure 28 Material SAE1020 Strain

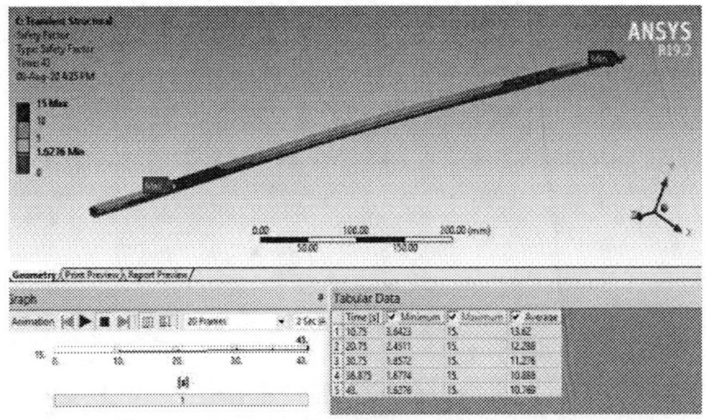

Figure 29 Material SAE1020 Safety factor

7.3 Helical twisted ropeway transient analysis results:

7.3.1 Material SAE1020

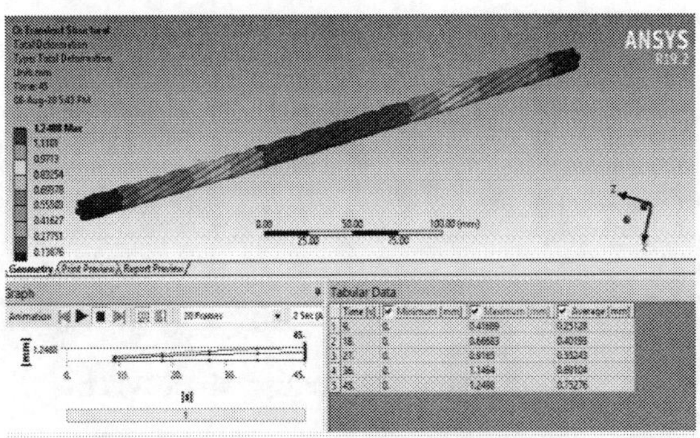

Figure 30 Material SAE1020 Deformation

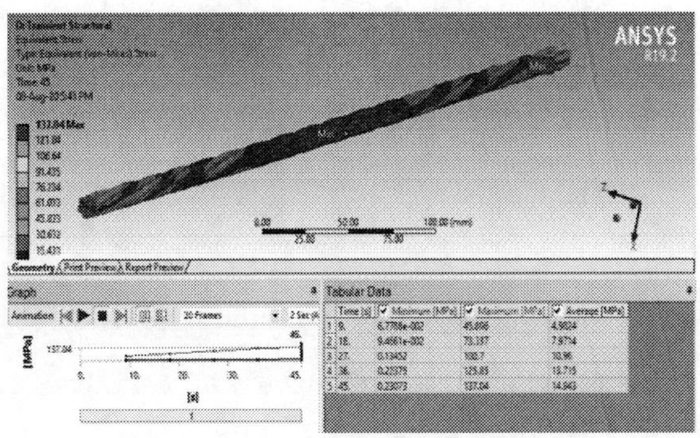

Figure 31 Material SAE1020 Stress

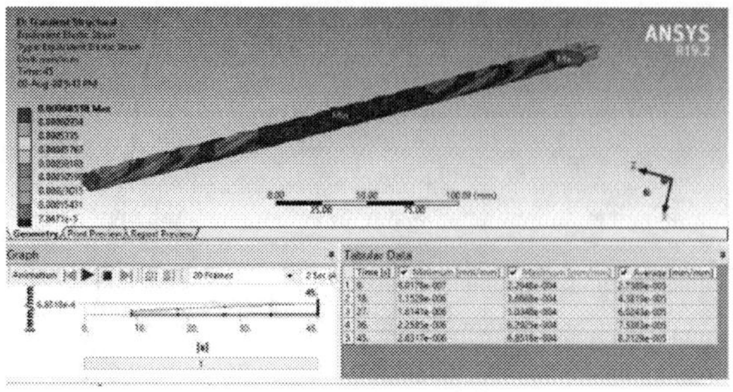

Figure 32 Material SAE1020 Strain

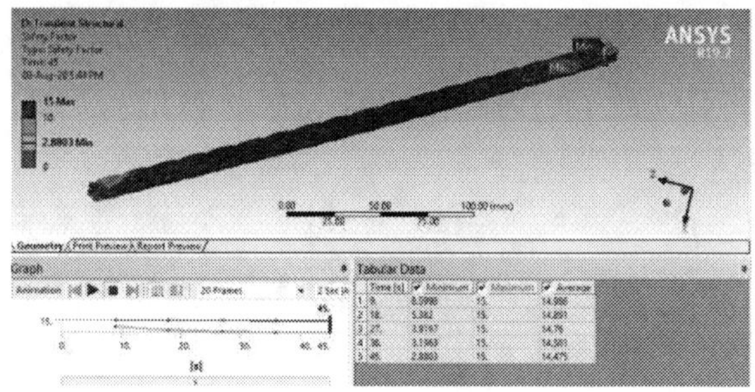

Figure 33 Material SAE1020 Safety factor

8. Results and discussion of static analysis:

8.1 Following table shows the static analysis results of straight ropeway string for different materials

Table 4 Static analysis results of straight ropeway string

Material	SAE 1020	SAE 1137	AL7475	AL7075
Deformation (mm)	6.401	6.4014	18.333	17.975
Stress (Mpa)	260.51	260.5	261.97	261.9
Strain	0.0013025	0.0013025	0.0037265	0.0036537
Safety factor	1.5151	1.9193	1.7178	1.9201

8.2 Following table shows the static analysis results of helical twisted ropeway string for different materials

Table 5 Static analysis results of helical twisted ropeway string

Material	SAE 1020	SAE 1137	AL7475	AL7075
Deformation (mm)	2.3689	2.3689	6.8794	6.745
Stress (Mpa)	257.85	257.85	255.12	255.21
Strain	0.0012893	0.0012893	0.003629	0.0035581
Safety factor	1.5307	1.9391	1.7639	1.9716

9. Results and discussion of modal analysis:

9.1 Following table shows the modal analysis results of straight ropeway string for different materials

Table 6 Modal analysis of straight ropeway string

Material	SAE 1020	SAE 1137	AL7475	AL7075
Mode1 (Hz)	56.518	56.772	56.449	55.896
Mode2 (Hz)	74.643	74.977	74.269	73.54
Mode3 (Hz)	145.68	146.33	145.72	144.29

Mode4 (Hz)	173.77	174.54	173.27	171.57

9.2 Following table shows the modal analysis results of helical twisted ropeway string for different materials

Table 7 Modal analysis of helical twisted ropeway string

Material	SAE 1020	SAE 1137	AL7475	AL7075
Mode1 (Hz)	94.525	94.948	92.863	93.783
Mode2 (Hz)	95.155	95.981	93.464	94.39
Mode3 (Hz)	206.47	207.4	203.23	205.24
Mode4 (Hz)	208.21	209.15	204.9	206.93

10. Conclusion:

In this thesis both straight and helical twisted ropeway strings were designed and analyzed with static and dynamic boundary conditions with 4 different materials, from analysis results it is showing that 1m straight rope way string straight can withstand up to 430 N only when it made up of SAE 1020, at this boundary condition the minimum safety factor is near to 1.5, so that this is the maximum limit which the rope way can bear, in similar case helical twisted rope string can withstand up to 450N, it means by using helical twisted rope string the strength nearly increases up to 4 to 5% approximately with same material properties.
When comparing all other materials results SAE 1137 has better strength values compare to SAE 1020 and this material can withstand more load and it can produce more durability, in same way among AL7075 has highest strength compare to AL7475, by comparing 4 materials results, al7075 has taken 1st position and then SAE 1137 has taken 2nd place in terms strength properties.

By knowing only static analysis it is not the way to decide an optimum rope structure or material for it, so that dynamic analysis is also carried on for each material and calculated natural frequency results, from the results SAE 1137 & AL7075 has better natural frequency values.

Finally thesis can concluded with helical twisted with SAE 1137 or AL7075 materials are more optimum, in this case each material has their own advantages and limitations, so that according to requirement it can possible to choose required material.

11. References:

[1] A.J. Wallis Tayler (1911) *Aerial or wire ropeways –Their construction and management*, Crosby Lockwood and Son, London, 1911

[2] Walter G. Booth (1965) *The Design and Application of Aerial Ropeways*, Master of engineering thesis, M C University, 1965

[3] James M. W. and Brown john (1998) *Dynamics of an aerial cableway system, (Elsevier) Engineering structures*, Vol 20 pp.826-836

[4] Jacques Dubuisson, Michel Cantin (1993) Slipping resistance of monocable aerial ropeway carrier grips, OITAF congress in Barcelona

[5] Andreas Pichler(1999) *Dynamic Behavior of the Grip and the Terminal Equipment of a Detachable Monocable Ropeway at the Terminal Entry*

[6] Mark Löhr, *Adams simulation for ropeway technology*, Thesis at Institute for Material Handling, Technical University of Munich.

[7] Stephan Liedl, *Motions and forces in the rope system of aerial ropeways during operation,* Thesis at Institute for Material Handling, Technical University of Munich.

[8] Alexander BorisoffKazakoff (2012) *Advances in engineering software for lift transportation systems,* Journal of Theoretical and Applied Mechanics, Vol 42 pp.3-22

[9] Marta Knawa and Danuta Bryja (2009) *Modeling problems of steeply inclined cableway subjected to moving load,* PAMM Proc. Appl. Math. Mech. Vol 9,pp.263 – 264 .

[10] Danuta Bryja and Marta Knawa (2011) *Computational model of an inclined aerial ropeway and numerical method for analyzing nonlinear cable-car interaction,* (Elsevier) International Computers and Structures 89, pp.1895–1905.

IOP Conf. Series: Materials Science and Engineering **998** (2020) 012069 doi:10.1088/1757-899X/998/1/012069

Consequence modeling of Ethyl Acetate storage tank

Harisivasri Phanindra K[1*], M. Sreenivasa Rao[2], S.P.Sivapirakasam[3] and Muralimohan Cheepu[4]

[1]Department of Mechanical Engineering, Malla Reddy Engineering College, Maisammaguda, Secunderabad, Telangana State-500100, India
[2]Department of Mechanical Engineering, Godavari Institute of Engineering and Technology, Rajahmundry-533296, India
[3]Department of Mechanical Engineering, National Institute of Technology, Tiruchirappalli-620015, India
[4]Department of Materials System Engineering, Pukyong National University, Busan 48510, Republic of Korea

*E-mail: khssphanindra@gmail.com

Abstract. Consequence modeling of ethyl acetate 40 KL aboveground storage tank was carried out under different scenarios. The toxic concentration of ethyl acetate beyond IDLH (i.e. 2000 ppm) covered up to 23 meters from the release source when there was a leak that occurred 2 inches from the bottom of the tank. The flammable vapor cloud enveloped up to 13 meters from the release source when the concentration of ethyl acetate was beyond 60% of LEL (i.e.13080 ppm). The mass of the fire ball was considered as 100% in order to get the worst case scenario during BLEVE. Based on the mass of the methyl acetate the diameter and duration of the fire ball was 190 meters and 12 seconds respectively during BELVE.

1. Introduction

The risks associated with the pharmaceutical manufacturing industries are proportionate to their fast growth and development in India. These industries have high probability for incidents due to the use of highly flammable and toxic materials in the process [1]. Some of the chemicals which are highly flammable include ethyl acetate, ethanol, methanol, toluene etc.

In general, the flammable liquids will be stored in aboveground or underground storage tanks at a dedicated location according to the Indian Petroleum Rules, 2020. These chemicals will be transferred to day tanks and to process area from the local day tanks. The release of these flammable liquids during the process can cause fire and thermal radiation. The damage to people and property can be controlled with effective emergency preparedness. The consequence modeling is required in order to estimate the potential for damage or injury from a specific incident [1,2,3].

Several softwares are available for the consequence modeling of hazardous chemicals. In this study, ALOHA (Areal locations of Hazardous Atmosphere) software was used for modeling the consequence [4]. This software was developed by the Environmental Protection Agency (EPA), USA.

There are several consequence modeling studies are available for different chemicals [1,2,3] such as LPG, Ammonia, Toluene etc. under different conditions. However, there were no studies reported on Ethyl acetate. Ethyl acetate is a solvent and is suitable for food applications, and is a commonly

IOP Conf. Series: Materials Science and Engineering **998** (2020) 012069 doi:10.1088/1757-899X/998/1/012069

used chemical in the pharmaceutical manufacturing process. According to the Indian Manufacture, Storage and Import of Hazardous Chemicals Rules 1989 (Amended in 2000) Ethyl acetate is a hazardous chemical and categorized under "very highly flammable liquids". So, it is essential to understand the consequence analysis of Ethyl acetate. In this research paper, consequence modeling of the ethyl acetate was carried out for the 40 KL aboveground storage tank.

2. Materials and Methods

Ethyl acetate solvent was considered for the consequence modeling. Ethyl acetate is highly flammable and low toxic. When exposed to Ethyl acetate at 400 PPM, it will cause irritation to eyes and nose. IDLH (immediately dangerous to life or health) value of Ethyl acetate is 2000 ppm. However, it is difficult to obtain high exposure concentrations at normal temperature. The following consequence models were obtained: 1) Toxic area of vapor cloud, 2) Flammable area of vapor cloud, 3) Thermal radiation from pool fire and 4) Thermal radiation from BLEVE (Boiling Liquid Expanding Vapor Explosion).

The consequence modeling in terms of affected distance from source was determined using the ALOHA 5.4.7 software [5]. The various factors required for the consequence modeling by the aid of ALOHA includes geographical conditions, atmospheric data, and chemical data and source details.

In this work, we have considered the tank's total volume to be 42 KL containing liquid Ethyl acetate of 40 KL with L/D ratio of 2 [6]. The leak size was considered to be 2 inch from bottom of the tank due to rupture of the pipe line [7].

The incident location considered is in Visakhapatnam Pharma SEZ (Special Economic Zone) area. The atmospheric data of Visakhapatnam was considered from the reliable sources [8], average temperature of air was 35°C, average wind speed was around 4.3 m/s, and relative humidity was 70%. D-stability class was considered during the dispersion modeling in order to consider the neutral conditions based on the recommendation from ALOHA.

3. Results and Discussion

Consequence models were obtained for Ethyl acetate in different scenarios such as toxic area of vapor cloud, flammable area of vapor cloud, thermal radiation from pool fire and thermal radiation from BLEVE. Table 1 summarizes the consequence modeling in terms of affected distance from the source for ethyl acetate as per the assumptions stated in the section 2 of this paper.

Table 1. Summary of the consequence modeling results of Ethyl acetate

S.No.	Effect/ Scenario		Effected distance from source (in m)
1	Toxic area of vapor cloud	400 ppm	96
2		2000 ppm	23
3	Flammable area of vapor cloud	60% LEL	13
4		10% LEL	31
5	Thermal radiation level during pool fire	10 kW/m^2	13
6		5 kW/m^2	16
7		2 kW/m^2	21
8	Thermal radiation - BLEVE	10 kW/m^2	280
9		5 kW/m^2	403
10		2 kW/m^2	634

3.1. Toxic area of vapor cloud

Figure 1 shows the threat zones of toxic area of vapor cloud for Ethyl acetate which obtained from ALOHA software. It is obvious that the distance is indirectly proportional to concentration, higher the concentration lower the distance covered. The orange zone covers 23 to 96 meters from the release source when the concentration of Ethyl acetate is beyond 400 ppm and less than IDLH (i.e.2000 ppm). When people are exposed in orange zone, they can feel irritation to eyes and nose. The red zone covers up to 23 meters from the release source when the concentration of Ethyl acetate was beyond IDLH (i.e. 2000 ppm). The red zone was not drawn by ALOHA software because effects of near-field patchiness make dispersion predictions less reliable for short distances. However, the value can obtained from ALOHA program separately. When people are exposed in red zone, it can lead to life-threatening health effects or death.

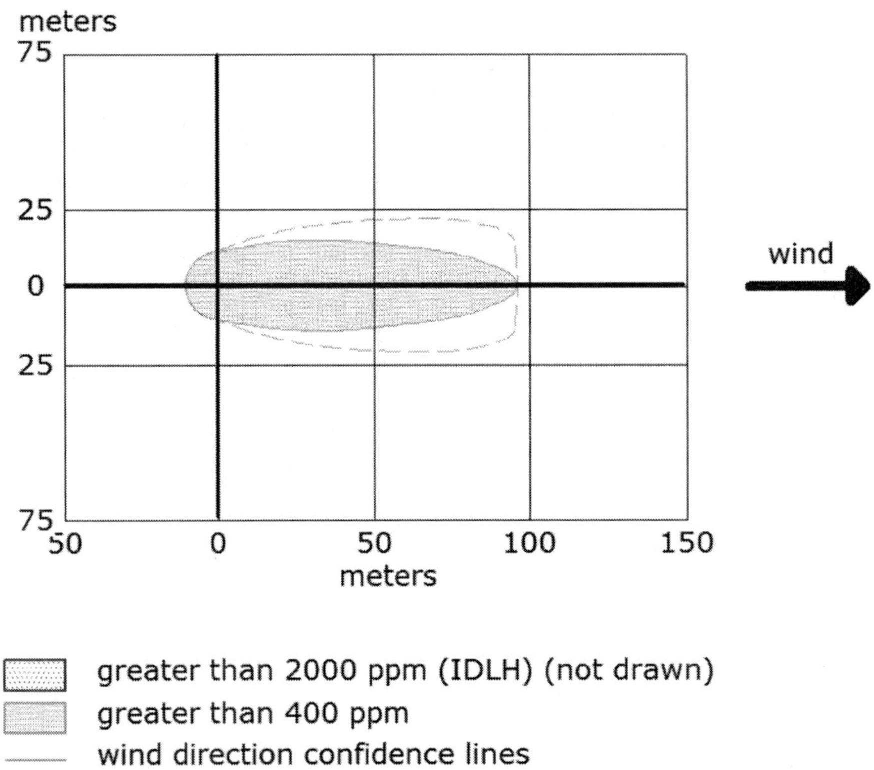

Figure 1. Threat zones of toxic area of vapor cloud for ethyl acetate from ALOHA

3.2. Flammable area of vapor cloud

Figure 2 shows the threat zones of toxic area of flammable area of cloud for Ethyl acetate which was obtained from ALOHA software. It is obvious that the distance is indirectly proportional to concentration, higher the concentration lower the distance covered. The yellow zone covers 13 to 31 meters from the release source when the concentration of Ethyl acetate is beyond 10% of LEL (i.e. 2180 ppm) and 60% of LEL (i.e.13080 ppm). All works to be suspended in this zone during the emergency to minimize heat/spark/flame sources. However, emergency activities can be carried out with necessary precautions.

The red zone covers up to 13 meters from the release source when the concentration of Ethyl acetate beyond 60% of LEL (i.e.13080 ppm). All works to be suspended in this zone during the emergency to minimize heat/spark/flame sources. As far possible emergency activities (i.e. firefighting) is to be carried out from beyond the red zone. Flammable area of cloud zones was also not drawn by ALOHA software because effects of near-field patchiness make dispersion predictions less reliable for short distances.

Threat Modeled: Flammable Area of Vapor Cloud
Model Run: Gaussian
Red : 13 meters --- (13080 ppm = 60% LEL = Flame Pockets)
Note: Threat zone was not drawn because effects of near-field patchiness make dispersion predictions less reliable for short distances.
Yellow: 31 meters --- (2180 ppm = 10% LEL)
Note: Threat zone was not drawn because effects of near-field patchiness make dispersion predictions less reliable for short distances.

Figure 2. Threat zones of flammable area of vapor cloud for Ethyl acetate from ALOHA

3.3. Thermal radiation from pool fire

Figure 3 shows the threat zones of thermal radiation during pool fire for Ethyl acetate which was obtained from ALOHA software. It is obvious that the distance is indirectly proportional to concentration, higher the concentration lower the distance covered. The yellow zone covers 16 to 21 meters from the release source when the heat radiation from pool fire of Ethyl acetate is beyond 2 kW/m^2 and less than the 5 kW/m^2. When people are exposed to thermal radiation from pool fire in yellow zone, they can feel pain within 60 seconds.

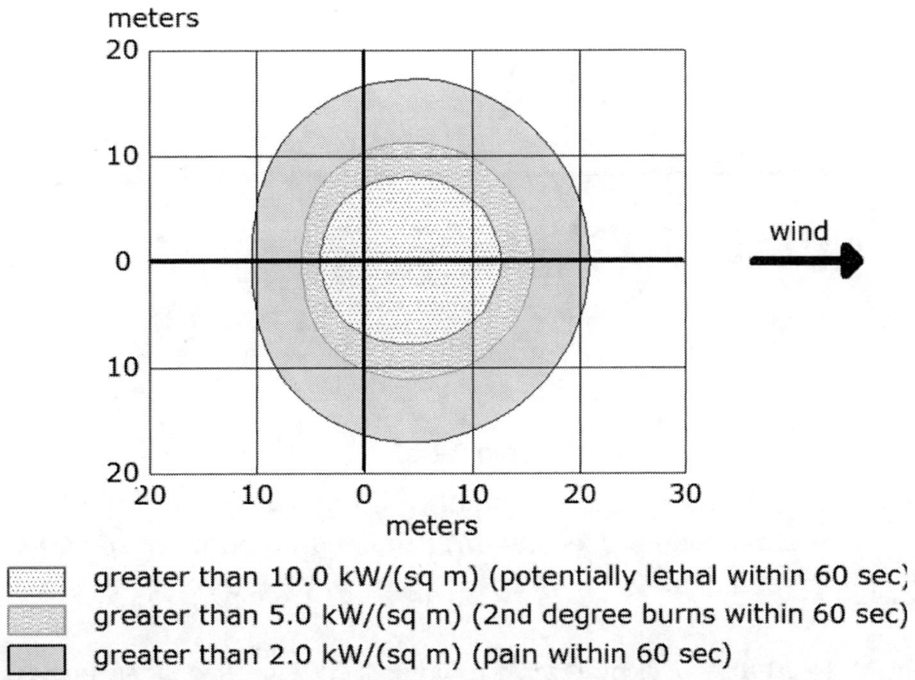

Figure 3. Threat zones of thermal radiation during pool fire for ethyl acetate from ALOHA

The orange zone covers from 13 to 16 meters from the release source when the heat radiation from pool fire of Ethyl acetate is beyond 5 kW/m^2 and less than the 10 kW/m^2. It can cause second degree burns within 60 seconds when people are exposed to thermal radiation from pool fire in orange zone. The red zone covers up to13 meters from the release source when the heat radiation from pool fire of Ethyl acetate is beyond 10 kW/m^2. When people are exposed to thermal radiation from pool fire in red zone, it can lead to lethal within 60 seconds.

3.4. Thermal radiation from BLEVE

BLEVE is a kind of explosion which caused due to the rupture of pressurized liquid in a tank. The pressure in tanks reaches abnormal condition when Methyl acetate temperature reaches beyond the boiling point. This situation can arise due to thermal radiation from pool fire from same tank or material from adjacent tank and BLEVE from other tanks located nearby. The mass of the fire ball was considered as 100% in order to get the worst case scenario. Based on the mass of the Methyl acetate the diameter and duration of the fire ball was 190 meters and 12 seconds respectively, these values were computed from ALOHA software modeling.

Figure 4 shows the threat zones of thermal radiation during BLEVE for Ethyl acetate which obtained from ALOHA software. It is obvious that the distance is indirectly proportional to concentration, higher the concentration lower the distance covered. The yellow zone covers 403 to 634 meters from the release source when the heat radiation from BLEVE of Ethyl acetate is beyond 2 kW/m^2 and less than the 5 kW/m^2. When people are exposed to thermal radiation from BLEVE in yellow zone, they can feel pain within 60 seconds.

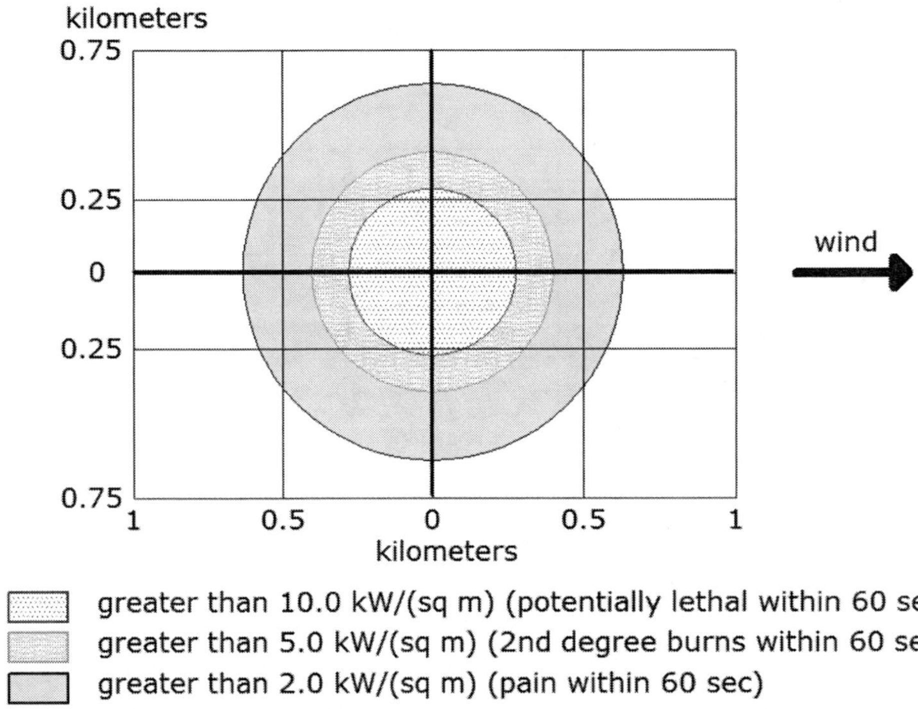

Figure 4. Threat zones of thermal radiation during BLEVE for ethyl acetate from ALOHA

The orange zone covers 280 to 403 meters from the release source when the heat radiation from BLEVE of Ethyl acetate was beyond 5 kW/m^2 and less than the 10 kW/m^2. It can cause second degree burns within 60 seconds when people are exposed to thermal radiation from BLEVE in orange zone. The red zone covers up to 280 meters from the release source when the heat radiation from BLEVE of Ethyl acetate was beyond 10 kW/m^2. When people are exposed to thermal radiation from BLEVE in red zone, it can lead to lethal within 60 seconds.

The data obtained from ALOHA software is useful for planning the emergency preparedness of Ethyl acetate storage tank. It is always recommended to prevent the chemical release or fire incidents from the storage facility. The following recommendation will be useful for prevention of chemical release and fire incidents during the storage of Ethyl acetate which are not limited:

- All the electrical fitting must be flame proof according to the hazardous area classification
- Ensure the electrical continuity by bonding and earthing to the storage tank
- Ensure the electrical continuity by bonding all the pipe line flange joins
- Ensure the electrical continuity by bonding and earthing while unloading or loading operations from the taker
- The minimum safe distance between storage vessels must be maintained
- Regular inspection to be carried out for the storage tank, pipelines, electrical wires and fitting etc.
- Suitable bund wall to be provided for the storage tank
- Notice board on "no smoking/mobile/necked flame" to be displayed
- Possibilities for nitrogen interstation to be explored for the storage tank
- Flame arrestor to be provided the vent line
- Train the workforces related to hazard and precautions while handling Ethyl acetate

4. Conclusions

Consequence modeling of Ethyl acetate in different scenarios was computed using ALOHA software. The orange zone for toxic cloud area covers 23 to 96 meters from the release source which cases the irritation to eyes and nose. The red zone for toxic cloud area covers up to 23 meters, it can lead to life-threatening health effects or death. The red zone during thermal radiation from BELVE covers up to 280 meters from storage tank, it will be lethal for humans within 60 seconds for personnel located in this zone. The red zone during the flammable area of vapour cloud covers up to 13 meters from the storage tank, as far possible emergency activities (i.e. firefighting) is to be carried out from beyond this red zone. The data obtained from this research work will be useful for planning the emergency preparedness of Ethyl acetate storage tank.

5. Acknowledgement

The first author and second author thank the management of Godavari Institute of Engineering & Technology (A), Rajahmundry for their continuous encouragement towards the research work.

References

[1] Haghnazarloo H, Parvini M and Lotfollahi MN 2015 *J. Loss. Prev. Process. Ind.* **37** 11.

[2] Tseng JM, Su TS and Kuo CY 2012 *Procedia. Eng.* **45** 384.

[3] Khanmohamadi M, Bagheri M, Khademi N and Ghannadpour SF 2018 *Saf. Sci.* **110** 230.

[4] Jones R, Lehr W, Simecek-Beatty D and Reynolds M 2013 ALOHA® (Areal Locations of Hazardous Atmospheres) 5.4. 4: *Technical Documentation.*

[5] Jabbari M, Atabi F and Ghorbani R 2020 *J. Loss. Prev. Process. Ind.* **104** 139.

[6] Tipasri W and Wongwuttanasatian T 2019 *Energy. Procedia.* **156** 254.

[7] Center for Chemical Process Safety 2010 Guidelines for Chemical Process Quantitative Risk Analysis, New York.

[8] Police S, Sahu SK and Pandit GG 2016 Atmos. Pollut. Res. **7** 725.

Influence of rotational speed on the dissimilar friction welding of heat-treated aluminum alloys

G. Ravi Chandra[1], S. Venukumar[1], Muralimohan Cheepu[2]*

[1]Department of Mechanical Engineering, Vardhaman College of Engineering, Telangana 501218, India
[2]Super-TIG Welding Co., Limited, Busan 48547, Republic of Korea

*Corresponding Email: muralicheepu@gmail.com

Abstract. Union of dissimilar alloys by means of welding processes is often challenging than union of like material or alloys. The trivial distinction in mechanical, metallurgical and formation of alloys lead in major changes in the welded joints. Friction welding can be implemented effectively to unite a much wide variety of dissimilar materials. It efficiently resolves so many complex issues in front of contemporary manufacturing, like recuperating quality, dipping costs and safeguarding the environs. It produces the high strength joints compared to fusion welding process. The properties of the welded joints contingent on the right assortment of the welding parameters. In this investigation , the effect of rotational speed on mechanical properties was assessed by tensile, microhardness tests and weld flash appearance of friction welded joint of aluminium alloy 6063-O and aluminium alloy 6063-T6.

1. Introduction

The joining of dissimilar alloys demanding for the usage of various materials in the manufacturing industries. The recent technologies always intending to ease the components by combining various parts made by different materials. In order to obtain the multiplex components it is needed to join the parts effectively using suitable joining methods [1-2]. The joining of same materials leads in successful joint interface with solid weld integrity. Whereas, the union of dissimilar materials lacks in formation of strong weld interface and joining strength [3-5]. However, it is depending the type of dissimilar couples and their elemental composition. The elemental composition has much influence on the development of resulting phases that can deteriorate mechanical properties of welded joints [6-9]. The formation of the brittle phases and their position in the weld region is depending the type of welding processes applied for the joining. The severity of the brittle phases along with their size and amount varies greatly with the conventional welding methods compared to solid state welding methods [10-13]. The amount heat input into the weld is one of the important criterion to deal with the dissimilar welds. It is a known fact that the arc welding methods have high heat input and much heat energy stored in the welds during welding. The large amount of heat for a long duration in the welds until it cools down from the molten phases, welds were experienced to a large amount of thermal stresses and microstructural changes [14-18].

The dissimilar combinations of welded joints produced under arc welding processes experienced for the development of various phases in the fusion region. The dissimilar combinations of stainless

IOP Conf. Series: Materials Science and Engineering 998 (2020) 012070 doi:10.1088/1757-899X/998/1/012070

steel to aluminum using arc joining leads in the formation of Fe-Al-Zn secondary phases alongside the interface. The addition of Zn based filler might have beneficial for achieving strong welds, whereas the Zn also involved to make the intermetallic compound [19].The same combination of materials were joined by solid state welding methods resulted in the formation of very few intermetallics with excellent mechanical properties [20-22].There are many other dissimilar combinations were successfully joined using solid state welding methods with the excellent mechanical strength and strong metallurgical properties [23-30]. In particular, some of the alloys consisting of various types of heat treated alloys within the same materials with variation in their physical properties. the joining of these materials have different strategy that the degradation of the mechanical strength after welding due to reheating the alloys due to heat produced by welding processes [31-33]. The effect of welding processes also has similar effect on these alloys for the degradation of mechanical strength of the joints. The solid state welding methods were resulted in the producing of efficient welds over the arc welding methods to remain its original properties after welding [34-36].

Some of the arc welding methods were recently developed to produce dissimilar welds using cold metal transfer method. However, this process is limited to the thickness of metal and some of the geometries [37-39]. To weld these heat treated alloys friction stir and friction based welding methods were most suitable and easy to produce the strong welds [40-47]. In the present investigation, aluminum alloys with two different heat treated conditions such as AA6063-O and AA6063-T6 welded using friction welding. The welds were formed by friction welding under effect of rotational speed in order to keep the properties of the alloys constant with strong weld strength. The metallurgical and mechanical properties of the joints were studied using tensile and microhardness tests. The strength of welds were identified according to the rotational speed and structural changes of weld flash have been studied.

2. Experimental procedure

The materials put to use for present experiment were rods of 12 mm diameter AA6063-O and AA6063-T6. The sample for uniting has been procured and made to 70 mm span with uniform profile. Precedent to friction welding, the abutment faces of the original materials were make ready using #400 and #800 grit emery sheets. The specimens were cleaned with running water followed by distilled water. After they were wiped by means of acetone to take away the organic adulterant. Among the friction welding parameters, rotational speed was varied; this parameter was delineate to have noteworthy outcome on the joint properties and pursuance. The former variables those were remained without change were upset pressure, upset time, friction pressure and burn off length. For the period of friction welding, AA6063-T6 rod was kept fastened, and AA6063-O was used as the rotating specimen which was inserted in the spindle. A downward force is applied to maintain the contact. Individual part is placed in a motionless fix. The next piece is placed within the revolving spindle, which be after that reduced to a speed of prescribed revolving pace. When the desired pace is achieved already defined axial force will be applied by moving the spindle downwards. Such settings were vindicated for a persistent period. In the next step of load is imposed till the needed temperatures and material state remain. In the course of this juncture these materials are plasticized (develop into workable).

The amount of "burn off length" activates the end point when the element gets to intended desired span. Revolving speed is halted. After that axial force is functional to generate "forge pressure" to finishing the weld. This supply molecular bonding and grain refinement in the course of t weld region. The welds were cut for microstructural analysis and for the evaluation of mechanical properties. The welds were performed under microhardness measurements to analyse the strength variations in the various zones. The joint properties were evaluated by using tensile tests and the fracture surfaces characterised using scanning electron microscope. The axial reduction as a weld blaze was bulged around weld boundary and it was measured for analysis using diameter and thickness for the different welding conditions. The tensile sample was set according to ASTM benchmark to test the joints strength. The failure joints were cut at the neck of the tensile tested specimens and carefully stored and used scanning electron microscope to observe fracture phenomenon and characteristics.

3. Results and discussion

The friction welds joined between heat treated dissimilar aluminum alloys, which have differences in their physical and metallurgical properties. In friction welding, due to the absence of melting the base materials no evidence of severe intermetallic phases. However, joint quality and the process parameters and welding method optimization are major concern in order to obtain the reliable welds. The selection of process parameters are depending the type of dissimilar combination of materials and their mechanical properties. The development of weld twinkle near boundary from both substrates is symmetry for the similar materials. Whereas it is varying with the dissimilar materials in their size and diameter due to their hard and soft nature of the properties. Joining of similar materials but with different heat treatment conditions is quite difficult to find the welding conditions due to the difference in their properties. In order to weld the ductile materials the selection of rotational speed is one of the solutions to obtain the strong welds. The rotational velocity and the force on the substrates have influence to deform the faying surface with extruded weld flash. The amount of weld flash is one of the significance to estimate the quality of the welds. The amount of weld zone is steadily augmented by the mounting of rotational speed from the 710 rpm to 1120 rpm. The amount of weld flash is quite higher for the aluminum alloys compared to other alloys due to their soft nature of the property at high temperature. Figure 1 shows the axial reduction of the welds at diverse rotational speed. Maximum axial shortening of 16 mm was occurred for the welds produced under the rotational speed of 1210 rpm. The axial shortening increased from 8.5 mm to 16 mm as the increasing of rotational speed. The axial shortening is almost doubled with the rotational speed from 710 rpm to 1120 rpm. Even though small amount of increment in the rotational speed there is a large amount of axial shortening in the welds. Therefore, the interval of increasing the rotational speed was decided smaller and observed the continuous change in the axial shortening. Based on the weld flash formation and axial shortening, it is easy to estimate the optimal welding conditions. It is observed that the axial shortening of 12 mm reasonable amount for the aluminum alloys and expected that condition is optimal and the further trails were not performed. The weld flash at the 1120 rpm tends to produce the cracks and changes the circular shape to irregular shape. It is not a common phenomenon in the formation of weld flash within the optimal range. The large amount of decrease in axial shortening is not beneficial and economical to the production, due the loss of material and accuracy of the components. Therefore, the small amount of axial shortening is recommended to avoid the material loss and improve the joint quality.

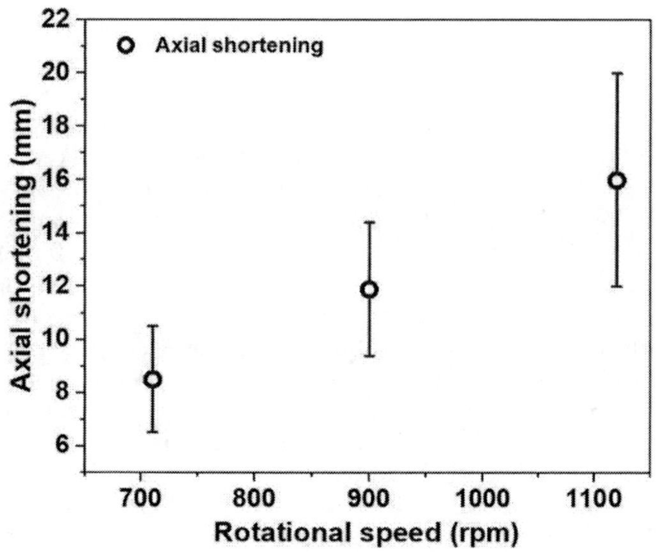

Figure 1. Axial shortening in the welds according to the rotational speed.

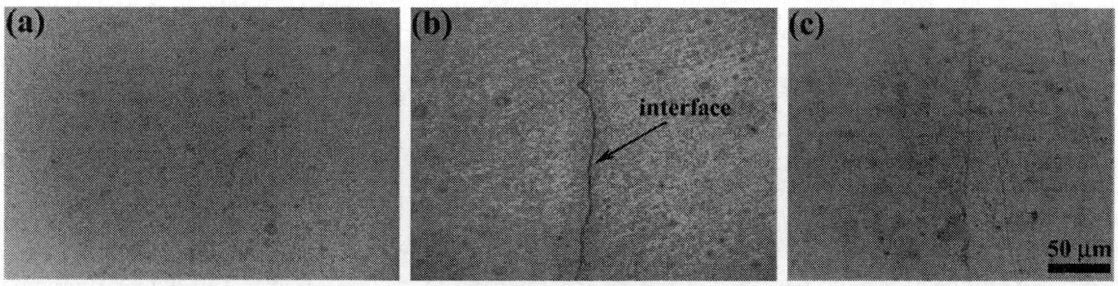

Figure 2. Optical microstructures of the welds showing the weld interface at different rotational speed (a) 710 rpm (b) 900 rpm and (c) 1120 rpm.

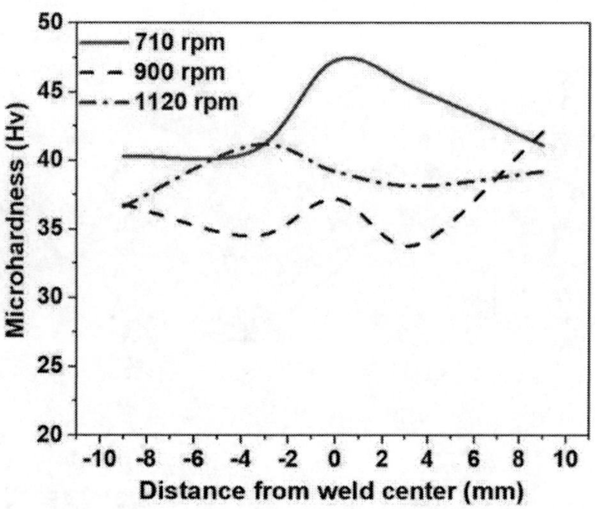

Figure 3. Microhardness of the welds at three different rotational speed.

Figure 2 elaborates the optical microstructures of the friction welds produced under different rotational speeds. The weld interface of the 710 rpm welds exhibited the nonlinear shape with the weaker bonding. The interface line is not formed strongly to obtain the required metallurgical bond between the two alloys. The metallurgical reactions and the bonding lines are subject to the correct selection of the welding situation. Based on the weld interface and its linear shape it is analysed that the rotational speed is not enough to make the strong bonding line between the alloys. The microstructure of the alloys adjacent to the interface is not changed greatly with the grain size or deformed regions. Figure 2b exhibit the microstructure of the welds at 900 rpm rotational speed with the clear interface between two substrates. The interface is linear and formed strongly with the metallurgical bonding without any defects along the weld interface. The required amount of heat and pressured is sufficiently obtained by the rotational speed of 900 rpm. The microstructure at the interface of the substrates shows clearly the formation of fine equi-axed grains. The presence of equi-axed grains region is narrow and the coarser grains region is slightly wider, which is away from the interface. The weld interface at the 1120 rpm rotational speed showed that the incomplete formation of the weld interface with a weaker metallurgical bonding (see Fig. 2c). The excess amount of heat and pressure due to high rotational speed has caused for the diminishing of metallurgical bonding at weld interface. Moreover, the interface microstructure also not formed clearly with the equi-axed grains and heat affected zone. The microstructural analysis confirmed the low or higher level of rotational speed causes the weaker metallurgical bonding.

Figure 3 elucidates microhardness of respective welds at three unlike rotational speed. The highest hardness was obtained for the welds of 700 rpm rotational speed compared to other welds. The hardness at the weld interface and adjacent regions have highest hardness. Even though the microstructures of the welds did not exhibit the clear view of the weld interface, the hardness results confirmed that the region along the interface and the dynamic recrystallization zone. The microstructures of the aluminum alloys neither are nor clear after experiencing the severe plastic deformation. The microhardness of the welds of non-identical zones of heat affected zone and the weld zone were presented in figure 4. The hardness of welds decreased gradually from the 700 rpm to 1120 rpm of rotational speed. Figure 5 shows the tensile strength of the welds at different rotational speed. The highest strength achieved at the rotational speed of 710 rpm contrast to the supplementary welding conditions. The strength of the welds progressively changed with rotational speed as like the microhardness. The lower rotational speeds have better results to achieve the friction welds between aluminum alloys.

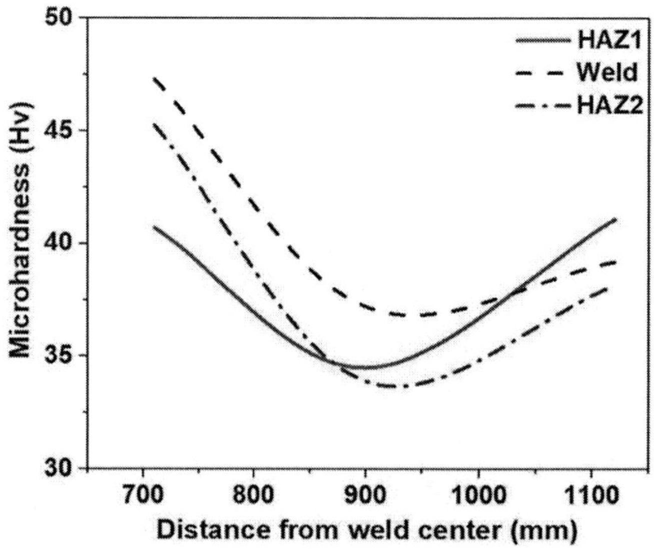

Figure 4. Microhardness of the welds at different zones.

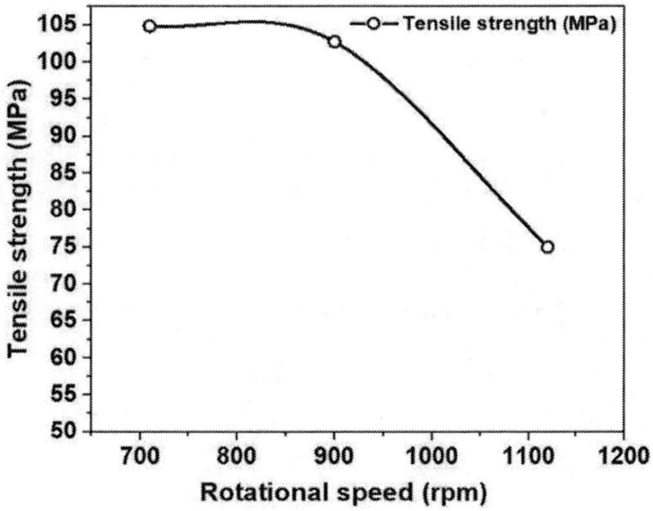

Figure 5. Tensile strength of the welds according to the rotational speed.

4. Conclusions

Friction welding of dissimilar aluminum alloys at heat treatment conditions of 6063-O and 6063-T6 were joined successfully using different rotational speed. The following observations were concluded based on the experimental results.

- The process parameter of rotational speed range was identified to weld the aluminum 6063 alloy.
- The axial shortening of the friction welds increased with the incremental of rotational speed along with the weld flash diameter.
- Microhardness of the welds at lower rotational speed achieved higher than the maximum rotational speed. The hardness at the dynamic recrystallization zone is superior to the weld interface.
- The highest tensile strength achieved for the welds at the rotational speed of 710 rpm with the base metal failure.
- The linear weld interface and the fine equi-axed region was obtained in the welds along with the curved shape of the interface at the periphery regions.

References

[1] Murakami T, Kazuhiro N, Hongjun T and Masao U 2003 Dissimilar metal joining of aluminum to steel by MIG arc brazing using flux cored wire *ISIJ international.* **43**, 1596-602.

[2] Muralimohan C H, Ashfaq M, Ashiri R, Muthupandi V and Sivaprasad K 2016 Analysis and characterization of the role of Ni interlayer in the friction welding of titanium and 304 austenitic stainless steel *Metall. Mater. Trans. A.* **47** 347-59

[3] Cheepu M M, Muthupandi V and Loganathan S 2012 Friction welding of titanium to 304 stainless steel with electroplated nickel interlayer *Mater. Sci. Forum.* **710** 620-25

[4] Cheepu M, Venkateswarlu D, Rao P N, Muthupandi V, Sivaprasad K and Che W S 2019 Microstructure characterization of superalloy 718 during dissimilar rotary friction welding *Mater. Sci. Forum.***969** 211-7.

[5] Cheepu M, Muthupandi V and Che W S 2019 Interface microstructural characterization of titanium to stainless steel dissimilar friction welds. *TMS 2019 148th Annual Meeting &*

Exhibition Supplemental Proceedings, pp. 259-268. Springer, Cham, 2019.
[6] Venukumar S, Cheepu M, BabuT V and Venkateswarlu D 2019 TIG arc welding-brazing of dissimilar metals-an overview *Mater. Sci. Forum.* **969** 768-74.
[7] Cheepu M and Che W S 2020 Influence of friction pressure on microstructure and joining phenomena of dissimilar joints *Trans. Indian. Inst. Met.***73** 1455-60.
[8] Anuradha M, Das V C, Venkateswarlu D and Cheepu M 2019 Microstructure characterization in dissimilar TIG welds of Inconel alloy 718 and high strength tensile steel *Mater. Sci. Forum.***969**496-501.
[9] Anuradha M, Das V C, Venkateswarlu D and Cheepu M 2019 Parameter optimization for laser welding of high strength dissimilar materials*Mater. Sci. Forum.***969**558-64.
[10] Shiva A, Cheepu M, Kantumuchu V C, Kumar K R, Venkateswarlu D, SrinivasB and JeromeS 2018 Microstructure characterization of Al-TiC surface composite fabricated by friction stir processing *IOP Conf. Ser. Mater. Sci. Eng.***330**012060.
[11] Muralimohan C H and Muthupandi V 2013 Friction welding of type 304 stainless steel to CP titanium using nickel interlayer *Adv. Mater. Res.* **794** 351-57
[12] Cheepu M, Muthupandi V, Srinivas B and Sivaprasad K 2018 *Techno-Societal 2016, International Conference on Advanced Technologies for Societal Applications ICATSA 2016* ed P M Pawar, B P Ronge, R Balasubramaniam and S Seshabhattar (Cham: Springer) Development of a friction welded bimetallic joints between titanium and 304 austenitic stainless steel 709-17 Chapter 73. https://doi.org/10.1007/978-3-319-53556-2_73
[13] Muralimohan C H, Muthupandi V and Sivaprasad K 2014 Properties of friction welding titanium-stainless steel joints with a nickel interlayer *Procedia. Mater. Sci.* **5** 1120-29
[14] Cheepu M, Ashfaq M and Muthupandi V 2017 A new approach for using interlayer and analysis of the friction welding of titanium to stainless steel *Trans. Indian. Inst. Met.* **70** 2591-600
[15] Cheepu M, Muthupandi V and Che W S 2018 Improving mechanical properties of dissimilar material friction welds *Appl. Mech. Mater.* **877** 157-62
[16] Cheepu M, Haribabu S, Ramachandraiah T, Srinivas B, Venkateswarulu D, Karna S, Alapati S and Che W S 2018 Fabrication and analysis of accumulative roll bonding process between magnesium and aluminum multi-layers *Appl. Mech. Mater.* **877** 183-89
[17] Dong H, Liqun Y, Chuang D and Sindo K 2010 Arc joining of aluminum alloy to stainless steel with flux-cored Zn-based filler metal *Mater. Sci. Eng. A.* **527**7151-54.
[18] Matsuda, T, Hironobu A, Tomokazu S, Ryo Y, Hisashi H, Shozo O and Akio H 2019 High-frequency linear friction welding of aluminum alloys to stainless steel *J. Mater. Proces. Technol.* **269** 45-51.
[19] Anuradha M, Das V C, SusilaP, Cheepu M and VenkateswarluD 2020 Effect of welding parameters on TIG welding of Inconel 718 to AISI 4140 steel *Trans. Indian. Inst. Met.* **73** 1515-20.
[20] Muralimohan C H, Muthupandi V and Sivaprasad K 2014 The influence of aluminium intermediate layer in dissimilar friction welds *Inter. J. Mater. Res.* **105** 350-57
[21] Muralimohan C H, Haribabu S, Reddy Y H, Muthupandi V and Sivaprasad K 2014 Evaluation of microstructures and mechanical properties of dissimilar materials by friction welding *Procedia. Mater. Sci.* **5** 1107-13
[22] Muralimohan C H, Haribabu S, Reddy Y H, Muthupandi V and Sivaprasad K 2015 Joining of AISI 1040 steel to 6082-T6 aluminium alloy by friction welding *J. Adv. Mech. Eng. Sci.* **1** 57-64
[23] Cheepu M and CheW-S 2019 Characterization of microstructure and interface reactions in friction welded bimetallic joints of titanium to 304 stainless steel using nickel interlayer *Trans. Indian. Inst. Met.***72**1597-601.
[24] Haribabu S, CheepuM, TammineniL, GurasalaN K, DevuriV and Kantumuchu V C 2019 Dissimilar friction welding of AISI 304 austenitic stainless steel and AISI D3 tool steel:

IOP Conf. Series: Materials Science and Engineering 998 (2020) 012070 doi:10.1088/1757-899X/998/1/012070

Mechanical properties and microstructural characterization. *Advances in Materials and Metallurgy*, pp. 271-281. Springer, Singapore, 2019.

[25] Cheepu M, MuthupandiV, VenkateswarluD, SrinivasB and CheW-S 2017 Interfacial microstructures and characterization of the titanium—stainless steel friction welds using interlayer technique. *International Conference on Physics and Mechanics of New Materials and Their Applications*, pp. 267-283. Springer, Cham, 2017.

[26] Cheepu M and Che W S 2019 Friction welding of titanium to stainless steel using Al interlayer *Trans. Indian. Inst. Met.* **72** 1563-8.

[27] Cheepu M and Che W S 2020 Characterization of interfacial microstructure in friction welds between Inconel 718 and SM45C steel *Trans. Indian. Inst. Met.***73**1567-71.

[28] Anuradha M, Das V C, SusilaP, Cheepu M and VenkateswarluD 2020 Microstructure and mechanical properties for the dissimilar joining of Inconel 718 alloy to high-strength steel by TIG welding*Trans. Indian. Inst. Met.***73**1521-1525.

[29] Cheepu M and Susila P 2020 Growth rate of intermetallics in aluminum to copper dissimilar welding *Trans. Indian. Inst. Met.***73**1509-14.

[30] Cheepu M and Che W S 2019 Effect of burn-off length on the properties of friction welded dissimilar steel bars*J. Weld. Join.* **37**, 40-5.

[31] Cheepu M, Venkateswarlu D, Mahapatra M M and Che W S 2017 Influence of heat treatment conditions of Al-Cu aluminum alloy on mechanical properties of the friction stir welded joints *Korean Welding and Joining Society* **11** 264-264. http://www.dbpia.co.kr/Journal/ArticleDetail/NODE07278590

[32] Venkateswarlu D, CheepuM, Rao P N, Kumaran S S and Srinivasan N 2019 Characterization of microstructure and mechanical properties of AA2219-O and T6 friction stir welds*Mater. Sci. Forum.***969** 205-10.

[33] Srinivas B, CheepuM, SivaprasadK and MuthupandiV 2018 Effect of gaussian beam on microstructural and mechanical properties of dissimilar laser welding of AA5083 and AA6061 alloys *IOP Conf. Ser. Mater. Sci. Eng.***330**, 012066.

[34] Srinivas B, Krishna N M, CheepuM, SivaprasadK and MuthupandiV Studies on post weld heat treatment of dissimilar aluminum alloys by laser beam welding technique *IOP Conf. Ser. Mater. Sci. Eng.***330** 012079.

[35] Karna S, CheepuM, VenkateswaruluD and SrikanthV 2018 Recent developments and research progress on friction stir welding of titanium alloys: an overview *IOP Conf. Ser. Mater. Sci. Eng.***330**012068.

[36] Cheepu M, SrinivasB, AbhishekN, RamachandraiahT, KarnaS, VenkateswarluD, Alapati S and Che W S 2018 Dissimilar joining of stainless steel and 5083 aluminum alloy sheets by gas tungsten arc welding-brazing process *IOP Conf. Ser. Mater. Sci. Eng.***330**012048.

[37] Ahsan Md RU, CheepuM, AshiriR, KimT-H, JeongC and ParkY-D 2017 Mechanisms of weld pool flow and slag formation location in cold metal transfer (CMT) gas metal arc welding (GMAW) *Weld. World.* **61** 1275-85.

[38] Venukumar S, CheepuM, BabuT V and VenkateswarluD 2019 Cold metal transfer (CMT) welding of dissimilar materials: An overview *Mater. Sci. Forum.***969** 685-90.

[39] Krishnaja D, CheepuM and VenkateswarluD 2018 A review of research progress on dissimilar laser weld-brazing of automotive applications *IOP Conf. Ser. Mater. Sci. Eng.***330**012073.

[40] Avula D, DevuriV, CheepuM and Dwivedi D K 2018 Tensile properties of friction stir welded joints of AA 2024-T6 alloy at different welding speeds*IOP Conf. Ser. Mater. Sci. Eng.***330**012081.

[41] Venkateswarulu D, Cheepu M, Krishnaja D and Muthukumaran S 2018 Influence of water cooling and post-weld ageing on mechanical and microstructural properties of the friction-stir welded 6061 aluminium alloy joints *Appl. Mech. Mater.* **877** 163-76

[42] Devireddy K, Devuri V, Cheepu M and Kumar B K 2018 Analysis of the influence of friction stir processing on gas tungsten arc welding of 2024 aluminum alloy weld zone *Int. J. Mech.*

Prod. Eng. Res. Dev. **8** 243-52.

[43] Venkateswarlu D, CheepuM and MahapatraM M 2018 Analysing the friction stir welded joints of AA2219 Al-Cu alloy in different heat-treated-state*IOP Conf. Ser. Mater. Sci. Eng.***330**012074.

[44] Cheepu M and SusilaP 2020 Interface microstructure characteristics of friction-welded joint of titanium to stainless steel with interlayer*Trans. Indian. Inst. Met.***73**1497-501.

[45] Kong Y S, CheepuM and ParkY W 2020 Effect of heating time on thermomechanical behavior of friction-welded A105 bar to A312 pipe joints*Trans. Indian. Inst. Met.***73**1433-8.

[46] Kong Y-S, CheepuM and KimD-G 2020 Microstructure and mechanical properties of friction-welded and post-heat-treated Inconel 718*Trans. Indian. Inst. Met.***73**1449-53.

[47] Cheepu M, Cheepu H, A Karpagaraj and Che W S 2020 Influence of joint interface on mechanical properties in dissimilar friction welds *Adv. Mater. Process. Technol.* 1-13.https://doi.org/10.1080/2374068X.2020.1832413

AUTHOR INDEX

Institute of Physics
Dirac House, Temple Back
Bristol BS1 6BE UK

CURRAN ASSOCIATES INC.
proceedings
.com

ISSN: 1757-8981
ISBN 978-1-7138-4802-8
9 781713 848028